Technical Mathematics

Technical Mathematics

John C. Peterson

Chattanooga State Technical Community College

Delmar Publishers Inc.®

I(T)P™

Cover photo courtesy of the Jacksonville Chamber of Commerce, Jacksonville, Florida.

DELMAR STAFF

Publisher: Michael A. McDermott
Developmental Editor: Mary E. Clyne
Project Editor: Cynthia Lassonde
Senior Production Supervisor: Larry Main
Art Coordinator: Brian Yacur
Senior Design Supervisor: Susan C. Mathews

For information, address Delmar Publishers Inc.
3 Columbia Circle, Box 15-015
Albany, NY 12212-5015

Printed in the United States of America
Published simultaneously in Canada by Nelson Canada
a division of The Thomson Corporation

Library of Congress Cataloging in Publication Data
Peterson, John C. (John Charles), 1939–
    Technical mathematics/John C. Peterson.
        p.   cm.
    Includes index.
    ISBN 0-8273-4575-5
    1. Mathematics.   I. Title.
QA39.2.P49   1994
510—dc20   92-32234
                                                                                   CIP

Contents

Preface xiii

CHAPTER 1   The Real Number System 1

1.1   Some Sets of Numbers 2

1.2   Basic Laws of Real Numbers 7

1.3   Basic Operations with Real Numbers 13

1.4   Laws of Exponents 23

1.5   Significant Digits and Rounding Off 29

1.6   Scientific Notation 36

1.7   Roots 42

        Chapter 1 Review 48

        Chapter 1 Test 49

CHAPTER 2   Algebraic Concepts and Operations 51

2.1   Addition and Subtraction 52

2.2   Multiplication 56

2.3   Division 62

2.4   Solving Equations 69

2.5   Applications of Equations 76

        Chapter 2 Review 85

        Chapter 2 Test 87

CHAPTER 3   Geometry 88

3.1   Lines and Angles 89

3.2	Triangles	97
3.3	Other Polygons	104
3.4	Circles	110
3.5	Geometric Solids	117
	Chapter 3 Review	127
	Chapter 3 Test	130

CHAPTER 4 Functions and Graphs 132

4.1	Relations and Functions	133
4.2	Operations on Functions; Composite Functions	143
4.3	Rectangular Coordinates	148
4.4	Graphs	152
4.5	Graphing Calculators and Computer-Aided Graphing	159
4.6	Using Graphs to Solve Equations	163
	Chapter 4 Review	172
	Chapter 4 Test	174

CHAPTER 5 Systems of Linear Equations and Determinants 175

5.1	Linear Equations	176
5.2	Graphical and Algebraic Methods for Solving Two Linear Equations in Two Variables	180
5.3	Algebraic Methods for Solving Three Linear Equations in Three Variables	188
5.4	Determinants	195
5.5	Using Cramer's Rule to Solve Systems of Linear Equations	205
	Chapter 5 Review	212
	Chapter 5 Test	215

CHAPTER 6 Ratio, Proportion, and Variation 216

6.1	Ratio and Proportion	217

6.2   Similar Geometric Shapes 222

6.3   Direct and Inverse Variation 226

6.4   Joint and Combined Variation 231

       Chapter 6 Review 235

       Chapter 6 Test 237

CHAPTER 7   Factoring and Algebraic Fractions 238

7.1   Special Products 239

7.2   Factoring 245

7.3   Factoring Trinomials 249

7.4   Fractions 256

7.5   Multiplication and Division of Fractions 260

7.6   Addition and Subtraction of Fractions 264

       Chapter 7 Review 273

       Chapter 7 Test 274

CHAPTER 8   Fractional and Quadratic Equations 275

8.1   Fractional Equations 276

8.2   Quadratic Equations and Factoring 281

8.3   Completing the Square 289

8.4   The Quadratic Formula 293

       Chapter 8 Review 299

       Chapter 8 Test 301

CHAPTER 9   Trigonometric Functions 302

9.1   Angles, Angle Measure, and Trigonometric Functions 303

9.2   Values of the Trigonometric Functions 308

9.3   The Right Triangle 315

9.4   Trigonometric Functions of any Angle 322

9.5   Inverse Trigonometric Functions 327

9.6   Applications of Trigonometry 331

Chapter 9 Review 336

Chapter 9 Test 340

CHAPTER 10   Vectors and Trigonometric Functions 341

10.1 Introduction to Vectors 342

10.2 Adding and Subtracting Vectors 350

10.3 Applications of Vectors 358

10.4 Oblique Triangles: Law of Sines 362

10.5 Oblique Triangles: Law of Cosines 369

Chapter 10 Review 373

Chapter 10 Test 375

CHAPTER 11   Graphs of Trigonometric Functions 376

11.1 Sine and Cosine Curves: Amplitude and Period 377

11.2 Sine and Cosine Curves: Displacement or Phase Shift 382

11.3 Composite Sine and Cosine Curves 385

11.4 Graphs of Other Trigonometric Functions 391

11.5 Applications of Trigonometric Graphs 395

11.6 Parametric Equations 400

11.7 Polar Coordinates 404

Chapter 11 Review 410

Chapter 11 Test 411

CHAPTER 12   Exponents and Radicals 413

12.1 Fractional Exponents 414

12.2 Laws of Radicals 418

12.3 Basic Operations with Radicals 424

12.4 Equations with Radicals 430

Chapter 12 Review 435

Chapter 12 Test 437

CHAPTER 13 Exponential and Logarithmic Functions 438

13.1 Exponential Functions 439

13.2 The Exponential Function e^x 443

13.3 Logarithmic Functions 449

13.4 Properties of Logarithms 453

13.5 Exponential and Logarithmic Equations 457

13.6 Graphs Using Semilogarithmic and Logarithmic Paper 463

Chapter 13 Review 467

Chapter 13 Test 469

CHAPTER 14 Complex Numbers 471

14.1 Imaginary and Complex Numbers 472

14.2 Operations with Complex Numbers 477

14.3 Graphing Complex Numbers; Polar Form of a Complex Number 486

14.4 Exponential Form of a Complex Number 492

14.5 Operations in Polar Form; DeMoivre's Formula 497

14.6 Complex Numbers in AC Circuits 503

Chapter 14 Review 509

Chapter 14 Test 510

CHAPTER 15 An Introduction to Plane Analytic Geometry 511

15.1 Basic Definitions and Straight Lines 512

15.2 The Circle 519

15.3 The Parabola 525

15.4 The Ellipse 532

15.5 The Hyperbola 538

15.6 Translation of Axes 543

15.7 Rotation of Axes; the General Second-Degree Equation 548

15.8 Conic Sections in Polar Coordinates 553

     Chapter 15 Review 557

     Chapter 15 Test 559

CHAPTER 16 Systems of Equations and Inequalities 560

16.1 Solutions of Nonlinear Systems of Equations 561

16.2 Properties of Inequalities; Linear Inequalities 566

16.3 Nonlinear Inequalities 572

16.4 Inequalities in Two Variables 577

16.5 Systems of Inequalities; Linear Programming 580

     Chapter 16 Review 586

     Chapter 16 Test 588

CHAPTER 17 Matrices 589

17.1 Matrices 590

17.2 Multiplication of Matrices 598

17.3 Inverses of Matrices 604

17.4 Matrices and Linear Equations 609

     Chapter 17 Review 614

     Chapter 17 Test 616

CHAPTER 18 Higher Degree Equations 617

18.1 The Remainder and Factor Theorems 618

18.2 Roots of an Equation 624

18.3 Rational Roots 628

18.4 Irrational Roots 636

18.5 Rational Functions 640

Chapter 18 Review 646

Chapter 18 Test 647

CHAPTER 19 Sequences, Series, and the Binomial Formula 648

19.1 Sequences 649

19.2 Arithmetic and Geometric Sequences 652

19.3 Series...658

19.4 Infinite Geometric Series 665

19.5 The Binomial Theorem 670

Chapter 19 Review 677

Chapter 19 Test 679

CHAPTER 20 Trigonometric Formulas, Identities, and Equations 680

20.1 Basic Identities..681

20.2 The Sum and Difference Identities 685

20.3 The Double- and Half-Angle Identities 691

20.4 Trigonometric Equations 699

Chapter 20 Review 704

Chapter 20 Test 705

CHAPTER 21 Statistics and Empirical Methods 707

21.1 Probability...708

21.2 Measures of Central Tendency 712

21.3 Measures of Dispersion 720

21.4 Fitting a Line to Data 725

21.5 Fitting Nonlinear Curves to Data......................................731

Chapter 21 Review 736

Chapter 21 Test 737

APPENDIX A The Electronic Hand-Held Calculator 739

    A.1   Introduction...739

    A.2   Basic Operations with a Calculator 741

    A.3   Some Special Calculator Keys.............................747

    A.4   Graphing Calculators and Computer-Aided Graphing 750

APPENDIX B The Metric System 752

ANSWERS TO ODD-NUMBERED EXERCISES 762

COMPUTER PROGRAMS 818

INDEX OF APPLICATIONS 840

INDEX 842

Preface

Technical Mathematics is a student-oriented textbook, designed to be easily read and understood by students who are preparing for technical or scientific careers. Mathematics can be a difficult subject to read or study, especially when the math terms become tangled with the technical terms used in the applications. This text offers clear, readable developments of the math concepts. But learning mathematics takes more than reading—it requires extensive practice. Therefore, *Technical Mathematics* provides plentiful opportunities to practice problem solving. Together, these two essential elements have formed our constant focus in the development of this text: thorough, uncomplicated discussions followed by examples and problem sets that draw on real-world applications of math

Organization and Approach

The first three chapters of this text form a thorough introduction to the real number system, algebra, and geometry. Much of the material in these chapters might be familiar to your students and can be covered quickly or skipped entirely. The remaining chapters focus on the precalculus areas of mathematics through trigonometry, analytic geometry, and introductory statistics.

This text offers an integrated presentation of algebra, geometry, trigonometry, and analytic geometry. Practical applications of mathematics in the technical and scientific areas are provided throughout the text. Students also receive ample opportunities to solve problems using scientific calculators, graphing calculators, and microcomputers.

This textbook strives to make the reading and comprehension of its contents easier for students. The uncomplicated writing style makes difficult discussions easier to grasp. Each important concept is supported by numerous examples, applications, and practice problems

Help for Teaching and Learning

Well-written, carefully illustrated discussions form the foundation of any textbook; but your students will form their math skills with *practice*. We have developed a package of learning features that will inspire students to practice hard, and this will make their efforts, and yours, effective.

Problem Solving

Students often have trouble assessing a real situation and sorting out the pertinent information. Our approach to problem solving encourages students to visualize problems, organize information, and develop their intuitive abilities. The applications, examples, and problems are intended to teach students how to interpret real-world situations. Many problems are accompanied by illustrations and photos intended to guide students in analyzing information.

Problem solving is formally introduced in Sections 2.4 and 2.5, and is developed throughout the book. Hints, Notes, and Cautions are offered to hone problem-solving techniques, and many boxed features include guidelines for solving certain types of problems.

Visualizing Mathematics

For students to develop good problem-solving skills, they must learn to visualize mathematical problems. This text offers a unique feature that will help students to see the important information in a problem. Each chapter opens with a photograph that demonstrates mathematics at work in the real world. Later in the chapter, this photograph is analyzed in an example and used to solve a problem based on the picture. A sketch placed next to the photo extracts the important mathematical elements. Where possible, this sketch is imposed over the photograph to show students how to cull the important elements from a real situation, and how to create a sketch that will help to solve a problem.

Using Technology in Problem Solving

This text emphasizes the wise use of technologies—scientific calculators, graphing calculators, and microcomputers—because they can take the drudgery out of repetitive computations and can help students learn faster. Reducing the drudgery can help students focus on the important aspects of the operations they are performing. However, students must be aware that not all solutions are aided by technology. They must also learn how to determine whether a calculator or computer has given a correct answer, a reasonable answer, or a wrong answer. Rather than indicate that only certain problems should be solved with a calculator or computer, the author has assumed that any problem can be solved with the help of one of these tools. Calculator keystroke sequences appear frequently throughout the text to help students learn to use these important tools efficiently. Some examples and exercises may be completed using a graphing calculator or a computer graphing program. These instances are indicated by the placement of a graphing calculator

icon . Problems that require students to do some computer programming are

indicated by a computer icon . Answers to these exercises are given in GW-

BASIC. The *Computer Programs* section provides the debugged programs, along with additional problems to work using these programs. The use of computer programs throughout the book also allows students to explore what can and does happen when small changes are made in a problem or in generating recursive answers. For those students who do not wish to write the programs, a disk is available that contains all the programs created in the exercises.

Examples

The book contains frequent, realistic examples that show how to apply mathematical concepts. Examples cover both routine mathematical manipulations and applications. Approximately 20 percent of these examples have been selected from trade, vocational, technical, and industrial areas to demonstrate how mathematics is applied to all fields of technology.

Exercises

Ample practice is provided in more than 5,200 exercises, many of which show the great variety of technical and scientific applications for math.

Methods for solving most of the exercises can be found by reviewing the worked examples. Answers to the odd-numbered questions appear in the back of the book. A separate solutions manual for all problems is available to instructors, with a manual of selected solutions available to students.

Applications

Most technical math courses are designed to prepare students to solve applied problems in their chosen technical fields. For this reason, we have taken care to present many exercises and examples that show how to apply math in a variety of technical fields. These applications are highlighted in the text with special headings that indicate the applied field. The text also offers an index of the applications by field.

Pedagogical Features

Throughout the book, you will find rules, formulas, guidelines, and hints that are boxed for easy identification. The boxes make the material easy to locate, and also make the book a valuable reference after the course is finished.

Five icons appear frequently in the margins:

 The graphing calculator icon indicates that a certain example or exercise may be used with a graphing calculator. However, students may prefer to use a computer graphing program.

 The computer icon indicates that a certain problem requires the student to do some computer programming.

 Many years of teaching experience have shown the author where students make common errors. These places are indicated by the "CAUTION" icon. This icon not only points out the potential errors, but how to avoid them.

 The "HINT" icon points out a valuable technique or learning hint that students can use to solve problems.

≡ **Note** The "NOTE" icon refreshes students' memories about a concept, or points out interesting or unusual ideas. These notes highlight ideas that might easily be overlooked, alternative ways to solve a problem, or a possible shortcut.

Chapter Review

Each chapter concludes with a list of the important terms covered in the chapter, a generous set of review exercises, and a chapter test. The review exercises include both routine computations and applications. Many of these exercises draw on more than one section of the chapter, requiring students to think harder and synthesize their learning. The chapter test gives students an idea of the types of questions they might expect for a fifty-minute exam.

Supplements

We have carried our developmental themes of straightforward discussions and plenty of practice into our supplements package. This flexible package will allow you to emphasize the teaching strategies your course demands, and give your students even more problem-solving experience.

The *Solutions Manual* contains fully worked solutions for every text problem. The *Crossover Guide* will help you make the switch from your current text by showing you exactly how the topic coverage in *Technical Mathematics* aligns with your text. The *Student Solutions Manual* contains fully worked solutions for every other odd problem. All of these solutions have been thoroughly examined for accuracy.

A *Computerized Test Bank* is available on 3 1/2-inch disks. Illustrations are included on the disks for maximum flexibility. A printed version is also offered.

Computer Software is available in two forms. A disk is available containing all of the debugged computer programs that appear in the *Computer Programs* section. In addition, a tutorial program is available to help teach problem-solving skills.

A package of *two-color transparencies* and *transparency masters* is available.

Acknowledgments

Anyone who writes a book enjoys the direct and indirect assistance of many people. I would especially like to thank the late Jeff Jones, the first accuracy checker for this text. He worked every example exercise and provided valuable editorial suggestions. Rosanna Templin and Jessica Daugherty typed most of the manuscript and solutions. Melissa Savage also helped to type and scan some material. Not only did they do this with speed and efficiency, but their experience with this difficult material made my job so much easier. Special thanks go to Alan Herweyer for preparing the *Solutions Manual,* and to Darryl Sheppard for debugging the computer programs. Miss Arden Brugger of the Jacksonville Public Library provided useful data on the Dame Point Freeway Bridge, shown on the cover.

The author and publisher gratefully acknowledge the assistance of many talented reviewers and contributors. Allan Hymers, of Scarborough, Ontario, contributed several hundred problems and applications. The following reviewers also submitted applications for the text:

Calvin Holt, Franklin, Va.

Roberta Laine, West Bend, Wis.

Frank Weeks, Gresham, Oreg.

Natalie Woodrow, Sweetwater, Tex.

The following instructors provided valuable criticism during several review stages:

Haya Adner, Queensborough Community College, Bayside, N.Y.
Doris Bratcher, Northwest Iowa Technical College, Iowa
Granville Brown, Grand Rapids Community College, Grand Rapids, Mich.
Ronald Bukowski, Johnson Technical Institute, Scranton, Pa.
Bill Ferguson, Columbus State Community College, Columbus, Ohio
Marion Graziano, Montgomery County Community College, Blue Bell, Pa.
Henry Hosek, Purdue University, Calumet, Ind.
David Hutchinson, Mohawk College of Technical and Applied Arts, Hamilton, Ont.
Allan Hymers, Centennial College, Scarborough, Ont.
Stanley Koper, Triangle Tech, Belle Vernon, Pa.
Larry Lance, Columbus State Community College, Columbus, Ohio
Robert Opel, Waukesha County Technical College, Waukesha, Wis.
Steven B. Ottman, Southeast Community College, Nebraska
Gordon Schlafmann, Western Wisconsin Technical College
Cathy Vollstedt, North Central Technical College, Schofield, Wis.
Anita Walker, Asheville, N.C.

During several rigorous accuracy checks, the following professionals examined the text and solutions in detail:

C. Lea Campbell, Lamar University, Port Arthur, Tex.
Gerry East, Tulane University, New Orleans, La.
Diane Ferris, Portland Community College, Portland, Oreg.
Alan Herweyer, Chattanooga State Technical College, Chattanooga, Tenn.
Barbara Karp, Newcastle, Ind.
Amy Relyea, Miami University, Oxford, Ohio
Gordon Schlafmann, Western Wisconsin Technical College
Randall Sowell, Central Virginia Community College, Lynchburg, Va.
George Wyatt, Rensselaer Polytechnic Institute, Troy, N.Y.
John Wisthoff, Anne Arundel Community College, Md.

Several enthusiastic instructors generously agreed to test the final pages in their classrooms:

C. Lea Campbell, Lamar University, Port Arthur, Tex.
Bill Ferguson, Columbus State Community College, Columbus, Ohio
Marion Graziano, Montgomery County Community College, Blue Bell, Pa.
Henry Hosek, Purdue University, Calumet, Ind.
Robert Opel, Waukesha County Technical College, Waukesha, Wis.
Arthur J. Varie, Trumbull County JVS, Warren, Ohio

No one could complete an effort such as this without the support, encouragement, and tolerance of an understanding family. I want to thank my wife, Dr. Marla Peterson, and our son, Matt, for giving me the time and understanding to complete this work.

John C. Peterson

1

The Real Number System

In order to work mathematics, we need to be able to add, subtract, multiply, and divide real numbers. In Section 1.3, we will see how to subtract real numbers to determine the thickness of this pipe.

Courtesy of Ruby Gold

Every technician must be able to use mathematics. The science of mathematics is a universal language that helps technicians communicate and do their work. Advances in technology require that technicians know more and more mathematics. Technicians must be able to solve mathematical problems quickly and accurately.

The mathematics that are developed in this text provide the foundation for work in almost any field of technology. Perhaps, as important as the actual mathematics is the ability to use calculators and computers to help work mathematics problems. As you use this book you should learn how to use calculators and computers to help you with mathematics. Basic calculator instructions are in Appendix A.

≡ 1.1
SOME SETS OF NUMBERS

Fundamental to our work in mathematics is an understanding of the different types of numbers that we will use. One of the first sets of numbers that you learned was the **natural numbers**. These numbers are 1, 2, 3, 4, and so forth. These were also the first numbers invented by people. Later, people discovered a need for the number zero (0). When 0 was added to the natural numbers, a new set of numbers, called the **whole numbers**, was formed.

Integers

People continued to use the whole numbers until they discovered that they were not able to solve some problems with the set of whole numbers. Problems such as $3 - 5$ required that a new set of numbers be invented: the **integers**. The integers included the natural numbers, zero, and a negative value for each of the natural numbers. The set of integers consists of the numbers

$$\ldots, -4, -3, -2, -1, 0, 1, 2, 3, \ldots$$

The negative numbers are called the **negative integers** and the natural numbers are called the **positive integers**. The number 0 is neither positive nor negative. Positive numbers can be written with a plus (+) sign in front as +1, +2, +3, ... but this is not necessary.

≡ Note

Three points of ellipsis (...) in a series of numbers means that some numbers have been omitted. They are placed wherever the missing numbers would be.

EXAMPLE 1.1

The numbers 1, 14, 973, and 8,436 are all positive integers. They are also natural numbers.

The numbers -2, -18, and $-4,392$ are all negative integers, but not natural numbers.

Rational Numbers

The rational numbers were created to indicate when a part of something was needed. The rational numbers are used to represent division of one integer by a positive or negative integer. There are both positive and negative rational numbers. Rational numbers can be expressed in more than one way. The numbers $\frac{4}{2}$, $\frac{6}{3}$, $\frac{2}{1}$, and $\frac{142}{71}$ are all different ways of writing 2.

EXAMPLE 1.2

Some positive rational numbers are 14, $\frac{7}{3}$, $\frac{2}{5}$, $\frac{271}{18}$, $\frac{4}{2}$, and 973.

Negative rational numbers include $-\frac{2}{3}$, -5, $-\frac{9}{7}$, $-\frac{5}{23}$ and $\frac{-14}{2}$.

Zero (0) is a rational number that is neither positive nor negative.

Irrational Numbers

At one time people thought that everything could be expressed as a rational number and that every problem would have a rational number as an answer. In fact, there is a story that the first person who proved that all numbers are not rational was executed because of this proof. Perhaps because it is not rational to execute a person just because they discovered a new group of numbers, these new numbers were called the **irrational numbers**. Some of the irrational numbers are $\sqrt{2}$, $\sqrt{5}$, π, and $-\sqrt{17}$. While we will not prove it here, it is not possible to represent any of these numbers by the division of two integers.

Real Numbers

When the rational numbers are combined with the irrational numbers, a new group of numbers is formed—the **real numbers**. Most of our work in this text uses the real numbers. Later we will introduce a new group of numbers, called the imaginary numbers, and combine the real and imaginary numbers into another group, the complex numbers.

At times it is convenient to represent the real numbers on a line called the **number line**. The usual number line is a horizontal line that has been marked in equally spaced intervals. One of these marks is called the **origin** and is indicated by the number zero (0). The marks to the right are labeled using the positive integers. The negative integers are used to designate the marks to the left of the origin. A typical number line is shown in Figure 1.1.

$-6\ -5\ -4\ -3\ -2\ -1\ \ 0\ \ 1\ \ 2\ \ 3\ \ 4\ \ 5\ \ 6$

FIGURE 1.1

The other real numbers are located between the integers.

The fact that a number line does not have to be horizontal can be demonstrated by a thermometer. Most wall thermometers hang vertically with the positive numbers above the negative numbers and this represents a vertical number line. But the fact that some thermometers are circular shows that nothing requires a number line to be a straight line.

Decimals

Each real number can be represented by a decimal number. The rational numbers are represented by repeating decimals. Irrational numbers are represented by non-repeating decimals. The decimal representations allow us to position these numbers accurately on the number line.

EXAMPLE 1.3

Repeating decimals include $\frac{1}{2} = 0.5000\ldots$ (repeats 0), $\frac{13}{4} = 3.2500\ldots$ (repeats 0), $-\frac{4}{3} = -1.333\ldots$ (repeats 3), and $\frac{-41}{11} = -3.727272\ldots$ (repeats 72). These are sometimes represented with a bar over the repeating digits. Thus $\frac{1}{2} = 0.5\overline{0}$, $\frac{13}{4} = 3.25\overline{0}$, $-\frac{4}{3} = -1.\overline{3}$, and $\frac{-41}{11} = -3.\overline{72}$. Numbers that repeat zero are called **terminating decimals** and are usually written without indicating the repeating zero. When this is done, we have $\frac{1}{2} = 0.5$ and $\frac{13}{4} = 3.25$.

Irrational numbers are represented by nonrepeating decimals. For example, here are the decimal representations of four irrational numbers:

$$\sqrt{2} = 1.414213\ldots \qquad\qquad \sqrt{7} = 2.6457513\ldots$$

$$\pi = 3.14159265359\ldots \qquad -\frac{\sqrt{15}}{2} = -1.93649167\ldots$$

≡ **Note**

FIGURE 1.2

Here, the three points of ellipsis are placed at the end of a number to indicate that not all of the digits have been shown.

The location of each of the numbers in Example 1.3 is shown in Figure 1.2.

Absolute Values

Knowing the different kinds of numbers can be useful. However, what is more important is how the numbers relate to each other and how we can use them. One of the basic ideas is the distance of a number from zero or, on a number line, from the origin. The number 3 is three units from zero. So is the number -3. The number $-\frac{8}{5}$ is $\frac{8}{5}$ units from zero. The distance of a number from zero is called the **absolute value** of the number. The absolute value of a number is indicated by placing the number between vertical bars, $|\ \ |$.

EXAMPLE 1.4

The absolute value of 4 is 4, the absolute value of -9 is 9. These, and other absolute values, are written as

$$|4| = 4$$
$$|-9| = 9$$
$$\left|\frac{4}{3}\right| = \frac{4}{3}$$
$$\left|-\frac{18}{7}\right| = \frac{18}{7}$$
$$\left|-\sqrt{7}\right| = \sqrt{7}$$
$$|0| = 0$$

Inequalities

Some other important ideas include how numbers relate to each other. The relationship most often used is equality, or when two different numerals represent the same amount. Since $\frac{6}{3}$ and 2 both represent the same amount, we say they are equal, use the equal sign ($=$), and indicate this by writing $\frac{6}{3} = 2$.

Other relationships are concerned with numbers that are **not equal to** each other. To indicate that two numbers do not represent the same amount we use the "not equal to" symbol, $\neq$. Thus, $3 \neq 4$ and $-5 \neq 17$ are two examples showing how to use the "not equal to" symbol.

The $\neq$ sign is one example of a **sign of inequality**. The signs of inequality are often used to indicate which of two numbers is the larger. A number is larger than another if it is located farther to the right on the number line. In this case we say that the number that is farther to the right on the number line **is greater than** the other number. The symbol $>$ is used to indicate that one number is greater than another. Another definition says that when the second number is subtracted from the first and the answer is positive, then the first number is greater than the second.

EXAMPLE 1.5

$7 > 5$, because $7 - 5 = 2$, a positive number
$4 > -2$, because $4 - (-2) = 6$, a positive number
$-3 > -10$, because $-3 - (-10) = 7$, a positive number

If the first number is to the left of the second number on the number line, or the answer is negative when the second number is subtracted from the first, then the first number **is less than** the second. We use the symbol $<$.

EXAMPLE 1.6

$5 < 7$, because $5 - 7 = -2$, a negative number
$-2 < 4$, because $-2 - 4 = -6$, a negative number
$-10 < -3$, because $-10 - (-3) = -7$, a negative number

There are two other inequality symbols. The symbol $\leq$ means "is less than or equal to" and the symbol $\geq$ indicates "is greater than or equal to."

Reciprocals

Every number, except zero, has a **reciprocal**. The reciprocal of a number is equal to 1 divided by that number. If the number is a fraction, then the reciprocal is 1 divided by the fraction or $\dfrac{1}{\text{fraction}}$. Thus, the reciprocal of $\dfrac{4}{3}$ is $\dfrac{1}{\frac{4}{3}} = \dfrac{3}{4}$.

EXAMPLE 1.7

The reciprocal of 9 is $\frac{1}{9}$.

The reciprocal of $\frac{2}{3}$ is $\dfrac{1}{\frac{2}{3}} = \dfrac{3}{2}$.

The reciprocal of $\sqrt{7}$ is $\dfrac{1}{\sqrt{7}}$.

The reciprocal of $\dfrac{-5}{2}$ is $\dfrac{1}{\frac{-5}{2}} = \dfrac{-2}{5}$.

≡ **Note**   The product of a number and its reciprocal is 1.

In this section we have introduced some of the different sets of numbers and ways to indicate the relationship between two numbers. In the next section we will review the basic operations on, and the basic laws of, the real numbers.

Exercise Set 1.1

In Exercises 1–4, indicate all the sets of numbers to which each number belongs. Remember, a number can belong to more than one set of numbers.

1. 15

2. $\frac{-2}{3}$

3. $\frac{-\sqrt{7}}{8}$

4. 0

Find the absolute value of the numbers in Exercises 5–8.

5. 15

6. $\frac{-2}{3}$

7. $\frac{-\sqrt{7}}{8}$

8. 0

Locate each of the numbers in Exercises 9–12 on a number line.

9. $\frac{4}{7}$

10. 2.5

11. $-\frac{8}{3}$

12. $\frac{-\pi}{3}$

In Exercises 13–20, insert the correct sign of inequality ($<$ or $>$) between the given pairs of numbers.

13. $2, 3$

14. $5, 3$

15. $-4, 7$

16. $9, -7$

17. $-3, -8$

18. $-15, -7$

19. $\frac{-2}{3}, -\frac{1}{2}$

20. $0.7, 0.5$

Find the reciprocals of the numbers in Exercises 21–24.

21. -5

22. $\frac{1}{2}$

23. $\frac{17}{3}$

24. $\frac{-2}{\pi}$

Solve Exercises 25–29.

25. List the following numbers in numerical order from smallest to largest: $-5, -\frac{2}{3}, |-8|, \frac{16}{3}, -|4|, \frac{-1}{3}$

26. List the following numbers in numerical order from smallest to largest: $\pi, \frac{22}{7}, -\sqrt{2}, -\sqrt{7}, \sqrt{7} - \sqrt{2}, \frac{7}{5}, -\left|\frac{7}{4}\right|$

27. *Automotive technician* Front end shims come in $\frac{1}{32}$-, $\frac{1}{16}$-, and $\frac{1}{8}$-inch thicknesses.
  (a) Which is the smallest shim?
  (b) Which is the largest shim?
  (c) What is the decimal size of the $\frac{1}{16}$-inch shim?

28. *Automotive technician* The service manager records how long each technician has worked on a job in hours written as a decimal. If you spent $\frac{3}{4}$ h to change a front engine mount, $1\frac{1}{2}$ h to replace the engine oil pan, and $2\frac{1}{4}$ h to replace the clutch, how would the service manager record these times?

29. *Electricity* The voltage across an element with respect to the ground is -19.4 V initially and then it changes to -16.8 V. Determine the absolute value of the change in the voltage.

≡ 1.2
BASIC LAWS OF REAL NUMBERS

Knowing the different kinds of numbers is important. But, it is even more important that you can use these numbers. In order to use them, there are some basic laws of the real numbers that you should know. Two of these laws apply to addition and multiplication.

Commutative Laws

The first laws state that it makes no difference in which order two numbers are added or multiplied. These are called the **commutative laws**. For example, the commutative law for addition would state that $5 + 2 = 2 + 5$ and also $\frac{17}{4} + \frac{-3}{7} = \frac{-3}{7} + \frac{17}{4}$. In the same way, the commutative law for multiplication guarantees that $4 \times 8 = 8 \times 4$ and also that $\frac{-\pi}{3} \times \frac{6}{\sqrt{2}} = \frac{6}{\sqrt{2}} \times \frac{-\pi}{3}$.

Commutative Laws for Addition and Multiplication

Symbolically, the commutative laws would be written as follows:
Commutative law for addition: $a + b = b + a$.
Commutative law for multiplication: $a \times b = b \times a$.

 Caution Subtraction and division are **not** commutative. Thus,

$$7 - 4 \neq 4 - 7 \qquad \text{and} \qquad 12 \div 3 \neq 3 \div 12$$

Associative Laws

The second laws state that when you have a group of three numbers it makes no difference which two are added or multiplied first. These are called the **associative laws**. For example, the associative law for addition would state that $(4 + 2) + 8 = 4 + (2 + 8)$ or $\left(\frac{6}{5} + \frac{-2}{3}\right) + \sqrt{3} = \frac{6}{5} + \left(\frac{-2}{3} + \sqrt{3}\right)$. Similarly, the associative law for multiplication guarantees that $(6 \times 7) \times 2 = 6 \times (7 \times 2)$ and $\left(\frac{\pi}{4} \times \sqrt{5}\right) \times \frac{1}{\sqrt{7}} = \frac{\pi}{4} \times \left(\sqrt{5} \times \frac{1}{\sqrt{7}}\right)$

Associative Laws for Addition and Multiplication

Symbolically, the associative laws would be written as follows:
Associative law for addition: $(a + b) + c = a + (b + c)$.
Associative law for multiplication: $(ab)c = a(bc)$.

 Caution Subtraction and division are **not** associative. Thus,

$$15 - (7 - 3) \neq (15 - 7) - 3 \qquad \text{and} \qquad 18 \div (6 \div 3) \neq (18 \div 6) \div 3$$

Distributive Law

A third law combines addition and multiplication and states that multiplication distributes over addition. This law, called the **distributive law**, means that if you have a problem such as $12(13+7)$ you could evaluate it two ways. One way would be to first add the numbers inside the parentheses and get $12(13+7) = 12 \times 20 = 240$. The other method would first distribute the multiplication and then add those products. For this same problem, you would get $12(13+7) = (12 \times 13) + (12 \times 7) = 156 + 84 = 240$.

Distributive Law

Symbolically, the distributive law is written as

$$a(b+c) = ab + ac.$$

Order of Operations

Now, let's consider the order of operations. The order of operations is very important in mathematics. Mathematicians have made some basic agreements that some operations are to be performed before others. By these agreements, operations such as addition, subtraction, multiplication, division, and raising to powers are to be performed in the following order: (Remember, not all operations are in every problem.)

Order of Operations

Perform the operations in a problem in the following order.

1. Operations inside parentheses or above or below a fraction bar. Always start with the innermost parentheses and work outward.
2. Raising to a power.
3. Multiplications and divisions in the order in which they appear from left to right.
4. Additions and subtractions in the order in which they appear from left to right.

Hint

Some people use the acronym "**P**lease **E**xcuse **M**y **D**ear **A**unt **S**ally" to help remember the order of operations. Here the **P** in "Please" stands for parentheses; the **E** for exponents (raise to a power), **M** and **D** for multiplication and division, and the **A** in "Aunt" and **S** in "Sally" for addition and subtraction.

Computers and most scientific calculators are programmed to perform the operations according to the previous rules. The logic used by these calculators and computers is called **algebraic logic** or the **algebraic operating system**.

Checking: Does Your Calculator Use Algebraic Logic?

Work this problem on your calculator: $3 + 5 \times 7$. If your calculator gives the answer 38, then it uses algebraic logic.

If your calculator gives the answer 56, then it does not use algebraic logic.

If your calculator gives an answer other than 38 or 56, then you probably made a mistake. Try again.

EXAMPLE 1.8

We will consider the problem $6 + 10 \div 2$ in two different ways. Following the rules, we should do the division first:

$$6 + 10 \div 2 = 6 + 5 = 11$$

Notice that $6 + 10 \div 2$ is performed as if it were $6 + (10 \div 2)$. If you wanted to indicate that the addition should be performed first, then parentheses would have to be added:

$$(6 + 10) \div 2 = 16 \div 2 = 8$$

As you can see, the answers 8 and 11 are not the same. This shows the importance of performing the operations in the correct order.

≡ **Note** $(6 + 10) \div 2$ is sometimes written as $\frac{6+10}{2}$. Here we perform the operation above the fraction bar (order of operation #1) before we perform the division. Thus, $\frac{6+10}{2} = \frac{16}{2} = 8$.

EXAMPLE 1.9

Solve $8 + 9 \times 2 + 16 - 12 \div 4$

Solution Orders of operations #1 and #2 do not apply because there are no parentheses, fractions, or exponents. According to order of operation #3, we should perform the multiplication and division in order from left to right. Thus, if we use parentheses to indicate what to do first, we get

$$8 + 9 \times 2 + 16 - 12 \div 4 = 8 + (9 \times 2) + 16 - (12 \div 4)$$
$$= 8 + 18 + 16 - 3$$
$$= 39$$

EXAMPLE 1.10

Solve $6 + 12 \div 4 \times 5 - 4 \times 3$

Solution We begin by performing the multiplication and division operations in order from left to right. First we perform $12 \div 4$ and then multiply this answer by

EXAMPLE 1.10 (Cont.)

5 as shown.

$$6 + 12 \div 4 \times 5 - 4 \times 3 = 6 + 3 \times 5 - 4 \times 3$$
$$= 6 + 15 - 4 \times 3$$
$$= 6 + 15 - 12$$
$$= 21 - 12$$
$$= 9$$

EXAMPLE 1.11

Solve $28 - (26 - (3 - (4 - 3)))$

Solution We begin with the innermost set of parentheses and perform the operations inside of them. So, we begin by working $4 - 3$.

$$28 - (26 - (3 - (4 - 3))) = 28 - (26 - (3 - 1))$$
$$= 28 - (26 - 2)$$
$$= 28 - 24$$
$$= 4$$

Notice that at each step we performed the operations in the innermost parentheses.

Sometimes different grouping symbols are used to help a person see symbols that go together. Example 1.11 might have been written as follows

$$28 - \{26 - [3 - (4 - 3)]\}$$

You would first work the problem inside the parentheses ( ), then inside the brackets [ ], and finally, inside the braces { }.

Identity Elements

Two numbers that are very important are the **identity elements**. There is an identity element for addition and one for multiplication. The identity element for addition is zero (0) and for multiplication is one (1). When you add the identity element for addition to a number, the answer does not change. So, $4 + 0 = 4$; $\frac{-3}{8} + 0 = \frac{-3}{8}$; and $0 + \frac{\sqrt{2}}{5} = \frac{\sqrt{2}}{5}$. In the same way, when you multiply a number by the identity element for multiplication, the answer does not change. For example, $17 \times 1 = 17$; $1 \times \sqrt{5} = \sqrt{5}$; and $\frac{-9}{7} \times 1 = \frac{-9}{7}$.

Identity Elements

Symbolically, these are written as follows:

Identity element for addition is 0: $a + 0 = a$.

Identity element for multiplication is 1: $a \times 1 = a$.

Inverses

Now that we have identity elements for addition and multiplication we can determine the inverses of each number. Every number, except zero, has two inverses—an inverse for addition and an inverse for multiplication. Zero has only an inverse for addition.

The inverse for addition, or **additive inverse**, of a number is the number that gives a sum of zero when added to the original number. Thus, the additive inverse of 5 is -5 because $5 + (-5) = 0$. (This also means that 5 is the additive inverse of -5.) The additive inverse of $\frac{-8}{7}$ is $\frac{8}{7}$ because $\frac{-8}{7} + \frac{8}{7} = 0$ and the additive inverse of $-\sqrt{5}$ is $\sqrt{5}$ because $-\sqrt{5} + \sqrt{5} = 0$.

≡ Note The additive inverse of a number b is $-b$ because $b + -b = 0$. If b is a negative number, such as -8, then $-b$ is positive, $-(-8) = 8$. The additive inverse of zero is zero because $0 + 0 = 0$.

💡 Hint To enter a negative number on a calculator you need to press a special key. On some calculators it is the $\boxed{+/-}$ or $\boxed{\text{CHS}}$ key and it is pressed *after* the number is entered. Thus, -25 would be entered as $25 \boxed{+/-}$. Some calculators use a $\boxed{(-)}$ key and it is pressed *before* the number is entered. Here, -35 would be entered as $\boxed{(-)}$ 35.

The other inverse that each number (except zero) has is an inverse for multiplication. The **multiplicative inverse** of a number is the number that will give a product of 1 when multiplied by the original number. So, the multiplicative inverse of 5 is $\frac{1}{5}$ because $5 \times \frac{1}{5} = 1$. (This means that the multiplicative inverse of $\frac{1}{5}$ is 5.) The multiplicative inverse of $\frac{-3}{8}$ is $\frac{8}{-3}$ because $\frac{-3}{8} \times \frac{8}{-3} = 1$ and the multiplicative inverse of $-\frac{\sqrt{2}}{3}$ is $-\frac{3}{\sqrt{2}}$ because $-\frac{\sqrt{2}}{3} \times \left(-\frac{3}{\sqrt{2}}\right) = 1$

The multiplicative inverse of a number b is $\frac{1}{b}$ because $b \times \frac{1}{b} = 1$ provided that $b \neq 0$. Notice that a number and its multiplicative inverse have the same sign. Zero does not have a multiplicative inverse because the product of 0 and any other number is 0. The fact that $b \neq 0$ means that the denominator of a fraction, and the divisor in a division problem, cannot be 0. Since $\frac{1}{0}$ is not a number, $0 \times \frac{1}{0}$ is not defined. In particular, $0 \times \frac{1}{0} \neq 0$ and $0 \times \frac{1}{0} \neq 1$.

≡ Note You might have already noticed that the multiplicative inverse and the reciprocal of a number are the same.

Inverse Elements

Symbolically, the inverse elements are written as:

Inverse element for addition of the number b is $-b$.

Inverse element for multiplication, or reciprocal, of the number b, $b \neq 0$, is $\frac{1}{b}$.

There are two other properties of zero that are important to remember. The product of any number multiplied by zero is zero. The quotient of zero divided by any number (except zero) is also zero. For example, $7 \times 0 = 0$ and $0 \div 9 = 0$. In symbols we would write these as follows:

Properties of 0

$b \times 0 = 0$

$0 \div b = 0$ (or $\dfrac{0}{b} = 0$) if $b \neq 0$

Exercise Set 1.2

In Exercises 1–12, determine which of the basic laws of real numbers is being used.

1. $4 + 3 = 3 + 4$

2. $8(4 \cdot 6) = (8 \cdot 4)6$

3. $9 \cdot 7 = 7 \cdot 9$

4. $4(3 + 5) = 4 \cdot 3 + 4 \cdot 5$

5. $(9 + 3) + 6 = 9 + (3 + 6)$

6. $(\pi + \sqrt{2})\sqrt{5} = \pi\sqrt{5} + \sqrt{2}\sqrt{5}$

7. $1 \times 81 = 81$

8. $\dfrac{3}{16} \times \dfrac{16}{3} = 1$

9. $-5 + 5 = 0$

10. $19 - 5 = -5 + 19$

11. $(\sqrt{3} \cdot \sqrt{4})\sqrt{5} = \sqrt{3}(\sqrt{4} \cdot \sqrt{5})$

12. $16 = 0 + 16$

Name the additive inverse of each of the numbers in Exercises 13–16.

13. 91

14. -8

15. $\sqrt{2}$

16. $\dfrac{1}{3}$

Name the multiplicative inverse of each of the numbers in Exercises 17–20.

17. $\dfrac{1}{2}$

18. -5

19. $\dfrac{\sqrt{2}}{2}$

20. $-\dfrac{3}{7}$

Use the correct order of operations to solve Exercises 21–30.

21. $16 - 8 \div 4$

22. $16 \div 8 + 2$

23. $24 + 3 - 10 \div 5 \times 8 + 2$

24. $13 \times 7 - 26 \div 5 + 5$

25. $(7 - 2) - (3 + 8 - 7)$

26. $7 - 2 - 3 + 8 - 7$

27. $7 \times 3 + 5 \times 2$

28. $6 \times 4 - 3 \times 5$

29. $\{-[5 - (8 - 4) + (3 - 7)] - (4 - 2)\}$

30. $(14 + 3(8 - 6) + 4(9 - 5))$

≡ 1.3
BASIC OPERATIONS WITH REAL NUMBERS

The ability to work with numbers is very important in mathematics. In this book you will be using calculators and computers to help solve problems. Yet, in many ways this makes it more important that you have good arithmetic skills. Both the calculator and the computer will only give the correct answer if you correctly tell the machine what to do. This section provides a brief review of arithmetic with real numbers. The first group of rules will use integers in the examples. Later in this section, we will examine the arithmetic of rational numbers.

Addition of Integers

Rule 1

To add two real numbers with the same signs, add the numbers and give to the sum the sign of the original numbers.

EXAMPLE 1.12

$+8 + (+7)$

Solution $8 + 7 = 15$

8 and 7 both have a + sign.
So, $+8 + (+7) = +15$.

EXAMPLE 1.13

$-9 + (-24)$

Solution $9 + 24 = 33$

9 and 24 both have a − sign.
So, $-9 + (-24) = -33$.

Rule 2

To add two real numbers with different signs, take the absolute value of both numbers, subtract the smaller absolute value from the larger, and give to the answer the sign of the number with the larger absolute value.

EXAMPLE 1.14

$-8 + (+5)$

Solution $|-8| = 8$

$|+5| = 5$

$8 > 5$, so the answer will be negative.
$8 - 5 = 3$ and so, $-8 + (+5) = -3$.

EXAMPLE 1.15

$+14 + (-9)$

Solution $|+14| = 14$

$|-9| = 9$

$14 > 9$ so the answer will be positive.

$14 - 9 = 5$ and so $+14 + (-9) = +5$.

≡ **Note**   Remember that positive numbers are often written without a + sign.

EXAMPLE 1.16

$-19 + (-18) + 23$

Solution $-19 + (-18) + 23 = -37 + 23$, because $-19 + (-18) = -37$

$= -14$

Application

EXAMPLE 1.17

When you started your car one morning, the temperature was $-16°$C. Later you hear someone say that the temperature has gone up $24°$C. What is the latest temperature?

Solution   To find the new temperature, we add the increase to the morning temperature:

$$-16 + 24 = 8$$

The new temperature is $8°$C.

Application

EXAMPLE 1.18

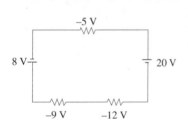

FIGURE 1.3

Figure 1.3 shows a closed circuit with two voltage sources and three resistors. The sum of voltages in such a loop must be 0. Check the voltage drops given in the figure to determine if the sum is 0.

Solution   The voltage drops in Figure 1.3 are 8 V, -5 V, 20 V, -12 V, and -9 V. The sum of the voltage drops is

$$8 + (-5) + 20 + (-12) + (-9) = 3 + 20 + (-12) + (-9)$$

$$= 23 + (-12) + (-9)$$

$$= 11 + (-9)$$

$$= 2$$

A sum of 2 V indicates an error, probably in one of the measurements.

Subtraction of Integers

> **Rule 3**
>
> To subtract one real number from another, change the sign of the number being subtracted and then add according to Rule 2.

EXAMPLE 1.19

$(-6) - (-9)$

Solution $(-6) - (-9) = (-6) + (+9)$ Change -9 to $+9$ and add.

$$= 3$$

EXAMPLE 1.20

$(-13) - (+7)$

Solution $(-13) - (+7) = (-13) + (-7)$ Change $+7$ to -7 and add.

$$= -20$$

 Caution   Do not use the $\boxed{-}$ key to enter a negative number into your calculator. The $\boxed{-}$ key is used only for subtraction.

EXAMPLE 1.21

$14 - 29$

Solution $14 - 29 = 14 + (-29)$ Change 29 to -29 and add.

$$= -15$$

Application

EXAMPLE 1.22

When performing a front-end alignment, a technician must work with the included angle, the steering axis inclination, and the camber. To find the steering axis angle, you subtract the camber angle from the included angle. If the included angle is $5°$ and the camber angle is $-2°$, what is the steering axis?

Solution   Subtracting the camber angle from the included angle, we have

$$5° - (-2°) = 5° + 2°$$

$$= 7°$$

The steering axis angle is $7°$.

Multiplication and Division of Integers

> **Rule 4**
>
> The product (or quotient) of two real numbers with the same sign is the product (or quotient) of their absolute values.

EXAMPLE 1.23

$(-8)(-9)$

Solution $(-8)(-9) = |-8| \times |-9|$

$$= 8 \times 9$$
$$= 72$$

EXAMPLE 1.24

$(-27) \div (-3)$

Solution $(-27) \div (-3) = |-27| \div |-3|$

$$= 27 \div 3$$
$$= 9$$

> **Rule 5**
>
> The product (or quotient) of two real numbers with different signs is the additive inverse of the product (or quotient) of their absolute values.

EXAMPLE 1.25

$(-8)(9)$

Solution $|-8| = 8$ $|9| = 9$ $8 \times 9 = 72$

The additive inverse of 72 is -72.
So, $(-8)(9) = -72$.

EXAMPLE 1.26

$81 \div -3$

Solution $|81| = 81$ $|-3| = 3$ $81 \div 3 = 27$

The additive inverse of 27 is -27 and so $81 \div (-3) = -27$.

Application

EXAMPLE 1.27

The total cost of a car, including finance charges, is \$9,216. This is to be repaid in 48 equal payments. How much is each payment?

EXAMPLE 1.27 (Cont.)

Solution To find the size of each payment, we need to divide the total cost by the number of payments.

$$9,216 \div 48 = 192$$

Each payment is $192.

Application

EXAMPLE 1.28

A metal plate contracts (shrinks) 0.2 mm for each degree below 60°F. If the plate is 42 mm wide at 60°F, what is its width at 35°F?

Solution The temperature change from 60°F to 35°F is $60° - 35° = 25°$F. A decrease, or shrinkage, of 0.2 mm can be written as -0.2. So, for a 25° change in temperature, the total shrinkage is

$$-0.2 \times 25 = -5.0$$

So, the plate will shrink 5 mm and, at 35°F, its width is

$$42 \text{ mm} - 5 \text{ mm} = 37 \text{ mm}$$

Arithmetic of Rational Numbers

We will now look at the arithmetic of rational numbers. Remember, the numerator of a rational number is the top number and the denominator is the bottom number. In the rational number $\frac{-5}{8}$, -5 is the numerator and 8 is the denominator.

Sometimes a rational number is written as a mixed number that combines an integer and a rational number. Examples of mixed numbers are $2\frac{1}{2}$ and $-4\frac{2}{3}$. While it is not shown, the two parts of a mixed number are being added. Thus, $2\frac{1}{2}$ means $2 + \frac{1}{2}$ and $-4\frac{2}{3}$ means $-(4 + \frac{2}{3})$.

Every rational number can be written with either one negative sign or all positive signs. If a rational number is written with more than one negative sign you can rewrite it by reducing the number of negative signs by twos.

EXAMPLE 1.29

$-\frac{-3}{-8}$ has three negative signs. You can delete any two of these and get $\frac{-3}{8}$ or $-\frac{3}{8}$ or $\frac{3}{-8}$ depending on which two negative signs you delete. All three of these are equivalent to the original number.

EXAMPLE 1.30

$-\frac{9}{-5}$ has two negative signs. If you delete both you get $\frac{9}{5}$, which is equivalent to the original number.

> **Rule 6**
>
> To add (or subtract) two rational numbers, change both denominators to the same positive integer (the **common denominator**), add (or subtract) the numerators, and place the result over the common denominator.

EXAMPLE 1.31

$\frac{2}{3} + \frac{-5}{6}$

Solution The denominators are 3 and 6. A common denominator of 3 and 6 is 6, so $\frac{2}{3} = \frac{4}{6}$. The problem then becomes $\frac{2}{3} + \frac{-5}{6} = \frac{4}{6} + \frac{-5}{6} = \frac{4+(-5)}{6} = \frac{-1}{6}$. So, $\frac{2}{3} + \frac{-5}{6} = \frac{-1}{6}$.

EXAMPLE 1.32

$\frac{7}{-8} + \frac{-5}{6}$

Solution The denominators are -8 and 6. A common denominator of -8 and 6 is 24. So, $\frac{7}{-8} = \frac{-21}{24}$ and $\frac{-5}{6} = \frac{-20}{24}$. Thus,

$$\frac{7}{-8} + \frac{-5}{6} = \frac{-21}{24} + \frac{-20}{24}$$

$$= \frac{-21 + (-20)}{24}$$

$$= \frac{-41}{24}$$

EXAMPLE 1.33

$\frac{3}{8} + \left(-1\frac{5}{16}\right)$

Solution A common denominator is 16. Thus, $\frac{3}{8} = \frac{6}{16}$ and $-1\frac{5}{16} = \frac{-21}{16}$. The example then becomes $\frac{6}{16} + \frac{-21}{16} = \frac{-15}{16}$.

EXAMPLE 1.34

$\frac{-5}{7} - \frac{-8}{3}$

Solution This is a subtraction problem. We first change it to an addition problem using Rule 3: $\frac{-5}{7} - \frac{-8}{3} = \frac{-5}{7} + \frac{8}{3}$. A common denominator of 7 and 3 is 21. $\frac{-5}{7} = \frac{-15}{21}$ and $\frac{8}{3} = \frac{56}{21}$. Thus,

$$\frac{-5}{7} - \frac{-8}{3} = \frac{-15}{21} + \frac{56}{21}$$

$$= \frac{-15 + 56}{21}$$

$$= \frac{41}{21}$$

Application

| EXAMPLE 1.35 | The circular pipe shown in Figure 1.4a has an inside radius of 1.35 cm and an outside radius of 1.575 cm. What is the thickness of the pipe? |

FIGURE 1.4

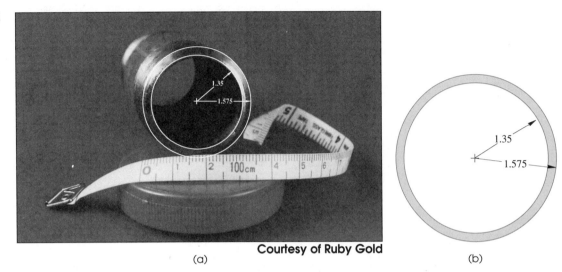

Courtesy of Ruby Gold

(a)   (b)

Solution   An outline of the pipe is shown in Figure 1.4b. To find the thickness, we need to subtract the inner radius from the outer radius, or $1.575 - 1.35$.

$$\begin{array}{r} 1.575 \\ -1.35 \\ \hline 0.225 \end{array}$$

The pipe is 0.225 cm thick.

Rule 7

To multiply two rational numbers, multiply the numerators and multiply the denominators.

| EXAMPLE 1.36 | $\dfrac{3}{4} \times \dfrac{-5}{8}$ |
| | **Solution** $\dfrac{3}{4} \times \dfrac{-5}{8} = \dfrac{3 \times (-5)}{4 \times 8} = \dfrac{-15}{32}$ |

| EXAMPLE 1.37 | $\left(\dfrac{-7}{8}\right)(-9)$ |
| | **Solution** $\left(\dfrac{-7}{8}\right)(-9) = \dfrac{-7}{8} \times \dfrac{-9}{1} = \dfrac{(-7)(-9)}{8 \times 1} = \dfrac{63}{8}$ |

Hint

Sometimes it is possible to "cancel" factors that appear in the numerators and denominators *before* you multiply the numbers. For example, in $\frac{5}{12} \times \frac{8}{25} = \frac{5\times 8}{12\times 25}$, the numerator and denominator have a factor of 5×4 in common. Thus, we can write

$$\frac{5}{12} \times \frac{8}{25} = \frac{5 \times 8}{12 \times 25} = \frac{\cancel{5} \times \cancel{4} \times 2}{3 \times \cancel{4} \times \cancel{5} \times 5} = \frac{2}{3 \times 5} = \frac{2}{15}$$

Application

EXAMPLE 1.38

An auto shop charges \$56 for each hour they work on your car. How much will they charge if they work on it for $3\frac{1}{4}$ hours?

Solution   We need to multiply $56 \times 3\frac{1}{4}$. We begin by changing 56 and $3\frac{1}{4}$ to fractions. We write 56 as $\frac{56}{1}$ and $3\frac{1}{4}$ as $\frac{13}{4}$.

$$56 \times 3\frac{1}{4} = \frac{56}{1} \times \frac{13}{4}$$
$$= \frac{56 \times 13}{1 \times 4} \qquad \text{or} \qquad \frac{\cancel{4} \times 14 \times 13}{1 \times \cancel{4}} = \frac{14 \times 13}{1}$$
$$= \frac{728}{4} = 182 \qquad\qquad\qquad = 182$$

They will charge \$182.

Application

EXAMPLE 1.39

To change 14°F to its equivalent temperature in degrees Celsius, you need to compute $\frac{5}{9}(14° - 32°)$. What is this temperature?

Solution $\frac{5}{9}(14° - 32°) = \frac{5}{9}(-18°)$
$$= \frac{5}{9} \times \frac{-18}{1}$$
$$= \frac{5 \times (-18)}{9 \times 1} \qquad \text{or} \qquad \frac{5 \times (-2) \times \cancel{9}}{\cancel{9} \times 1} = \frac{-10}{1}$$
$$= \frac{-90}{9} = -10 \qquad\qquad\qquad = -10$$

So, $-10°C$ is the same as 14°F.

Rule 8

To divide one rational number by another, multiply the first number by the reciprocal of the second.

EXAMPLE 1.40

$\frac{-5}{8} \div \frac{2}{3}$

EXAMPLE 1.40 (Cont.)

Solution   The reciprocal of $\frac{2}{3}$ is $\frac{3}{2}$ so

$$\frac{-5}{8} \div \frac{2}{3} = \frac{-5}{8} \times \frac{3}{2}$$

$$= \frac{(-5)(3)}{8 \times 2}$$

$$= \frac{-15}{16}$$

EXAMPLE 1.41

$$\frac{-7}{8} \div \frac{-9}{5}$$

Solution   The reciprocal of $\frac{-9}{5}$ is $\frac{5}{-9}$ $\left(\text{or } \frac{-5}{9}\right)$ and so

$$\frac{-7}{8} \div \frac{-9}{5} = \frac{-7}{8} \times \frac{-5}{9}$$

$$= \frac{(-7)(-5)}{8 \times 9}$$

$$= \frac{35}{72}$$

Application

EXAMPLE 1.42

A tank of fuel weighs $185\frac{1}{6}$ pounds. If a gallon of gasoline weighs $7\frac{1}{3}$ pounds, how many gallons are in the tank?

Solution   We need to divide the weight of the tank by the number of pounds for one gallon.

$$185\frac{1}{6} \div 7\frac{1}{3} = \frac{1,111}{6} \div \frac{22}{3}$$

$$= \frac{1,111}{6} \times \frac{3}{22}$$

$$= \frac{1,111 \times 3}{6 \times 22}$$

$$= \frac{3,333}{132}$$

$$= 25\frac{33}{132} = 25\frac{1}{4}$$

The tank contains $25\frac{1}{4}$ gallons of gasoline.

Exercise Set 1.3

Perform the indicated operation in Exercises 1–40.

1. $+27 + (+23)$

2. $8 + (-19)$

3. $27 + (-13)$

4. $-9 + (-8)$

5. $7 - 16$

6. $29 - (-8)$

7. $-8 - 16$

8. $-25 - (-13)$

9. $-37 - (-49)$

10. $(-2)(6)$

11. $(-3)(-5)$

12. $(7)(-8)$

13. $-38 \div 4$

14. $-45 \div (-9)$

15. $\frac{3}{4} + \frac{-5}{8}$

16. $-1\frac{3}{4} + \frac{-2}{3}$

17. $\frac{-9}{5} + \frac{7}{3}$

18. $\frac{-2}{3} + \frac{5}{6}$

19. $\frac{2}{5} + \frac{-1}{4}$

20. $\frac{-4}{5} + \frac{-5}{6}$

21. $\frac{3}{8} - \frac{-1}{4}$

22. $1\frac{1}{3} - \frac{-5}{6}$

23. $\frac{-9}{10} - \frac{2}{3}$

24. $-\frac{5}{16} - \frac{-3}{8}$

25. $\frac{5}{32} - \frac{1}{8}$

26. $-\frac{7}{3} - \frac{-6}{7}$

27. $\frac{-2}{3} \times \frac{4}{5}$

28. $\frac{3}{4} \times \frac{-5}{8}$

29. $\frac{-1}{8} \times \frac{-3}{4}$

30. $\frac{9}{16} \times \frac{1}{2}$

31. $-\frac{4}{3} \times \frac{5}{2}$

32. $\frac{-9}{5} \times \frac{-3}{8}$

33. $-\frac{3}{4} \div \frac{-5}{8}$

34. $1\frac{3}{4} \div \frac{-2}{3}$

35. $-\frac{3}{5} \div 4$

36. $-\frac{3}{8} \div \frac{1}{4}$

37. $-\frac{7}{5} \div \left(-\frac{5}{7}\right)$

38. $\frac{2}{3} \div \left(-\frac{7}{3}\right)$

39. $\left(-2 + \frac{2}{3}\right) \times \frac{-1}{2} - \left[\frac{3}{2} \div (-3)\right] + \frac{7}{3}$

40. $\frac{4}{3} \times \frac{-7}{8} \times \frac{3}{-5} \div \frac{4}{5} + \frac{-5}{8} - \frac{7}{8}$

Solve Exercises 41–49.

41. *Recreation* A ski resort received $17\frac{1}{2}''$ of snow in December, $15\frac{7}{8}''$ in January, $29\frac{3}{4}''$ in February, and $15\frac{3}{8}''$ in March. How much snow did the resort get during these four months?

42. *Auto technician* The toe-in reading is $\frac{5}{32}$ on one front wheel and $\frac{3}{16}$ on the other. What is the total toe-in?

43. *Electricity* When two or more voltages are connected in series, the total voltage is the sum of the separate voltages. When the voltages are connected in the same direction, they are all considered to be positive. When the voltages are connected in the opposite direction, one direction is considered positive and the other negative. Find the total voltage of the batteries in Figure 1.5.

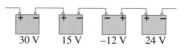

30 V   15 V   −12 V   24 V

FIGURE 1.5

44. *Construction* A planer removed $\frac{1}{5}$ inch from a $1\frac{5}{8}$-inch board. Find the thickness of the finished board.

45. *Recreation* A $14\frac{1}{2}$-mi race has 3 checkpoints. The first checkpoint is $4\frac{5}{8}$ mi from the starting point. The second checkpoint is $3\frac{1}{5}$ mi from the first checkpoint. The third checkpoint is $3\frac{3}{8}$ mi from the second checkpoint.

(a) How many miles is it from the starting point to the second checkpoint?

(b) How many miles is it from the starting point to the third checkpoint?

(c) How many miles is it from the third checkpoint to the finish line?

46. *Automotive technology* An automobile traveled $427\frac{1}{5}$ mi on $13\frac{3}{4}$ gal of gasoline. How many miles did it get per gallon?

47. *Electrical technician* A certain job requires 37 pieces of electrical wire that are each $2'3\frac{1}{2}''$ long. What is the total length of wire that is needed?

48. *Construction* A 4-ft by 8-ft sheet of plywood is cut into strips, each $1\frac{5}{16}$ in. by 4 ft. The saw cut is $\frac{1}{8}$-in. wide. How many strips can be cut from the sheet?

49. *Construction* Find the length indicated by the question mark in Figure 1.6.

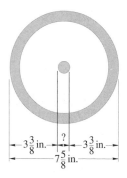

FIGURE 1.6

≡ 1.4
LAWS OF EXPONENTS

We have introduced the different types of numbers, the basic laws of arithmetic, and the basic operations with real numbers. In this section, we will learn some shorthand notation used in mathematics that will help us with the work in future sections.

Exponents

It is often necessary to multiply a number, b, by itself several times. The notation b^n is used to indicate that a total of n bs are being multiplied. The number b is called the **base**, and n is called the **exponent**. We say that b^n is the "nth power of b" or "b to the nth power." When the exponent is 1, the 1 is often not written. Thus, $19^1 = 19$.

EXAMPLE 1.43

$$3^4 = 3 \times 3 \times 3 \times 3$$
$$= 81$$

The base is 3 and the exponent is 4. This is 3 to the fourth power or the fourth power of 3.

$$7^5 = 7 \times 7 \times 7 \times 7 \times 7$$
$$= 16{,}807$$

The base is 7 and the exponent is 5. This is 7 to the fifth power.

Notice how much easier it is to write 7^5 than 16,807.

$$9^2 = 9 \times 9$$
$$= 81$$

The base is 9, the exponent 2. This is the second power of 9 or, as it is more frequently called, 9 squared.

$$\left(\frac{5}{4}\right)^3 = \frac{5}{4} \times \frac{5}{4} \times \frac{5}{4}$$
$$= \frac{125}{64}$$

The base is $\frac{5}{4}$, the exponent is 3. This is the third power of $\frac{5}{4}$ or $\frac{5}{4}$ cubed.

$$(-3)^4 = (-3)(-3)(-3)(-3)$$
$$= 81$$

The base is -3, the exponent 4.

At the present time the exponent must be an integer. Later we will learn how to use and understand expressions where the exponent is any real number.

Rules of Exponents

As with all of the operations we have studied so far, there are some basic rules of exponents.

Rule 1 for Exponents	$b^m b^n = b^{m+n}$

This rule states that to multiply two or more powers of the same base, you add the exponents.

EXAMPLE 1.44

$$3^5 \cdot 3^2 = 3^{5+2} = 3^7 \text{ because } 3^5 \cdot 3^2 = (3 \times 3 \times 3 \times 3 \times 3)(3 \times 3)$$
$$= 3 \times 3 \times 3 \times 3 \times 3 \times 3 \times 3$$
$$= 3^7$$

EXAMPLE 1.45

$$x^3 \cdot x^4 \cdot x = x^3 \cdot x^4 \cdot x^1 = x^{3+4+1} = x^8$$

Rule 2 for Exponents	$(b^m)^n = b^{mn}$

This means that in order to take a power of a power you multiply the exponents.

EXAMPLE 1.46

$$(3^2)^4 = 3^{2 \cdot 4} = 3^8 \text{ because } (3^2)^4 = (3^2)(3^2)(3^2)(3^2)$$
$$= (3 \times 3)(3 \times 3)(3 \times 3)(3 \times 3)$$
$$= 3 \times 3 \times 3 \times 3 \times 3 \times 3 \times 3 \times 3$$
$$= 3^8$$

EXAMPLE 1.47

$$(x^4)^5 = x^{4 \cdot 5} = x^{20}$$

Rule 3 for Exponents	$(ab)^n = a^n b^n$

EXAMPLE 1.48

$$(2 \cdot 5)^3 = (2 \cdot 5)(2 \cdot 5)(2 \cdot 5)$$
$$= (2 \cdot 2 \cdot 2)(5 \cdot 5 \cdot 5)$$
$$= 2^3 \cdot 5^3$$

EXAMPLE 1.49

$$(xy)^5 = x^5 y^5$$

Rule 4a for Exponents	$\left(\dfrac{a}{b}\right)^m = \dfrac{a^m}{b^m}$, if $b \neq 0$

This says that to raise a quotient to a power, both the numerator and denominator are raised to that power.

EXAMPLE 1.50

$$\left(\frac{2}{5}\right)^4 = \frac{2^4}{5^4} \text{ because } \left(\frac{2}{5}\right)^4 = \left(\frac{2}{5}\right)\left(\frac{2}{5}\right)\left(\frac{2}{5}\right)\left(\frac{2}{5}\right)$$

$$= \frac{2 \times 2 \times 2 \times 2}{5 \times 5 \times 5 \times 5}$$

$$= \frac{2^4}{5^4}$$

Rule 4b for Exponents	$\left(\dfrac{1}{b}\right)^m = \dfrac{1}{b^m}$, if $b \neq 0$
This is a special case of Rule 4a.	

EXAMPLE 1.51

$$\left(\frac{1}{3}\right)^4 = \frac{1}{3^4}$$

Rule 5 for Exponents	$\dfrac{b^m}{b^n} = b^{m-n}$, if $b \neq 0$

To divide two powers with the same base, you subtract the exponent of the denominator from the exponent of the numerator.

There are really three cases to this rule depending on which is larger, m or n, or if they are the same. Let's look first at when $m > n$.

EXAMPLE 1.52

$$\frac{5^7}{5^4} = 5^{7-4} = 5^3 \text{ because } \frac{5^7}{5^4} = \frac{5 \times 5 \times 5 \times 5 \times 5 \times 5 \times 5}{5 \times 5 \times 5 \times 5}$$

$$= 5 \times 5 \times 5$$

$$= 5^3$$

EXAMPLE 1.53

$$\frac{x^9}{x^4} = \frac{x \cdot x \cdot x \cdot x \cdot x \cdot x \cdot x \cdot x \cdot x}{x \cdot x \cdot x \cdot x} = x^5$$

Next, let's look at when $m = n$. Then we have $\dfrac{b^n}{b^n} = b^{n-n} = b^0$. But, $\dfrac{b^n}{b^n} = 1$ and so $b^0 = 1$.

EXAMPLE 1.54

$$\frac{5^2}{5^2} = 5^{2-2} = 5^0 = 1$$

and $\quad \dfrac{5^2}{5^2} = \dfrac{5 \times 5}{5 \times 5} = \dfrac{25}{25} = 1$

Finally, let's look at when $m < n$.

EXAMPLE 1.55

$$\frac{6^3}{6^7} = 6^{3-7} = 6^{-4}$$

But, $\quad \dfrac{6^3}{6^7} = \dfrac{6 \times 6 \times 6}{6 \times 6 \times 6 \times 6 \times 6 \times 6 \times 6} = \dfrac{1}{6 \times 6 \times 6 \times 6} = \dfrac{1}{6^4}$

EXAMPLE 1.56

$$\frac{x^5}{x^7} = x^{5-7} = x^{-2}$$

and $\quad \dfrac{x^5}{x^7} = \dfrac{x \cdot x \cdot x \cdot x \cdot x}{x \cdot x \cdot x \cdot x \cdot x \cdot x \cdot x} = \dfrac{1}{x \cdot x} = \dfrac{1}{x^2}$

Examples 1.54–1.56 have set the foundation for two more rules. From Example 1.54 we can see that

Rule 6 for Exponents	$b^0 = 1$, if $b \neq 0$

Any nonzero number raised to the zero power has a value of one.
From Examples 1.55–1.56, we have

Rule 7 for Exponents	$b^{-n} = \dfrac{1}{b^n}$, if $b \neq 0$

This rule says that any number raised to a negative exponent is the same as the reciprocal of the number raised to that positive power.

≡ **Note**   Since $n = -(-n)$, it sometimes helps to think of this as

$$b^n = b^{-(-n)} = \frac{1}{b^{-n}}$$

EXAMPLE 1.57

$$\left(x^2 y^3 z w^0\right)^{-4} = \frac{1}{\left(x^2 y^3 z w^0\right)^4} = \frac{1}{x^8 y^{12} z^4}$$

EXAMPLE 1.58

We will use the previous note $\left(\text{that } b^n = \dfrac{1}{b^{-n}}\right)$ to solve this example.

$$\left(\frac{a^2 b}{x^3 w^2}\right)^{-5} = \frac{\left(a^2 b\right)^{-5}}{\left(x^3 w^2\right)^{-5}} = \frac{\left(x^3 w^2\right)^5}{\left(a^2 b\right)^5}$$

$$= \frac{x^{15} w^{10}}{a^{10} b^5}$$

Application

EXAMPLE 1.59

The total resistance, R, of a series-parallel circuit is given by the formula

$$R = \left(\frac{1}{R_1} + \frac{1}{R_2}\right)^{-1} + R_3$$

Find R when $R_1 = 0.75\ \Omega$, $R_2 = 0.50\ \Omega$, and $R_3 = 0.60\ \Omega$.

Solution   We begin by adding the terms in the parentheses.

$$R = \left(\frac{1}{R_1} + \frac{1}{R_2}\right)^{-1} + R_3$$

$$= \left(\frac{1}{0.75} + \frac{1}{0.50}\right)^{-1} + 0.60$$

$$= \left(\frac{0.50}{(0.75)(0.50)} + \frac{0.75}{(0.75)(0.50)}\right)^{-1} + 0.60$$

$$= \left(\frac{0.50 + 0.75}{(0.75)(0.50)}\right)^{-1} + 0.60$$

$$= \left(\frac{1.25}{(0.75)(0.50)}\right)^{-1} + 0.60$$

$$= \frac{(0.75)(0.50)}{1.25} + 0.60$$

$$= 0.3 + 0.60$$

$$= 0.9$$

The total resistance is $0.9\ \Omega$.

≡ **Note**   Remember when we discussed the order of operations in Section 1.2, that order #2 was raising to a power. Thus, unless grouping symbols indicate otherwise, perform all operations of raising to a power before multiplication, division, addition, or subtraction.

Exercise Set 1.4

Evaluate Exercises 1–8.

1. 5^3

2. 3.8^0

3. $\left(\dfrac{2}{3}\right)^{-1}$

4. $\left(\dfrac{3}{5}\right)^{-2}$

5. $(-4)^2$

6. $(-5)^4$

7. $\dfrac{7}{7^3}$

8. -3^2

In Exercises 9–42, perform the indicated operations. Leave all answers in terms of positive exponents.

9. $3^2 \cdot 3^4$

10. $d^8 d^5$

11. $2^4 \cdot 2^3 \cdot 2^5$

12. $f^3 f^4 f^1$

13. $\dfrac{2^5}{2^3}$

14. $\dfrac{5^{14}}{5^3}$

15. $(2^3)^2$

16. $(5^7)^3$

17. $(x^4)^5$

18. $(xy^3)^4$

19. $(a^{-2}b)^{-3}$

20. $\left(\dfrac{2}{3}\right)^4$

21. $\left(\dfrac{x}{4}\right)^3$

22. $\left(\dfrac{a}{b^3}\right)^5$

23. $\left(\dfrac{a^2 b}{c^3}\right)^4$

24. 4^{-3}

25. x^{-7}

26. $\dfrac{1}{p^{-5}}$

27. $\left(\dfrac{1}{5}\right)^3$

28. $\dfrac{x^4}{x^2}$

29. $\dfrac{7^3}{7^8}$

30. $\dfrac{5^2}{5^{10}}$

31. $\dfrac{a^2 y^3}{a^5 y^7}$

32. $\dfrac{x^4 y b^2}{xy^3 b^5}$

33. $\dfrac{a^2 p^5 y^3}{a^6 p^5 y}$

34. $\dfrac{p^3 q^4 r^2}{p^4 r^2}$

35. $(pr^2)^{-1}$

36. $\left(\dfrac{2x^2}{y}\right)^{-1}$

37. $\left(\dfrac{4y^3}{5^2}\right)^{-1}$

38. $\left(\dfrac{4x^3}{y^2}\right)^{-2}$

39. $\left(\dfrac{2b^2}{y^5}\right)^{-3}$

40. $(-8pr^2)^{-3}$

41. $(-b^4)^6$

42. $ap^2(-a^2 p^3)^2$

Solve Exercise 43.

43. *Electricity*   The impedance in an *RC* circuit is given by the expression

$$Z_{RC} = \sqrt{R^2 + \left((2\pi fC)^{-1}\right)^2}$$

Determine the impedance if $R = 40\ \Omega, f = 60$ Hz, and $C = 8 \times 10^{-5}$ F.

≡ **1.5**
SIGNIFICANT DIGITS AND ROUNDING OFF

A great deal of technical work deals with measurements. Problems in this book are in both the metric system and the customary (or English) system of measurement. If you are not familiar with these two systems of measurement, you should read Appendix B.

No measurement deals with exact numbers. For example, suppose an automotive technician says that the outside diameter of a valve stem is 7.1 mm. While the diameter may be exactly 7.1 mm, it is more than likely to be a little more or less than 7.1 mm.

The amount of precision in such a measurement depends on the measuring instrument and the person doing the measuring. In mathematical terms, we have the following definitions of precision, significant digits, and accuracy.

Precision, Significant Digits, and Accuracy

The **precision** of a measurement is indicated by the position of the last significant digit relative to the decimal point.

The **significant digits** are those that are determined by measurement.

Accuracy refers to the number of significant digits.

A number with five significant digits, such as 7.1043, is more accurate than a number with four significant digits, such as 7.104.

Guidelines for Determining Which Digits are Significant

1. All nonzero digits are significant.
2. Zero digits that lie between significant digits are significant. For example, 307 has three significant digits.
3. Zero digits that lie to the right of both the decimal point and the last nonzero digit are significant. For example, .860 has three significant digits.
4. Zeros at the beginning of a decimal fraction are not significant. For example, both .045 and 0.045 have two significant digits since the zeros serve only to locate the decimal point.
5. Zeros written at the end of a whole number are significant only if there is a "tilde" (˜) or a "bar" (⁻) over the last significant digit. For example, $980,\tilde{0}00$ and $980,\bar{0}00$ both have four significant digits.

Perhaps a few words about notation are needed here. A number such as 15,340,000 is written in the metric system by using spaces instead of commas, as 15 340 000. The spaces are also used for numbers smaller than one, for example, 0.00002471 would be written in the metric system as 0.000 024 71. A four-digit number in the metric system does not need to be written with the space, just as we do not always use a comma in such a number. Thus, 2400 and 2 400 are both correct in the metric system and 2,400 and 2400 are both correct in the English (or standard) system. As a technician, you may encounter both metric and standard or English notations and should recognize a number in either system.

Concept of Error

Let us now return to the earlier measurement of the outside diameter of a valve stem. If the person who measured this diameter had used an instrument that measured in thousandths of millimeters and the diameter had measured 7.1 mm to the nearest thousandth, then the measurement should have been written 7.100 mm.

Since we have spent so much time describing approximate numbers and the method with which to indicate how carefully they were measured, the question might arise as to what is an exact number. **Exact numbers** result from some definition or from counting. For example, the number of spark plugs in an 8-cylinder vehicle is an exact number; the length of a board is an approximate number.

Related to the ideas of accuracy and precision are the ideas of absolute, relative, and percent error. The **absolute error** is the true value subtracted from the approximate value of a number. The absolute error can be either positive or negative depending on whether the true value is smaller or larger than the approximate value.

The **relative error** is the ratio of the absolute error to the true value. Relative error is usually expressed as a percent. When relative error is written as a percent it is often called **percent error**.

Three Types of Errors

Absolute error = approximate value − true value

$$\text{Relative error} = \frac{\text{absolute error}}{\text{true value}}.$$

Percent error = Relative error × 100%

EXAMPLE 1.60

The outside diameter of a valve stem was measured as 7.127 mm when the actual diameter was 7.134 6 mm. What are the absolute, relative, and percent errors?

Solution   Absolute error = approximate value − true value

$$= 7.127 \text{ mm} - 7.134\,6 \text{ mm}$$

$$= -0.007\,6 \text{ mm}$$

$$\text{Relative error} = \frac{\text{absolute error}}{\text{true value}}$$

$$= \frac{-0.007\,6 \text{ mm}}{7.134\,6 \text{ mm}}$$

$$= -0.001\,065 \text{ mm}$$

$$\text{Percent error} = \text{Relative error} \times 100\%$$

$$= -0.001\,065 \times 100\%$$

$$= -0.106\,5\%$$

≡ Note   When computing with approximate numbers, the answer will depend on the accuracy or precision of those numbers.

Working With Approximate Numbers

When you add or subtract approximate numbers, the result is only as precise as the least precise number.

EXAMPLE 1.61

Add 182.7, 43.69, 2,470.765, and 0.32, and express the answer to the correct precision.

Solution   First, we will add the numbers:

$$
\begin{array}{r}
182.7 \\
43.69 \\
2,470.765 \\
+ \quad 0.32 \\
\hline
2,697.475
\end{array}
$$

These four numbers have 182.7 as their least precise number, so the sum can only be significant to the tenths place. The answer would be rounded off to 2,697.5.

Application

EXAMPLE 1.62

The masses of 5 pieces of metal plate are 16.63 kg, 738.6 kg, 4.314 kg, 21.645 kg, and 0.875 2 kg. Find the total mass of the four pieces correct to 4 significant digits.

Solution   These five numbers have 738.6 as their least precise number, so the sum can only be significant to the tenths place. First, we will add the numbers:

$$
\begin{array}{r}
16.63 \\
738.6 \\
4.314 \\
21.645 \\
+ \quad 0.875\,2 \\
\hline
782.064\,2
\end{array}
$$

The answer would be rounded off to 782.1 kg.

When multiplying or dividing approximate numbers, the errors are enlarged. For this reason, the result is only as accurate as the least accurate number.

EXAMPLE 1.63

If a rectangle measures 42.37 mm long and 5.81 mm wide, it has an area of

$$
\begin{array}{r}
42.37 \text{ mm} \\
\times \quad 5.81 \text{ mm} \\
\hline
246.169\,7 \text{ mm}^2
\end{array}
$$

| **EXAMPLE 1.63 (Cont.)** | Since 42.37 has four significant digits and 5.81 has only three significant digits, the product is rounded off to three significant digits and the area is 246 mm^2. |

Rounding off Numbers

In general, to round off a number to a certain number of significant digits, examine the digit in the next place to the right. If this digit is less than 5 (0, 1, 2, 3, or 4) then accept the digit in the last place to the left of this digit. If the digit is 5 or more (5, 6, 7, 8, or 9) then increase the digit in the last place by one.

| **EXAMPLE 1.64** | Round off 83.427 to the nearest hundredth.

Solution The digit 2 is in the hundredths place. The next digit to the right is a 7. This is more than 5, so the 2 is increased by 1 (to 3) and the 7 is dropped. Thus, 83.427 rounded to the nearest hundredth is 83.43. |

| **EXAMPLE 1.65** | Round 5.2348 to the nearest hundredth.

Solution Here the digit in the hundredths place is a 3. The next digit to the right is a 4. Since 4 is less than 5, the 4 and all other digits to its right are dropped. So, 5.2348 rounded off to the nearest hundredth is 5.23. |

In many scientific and engineering applications a different round-off rule is used when the test digit is 5. This rule is known as either the **Odd-Five Rule** or the **Round-to-the-Even Rule**.

The Odd-Five Rule states that if the test digit is a 5 and it is the last nonzero digit in a number, then add 1 to the round off digit if it is odd (1, 3, 5, 7, or 9) or retain the original round-off digit if it is even (0, 2, 4, 6, or 8).

| **EXAMPLE 1.66** | Round off 43.725 to the nearest hundredth by the Odd-Five Rule.

Solution We first note that the test digit is 5 and that it is the last nonzero digit. Next, because the round-off digit 2 is even, it is retained and the digit 5 is dropped. Thus, 43.725 would round off to the nearest hundredth to 43.72, by the Odd-Five Rule. |

| **EXAMPLE 1.67** | Round off 153.835 to the nearest hundredth by using the Odd-Five Rule.

Solution Again the test digit is 5 and it is the last nonzero digit, but here the round-off digit is 3, an odd number. Using the Odd-Five Rule, we add 1 to the 3 and drop the 5, and the number is rounded off to 153.84. |

≡ **Note** Unless it is specified, we will **not** use the Odd-Five Rule in this book.

Caution

A precaution about significant digits and electronic calculators: Most calculators display 8 or 10 digits and store 2 or 3 more for rounding off purposes. However, this is not always the case. Some calculators employ a method called **truncation** in which any digits not displayed are discarded. Thus, 489.781 truncated to tenths is 489.7. Rounded off to tenths it would be 489.8. This can result in different answers when two different calculators are used.

Estimation

As we mentioned in the introduction to Section 1.3, you will most likely be using calculators and computers in your work as a technician. You must be able to recognize when an answer provided by these tools is reasonable. The ability to estimate answers is very important in your ability to recognize when answers are reasonable. Your skill at rounding off numbers will help you estimate answers to problems.

In addition and subtraction, round off each number to one or two significant digits and add (or subtract) these rounded off figures.

EXAMPLE 1.68

Estimate the sum of 4.7 + 8.6 + 9.2 + 4.1.

Solution   The numbers in this sum, 4.7 + 8.6 + 9.2 + 4.1, would be rounded off to 5 + 9 + 9 + 4 = 27. The actual sum of the original four numbers is 26.6, which rounds off to 27.

EXAMPLE 1.69

Estimate the sum of 93.74 + 182.7 + 14,325 + 83.43.

Solution   Rounding off each number to the nearest hundred, you get the estimate 100 + 200 + 14,300 + 100 = 14,700. The actual total is 14,684.87.

We will learn how to estimate products and quotients in Section 1.6, which involves scientific notation.

Rounding-Off with a Calculator

Most calculators display from 8 to 10 digits. It is often helpful to set your calculator so that it will only display a certain number of digits to the right of the decimal point. For this, you use the $\boxed{\text{Fix}}$ key, or mode, of the calculator. So, if you want your answer to show 3 decimal places to the right of the decimal point, set your calculator mode to $\boxed{\text{Fix}}$ 3. The next example shows how to get the answer of a sum to the correct amount of precision.

EXAMPLE 1.70

Use your calculator to add 182.7, 43.69, 2,470.765, and 0.32. Express the answer to the correct number of decimal places.

Solution   We worked this example earlier in Example 1.61, and found the answer to be 2,697.5. Now, let's use our calculator. The least precise number is 182.7. Set the calculator to $\boxed{\text{Fix}}$ 1. This will cause the calculator to display our answer with

EXAMPLE 1.70 (Cont.)	only one digit to the right of the decimal point. Now we enter each number that we want to add.

	PRESS	DISPLAY
	182.7 $\boxed{+}$	182.7
	43.69 $\boxed{+}$	226.4
	2470.765 $\boxed{+}$	2697.2
	0.32 $\boxed{=}$	2697.5

Once again, we get the answer of 2,697.5.

Exercise Set 1.5

In Exercises 1–6, determine whether the given numbers are exact or approximate.

1. There are 27 students in class.

2. The car traveled at 88 km/h.

3. A calculator is 148 mm by 79 mm by 35 mm.

4. She bought 18 bolts for $2.79.

5. Of all the people working in the United States, 7.9% are technicians.

6. A sheet of plywood is 1 200 mm by 2 400 mm.

Give the number of significant digits for the numbers in Exercises 7–14.

7. 6.05

8. 4,030

9. 12.0

10. 0.432

11. 4,000

12. 0.00290

13. 70.06

14. 160.070

Which numbers in Exercises 15–23 are (a) more accurate and (b) more precise. (It is possible for both numbers to be accurate or for both to be precise.)

15. 6.05; 2.8

16. 0.027; 6.324

17. 0.027; 5.01

18. 19,020; 29,000

19. 27,0$\overline{0}$0; 37,800

20. 6,000; 0.003

21. 0.2; 86

22. 3.05; 305.00

23. 140.070; 140,070

Round off each number in Exercises 24–31 to (a) one, (b) two, and (c) three significant digits.

24. 4.362

25. 14.37

26. 4.065

27. 7.035

28. 0.006155

29. 403.2

30. 0.03725

31. 305.4

In Exercises 32–39, round off each of the numbers to (a) one, (b) two, and (c) three significant digits using the Odd-Five Rule. (These are the same numbers you rounded off in Exercises 24–31.)

32. 4.362

33. 14.37

34. 4.065

35. 7.035

36. 0.006155

37. 403.2

38. 0.03725

39. 305.4

Round off each number in Exercises 40–47 to the nearest (a) ten, (b) tenth, and (c) thousandth.

40. 25.3345

41. 89.8992

42. 125.3755

43. 237.3017

44. 96.99854

45. 437.9975

46. 12.3405

47. 78.6705

Solve Exercises 48–49.

48. *Construction* The actual length of an I-beam is 12.445 m. An engineer measures the beam as 12.45 m. What are the absolute, relative, and percent errors in this measurement?

49. *Computer science* A microcomputer chip is supposed to measure 24 mm long, 8 mm wide, and 3 mm thick. One chip, when measured with a micrometer, is 23.72 mm long, 8.35 mm wide, and 2.98 mm thick. What are the absolute, relative, and percent errors of each of these measurements?

Estimate the answers to Exercises 50–53, then work the exercises and round off the answers using the rules for approximate numbers.

50. $4.31 + 2.015 + 18.35$

51. $97.83 - 4.378 + 5.92$

52. $243.7 + 85.37 - 62.105 + 143.8$

53. $1,342.8 + 85.32 + 173.54 + 16$

Work Exercises 54–57 and round off the answer using the rules for approximate numbers.

54. 14.3×5.7

55. 20.4×50.1

56. 0.0034×2.50

57. 3.4×5.00

Solve Exercises 58–61.

58. *Automotive technology* The 2 front parking lamps of a car each draw a current of 0.417 A (amperes). The 2 tail lamps each draw a current of 0.457 A. The license plate lamp draws 0.736 A. Find the total current drawn by the 5 lamps. Give your answer correct to 3 significant digits.

59. *Electricity* An inductor has 37 layers of wire. Each layer contains 132 turns. The average length of a turn is 0.072 m. Find the length of the wire in the coil. Give your answer correct to 5 significant digits.

60. *Construction* A pile of lumber has 379 pieces with an average length of 11 ft $10\frac{1}{8}$ in. Find the total length of the lumber, in inches, correct to **(a)** three decimal places and **(b)** three significant digits.

61. *Metalworking* A milling machine cutter with 18 teeth removes 0.086 mm of steel per tooth. The cutter makes 597.3 revolutions. What length of material is removed? Round off your answer to 4 significant digits.

≡ 1.6
SCIENTIFIC NOTATION

In scientific and technical work it is often necessary to work with very large or very small numbers. For example, the star Sirius is approximately 8 220 000 000 000 000 km from earth. On the other hand, the radius of an electron is about 0.000 000 000 000 002 82 m.

Writing Numbers in Scientific Notation

Writing numbers with all these zeros is very time consuming and increases the chance of making an error by either omitting a zero or by using too many zeros. In order to save time and reduce the chance of making a mistake, scientists adopted a method for abbreviating numerals. This method is called **scientific notation**.

Expressing a Number in Scientific Notation

To express any number in scientific notation, write the number as a product of a power of ten and a number between 1 and 10.

EXAMPLE 1.71

$$2,400 = 2.4 \times 1,000 = 2.4 \times 10^3$$

$$38,900,000 = 3.89 \times 10,000,000 = 3.89 \times 10^7$$

$$4,070,000,000,000 = 4.07 \times 1,000,000,000,000 = 4.07 \times 10^{12}$$

$$100,000 = 1 \times 10^5 \quad \text{or} \quad 10^5$$

$$0.036 = 3.6 \times \frac{1}{100} = 3.6 \times 10^{-2}$$

$$0.000\,000\,403 = 4.03 \times \frac{1}{10\,000\,000} = 4.03 \times 10^{-7}$$

$$8.2 = 8.2 \times 10^0 \text{ or } 8.2$$

Perhaps the easiest way to remember how to express a number in scientific notation is to follow these two guidelines:

Guidelines for Expressing a Number in Scientific Notation

1. Move the decimal point to just after the first nonzero digit.
2. Then multiply by 10^n if you shifted the decimal point n places to the left or by 10^{-n} if you shifted the decimal point n places to the right.

EXAMPLE 1.72

Express 20 500 in scientific notation.

EXAMPLE 1.72 (Cont.)

Solution $20\,500 = 2.05 \times 10^4$

Move decimal point
four places to the left.

EXAMPLE 1.73

Express 0.000 037 in scientific notation.

Solution $0.000\,037 = 3.7 \times 10^{-5}$

Move decimal point
five places to the right.

Any zeros that are significant should be included when the number is written in scientific notation. For example, $0.002\,30 = 2.30 \times 10^{-3}$ and $847\,\tilde{0}00 = 8.470 \times 10^5$.

To change a number from scientific notation to ordinary notation you just reverse the process. To change 8.37×10^6 to ordinary notation you would move the decimal point six places to the right. So, $8.37 \times 10^6 = 8\,370\,000$. A number like 4.61×10^{-4} is changed by moving the decimal point four places to the left. Then $4.61 \times 10^{-4} = 0.000\,461$.

Obviously scientific notation relies heavily on exponents. This will give us the first chance to use the rules for exponents that we learned in Section 1.4. Also, many calculators use scientific notation to express some numbers. We will now compute products and quotients using scientific notation.

Scientific Notation on a Calculator

A calculator or a computer will automatically put a number in scientific notation if the number contains too many digits to show on the screen. For example, if you multiply 250,000 and 19,700,000, the calculator displays the answer as

$$4.925\text{E}12$$

$$\text{or} \quad 4.925 \quad 12$$

which should be interpreted as 4.925×10^{12}.

Similarly, the answer to $0.000\,025 \div 800\,000$ would be displayed as

$$3.125\text{E}{-}11$$

$$\text{or} \quad 3.125 \quad -11$$

which is interpreted as 3.125×10^{-11}.

 Note

Some calculators place a space between the number and the exponent while other calculators use the letter "E."

To enter the number in scientific notation on a calculator, you need to use the Enter Exponent key, labeled as $\boxed{\text{EE}}$, $\boxed{\text{EEX}}$, or $\boxed{\text{EXP}}$.

EXAMPLE 1.74

Enter 4.37×10^6 on a calculator.

Solution

PRESS		DISPLAY		
4.37 $\boxed{\text{EE}}$	4.37	00	or	4.37E
6	4.37	06	or	4.37E6

EXAMPLE 1.75

Enter 9.87×10^{-15} on a calculator.

Solution   The procedure depends on the type of calculator. With some models, you use this method:

PRESS	DISPLAY	
9.87 $\boxed{\text{EE}}$	9.87	00
15 $\boxed{+/-}$	9.87	-15

With other calculators, such as the Casio® *fx-7700G* or *TI®–81* graphics calculator, you follow these steps:

PRESS	DISPLAY
9.87 $\boxed{\text{EE}}$	9.87E
$\boxed{(-)}$ 15	9.87E-15

Special care has to be taken when entering a power of 10. The next example shows a correct procedure.

EXAMPLE 1.76

Enter 10^{25} into a calculator.

Solution

PRESS		DISPLAY		
1 $\boxed{\text{EE}}$	1	00	or	1E
25	1	25	or	1E25

Caution   Remember that powers of 10, such as 10^{25}, really represent 1×10^{25}. Thus, you enter 1 $\boxed{\text{EE}}$ 25 and not 10 $\boxed{\text{EE}}$ 25. However, on a *TI–81*, $\boxed{\text{EE}}$ 25 will work.

Products and Quotients

The product of $87\,000\,000 \times 470\,000\,000\,000$ can be simplified with scientific notation. Change each number to scientific notation and then multiply.

$$
\begin{aligned}
87\,000\,000 \times 470\,000\,000\,000 &= 8.7 \times 10^7 \times 4.7 \times 10^{11} \\
&= (8.7 \times 4.7) \times \left(10^7 \times 10^{11}\right) \\
&= 40.89 \times 10^{7+11} \\
&= 40.89 \times 10^{18} \\
&= 4.089 \times 10^{19}
\end{aligned}
$$

Notice that we had to change our first answer $\left(40.89 \times 10^{18}\right)$ to scientific notation because 40.89 is not between 1 and 10.

Similarly, a large and small number can be multiplied more easily by using scientific notation, as shown in the next example.

EXAMPLE 1.77

Solve $4\,100\,000\,000 \times 0.000\,002\,4$.

Solution
$$
\begin{aligned}
4\,100\,000\,000 \times 0.000\,002\,4 &= 4.1 \times 10^9 \times 2.4 \times 10^{-6} \\
&= (4.1 \times 2.4) \times \left(10^9 \times 10^{-6}\right) \\
&= 9.84 \times 10^3 \\
&= 9\,840
\end{aligned}
$$

Division can also be simplified using scientific notation.

EXAMPLE 1.78

Solve $0.000\,000\,036 \div 0.000\,012$.

$$
\begin{aligned}
0.000\,000\,036 \div 0.000\,012 &= 3.6 \times 10^{-8} \div 1.2 \times 10^{-5} \\
&= \frac{3.6 \times 10^{-8}}{1.2 \times 10^{-5}} \\
&= \frac{3.6}{1.2} \times 10^{-8+5} \\
&= 3 \times 10^{-8+5} \\
&= 3 \times 10^{-3}
\end{aligned}
$$

EXAMPLE 1.79

Use a calculator to solve $4\,200\,000\,000 \div 0.000\,000\,025$.

EXAMPLE 1.79 (Cont.)

Solution First, we convert each of these numbers to scientific notation.

$$4\,200\,000\,000 = 4.2 \times 10^9$$

$$0.000\,000\,025 = 2.5 \times 10^{-8}$$

Now, we use our calculator.

PRESS	DISPLAY	
4.2 EE	4.2	00
9 ÷	4.2	09
2.5 EE	2.5	00
8 +/−	2.5	−08
=	1.68	17

So, the quotient is 1.68×10^{17}.

Application

EXAMPLE 1.80

The alternating current reactance of a circuit, X_L, is given in ohms (Ω) by the formula

$$X_L = 2\pi fL$$

where f is the frequency of the alternating current in hertz (Hz), and L is the inductance of the circuit or inductor in henrys (H). Compute the inductive reactance when $f = 10\,000$ Hz and $L = 0.000\,006\,5$ H.

Solution We will use a calculator to compute this product. First, we convert 10 000 Hz and 0.000 006 5 H to scientific notation.

$$10\,000 = 1 \times 10^4$$

$$0.000\,006\,5 = 6.5 \times 10^{-6}$$

Now, we use our calculator:

PRESS	DISPLAY	
2 ×	2	
π ×	6.283185307	
1 EE	1	00
4	1	04
×	62831.85307	
6.5 EE	6.5	00
6 +/−	6.5	−06
=	0.408407045	

So, the reactance is about 0.41 Ω.

Using Scientific Notation for Estimates

Scientific notation can also be used to help estimate products and quotients.

EXAMPLE 1.81

Estimate the product of $362 \times 2{,}165 \times 82$.

Solution First, round off each number so that it has only one nonzero digit. Thus, 362 is rounded off to 400, 2,165 to 2,000, and 82 to 80. Express each of these in scientific notation and multiply.

$$400 \times 2{,}000 \times 80 = 4 \times 10^2 \times 2 \times 10^3 \times 8 \times 10^1$$
$$= (4 \times 2 \times 8) \times (10^2 \times 10^3 \times 10^1)$$
$$= 64 \times 10^6$$
$$= 64{,}000{,}000$$

The actual product of $362 \times 2{,}165 \times 82$ is 64,265,860.

Much of the time the estimated product (or quotient) will not be this close to the actual product (or quotient). This method of estimation is best used to make sure that you are placing the decimal point in the correct place.

For example, $2\,965 \times 650 \times 24 = 46\,254\,000$. Using the estimation procedure previously described you would get

$$3\,000 \times 700 \times 20 = 3 \times 10^3 \times 7 \times 10^2 \times 2 \times 10$$
$$= (3 \times 7 \times 2) \times (10^3 \times 10^2 \times 10)$$
$$= 42 \times 10^6 = 42\,000\,000$$

As you can see, $46\,254\,000$ is quite a bit larger than $42\,000\,000$. Yet, the estimation procedure indicated that the correct answer would have 8 digits to the left of the decimal point.

Exercise Set 1.6

In Exercises 1–10, change each of the numbers to scientific notation.

1. $42\,000$
2. $370\,000\,000$
3. $0.000\,38$
4. $0.000\,007\,5$
5. $9\,807\,0\tilde{0}0\,000$
6. $87\,0\tilde{0}0\,000$
7. $0.000\,097\,0$
8. 0.400
9. 4.3
10. 2.07

In Exercises 11–18, change each number from scientific notation to ordinary notation.

11. 4.5×10^3
12. 3.7×10^5
13. 4.05×10^7
14. 3.05×10^8
15. 6.3×10^{-5}
16. 1.87×10^{-8}
17. 7.2×10
18. 9.6×10^{-1}

In Exercises 19–40, perform the indicated calculations by using scientific notation.

19. $760\,000 \times 20\,400\,000\,000$

20. $(43\,200\,000)(850\,000\,000)$

21. $0.000\,035 \times 0.000\,000\,76$

22. $(0.000\,42)(0.000\,075)$

23. $(840\,000\,000)(0.000\,35)$

24. $(0.000\,0042)(23\,000)$

25. $(70\,400)(0.000\,003\,2)$

26. $(0.000\,302)(4\,370\,000\,000)$

27. $(28\,800\,000\,000) \div (240\,000)$

28. $55\,500\,000\,000 \div 370\,000$

29. $375\,000 \div 150\,000\,000$

30. $79\,800 \div 840\,000\,000$

31. $0.003\,2 \div 0.000\,000\,16$

32. $0.000\,48 \div 0.000\,000\,3$

33. $0.000\,000\,36 \div 0.000\,2$

34. $0.000\,000\,009\,8 \div 0.000\,014$

35. $(8\,760\,000)(245\,000\,000)(6\,4\tilde{0}0\,000\,000)$

36. $(4\,360\,000)(625\,000\,000)(38\,700\,000\,000)$

37. $\dfrac{(250\,200)(630\,000\,000)}{0.000\,000\,006\,3}$

38. $\dfrac{(25\,200)(8\,0\tilde{0}0\,000)}{3\,970\,000\,000}$

39. $\dfrac{0.000\,005\,2 \times 480\,000\,000}{0.000\,000\,000\,006\,4}$

40. $\dfrac{96\,000\,000 \times 810\,000}{243\,000\,000\,000\,000}$

Solve Exercises 41–48.

41. Estimate the computation of $\dfrac{325 \times 86.5}{43.1}$.

42. Estimate the computation of $\dfrac{453 \times 672.4}{3.81 \times 42.3}$.

43. *Physics*   An electron moves at the rate of $300\,000\,000\,000$ mm/s (millimeters per second). How many millimeters will it move in $0.000\,003\,7$ s?

44. *Astronomy*   The speed of light is approximately $300\,000$ km/s. How many minutes will it take for a ray of light to reach the earth from the sun, if the sun is $150\,000\,000$ km from earth?

45. *Physics*   One atomic mass unit (amu) is $1.660\,6 \times 10^{-27}$ kg. If 1 carbon atom has 12 atomic mass units, what is the mass of $14\,000\,000$ carbon atoms?

46. *Physics*   One iron atom contains 55 amu. What is the mass in kilograms of $230\,000\,000$ iron atoms?

47. *Physics*   The rest mass of one electron is 9.1095×10^{-31} kg and the rest mass of one neutron is 1.6750×10^{-27} kg. Which has the larger mass—one electron or one neutron? How many times heavier is it?

48. *Electronics*   Compute the inductive reactance when $f = 10{,}000{,}000$ Hz and $L = 0.015$ H. (See Example 1.80.)

≡ 1.7
ROOTS

In Section 1.4 we have learned how to take powers of numbers and the different operations with exponents. In this section we will look at the reverse process of taking exponents. Remember that 3^4 meant $3 \times 3 \times 3 \times 3 = 81$. Now, suppose you were asked to find the number (or numbers) that you could raise to the 4th power and get 81. This number (or numbers) would be called the fourth root (or roots) of 81. We use the symbol $\sqrt[4]{81}$ to indicate the fourth root of 81.

In general, the symbol $\sqrt[n]{b}$ is used to indicate the *n*th root of a number *b*. When $n = 2$ it is not necessary to put the 2 in the symbol. Thus, $\sqrt{9}$ means $\sqrt[2]{9}$. The $\sqrt{}$ sign is called the **root** or **radical sign**.

EXAMPLE 1.82

$\sqrt{16}$ This is the second root of 16 or, as it is more commonly called, the square root of 16.

EXAMPLE 1.82 (Cont.)

$\sqrt[3]{8}$ This is the third root of 8 or the cube root of 8.

$\sqrt[6]{17}$ This is the sixth root of 17.

The solution to $\sqrt{16}$, or to the question "What number squared is 16?", can readily be seen to be either 4 or -4. This presents a problem. Do you write both as the solution to $\sqrt{16}$? Again, mathematicians have decided on a way to interpret $\sqrt{16}$. They have decided that $\sqrt{16} = 4$ and not -4. The symbol $-\sqrt{16}$ is used when they want the value -4.

What we want to find as the solution to $\sqrt[n]{b}$ is the **principal nth root of b**.

Principal nth Root of b

The principal nth root of b is positive if b is positive.

The principal nth root of b is negative if b is negative and n is odd.

EXAMPLE 1.83

$$\sqrt{16} = 4 \text{ and not } -4$$

$$\sqrt[3]{8} = 2 \text{ since } 2 \times 2 \times 2 = 8$$

$$\sqrt[3]{-8} = -2 \text{ since } -2 \times -2 \times -2 = -8$$

≡ Note

If you want the negative number that is the square root of b then you must write $-\sqrt{b}$. Thus, $-\sqrt{16} = -4$.

We will see later that there are n different numbers that, when raised to the nth power, will give the number b. Thus, there are n different numbers that will satisfy the definition of $\sqrt[n]{b}$. In most cases we will limit ourselves to just one of these—the principal nth root.

Rules of Roots

There are some rules of roots that are very similar to the rules of exponents.

Rule 1 of Roots $\qquad\qquad\qquad\qquad \sqrt[n]{ab} = \sqrt[n]{a} \cdot \sqrt[n]{b}$

This rule says that the root of a product is the same as the product of the root of each factor. Compare this to Rule 3 for exponents.

EXAMPLE 1.84

$$\sqrt{36} = \sqrt{4 \cdot 9} = \sqrt{4} \cdot \sqrt{9} = 2 \cdot 3 = 6$$

$$\sqrt{75} = \sqrt{25 \cdot 3} = \sqrt{25} \cdot \sqrt{3} = 5\sqrt{3}$$

$$\sqrt[3]{40} = \sqrt[3]{8 \cdot 5} = \sqrt[3]{8}\sqrt[3]{5} = 2\sqrt[3]{5}$$

Rule 2 of Roots	$\sqrt[n]{\dfrac{a}{b}} = \dfrac{\sqrt[n]{a}}{\sqrt[n]{b}}$

In words, this says that the root of a quotient is the root of the numerator divided by the root of the denominator.

EXAMPLE 1.85

$$\sqrt{\frac{9}{4}} = \frac{\sqrt{9}}{\sqrt{4}} = \frac{3}{2}$$

$$\sqrt[3]{\frac{5}{8}} = \frac{\sqrt[3]{5}}{\sqrt[3]{8}} = \frac{\sqrt[3]{5}}{2}$$

Rule 3 of Roots	$\left(\sqrt[n]{b}\right)^n = b$

This rule states that when the root of a number is raised to the same power as the root, then the result is the original number.

EXAMPLE 1.86

$$\left(\sqrt[3]{5}\right)^3 = 5$$

$$\left(\sqrt[11]{\frac{7}{3}}\right)^{11} = \frac{7}{3}$$

Fractional Exponents

Rule 4 of Roots	$\sqrt[n]{b} = b^{1/n}$

≡ **Note**   Rule 4 is very important. It means that every radical or root can be written as a fractional exponent. This means that all the rules of exponents will hold for roots. This causes Rule 3 to make even more sense because $\left(\sqrt[n]{b}\right)^n = \left(b^{1/n}\right)^n = b^{n/n} = b^1 = b$.

EXAMPLE 1.87

$$27^{1/3} = \sqrt[3]{27} = 3$$

$$16^{1/4} = \sqrt[4]{16} = 2$$

Application

EXAMPLE 1.88

According to *Blasius' formula*, the friction factor f of flow in a smooth pipe is

$$f = \frac{0.316}{\sqrt[4]{Re}}$$

where Re is *Reynold's number,* a constant that depends on the average velocity of the fluid flow, the diameter of the pipe, and the kinematic viscosity of the fluid. Calculate f when Re $= 7{,}340$.

Solution   We will show how to use a calculator to solve this problem. Remember, we are calculating

$$\frac{0.316}{\sqrt[4]{7{,}340}}$$

and also remember that $\sqrt[4]{7{,}340} = 7{,}340^{1/4}$.

   Now, we use our calculator:

PRESS	DISPLAY
.316 $\div$	0.316
7340 $\boxed{x^y}$	7340
$\boxed{(}$ 1 $\boxed{\div}$ 4 $\boxed{)}$	0.25
$\boxed{=}$	0.034139964

So, the friction factor is about 0.0341.

Application

EXAMPLE 1.89

For a large rectangular orifice of width b and depth $h_2 - h_1$, the discharge, Q, is

$$Q = \frac{2}{3} C_Q b \left(h_2^{2/3} - h_1^{2/3} \right) \sqrt{2g}$$

where C_Q is the coefficient of discharge. Determine the discharge if $C_Q = 0.96$, $b = 0.200$ m, $h_2 = 0.512$ m, $h_1 = 0.385$ m, and $g = 9.81$ m/s^2.

Solution   We will show how to use a calculator to solve this problem. Remember, after the values above are substituted, we are calculating $\frac{2}{3}(0.96)(0.200)(0.512^{2/3} - 0.385^{2/3})\sqrt{2(9.81)}$

PRESS	DISPLAY
2 $\boxed{\div}$ 3 $\boxed{\times}$	0.666666666
.96 $\boxed{\times}$ .2 $\boxed{\times}$	0.128
$\boxed{(}$.512 $\boxed{x^y}$ $\boxed{(}$ 2 $\boxed{\div}$ 3 $\boxed{)}$	0.666666666
$\boxed{-}$	0.64
.385 $\boxed{x^y}$ $\boxed{(}$ 2 $\boxed{\div}$ 3 $\boxed{)}$	0.666666666
$\boxed{)}$	0.110774835
$\boxed{\times}$ $\boxed{(}$ $\boxed{(}$ 2 $\boxed{\times}$ 9.81 $\boxed{)}$	19.62
$\boxed{\sqrt{}}$ $\boxed{)}$ $\boxed{=}$	0.06280592

So, the discharge is about 0.063 m^3/s.

Rule 4 also means that we can have numbers such as $4^{3/2} = \left(4^{1/2}\right)^3 = 2^3 = 8$. This also allows us to find some higher degree roots by simplifying them into combinations of similar roots. Thus $\sqrt[mn]{b} = \sqrt[m]{\sqrt[n]{b}}$ because, by the rules of exponents, $\sqrt[mn]{b} = b^{1/mn} = \left(b^{1/n}\right)^{1/m} = \sqrt[m]{\sqrt[n]{b}}$. It may not look as if this has been simplified. However, it has been, as Example 1.90 shows.

EXAMPLE 1.90

$$\sqrt[6]{64} = \sqrt[3]{\sqrt{64}} = \sqrt[3]{8} = 2$$

$$\sqrt[6]{729} = \sqrt[3]{\sqrt{729}} = \sqrt[3]{27} = 3$$

$$\sqrt[8]{256} = \sqrt[4]{\sqrt{256}} = \sqrt[4]{16} = 2$$

You will remember that in defining the principal nth root of a number, we said that if b was negative then n had to be an odd number. This means that, as yet, we have no definition for numbers such as $\sqrt{-1}$, $\sqrt{-9}$, or $\sqrt[4]{-16}$. In general, these are new types of numbers called imaginary numbers. They do not belong to the set of real numbers.

The special symbol j is used to represent $\sqrt{-1}$. (Some areas use i for $\sqrt{-1}$.) This allows us to represent numbers such as $\sqrt{-9}$ in terms of j, since $\sqrt{-9} = \sqrt{(9)(-1)} = \sqrt{9}\sqrt{-1} = 3j$. These are called imaginary numbers and we will study them in a later chapter. For the present time we need to remember that we cannot take an even root of a negative number.

Exercise Set 1.7

Find the principal *n*th root of each real number in Exercises 1–16.

1. $\sqrt{25}$
2. $\sqrt{36}$
3. $\sqrt{144}$
4. $\sqrt{121}$
5. $\sqrt[3]{8}$
6. $\sqrt[3]{-64}$
7. $\sqrt[3]{-27}$

8. $\sqrt[4]{81}$
9. $\sqrt[4]{16}$
10. $\sqrt[5]{-243}$
11. $\sqrt[4]{\dfrac{16}{81}}$
12. $\sqrt[3]{\dfrac{-27}{1,000}}$

13. $\sqrt{0.04}$
14. $\sqrt{0.25}$
15. $\sqrt[3]{-0.001}$
16. $\sqrt[3]{0.125}$

Simplify each of the real numbers in Exercises 17–48. Use the rules for roots and the rules for exponents.

17. $\sqrt{3^2}$
18. $\sqrt[3]{5^3}$
19. $\sqrt[4]{8.32^4}$
20. $\sqrt[6]{7.91^6}$

21. $\sqrt[3]{5}\,\sqrt[3]{25}$
22. $\sqrt[5]{-3}\,\sqrt[5]{81}$
23. $\sqrt[4]{8}\,\sqrt[4]{9}$
24. $\sqrt[6]{12}\,\sqrt[6]{48}$

25. $\dfrac{\sqrt{75}}{\sqrt{3}}$
26. $\dfrac{\sqrt{112}}{\sqrt{7}}$

27. $\dfrac{\sqrt[3]{5}}{\sqrt[3]{40}}$

28. $\dfrac{\sqrt[3]{11}}{\sqrt[3]{297}}$

29. $\sqrt[3]{2^3} + \sqrt[4]{5^4}$

30. $\sqrt[3]{3^2} - \sqrt[3]{2^3}$

31. $\sqrt[3]{\left(\dfrac{2}{3}\right)^3} - \sqrt[4]{\left(\dfrac{1}{3}\right)^4}$

32. $\sqrt[5]{\left(\dfrac{3}{4}\right)^5} + \sqrt[3]{\left(\dfrac{5}{4}\right)^3}$

33. $\sqrt{5^{2/3}}$

34. $\sqrt{7^{2/5}}$

35. $\sqrt[3]{16^{3/4}}$

36. $\sqrt[4]{27^{4/3}}$

37. $16^{3/4}$

38. $(-27)^{2/3}$

39. $(-8)^{2/3}$

40. $25^{3/2}$

41. $8^{-2/3}$

42. $9^{-3/2}$

43. $(-27)^{4/3}$

44. $(-32)^{3/5}$

45. $\sqrt{\dfrac{81}{(8)(0.01)}}$

46. $\sqrt[3]{\dfrac{(27)(0.008)^2}{0.027}}$

47. $\sqrt[3]{\dfrac{(0.125)^3\sqrt{144}}{\dfrac{3}{2}}}$

48. $\sqrt{\dfrac{64}{(0.25)(0.16)}}$

Solve Exercises 49–52.

49. *Electronics*   A wave consists of many crests and troughs. The distance between any two crests is called the wavelength. The frequency of a wave is the number of waves that pass a given point in a unit of time. The relationship between the wavelength λ of a wave, the frequency f in hertz (Hz), and the speed c at which the wave is propagated is given by the formula $\lambda = \dfrac{c}{f}$. λ and c must be in the same units of length, that is, if λ is given in meters, then c must be in meters per second (m/s) or if λ is given in feet, then c must be in feet per second (ft/s). What is the wavelength in meters of a radio wave with a frequency of 143.5 megahertz (MHz)(143.5 MHz = 143 500 000 Hz) if $c = 300\,000\,000$ m/s?

50. *Wastewater technology*   The minimum retention time (in days) of a certain waste-handling system is given by the expression below. Evaluate the given expression.

$$\dfrac{1}{0.35\left[1 - \sqrt{\dfrac{8{,}100}{8{,}100 + 121{,}000}}\right] - 0.045}$$

51. *Acoustics*   The speed of sound of a longitudinal sound wave in a liquid, v_L, is given by the formula

$$v_L = \sqrt{\dfrac{K}{\rho}}$$

where K is the bulk modulus and ρ is the mass density of the liquid. Determine the speed of sound in water if $K = 2.1 \times 10^9$ N/m^2 and $\rho = 1\,000$ kg/m^3.

52. *Wastewater technology*   For a large rectangular orifice of depth $h_2 - h_1$, the average discharge velocity is

$$\bar{v} = \dfrac{2C_v(h_2^{2/3} - h_1^{2/3})\sqrt{2g}}{3(h_2 - h_1)}$$

Find $\bar{v}$ (in meters per second) when C_v, the coefficient of velocity, is 0.96, $h_2 = 0.512$ m, $h_1 = 0.385$ m, and $g = 9.81$ m/s^2.

≡ CHAPTER 1 REVIEW

Important Terms and Concepts

Absolute value
Accuracy
Associative laws
Commutative laws
Decimals
Distributive law
Exact numbers
Exponents
  Fractional
  Rules for
Identity elements
Inequalities
Integers
  Negative
  Positive

Zero
Inverse elements
Irrational numbers
Natural numbers
Order of operations
Precision
Rational numbers
Real numbers
Reciprocal
Roots
Rounding off
Scientific notation
Significant digits
Zero

Review Exercises

1. To which sets of numbers do each of the following numbers belong?
   (a) -24
   (b) $\dfrac{15}{7}$
   (c) $\sqrt[3]{-17}$

2. Give the absolute value of each of these numbers.
   (a) $\sqrt{42}$
   (b) -16
   (c) $\dfrac{-5}{8}$

3. Give the reciprocal of each of these numbers.
   (a) $\dfrac{2}{3}$

   (b) -8
   (c) $\dfrac{-1}{5}$

4. Give the additive inverse of each of these numbers.
   (a) -5
   (b) 17
   (c) $4\sqrt{2}$

5. In each part, identify the basic law of real numbers that is being used.
   (a) $14 + 7 = 7 + 14$
   (b) $6(8 + 2) = 6 \cdot 8 + 6 \cdot 2$
   (c) $14 = 0 + 14$
   (d) $\dfrac{2}{3} \times \dfrac{3}{2} = 1$

In Exercises 6–20, perform the indicated operation.

6. $16 + 48$

7. $37 + (-81)$

8. $95 - 42$

9. $37 - (-61)$

10. $4 \times (-8)$

11. $\dfrac{2}{3} + \dfrac{-5}{6}$

12. $\dfrac{4}{5} - \dfrac{5}{6}$

13. $-9 \div 3$

14. $12 \div (-4)$

15. $\dfrac{2}{3} \times \dfrac{-3}{5}$

16. $3\dfrac{3}{4} \times \left(-4\dfrac{1}{3}\right)$

17. $\dfrac{2}{3} \div \dfrac{1}{4}$

18. $\dfrac{1}{5} \div \dfrac{-2}{15}$

19. $2\dfrac{1}{2} \div 3\dfrac{1}{4}$

20. $-4\dfrac{1}{3} \div \left(-3\dfrac{1}{6}\right)$

Evaluate each number in Exercises 21–26.

21. 2^5

22. $(-3)^4$

23. $(-4)^3$

24. $8^{1/3}$

25. $4^{1/2}$

26. $(-64)^{1/3}$

In Exercises 27–38, perform the indicated operations. Leave all answers in terms of positive exponents.

27. $2^5 \cdot 2^3$

28. $3^5 \cdot 3^4$

29. $2^{-3} \cdot 2^5$

30. $16^{-4} \cdot 16^{-3}$

31. $\left(4^3\right)^5$

32. $\left(2^{-3}\right)^4$

33. $\left(4^{1/3}\right)^3$

34. $\left(5^{1/4}\right)^{2/3}$

35. $\dfrac{a^2 b^3}{ab^4}$

36. $\dfrac{x^2 y^3 z}{x^3 yz}$

37. $\left(ax^2\right)^{-2}$

38. $\dfrac{\left(by^{-2}\right)^2}{\left(cy^{-4}\right)^{1/4}}$

Which of the pairs of numbers in Exercises 39–42 is (a) more accurate and which is (b) more precise?

39. 2.37; 42,000

40. 0.002; 2.02

41. 7; 0.7

42. 1200; 0.0021

In Exercises 43–46, round off each number to (a) two significant digits, (b) three significant digits, (c) the nearest tenth, and (d) the nearest hundredth.

43. 7.351

44. 18.2874

45. 2.0528

46. 4.028476

Change each of the numbers in Exercises 47–50 to scientific notation.

47. 371 000 000 000

48. 2 540 000 000 000 000

49. 0.000 000 000 024

50. 0.000 000 000 000 000 000 049 1

Find the principal root of the real numbers in Exercises 51–54.

51. $\sqrt{144}$

52. $\sqrt[3]{-64}$

53. $\sqrt[4]{625}$

54. $\sqrt{\dfrac{36}{121}}$

≡ CHAPTER 1 TEST

1. Give the absolute value of each of the following:

   (a) $\dfrac{4}{3}$

   (b) $\dfrac{1}{2} - \dfrac{5}{8}$

   (c) -6

2. What is the reciprocal of $-\dfrac{3}{7}$?

3. What is the additive inverse of $-\dfrac{5}{8}$?

In Exercises 4–21, perform the indicated operation.

4. $35 + 76$

5. $-47 - 65$

6. -5×14

7. $\dfrac{5}{3} + 5\dfrac{2}{3}$

8. $\dfrac{7}{4} - (-\dfrac{3}{5})$

9. $-4\dfrac{1}{2} \times 2\dfrac{1}{3}$

10. $\dfrac{-5}{7} \div \dfrac{15}{28}$

11. $-2\dfrac{1}{3} \div -4\dfrac{5}{6}$

12. Evaluate $(-5)^3$.

13. $4^{3/2}$

14. $2^6 \cdot 2^{-4}$

15. $\left(5^{1/4}\right)^8$

16. $3^{5/2} \div 3^{-3/2}$

17. $\dfrac{a^3 b^5}{a^4 b^2}$

18. Which is more accurate: 4.516 or 37 000?

19. Which is more precise: 0.000 51 or 51.02?

20. Write 47,500,000,000,000 in scientific notation.

21. What is the principal root of $\sqrt{\dfrac{49}{25}}$?

2

Algebraic Concepts and Operations

In order to operate correctly, a computer relies on instructions given to it by a programmer. The computer programmer must understand algebra in order to tell the computer what to do.

Courtesy of Ruby Gold

Until now we have spent most of our time working with real numbers. We have learned the basic operations with real numbers, how to take powers, roots, and absolute values, and how to write real numbers in scientific notation. Now it is time to begin our work in algebra.

In this chapter, we will begin by looking at the operations of addition, subtraction, multiplication, and division of algebraic expressions. Once these four operations have been learned, we will use them to learn how to solve equations. Finally, we will use our equation-solving skills to solve many problems like those encountered in technical areas.

☰ 2.1
ADDITION AND SUBTRACTION

One of the first things that everyone notices about algebra is that it uses letters and other symbols to represent numbers. Algebra also uses letters and other symbols to represent unknown amounts in equations and inequalities. In the first chapter we used letters to help present some of the rules. For example, we said that $b^{-n} = \dfrac{1}{b^n}$, if $b \neq 0$.

Variables

When a letter or other symbol is used to indicate something that can be assigned any value from a given or implied set of numbers, it is called a **variable.** We have already used letters as variables in this book. For example, we used a, b, and c as variables when we said that $a(b + c) = ab + ac$. We used b and n as variables when we said that $b^{-n} = \dfrac{1}{b^n}$, if $b \neq 0$. Another place where variables are used is on a calculator. For example, the $\boxed{x^2}$, $\boxed{\sqrt{x}}$, $\boxed{1/x}$, and $\boxed{y^x}$ keys on a calculator all use the letter x as a variable and one uses the letter y.

Constants

Letters or other symbols may also be used to designate fixed, but unspecified, numbers called **constants.** One constant that you are already familiar with is π, which represents the number $3.14159265\ldots$. (The $\ldots$ at the end of this number means that there are more digits after the 5.) Usually letters near the beginning of the alphabet, like a, b, c, d, are used to represent constants. Letters near the end of the alphabet, such as w, x, y, and z, are used to indicate variables. Thus, in the expression $ax + b$, a and b represent constants and x is a variable.

Algebraic Expressions

The term **algebraic expression** is used for any combination of variables and constants that is formed using a finite number of operations. Examples of algebraic expressions include ax^2, $ax^2 + bx + c$, $\dfrac{x^3 + \sqrt{y}}{ax^2 - by}$, $\dfrac{4x^{-3} + 7x^2}{2x - 3w}$, and $1.8C + 32$.

Algebraic Terms

If an algebraic expression consists of parts connected by plus or minus signs, it is called an **algebraic sum.** Each of the parts of an algebraic sum, together with the sign preceding it, is called an **algebraic term.**

EXAMPLE 2.1

What are the algebraic terms of the algebraic sum $5x^7 + 2x^3y - \dfrac{4x}{y^2}$?

Solution This algebraic sum has three algebraic terms: $5x^7$, $2x^3y$, and $-\dfrac{4x}{y^2}$.

An algebraic term consists of several **factors.** The term $5xy$ has the individual factors of 5, x, and y. Other factors are products of two or more of the individual factors. Thus, $5x$, $5y$, xy, and $5xy$ are also factors of $5xy$.

EXAMPLE 2.2

What are the factors of $\dfrac{6x^2}{y}$?

Solution The individual factors are 2, 3, x, and $\dfrac{1}{y}$. Other factors are 6, $2x$, $3x$,

$6x$, x^2, $2x^2$, $3x^2$, $6x^2$, $\dfrac{2}{y}$, $\dfrac{3}{y}$, $\dfrac{6}{y}$, $\dfrac{x}{y}$, $\dfrac{2x}{y}$, $\dfrac{3x}{y}$, $\dfrac{6x}{y}$, $\dfrac{x^2}{y}$, $\dfrac{2x^2}{y}$, $\dfrac{3x^2}{y}$, and $\dfrac{6x^2}{y}$.

EXAMPLE 2.3

What are the factors of $\dfrac{-10\,(x+y)}{z}$?

Solution The individual factors are -1, 2, 5, $x + y$, and $\dfrac{1}{z}$. The other factors are products of these factors.

Each term has two parts. One part is the coefficient and the other part contains the variables. The **coefficient** is the product of all of the constants. Normally the coefficient is written at the front of the term. A variable with no visible coefficient, such as x or y, is understood to have a coefficient of 1. The coefficient of the term $4x^2$ is 4; of $18y^3z$ is 18; of $\dfrac{w^2}{3}$ is $\dfrac{1}{3}$; and of yx^2 is 1. The coefficient of $4\pi bx$ is $4\pi b$ if we assume that b is a constant.

Terms that involve exactly the same variables raised to exactly the same power are called **like terms.** For example, $7x^2y$ and $4x^2y$ are like terms because both contain the same variables (x and y), the variable x is raised to the second power in each term and the y is raised to the first power. $4xyz^3$ and $5xyz^2$ are not like terms even though they both contain the variables x, y, and z, because one term has the variable z raised to the third power and in the other term it is raised to the second power.

Monomials and Multinomials

An algebraic expression with only one term is a **monomial.** An algebraic sum with exactly two terms is a **binomial;** an algebraic sum with three terms is a **trinomial;** and an algebraic sum with two or more terms is a **multinomial.** As you can see, binomials and trinomials are special kinds of multinomials.

EXAMPLE 2.4

$4x^2$, $\frac{7}{5}xy$, and $-8xz^3\sqrt{w}$ are all monomials.

$ax^2 + by$, $\sqrt{3}x^3z - 2w$, and $4x^2 + 2$ are binomials.

$3x^3 - 2x + 1$ and $2xy + 3yz - 4\sqrt{2y}$ are trinomials.

EXAMPLE 2.4 (Cont.)

$16x^3 - 7\sqrt{2}y^2 + \frac{2}{3}xy - 7^5x^3y$ is a multinomial with four terms. In this last expression the coefficient of the first term is 16, the second term has a coefficient of $-7\sqrt{2}$, the third $\frac{2}{3}$, and the fourth -7^5 (or $-16,807$).

Adding and Subtracting Multinomials

In order to add or subtract monomials, we add or subtract the coefficients of like terms. So, $3x^2 + 5x^2 = 8x^2$. We can justify this with the distributive property, since $3x^2 + 5x^2 = (3 + 5)x^2 = 8x^2$.

≡ **Note**   Remember, you can only add or subtract like terms.

So,

$$3x^2 + 2xy - y^3 - 8xy = 3x^2 - 6xy - y^3$$

because the only two like terms are $2xy$ and $-8xy$. When these are added, we get $-6xy$. None of the other terms can be combined because they are not like terms.

EXAMPLE 2.5

$$(3x^2 + 2yx - 5y) + (-6xy + 6x^2 - 7y)$$
$$= (3x^2 + 6x^2) + (2yx - 6xy) + (-5y - 7y)$$
$$= 9x^2 - 4yx - 12y$$

Notice that $2yx$ and $-6xy$ are like terms even though the variables are written in a different order. This is an application of the commutative law introduced in Section 1.2.

EXAMPLE 2.6

$$(4y - 3z + 4xy) - (7y - 3z - 6xy)$$
$$= 4y - 3z + 4xy - 7y + 3z + 6xy$$
$$= (4y - 7y) + (-3z + 3z) + (4xy + 6xy)$$
$$= -3y + 10xy$$

Notice that when you have a negative sign in front of a parenthesis it means to multiply *each* term inside the parentheses by -1. So,

$$-(7y - 3z - 6xy) = -7y + 3z + 6xy$$

As mentioned in Section 1.2, there are several different symbols for grouping. These symbols are parentheses ( ), brackets [ ], and braces { }. When it is necessary to simplify an expression, you should start removing grouping symbols from the inside. For example, the expression

$$5\{2x - [3x + (4x - 5) + 2] - 3x\}$$

would be simplified by first removing the parentheses and combining any like terms that are within the brackets.

$$5\{2x - [3x + (4x - 5) + 2] - 3x\} = 5\{2x - [3x + 4x - 5 + 2] - 3x\}$$
$$= 5\{2x - [7x - 3] - 3x\}$$

Next, remove the brackets. Do not forget the negative sign in front of the left bracket [. This indicates that each term inside the brackets is to be multiplied by -1 when the brackets are removed.

$$= 5\{2x - [7x - 3] - 3x\}$$
$$= 5\{2x - 7x + 3 - 3x\}$$
$$= 5\{-8x + 3\}$$

Now, remove the braces. The 5 in front of the left brace { means that each term within the braces should be multiplied by 5.

$$= 5\{-8x + 3\}$$
$$= -40x + 15$$

Application

EXAMPLE 2.7

The total resistance R in a parallel circuit with two resistances, R_1 and R_2, is given by the formula $\dfrac{1}{R} = \dfrac{1}{R_1} + \dfrac{1}{R_2}$. Simplify the right-hand side of this equation.

Solution In order to simplify these two fractions, we need to have a common denominator. For this fraction, the common denominator is $R_1 R_2$. Multiplying the right-hand side of the equation by $\dfrac{R_1 R_2}{R_1 R_2}$ produces

$$\frac{1}{R} = \left(\frac{R_1 R_2}{R_1 R_2}\right)\left(\frac{1}{R_1} + \frac{1}{R_2}\right)$$
$$= \left(\frac{R_1 R_2}{R_1 R_2}\right)\left(\frac{1}{R_1}\right) + \left(\frac{R_1 R_2}{R_1 R_2}\right)\left(\frac{1}{R_2}\right)$$
$$= \frac{R_1 R_2}{R_1^2 R_2} + \frac{R_1 R_2}{R_1 R_2^2}$$
$$= \frac{R_2}{R_1 R_2} + \frac{R_1}{R_1 R_2}$$
$$= \frac{R_2 + R_1}{R_1 R_2}$$

Exercise Set 2.1

In Exercises 1–50, simplify each algebraic expression.

1. $4x + 7x$

2. $5y - 2y$

3. $3z - z$

4. $7w + 4w - w$

5. $8x + 9x^2 - 2x$

6. $11y - 7y + 6y^2$

7. $10w + w^2 - 8w^2$

8. $y^2 - 6y^2 + 4y$

9. $ax^2 + a^2x + ax^2$

10. $by - by^2 + by$

11. $7xy^2 - 5x^2y + 4xy^2$

12. $12wz - 8w^2z + 6w^2z$

13. $(a + 6b) - (a - 6b)$

14. $(x - 7y) - (7y - x)$

15. $(2a^2 + 3b) + (2b + 4a)$

16. $(7c^2 - 8d) + (6d - 8c)$

17. $(4x^2 + 3x) - (2x^2 - 3x)$

18. $(3y^2 - 4x) - (4y + 2x)$

19. $2(6y^2 + 7x)$

20. $5(3a + 4b)$

21. $-3(4b - 2c)$

22. $-2(-6b + 3a)$

23. $4(a + b) + 3(b + a)$

24. $2(c + d) + 8(d + c)$

25. $3(x + y) - 2(x + y)$

26. $3(x^2 - y) - 2(y + x^2)$

27. $2(a + b + c) + 3(a + b - c)$

28. $4(x - y + z) + 2(x + y - z)$

29. $3[2(x + y)]$

30. $4[3(x - y)]$

31. $3(a + b) + 4(a + b) - 2(a + b)$

32. $2(x + y) - 3(x + y) - 4(x + y)$

33. $2(a+b+c)+3(a-b+c)+(a-b-c)$

34. $3(x-y+z)-2(x+y-z)+4(-x-y+z)$

35. $2(x+3y) - 3(x-2y) + 5(2x-y)$

36. $3(y - 2a) - (a + 3y) + 4(a - 3y)$

37. $3(x + y - z) - 2(3x + 2y - z) - 3(x - y + 4z)$

38. $5(x - y + z) - (y - x + z) + 2(x - 2y + z)$

39. $(x + y) - 3(x - z) + 4(y + 4z) - 2(x + y - 3z)$

40. $(a + b) - 2(b - c) + 4(c + 2d) - 5(d + 2c - 3b - a)$

41. $x + [3x + 2(x + y)]$

42. $x + [5y + 3(y - x)]$

43. $y - [2z - 3(y + z) + y]$

44. $2w - [4z - 5(z + w) + 2w]$

45. $[2x+3(x+y) - 2(x-y)+y] - 2x$

46. $[4a-8(a+b)+2(a-b)+3a]-3b$

47. $-\{-[2a - (3b + a)]\}$

48. $-\{-3[4x - (5x - 4y)]\}$

49. $5a-2\{4[a+2(4a+b)-b]+a\}-a$

50. $7x - 3\{-[x + 2(x - y) - y] + 2x\} - 3y$

Solve the word problems in Exercises 51–52.

51. *Civil engineering* In order to determine the cost of widening a highway, several cost comparisons were used. These led to the following expression for determining the total cost: $p + \frac{1}{2}p + \frac{2}{3}p$. Simplify by combining like terms.

52. *Construction* A concrete mix has the following ingredients by volume: 1 part cement, 2 parts water, 3 parts aggregate, and 3 parts sand. These are used to produce the expression $x + 2x + 3x + 3x$. This expression is then used to determine the amount of each ingredient for a specific amount of concrete. Simplify by combining like terms.

≡ 2.2
MULTIPLICATION

In Section 1.4 we learned some rules for exponents. The very first rule stated that $b^m b^n = b^{m+n}$. We will use this rule as a foundation for learning how to multiply multinomials. We will also use the rule for multiplying real numbers from Section 1.3.

Multiplying Monomials

First, we will find the product of two or more monomials. When multiplying monomials, first multiply the numerical coefficients to get the numerical coefficient of the product. Then, multiply the remaining factors using the rules of exponents.

EXAMPLE 2.8

$$(3x^3)(-7bx^4) = -21bx^{3+4} = -21bx^7$$
$$(4ax^2)(7bx^3) = 28abx^{2+3} = 28abx^5$$
$$(-5xy^2z^3)(3x^2y^2) = -15x^3y^4z^3$$
$$\left(\sqrt{2}x^3y\right)\left(\sqrt{3}xy^4\right) = \sqrt{6}x^4y^5$$

To multiply a monomial and a multinomial, use the distributive property. Distribute the monomial over the multinomial. Thus, by the distributive property $a(b+c) = ab+ac$. Each term of the multinomial will be multiplied by the monomial.

EXAMPLE 2.9

$$4x^2y(3xy^2 + 2x^3y) = (4x^2y)(3xy^2) + (4x^2y)(2x^3y)$$
$$= 12x^3y^3 + 8x^5y^2$$

EXAMPLE 2.10

$$-7ab^2(2ax^3 - 5abx + 3b^3) = (-7ab^2)(2ax^3) + (-7ab^2)(-5abx)$$
$$+ (-7ab^2)(3b^3)$$
$$= -14a^2b^2x^3 + 35a^2b^3x - 21ab^5$$

As you gain experience, you will find that you may not always need to write all the steps. The more you practice, the better you will become, until you may be able to write the correct answer without writing the middle expression.

Multiplying Multinomials

To multiply one multinomial by another, multiply each term in one multinomial by each term in the other multinomial. Again, use the distributive property to help in the multiplication.

EXAMPLE 2.11

$$(x^2 + b)(x + c) = x^2(x + c) + b(x + c)$$
$$= x^3 + cx^2 + bx + bc$$

EXAMPLE 2.12

$$(y^2 + 2f)(y^2 - 3f) = y^2(y^2 - 3f) + 2f(y^2 - 3f)$$
$$= y^4 - 3y^2f + 2y^2f - 6f^2$$
$$= y^4 - y^2f - 6f^2$$

The preceding two examples show multiplication of two binomials. The same procedure is used for multiplying any two multinomials. The next example shows how to multiply two trinomials. This example uses an extended version of the distributive property, where $a(b + c + d) = ab + ac + ad$.

EXAMPLE 2.13

$$(2x + 3y + z)(4x - 3y - z) = 2x(4x - 3y - z) + 3y(4x - 3y - z)$$
$$+ z(4x - 3y - z)$$
$$= (8x^2 - 6xy - 2xz)$$
$$+ (12xy - 9y^2 - 3yz)$$
$$+ (4xz - 3yz - z^2)$$
$$= 8x^2 + 6xy + 2xz - 9y^2 - 6yz - z^2$$

Another method is to multiply in the same manner that you used with real numbers. This is shown in Examples 2.14 and 2.15. When using this method, like terms are put in the same column.

EXAMPLE 2.14

$(2x + 5)(3x - 4)$

Solution

$$
\begin{array}{r}
2x + 5 \\
\times \quad 3x - 4 \\
\hline
-8x - 20 \\
6x^2 + 15x \quad\quad \\
\hline
6x^2 + 7x - 20
\end{array}
$$
Add the like terms $15x$ and $-8x$.

EXAMPLE 2.15

$(x^3 + 2x^2 - 3x + 7)(5x^2 - 4x + 2)$

Solution

$$
\begin{array}{r}
x^3 + 2x^2 - 3x + 7 \\
\times \quad\quad 5x^2 - 4x + 2 \\
\hline
2x^3 + 4x^2 - 6x + 14 \\
-4x^4 - 8x^3 + 12x^2 - 28x \quad\quad \\
5x^5 + 10x^4 - 15x^3 + 35x^2 \quad\quad\quad\quad \\
\hline
5x^5 + 6x^4 - 21x^3 + 51x^2 - 34x + 14
\end{array}
$$

The FOIL Method for Multiplying Binomials

There will be many times when you will need to multiply two binomials. Because this happens so often, there are a few patterns that we should notice. The first pattern, called the **FOIL method,** is one that many people use to remember how to multiply two binomials. The letters for FOIL come from the first letters for the words First, Outside, Inside, and Last.

 Hint

Use the FOIL method to help you remember how to multiply binomials. FOIL indicates the order in which terms of two binomials could be multiplied.

>   **F**   suggests the product of the **F**irst terms.
>   **O**   suggests the product of the **O**utside terms.
>   **I**   suggests the product of the **I**nside terms.
>   **L**   suggests the product of the **L**ast terms.

EXAMPLE 2.16

Use the FOIL method to multiply $(x + 4)(x + 3)$.

Solution

$(x + 4)(x + 3)$ x^2 the product of the **F**irst terms
       F

$(x + 4)(x + 3)$ $3x$ the **O**utside terms
       O

$(x + 4)(x + 3)$ $4x$ the **I**nside terms
       I

$(x + 4)(x + 3)$ 12 the **L**ast terms
       L

Adding and combining the terms in the right-hand column

$$x^2 + 3x + 4x + 12 = x^2 + 7x + 12$$

gives the correct answer.

This example looks much longer than it really is. Try it! You will find that this will help "foil" errors caused by not multiplying all the terms.

The Special Product $(a + b)(a - b)$

One special product is shown by this example:

$$(x + 4)(x - 4) = x^2 - 4x + 4x - 16 = x^2 - 16$$

Notice that the two middle terms are additive inverses of each other and their sum is zero.

Let's look at the general case of this type of problem.

$$(a + b)(a - b) = a^2 - ab + ab - b^2 = a^2 - b^2$$

What is so special about this problem? Look at the two factors in the original problem. One factor is $a + b$ and the other is $a - b$. Notice that each factor has the same first term a, and that the second terms are additive inverses of each other, b and $-b$. Whenever you have a product of two binomials in the form $a + b$ and $a - b$, the answer is $a^2 - b^2$.

Difference of Squares	$(a+b)(a-b) = a^2 - b^2$

EXAMPLE 2.17

$$(x+7)(x-7) = x^2 - 49 \qquad \text{Note that } 49 = 7^2.$$

$$(3a+b)(3a-b) = 9a^2 - b^2 \qquad \text{Note that } (3a)^2 = 9a^2.$$

$$(\frac{4}{3}x - 2y)(\frac{4}{3}x + 2y) = \frac{16}{9}x^2 - 4y^2$$

The Special Product $(a + b)^2$

The final special product of binomials is a perfect square: $(a+b)^2$. Remember that $(a+b)^2 = (a+b)(a+b)$. If we multiply these we get

$$(a+b)^2 = (a+b)(a+b) = a^2 + ab + ab + b^2 = a^2 + 2ab + b^2$$

The answer contains the squares of the first and last terms. The middle term in the answer is twice the product of the two terms, a and b. So, when you have a square of a binomial you get $(a+b)^2 = a^2 + 2ab + b^2$ or $(a-b)^2 = a^2 - 2ab + b^2$.

Square of a Binomial	$(a+b)^2 = a^2 + 2ab + b^2$
	$(a-b)^2 = a^2 - 2ab + b^2$

EXAMPLE 2.18

$$(x+3y)^2 = x^2 + 6xy + 9y^2$$

$$(t+7)^2 = t^2 + 14t + 49$$

$$(4a-b)^2 = 16a^2 - 8ab + b^2$$

$$\left(\frac{ax}{2} - \frac{3c}{4}\right)^2 = \frac{a^2x^2}{4} - \frac{3}{4}acx + \frac{9c^2}{16} \qquad \text{Note that } 2\left(\frac{ax}{2}\right)\left(\frac{3c}{4}\right) = \frac{3}{4}acx.$$

Application

EXAMPLE 2.19

A physics equation that describes an elastic collision is $m_1(v_a - v_b)(v_a + v_b) = m_2(v_c - v_d)(v_c + v_d)$. Perform the indicated multiplications.

Solution We will begin by multiplying the expressions in parentheses.

$$m_1(v_a - v_b)(v_a + v_b) = m_2(v_c - v_d)(v_c + v_d)$$

$$m_1(v_a^2 - v_b^2) = m_2(v_c^2 - v_d^2)$$

Next, we shall multiply through the left-hand side by m_1 and the right-hand side by m_2, producing

$$m_1 v_a^2 - m_1 v_b^2 = m_2 v_c^2 - m_2 v_d^2$$

| **EXAMPLE 2.19 (Cont.)** | We cannot simplify this equation any more, because none of the remaining terms are like terms. |

∎

Exercise Set 2.2

In Exercises 1–60, perform the indicated multiplication.

1. $(a^2x)(ax^2)$

2. $(by^2)(b^2y)$

3. $(3ax)(2ax^2)$

4. $(5by)(3b^2y)$

5. $(2xw^2z)(-3x^2w)$

6. $(-4ya^2b)(6y^2b)$

7. $(3x)(4ax)(-2x^2b)$

8. $(4y)(3y^2b)(-5by^2)$

9. $2(5y-6)$

10. $4(3x-5)$

11. $-5(4w-7)$

12. $-3(8+5p)$

13. $3x(7y+4)$

14. $6x(8y-7)$

15. $-5t(-3+t)$

16. $-3n(2n-5)$

17. $\dfrac{1}{2}a(4a-2)$

18. $\dfrac{1}{3}x(-21x-15)$

19. $2x(3x^2-x+4)$

20. $3y(4y^2-5y-7)$

21. $4y^2(-5y^2+2y-5+3y^{-1}-6y^{-2})$

22. $5p^2(-4p^3-3p+2+p^{-1}-7p^{-2})$

23. $(a+b)(a+c)$

24. $(s+t)(s+2t)$

25. $(x+5)(x^2+6)$

26. $(y+3)(y^2+7)$

27. $(2x+y)(3x-y)$

28. $(4a+b)(8a-b)$

29. $(2a-b)(3a-2b)$

30. $(4p+q)(3p-2q)$

31. $(b-1)(2b+5)$

32. $(4x-1)(3x-2)$

33. $(7a^2b+3c)(8a^2b-3c)$

34. $(6p^2r+2t)(5p^2r+4t)$

35. $(x+4)(x-4)$

36. $(a+8)(a-8)$

37. $(p-6)(p+6)$

38. $(b-10)(b+10)$

39. $(ax+2)(ax-2)$

40. $(xy-3)(xy+3)$

41. $(2r^2+3x)(2r^2-3x)$

42. $(4p^3-7d)(4p^3+7d)$

43. $(5a^2x^3-4d)(5a^2x^3+4d)$

44. $(3p^2st-\dfrac{11}{3}w^3)(3p^2st+\dfrac{11}{3}w^3)$

45. $\left(\dfrac{2}{3}pa^2f+\dfrac{3}{4}tb^3\right)$
$\times\left(-\dfrac{2}{3}pa^2f+\dfrac{3}{4}tb^3\right)$

46. $\left(\dfrac{\sqrt{3}}{2}+\dfrac{7}{5}t^2u\right)$
$\times\left(-\dfrac{\sqrt{3}}{2}+\dfrac{7}{5}t^2u\right)$

47. $(x+y)^2$

48. $(p+r)^2$

49. $(x-5)^2$

50. $(b-7)^2$

51. $(a+3)^2$

52. $(w+5)^2$

53. $(2a+b)^2$

54. $(3c+d)^2$

55. $(3x-2y)^2$

56. $(5a-6f)^2$

57. $4x(x+4)(3x-2)$

58. $5y(y-6)(2y+3)$

59. $(x+y-z)(x-y+z)$

60. $(a+b+c)(a-b-c)$

Solve Exercises 61 and 62.

61. *Computer science*   The amount of time required to test a computer "chip" with n cells is given by the expression $2[n(n+2)+n]$. Simplify this expression. (In this problem, asking you to simplify means that you are to perform the indicated multiplications and add the like terms.)

62. *Aeronautical engineering*   An aircraft uses its radar to measure the direct echo range R to another object. If x represents the distance to the ground echo point, then the following expression results:

$$(2R-x)^2 - x^2 - R^2$$

Simplify this expression.

≡ 2.3
DIVISION

To find the quotient when one multinomial is divided by another, we will again need to use the rules of exponents. Rule 5 in Section 1.4 states that $\frac{b^m}{b^n} = b^{m-n}$, if $b \neq 0$. This is a very helpful rule in the division of algebraic expressions.

Remember that division can be indicated by $x \div y$, $\frac{x}{y}$, or x/y. You must be very careful, when reading algebra problems and writing your answers, to watch how terms are grouped. For example, in $a^2 + b/c + 3$ the only division is b divided by c. This is equivalent to $a^2 + \frac{b}{c} + 3$. Now, consider $(a^2 + b)/c + 3$. The parentheses indicate that the entire expression $a^2 + b$ is being divided by c. So $(a^2 + b)/c + 3 = \left(\frac{a^2 + b}{c} \right) + 3$.

This could also be written without parentheses as $\frac{a^2 + b}{c} + 3$. Finally, $(a^2 + b)/(c + 3)$ means that the expression $a^2 + b$ is being divided by the expression $c + 3$. So, $(a^2 + b)/(c + 3) = \frac{a^2 + b}{c + 3}$.

Dividing a Monomial by a Monomial

We will begin by dividing a monomial by a monomial. Consider $16x^5 \div 8x^2$. Using the laws of exponents, we have $x^5 \div x^2 = x^{5-2} = x^3$. If there are numerical coefficients, they are divided separately. So, $16x^5 \div 8x^2 = \frac{16}{8} \frac{x^5}{x^2} = 2x^{5-2} = 2x^3$.

Each variable should be treated separately. For example, $24x^3y^2 \div 8xy^2 = \frac{24}{8} \frac{x^3}{x} \frac{y^2}{y^2} = 3x^{3-1}y^{2-2} = 3x^2y^0 = 3x^2$.

EXAMPLE 2.20

Divide $6w^2xy^4z$ by $9wx^3y^2z$.

Solution

$$6w^2xy^4z \div 9wx^3y^2z = \frac{6w^2xy^4z}{9wx^3y^2z}$$

$$= \frac{6}{9} w^{2-1}x^{1-3}y^{4-2}z^{1-1}$$

$$= \frac{2}{3} wx^{-2}y^2z^0$$

$$= \frac{2wy^2}{3x^2}$$

EXAMPLE 2.21

Divide $35a^2b^3c^{-2}$ by $-7a^{-3}bc^2$.

Solution

$$35a^2b^3c^{-2} \div -7a^{-3}bc^2 = \frac{35a^2b^3c^{-2}}{-7a^{-3}bc^2}$$

$$= \frac{35}{-7} a^{2-(-3)}b^{3-1}c^{-2-2}$$

$$= -5a^5b^2c^{-4}$$

EXAMPLE 2.21 (Cont.)

$$= \frac{-5a^5b^2}{c^4}$$

Caution

Be careful! A typical error here is cancelling noncommon factors. Do not make either of the following mistakes:

$$\frac{\cancel{x} + y}{\cancel{x}} = 1 + y$$

or

$$\frac{x + \cancel{4}}{\cancel{4}} = x + 1$$

These are *not correct*, as you can see by substituting some values for x or y.

Dividing a Multinomial by a Monomial

In order to divide a multinomial by a monomial, you should divide each term of the multinomial by the monomial. You may remember that $\frac{a+b}{c} = \frac{a}{c} + \frac{b}{c}$. (This is part of Rule 6 in Section 1.3.)

EXAMPLE 2.22

Divide $(15x^2 + 9x)$ by $6x$.

Solution

$$(15x^2 + 9x) \div 6x = \frac{15x^2 + 9x}{6x}$$

$$= \frac{15x^2}{6x} + \frac{9x}{6x}$$

$$= \frac{5}{2}x + \frac{3}{2}$$

EXAMPLE 2.23

Divide $(4r^2st^3 - 8rs^3)$ by $-2rst^2$.

Solution

$$(4r^2st^3 - 8rs^3) \div -2rst^2 = \frac{4r^2st^3 - 8rs^3}{-2rst^2}$$

$$= \frac{4r^2st^3}{-2rst^2} + \frac{-8rs^3}{-2rst^2}$$

$$= -2rt + 4\frac{s^2}{t^2}$$

Remember to divide each term in the multinomial by the monomial.

EXAMPLE 2.24

Divide $(6xy - 4xy^2 + 9x^2y - 12x^2y^2)$ by $3xy$.

EXAMPLE 2.24 (Cont.)

Solution $\dfrac{6xy - 4xy^2 + 9x^2y - 12x^2y^2}{3xy} = \dfrac{6xy}{3xy} - \dfrac{4xy^2}{3xy} + \dfrac{9x^2y}{3xy} - \dfrac{12x^2y^2}{3xy}$

$$= 2 - \frac{4}{3}y + 3x - 4xy$$

Dividing a Multinomial by a Multinomial

The next step in the division process is to divide a multinomial by a multinomial. This is not as easy as dividing by a monomial, so we will learn it in two stages. In the first stage we will learn to divide a polynomial by a polynomial. In stage two we will apply what we learned in stage one.

Before we can learn to divide a polynomial by a polynomial we need to have a definition of a polynomial. A polynomial is a special type of multinomial. An algebraic sum is a **polynomial in x** if each term is of the form ax^n, where n is a non-negative integer. One example of a polynomial in x is $7x^3 - \sqrt{3}x^2 + \frac{1}{2}x - 2$. (Remember that -2 can be written as $-2x^0$.) Another polynomial in x is $7x^{15} + 2x^3 - 5x$.

Not every polynomial is a polynomial in x. For example, $4y^3 - y^2$ is a polynomial in y and $\frac{-\sqrt{3}}{2}$ is a polynomial that is a constant. The multinomial $4x^3 + 5x^2 - 3 + x^{-3}$ is *not* a polynomial because the exponent on the last term is -3, which is not a non-negative integer. The multinomial $6x + \sqrt{7x} = 6x + \sqrt{7}x^{1/2}$ is not a polynomial in x because the exponent on the variable in the last term is $\frac{1}{2}$, which is not an integer.

We will now examine how to divide one polynomial by another. Study Example 2.15. Make sure you read the comments in Example 2.15, because they explain the method we use to divide polynomials.

EXAMPLE 2.25

Divide $6x^2 - 4 - 4x + 8x^4$ by $2x + 1$.

Solution

1. Write both the dividend and the divisor in decreasing order of powers.
The dividend is $6x^2 - 4 - 4x + 8x^4$. The largest power is 4, the next largest is 2, and so the dividend should be written as $8x^4 + 6x^2 - 4x - 4$. The divisor, $2x + 1$, is already in descending order of the powers.

2. Are there any missing terms in the dividend? If so, write them with a coefficient of 0.
Here the dividend does not have an x^3 term. Rewrite the dividend as $8x^4 + 0x^3 + 6x^2 - 4x - 4$.

3. Set up the problem just as you would any long division problem.

$$2x + 1 \,\overline{)\, 8x^4 + 0x^3 + 6x^2 - 4x - 4}$$

4. Divide the first term in the dividend by the first term in the divisor.
In this example the first term in the dividend is $8x^4$ and the first term in the divisor is $2x$. The result of this division, $4x^3$, is written above the dividend directly over the term with the same power.

$$\begin{array}{r} 4x^3 \\ 2x + 1 \,\overline{)\, 8x^4 + 0x^3 + 6x^2 - 4x - 4} \end{array}$$

EXAMPLE 2.25 (Cont.)

5. Multiply the divisor by this first term of the quotient. Write the product below the dividend with like terms under the like terms in the dividend. Subtract this product from the dividend. We will call this difference the "new dividend."

$$
\begin{array}{r}
4x^3 \\
2x+1 \overline{\smash{)}\ 8x^4 + 0x^3 + 6x^2 - 4x - 4} \\
\underline{8x^4 + 4x^3 } \\
-4x^3 + 6x^2 - 4x - 4 \qquad \text{new dividend}
\end{array}
$$

Caution

Be careful doing the subtraction. A common error would be to get $+4x^3$ for the x^3 term instead of $-4x^3$.

6. Repeat the last two steps until the power of the new dividend is less than the power of the divisor. Each time the last two steps are repeated, you should divide the first term of the new dividend by the first term of the divisor as shown here.

A.
$$
\begin{array}{r}
4x^3 - 2x^2 \\
2x+1 \overline{\smash{)}\ 8x^4 + 0x^3 + 6x^2 - 4x - 4} \\
\underline{8x^4 + 4x^3 } \\
-4x^3 + 6x^2 - 4x - 4 \\
\underline{-4x^3 - 2x^2 } \\
8x^2 - 4x - 4 \qquad \text{new dividend}
\end{array}
$$

B.
$$
\begin{array}{r}
4x^3 \ -2x^2 \ +4x \\
2x+1 \overline{\smash{)}\ 8x^4 + 0x^3 + 6x^2 - 4x - 4} \\
\underline{8x^4 + 4x^3 } \\
-4x^3 + 6x^2 - 4x - 4 \\
\underline{-4x^3 - 2x^2 } \\
8x^2 - 4x - 4 \\
\underline{8x^2 + 4x } \\
-8x - 4 \qquad \text{new dividend}
\end{array}
$$

C.
$$
\begin{array}{r}
4x^3 - 2x^2 + 4x - 4 \\
2x+1 \overline{\smash{)}\ 8x^4 + 0x^3 + 6x^2 - 4x - 4} \\
\underline{8x^4 + 4x^3 } \\
-4x^3 + 6x^2 - 4x - 4 \\
\underline{-4x^3 - 2x^2 } \\
8x^2 - 4x - 4 \\
\underline{8x^2 + 4x } \\
-8x - 4 \\
\underline{-8x - 4} \\
0
\end{array}
$$

Stop! This is really $0x^0$ and its power, 0, is less than the power of the first term in the divisor, 1. So,

EXAMPLE 2.25 (Cont.)

$$(8x^4 + 6x^2 - 4x - 4) \div (2x + 1) = 4x^3 - 2x^2 + 4x - 4$$

When you divide one polynomial by another, your work should look like the final step, C. The other steps were given to help you see what we were doing. In the next example we will show it all in one step.

EXAMPLE 2.26

Divide $4x^5 - 2x^3 + 6x^2 - 8$ by $4x^2 + 2$.

Solution   The dividend and divisor are already in descending order. The dividend, $4x^5 - 2x^3 + 6x^2 - 8$, does not have an x^4 or an x^1 term. We should include those terms with coefficients of zero. The dividend now becomes $4x^5 + 0x^4 - 2x^3 + 6x^2 + 0x - 8$. We will now set the problem up in the standard long division format and divide.

$$
\begin{array}{r}
x^3 + 0x^2 - x + \dfrac{3}{2} \\
4x^2 + 2 \overline{)\, 4x^5 + 0x^4 - 2x^3 + 6x^2 + 0x - 8} \\
\underline{4x^5 \qquad\quad + 2x^3} \\
0x^4 - 4x^3 + 6x^2 + 0x - 8 \\
\underline{0x^4 \qquad\quad + 0x^2} \\
-4x^3 + 6x^2 + 0x - 8 \\
\underline{-4x^3 \qquad\quad - 2x} \\
6x^2 + 2x - 8 \\
\underline{6x^2 \qquad + 3} \\
2x - 11
\end{array}
$$

The power of the $2x$ term, 1, is less than the power of the x^2 term of the divisor, 2. This expression, $2x - 11$, is the remainder.

The remainder is written as $\dfrac{2x - 11}{4x^2 + 2}$. So,

$$(4x^5 - 2x^3 + 6x^2 - 8) \div (4x^2 + 2)$$

$$= x^3 + 0x^2 - x + \frac{3}{2} + \frac{2x - 11}{4x^2 + 2}$$

$$= x^3 - x + \frac{3}{2} + \frac{2x - 11}{4x^2 + 2}$$

There is a faster method of division that can be used with some polynomials. This method is known as synthetic division. We will learn about synthetic division in Chapter 16.

EXAMPLE 2.27

Divide $3x^4 - 6x^3 + x^2 + 12x$ by $3x^2 - 2$.

EXAMPLE 2.27 (Cont.) | **Solution**

$$
3x^2 - 2 \overline{\smash{\big)}\
\begin{array}{l}
 x^2 - 2x + 1 \\
3x^4 - 6x^3 + x^2 + 12x + 0 \\
\underline{3x^4 - 2x^2 } \\
 -6x^3 + 3x^2 + 12x + 0 \\
\underline{ -6x^3 + 4x } \\
 3x^2 + 8x + 0 \\
\underline{ 3x^2 - 2} \\
 8x + 2
\end{array}
}
$$

The solution to $(3x^4 - 6x^3 + x^2 + 12x) \div (3x^2 - 2)$ is $x^2 - 2x + 1 + \dfrac{8x + 2}{3x^2 - 2}$.

In the next two examples, we will look at a division problem in which both the dividend and the divisor have more than one variable. The method used will be very similar to the method for dividing one polynomial by another.

EXAMPLE 2.28 | Divide $27x^6 - 8y^6$ by $3x^2 - 2y^2$.

Solution

$$
3x^2 - 2y^2 \overline{\smash{\big)}\
\begin{array}{l}
 9x^4 + 6x^2 y^2 + 4y^4 \\
27x^6 - 8y^6 \\
\underline{27x^6 - 18x^4 y^2 } \\
 18x^4 y^2 - 8y^6 \\
\underline{ 18x^4 y^2 - 12x^2 y^4 } \\
 12x^2 y^4 - 8y^6 \\
\underline{ 12x^2 y^4 - 8y^6} \\
 0
\end{array}
}
$$

So, $(27x^6 - 8y^6) \div (3x^2 - 2y^2) = 9x^4 + 6x^2 y^2 + 4y^4$.

EXAMPLE 2.29 | Divide $2a^3 - 3a^2 b + 4ab^2 - 12b^3$ by $a - 2b$.

Solution

$$
a - 2b \overline{\smash{\big)}\
\begin{array}{l}
 2a^2 + ab + 6b^2 \\
2a^3 - 3a^2 b + 4ab^2 - 12b^3 \\
\underline{2a^3 - 4a^2 b } \\
 a^2 b + 4ab^2 - 12b^3 \\
\underline{ a^2 b - 2ab^2 } \\
 6ab^2 - 12b^3 \\
\underline{ 6ab^2 - 12b^3} \\
 0
\end{array}
}
$$

So, $(2a^3 - 3a^2 b + 4ab^2 - 12b^3) \div (a - 2b) = 2a^2 + ab + 6b^2$.

Exercise Set 2.3

In Exercises 1–86, perform the indicated division.

1. x^7 by x^3

2. y^8 by y^6

3. $2x^6$ by x^4

4. $3w^4$ by w^2

5. $12y^5$ by $4y^3$

6. $15a^7$ by $3a^4$

7. $-45ab^2$ by $15ab$

8. $-55xy^3$ by $-11xy$

9. $33xy^2z$ by $3xyz$

10. $65x^2yz$ by $5xyz$

11. $96a^2xy^3$ by $-16axy^2$

12. $105b^3yw^2$ by $-15b^2yw$

13. $144c^3d^2f$ by $8cf$

14. $162x^2yz^3$ by $9x^2z^2$

15. $9np^3$ by $-15n^3p^2$

16. $15rs^2t$ by $-27r^2st^3$

17. $8abcdx^2y$ by $14adxy^2$

18. $9efg^2hr$ by $24e^2fh^3r$

19. $2a^3 + a^2$ by a

20. $4x^4 - x^3$ by x^2

21. $36b^4 - 18b^2$ by $9b$

22. $49y^5 + 35y^3$ by $7y^2$

23. $42x^2 + 28x$ by 7

24. $56z^6 - 48z^3$ by 8

25. $34x^5 - 51x^2$ by $17x^2$

26. $105w^6 + 63w^4$ by $21w^2$

27. $24x^6 - 8x^4$ by $-4x^3$

28. $42y^7 - 24y^5$ by $6y^4$

29. $5x^2y + 5xy^2$ by xy

30. $7a^2b - 7ab^2$ by ab

31. $10x^2y + 15xy^2$ by $5xy$

32. $25p^2q - 15pq^2$ by $5pq$

33. $ap^2q - 2pq$ by pq

34. $bx^2w + 3xw$ by xw

35. $a^2bc + abc$ by abc

36. $x^3yz - xyz$ by xyz

37. $9x^2y^2z - 3xyz^2$ by $-3xyz$

38. $12a^2b^2c + 4abc^2$ by $-4abc$

39. $b^3x^2 + b^3$ by $-b$

40. $c^5y^3 - cy^2$ by $-c$

41. $x^2y + xy - xy^2$ by xy

42. $ab^2 - ab + a^2b$ by ab

43. $18x^3y^2z - 24x^2y^3z$ by $-12x^2yz$

44. $36a^4b^2c - 27a^2b^4c^2$ by $-27a^2bc$

45. $x^2 + 7x + 12$ by $x + 3$

46. $x^2 + x - 12$ by $x + 4$

47. $x^2 - 3x + 2$ by $x - 2$

48. $x^2 - 2x - 15$ by $x - 5$

49. $x^2 + x - 2$ by $x + 2$

50. $x^2 + x - 6$ by $x + 3$

51. $6a^2 + 17a + 7$ by $3a + 7$

52. $4b^2 + 10b - 6$ by $4b - 2$

53. $8y^2 - 8y - 6$ by $2y - 3$

54. $12t^2 + t - 6$ by $4t + 3$

55. $x^3 - 5x + 2$ by $x^2 + 2x - 1$

56. $d^3 + d^2 - 3d + 2$ by $d^2 + d + 2$

57. $6a^3 - 7a^2 + 10a - 4$ by $3a^2 - 2a + 4$

58. $9y^3 - 16y + 8$ by $3y^2 + 3y - 4$

59. $4x^3 - 3x + 4$ by $2x - 1$

60. $7p^3 + 2p - 5$ by $3p + 2$

61. $r^3 - r^2 - 6r + 5$ by $r + 2$

62. $2c^3 - 3c^2 + c - 4$ by $c - 2$

63. $x^4 - 81$ by $x - 3$

64. $y^4 - 81$ by $y + 3$

65. $23x^2 - 5x^4 + 12x^5 - 12 - 14x^3 + 8x$ by $3x^2 - 2 + x$

66. $10a + 21a^3 - 35 + 6a^5 - 25a^4 + 21a^7 - 14a^6$ by $2a + 7a^3 - 7$

67. $x^2 - y^2$ by $x - y$

68. $a^2 - b^2$ by $a + b$

69. $w^3 - z^3$ by $w - z$

70. $x^3 + y^3$ by $x + y$

71. $x^3 + xy^2 + x^2y + y^3$ by $x + y$

72. $a^3 - a^2b + ab^2 - b^3$ by $a - b$

73. $c^3d^3 - 8$ by $cd - 2$

74. $e^3f^3 + 27$ by $ef + 3$

75. $x^2 - 2xy + y^2$ by $x - y$

76. $a^2 + 6ab + 9b^2$ by $a + 3b$

77. $5p^3r - 10p^2 + 15pr^2 - p^2r^2 + 2pr - 3r^3$ by $5p - r$

78. $8x^3 - 12x^2y + 6xy^2 - y^3$ by $2x - y$

79. $a^2 - 2ad - 3d^2 + 3a - 13d - 8$ by $a - 3d - 1$

80. $2x^2 - xy - 6y^2 + x + 19y - 15$ by $2x + 3y - 5$

81. $a^2f - af^2$ by $a - f$

82. $d^3 - 2dm^2 + 2m^3 - d^2m$ by $d - m$

83. $a^2 - b^2 + 2bc - c^2$ by $a - b + c$

84. $e^2 + 2eh - f^2 + h^2$ by $e + f + h$

85. $a^4 + 2a^2 - a + 2$ by $a^2 + a + 2$

86. $x^6 - x^4 + 2x^2 - 1$ by $x^3 - x + 1$

Solve Exercise 87.

87. *Electronics*   The expression for the total resistance of three resistances in a parallel electrical circuit is

$$\frac{R_1R_2R_3}{R_2R_3 + R_1R_3 + R_1R_2}$$

Determine the reciprocal of this expression, and then perform the indicated division.

≡ 2.4
SOLVING EQUATIONS

Until now we have worked with real numbers and learned how to add, subtract, multiply, and divide algebraic expressions. In these next two sections we are going to begin learning how to use these skills to solve problems.

The ability to solve a problem often depends on the ability to write an equation for a problem and then solve that equation. We will use this section to learn how to solve an equation. In the next section we will start to learn how to write an equation for a problem.

Equations ▬▬▬▬▬▬▬▬▬▬▬▬▬▬▬▬▬▬▬▬▬▬▬▬▬▬▬▬▬▬▬▬▬

An **equation** is an algebraic statement. It asserts that two algebraic expressions are equal. The two algebraic expressions are called the left-hand side (or left side) of the equation and the right-hand side (or right side) of the equation. As the name indicates, the left-hand side is to the left of the equal sign and the right-hand side is to the right of the equal sign.

$$\underbrace{4x^3 + 2x - \sqrt{3}}_{\text{left-hand side}} \quad = \quad \underbrace{7x^2 - 5x}_{\text{right-hand side}}$$

Some equations are true for all values of their variables. These equations are called **identities.** For example, $(x + 2)(x - 2) = x^2 - 4$ is an identity. An equation that is true for only some values of the variables and not true for the other values is a **conditional equation.** Examples of conditional equations are $4x = -24$ and $y^2 = 25$. The first is true only when $x = -6$. The second is true only when $y = 5$ or $y = -5$.

Any value that can be substituted for the variable and that makes an equation true is called a **solution** or **root** of the equation. So, 5 is a solution of $2x^2 - 5x - 5 = 20$, since $2(5^2) - 5(5) - 5 = 20$. The root of $x + 7 = 10$ is 3, and this is often indicated by the phrase, "3 satisfies the equation $x + 7 = 10$." To solve an equation means to find all of its solutions or roots.

Equivalent Equations ▬▬▬▬▬▬▬▬▬▬▬▬▬▬▬▬▬▬▬▬▬▬▬▬▬▬▬

The purpose of this section is to help you learn how to find the solutions or roots of an equation. The techniques you learn here will be used whenever you need to solve an equation. Two equations are **equivalent** if they have exactly the same roots. There is a series of five operations that will allow you to change an equation into an equivalent equation.

To change an equation into an equivalent equation you can use any of the following five operations.

> **Operations for Changing Equations into Equivalent Equations**
>
> 1. Add or subtract the same algebraic expression or amount to both sides of the equation.
> 2. Multiply or divide both sides of the equation by the same algebraic expression, provided the expression does not equal zero.
> 3. Combine like terms on either side of the equation.
> 4. Replace one side (or both sides) with an identity.
> 5. Interchange the two sides of the equation.

In order to solve an equation you will generally have to use a combination of these five operations. You may have to use the same operation more than once. The following examples show how these operations are used. The first four examples use one operation and the other examples use a combination of operations.

EXAMPLE 2.30

Solve $x + 7 = 15$.

Solution
$$x + 7 = 15$$
$$(x + 7) - 7 = 15 - 7 \quad \text{Subtract 7 from both sides (Operation \#1).}$$
$$x = 8$$

Substituting this value in the original equation verifies that 8 is the solution, since $8 + 7 = 15$.

EXAMPLE 2.31

Solve $3x = -15$.

Solution
$$3x = -15$$
$$\frac{1}{3}(3x) = \frac{1}{3}(-15) \quad \text{Multiply both sides by } \tfrac{1}{3} \text{, the reciprocal of 3}$$
$$\text{(Operation \#2).}$$
$$x = -5$$

Checking the solution in the original equation, we see that -5 satisfies the equation, since $3(-5) = -15$.

EXAMPLE 2.32

Solve $5x - 3x - x = 20 - 12$.

Solution
$$5x - 3x - x = 20 - 12$$
$$x = 8 \qquad \text{Combine like terms (Operation \#3).}$$

Checking the root in the original, we see that $5 \times 8 - 3 \times 8 - 8 = 40 - 24 - 8 = 8$. This satisfies the original equation.

EXAMPLE 2.33

Solve $x^2 + 6x + 9 = 0$.

Solution $\quad x^2 + 6x + 9 = 0$

$\qquad\qquad (x + 3)^2 = 0 \quad$ Replace $x^2 + 6x + 9$ with $(x + 3)^2$ (Operation #4).

We will stop this example here. To continue solving this problem takes some more mathematics, which we will learn later. The important thing to notice is that we used an identity to rewrite one side of the equation.

The next four examples use combinations of the operations.

EXAMPLE 2.34

Solve $3x + 27 = 6x$.

Solution $\qquad\qquad 3x + 27 = 6x$

$$(3x + 27) - 3x = 6x - 3x \qquad \text{Subtract } 3x \text{ (Operation #1).}$$

$$27 = 3x \qquad \text{Combine like terms (Operation #3).}$$

$$3x = 27 \qquad \text{Switch sides (Operation #5).}$$

$$\tfrac{1}{3}(3x) = \tfrac{1}{3}(27) \qquad \text{Multiply by } \tfrac{1}{3} \text{ (Operation #2).}$$

$$x = 9$$

To check, replace the x in the original equation with the value we found, 9. The left-hand side is $3(9) + 27 = 27 + 27 = 54$ and the right-hand side is $6(9) = 54$. Since both sides give the same value, 9 is a solution.

EXAMPLE 2.35

Solve $7y + 6 = 216 - 3y$.

Solution $\qquad\qquad 7y + 6 = 216 - 3y$

$$7y + 6 + 3y = 216 - 3y + 3y \qquad \text{Add } 3y \text{ (Operation #1).}$$

$$10y + 6 = 216 \qquad \text{Combine terms (Operation #3).}$$

$$10y + 6 - 6 = 216 - 6 \qquad \text{Subtract 6 (Operation #1).}$$

$$10y = 210 \qquad \text{Combine terms.}$$

$$\frac{10y}{10} = \frac{210}{10} \qquad \text{Divide by 10 (Operation #2).}$$

$$y = 21$$

Check this answer in the original equation. When you replace y with 21 do you get the same value on both the left-hand and right-hand sides of the equation? You should.

EXAMPLE 2.35 (Cont.)

Notice that in the next-to-last step we divided both sides by 10. We could have multiplied both sides by $\frac{1}{10}$ and arrived at the same result. Do you understand why? Because division by 10 is the same as multiplication by $\frac{1}{10}$.

EXAMPLE 2.36

Solve $3(4z - 8) + 2(3z + 5) = 4(z + 7)$.

Solution $3(4z - 8) + 2(3z + 5) = 4(z + 7)$

$12z - 24 + 6z + 10 = 4z + 28$	Remove parentheses by distribution.
$18z - 14 = 4z + 28$	Combine terms.
$14z - 14 = 28$	Subtract $4z$.
$14z = 42$	Add 14.
$z = 3$	Divide by 14 (or multiply by $\frac{1}{14}$).

Check this in the original equation. You should get 40 on each side of the equation.

EXAMPLE 2.37

Solve $\frac{n}{2} + 5 = \frac{2n}{3}$.

Solution
$$\frac{n}{2} + 5 = \frac{2n}{3}$$

$$6\left(\frac{n}{2} + 5\right) = 6\left(\frac{2n}{3}\right) \qquad \text{Multiply by 6.} \left(6 \text{ is a common denominator of } \frac{n}{2} \text{ and } \frac{2n}{3}.\right)$$

$$3n + 30 = 4n$$

$$30 = n \qquad \text{Subtract } 3n.$$

Because most people like to give the variable first and then the value of that variable, you could now use Operation #5 and rewrite this as $n = 30$.

Check by substituting 30 for n in the original equation. The left side is $\frac{30}{2} + 5 = 15 + 5 = 20$. The right side is $\frac{2(30)}{3} = \frac{60}{3} = 20$. So, $n = 30$ satisfies the original equation.

Hints for Solving Problems

The last four examples used a combination of the five operations used to make equivalent equations. Each of the examples showed how to use one or more of the hints for solving problems. As you continue through this book, you will get some more hints to help you become a better problem-solver.

> **Hints for Solving Problems**
>
> 1. Eliminate fractions. Multiply both sides of the equation by a common denominator.
> 2. Remove grouping symbols. Perform the indicated multiplications and then combine terms to remove parentheses, brackets, and braces.
> 3. Combine like terms whenever possible.
> 4. Get all terms containing the variable on one side of the equation. All other terms should be placed on the other side of the equation.
> 5. Check your answer in the original equation.

Here are two more examples. See if you can work each problem before studying the solution given.

EXAMPLE 2.38

Solve $\dfrac{3p+7}{4} - \dfrac{2(p-5)}{3} = \dfrac{p-3}{2} + 6.$

Solution $\dfrac{3p+7}{4} - \dfrac{2(p-5)}{3} = \dfrac{p-3}{2} + 6$

$12\left[\dfrac{3p+7}{4} - \dfrac{2(p-5)}{3}\right] = 12\left[\dfrac{p-3}{2} + 6\right]$ Multiply both sides by 12, a common denominator of 2, 3, and 4.

$12\left[\dfrac{3p+7}{4}\right] - 12\left[\dfrac{2(p-5)}{3}\right] = 12\left[\dfrac{p-3}{2}\right] + 12(6)$ Distribute the 12.

$3[3p+7] - 4[2(p-5)] = 6[p-3] + 12(6)$

$9p + 21 - 4[2p - 10] = 6p - 18 + 72$ Remove grouping symbols.

$9p + 21 - 8p + 40 = 6p - 18 + 72$

$p + 61 = 6p + 54$ Combine terms.

$7 = 5p$ Use hint #4.

$\dfrac{7}{5} = p$ Divide by 5.

The solution is $\frac{7}{5}$, or 1.4. Check. You should get 5.2 on both sides of the equation when you evaluate each side with $p = 1.4$. This would be a good time to practice using your calculator to see if you get 5.2 on each side of the equation.

EXAMPLE 2.39

Solve $\dfrac{3w+a}{2b} - \dfrac{4(w+a)}{b} = \dfrac{2w-6}{3b} + \dfrac{2+6a}{b}$ for w.

EXAMPLE 2.39 (Cont.)

Solution   There are several letters in this equation, a, b, and w, but you are asked to solve for w. This means that w is the variable and that a and b should be treated as constants.

$$\frac{3w+a}{2b} - \frac{4(w+a)}{b} = \frac{2w-6}{3b} + \frac{2+6a}{b}$$

$$6b\left[\frac{3w+a}{2b} - \frac{4(w+a)}{b}\right] = 6b\left[\frac{2w-6}{3b} + \frac{2+6a}{b}\right] \qquad \text{Multiply both sides by a common denominator, } 6b.$$

$$6b\left[\frac{3w+a}{2b}\right] - 6b\left[\frac{4(w+a)}{b}\right] = 6b\left[\frac{2w-6}{3b}\right] + 6b\left[\frac{2+6a}{b}\right]$$

$$3[3w+a] - 6[4(w+a)] = 2[2w-6] + 6[2+6a]$$

$$9w + 3a - 24w - 24a = 4w - 12 + 12 + 36a \qquad \text{Remove grouping.}$$

$$-15w - 21a = 4w + 36a \qquad \text{Combine like terms.}$$

$$-15w - 4w = 36a + 21a \qquad \text{Hint \#4.}$$

$$-19w = 57a \qquad \text{Combine like terms.}$$

$$w = \frac{57a}{-19} \qquad \text{Divide both sides by } -19.$$

$$w = -3a$$

Check. Substitute $-3a$ for w in the original equation. When you simplify each side, you get $\frac{4a}{b}$. While it is easier to substitute for w at a later step, you might have made a mistake getting to that step. Sometimes you will even make an error when you copy the problem onto your paper. Always go back to the original problem to check your work.

Exercise Set 2.4

Solve the equations in Exercises 1–80.

1. $x - 7 = 32$
2. $y - 8 = 41$
3. $a + 13 = 25$
4. $b + 21 = 34$
5. $25 + c = 10$
6. $28 + d = 12$
7. $4.3 + w = 8.7$
8. $5.1 + z = 9.1$

9. $4x = 18$
10. $5y = 12$
11. $-3w = 24$
12. $6z = -42$
13. $12a = 18$
14. $15b = -25$
15. $21c = -14$
16. $24d = 16$

17. $\frac{p}{3} = 5$
18. $\frac{r}{5} = 4$
19. $\frac{t}{4} = -6$
20. $\frac{s}{-3} = -5$
21. $4a + 3 = 11$
22. $3b + 4 = 16$

23. $7 - 8d = 39$

24. $9 - 7c = 44$

25. $4x - 3 = -37$

26. $5y - 4 = -41$

27. $2.3w + 4.1 = 13.3$

28. $3.5z + 5.2 = 22.7$

29. $2x + 5x = 28$

30. $3y + 8y = 121$

31. $3x = 7 - 10$

32. $4x = 16 - 24$

33. $3a + 2(a + 5) = 45$

34. $4b + 3(7 + b) = 56$

35. $4(6 + c) - 5 = 21$

36. $5(7 + d) + 4 = 31$

37. $2(p - 4) + 3p = 16$

38. $7(n - 5) + 4n = 16$

39. $3x = 2x + 5$

40. $4y = 3y + 7$

41. $4w = 6w + 12$

42. $7z = 10z + 42$

43. $9a = 54 + 3a$

44. $8b = 55 + 3b$

45. $\dfrac{5x}{2} = \dfrac{4x}{3} - 7$

46. $\dfrac{3y}{7} = \dfrac{2y}{3} + 4$

47. $\dfrac{6p}{5} = \dfrac{3p}{2} + 4$

48. $\dfrac{5z}{3} = \dfrac{4z}{5} - 3$

49. $8n - 4 = 5n + 14$

50. $9p - 5 = 6p + 37$

51. $7r + 3 = 11r - 21$

52. $8s + 7 = 15s - 56$

53. $9t + 6 = 3t - 5$

54. $11u - 4 = 6u + 5$

55. $\dfrac{6x - 3}{2} = \dfrac{7x + 2}{3}$

56. $\dfrac{4r - 3}{3} = \dfrac{5r + 2}{2}$

57. $\dfrac{3t + 4}{4} = \dfrac{2t - 5}{2}$

58. $\dfrac{6a - 5}{3} = \dfrac{7a + 5}{6}$

59. $3(x + 5) = 2x - 3$

60. $2(y - 3) = 4 + 3y$

61. $5(w - 7) = 2w + 4$

62. $6(z + 5) = 2z - 9$

63. $\dfrac{x}{2} + \dfrac{x}{3} - \dfrac{x}{4} = 2$

64. $\dfrac{p}{2} - \dfrac{p}{3} - \dfrac{p}{4} = 3$

65. $\dfrac{4(a - 3)}{5} = \dfrac{3(a + 2)}{4}$

66. $\dfrac{5(b + 4)}{3} = \dfrac{4(b - 5)}{5}$

67. Solve $ax + b = 3ax$ for x

68. Solve $2by = 6 + 4by$ for y

69. Solve $ax - 3a + x = 5a$ for a

70. Solve $2(by - c) = 3\left(\dfrac{y}{2} - c\right)$ for y

71. $\dfrac{3}{x} + \dfrac{4}{x} = 3$

72. $\dfrac{5}{y} - \dfrac{3}{y} = 6$

73. $\dfrac{3}{4p} + \dfrac{1}{p} = \dfrac{7}{4}$

74. $\dfrac{6}{5q} - \dfrac{2}{q} = \dfrac{6}{5}$

75. $\dfrac{1}{x + 1} - \dfrac{2}{x - 1} = 0$

76. $\dfrac{3}{x + 2} - \dfrac{4}{x - 2} = 0$

77. $\dfrac{3}{2x} = \dfrac{1}{x + 5}$

78. $\dfrac{4}{3x} = \dfrac{2}{x + 1}$

79. $\dfrac{2x + 1}{2x - 1} = \dfrac{x - 1}{x - 3}$

80. $\dfrac{2x + 3}{2x + 5} = \dfrac{5x + 4}{5x + 2}$

Solve Exercises 81 and 82.

81. *Meteorology* The formula for converting Celsius temperatures to Fahrenheit temperatures is $F = \frac{9}{5}C + 32$, where F represents the Fahrenheit temperature and C, the Celsius temperature. Find the formula for converting Fahrenheit temperatures to Celsius by solving this equation for C.

82. *Physics* The velocity v of a falling object after t seconds is given by the formula $v = v_0 + at$, where v_0 represents the initial velocity and a is the acceleration due to gravity.

a. Solve this equation for t in order to determine the length of time that it takes the object to reach some velocity v.

b. Use your answer to part a to determine how long it takes an object to strike the ground, if its initial velocity is 12 m/s, its final velocity is 97 m/s, and the acceleration is 9.8 m/s^2.

≡ 2.5
APPLICATIONS OF EQUATIONS

In Section 2.4 we learned how to solve some equations. But, the ability to solve equations is helpful only if you are able to take a problem and write it in the form of one or more equations. Then, once you have the equations, you can solve them to find the answer to your original problem.

Most problems you have to solve at work will be verbal problems. Someone will tell you about a problem they want you to solve or they will write part, or all, of the problem. You will have to first take this verbal problem and organize it so that it is easier to understand. Then, you will look at the problem and decide what information is important and what is not. You should then take the important information and express it in one or more equations. Once you have written the equations, the most difficult part is over. All that is left is to solve the equations and check your answers. In this section, we will focus on taking written information and writing it as an equation.

Seven Suggestions to Help Solve Word Problems ▬▬▬▬

Verbal, or word problems, give several numerical relationships and then ask some questions about them. You must be able to translate the word problem into equations. The fewer equations you need, the easier it will be to solve the problem. Here are some suggestions and examples to help you.

Suggestions for Solving Word Problems

1. Read the problem carefully. Make sure you understand what the problem is asking. You may need to read the problem several times to fully understand it.
2. Clearly identify the unknown quantities. Identify each unknown quantity with a letter (or variable). Write down what each letter stands for. Use a letter for an unknown quantity that makes sense to you. For example, you might use d for distance or t for time.
3. If possible, represent all the unknowns in terms of just one variable.
4. Make a sketch (if possible) if it makes the problem clearer.
5. Analyze the problem carefully. Try to write one equation that shows how all the unknowns and knowns are related. If it is not possible to write one equation, use more. (Later we will learn how to solve several equations that show how given unknowns are related.)
6. Solve the equation.
7. Check your answer in the original problem.

EXAMPLE 2.40

In order to get an A in a word processing class, one teacher requires that you type an average of 85 words per minute for five different timings. Bill had speeds of 77, 78, 87, and 91 words per minute on his first four timings. How fast must he type on the next test in order to get an A?

EXAMPLE 2.40 (Cont.)

Solution   The unknown quantity is the typing speed on the next timed typing. We will let s represent the unknown typing speed. The average for the five timings is the sum of these five timings divided by five or $\dfrac{77 + 78 + 87 + 91 + s}{5}$. This average must be 85, and so we have the equation

$$\frac{77 + 78 + 87 + 91 + s}{5} = 85$$

$$\frac{333 + s}{5} = 85 \qquad \text{Combine terms.}$$

$$5\left(\frac{333 + s}{5}\right) = 5 \times 85 \qquad \text{Multiply by 5.}$$

$$333 + s = 425$$

$$333 + s - 333 = 425 - 333 \qquad \text{Subtract 333.}$$

$$s = 92$$

Bill must type 92 words per minute to have an average of 85 words per minute for the five timings, in order to get an A in the class.

EXAMPLE 2.41

At the end of a model year, a dealer advertises that the list prices of last year's models are 15% off. What was the original price of a car that is on sale for $7,990?

Solution   Suppose the original price of the car was p dollars. The discounted amount is 15% of the original price, or $0.15p$. The sale price of $7,990 is the original price p less the discounted amount, $0.15p$. This can be written as

$$p - 0.15p = 7,990$$

$$0.85p = 7,990 \qquad \text{Combine terms.}$$

$$p = \frac{7,990}{0.85} \qquad \text{Divide by 0.85.}$$

$$p = 9,400$$

The original price was $9,400. You can check your answer by taking 15% of $9,400 $(0.15 \times 9,400 = 1,410)$ and subtracting this from the original price $(9,400 - 1,410 = 7,990)$. Is this the sale price?

We will now look at some types of problems that have applications to technology. The problems selected provide a variety of examples.

Uniform Motion Problems

The distance an object travels is governed by the rate it is traveling and the time that it travels. This is often expressed with the formula

$$d = rt$$

where d is the distance traveled, r is the rate of travel, and t is the time spent traveling.

Application

EXAMPLE 2.42

If an airplane flies 785 km/h for 4.5 h, how far does it travel?

Solution   We use the formula $d = rt$. We are given the rate r as 785 km/h and the time t as 4.5 h. So,

$$d = rt$$
$$= 785(4.5)$$
$$= 3\,532.5 \text{ km}$$

The plane will travel 3 532.5 km in 4.5 h.

It sometimes helps to include the units as part of your work. In Example 2.42, we could have written

$$d = rt$$
$$= 785\,\frac{\text{km}}{\text{h}} \times 4.5\,\text{h}$$
$$= 3\,532.5\,\frac{\text{km} \cdot \cancel{\text{h}}}{\cancel{\text{h}}}$$
$$= 3\,532.5\,\text{km}$$

This same formula, $d = rt$, can be used to find any one of the three values, distance, rate, or time, provided that we know the other two.

Application

EXAMPLE 2.43

If a plane travels 1,150 mi in 2.5 h, what is its average rate of speed?

Solution   Again, we use the formula $d = rt$. This time we have the distance d as 1,150 mi and the time t as 2.5 h. So,

$$d = rt$$
$$1{,}150 = r(2.5)$$
$$\frac{1{,}150}{2.5} = r$$
$$460 = r$$

The plane averaged 460 mph.

Another motion problem is demonstrated in Example 2.44.

Application

EXAMPLE 2.44

A car traveling on an interstate highway leaves a rest stop at 3 p.m., traveling at 75 km/h. Another car leaves the same rest stop 15 min later, headed in the same direction. If the second car travels at 100 km/h, how long will it be before it overtakes the first car?

Solution   Notice that the problem asks how long it will be before the cars meet, but does not say if this should be how much time or how many km. Let's find both.

The distance formula says that $d = rt$. If t is the amount of time (in hours) since the first car left, then for the first car the distance it has traveled, d, is $d = 75t$.

The distance traveled by the second car will be the same. The rate 100 km/h and the time will be different from those of the first car. The second car left 15 min or $\frac{1}{4}$ h later. (Notice that we had to give this time in hours.) So, the time the second car traveled is $t - \frac{1}{4}$ h. Thus, we have $d = 100\left(t - \frac{1}{4}\right)$.

Both cars traveled the same distance, which means that

$$75t = 100\left(t - \frac{1}{4}\right)$$

$$75t = 100t - 25$$

$$75t - 100t = 100t - 25 - 100t \qquad \text{Subtract } 100t.$$

$$-25t = -25 \qquad \text{Combine terms.}$$

$$t = 1 \qquad \text{Divide by } -25.$$

The second car will catch the first 1 h after the first car left. This will be at 4 p.m., or 45 min after the second car leaves. The distance traveled by each car is $75(1) = 75$ km.

Work Problems

Work problems provide a different type of challenge. In these problems, we want to determine the amount of work each person or machine can do in a given amount of time.

Application

EXAMPLE 2.45

One printer can complete a certain job in 3 hours. Another printer can do the same job in 2 hours. How long would it take if both printers work on the job?

Solution   We will let h stand for the hours it takes both machines to complete the job. In one hour the two machines can complete $\frac{1}{h}$ of the job. The first printer does $\frac{1}{3}$ of the job in one hour when it works alone. The second printer does $\frac{1}{2}$ of the job

EXAMPLE 2.45 (Cont.)

in one hour when it works alone, but together they complete $\frac{1}{h}$ of the job in one hour. So, $\frac{1}{h} = \frac{1}{3} + \frac{1}{2}$. A common denominator is $6h$. Multiplying by $6h$ we get

$$6 = 2h + 3h$$

$$6 = 5h$$

$$\frac{6}{5} = h$$

They can complete the job together in $\frac{6}{5}$ hours or in 1 hour 12 minutes.

Mixture Problems

In a mixture problem, two quantities are combined, or mixed, to produce a third quantity. Each quantity has a different percentage of some ingredient. The problems often ask for the percent of the ingredient in the third quantity.

Application

EXAMPLE 2.46

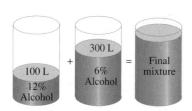

FIGURE 2.1a

If 100 L of gasohol containing 12% alcohol is mixed with 300 L that contains 6% alcohol, how much alcohol is in the final mixture? What is the percent of alcohol in the final mixture?

Solution   Figure 2.1a shows the two given quantities on the left and the final mixture on the right. The first quantity has $0.12 \times 100 = 12$ L of alcohol. The second quantity has $0.06 \times 300 = 18$ L of alcohol.

The total amount of gasohol is $100 + 300 = 400$ L. The total amount of alcohol is $12 + 18 = 30$ L. The alcohol is $\frac{30}{400} = 7.5\%$ of the gasohol. (See Figure 2.1b.)

Application

EXAMPLE 2.47

FIGURE 2.1b

A 15-L cooling system contains a solution of 20% antifreeze and 80% water. In order to get the most protection, the solution should contain 60% antifreeze and 40% water. How much of the 20% solution must be replaced with pure antifreeze to get the best protection?

Solution   Let $n =$ the number of liters of the original solution that will be replaced with antifreeze. The amount of original solution that is not replaced is $(15 - n)$ L. You know that you will end up with 60% of 15 L, or 9 L of antifreeze. Thus, you can write

antifreeze left + antifreeze added $= 9$ L

EXAMPLE 2.47 (Cont.)

The antifreeze that is left is 20% of $15 - n$, so we can rewrite this equation as

$$20\%(15 - n) + n = 9$$

$$0.2(15 - n) + n = 5$$

$$3 - 0.2n + n = 9$$

$$3 + 0.8n = 9$$

$$0.8n = 6$$

$$n = \frac{6}{0.8}$$

$$n = 7.5$$

So, 7.5 L of the 20% solution must be replaced with pure antifreeze.

Statics Problems

The **moment of force,** or **torque,** about a point O is the product of the force F and the perpendicular distance ℓ from the force to the point. The distance ℓ is also called the **moment arm** of the force about a point. The Greek letter tau, τ, is used to represent torque. Thus, we have the equation

$$\tau = F\ell$$

A torque that produces a counterclockwise rotation is considered positive, as shown in Figure 2.2a. A torque that produces a clockwise rotation is considered negative. (See Figure 2.2b.)

An object is in rotational equilibrium when the sum of all the torques acting on the object about any point is zero.

The **center of gravity** of an object is the point where the object's entire weight is regarded as being concentrated. If an object is hung from its center of gravity, it will not rotate. An object can also be balanced on its center of gravity without rotating. But, what is perhaps most important in studying the equilibrium of an object, is that its weight is considered to be a downward force from its center of gravity.

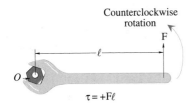

Counterclockwise rotation

$\tau = +F\ell$

FIGURE 2.2a

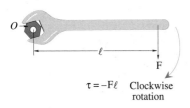

$\tau = -F\ell$ Clockwise rotation

FIGURE 2.2b

Application

EXAMPLE 2.48

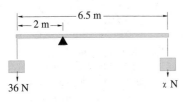

6.5 m

2 m

36 N

x N

FIGURE 2.3

A 6.5-m rigid rod of negligible weight has a 36-N (newton) weight hung from one end. An unknown weight is hung from the other end. If the rod is balanced at a point 2 m from the end with the 36-N weight, how much is the unknown weight? (See Figure 2.3.)

Solution   Since the rod and weights are balanced, the torques of the two weights must be equal. On the left side the moment arm is 2 m and the force is 36 N. The torque on the left-hand side, τ_1, is $\tau_1 = 36$ N $\times$ 2 m $= 72$ N $\cdot$ m. The moment arm on the right is 6.5 m $-$ 2 m $=$ 4.5 m. We are to find the force, x. So, the torque on the right-hand side is $\tau_2 = (x$ N$)(4.5$ m$) = 4.5x$ N $\cdot$ m. Since $\tau_1 = \tau_2$ then

EXAMPLE 2.48 (Cont.)

72 N · m = 4.5x N · m or 72 = 4.5x. Solving this, we get $x = \frac{72}{4.5} = 16$ N. The unknown weight is 16 N.

Application

EXAMPLE 2.49

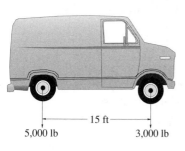

— 15 ft —

5,000 lb   3,000 lb

FIGURE 2.4

The front wheels of a truck together support 3,000 lb. Its rear wheels together support 5,000 lb. If the axles are 15 ft apart, where is the center of gravity? (See Figure 2.4.)

Solution   Let d represent the distance from the front axle to the center of gravity. Then $15 - d$ is the distance from the center of gravity to the rear axle. The torque on the right is $\tau_1 = 3,000d$ ft · lb. The torque on the left is $\tau_2 = -5,000(15 - d)$ ft · lb. (Notice that τ_2 is a counterclockwise rotation around the center of gravity and so it is positive. τ_1 is a clockwise rotation around the center of gravity and so it is negative.) Since the truck is in rotational equilibrium,

$$\tau_1 + \tau_2 = 0$$

$$3,000d - 5,000(15 - d) = 0$$

$$3,000d - 75,000 + 5,000d = 0$$

$$-75,000 + 8,000d = 0$$

$$8,000d = 75,000$$

$$d = \frac{75,000}{8,000}$$

$$d = 9.375$$

The center of gravity is 9.375 ft or 9′4.5″ behind the front axle.

Exercise Set 2.5

Solve Exercises 1–30.

1. On your first three mathematics exams you received scores of 79, 85, and 74. What do you need to get on the next exam in order to have an 80 average?

2. There are three parts to a state certification exam. You must pass all three parts with an average of 75 and you cannot get below 60 on any one part. Judy's first two scores were 65 and 72. What must she get on the last part in order to pass?

3. If Raphael got scores of 85 and 82 on the first two parts of the exam described in Exercise 2, what must he get on the third part in order to pass the exam?

4. *Business*   A resort promised that the temperature would average 72°F during your four-day vacation or you would get your money back. The first three days the average temperatures were 69°, 73°, and 68°. How warm does it have to get today for the resort to be able to keep your money?

5. *Business*   A discount store sells personal computers for $920. This price is 80% of the price at a wholesale store. What is the wholesale price? How much will you save at the discount store?

6. *Business*   The personal computer discount store makes a profit of 15% based on its cost for the computer in Exercise 5. What does the computer cost the discount store?

7. An insurance company gives you $1,839 to replace a stolen car. At the time they tell you that the car was only worth 30% of its original cost. What was the original cost?

8. *Automotive*   Because of inflation, the price of automotive parts increased by 3% in January and by another 4% in May. What was last year's price of parts that cost $227.63 after the May increase?

9. *Finance*   Sally invested $4,500. Part of it was invested at 7.5% and part of it at 6%. After one year her interest was $303. How much was invested at each rate?

10. *Finance*   To help pay for his education, José worked and invested his money. Altogether he was able to invest $6,200. Some money he invested at 8.2% and the rest at 7.25%. He was able to earn $482.75 in interest during the year. How much did he invest at each rate?

11. *Business*   A factory pays time and a half for all hours over 40 hours per week. Juannita makes $8.50 an hour and one week brought home $429.25. How many hours of overtime did she work?

12. *Business*   Mladen is paid $8.20 an hour. He also gets paid time-and-a-half overtime when he works more than 40 hours a week if it is during the week. He gets double time if the extra hours are on the weekend. One week Mladen brought home $524.80 for 54 hours of work. How many hours did he work on the weekend?

13. *Transportation*   A freight train leaves Chicago traveling at an average rate of 38 mph. How far will it travel in 7 h?

14. *Transportation*   How long does it take the train in Exercise 13 to travel 475 mi?

15. *Environmental science*   An oil tanker has hit a reef and a hole has been knocked in the side of the tanker. Oil is leaking out of the tanker and forming an oil slick. The oil slick is moving toward a beach 380 km away at a rate of 12 km/d (kilometers per day). The day after the oil spill, cleanup ships leave a dock at the beach and head directly toward the oil slick at a rate of 80 km/d. How far will the ships be from the beach when they reach the oil slick?

16. A school group is traveling together on a school bus. The bus leaves a rest stop on an interstate highway at 1:00 p.m. and travels at a rate of 60 mph. One of the students did not get back on the bus before it left. A highway patrol car leaves the rest stop with the student at 1:30 p.m. and tries to catch the bus. If the patrol car averages 80 mph, at what time will it catch the bus? How far from the rest stop will the bus have traveled?

17. *Machine technology*   The worker on machine A can complete a certain job in 6 h. The worker on machine B can do the same job in 4 h. In a rush situation, how long would it take both machines to do the job?

18. *Construction*   Manuel can build a house in 45 days. With Errol's help they can build the house in 30 days. How long would it take Errol to build the same house by himself?

19. A tank can be filled by Pipe A in 4 h. Pipe B fills the same tank in 2 h. How long will it take to fill the tank if both pipes are used at the same time?

20. *Energy*   A solar collector can generate 70 kJ (kilojoules) in 12 min. A second solar collector can generate the same amount of kJ in 4 min. How long will it take the two of them working together to generate 70 kJ?

21. *Chemistry*   To generate hydrogen in a chemistry laboratory, a 40% solution of sulfuric acid is needed. You have 50 mL of an 86% solution of sulfuric acid. How many mL of water should be added to dilute the solution to the required 40% sulfuric acid level?

22. *Petroleum engineering*   A petroleum distributor has two gasohol storage tanks. One tank contains 8% alcohol and the other 14% alcohol. An order is received for 1 000 000 L of gasohol containing 9% alcohol. How many liters from each tank should be used to fill this order?

23. *Chemistry*   A copper alloy that is 35% copper is to be combined with an alloy that is 75% copper. The result will be 750 kg of an alloy that is 60% copper. How many kg of each alloy should be used?

24. *Chemistry*   How many pounds of an alloy of 20% silver must be melted with 50 lb of an alloy with 30% silver in order to get an alloy of 27% silver?

25. *Civil engineering*   A horizontal beam of negligible weight is 20 ft long and supported by columns at each end. A weight of 850 lb is placed at one spot on the beam. The force on one end is 500 lb. What is the force on the other end? Where is the weight located?

26. *Transportation engineering*   A loaded truck weighs 140 000 N. The front wheels of the truck support 60 000 N and the rear wheels support the rest. Where is the center of gravity if the axles are 4.2 m apart?

27. *Physics*   Locate the center of gravity of the beam in Figure 2.5 if the beam is 12 ft long.

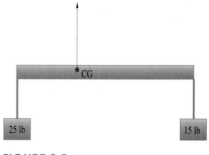

FIGURE 2.5

28. *Machine technology*   Locate the center of gravity of the machine part in Figure 2.6 if it is all constructed from the same material.

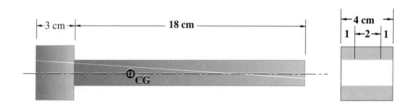

29. *Machine technology*   Locate the center of gravity of the machine part in Figure 2.7 if it is all made of the same metal. (Hint: The volume of a cylinder is $\pi r^2 h$ or $\frac{\pi}{4} d^2 h$, where r is the radius, d is the diameter of the circular bottom, and h is the height. This machine part is made of two circular cylinders.)

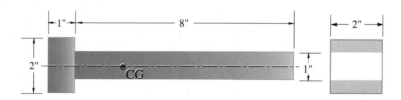

30. *Physics*   The center of gravity of the object in Figure 2.8 is 10 cm from the left side. What is the thickness of the right side?

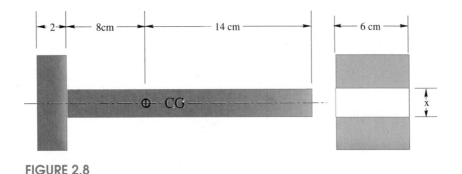

FIGURE 2.8

≡ **CHAPTER 2 REVIEW**

Important Terms and Concepts

Algebraic expression

Algebraic sum

Algebraic term

Binomial

Center of gravity

Coefficient

Conditional equation

Constant

Equation

Equivalent equations

Factors

FOIL method for multiplying binomials

Identities

Like terms

Mixture problems

Moment arm

Moment of force

Monomial

Multinomial

Polynomial

Root

Solution

Solving word problems

Statics problems

Torque

Trinomial

Uniform motion problems

Variable

Work problems

Review Exercises

Simplify the algebraic expressions in Exercises 1–48.

1. $8y - 5y$

2. $4z + 15z$

3. $7x - 4x + 2x - 8$

4. $-9a + 4a - 3a + 2$

5. $(2x^2 + 3x + 4) + (5x^2 - 3x + 7)$

6. $(3y^2 - 4y - 3) + (5y - 3y^2 + 6)$

7. $2(8x + 4)$

8. $-3(4a - 2)$

9. $-(3x - 1)$

10. $-(2z + 5)$

11. $(4x^2 + 3x) - (2x - 5x^2 + 2)$

12. $(7y^2 + 6y) - (6y - 7y^2) + 2y - 5$

13. $2(a + b) - 3(a - b) + 4(a + b)$

14. $6(c - d) - 4(d - c) + 2(c + d)$

15. $(ax^2)(a^2x)$

16. $(cy^3)(dy)$

17. $(9ax^2)(3x)$

18. $(6cy^2z)(2cz^3)$

19. $4(5x - 6)$

20. $3(12y - 5)$

21. $2x(4x - 5)$

22. $3a(6a + a^2)$

23. $(a + 4)(a - 4)$

24. $(x - 9)(x + 9)$

25. $(2a - b)(3a - b)$

26. $(4x + 1)(3x - 7)$

27. $(3x^2 + 2)(2x^2 - 3)$

28. $(4a^3 + 2)(6a - 3)$

29. $(x + 2)^2$

30. $(3 - y)^2$

31. $5x(3x - 4)(2x + 1)$

32. $6a(4a + 3)(a - 2)$

33. $a^5 \div a^2$

34. $x^7 \div x^3$

35. $8a^2 \div 2a$

36. $27b^3 \div 3b$

37. $45a^2x^3 \div (-5ax)$

38. $52b^4c^2 \div (-4b^2)$

39. $(36x^2 - 16x) \div 2x$

40. $(39b^3 + 52b^5) \div 13b^2$

41. $(x^2 - x - 12) \div (x + 3)$

42. $(x^2 - x - 30) \div (x - 6)$

43. $(x^3 - 27) \div (x - 3)$

44. $(8a^3 + 64) \div (2a + 4)$

45. $(x^2 - y^2) \div (x + y)$

46. $(y^2 - a^2) \div (y + a)$

47. $(x^3 - 2x^2y + 2y^3 - xy^2) \div (x^2 - y^2)$

48. $(a^3 - b^3 + 2ba + 3a^2b - 3ab^2) \div (a - b)$

Solve each of the equations in Exercises 49–68.

49. $x + 9 = 47$

50. $y - 19 = -32$

51. $2x = 15$

52. $-3y = 14$

53. $\frac{x}{4} = 9$

54. $\frac{y}{3} = -7$

55. $4x - 3 = 17$

56. $7 + 8y = 23$

57. $3.4a - 7.1 = 8.2$

58. $6.2b + 19.1 = 59.4$

59. $4x + 3 = 2x$

60. $7a - 2 = 2a$

61. $4b + 2 = 3b - 5$

62. $7c + 9 = 12c - 4$

63. $\dfrac{4(x - 3)}{3} = \dfrac{5(x + 4)}{2}$

64. $\dfrac{3(y - 7)}{5} = \dfrac{5(y + 4)}{2}$

65. $\dfrac{2}{a} - \dfrac{3}{a} = 5$

66. $\dfrac{4}{x} + \dfrac{5}{x} = \dfrac{1}{8}$

67. $\dfrac{3}{2a} = \dfrac{3}{a + 2}$

68. $\dfrac{9}{4b} = \dfrac{12}{b + 4}$

Solve Exercises 69–75.

69. A student received grades of 68, 70, and 74 on the first three exams. What does the student need to achieve on the next exam to have an average of 72?

70. *Business*   The price of a new television was $342.93. This price included 6.5% sales tax. How much was the television before the tax was added?

71. *Transportation*   An airplane leaves New York City and flies at a rate of 755 km/h. How long does it take to fly 2 718 km?

72. *Space technology*   A damaged space satellite passes over Houston at midnight and is traveling at a rate of 330 km/h. A space shuttle is attempting to overtake the satellite so it can be brought on board for repairs. The shuttle passes over Houston at 1:45 a.m., traveling in the same direction and orbit as the satellite and moving at 430 km/h. At what time does the shuttle overtake the satellite?

73. *Chemistry*   From 50 kg of solder that is half lead and half zinc, 10 kg of solder is removed and 15 kg of lead is added. How much lead is in the final mixture? What percent of the final mixture is lead?

74. *Physics*   A horizontal bar of negligible weight is supported by two columns that are 8 m apart. A load of 460 N is applied to the bar at a point x m from one end. The force on that end is 320 N. What is the force on the other end? Where was the load applied? (See Figure 2.9.)

460 N

FIGURE 2.9

75. *Physics*   A 490 N beam is suspended from two cables 10 m apart. A 2 156 N mass is placed 4 m from one end of the beam. How much tension is on each cable? (See Figure 2.10.)

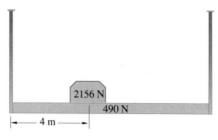

FIGURE 2.10

☰ CHAPTER 2 TEST

1. Simplify $2x + 5x^2 - 5x$.

2. Simplify $(4a^3 - 2b) - (3b + a^3)$.

In Exercises 3–8, perform the indicated operation and simplify your answer.

3. $4x + [3(x + y - 2) - 5(x - y)]$

4. $(4xy^3z)(\frac{1}{2}xy^{-2}z^2)$

5. $(2b - 3)(2b + 3)$

6. $(x^3 + 3x) \div x$

7. $(6x^5 + 4x^3 - 1) \div 2x^2$

8. $(3x^3 - 2x^2 + x - 3) \div (x + 2)$

9. $\dfrac{y}{3y + 2} - \dfrac{4}{y - 1}$

10. Solve for x: $5x - 8 = 3x$.

11. Solve for x: $\dfrac{7x + 3}{2} - \dfrac{9x - 12}{4} = 8$.

12. The price of a certain graphing calculator is $74.85. This price includes a 7% sales tax. How much was the calculator before the tax was added?

13. Your automobile's cooling system, including the heater and coolant reserve system, has a capacity of 9 qt of coolant. You know that the system currently contains 50% antifreeze and 50% water. You want to remove some of the solution and replace it with antifreeze, so that the final mixture is 60% antifreeze and 40% water. How much solution should you remove?

3

Geometry

Some oven hoods are made by combining prisms and the frustum of a pyramid. In Example 3.27, we shall see how to determine the amount of metal it takes to make an oven hood like the one shown here.

Courtesy of Best (USA) Inc.

Yⁿou have been working with geometry and geometric ideas most of your life. Geometry deals with the properties and measurements of lines, angles, plane figures, and solid figures. Geometry is a very important tool in technical mathematics. In this chapter we explore the basic geometric properties and measurements needed in technology.

In Chapter 4, we will begin to combine geometry and algebra. In this chapter, we will focus on the geometric ideas and skills that you will need in the following chapters and in many technical jobs.

☰ **3.1**
LINES AND ANGLES

Let's begin by discussing lines, segments, and rays. They are the building blocks for the remainder of this section. We will then look at two ways in which angles are measured. Finally, we will give a few areas in which angles and their measures are used.

Lines, Segments, and Rays

The basic parts of geometry are lines, angles, planes, surfaces, and the figures that they form. In this section, we will focus on the first two: lines and angles. A **line segment,** or **segment,** is a portion of a straight line between two endpoints. It is named by its endpoints, and a bar is placed over the endpoints' names. For example, the segment joining the points A and B, written $\overline{AB}$, is shown in Figure 3.1a. This segment could also be named $\overline{BA}$. A **ray** or **half-line** is the portion of a line that lies on one side of a point and includes the point. For example, in Figure 3.1b, ray $\overrightarrow{CD}$ begins at point C and passes through D. The beginning point of the ray is called the ray's **endpoint.** This second point used to name the ray can be any point on the ray other than the endpoint. Thus, in Figure 3.1b, $\overrightarrow{CD} = \overrightarrow{CE}$.

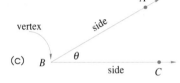

(a) line segment $\overline{AB}$ or $\overline{BA}$

(b) ray, or half-line $\overrightarrow{CD}$

(c)

FIGURE 3.1

Angles

An **angle** is formed by two rays that have the same endpoint. This common endpoint is called the **vertex** of the angle and the two rays are called the **sides** of the angle. The angle in Figure 3.1c has its vertex at B and the two rays that form the sides are $\overrightarrow{BA}$ and $\overrightarrow{BC}$. This angle has several possible names, including $\angle B$, $\angle ABC$, $\angle CBA$, and $\angle \theta$, where the symbol $\angle$ means angle.

Another way to think of an angle is to think of it as generated by moving a ray from an initial position to a terminal position. One revolution is the amount a ray would move to return to its original position. If this generated angle was placed inside a circle with its vertex at the center of the circle, then you would have a figure something like the one shown in Figure 3.2.

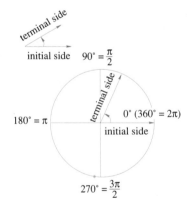

FIGURE 3.2

Degrees and Radians

Angles are measured using several different systems. The two most common units are **degrees** and **radians.** We will use both in this book. Look again at Figure 3.2. The end of the initial side is marked $0°$ ($360° = 2\pi$). This indicates that if the terminal side were to rotate completely around the circle and stop at the initial side, the size of the angle would be 360 degrees ($360°$) or 2π radians (2π rad). Later we will see a relationship between the radian and the distance around a circle.

Other important angle measures are given in relation to the distance around the circle. One-fourth of the way around the circle is $90°$ or $\frac{\pi}{2}$ rad. Halfway around the circle is $180°$ or π rad and three-fourths of the way is $270°$ or $\frac{3\pi}{2}$ rad.

☰ **Note**   If there is no degree symbol (°) after an angle measure, then assume that it is a radian measure.

Each degree is divided into 60 minutes and each minute into 60 seconds. The symbol $'$ is used for minutes and $''$ for seconds. So, $60' = 1°$ and $60'' = 1'$. Recently the use of calculators has caused decimal values for degrees to become more popular. Thus, in decimal degrees, $15' = \left(\frac{15}{60}\right)° = 0.25°$. The increased use of calculators should cause decimal values for degrees to become even more widely used.

There are some other widely used angles besides the ones shown in Figure 3.2. These are the $30° = \frac{\pi}{6}$, $45° = \frac{\pi}{4}$, and $60° = \frac{\pi}{3}$ angles. There are times when you are going to need to convert from degrees to radians or from radians to degrees. Since $180° = \pi$ rad, then

$$1 \text{ rad} = \frac{180°}{\pi} \approx \frac{180°}{3.1416} \approx 57.296°$$

(The symbol $\approx$ means "is approximately equal to.")

and you also have

$$1° = \frac{\pi}{180°} \approx \frac{3.1416}{180°} \approx 0.01745 \text{ rad}$$

Therefore, we have the following conversion formulas.

Conversion Between Degrees and Radians

To change from radians to degrees, multiply the number of radians by $\frac{180°}{\pi}$, or 57.296°.

To change from degrees to radians, multiply by $\frac{\pi}{180°}$, or 0.01745 (or divide by 57.296).

EXAMPLE 3.1

Change **(a)** 1.89 and **(b)** $\frac{5\pi}{6}$ to degrees.

Solutions   Since there is no degree symbol, these must be in radians.

(a) $1.89 \text{ rad} \approx 1.89 \text{ rad} \times \dfrac{57.296°}{1 \text{ rad}} = 108.29°$

(b) $\dfrac{5\pi}{6} = \dfrac{5\pi}{6} \times \dfrac{180°}{\pi} = 150°$

EXAMPLE 3.2

Change **(a)** 120° and **(b)** 82.5° to radians.

Solutions

(a) $120° = 120° \times \dfrac{\pi}{180°} \text{ rad} = \dfrac{2\pi}{3} \text{ rad}$

Notice here that the answer is left in terms of π and we have $120° = \dfrac{2\pi}{3}$, the exact answer. You may leave the answer in terms of π or compute a decimal approximation and see that $120° \approx 2.094$ rad.

EXAMPLE 3.2 (Cont.)

(b) $82.5° \approx 82.5° \times 0.01745 \text{ rad} = 1.439625 \approx 1.44 \text{ rad}$

or $\quad 82.5° = 82.5° \div 57.296 \approx 1.440 \text{ rad}$

or $\quad 82.5° = 82.5° \times \dfrac{\pi}{180°} = \dfrac{11\pi}{24} \approx 1.440 \text{ rad}$

 Hint

Some people find that it is easier to convert between degrees and radians by using the proportion

$$\frac{D}{180°} = \frac{R}{\pi}$$

where D represents a known or unknown number of degrees and R represents a known or unknown number of radians.

EXAMPLE 3.3

Use the proportion $\dfrac{D}{180°} = \dfrac{R}{\pi}$ to change **(a)** $120°$ to radians and **(b)** $\dfrac{5\pi}{6}$ to degrees.

Solutions

(a) Since we are asked to convert $120°$ to radians, we substitute $120°$ for D in the proportion and solve it for R as follows:

$$\frac{120°}{180°} = \frac{R}{\pi}$$

$$R = \frac{120°}{180°}\pi$$

$$= \frac{2}{3}\pi$$

Thus, $120° = \dfrac{2}{3}\pi = \dfrac{2\pi}{3}$ radians.

(b) Here we want to convert $\dfrac{5\pi}{6}$ to degrees, so we substitute $\dfrac{5\pi}{6}$ for R in the proportion, and obtain

$$\frac{D}{180°} = \frac{R}{\pi}$$

$$\frac{D}{180°} = \frac{5\pi/6}{\pi}$$

$$D = \frac{5\pi}{6} \cdot \frac{180°}{\pi}$$

$$= 150°$$

As in Example 3.1(b), we obtain $\dfrac{5\pi}{6} = 150°$.

An easier method to work these problems would be to use a calculator that has a conversion factor built in. (Introductory material on calculators is in Appendix A.) Some calculators have a $\boxed{\text{2nd}}$ $\boxed{\text{D·R}}$, $\boxed{\text{2nd}}$ $\boxed{\text{D↔R}}$, or $\boxed{\text{2nd}}$ $\boxed{\text{→RAD}}$ key. This indicates that this key can be used to change from degrees to radians. In Example 3.4, we will use a calculator to work the same problems that were in Example 3.2.

EXAMPLE 3.4

Use a calculator to change **(a)** 120° and **(b)** 82.5° to radians.

Solutions

		PRESS	DISPLAY
(a)		120 ⎡2nd⎤ ⎡D·R⎤	2.0943951
(b)		82.5 ⎡2nd⎤ ⎡D·R⎤	1.4398966

Our answer to **(b)** is the same as in Example 3.2. But the answer for **(a)** looks different. In Example 3.2(a), we obtained $\frac{2\pi}{3}$, but the calculator gave an answer of 2.0943951. However, a little work on your calculator will show you that these are approximately equal.

Caution

Graphing calculators, such as the Texas Instruments *TI-81* and the Casio *fx-7700G*, follow different procedures than those just shown. Consult the owner's manual for your calculator to find the exact steps you should use.

The calculator can also be used to convert from radians to degrees, but now you may need to use the ⎡INV⎤ key. This is the INVerse key. By using it before you use another function key, you reverse the operation of that function or key. In this example, we will need to use both the ⎡INV⎤ and the ⎡2nd⎤ keys. These can be used in either order; however, we will always use ⎡INV⎤ and then the ⎡2nd⎤. Some calculators have a ⎡2nd⎤ ⎡→DEG⎤ key that is used to convert radians to degrees. We will now use a calculator to work the same problems you saw in Example 3.1.

EXAMPLE 3.5

Use a calculator to change **(a)** 1.89 and **(b)** $\frac{5\pi}{6}$ to degrees.

Solutions

		PRESS	DISPLAY
(a)		1.89 ⎡INV⎤ ⎡2nd⎤ ⎡D·R⎤	108.28902
(b)		5 ⎡×⎤ ⎡2nd⎤ ⎡π⎤	3.1415927
		⎡÷⎤	15.707963
		6 ⎡=⎤	2.6179939
		⎡INV⎤ ⎡2nd⎤ ⎡D·R⎤	150

Part **(b)** of Example 3.4 was more complicated because we had to first convert $\frac{5\pi}{6}$ rad to its decimal approximation and then convert that approximation to degrees.

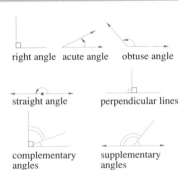

right angle   acute angle   obtuse angle

straight angle   perpendicular lines

complementary angles   supplementary angles

FIGURE 3.3

Types of Angles

Figure 3.3 shows four basic types of angles. A **right angle** has a measure of 90° or $\frac{\pi}{2}$ rad. It is usually indicated by placing a small square at the vertex of the angle. An **acute angle** measures between 0° and 90° (0 and $\frac{\pi}{2}$ rad) and an **obtuse angle** measures between 90° and 180° ($\frac{\pi}{2}$ and π rad). A **straight angle** has a measure of 180° or π rad. If two lines meet and form a right angle, then the lines are

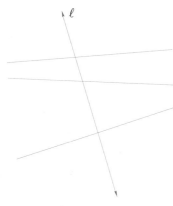

FIGURE 3.4

FIGURE 3.5

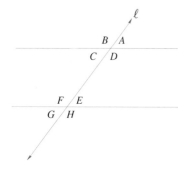

FIGURE 3.6

perpendicular. Two angles are **supplementary** if their sum is a straight angle (180° or π rad) and are **complementary** if their sum is a right angle (90° or $\frac{\pi}{2}$ rad).

Figure 3.4 shows two intersecting lines. Angles A and C are **vertical angles.** Angles B and D are also vertical angles. As you can probably see, vertical angles are the same size. In geometry, when two objects are the same size we say that they are **congruent.** Two 37° angles are said to be congruent. Two segments that are 3 cm long are congruent. So, since vertical angles are the same size, we can say that vertical angles are congruent.

Two angles that have the same vertex and a common side are called **adjacent angles.** In Figure 3.4, $\angle A$ and $\angle B$ are adjacent angles. They are also supplementary angles. There are three other pairs of adjacent angles in Figure 3.4. Can you find them?

A **transversal** is a line that intersects, or crosses, two or more lines. For example, in Figure 3.5, line l is a transversal of the other three lines. In Figure 3.6, line ℓ is a transversal of the other two lines. Angles A, B, G, and H are called **exterior angles** and angles C, D, E, and F are **interior angles.** Two angles that are not adjacent but are on different sides of the transversal are called **alternate angles.** Angles A and G are both exterior angles and alternate angles, so we say that they are **alternate exterior angles.** So are angles B and H. When you refer to alternate exterior angles, you are referring to a pair of angles. Another pair of angles is called the **alternate interior angles,** such as the pair $\angle C$ and $\angle E$ or the pair $\angle D$ and $\angle F$.

If there are two nonadjacent angles on the same side of the transversal, and one is an interior angle and the other is an exterior angle, then they are **corresponding angles.** In Figure 3.6, $\angle A$ and $\angle E$ are corresponding angles. So are $\angle D$ and $\angle H$. On the other side of the transversal, $\angle B$ and $\angle F$ form one set of corresponding angles and $\angle C$ and $\angle G$ form the other.

If a transversal intersects two parallel lines, as in Figure 3.6, then some interesting things are true. There are several theorems in geometry about parallel lines and a transversal. These theorems are summarized in the following theorem.

Theorem

If two parallel lines are intersected by a transversal, then the

(a) alternate interior angles are congruent,

(b) alternate exterior angles are congruent, and

(c) corresponding angles are congruent.

One pair of alternate interior angles in Figure 3.6 is $\angle C$ and $\angle E$. Another pair is $\angle D$ and $\angle F$. Since alternate interior angles are congruent, $\angle C \cong \angle E$ and $\angle D \cong \angle F$. (The symbol $\cong$ means "is congruent to.") The alternate exterior angles in Figure 3.6 are also congruent, so $\angle A \cong \angle G$ and $\angle B \cong \angle H$.

In Figure 3.6, $\angle A \cong \angle E$ and $\angle C \cong \angle G$, because they are corresponding angles. We also know that $\angle A \cong \angle C$, because they are vertical angles. So, $\angle A \cong \angle C \cong \angle E \cong \angle G$. In the same way we can show that $\angle B \cong \angle D \cong \angle F \cong \angle H$.

Application

EXAMPLE 3.6

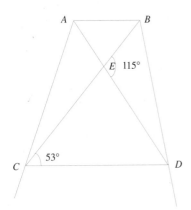

FIGURE 3.7

Some structures, such as the Eiffel Tower in Paris, France, or the Sunsphere in Knoxville, Tennessee, use a crossed girder design. A portion of a similar girder arrangement is shown in Figure 3.7, with $\overline{AB}$ parallel to $\overline{CD}$. If $\angle DCB = 53°$ and $\angle DEB = 115°$, what are the sizes of **(a)** $\angle ABC$, **(b)** $\angle CED$, and **(c)** $\angle AEC$?

Solutions

(a) $\angle ABC$ is an alternate interior angle of $\angle BCD$. Since $\angle BCD = 53°$, $\angle ABC = 53°$.

(b) $\angle CED$ is a supplementary angle of $\angle DEB$. We are given $\angle DEB = 115°$. So, $\angle CED = 180° - 115° = 65°$.

(c) $\angle AEC$ and $\angle DEB$ are vertical angles. So, $\angle AEC = \angle DEB = 115°$.

Another theorem says that if parallel lines are intersected by two transversals, then the corresponding segments are proportional. **Corresponding segments** are the parts of the transversals between the same parallel lines. In Figure 3.8, lines P_1, P_2 and P_3 are parallel lines intersected by transversals ℓ_1 and ℓ_2. The points of intersection have been marked with capital letters. Segments $\overline{AB}$ and $\overline{DE}$ are corresponding segments, as are $\overline{BC}$ and $\overline{EF}$. Likewise, $\overline{AC}$ and $\overline{DF}$ are corresponding segments. Therefore, we have the proportion

$$\frac{AB}{DE} = \frac{BC}{EF} = \frac{AC}{DF}$$

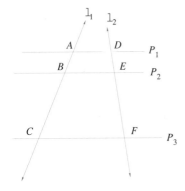

FIGURE 3.8

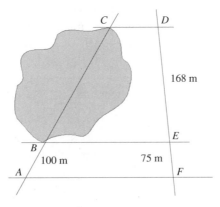

FIGURE 3.9 `

Application

EXAMPLE 3.7

Find the distance across the lake in Figure 3.9.

Solution   You carefully use stakes to mark off the segments $\overline{AF}$, $\overline{BE}$, and $\overline{CD}$. The segments are parallel. A, B, C are in a line, as are D, E, and F. We know

EXAMPLE 3.7 (Cont.)

that corresponding segments are proportional, so $\dfrac{AB}{EF} = \dfrac{BC}{DE}$. Using the lengths in Figure 3.9, $\dfrac{100}{75} = \dfrac{BC}{168}$. Solving this, we get $BC = \dfrac{100 \times 168}{75} = 224$. It is 224 m across the lake.

Exercise Set 3.1

Convert each of the angle measures in Exercises 1–14 from degrees to radians or from radians to degrees without using a calculator. (You may leave your answers in terms of π.) When you have finished, check your work with a calculator.

1. $15°$

2. $75°$

3. $210°$

4. $10°45'$

5. $85.4°$

6. $48.6°$

7. $163.5°$

8. $242°35'$

9. $\dfrac{4\pi}{3}$

10. $\dfrac{\pi}{6}$

11. 1.3π

12. 2.15

13. 0.25

14. 1.1

Solve Exercises 15 and 16.

15. Find the supplement of a $35°$ angle in **(a)** degrees and **(b)** radians.

16. Find the complement of a $65°$ angle is **(a)** degrees and **(b)** radians.

In Exercises 17–22, find the measures of angles A and B. In Exercises 19–23, lines ℓ_1 and ℓ_2 are parallel.

17.

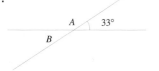

18.

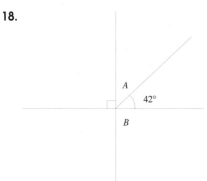

19.

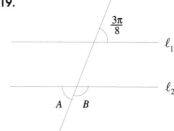

20.

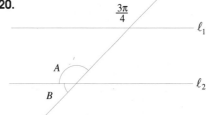

21.

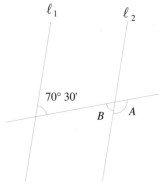

22.

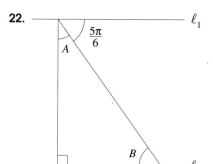

Solve Exercises 23–28.

23. What is the distance from A to B in Figure 3.10 if lines ℓ_1, ℓ_2, and ℓ_3 are parallel?

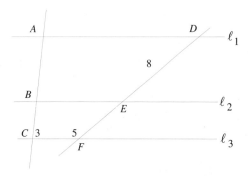

FIGURE 3.10

24. In Figure 3.11, if lines ℓ_1, ℓ_2, and ℓ_3 are parallel, what is the distance from A to B?

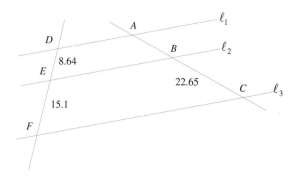

FIGURE 3.11

25. In Figure 3.12, angles A and B are the same size. Find the sizes of angles A and B.

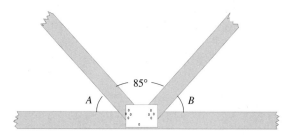

FIGURE 3.12

26. *Civil engineering*   Figure 3.13 shows part of a bridge that is made of suspended cables hung from a girder to the deck of the bridge. What is the length from A to B?

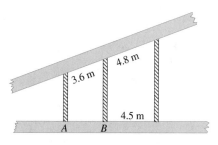

FIGURE 3.13

27. *Electronics*   The angular frequency ω of an ac current is $\omega = 2\pi f$ rad/s, where f is the frequency. If $f = 60$ Hz for ordinary house current, what is the angular frequency?

28. *Mechanical technology*   An angular velocity ω of a rigid object, such as a drive shaft, pulley, or wheel, is related to the angle θ through which the body rotates in a period of time t by the formula $\theta = \omega t$ or $\omega = \dfrac{\theta}{t}$.

(a) Find the angular velocity of a gear wheel that rotates $285°$ in 0.6 s.

(b) Find the angular velocity of a wheel that rotates $\dfrac{11\pi}{16}$ rad in 0.9 s.

≡ 3.2
TRIANGLES

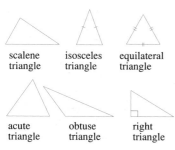

scalene triangle   isosceles triangle   equilateral triangle

acute triangle   obtuse triangle   right triangle

FIGURE 3.14

A **polygon** is a figure in a plane that is formed by three or more line segments, called **sides,** joined at their endpoints. The endpoints are called **vertices** and one endpoint is called a **vertex.** In this section, we will be concerned only with the one type of polygon called a triangle.

A **triangle** is a polygon that has exactly three sides. (See Figure 3.14.) Triangles are named according to a property of their sides or their angles. When classified according to its sides, a triangle is **scalene** if none of the sides are the same length, **isosceles** if two sides are the same length, and **equilateral** if all three sides are the same length. When sides are the same length, single, double, or triple marks are used to show which sides are congruent.

An **acute triangle** has three acute angles, an **obtuse triangle** has one obtuse angle, and a **right triangle** has one right angle. (See Figure 3.14.) In an equilateral triangle, all three angles are congruent. In an isosceles triangle the angles opposite the congruent sides are congruent.

The sum of the three angles of a triangle is $180°$, or π rad. Since all three angles in an equilateral triangle are congruent, the size of each angle must be $\dfrac{180°}{3} = 60°$ or $\dfrac{\pi}{3}$ rad. Since a right angle is $90°$, a right triangle must have two acute angles, because the other two angles, when added together, can have only a total of $90°$. This means that each of the angles must be less than $90°$. In a similar manner, an obtuse triangle must have two acute angles.

EXAMPLE 3.8

Find the size of the third angle of a triangle if the other two angles measure $37°$ and $96°$.

Solution   If we call the unknown angle $\angle A$, then

$$\angle A = 180° - 37° - 96° = 47°$$

The third angle of this triangle is $47°$.

The **perimeter** of a triangle is the distance around the triangle. To find the perimeter, you add the lengths of the three sides. If the lengths of the sides are a, b, and c, then the perimeter P is $P = a + b + c$.

Application

EXAMPLE 3.9

A triangular piece of land measures 42 m, 36.2 m, and 58.7 m on the three sides. What is the perimeter of this plot of land?

Solution $P = a + b + c$

$$= 42 + 36.2 + 58.7$$

$$= 136.9$$

The perimeter is 136.9 m.

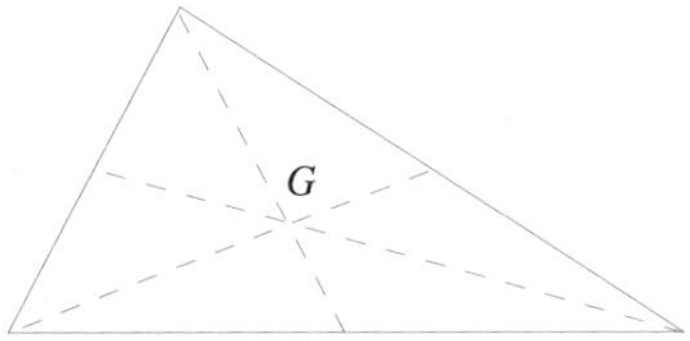

FIGURE 3.15

A segment drawn from a vertex to the middle (or midpoint) of the opposite side is called a **median.** If you draw all three medians of a triangle, they meet in a common point, G, called the **centroid,** as shown in Figure 3.15. The centroid is the center of gravity of a triangle.

A segment drawn from a vertex and perpendicular to the opposite side is an **altitude** of a triangle. The opposite side is called the **base.** (See Figure 3.16.) When solving some problems, it is sometimes necessary to extend the base so that it will intersect the altitude.

The area of a triangle is found by the product of $\frac{1}{2}$ and the lengths of the base and the altitude. If b is the length of the base and h the length of the altitude, then the area A is given by the formula

$$A = \frac{1}{2}bh$$

If the base and altitude are in the same units, then the area is in square units.

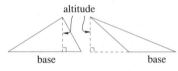

FIGURE 3.16

EXAMPLE 3.10

Find the area of a triangle with a base of 1.2 m and an altitude of 4.5 m.

Solution $A = \frac{1}{2}bh$

$$= \frac{1}{2}(1.2)(4.5)$$

$$= 2.7$$

The area is 2.7 m^2.

Sometimes the length of the altitude is not known. It is possible to find the area using **Hero's formula,** also referred to as **Heron's formula.** Instead of the altitude, you will need the lengths of the three sides and the semiperimeter. If we let a, b, and c represent the lengths of the sides, then the **semiperimeter,** s, is found using the formula $s = \dfrac{a + b + c}{2}$. We can find the area from Hero's formula:

$$A = \sqrt{s(s - a)(s - b)(s - c)}$$

Application

EXAMPLE 3.11

A triangular piece of land measures $51'9''$, $47'3''$, and $82'6''$ on the three sides. What is the area of the lot of land?

Solution   We will let $a = 51'9'' = 51.75$ ft, $b = 47'3'' = 47.25$ ft, and $c = 82'6'' = 82.5$ ft.

$$s = \frac{a+b+c}{2} = \frac{51.75 + 47.25 + 82.5}{2} = \frac{181.5}{2} = 90.75$$

$$A = \sqrt{s(s-a)(s-b)(s-c)}$$

$$= \sqrt{90.75\,(90.75 - 51.75)\,(90.75 - 47.25)\,(90.75 - 82.5)}$$

$$= \sqrt{90.75\,(39)\,(43.5)\,(8.25)}$$

$$= \sqrt{1,270,148.3}$$

$$\approx 1,127.0086$$

The area is about $1,127$ ft^2.

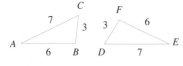

FIGURE 3.17

Two triangles are congruent if the corresponding angles and sides of each triangle are congruent. For example, in Figure 3.17, sides $\overline{AB}$ and $\overline{EF}$ are both 6 units long, $\overline{BC}$ and $\overline{DF}$ are both 3 units long, and $\overline{AC}$ and $\overline{DE}$ are 7 units long. If you measure the angles you will find that $\angle A \cong \angle E$, $\angle B \cong \angle F$ and $\angle C \cong \angle D$. As you can see, **congruent triangles** have the same size and shape.

We indicate that the triangles in Figure 3.17 are congruent by writing $\triangle ABC \cong \triangle EFD$. When we show that two triangles are congruent, we write it so the corresponding vertices are in the same order. Thus, writing $\triangle ABC \cong \triangle EFD$ indicates that $\angle A \cong \angle E$, $\angle B \cong \angle F$, and $\angle C \cong \angle D$.

Two triangles that have the same shape are said to be **similar triangles.** In similar triangles, the corresponding angles are congruent. In Figure 3.18, $\angle A \cong \angle D$, $\angle B \cong \angle E$, and $\angle C \cong \angle F$. Since the corresponding angles are the same size, the triangles are similar. Congruent triangles are a special case of similar triangles in which the corresponding sides are the same length. In all similar triangles, the corresponding sides are proportional. In Figure 3.18, the sides of the larger triangle are three times as large as those of the smaller triangle. Thus, the ratios of the corresponding sides of the larger triangle to the smaller triangle are $\frac{12}{4} = \frac{9}{3} = \frac{6}{2} = \frac{3}{1}$.

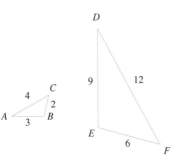

FIGURE 3.18

Application

EXAMPLE 3.12

A television tower casts a shadow that is 150 m long. At the same time, a vertical pole that is 1.6 m high casts a shadow 1.2 m long. How high is the television tower? (See Figure 3.19.)

EXAMPLE 3.12 (Cont.)

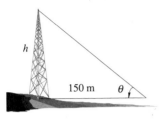

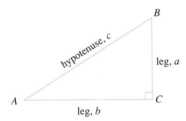

FIGURE 3.19

Solution   We have two similar triangles, so the corresponding sides are proportional. If we call the unknown height h, then we have

$$\frac{h}{1.6} = \frac{150}{1.2}$$

$$h = \frac{150\,(1.6)}{1.2}$$

$$= 200$$

The height of the tower is 200 m.

One of the most valuable theorems in geometry involves the right triangle. In a right triangle, the side opposite the right angle is called the **hypotenuse,** as shown in Figure 3.20. The other two sides áre the **legs.** If the lengths of the two legs are a and b and the length of the hypotenuse is c then the **Pythagorean theorem** is stated as follows.

FIGURE 3.20

Pythagorean Theorem

$\triangle ABC$ is a right triangle with a hypotenuse of length c and legs of lengths a and b, if and only if

$$a^2 + b^2 = c^2$$

EXAMPLE 3.13

A right triangle has legs of length 6 in. and 8 in. What is the length of the hypotenuse?

Solution   If we let $a = 6$ and $b = 8$, then, by using the Pythagorean theorem, we have

$$c^2 = a^2 + b^2$$

$$= 6^2 + 8^2$$

$$= 36 + 64$$

$$= 100$$

$$c = \sqrt{100} = 10$$

The hypotenuse is 10 in. long.

EXAMPLE 3.14

The hypotenuse of a right triangle is 7.3 cm long. One leg is 4.8 cm long. What is the length of the other leg?

EXAMPLE 3.14 (Cont.)

Solution   We are given $c = 7.3$ and $a = 4.8$, and so, by using the Pythagorean theorem, we obtain

$$c^2 = a^2 + b^2$$

$$(7.3)^2 = (4.8)^2 + b^2$$

$$(7.3)^2 - (4.8)^2 = b^2$$

or $$b^2 = (7.3)^2 - (4.8)^2$$

$$= 53.29 - 23.04$$

$$= 30.25$$

$$b = \sqrt{30.25} = 5.5$$

The length of the remaining side is 5.5 cm.

Application

EXAMPLE 3.15

In Example 3.12 we found the height of a television tower. If a cable could be strung in a straight line from the top of the tower to a point on the ground 130 m from the base of the tower, how long would the cable have to be?

Solution   The cable, tower, and ground form a right triangle. We know that the height of the tower is 200 m. This is one leg of the triangle. The other leg is 130 m. The cable will form the hypotenuse. So,

$$c^2 = a^2 + b^2$$

$$c^2 = 130^2 + 200^2$$

$$= 16\,900 + 40\,000$$

$$= 56\,900$$

$$c = \sqrt{56\,900} \approx 238.54$$

Since the given data, 200 m and 130 m, have three significant digits, the cable would be about 239 m long.

The last example might not have been exactly realistic. It failed to account for any sag in the cable or any additional cable that would be needed to fasten it at each end. Later, we will find some methods to get more accurate answers to this problem.

Remember that the Pythagorean theorem can be used only with right triangles. The following exercises provide some additional examples that use the Pythagorean theorem.

Exercise Set 3.2

Find the indicated variables in Exercises 1–8. Identical marks on segments or angles indicate that they are congruent.

1.

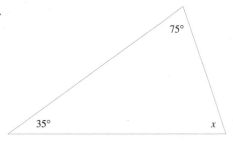

75°

35°   x

2.

122°20′

x   15°30′

3.

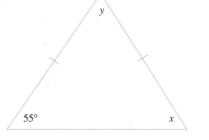

y

55°   x

4.

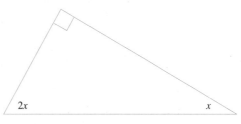

2x   x

5.

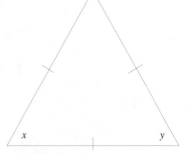

x   y

6.

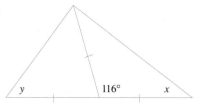

y   116°   x

7.

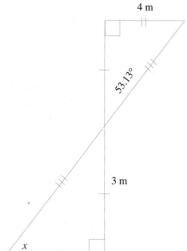

4 m

53.13°

3 m

x

8.

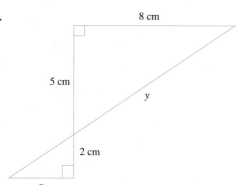

8 cm

5 cm

y

2 cm

x

Find the area and perimeter of △*ABC* in Exercises 9–12.

9.

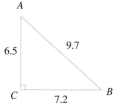

10.

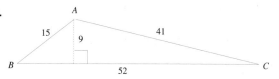

11.

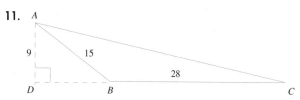

12.

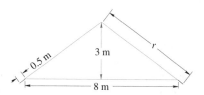

Solve Exercises 13–24.

13. *Landscape architecture*   A triangular piece of land measures 23.2 m, 47.6 m, and 62.5 m. **(a)** How much fencing is needed to enclose this land? **(b)** How many square meters of sod would be needed to sod the entire piece of land?

14. The shadow of a building is $123'6''$. At the same time, the shadow of a yard stick is $2'3''$. What is the height of the building?

15. A ladder 15 m long reaches the top of a building when its foot is 5 m from the building. How high is the building?

16. *Mechanical engineering*   A triangular metal plate was made by cutting a rectangular plate along a diagonal. If the rectangular piece was 23 cm long and 16 cm high, what is the area of the plate?

17. *Transportation engineering*   A traffic light support is to be suspended parallel to the ground. It reaches diagonally across the intersection of two perpendicular streets. One street is 45 ft wide and the other is 62 ft wide. Determine the length of the support.

18. *Construction*   A house is 8 m wide and the ridge is 3 m higher than the side walls. If the rafters, r, extend 0.5 m beyond the sides of the house, how long are the rafters? (See Figure 3.21.)

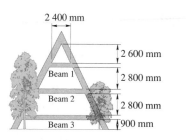

FIGURE 3.21

19. *Construction*   Several cross beams are placed across an A-frame house. The highest beam is placed 2 600 mm from the ridge and the top of this beam is 2 400 mm long. The top of the next beam is placed 2 800 mm below the bottom of the top beam and the top of the third is 2 800 mm below the bottom of the second. The bottom of the third beam is 900 mm above the ground. Each beam is 300 mm thick. **(a)** How long are the tops of the second and third beams? **(b)** How far apart is the base of the A-frame? (See Figure 3.22.)

FIGURE 3.22

20. *Construction* An antenna 175 m high is supported by cables positioned around the antenna. One cable is 50 m from the base of the antenna, one is 75 m, and the third is 100 m. How long is each cable?

21. *Electrical engineering* A 30 m length of conduit is bent as shown in Figure 3.23. What is the length of the offset, x?

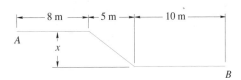

FIGURE 3.23

22. *Electrical engineering* If it were possible to run a straight conduit from A to B in Figure 3.23, how much conduit would be saved?

23. *Construction* A gas pipeline was constructed across a ravine by going down one side, across the bottom, and up the other side, as shown in Figure 3.24. **(a)** How much pipe was used to get from A to B? **(b)** How much would have been needed if it would have been possible to go directly from A to B?

FIGURE 3.24

24. *Physics* The effect of a moving load on the stress beam is shown in the influence diagram of Figure 3.25. If the distance from A to B is 24 ft, find the lengths of $\overline{BC}$ and $\overline{CD}$. (Assume that $\overline{AF}$ is parallel to $\overline{EG}$.)

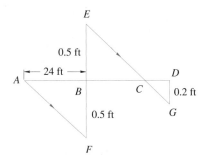

FIGURE 3.25

≡ 3.3
OTHER POLYGONS

In Section 3.2 we looked at the simplest type of polygon, the triangle. In this section, we will look at polygons that have more than three sides.

After the triangle, the most commonly used polygons are those with four sides. A polygon with exactly four sides is a **quadrilateral.** Other polygons that we will use are the **pentagon,** which has exactly five sides, the **hexagon** with six sides, and the **octagon** with eight sides. Any polygon that has all congruent sides and congruent angles is a **regular polygon.** For example, equilateral is another name for a regular triangle.

Quadrilaterals

All quadrilaterals have four sides. Like triangles, different names are given to the quadrilaterals with special properties. The properties deal with the lengths of the sides, if the sides are parallel, and the sizes of the angles. Among the special types of quadrilaterals are the ones shown in Figure 3.26.

The **kite** has two pairs of adjacent sides congruent. A **trapezoid** has at least one pair of opposite sides parallel. If a trapezoid also has a pair of congruent sides that are not parallel, it is an **isosceles trapezoid** and has the congruent angles indicated in Figure 3.26. A **parallelogram** is a quadrilateral with both pairs of opposite sides parallel. As a result, the opposite sides are also congruent.

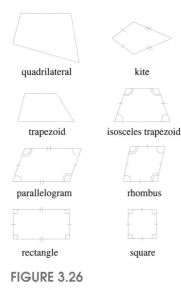

quadrilateral

kite

trapezoid

isosceles trapezoid

parallelogram

rhombus

rectangle

square

FIGURE 3.26

A **rhombus** is a parallelogram that has all four sides congruent. A **rectangle** is a parallelogram that has four right angles. A **square** is a rectangle that has four congruent sides. As you can see in Figure 3.26, a square is a special kind of rhombus.

To find the perimeter of a quadrilateral, you must add the lengths of the four sides. In the case of a rhombus or square, this is made easier by the fact that all four sides are the same length. So, if s is the length of one side of a rhombus or square, the perimeter is $4s$. A parallelogram, rectangle, and kite are not much more difficult, since each has two pairs of congruent sides. If the lengths of the sides that are not congruent are a and b, then the perimeter is $2(a + b)$.

In most cases, the area of a quadrilateral depends on the length of one or more sides and the distance between this side and its opposite side. If the height h is the distance between the sides and the length of the base is b, then the area of a rectangle, square, rhombus, or parallelogram can all be expressed as $A = bh$. In a trapezoid you must use the average length of the two parallel sides, $\frac{1}{2}(b_1 + b_2)$ and the height, h. So, for a trapezoid, $A = \frac{1}{2}h(b_1 + b_2)$. A kite is an exception, since its area can be found by multiplying $\frac{1}{2}$ by the lengths of the diagonals, d_1 and d_2. So, the area of a kite is $\frac{1}{2}d_1d_2$. All of this is summarized in Figure 3.27, where the parts of each figure are labeled and the formulas for perimeter and area of each polygon are given.

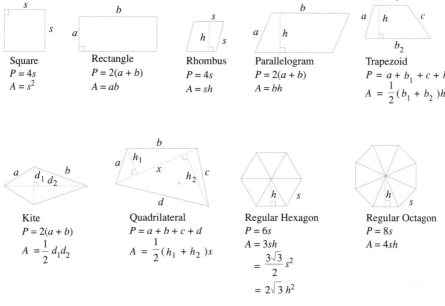

Square
$P = 4s$
$A = s^2$

Rectangle
$P = 2(a + b)$
$A = ab$

Rhombus
$P = 4s$
$A = sh$

Parallelogram
$P = 2(a + b)$
$A = bh$

Trapezoid
$P = a + b_1 + c + b_2$
$A = \frac{1}{2}(b_1 + b_2)h$

Kite
$P = 2(a + b)$
$A = \frac{1}{2}d_1d_2$

Quadrilateral
$P = a + b + c + d$
$A = \frac{1}{2}(h_1 + h_2)x$

Regular Hexagon
$P = 6s$
$A = 3sh$
$= \frac{3\sqrt{3}}{2}s^2$
$= 2\sqrt{3}h^2$

Regular Octagon
$P = 8s$
$A = 4sh$

FIGURE 3.27

Area and Perimeter of Other Polygons

Figure 3.27 shows the formulas for the perimeter and area of a general quadrilateral. Also included are the area and perimeter for a regular hexagon and a regular octagon. In general, with a regular polygon of n sides, you can divide it into n congruent triangles, each with height h and base with length s. The perimeter of the polygon is ns and the area is $\frac{1}{2}nsh$. In Figure 3.27, two additional formulas for the area of a regular hexagon are given. These come from the fact that it is divided into six equilateral triangles and by using the Pythagorean theorem.

EXAMPLE 3.16

Find the perimeter and area of a rectangle with a length of 16 m and a width of 9 m.

Solution $P = 2(a + b)$

$$= 2(16 + 9)$$

$$= 2(25) = 50$$

$$A = ab$$

$$= 16 \cdot 9$$

$$= 144$$

The perimeter is 50 meters and the area is 144 square meters. These answers are usually written as 50 m and 144 m^2.

EXAMPLE 3.17

A trapezoid has bases of 14 and 18 in. and a height of 7 in. What is its area?

Solution $A = \frac{1}{2}h(b_1 + b_2)$

$$= \frac{1}{2}(7)(14 + 18)$$

$$= \frac{1}{2}(7)(32)$$

$$= (7)\frac{1}{2}(32)$$

$$= 7 \cdot 16 = 112$$

The area is 112 square inches (often written as 112 in.^{2}).

EXAMPLE 3.18

A metal plate is made by cutting a rectangle out of an isosceles trapezoid, using the dimensions shown in Figure 3.28. **(a)** Find the area and **(b)** the perimeter of this metal plate.

EXAMPLE 3.18 (Cont.)

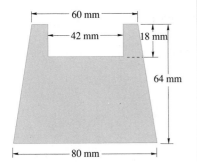

FIGURE 3.28

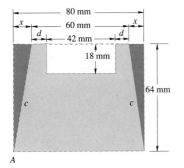

FIGURE 3.29

Solutions

(a) To find the area of this plate, we first find the area of the trapezoid and then of the rectangle. The area of the plate is the difference between the two.

$$A_{\text{trapezoid}} = \frac{1}{2}h(b_1 + b_2)$$

$$= \frac{1}{2}(64)(60 + 80)$$

$$= \frac{1}{2}(64)(140)$$

$$= 4{,}480$$

$$A_{\text{rectangle}} = bh$$

$$= (42)(18)$$

$$= 756$$

$$A_{\text{plate}} = A_{\text{trapezoid}} - A_{\text{rectangle}}$$

$$= 4{,}480 - 756$$

$$= 3{,}724$$

(b) We have all the measurements we need to find the perimeter except for the lengths of the two slanted sides of the trapezoid. For this we will use the Pythagorean theorem. Look at Figure 3.29. The plate has been drawn inside a rectangle. Because the plate was cut from an isosceles trapezoid, we know that the two lengths marked x are the same length. We also know that $2x + 60 = 80$, so $x = 10$ mm.

The slanted sides of the trapezoid are the hypotenuse c of the two shaded triangles in Figure 3.29. The legs of these triangles are 10 and 64 so

$$c^2 = 10^2 + 64^2$$

$$= 100 + 4\,096$$

$$= 4\,196$$

$$c = \sqrt{4\,196} \approx 64.78$$

If we assume that the rectangular cutout is centered, then

$$d = \left(\frac{60 - 42}{2}\right) = 9 \text{ mm}$$

The perimeter can be found by starting at point A and adding the lengths of the sides in a clockwise direction:

$$P = 64.78 + 9 + 18 + 42 + 18 + 9 + 64.78 + 80$$

$$= 305.56$$

The perimeter of this plate is approximately 305.56 mm.

Exercise Set 3.3

Find the perimeter and area of each polygon in Exercises 1–10. Use the formulas in Figure 3.27.

1.

15 cm
15 cm

2.

9 in.
25 in.

3.

$11\frac{1}{2}$ in.
$15\frac{1}{4}$ in.
33 in.

4.

23.4 mm
11.2 mm
24.9 mm

5.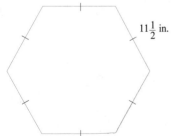

$11\frac{1}{2}$ in.

6.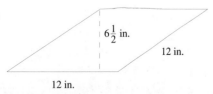

$6\frac{1}{2}$ in.
12 in.
12 in.

7.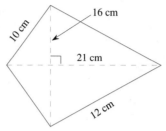

10 cm
16 cm
21 cm
12 cm

8.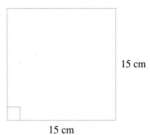

318 mm
213 mm
20.8 mm

9.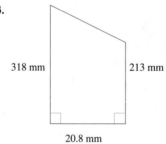

21.2 mm
15.7 mm
13.2 mm

10.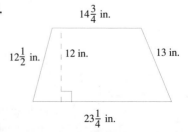

$14\frac{3}{4}$ in.
$12\frac{1}{2}$ in.
12 in.
13 in.
$23\frac{1}{4}$ in.

Solve Exercises 11–22.

11. *Construction*   Find the area of the L-shaped patio in Figure 3.30.

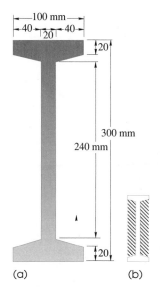

FIGURE 3.30

12. *Construction*   You plan to put a brick border around the patio in Figure 3.30. Each brick is 8 in. long and 4 in. wide. If the bricks are placed end to end, how many bricks are needed?

13. *Mechanical engineering*   What is the area of the cross-section of the I-beam in Figure 3.31a? (Hint: Think of it as a rectangle and subtract the areas of the two trapezoids. See Figure 3.31b.)

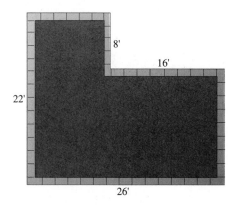

(a)   (b)

FIGURE 3.31

14. *Civil engineering*   Find the area of the concrete highway support shown in Figure 3.32.

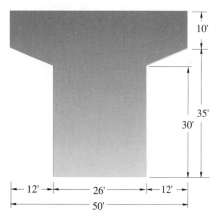

FIGURE 3.32

15. *Machine technology*   A hexagonal bolt measures $\frac{7}{8}$ '' across the short distance. (See Figure 3.33.) What are the perimeter and area of this bolt?

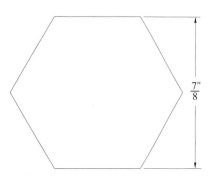

FIGURE 3.33

16. *Machine technology*   A hexagonal bolt measures 15 mm across the short distance. What are the perimeter and area of this bolt?

17. *Civil engineering*   A river bed is going to be constructed in the shape of a trapezoid. The height of the trapezoid is designed to contain flood waters between the dykes, which form the walls. The trapezoid has the dimensions given in Figure 3.34. (a) What is the cross-sectional area? (b) How many linear feet

of concrete are needed to surface the walls and the bottom?

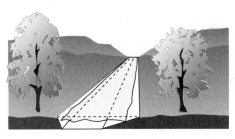

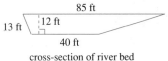

85 ft

13 ft   12 ft

40 ft

cross-section of river bed

FIGURE 3.34

18. *Interior design*   What will it cost to carpet the floor of a rectangular room that is $20'6''$ by $15'3''$ at \$6.75 a square yard? (There are 9 ft^2 in 1 yd^2.)

19. *Civil engineering*   A structural supporting member is made in the shape of an angle as shown in Figure 3.35. What is the cross-sectional area?

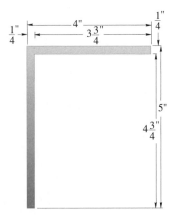

$\frac{1}{4}''$ 4'' $3\frac{3}{4}''$ $\frac{1}{4}''$

5''

$4\frac{3}{4}''$

FIGURE 3.35

20. *Architecture*   How many $9''$ square tiles will cover a floor $12'$ by $17'3''$?

21. *Civil engineering*   Find the area of the cross-section of the structural tee in Figure 3.36.

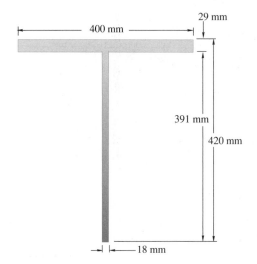

29 mm

400 mm

391 mm

420 mm

18 mm

FIGURE 3.36

22. *Architecture*   Find the area of the side of the house in Figure 3.37.

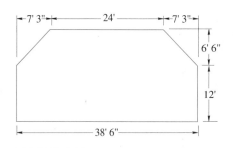

7' 3"   24'   7' 3"

6' 6"

12'

38' 6"

FIGURE 3.37

☰ 3.4
CIRCLES

Until now, the geometric figures we have studied have all been made of segments joined at their endpoints. A circle is a different type of geometric figure. Like the polygons, a circle is part of a plane. A **circle** is made up of the set of points that are all the same distance from a point called the **center,** as shown in Figure 3.38.

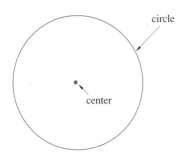

FIGURE 3.38

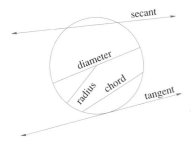

FIGURE 3.39

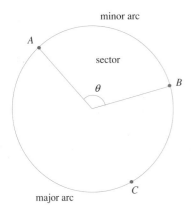

FIGURE 3.40

Parts of a Circle

There are many parts to a circle. Some of them are shown in Figure 3.39. The **radius** is a segment with one endpoint at the center and the other on the circle. A **chord** is any segment with both endpoints on the circle. A **diameter** is a chord through the center of the circle. A **secant** is a line that intersects a circle twice, and a **tangent** intersects the circle in exactly one point. A tangent is perpendicular to the radius at its point of tangency. The **circumference** of a circle is the name for the perimeter and is equal to the distance around the circle.

Figure 3.40 shows a few more parts of the circle that will be used later. A **central angle** is an angle with a vertex at the center of the circle. An **arc** is a section of a circle and is often described in terms of the size of its central angle. Thus, we might refer to a 20° arc or an arc of $\frac{\pi}{9}$ rad.

An arc with length equal to the radius is 1 rad. A central angle divides a circle into a **minor arc** and a **major arc.** We also may refer to an arc by its endpoints. The minor arc in Figure 3.40 is identified as $\overset{\frown}{AB}$. The major arc is identified as $\overset{\frown}{ACB}$, where A and B are the endpoints and C is any other point on the major arc. The length of an arc is denoted by placing an m in front of the name of the arc. Thus, $m\overset{\frown}{AB}$ is the length of $\overset{\frown}{AB}$. A **sector** is the region inside the circle and is bounded by a central angle and an arc.

Circumference and Area of Circles

The formulas for the circumference and area of a circle involve the use of the irrational number π. The circumference C of a circle is

$$C = \pi d$$

or $C = 2\pi r,$ (where $d = 2r$)

where d is the length of a diameter and r is the length of a radius.

The area of a circle is

$$A = \pi r^2$$

Some mechanics use the formula

$$A = 0.785d^2$$

Since $d = 2r$, you can see that $d^2 = 4r^2$ and so $r^2 = \frac{d^2}{4}$. The area of the circle can be written as $A = \pi \left(\frac{d^2}{4} \right) = \frac{\pi}{4} d^2$. Since $\frac{\pi}{4} \approx 0.785$, we have $A = 0.785d^2$.

Circumference and Area of a Circle

A circle with radius r and diameter d has a circumference C and an area A where

$$C = 2\pi r = \pi d$$

and $A = \pi r^2$

EXAMPLE 3.19

Find the circumference and area of a circle with **(a)** diameter 7.00 in. and **(b)** radius 8.3 mm.

Solution

(a) $d = 7.00$ in.

$$C = \pi d$$
$$= \pi 7$$
$$\approx 21.99 \text{ in.}$$
$$A = \pi r^2$$
$$= \pi \left(\frac{7}{2}\right)^2$$
$$\approx 38.48 \text{ in.}^2$$

(b) $r = 8.3$ mm

$$C = 2\pi r$$
$$= 2\pi(8.3)$$
$$\approx 52.2 \text{ mm}$$
$$A = \pi r^2$$
$$= \pi(8.3)^2$$
$$\approx 216.4 \text{ mm}^2$$

This example raises a question. What value should you use for π? If you use 3.14 or $3\frac{1}{7}$, your answers may differ slightly from those shown. They were obtained using a value programmed into a calculator and recalled by pressing the $\boxed{\pi}$ key, with the resulting display of 3.1415927. As you can see, $\pi \neq 3.14$. We use 3.14 as an approximation of π.

Arc Length

The arc length s is a direct result of the size of the angle that determines the arc. An angle of 2π rad is a complete circle and has an arc length equal to the circumference, $2\pi r$. An angle of π rad is half a circle, and so its arc length is πr. Similarly, an angle of 3 rad has an arc length of $3r$. In general, a central angle of θ rad has an arc length of θr. This gives us the following formula.

Arc Length

An arc formed from a circle of radius r and central angle of θ rad has an arc length s, where

$$s = \theta r$$

Similarly, the area of a sector of a circle can be derived from the formula for the area of a circle. A complete circle has an angle of 2π rad and an area of $\pi r^2 = \frac{1}{2}(2\pi)r^2$. A semicircle, or half a circle, has an angle of π rad and an area of $\frac{1}{2}\pi r^2$. In general, a central angle of θ rad forms a sector with area $\frac{1}{2}\theta r^2 = \frac{1}{2}r^2\theta$, as given in the following box.

Area of a Sector

A sector formed by a circle with radius r and a central angle of θ rad has an area, A, where

$$A = \frac{1}{2}r^2\theta$$

≡ **Note**   In both of the previous formulas, the central angle θ *must* be in radians.

Application

EXAMPLE 3.20

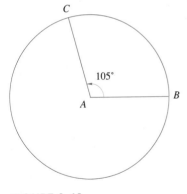

FIGURE 3.41

A landscaper is going to put some plastic edging around a pie-shaped flower bed. The flower bed is formed from a circle with radius 9.0 ft and a central angle of 105°. How much edging, in feet, will be needed?

Solution   A sketch of the flower bed is shown in Figure 3.41. The edging will go from C to A then to B and back to C around the arc of the circle. The length of edging needed is $CA + AB + m\widehat{BC} = 9.0 + 9.0 + m\widehat{BC} = 18.0 + m\widehat{BC}$. All we need to determine is the length of the arc $\widehat{BC}$.

To determine the length of the arc, we will use $s = \theta r$. Here, $r = 9.0$ and $\theta = 105°$. To use this formula, we must convert 105° to radians. Using the proportion $\dfrac{D}{180°} = \dfrac{R}{\pi}$, with $D = 105°$, we obtain $R = \dfrac{7\pi}{12}$. Thus,

$$m\widehat{BC} = r\theta$$

$$= 9.0\left(\frac{7\pi}{12}\right)$$

$$= \frac{21.0\pi}{4} \approx 16.5 \text{ ft}$$

So, the length of edging needed is $18.0 + 16.5 = 34.5$ ft.

Application

EXAMPLE 3.21

A machine shop is installing two pulleys with radii 570 mm and 130 mm, respectively. The pulleys, as shown in Figure 3.42, are 1 250 mm apart and the length AB is 1 170 mm. If $\angle AOP = 70°$, what is the length of the driving belt?

Solution   To solve this problem we must add the lengths of the straight sections of the belt, AB and CD, and the two arc lengths, s_1 and s_2. We are given $AB = 1\,170$ mm.

EXAMPLE 3.21 (Cont.)

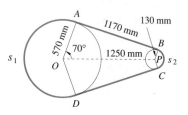

FIGURE 3.42

So, $CD = 1\,170$. To find s_1 we will use $s_1 = r\theta$. We are given $r = 570$ mm, $\theta = 360° - \angle AOD = 360° - 140° = 220° \approx 3.84$ rad. This means that

$$s_1 = r\theta$$
$$= 570(3.84)$$
$$= 2\,188.8 \text{ mm}$$

Now, $s_2 = r\theta$, where $r_2 = 130$ mm and $\theta_2 = 140° \approx 2.44$ rad. So,

$$s_2 = r_2\theta_2$$
$$= (130)(2.44)$$
$$= 317.2 \text{ mm}$$

The length of the belt is $1\,170 + 317.2 + 1\,170 + 2\,188.8 = 4\,846$ mm.

Application

EXAMPLE 3.22

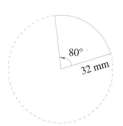

FIGURE 3.43

A pie-shaped piece is going to be cut out of a circular piece of metal. The radius of the circle is 32.0 mm and the central angle of the sector is 80°. What is the area of this sector? (See Figure 3.43.)

Solution We will use the formula $A = \frac{1}{2}r^2\theta$. We are given $r = 32.0$ mm and $\theta = 80° \approx 1.396$ rad, so

$$A = \frac{1}{2}(32.0)^2(1.396)$$
$$\approx 714.752$$

The area is about 714.8 mm^2.

Application

EXAMPLE 3.23

Two competing pizza companies sell pizza by the slice. At Checker's Pizza, a typical slice of pizza is a sector with a central angle of 60° formed from an 8″ radius pizza. A slice sells for $1.75. Pizza Plus makes each slice from a 10″ radius pizza with a central angle of 45°. A slice sells for $2.10. At which company do you get the most pizza for the price?

Solution At Checker's, a sector with $\theta = 60° = \frac{\pi}{3}$ and $r = 8″$ has an area $A = \frac{1}{2}r^2\theta = \frac{1}{2}(8^2)\frac{\pi}{3} = \frac{32\pi}{3}$ in.^{2} At $1.75 per slice, this is $1.75 \div \frac{32\pi}{3} \approx$ $0.052/in.^{2}

At Pizza Plus a sector with $\theta = 45° = \frac{\pi}{4}$ and $r = 10″$ has an area $A = \frac{1}{2}r^2\theta = \frac{1}{2}(10^2)\frac{\pi}{4} = \frac{25\pi}{2}$ in.^{2} At $2.10 per slice, this is $2.10 \div \frac{25\pi}{2} \approx$ $0.053/in.^{2}

Since the Checker's pizza is $0.052/in.^{2} and the Pizza Plus slice cost $0.053/in.^{2}, a slice from Checker's Pizza is the better bargain.

Exercise Set 3.4

Find the area and circumference of the circles in Exercises 1–8 with the given radius or diameter. (You may leave answers in terms of π.)

1. $r = 4$ cm
2. $d = 16$ in.
3. $r = 5$ in.
4. $d = 23$ mm

5. $r = 14.2$ mm
6. $r = 13\frac{1}{4}$ in.
7. $d = 24.20$ mm
8. $d = 23\frac{1}{2}$ in.

Solve Exercises 9–20.

9. *Interior design* **(a)** What is the area of a circular table top with a diameter of 48 in.? **(b)** How much metal edging would be needed to go around this table?

10. *Electricity* A coil of bell wire has 42 turns. The diameter of the coil is 0.5 m. How long is the wire on this coil?

11. *Industrial engineering* Two circular drums are to be riveted together, as shown in Figure 3.44, with the rivets spaced 75 mm apart. How many rivets will be needed?

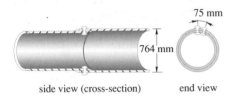

side view (cross-section)   end view

FIGURE 3.44

12. *Industrial design* Two pulleys each with a radius of $3'2''$ have their centers $11'9\frac{1}{4}''$ apart, as shown in Figure 3.45. What is the length of the belt needed for these pulleys?

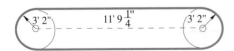

FIGURE 3.45

13. *Sheet metal technology* A sheet of copper has been cut in the shape of a sector of a circle. It is going to be rolled up to form a cone. **(a)** What is the arc length, $\widehat{AB}$, of the sector? **(b)** What is the area of this piece of copper? (See Figure 3.46.)

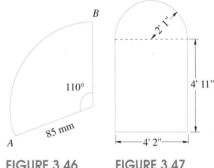

FIGURE 3.46   FIGURE 3.47

14. *Architecture* **(a)** Find the length of molding used around the window opening in Figure 3.47. **(b)** What is the area of the glass needed for this window?

15. *Architecture* The top of a stained glass window has the shape shown in Figure 3.48. The triangle is an equilateral triangle and the two arcs have their centers at the opposite vertices. That is, the arc from B to C is from a circle with center A. **(a)** What is the amount of molding needed for this window? **(b)** What is the area of the glass needed for this window?

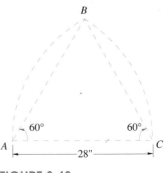

FIGURE 3.48

16. *Architecture* (a) Find the length of molding needed to go around the window molding in Figure 3.49. (b) What is the area of the glass needed for this window? (The radii for the arcs are 2 400 mm.)

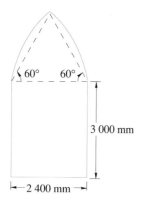

FIGURE 3.49

17. *Industrial design* (a) What is the area of the table top in Figure 3.50? (b) How much metal edging would be required for this table?

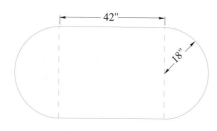

FIGURE 3.50

18. *Civil engineering* A cross-section of pipe is shown in Figure 3.51. What is the area of the cross-section?

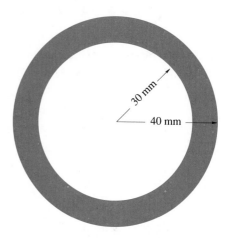

FIGURE 3.51

19. *Electronics* The resistance in a circuit can be reduced by 50% by replacing a wire with another wire that has twice the cross-sectional area. What is the diameter of a wire that will replace one that has a diameter of 8.42 mm?

20. *Space technology* A communications satellite is orbiting the Earth at a fixed altitude above the equator. If the radius of the Earth is 3,960 mi and the satellite is in direct communication with $\frac{1}{3}$ of the equator, what is the height h of the satellite? (See Figure 3.52.)

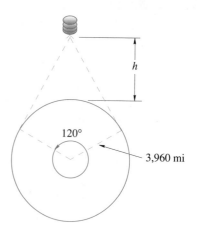

FIGURE 3.52

≡ 3.5
GEOMETRIC SOLIDS

The geometric figures we have looked at thus far have all been plane figures; that is, figures that can be drawn in two dimensions. But, we live in a three-dimensional world. Most of the objects we work with can be characterized as three-dimensional solids. The plane geometric figures we have studied are what the objects look like on one side, when they are taken apart, or "sliced" into cross-sections.

Cylinders

A **cylinder** is a solid whose ends, or **bases,** are parallel congruent plane figures arranged in such a manner that the segments connecting corresponding points on the bases are parallel. These segments are called **elements.** In the first cylinder in Figure 3.53, $\overline{AA'}$, $\overline{BB'}$, and $\overline{CC'}$ are all elements of the cylinder. A **circular cylinder** is a cylinder in which both bases are circles. A **right circular cylinder** is the most common type of cylinder and is formed when the bases are perpendicular to the elements. The **height** or **altitude** of a cylinder is a segment that is perpendicular to both bases.

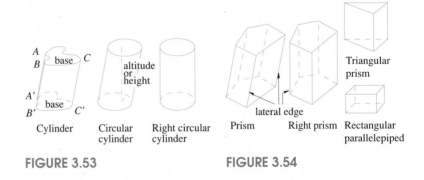

| Cylinder | Circular cylinder | Right circular cylinder | Prism | Right prism | Rectangular parallelepiped |

FIGURE 3.53   **FIGURE 3.54**

Prisms

As shown in Figure 3.54, a **prism** is a solid with ends, or **bases,** that are parallel congruent polygons and with sides, called **faces** or **lateral faces,** that are parallelograms. The segments that form the intersections of the lateral faces are called the **lateral edges.** The height, or altitude, of a prism is the distance between the bases. A **right prism** has its bases perpendicular to its lateral edges; hence, its faces are rectangles.

Prisms get their names from their bases. If the bases are regular polygons, then the prism is a **regular prism.** A **triangular prism** has triangles for bases and a **rectangular prism** has rectangles for bases. The most common prisms are the right rectangular prisms, which are called **rectangular parallelepipeds** and the right square prism, more commonly known as a **cube.**

There are two kinds of areas that are usually associated with any solid figure. The **lateral area** is the sum of the areas of all the sides. The **total surface area** is the lateral area plus the area of the bases. Because the lateral surface of a right prism or right cylinder can be unfolded to form a parallelogram if it is cut along an element, the lateral area L is found by multiplying the perimeter or circumference of the base times the height. The volume of a cylinder or prism is the area of the base B times the height.

> ### Lateral Area, Total Surface Area, and Volume of a Cylinder or Prism
>
> The lateral area, total surface area, and volume of a cylinder or prism are given by the following formulas
>
> | Solid | Lateral Area L | Total Surface Area T | Volume V |
> |---|---|---|---|
> | Prism | ph | $ph + 2B$ | Bh |
> | Cylinder | $2\pi rh$ | $2\pi r(r + h)$ | $\pi r^2 h$ |
>
> where p is the perimeter of a base of the prism, h is the height, r is the radius of a base of the cylinder, and B is the area of a base.

EXAMPLE 3.24

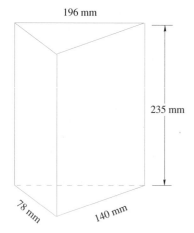

196 mm

235 mm

78 mm   140 mm

FIGURE 3.55

Find the (a) lateral area, (b) total surface area, and (c) volume of the triangular prism shown in Figure 3.55.

Solution

(a) Lateral area $L = ph$, where p is the perimeter of the base and h the height of the prism. We are given $h = 235$ and add the lengths of the triangle's sides to obtain $p = 78 + 140 + 196 = 414$.

$$L = ph$$
$$= 414 \times 235$$
$$= 97\,290$$

The lateral area is 97 290 mm².

(b) Total surface area $T = L + 2B$, where B is the area of a base. Using Hero's formula with $s = \dfrac{p}{2} = \dfrac{414}{2} = 207$, we get

$$B = \sqrt{s(s - a)(s - b)(s - c)}$$
$$= \sqrt{207(207 - 78)(207 - 140)(207 - 196)}$$
$$= \sqrt{207(129)(67)(11)}$$
$$= \sqrt{19\,680\,111}$$
$$\approx 4\,436.23$$
$$T = 2B + L$$
$$= 2(4\,436.23) + 97\,290$$
$$= 8\,872.46 + 97\,290$$
$$= 106\,162.46$$

The total surface area is 106 162.46 mm².

EXAMPLE 3.24 (Cont.)

(c) Volume $V = Bh$

$$= 4\,436.23 \times 235$$

$$= 1\,042\,514.1$$

The volume is $1\,042\,514.1$ mm^3.

Cones

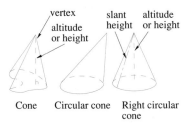

Cone   Circular cone   Right circular cone

FIGURE 3.56

A cone is formed by drawing segments from a plane figure, the **base,** to a point called the **vertex.** The vertex cannot be in the same plane as the base. The **altitude** is a segment from the vertex and perpendicular to the base. The most common cones are the **circular cone** and the **right circular cone.** Both have a base that is a circle. In a right circular cone, the altitude intersects the base at its center. The **slant height** of a right circular cone is a segment from the vertex to a point on the circumference. (See Figure 3.56.)

Pyramids

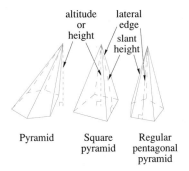

Pyramid   Square pyramid   Regular pentagonal pyramid

FIGURE 3.57

A **pyramid** is a special type of cone, which has a base that is a polygon. A typical pyramid and some of its parts are shown in Figure 3.57. Each side of a pyramid is a triangle and is called a **lateral face.** The lateral faces meet at the **lateral edges.** As with prisms, pyramids are classified according to the shape of their base. A **regular pyramid** has a regular polygon for a base and an altitude that is perpendicular to the base at its center. The slant height of a regular pyramid is the altitude of any of the lateral faces.

The volume V of a cone or pyramid is one-third the area of the base B times the height h, or $V = \frac{1}{3}Bh$. For the lateral areas we will only consider those of right circular cones and regular pyramids. The lateral area L is one-half the slant height s times the perimeter or circumference of the base. The total surface area is the lateral area plus the area of the base.

Lateral Area, Total Surface Area, and Volume of Cones or Pyramids

The lateral area, total surface area, and volume of a right circular cone or regular pyramid are given by the following formulas

Solid	Lateral Area L	Total Surface Area T	Volume V
Pyramid	$\frac{1}{2}ps$	$\frac{1}{2}ps + B$	$\frac{1}{3}Bh$
Cone	πrs	$\pi r(r + s)$	$\frac{1}{3}\pi r^2 h$

where p is the perimeter of a base of the pyramid, h is the height or altitude, s is the slant height, r is the radius of a base of the cone, and B is the area of a base.

Application

EXAMPLE 3.25

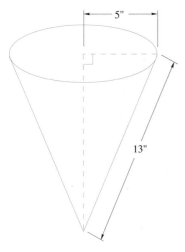

FIGURE 3.58

How many square inches of metal are needed to make a cone-shaped container like the one in Figure 3.58? What is the volume of this cone?

Solution   The amount of metal needed is the lateral area, L. We know that

$$L = \frac{1}{2}Cs$$

where C is the circumference of the base and s is the slant height of the cone.

$$C = 2\pi r \qquad \text{where } r \text{ is the radius of the base}$$

$$= 2\pi 5 = 10\pi$$

$$L = \frac{1}{2}(10\pi)(13)$$

$$= 65\pi \approx 204.2$$

It will take about 204.2 in.^{2} of metal.

$V = \frac{1}{3}\pi r^2 h$, where h is the height of the cone. The slant height, a radius, and the altitude form a right triangle with the slant height equal to the hypotenuse. So, $h^2 + 5^2 = 13^2$ or $h = 12$.

$$V = \frac{1}{3}\pi(5)^2(12)$$

$$= 100\pi \approx 314.2$$

The volume is about 314.2 in.^{3}

Frustums

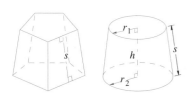

FIGURE 3.59

A **frustum** of a cone or a pyramid is formed by a plane parallel to the base that intersects the solid between the vertex and the base. We will limit our study to frustums of right circular cones or right regular pyramids, as shown in Figure 3.59. If the height of the frustum is h and the areas of the bases are B_1 and B_2 then the volume V is given by

$$V = \frac{h}{3}\left(B_1 + B_2 + \sqrt{B_1 B_2}\right)$$

The lateral area L is given using p_1 and p_2 as the perimeter of the bases and s, the slant height:

$$L = \frac{s}{2}(p_1 + p_2)$$

For the frustum of a cone with bases of radii r_1 and r_2, this becomes

$$L = \pi s(r_1 + r_2)$$

Application

EXAMPLE 3.26

The concrete base of a light pole is constructed in the form of the frustum of a square pyramid, as shown in Figure 3.60. What is the volume of the base for the light pole?

EXAMPLE 3.26 (Cont.)

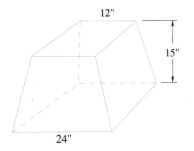

FIGURE 3.60

Solution The volume of a frustum of a square pyramid is given by the formula $V = \frac{h}{3}\left(B_1 + B_2 + \sqrt{B_1 B_2}\right)$. We must first find the area of the bases. If B_1 is the area of the top base, then $B_1 = 12^2 = 144$ in.^{2} and $B_2 = 24^2 = 576$ in.^{2} So,

$$V = \frac{h}{3}\left(B_1 + B_2 + \sqrt{B_1 B_2}\right)$$

$$= \frac{15}{3}\left(144 + 576 + \sqrt{144 \cdot 576}\right)$$

$$= \frac{15}{3}\left(144 + 576 + 288\right)$$

$$= 5{,}040$$

The volume is 5,040 in.^{3}

Application

EXAMPLE 3.27

The picture in Figure 3.61a shows the oven hood pictured at the beginning of the chapter. The hood was made by combining two prisms and the frustum of a square pyramid. In Figure 3.61b, a line drawing of the hood is shown with the measurements indicated. The bottom of the pyramid is 72 in. on each of the two longer sides and 36 in. on each of the two shorter sides. The top base measures 18 in. on each side. The slant height for the longer sides is 15 in. For the shorter sides, the slant height is 29.5 in. The top prism is 36 in. high and the bottom prism is 6 in. high. **(a)** How much metal will it take to form the outside of the part of the oven hood formed by the frustum of the pyramid? **(b)** How much metal will it take to form the outside of the entire oven hood?

FIGURE 3.61

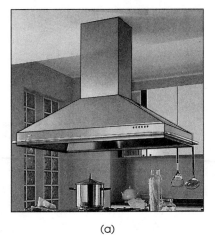

(a)

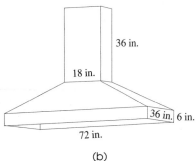

(b)

Solutions

(a) The amount of metal needed can be found by determining the lateral area L of this pyramid. Notice that this is a frustum of a rectangular pyramid and not

EXAMPLE 3.27 (Cont.)

FIGURE 3.61c

of a square pyramid. This means that we cannot use the above formula for the lateral area of the frustum of a pyramid. Since we cannot use the formula, we will think of this frustum as if it were made of four trapezoids, as shown in Figure 3.61c. These four trapezoids are two pairs of congruent trapezoids, so we need only find the areas of the left two trapezoids, marked T_1 and T_2, and double their total area to get the lateral surface area of the frustum in the oven hood.

The left trapezoid T_1, has an area L_1 of

$$L_1 = \frac{1}{2}(b_1 + b_2)h$$

$$= \frac{1}{2}(18 + 72)15$$

$$= 675$$

Trapezoid T_2 has an area of $L_2 = \frac{1}{2}(36 + 18)(29.5) = 796.5$ in.^{2} The total lateral surface area for the frustum and hence the amount of metal it will take to make this portion of the oven hood is $2(675 + 796.5) = 2,943$ in.^{2}

(b) To get the amount of metal it will take to form the outside of the entire oven hood, we add the answer to (a) to the lateral surface areas of the two prisms. The top prism has a square base that is 18 in. on each side and the prism is 36 in. high. So, it has a perimeter of $4 \times 18 = 72$ in. and its lateral area is $L_t = 36 \times 72 = 2,592$ in.^{2} The bottom prism is a rectangular prism with a perimeter of $72 + 36 + 72 + 36 = 216$ in. and a height of 6 in. Hence, its lateral surface area is $L_b = 6 \times 216 = 1,296$ in.^{2} The total amount of metal needed to form the outside of this oven hood is then $2,943 + 2,592 + 1,296 = 6,831$ in.^{2} ▪

Spheres

The final geometric solid that we will consider is the sphere. A **sphere** consists of all the points in space that are a fixed distance from some fixed point called the center. For a sphere with radius r, the volume and surface area are given by the following formulas.

Surface Area and Volume of a Sphere

A sphere with radius r has a surface area S and volume V where

$$S = 4\pi r^2$$

$$V = \frac{4}{3}\pi r^3$$

Application

EXAMPLE 3.28

A water tower, like the one in Figure 3.62, is in the shape of a sphere on top of a cylinder. Most of the water is stored in the sphere, which has a radius of 30 ft. How much water can the tower hold?

EXAMPLE 3.28 (Cont.)

FIGURE 3.62

Solution We want to determine the volume of a sphere with a radius of 30 ft. The formula for the volume of a sphere is $V = \frac{4}{3}\pi r^3$. Since $r = 30$ ft, we have

$$V = \frac{4}{3}\pi(30)^3$$

$$= \frac{4}{3}\pi(27{,}000)$$

$$= 36{,}000\pi$$

$$\approx 113{,}097.34$$

The water tower will hold approximately $113{,}097$ ft^3 of water.

A summary of all the formulas for the areas and volumes of the solid figures is given in Table 3.1.

TABLE 3.1 Areas and Volumes of Solid Figures

Solid	Lateral Area (L)	Total Surface Area (T)	Volume (V)
Rectangular prism	$2h(l+w)$	$2lw + 2lh + 2hw$ $= 2(lw + lh + hw)$	lwh
Cube	$4s^2$	$6s^2$	s^3
Prism	ph	$ph + 2B$	Bh
Cylinder	$2\pi rh$	$2\pi r(r+h)$	$\pi r^2 h$
Pyramid	$\frac{1}{2}ps$	$\frac{1}{2}ps + B$	$\frac{1}{3}Bh$
Cone	πrs	$\pi r(r+s)$	$\frac{1}{3}\pi r^2 h$
Frustum of pyramid	$\frac{s}{2}(p_1 + p_2)$	$\frac{s}{2}(p_1 + p_2) + B_1 + B_2$	$\frac{h}{3}\left(B_1 + B_2 + \sqrt{B_1 B_2}\right)$
Frustum of cone	$\pi s(r_1 + r_2)$	$\pi[r_1(r_1 + s) + r_2(r_2 + s)]$	$\frac{\pi h}{3}(r_1^2 + r_2^2 + r_1 r_2)$
Sphere	not applicable	$4\pi r^2$	$\frac{4}{3}\pi r^3$

Exercise Set 3.5

Find the lateral area, total surface area, and volume of each of the solids in Exercises 1–8.

1.

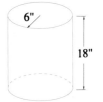

6"

18"

2.

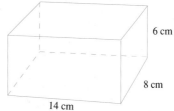

6 cm

8 cm

14 cm

3.

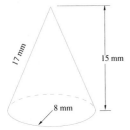

4.

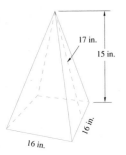

5.

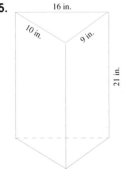

6.

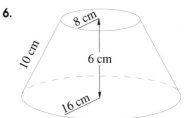

7.

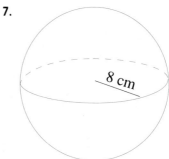

8.

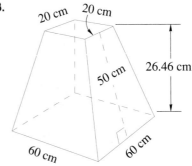

Solve Exercises 9–25.

9. *Civil engineering* The cross-section of a road is shown in Figure 3.63. Find the number of cubic yards of concrete it will take to pave 1 mi of this road. (There are 27 ft^3 in 1 yd^3 and 5,280 ft in 1 mi.)

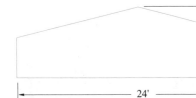

FIGURE 3.63

10. *Civil engineering* The cross-section of an I-beam is shown in Figure 3.64. **(a)** What is the volume of this beam if it is 10 m long? **(b)** How much paint, in cm^2, will be needed for this beam? **(c)** If 1 cm^3 of steel has a mass of 0.008 kg what is the mass of this beam?

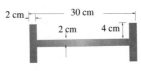

FIGURE 3.64

11. *Transportation* A railroad container car in the shape of a rectangular parallelepiped is 30 ft long, 10 ft wide and 12 ft high. **(a)** How much can the container car

11. *Transportation* A railroad container car in the shape of a rectangular parallelepiped is 30 ft long, 10 ft wide and 12 ft high. **(a)** How much can the container car hold? **(b)** How many square feet of aluminum were required to make the car?

12. *Energy* A cylindrical gas tank has a radius of 48 ft and a height of 140 ft. What are the volume and total surface area of the tank?

13. A cylindrical soup can has a diameter of 66 mm and a height of 95 mm. **(a)** How much soup can the can hold (in mm^3)? **(b)** How many square millimeters of paper are needed for the label if the ends overlap 5 mm?

14. *Mechanics* A cross-section of pipe is shown in Figure 3.65. If the pipe is 2 m long what is the volume of the material needed to make the pipe?

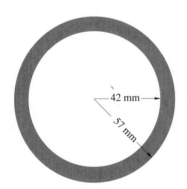

FIGURE 3.65

15. *Energy* A spherical fuel tank has a radius of 10 m. What is its volume?

16. *Sheet metal technology* A vent hood is made in the shape of a frustum of a square pyramid that is open at the top and bottom, as indicated in Figure 3.66. If the slant height is 730 mm, how much metal did it take to make this vent?

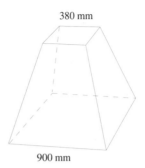

FIGURE 3.66

17. *Civil engineering* The concrete highway support in Figure 3.67 is 2 ft 6 in. thick. How many cubic yards of concrete are needed to make one support? (27 ft^3 = 1 yd^3.)

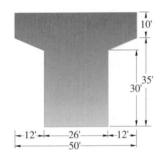

FIGURE 3.67

18. *Construction* How many cubic feet of dirt had to be excavated to dig a $\frac{1}{4}$-mi (1,320 ft) section of the river bed in Figure 3.68?

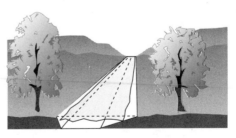

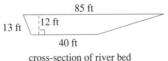

cross-section of river bed

FIGURE 3.68

19. *Sheet metal technology* What is the volume of the cone that was made from the piece of copper in Figure 3.69?

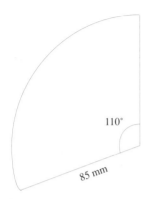

110°

85 mm

FIGURE 3.69

20. *Sheet metal technology* A manufacturer has an order for 500 tubs in the shape of a frustum of a cone that is 490 mm across at the top, 380 mm across at the bottom, and 260 mm deep. How much material will be needed for the sides of the tubs?

21. *Sheet metal technology* How many square inches of tin are required to make a funnel in the shape of a frustum of a cone that has a top and bottom with diameters of 3 in. and 8 in., respectively, and a slant height of 12 in.?

22. *Sheet metal technology* The funnel in Figure 3.70 has a diameter 3 cm at *B* and 1 cm at *A*. How much metal is needed to make this funnel?

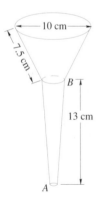

10 cm

7.5 cm

B

13 cm

A

FIGURE 3.70

23. *Agricultural technology* Figure 3.71 shows the top view of four grain elevators. When the elevators are filled, the grain will overflow to fill the space in the middle. The radius of each elevator is 3 m and the height is 10 m. What is the total volume that can be held by the four elevators and the space in the middle if the grain is leveled at the top?

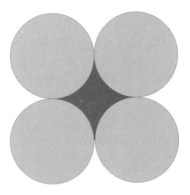

FIGURE 3.71

24. *Automotive technology* The piston displacement is the volume of the cylinder with a given bore (diameter) and piston stroke (height). **(a)** If the bore is 7 cm and the stroke is 8 cm, what is the displacement for this piston? **(b)** If this is a 6-cylinder engine, what is the total engine displacement?

25. *Construction* A pile of sand falls naturally into a cone. If a pile is 4 ft high and 12 ft in diameter, how many cubic feet of sand are there?

☰ CHAPTER 3 REVIEW

Important Terms and Concepts

Adjacent angles

Altitude

Angle
  Acute
  Obtuse
  Right

Arc

Arc length

Area

Central angle

Chord

Circumference

Complementary angle

Cone

Congruent angles

Congruent polygons

Corresponding angles

Cylinder

Degree

Diameter

Frustum

Hero's formula

Hexagon

Hypotenuse

Lateral angle

Line segment

Octagon

Parallel lines

Parallelogram

Pentagon

Perimeter

Perpendicular lines

Polygon

Prism

Pythagorean theorem

Quadrilateral

Radians

Radius

Rectangle

Rhombus

Similar polygons

Square

Supplementary angles

Tangent

Total surface area

Transversal

Trapezoid

Triangle
  Acute
  Equilateral
  Isosceles
  Obtuse
  Right
  Scalene

Vertex

Review Exercises

Convert each of the angle measures in Exercises 1–4 to either radians or degrees, without using a calculator. When you have finished, check your work by using a calculator.

1. $27°$

2. $212°$

3. 1.1π

4. 0.75

Solve Exercises 5 and 6.

5. What is the supplement of a $137°$ angle?

6. What is the complement of a $\frac{\pi}{6}$ angle?

In Exercises 7–8 find the measures of angles *A* and *B* if lines ℓ_1 and ℓ_2 are parallel.

7.

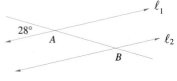

8.

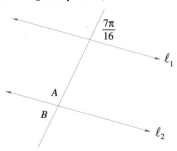

9. What is the distance from *A* to *B* in Figure 3.72, if line ℓ_1, ℓ_2, and ℓ_3 are parallel?

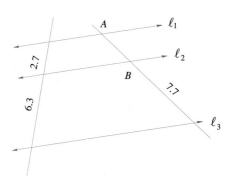

FIGURE 3.72

Find the variables indicated in Exercises 10–11.

10.

11.

Find the length of the missing side in the right triangles in Exercises 12–13.

12.

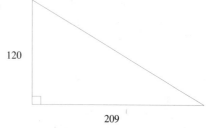

13.

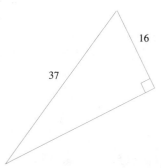

Find the area and perimeter or circumference of each of the figures in Exercises 14–21.

14.

15.

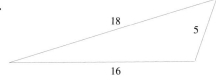

16.

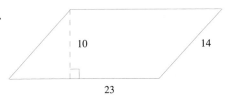

17.

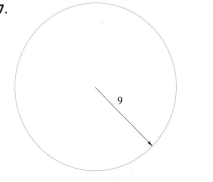

18.

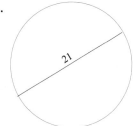

19.

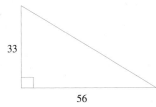

20.

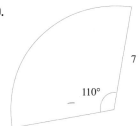

21.

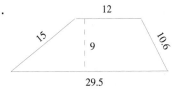

Find the lateral area, total surface area, and volume of each of the solid figures in Exercises 22–29.

22.

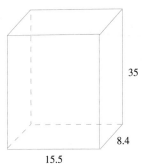

23.

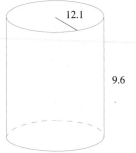

24.

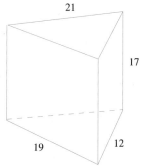

25.

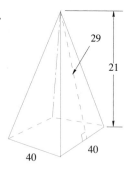

26.

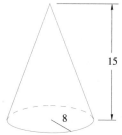

27.

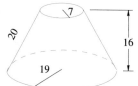
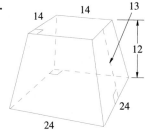

28.

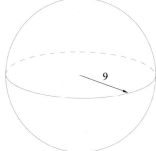

29.

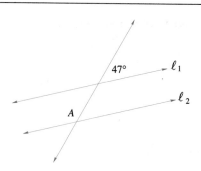

≡ CHAPTER 3 TEST

1. Convert 35° to radians.

2. Convert $\frac{7\pi}{15}$ to degrees.

3. What is the supplement of a 76° angle?

4. In Figure 3.73, lines ℓ_1 and ℓ_2 are parallel. What is the measure of angle A?

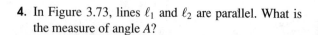

FIGURE 3.73

5. In Figure 3.74, what is the length of $\overline{AB}$, if ℓ_1, ℓ_2, and ℓ_3 are parallel?

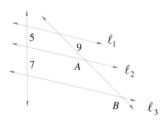

FIGURE 3.74

6. Determine the length of side a in Figure 3.75.

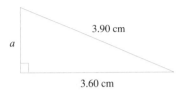

FIGURE 3.75

Find the perimeter of each of the figures in Exercises 7–8.

7.

36.4 cm

14.3 cm

8.

12 in.

5 in.   5 in.

4 in.   4 in.

Find the area of each of the figures in Exercises 9 and 10.

9.

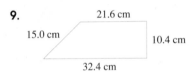

21.6 cm

15.0 cm

10.4 cm

32.4 cm

10.

9 in.

Solve Exercises 11–15.

11. A building casts a shadow of 120 m. At the same time, an antenna that is 14 m high casts a shadow of 11.6 m. How tall is the building?

12. An automobile tire has a diameter of 62 cm. How many revolutions must the tire make when the car travels 25 m in a straight line?

13. What is the volume of the box in Figure 3.76?

162 mm

CAKE

231 mm

52 mm

FIGURE 3.76

14. A spherical storage tank has a diameter of 35 ft. What is its volume?

15. The part of a cylindrical soup can that is covered by the label is 9.5 cm tall and has a diameter of 6.5 cm. What is the area of a label that covers the entire side of the can and that needs a 0.8 cm overlap to glue the ends of the label?

4

Functions and Graphs

Some technicians have to take water samples at a given distance from a source of pollution. In Sections 4.4 and 4.6, we will see how to use functions and their graphs to help analyze these samples and predict the effects of pollution.

Courtesy of Janet Essman, New York State Department of Environmental Conservation

In Chapter 2, we looked at algebraic equations. We learned some ways to solve equations, and we learned how we could take some information and write an equation that showed how that information was related.

Some of the equations involved variables that stood for things related to each other. One equation used variables to show the relationship between distance, rate, and time. Torque, force, and moment arm length were other relationships in which variables were used. In each of these last two relationships, once we knew two of the variables we could determine the third.

≡ 4.1
RELATIONS AND FUNCTIONS

In this section, we will learn to use a particular kind of relationship called a function. The key to the mathematical analysis of a technical problem is your ability to recognize the relationship between the variables that describe the problem. Such a relationship often takes the form of a formula that expresses one variable as a function of another variable. We will be studying the basic ideas of a function and, later in the chapter, show how we can draw a picture or graph of a function. These basic ideas will form the foundation for much of the work we will do later in this book.

Relations

We have used the word relationship very loosely. A **relation** in mathematics is used to represent a relationship between two numbers, variables, or objects. A relation is often written as a set of ordered pairs or by a rule that describes how the items are related. In mathematics this rule may be given as an equation, an inequality, or a system of equations or inequalities. A **function** is a special kind of relation.

EXAMPLE 4.1

$a = 5b + 1$ is a relation between a and b. Pick a value for b, say $b = 3$, then $a = 5(3) + 1 = 16$.

EXAMPLE 4.2

The equation $I = 2.54C$ expresses a relation between the two variables I and C. So, if you select a value for C, say $C = 5$, you would get $I = 2.54(5) = 12.7$. You could also pick a value for I, say $I = 8.89$. Then $8.89 = 2.54C$ and $C = \frac{8.89}{2.54} = 3.5$.

EXAMPLE 4.3

The equation $v = \frac{1}{2}t^2 + 9t + 1$ describes a relation between the two variables v and t. If you let $t = 4$, then you get $v = \frac{1}{2}(4^2) + 9(4) + 1 = 45$. Let $v = 21$, then $t = 2$ or $t = -20$ will both work.

EXAMPLE 4.4

$x^2 + y^2 = 64$ is a relation between x and y. Pick a value for x between -8 and 8. If you pick $x = 0$ then $0 + y^2 = 64$ or $y^2 = 64$, which means $y = \pm 8$. If $x = 1$, then $y = \pm\sqrt{63}$. If $x = -8$, then $y = 0$. In general, $y = \pm\sqrt{64 - x^2}$. Notice that you cannot select a number smaller than -8 or greater than 8 because that would require taking the square root of a negative number.

Ordered Pairs

In addition to defining a relation by an equation, we may also define it as a set of **ordered pairs**. For example, the solutions to $I = 2.54C$ are pairs of numbers. If we agree to write the ordered pairs as (C, I), where the first number is the C-value and the second number is the I-value, then these ordered pairs also define the relation between C and I. Some of these ordered pairs are $(1, 2.54)$, $(2, 5.08)$, $(2.5, 6.35)$, $(5, 12.7)$, $(3.5, 8.89)$, and $(-3, -7.62)$.

In technical work you often work with ordered pairs of numbers. An example of a set of ordered pairs of numbers is given in Table 4.1. In this table pressures in kilopascals (kPa) and the corresponding pressure in pounds per square inch (psi) are given. For each pressure in psi there is a corresponding pressure in kPa.

Ordered pairs can be represented in table form or with each corresponding ordered pair written between parentheses. If we use the ordered pairs from Table 4.1 we would write a kPa pressure as the first number and the corresponding psi pressure as the second number. The first four ordered pairs from Table 4.1 would be written as (140, 20), (145, 21), (155, 22), and (160, 23).

TABLE 4.1   Inflation Pressure Conversion Chart (kilopascals to psi)

kPa	psi	kPa	psi
140	20	215	31
145	21	220	32
155	22	230	33
160	23	235	34
165	24	240	35
170	25	250	36
180	26	275	40
185	27	310	45
190	28	345	50
200	29	380	55
205	30	415	60

Conversion: 6.9 kPa = 1 psi
1991 Buick Skyhawk Service Manual. Flint, Michigan: Service Department, Buick Division, General Motors Corporation, 1991. Figure 3 Inflation Pressure Conversion, p. 3E-2

Whenever you have a set of ordered pairs of numbers $(x,\ y)$ such that for each value of x there corresponds exactly one value of y, you have a function. You will notice that at the bottom of Table 4.1, the function is given in equation form, 6.9 kPa = 1 psi. We might rewrite this as $k = 6.9p$, where p represents a pressure in psi and k stands for a pressure in kPa. In this case, the ordered pairs are of the form $(k,\ p)$.

Domain and Range

For any relation, the set of all possible first component values is called the **domain** and the set of all second component values that can result from using values in the domain is called the **range**.

EXAMPLE 4.5

The set of ordered pairs

$$\{(2,\ 5),\ (3,\ 8),\ (4,\ 12),\ (6,\ -12)\}$$

EXAMPLE 4.5 (Cont.)

expresses a relation between the set of first components, or domain, $\{2, 3, 4, 6\}$, and the set of second components, or range, $\{5, 8, 12, -12\}$.

≡ **Note** The domain and range are important ideas that will help us better understand a relation, especially when we begin graphing, and they will help us to make better use of graphing calculators and computer graphing software.

EXAMPLE 4.6

(a) In Example 4.4 where $x^2 + y^2 = 64$, both the domain and range were the real numbers between, and including, -8 and 8.

(b) In Example 4.3, the equation $v = \frac{1}{2}t^2 + 9t + 1$ expressed how the first component v was obtained from the second component t. The domain contains all the real numbers. The range would be all real numbers v greater than or equal to -39.5, or $v \geq -39.5$.

(c) If $y = \sqrt{x^2 - 81}$, then the domain would be all real numbers such that $x^2 - 81 \geq 0$. That means that the domain is the values of x larger than or equal to 9 or less than or equal to -9, or $x \leq -9$ or $x \leq 9$. The range would be the nonnegative real numbers, or $y \geq 0$.

(d) If $y = \dfrac{5}{x - 3}$, the domain would be all real numbers except 3, because if $x = 3$, then $x - 3 = 0$ and $\dfrac{5}{x - 3}$ is not defined. (Remember, division by zero is not defined.) The range is all real numbers except 0, because a fraction is 0 only when the numerator is 0, but this numerator is always 5 (and so, never 0).

Functions

In the relation in Example 4.4, there are two values of y for each value of x, unless $x = 8$ or $x = -8$. If a relation between x and y gives only one value of y for each value of x, then we say that y is a **function of x**. Every function is a relation. However, as Example 4.4 demonstrated, not every relation is a function.

Function

The variable y is a function of x if a relation between x and y produces exactly one value of y for each value of x.

EXAMPLE 4.7

$y = 7x - 2$ is a function. Pick any value for x, say -4. Then $y = 7(-4) - 2 = -28 - 2 = -30$.

For each value of x there is only one value of y.

Independent and Dependent Variables

For the function $y = 7x - 2$, the variable x is called the **independent variable** and y is the **dependent variable**. These terms are used because the value of y depends upon the value selected for x. Since the value selected for x can be freely chosen, x is called the independent variable. These terms are arbitrary. After all, if we had written the function as $x = \frac{1}{3}(y + 5)$, then y would have been the independent variable and x the dependent variable.

Any time you have a function, you can represent it as a set of ordered pairs, as a table of values, as an equation, and, as we will see later in this chapter, as a graph. If the domain of a function is very large it may not be possible to show all possible ordered pairs or a complete table of values. In Examples 4.8 and 4.9, we show *some* of the possible ordered pairs and table values for two functions. Example 4.10 gives an example of some ordered pairs that do not form a function.

EXAMPLE 4.8

Equation: $y = 2x + 7$

Table:

x	-2	-1	0	$\frac{1}{2}$	1	2
y	3	5	7	8	9	11

Ordered pairs: $(-2, 3), (-1, 5), (0, 7), (\frac{1}{2}, 8), (1, 9), (2, 11), \ldots$

Domain: All real numbers

Range: All real numbers

EXAMPLE 4.9

Equation: $y = x^2 - 3$

Table:

x	-3	-2	-1	0	1	2	3
y	6	1	-2	-3	-2	1	6

Ordered pairs: $(-3, 6), (-2, 1), (-1, -2), (0, -3), (1, -2), (2, 1), (3, 6), \ldots$

Domain: All real numbers

Range: All real numbers greater than or equal to -3

EXAMPLE 4.10

The set of ordered pairs $(0, 1), (2, 5), (2, -8)$ is not a function because there are two different second values, 5 and -8, for the first number, 2.

A function is often identified by a letter or group of letters, such as $f, g, h, j, F, G, C,$ tan, ln, ABS, or $\cos^{-1}$. If x is the independent variable and y the dependent variable, then the number that corresponds to y is designated as $f(x), g(x), h(x), F(x),$ or $\tan(x)$ depending on how the function is identified.

The notation $f(x)$ is read "f of x" or "f at x." We say y is a function of x and write $y = f(x)$. Thus the function in Example 4.8 could be written as $f(x) = 2x + 7$ and the function in Example 4.9 as $g(x) = x^2 - 3$.

The independent variable does not have to be x. We often use functions of other variables. Thus, $f(y) = 5y + \frac{1}{2}$ is a function of y and $h(t) = \frac{4}{3}t^2 - 2t + 1$ is a function of t.

If you want to know the value of a function at a specific point, then function notation is very useful. For example, suppose you have the function $f(x) = 2x^2 - 3$ and you want to know what value of f corresponds to $x = 4$. You would then be asking for $f(4)$ and you should substitute 4 for each x in the function. The result is

$$f(4) = 2(4^2) - 3$$
$$= 2(16) - 3$$
$$= 29$$

The value of $f(4)$ is 29. This gives the ordered pair $(4, f(4))$ or $(4, 29)$.

EXAMPLE 4.11

If $f(x) = 3x^2 - 5x + 2$, find $f(-2)$.

Solution   Substitute -2 for each value of x, with the result
$$f(-2) = 3(-2)^2 - 5(-2) + 2$$
$$= 3(4) - (-10) + 2$$
$$= 12 + 10 + 2$$
$$= 24$$

So, $f(-2) = 24$.

EXAMPLE 4.12

If $d(t) = 88t - 2.5$, find $d(4.5)$.

Solution   Substitute 4.5 for the variable t, with the result
$$d(4.5) = 88(4.5) - 2.5$$
$$= 396 - 2.5$$
$$= 393.5$$

Thus, we have found that $d(4.5) = 393.5$.

EXAMPLE 4.13

If $g(x) = 3x^2 - 2x + 1$, then find $g(5 + h)$.

Solution

Replace each x with $5 + h$.
$$g(5 + h) = 3(5 + h)^2 - 2(5 + h) + 1$$
$$= 3(25 + 10h + h^2) - 2(5 + h) + 1$$
$$= (75 + 30h + 3h^2) - (10 + 2h) + 1$$
$$= 75 + 30h + 3h^2 - 10 - 2h + 1$$
$$= 66 + 28h + 3h^2$$

So, $g(5 + h) = 66 + 28h + 3h^2$.

Application

EXAMPLE 4.14

If 100 m of fencing is used to enclose a rectangular yard, then the resulting area of the fenced yard is given by the function

$$A(x) = x(50 - x)$$

where x is the length of the rectangle. **(a)** What is the area when the length is 10 m? **(b)** What is the area when the length is 30 m? **(c)** What restrictions must be placed on the domain x so that the problem makes physical sense?

Solution

(a) When the length is 10 m, we have $x = 10$, and the area is

$$A(10) = x(50 - x)$$
$$= 10(50 - 10)$$
$$= 10(40) = 400 \text{ m}^2$$

(b) Here $x = 30$ m, so the area is

$$A(30) = x(50 - x)$$
$$= 30(50 - 30)$$
$$= 30(20) = 600 \text{ m}^2$$

(c) The domain should be $0 < x < 50$. If $x \leq 0$ or $x \geq 50$, we would obtain a length or width that is either 0 or a negative number. This would produce an area that is either 0 or a negative number, which would not make physical sense.

Application

EXAMPLE 4.15

The water pressure on the base of a dam is a function of the depth of the water. The weight density of water is 9 800 newtons per square meter (9 800 N/m^2). The water pressure P beneath the surface of the water is $P(d) = 9\,800d$, where d represents the depth of the water in meters. What is the pressure at the base of a dam that is 20.5 m below the water's surface?

Solution   We know $P(d) = 9\,800d$ and we are given $d = 20.5$ m. Substituting, we obtain $P(d) = 9\,800(20.5) = 200\,900$. So, the pressure is 200 900 N/m^2 or 200 900 Pa.

Very often, we do not bother to write a formula as a function. For instance, in the last example, we had the function $P(d) = 9\,800d$. Most of the time we simply write this as $P = 9\,800d$.

Explicit and Implicit Functions

When a function is expressed in the form $y = f(x)$, it is called an **explicit function**. If the relationship between x and y is not of this form, we say that x and y are related implicitly or that it is an **implicit function**. Some implicit functions can be used more easily if they are written as explicit functions.

EXAMPLE 4.16

(a) $4x - y = 2$ is an implicit equation. If we solve for y, we get an explicit function, $y = 4x - 2$.

(b) $x^2 + y^2 = 64$ is an implicit equation. It defines two explicit functions, $y = \sqrt{64 - x^2}$ and $y = -\sqrt{64 - x^2}$.

(c) $x + 3y^2 - y = 7$ is an implicit function that defines two explicit functions,
$$y = \frac{1}{6} + \sqrt{\frac{85}{36} - \frac{x}{3}} \text{ and } y = \frac{1}{6} - \sqrt{\frac{85}{36} - \frac{x}{3}}.$$

(d) $x^3 - xy + 5y^4 = 7$ is an implicit function that does not define an explicit function.

Implicit functions can be written in functional notation. For example, $4x - y = 2$ could be written as a function of the two variables x and y, by writing $f(x, y) = 4x - y - 2$ and $x^2 + y^2 = 36$ as $g(x, y) = x^2 + y^2 - 36$. There is nothing that restricts a function to two variables. Thus, we could have the implicit function $x^2 + y^2 + z = 81$ and represent it as a function of three variables, or $h(x, y, z) = x^2 + y^2 + z - 81$.

Just as we learned that if $f(x) = 7x - 2$, then $f(-1) = 7(-1) - 2 = -9$, we can substitute values for the variables in functions of more than one variable.

EXAMPLE 4.17

Let $f(x, y, z) = 2x - \frac{1}{2}y + z^2 + 4$ and find $f(1, -2, 3)$.

Solution Here we substitute $x = 1$, $y = -2$, and $z = 3$, with the result

$$f(1, -2, 3) = 2(1) - \frac{1}{2}(-2) + 3^2 + 4$$

$$= 2 + 1 + 9 + 4$$

$$= 16$$

So, we have determined that $f(1, -2, 3) = 16$.

Inverse Functions

We often use a pair of functions that are natural opposites of each other. By a natural opposite, we mean that one of the functions "undoes" the result of the other function, much the same as subtraction "undoes" the result of addition and division "undoes" multiplication. Two functions with this relationship are called **inverse functions**.

To determine an inverse function of $y = f(x)$, you first solve for x. This gives an inverse relation for y. If this relation is also a function, say $g(y)$, then $g(y)$ is the

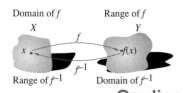

Domain of f Range of f
X Y

Range of f^{-1} Domain of f^{-1}

FIGURE 4.1

Caution

inverse function of $f(x)$. We use the notation $f^{-1}(x)$ to designate the inverse function of the function $f(x)$.

Figure 4.1 shows that the domain of f is the range of f^{-1} and the range of f is the domain of f^{-1}.

Do not confuse the -1 in f^{-1} with a negative exponent. The symbol f^{-1} does not represent $\frac{1}{f}$.

Guidelines for Finding f^{-1}

1. Solve the function $y = f(x)$ for x. Let $x = f^{-1}(y)$.
2. Exchange x and y to get $y = f^{-1}(x)$.
3. Check the domains and ranges: The domain of f and the range of f^{-1} should be the same, as should the domain of f^{-1} and the range of f.

EXAMPLE 4.18

Find the inverse function for each of the functions f: **(a)** $f(x) = 4x - 5$, **(b)** $g(x) = x^3 + 5$, and **(c)** $h(x) = x^2$.

Solution

(a) Let $y = f(x) = 4x - 5$. Solving for x, we get $x = \frac{y+5}{4}$ and we write $f^{-1}(y) = \frac{y+5}{4}$.

Since, by tradition, we usually let x represent the independent variable, we normally rewrite this as $f^{-1}(x) = \frac{x+5}{4}$.

(b) Let $y = g(x) = x^3 + 5$. Then $x = \sqrt[3]{y-5}$ and $g^{-1}(y) = \sqrt[3]{y-5}$ or, more traditionally, $g^{-1}(x) = \sqrt[3]{x-5}$.

(c) Let $y = h(x) = x^2$. Then $x = \pm\sqrt{y}$, but this is not the inverse function of $h(x)$. Why? Because the domain of h is all real numbers and the range of $x = \sqrt{y}$ is the nonnegative real numbers.

The range of an inverse function is the domain of the original function. Another way of saying this is that an inverse function "undoes" what the function did. In symbols, we would write $f^{-1}(f(x)) = x$. One way to see if you have the inverse of a function is to test some values in this equation.

To see what this means, let's look at the first function in Example 4.18: $f(x) = 4x - 5$. We said that $f^{-1}(x) = \frac{x+5}{4}$. If we try the value $x = 2$ in $f(x) = 4x - 5$, we get $f(2) = 4(2) - 5 = 3$. Now evaluate

$$f^{-1}(3) = \frac{3+5}{4}$$

$$= \frac{8}{4}$$

$$= 2$$

Good! We started and ended with 2, or $f^{-1}(f(2)) = f^{-1}(3) = 2$.
Try another number, say -10.

$$f(-10) = 4(-10) - 5$$

$$= -45$$

and $\qquad f^{-1}(-45) = \dfrac{-45 + 5}{4}$

$$= \dfrac{-40}{4}$$

$$= -10$$

Again it checks! $f^{-1}(f(-10)) = f^{-1}(-45) = -10$.

Look at the third function in Example 4.18: $y = h(x) = x^2$. We solved for x and obtained $x = \sqrt{y}$. Now check some values. $h(2) = 2^2 = 4$ and $\sqrt{4} = 2$, so this checks. Try a negative number, like -10. $h(-10) = (-10)^2 = 100$ and $\sqrt{100} = 10$. It does not work. What we hoped would be the inverse returned a value of 10 instead of -10.

How can we tell if a function has an inverse? We will explore that idea later in this chapter.

Exercise Set 4.1

Which of the relations in Exercises 1–8 are also functions of x?

1. $y = 10x + 2$
2. $y = 17 - 3x$
3. $y^2 = x^2 - 5$
4. $y = x^3 - 2$
5. $y = \sqrt{x} + 5$
6. $y = \sqrt[3]{x}$
7. $y^2 = 2x - 7$
8. $y = \pm\sqrt{x^2 - 5}$

Solve Exercises 9–15.

9. Is the set of ordered pairs $(0, 0), (1, 1), (2, 8), (3, 27)$, and $(4, 64)$ a function? Why or why not?

10. Is the set of ordered pairs $(-2, 4), (-1, 1), (0, 0)$, $(1, 1)$, and $(2, 4)$ a function? Why or why not?

11. Is the set of ordered pairs $(4, -2), (1, -1), (0, 0)$, $(1, 1)$, and $(4, 2)$ a function? Why or why not?

12. Does this table describe a function? Explain.

x	-40	-30	-20	-10	0	10	20	30	40
y	-40	-22	-4	14	32	50	68	86	104

13. Give the domain and range of the function defined by these ordered pairs: $(-3, -7), (-2, -5), (-1, -3)$, $(0, -1), (1, 1)$, and $(2, 3)$.

14. Give the domain and range of the function defined by these ordered pairs: $(-3, 3), (-2, -2), (-1, -5)$, $(0, -6), (1, -5), (2, -2), (3, 3)$.

15. Give the domain and range of the function defined by the values in this table.

x	-3	-2	-1	0	1	2	3
y	-25	-6	1	2	3	10	25

In Exercises 16–20, determine the domain and range of each function.

16. $y = x + 2$

17. $y = \dfrac{x}{x - 5}$

18. $y = \sqrt{x} - 2$

19. $y = x^2$

20. $y = \dfrac{15}{(x - 2)(x + 3)}$

In Exercises 21–25, given the function $f(x) = 3x - 2$, determine the following.

21. $f(0)$

22. $f(2)$

23. $f(-3)$

24. $f(b)$

25. $f(b + 3)$

In Exercises 26–30, given the function $g(x) = x^2 - 5x$, determine the following.

26. $g(0)$

27. $g(-3)$

28. $g(2)$

29. $g(-x)$

30. $g(x - 5)$

In Exercises 31–35, given the function $F(x) = \dfrac{2 - x}{x^2 + 2}$, determine the following.

31. $F(0)$

32. $F(2)$

33. $F(1.5)$

34. $F(2x)$

35. $\dfrac{F(2x)}{F(x)}$

Given the function $C(f) = \frac{5}{9}(f - 32)$, determine the following in Exercises 36–39.

36. $C(32)$

37. $C(98.6)$

38. $C(-40)$

39. $C(72)$

Which of the equations in Exercises 40–43 are in explicit form and which are in implicit form?

40. $y = 4x^2 + 5$

41. $y + 2x = 7$

42. $y = x^2y + x$

43. $y = x^3 - 5x + 3$

In Exercises 44–47, if $f(x, y) = 3x - 2y + 4$, then determine the indicated value.

44. $f(1, 0)$

45. $f(0, 1)$

46. $f(-1, 2)$

47. $f(2, -3)$

In Exercises 48–51, if $g(x, y) = x^2 - y^2 + 2xy$, determine the indicated value.

48. $g(-2, 0)$

49. $g(0, -2)$

50. $g(5, 4)$

51. $g(-3, 5)$

In Exercises 52–54, if $h(x, y, z) = 3x - 4y - 2z + xyz$, then determine the indicated value.

52. $h(-1, 2, 3)$

53. $h(2, -1, -3)$

54. $h(3, 2, -1)$

Find the inverses of each of the functions in Exercises 55–61.

55. $y = x - 5$

56. $2y = x + 7$

57. $y = 2x - 8$

58. $6y = 2x - 4$

59. $9y = 3x + 5$

60. $y^2 = x - 5$

61. $y = x^3 + 7$

Solve Exercises 62–64.

62. *Petroleum engineering*   The volume V of a cylindrical storage tank 40 ft high is a function of its radius and is described by $V = 40\pi r^2$. The total surface area S of the steel needed to construct this same tank is given by $S = 2\pi r(r + 40)$. Two tanks are designed. One has a radius of 20 ft and the other a radius of 30 ft. Compute the volume and total surface area of each tank.

63. *Dynamics*   The distance s that a free-falling object travels is a function of the time t since it started falling, and is described by the function $s(t) = 4.91t^2$, where s is measured in meters and t in seconds. **(a)** How far will an object fall in 1 s? **(b)** How far will it fall in 2 s? **(c)** How far will it fall in 5 s?

64. *Business*   The cost of renting a chain saw is $20 for the first 4 hours and $5 for each hour after 4. If $R(h)$ represents the rental charge for h hours, then $R(h) = 20 + 5(h - 4)$. Determine the cost of renting this chain saw for 7 hours.

≡ 4.2
OPERATIONS ON FUNCTIONS; COMPOSITE FUNCTIONS

This section introduces some operations on functions. Most of the operations involve the usual operations of addition, subtraction, multiplication, and division; but a new operation, called composition, will be introduced.

Operations on Functions ▬▬▬▬

Two functions, f and g, can be added, subtracted, multiplied, and divided according to the following definitions.

Basic Operations on Functions

If f and g are two functions, then their sum $(f + g)$, their difference $(f - g)$, their product $(f \cdot g)$, and their quotient $\left(f/g \text{ or } \dfrac{f}{g}\right)$ are defined by

$$(f + g)(x) = f(x) + g(x)$$

$$(f - g)(x) = f(x) - g(x)$$

$$(f \cdot g)(x) = f(x) \cdot g(x)$$

$$(f/g)(x) = \left(\frac{f}{g}\right)(x) = \frac{f(x)}{g(x)}$$

The domain of $f + g$, $f - g$, and $f \cdot g$ is the set of numbers that belong to both the domain of f and the domain of g. For the quotient function f/g, the domain also excludes all numbers where $g(x) = 0$.

EXAMPLE 4.19

Find each function, $(f + g)(x)$, $(f - g)(x)$, $(f \cdot g)(x)$, $(f/g)(x)$ and their domains, if $f(x) = x^2 - 16$ and $g(x) = x - 4$.

Solutions

$$\begin{aligned}
(f + g)(x) = f(x) + g(x) &= (x^2 - 16) + (x - 4) \\
&= x^2 - 16 + x - 4 \\
&= x^2 + x - 20
\end{aligned}$$

$$\begin{aligned}
(f - g)(x) = f(x) - g(x) &= (x^2 - 16) - (x - 4) \\
&= x^2 - 16 - x + 4 \\
&= x^2 - x - 12
\end{aligned}$$

$$\begin{aligned}
(f \cdot g)(x) = f(x) \cdot g(x) &= (x^2 - 16)(x - 4) \\
&= x^3 - 4x^2 - 16x + 64
\end{aligned}$$

$$\begin{aligned}
(f/g)(x) = \frac{f(x)}{g(x)} &= \frac{x^2 - 16}{x - 4} \\
&= \frac{(x + 4)(x - 4)}{x - 4} = x + 4
\end{aligned}$$

The domain of all the new functions, except for f/g, is the set of real numbers. Since $g(x) = 0$ when $x = 4$, the domain of f/g is all real numbers except $x = 4$.

Notice in Example 4.19 that the function $(f/g)(x) = x + 4$, if $x \neq 4$, and the function $h(x) = x + 4$ are identical except at $x = 4$. Because $(f/g)(x)$ is not defined when $x = 4$, its graph, as shown in Figure 4.2a, looks like the line $y = x + 4$ with a "hole" when $x = 4$. The graph of $h(x) = x + 4$ is shown in Figure 4.2b.

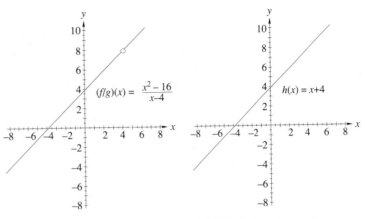

FIGURE 4.2a   FIGURE 4.2b

EXAMPLE 4.20

Find each function $(f + g)(x)$, $(f - g)(x)$, $(f \cdot g)(x)$ and $(f/g)(x)$ and their domains, if $f(x) = \sqrt{x + 7}$ and $g(x) = \sqrt{x - 5}$.

Solutions

$$(f + g)(x) = \sqrt{x + 7} + \sqrt{x - 5}$$

$$(f - g)(x) = \sqrt{x + 7} - \sqrt{x - 5}$$

$$(f \cdot g)(x) = \left(\sqrt{x + 7}\right)\left(\sqrt{x - 5}\right) = \sqrt{(x + 7)(x - 5)}$$

$$= \sqrt{x^2 + 2x - 35}$$

$$(f/g)(x) = \frac{\sqrt{x + 7}}{\sqrt{x - 5}} = \sqrt{\frac{x + 7}{x - 5}}$$

The domain of f is all real numbers $x \geq -7$. The domain of g is all real numbers $x \geq 5$. So, the domain of each new function is all real numbers $x \geq 5$ except for f/g; since $g(5) = 0$, the domain of f/g is all real numbers $x > 5$.

Composite Functions

There is another important way in which two functions f and g can be combined to form a new function. This new function is called the composite function. An example of a composite function is $y = (x - 7)^3$. If we let $u = g(x) = x - 7$, then $y = f(u) = u^3$. You could have also written this as $y = f(u) = f(g(x)) = (x - 7)^3$.

Composite Function

If f and g are two functions, the **composite function**, $f \circ g$, is defined by

$$(f \circ g)(x) = f(g(x))$$

where the domain of $f \circ g$ is all numbers x in the domain of g where $g(x)$ is in the domain of f.

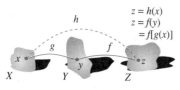

z = h(x)
z = f(y)
= f[g(x)]

FIGURE 4.3

The sketch in Figure 4.3 shows a function g, which assigns to each element x of set X, some element y of set Y. The figure also shows a function f that takes each element of set Y and assigns to it a value z of set Z. Using both f and g, an element x in X is assigned to an element z in Z. This process is a new function h, which takes an element x in X and assigns to it an element z in Z.

Caution

We do not perform composition in the same left-to-right direction as we do in reading. In the composition $f \circ g$, function g is performed first.

The definition of the domain is rather cumbersome, but it will make sense after a few examples.

EXAMPLE 4.21

If $f(x) = x^2$ and $g(x) = x - 4$, find **(a)** $(f \circ g)(x)$, **(b)** $(g \circ f)(x)$, **(c)** $(f \circ g)(5)$, and **(d)** $(g \circ f)(5)$

Solutions

(a) $(f \circ g)(x) = f(g(x)) = f(x - 4)$

$$= (x - 4)^2$$

(b) $(g \circ f)(x) = g(f(x)) = g(x^2) = x^2 - 4$

(c) $(f \circ g)(5) = (5 - 4)^2 = (1)^2 = 1$

(d) $(g \circ f)(5) = 5^2 - 4 = 25 - 4 = 21$

Because the domains of both of these functions are the entire set of real numbers, the domains of both $f \circ g$ and $g \circ f$ are all real numbers.

≡ Note Examples 4.21**(c)** and **(d)** show that in most cases $(f \circ g)(x) \neq (g \circ f)(x)$.

EXAMPLE 4.22

If $f(x) = x^2 - 9$ and $g(x) = \sqrt{x + 4}$, then find **(a)** $(f \circ g)(x)$, **(b)** $(g \circ f)(x)$, **(c)** $(f \circ g)(2)$, **(d)** $(g \circ f)(2)$, and **(e)** the domains of $f \circ g$ and $g \circ f$.

Solutions

(a) $(f \circ g)(x) = f(g(x)) = f\left(\sqrt{x + 4}\right) = \left(\sqrt{x + 4}\right)^2 - 9$

$$= (x + 4) - 9$$

$$= x - 5$$

(b) $(g \circ f)(x) = g(f(x)) = g(x^2 - 9) = \sqrt{(x^2 - 9) + 4}$

$$= \sqrt{x^2 - 5}$$

(c) $(f \circ g)(2) = 2 - 5 = -3$

(d) $(g \circ f)(2) = \sqrt{2^2 - 5} = \sqrt{-1}$, which is not a real number. Therefore, 2 is not in the domain of $g \circ f$.

EXAMPLE 4.22 (Cont.)

(e) The domain of f is all real numbers and the domain of g is all real numbers $x \geq -4$. The domain of $f \circ g$ will be all numbers $x \geq -4$. The domain of $g \circ f$ will be the real numbers x, where $f(x) \geq -4$ or where $x^2 - 9 \geq -4$. Simplifying this, we see that $x^2 \geq 5$, which further simplifies to $x \geq \sqrt{5}$ or $x \leq -\sqrt{5}$.

Application

EXAMPLE 4.23

An oil well in the Gulf of Mexico is leaking. There is no wind or current, so the oil is spreading on the water as a circle. At any time t, in minutes, after the leak began, the radius in meters of the circular oil slick is $r(t) = 12t$. If we know that the area of a circle of radius r is $A = \pi r^2$, (a) find the function $A(t)$, the area of the oil slick as a function of t, (b) determine the area of the slick, 15 min after the leak began, and (c) determine the area of the slick, 1 h after it began.

Solution

(a) We have $r(t) = 12t$ and $A(r) = \pi r^2$. Then $A(t) = (A \circ r)(t) = A(r(t)) = A(12t) = \pi(12t)^2 = 144\pi t^2$ m^2.

(b) When $t = 15$ min, $A(15) = 144\pi(15^2) = 32\,400\pi$ m^2.

(c) When $t = 1$ h, we must use 1 h $= 60$ min, because $A(t)$ is for t in minutes. $A(1\text{ h}) = A(60) = 144\pi(60^2) = 518\,400\pi$ m^2.

Exercise Set 4.2

In Exercises 1–18, let $f(x) = x^2 - 1$ and $g(x) = 3x + 5$. Determine the following.

1. $(f + g)(x)$
2. $(f + g)(4)$
3. $(f - g)(x)$
4. $(f - g)(4)$
5. $(g - f)(x)$
6. $(g - f)(5)$

7. $(f \cdot g)(x)$
8. $(f \cdot g)(5)$
9. $(f/g)(x)$
10. $(f/g)(-2)$
11. $(g/f)(x)$
12. $(g/f)(-2)$

13. The domains of f and g
14. The domains of f/g and g/f
15. $(f \circ g)(x)$
16. $(f \circ g)(-3)$
17. $(g \circ f)(x)$
18. $(g \circ f)(-3)$

In Exercises 19–36, let $f(x) = 3x - 1$ and $g(x) = 3x^2 + x$. Determine the following.

19. $(f + g)(x)$
20. $(f + g)(4)$
21. $(f - g)(x)$
22. $(f - g)(-2)$
23. $(g - f)(x)$
24. $(g - f)(-2)$

25. $(f \cdot g)(x)$
26. $(f \cdot g)(-1)$
27. $(f/g)(x)$
28. $(f/g)(-1)$
29. $(g/f)(x)$
30. $(g/f)(-1)$

31. $(f \circ g)(x)$
32. $(f \circ g)(5)$
33. $(g \circ f)(x)$
34. $(g \circ f)(5)$
35. The domains of f and g
36. The domains of f/g and g/f

Solve Exercises 37–42.

37. *Business* The cost of manufacturing an item is the sum of the fixed costs and the variable costs. Fixed costs are the costs the business must meet just to stay in business, such as rent and insurance. Variable costs are the costs, such as material, labor, and lighting, of producing just one item. Suppose the fixed cost, in dollars, of manufacturing computer chips is $F(n) = \$7,500$/week and the variable cost is \$15 per chip. **(a)** Find the cost function $C(n)$, in dollars, of manufacturing n computer chips per week. **(b)** How much will it cost to manufacture 100 chips? **(c)** How much will it cost to manufacture 1,000 chips?

38. *Business* In manufacturing a certain product, a company must pay a set-up cost of $S(n) = \dfrac{2,750,000}{n}$ dollars, where n is the number of units it produces in a production run, and with $1 \leq n \leq 25,000$. The fixed cost is $F(n) = \$12,500$/week and the variable cost is \$8/item. **(a)** Determine the cost function, $C(n)$, of setting up and manufacturing n items in a week. **(b)** What is the cost of manufacturing 100 items? **(c)** of manufacturing 1,000 items?

39. *Business* A company's profit is equal to the revenue it generates minus the cost of producing and selling the product. The revenue function for a certain product is $R(n) = 90n$ and its cost function is $C(n) = 30n + 275$, where n is the number of units sold. **(a)** If R and C are in dollars, determine the profit function, $P(n)$, for this product. **(b)** What is the profit if you sell 50 units?

40. *Business* The demand function $n = f(p)$ relates the number of units n that can be sold to the price p charged for each unit. The revenue is equal to the price per unit multiplied by the quantity sold. A manufacturer is currently selling its calculators for p dollars each. Based on past experience, the company believes that the weekly demand for the calculator is related to its price by the equation $f(p) = 750 - 4.2p$. **(a)** What is the revenue function? **(b)** How many calculators can they expect to sell if they charge \$100 for each calculator? **(c)** What is the revenue from the \$100 calculators? **(d)** How many calculators can they expect to sell at \$80 each? **(e)** What is the revenue for the \$80 calculator?

41. *Environmental science* The population P of a certain species of large fish depends on the number n of a smaller fish on which it feeds, with

$$P(n) = 300n^2 - 50n$$

The number of smaller fish depends on the amount a of its food supply, a type of plankton, with

$$n(a) = 7a + 4$$

Find the relationship between the population P of the larger fish and the amount a of plankton available.

42. *Meteorology* A spherical weather balloon is inflated so the radius r is changed at a constant rate of 3 cm/s. **(a)** Find an algebraic expression for $V(t)$, the volume of the balloon as a function of time t in seconds. **(b)** Determine the volume of the balloon after 10 s.

☰ 4.3

RECTANGULAR COORDINATES

In Chapter 1 we graphed some points on a number line. In this section we will expand our idea of graphing to a plane.

Why is a graph so important? A graph gives us a picture of an equation. By looking at a graph, we can often get a better idea of what we can expect an equation to do. The graphs will also help us find solutions to some of our problems.

In Section 1.1 we learned that we could represent the numbers by points on a line. To get a graph in a plane, we need two number lines. The two number lines are usually drawn perpendicular to each other and intersect at the number zero as shown in Figure 4.4. The point of intersection is called the **origin**. If one of the lines is horizontal; then the other is vertical. The horizontal number line is called the **x-axis** and the vertical number line the **y-axis**. This is called the **rectangular coordinate**

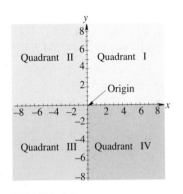

FIGURE 4.4

system or the **Cartesian coordinate system** in honor of the man who invented it, René Descartes.

On the x-axis, positive numbers are to the right of the origin and negative numbers to the left. On the y-axis, positive numbers are above the origin and negative values are below it. The two axes divide the plane into four regions called **quadrants**, with the first quadrant in the upper right section of the plane and the others numbered in a counterclockwise rotation around the origin. (See Figure 4.4.)

Suppose that P is a point in the plane. The coordinates of P can be determined by drawing a perpendicular line segment from P to the x-axis. If this perpendicular segment meets the x-axis at the value a then the x-coordinate of the point P is a. Now draw a perpendicular segment from P to the y-axis. It meets the y-axis at b, so the y-coordinate of P is b. The coordinates of P are the ordered pair (a, b).

≡ **Note**   The x-coordinate is always listed first. The order in which the coordinates are written is very important. In most cases, the point (a, b) is different than the point (b, a). (See Figure 4.5.)

EXAMPLE 4.24

The positions of A(2, 4), B(4, 2), C(−4, 5), D(7, 0), E(−4, −3), F(0, −3), and G(5, −4) are shown in Figure 4.6. Note that A(2, 4) and B(4, 2) are different points as are C(−4, 5) and G(5, −4).

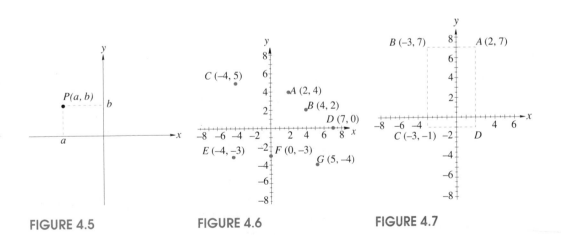

FIGURE 4.5   FIGURE 4.6   FIGURE 4.7

EXAMPLE 4.25

A(2, 7), B(−3, 7) and C(−3, −1) are three vertices of a rectangle as shown in Figure 4.7. What are the coordinates of the fourth vertex, D?

Solution   If we plot points A, B, and C, we can see that the missing vertex D will have the same x-coordinate as A, 2, and the same y-coordinate as C, −1. So, D has the coordinates (2, −1).

EXAMPLE 4.26

In Example 4.8, we found that $(-2, 3), (-1, 5), (0, 7), \left(\frac{1}{2}, 8\right), (1, 9)$, and $(2, 11)$ were ordered pairs for the equation $y = 2x + 7$. Plot these ordered pairs on a rectangular coordinate system.

Solution   The ordered pairs are plotted on the graph in Figure 4.8.

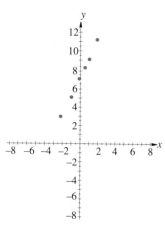

FIGURE 4.8

Application

EXAMPLE 4.27

The following table shows the results of a series of drillings used to determine the depth of the bedrock at a building site. These drillings were taken along a straight line down the middle of the lot, from the front to the back, where the building will be placed. In the table, x is the distance from the front of the parking lot and y is the corresponding depth. Both x and y are given in feet.

x	0	20	40	60	80	100	120	140	160
y	33	35	40	45	42	38	46	40	48

Plot these ordered pairs on a rectangular coordinate system. Connect the points in order to get an estimate of the profile of the bedrock.

Solution   The ordered pairs are plotted on the graph in Figure 4.9. The points have been connected in order. Note that this graph is "upside down." That is, the top of the dirt is along the x-axis, while the bedrock, which is below ground, is above the x-axis.

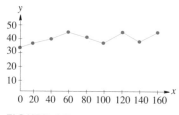

FIGURE 4.9

Exercise Set 4.3

In Exercises 1–10, plot the points on a rectangular coordinate system.

1. $(4, 5)$
2. $(1, 7)$
3. $(7, 1)$
4. $(-2, 4)$
5. $(-3, -5)$

6. $(6, -1)$
7. $(0, 0)$
8. $\left(-\frac{5}{2}, 3\right)$
9. $\left(2, -\frac{3}{2}\right)$
10. $(4.5, -1.5)$

Solve Exercises 11–22.

11. If $A(2, 5), B(-1, 5)$, and $C(2, -4)$ are three vertices of a rectangle, what are the coordinates of the fourth vertex?

12. In Example 4.9, we found that $(-3, 6), (-2, 1), (-1, -2), (0, -3), (1, -2), (2, 1)$, and $(3, 6)$ were ordered pairs of the equation $y = x^2 - 3$. Plot these ordered pairs on the same rectangular coordinate system.

13. Plot the ordered pairs $(-5, 3), (-3, 3), (0, 3), (1, 3), (4, 3)$, and $(6, 3)$. What do all of these points have in common?

14. Plot the ordered pairs $(-2, 6), (-2, 4), (-2, 1), (-2, -1), (-2, -3)$, and $(-2, -4)$. What do all of these points have in common?

15. *Automotive technology* Graph these ordered pairs from the conversion formula 6.9 kPa = 1 psi: $(138, 20), (172.5, 25), (207, 30), (241.5, 35), (276, 40), (345, 50), (414, 60)$. Use a ruler and draw the segment connecting $(138, 20)$ and $(414, 60)$. What seems to happen?

16. Where are all the points whose x-coordinates are 0?

17. Where are all the points whose y-coordinates are -2?

18. Where are all the points (x, y) for which $x > -3$?

19. Where are all the points (x, y) for which $x > 0$ and $y < 0$?

20. Where are all the points (x, y) for which $x > 1$ and $y < -2$?

21. *Automotive technology* If the antifreeze content of the coolant is increased, the boiling point of the coolant is also increased, as shown in the table below.

Percent antifreeze in coolant	0	10	20	30	40	50	60	70	80	90	100
Boiling temperature °F	210	212	214	218	222	228	236	246	258	271	330

(a) Plot these ordered pairs on a rectangular coordinate system.
(b) What will be the boiling point of the coolant if the system is filled with a recommended solution of 50% water and 50% antifreeze?
(c) What will be the boiling point of the coolant if the system is filled with a recommended solution of 40% water and 60% antifreeze?

22. *Automotive technology* The temperature-pressure relationship of the refrigerant R-12 is very important to maintain proper operation and for diagnosis. The table below indicates the pressure of R-12 at various temperatures.

Temperature °F	-20	-10	0	10	20	30	40	50	60	70	80
Atmospheric pressure (psi)	2.4	4.5	9.2	14.7	21.1	28.5	37.0	46.8	57.7	70.1	84.1

(a) Plot these ordered pairs on a rectangular coordinate system. Connect the ordered pairs in order to help you answer (b) and (c).
(b) One day the early morning temperature was 45°F. What was the pressure, in psi, when the temperature was 45°F?

(c) If a technician connected a pressure gauge to an air-conditioning system filled with R-12 on a 90°F summer day, what pressure, in psi, would the gauge indicate?

≡ 4.4
GRAPHS

In the last section, we learned how to graph a point on a plane. Earlier, we said that a function could be represented by a graph. In this section, we will see how we can use graphs to represent functions and we will use the computer to help us draw graphs.

> **Graph of an Equation**
>
> The graph of an equation in two variables x and y is formed by all the points $P(x, y)$ whose coordinates (x, y) satisfy the given equation.

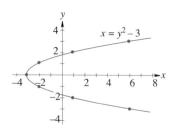

To graph an equation such as $x = y^2 - 3$, you can set up a table of values, plot the points in the table, and then connect the points smoothly. To fill the table, we will select values for y and use the equation to solve for the corresponding values for x. Because the ordered pairs have x listed first, we will list x as the top value in the table.

x	6	1	-2	-3	-2	1	6
y	-3	-2	-1	0	1	2	3

These points are plotted in Figure 4.10 and a smooth curve has been used to connect the points. Remember, this equation has an infinite number of possible ordered pairs. We have selected only seven of them. We must plot enough points to be confident that the points give an outline of the curve that is complete enough for us to tell what the actual curve looks like. Later, you will learn some ways to use mathematics to help determine when you have a sufficient number of points to sketch a curve.

FIGURE 4.10

In Figure 4.8 we plotted six ordered pairs for the equation $y = 2x + 7$: $(-2, 3)$, $(-1, 5)$, $(0, 7)$, $\left(\frac{1}{2}, 8\right)$, $(1, 9)$, and $(2, 11)$. If we connect these points, we see that we get the straight line in Figure 4.11. In fact, any equation of the form $Ax + By + C = 0$ where A, B, and C are constants and both A and B are not 0 is a linear equation whose graph is a straight line. The equation $y = 2x + 7$ is also a function and can be written as $f(x) = 2x + 7$.

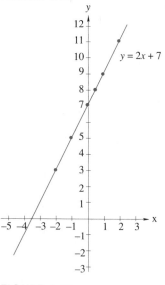

Intercepts

The graph of the equation $y = 2x + 7$ crosses both the x-axis and the y-axis. These points are called the **intercepts**. The graph crosses the y-axis at $(0, 7)$, so we say that the y-intercept is 7. It crosses the x-axis at $\left(-\frac{7}{2}, 0\right)$, so the x-intercept is $-\frac{7}{2}$.

FIGURE 4.11

Hint

Any time you have a linear equation, you will only need to plot two points in order to sketch the graph of that equation. The intercepts are often the easiest two points to determine and use when graphing a linear equation.

EXAMPLE 4.28

Plot the graph of $2x + 3y = 12$.

Solution This is a linear equation, so we will use two points to determine the line. If we find the intercepts, we get the points $(6, 0)$ and $(0, 4)$. These points and the line they determine are shown in Figure 4.12.

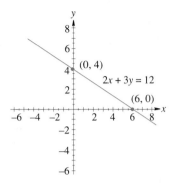

FIGURE 4.12

Slope

One important property of the graph of a straight line is its slope. The slope refers to the steepness or inclination of the line. The idea of the slope is so important that we will keep coming back to it. In fact, it is a major topic in calculus.

> **Slope**
>
> The **slope**, m, of a straight line is a measure of its steepness with respect to the x-axis. If (x_1, y_1) and (x_2, y_2) are two different points on a line, then the slope of the line is defined as
>
> $$m = \frac{y_2 - y_1}{x_2 - x_1}$$
>
> provided $x_2 - x_1 \neq 0$.

Hint

Some people like to remember the slope as the "rise over the run" or $\dfrac{\text{rise}}{\text{run}}$. Here "rise" means the vertical change and "run" the horizontal change.

EXAMPLE 4.29

What is the slope of the line $2x + 3y = 12$ in Example 4.28?

Solution We know that $(6, 0)$ and $(0, 4)$ are two points on this line. If we let $x_1 = 6, y_1 = 0, x_2 = 0$, and $y_2 = 4$, then

$$m = \frac{y_2 - y_1}{x_2 - x_1}$$

$$= \frac{4 - 0}{0 - 6} = \frac{4}{-6} = -\frac{2}{3}$$

The slope of this line is $m = -\frac{2}{3}$.

EXAMPLE 4.30

Graph $4y - 6x = 12$ and find the intercepts and the slope.

Solution This is a straight line. The intercepts are $(0, 3)$ and $(-2, 0)$. Plotting these two points and the line through them gives the graph in Figure 4.13. Letting $x_1 = 0, y_1 = 3, x_2 = -2$, and $y_2 = 0$, we determine that the slope is

$$m = \frac{0 - 3}{-2 - 0} = \frac{-3}{-2} = \frac{3}{2}$$

EXAMPLE 4.31

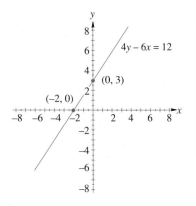

FIGURE 4.13

Graph $8x = 2 + y$, and find the intercepts and the slope.

Solution Again, this is a straight line. If we let $x = 0$, then we get $0 = 2 + y$ or $y = -2$, and so one intercept is $(0, -2)$. If we let $y = 0$, then $8x = 2$ and $x = \frac{1}{4}$. Thus, $\left(\frac{1}{4}, 0\right)$ is the other intercept. The slope is

$$m = \frac{-2 - 0}{0 - \frac{1}{4}} = \frac{-2}{-\frac{1}{4}} = 8$$

Plotting the two intercepts and drawing the line through them produces the line in Figure 4.14.

Look again at the last three examples. In Example 4.29, the slope of the line was $-\frac{2}{3}$. Notice that this line falls as the x-values get larger. In Example 4.30, the slope of the line was $\frac{3}{2}$. In Example 4.31, the slope of the line was 8. Both of these lines rose as the values of x got larger. Notice that the line in Example 4.31 was steeper than the line in Example 4.30.

In general, as the values of x increase, a straight line rises if it has a positive slope and falls if it has a negative slope. This is demonstrated in Figures 4.15 and 4.16.

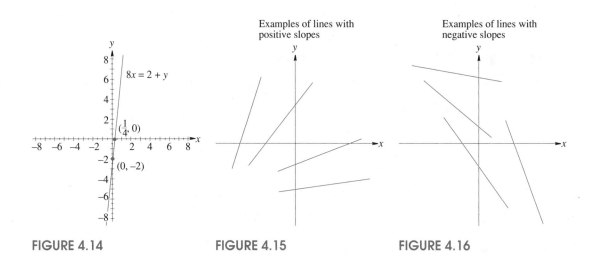

FIGURE 4.14   FIGURE 4.15   FIGURE 4.16

Graphs That Are Not Lines

Not every graph is a straight line. The next four examples show some graphs of functions that are not lines. This is followed by a test that tells how to use a graph to determine if it is a graph of a function.

EXAMPLE 4.32

Graph the function from Example 4.9, $y = x^2 - 3$.

Solution In Example 4.9, we found the ordered pairs in the following table.

x	-3	-2	-1	0	1	2	3
y	6	1	-2	-3	-2	1	6

If we plot these seven points and connect them with a smooth curve, we get the graph in Figure 4.17. Notice that this graph has a y-intercept at $y = -3$ and that it has two x-intercepts, when $x = \sqrt{3}$ and when $x = -\sqrt{3}$.

EXAMPLE 4.33

Graph the function $f(x) = x^2 - 2x$.

Solution Here is a table of some of the values for this function.

x	-3	-2	-1	0	1	2	3	4	5
$f(x)$	15	8	3	0	-1	0	3	8	15

These points and the curve connecting them are shown in Figure 4.18.

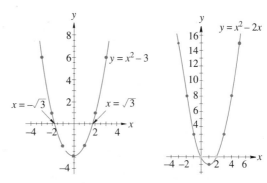

FIGURE 4.17   FIGURE 4.18

EXAMPLE 4.34

Graph the function $g(x) = \dfrac{x-1}{x}$.

Solution Note that this function is not defined when $x = 0$. A partial table of values for this function follows.

x	-5	-4	-3	-2	-1	-0.4	-0.1
$g(x)$	1.2	1.25	$1.3\overline{3}$	1.5	2	3.5	11

x	0	0.1	0.4	1	2	3	4	5
$g(x)$	DNE*	-9	-1.5	0	0.5	$1.6\overline{6}$	0.75	0.80

*DNE means Does Not Exist.

The graph is shown in Figure 4.19.

Another interesting graph is shown by Example 4.35.

EXAMPLE 4.35

Graph the function $f(x) = \sqrt{x + 4}$.

Solution   The function is only defined for $x \geq -4$. A partial table of values for this function follows. Its graph is shown in Figure 4.20.

x	-4	-3	-2	-1	0	1	2	3	4	5
y	0	1	$\sqrt{2} \approx$ 1.414	$\sqrt{3} \approx$ 1.732	2	$\sqrt{5} \approx$ 2.236	$\sqrt{6} \approx$ 2.449	$\sqrt{7} \approx$ 2.646	$\sqrt{8} \approx$ 2.828	3

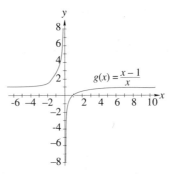

FIGURE 4.19

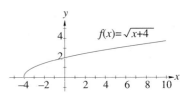

FIGURE 4.20

Application

EXAMPLE 4.36

The total cost to manufacture x items of a certain product is given by $C(x) = \dfrac{8}{x} + 0.5x$, where $x > 0$. Set up a partial table of values and sketch the graph of this function.

Solution   This function is only defined for $x > 0$. A partial table of values is shown. Its graph is shown in Figure 4.21.

x	1	2	3	4	5	6	7	8	9	10
y	8.5	5	$4\frac{1}{6}$	4	4.1	$4\frac{1}{3}$	$4\frac{9}{14}$	5	$5\frac{7}{18}$	5.8

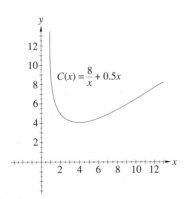

FIGURE 4.21

Vertical Line Test

A graph can provide us with an easy test to see if the equation that has been graphed is a function. This test is called the **vertical line test**.

> **Vertical Line Test**
>
> The vertical line test states that a graph is the graph of a function if no vertical line intersects the graph more than once.

To use the vertical line test, you look at a graph and determine if there are any places where a vertical line would intersect the graph more than once. If you cannot find any such places, then this is the graph of a function. If you can, then this is not the graph of a function. Figures 4.22a and 4.22b give some examples. The graph in Figure 4.22a is not the graph of a function, because the vertical line intersects the graph three times. On the other hand, the graph in Figure 4.22b is the graph of a function because no vertical line can intersect the graph more than once.

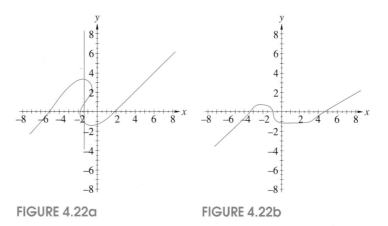

FIGURE 4.22a   FIGURE 4.22b

Application

EXAMPLE 4.37

The concentration in parts per million (ppm) of a certain pollutant m mi from a certain factory is given by

$$C(m) = \frac{50}{m^2 + 4}, m \geq 0$$

Set up a partial table of values and sketch the graph of this function.

Solution   This function is defined for values of $m \geq 0$. A partial table of values follows. The graph of the function is shown in Figure 4.23. Notice that this graph passes the vertical line test.

m	0	1	2	3	4	5	6
$C(m)$	12.5	10	6.25	3.85	2.5	1.72	1.25

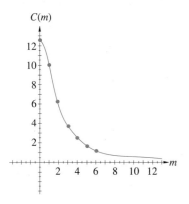

FIGURE 4.23

Exercise Set 4.4

For each of Exercises 1–10, (a) draw the graphs of the linear equations, (b) determine the intercepts and, (c) calculate the slope.

1. $y = x$
2. $y = 2x$
3. $y = -3x$
4. $y = \frac{1}{2}x - 3$
5. $y = -4x + 2$

6. $x + y = 2$
7. $x - y = 2$
8. $2x + y = 1$
9. $3x - 6y = 9$
10. $3x - 6y = -6$

In Exercises 11–20, set up a table of values and graph the given functions.

11. $y = x^2$
12. $y = x^2 + 3$
13. $y = x^2 - 2$
14. $y = x^2 + 2x + 1$
15. $y = x^2 - 6x + 9$
16. $y = -x^2 + 2$

17. $y = \frac{1}{x}$
18. $y = \frac{1}{x+3}$
19. $y = x^3$
20. $y = \sqrt[3]{x}$

In Exercises 21–26, set up a table of values and graph each of the equations.

21. $y^2 + x^2 = 25$
22. $y = \sqrt{25 - x^2}$
23. $y = |x + 2|$
24. $\frac{x^2}{4} + \frac{y^2}{9} = 1$

25. $\frac{x^2}{4} - \frac{y^2}{9} = 1$
26. $y = x^3 - x^2$

Which of the graphs in Exercises 27–30 are functions?

27.

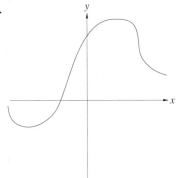

28.

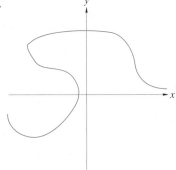

29.

30.

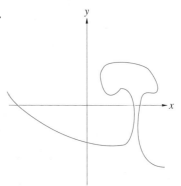

Solve Exercises 31–34.

31. *Ecology*   An ecologist is investigating the effects of air pollution from an industrial city in the plants surrounding the city. She estimates that the percentage of diseased plants is given by

$$p(k) = \frac{25}{2k+1}$$

   (a) Set up a table of values and graph the equation for $0 \le k \le 9$.
   (b) Does your graph pass the vertical line test?

32. *Ecology*   A second ecologist thinks that the function in the previous problem does not give the correct percentages. She thinks that the percentage of diseased plants k km from the city is given by

$$p(k) = \frac{25}{\sqrt{k+1}}$$

   Make a table of values and graph the function for $0 \le k \le 9$

33. *Business*   Based on past experience, a company decides that the weekly demand (in thousands) for a new microwaveable food product is $d(p) = -p^2 + 2.5p + 10$, where p is the price (in dollars) of the product.
   (a) What is the domain of this function?
   (b) Make a table of values for each $0.25 change in price of $0 < p \le \$2.00$.
   (c) Sketch a graph of this function.
   (d) What appears to be the price that will result in the most sales?

34. *Automotive technology*   In a chrome-electroplating process, the mass m in grams of the chrome plating increases according to the formula $m = 1 + 2^{t/2}$, where t is the time in minutes.
   (a) Set up a table of values and graph the equation for $1 \le t \le 15$.
   (b) How long does it take to form 100 g of plating?

≡ 4.5
GRAPHING CALCULATORS AND COMPUTER-AIDED GRAPHING

The introduction of microcomputers has allowed people to use inexpensive computers in their work for writing and for performing calculations quickly and accurately; but people discovered that words and numbers were not enough, and so computer graphics were introduced.

   Some of these graphing capabilities are now available on calculators. To some extent, the use of a graphing calculator or a computer removes some of the drudgery of plotting equations by hand. We will show you how you can use computer and calculator graphing capabilities to demonstrate some mathematical concepts and to relieve you of some work. However, we will also show you how to take advantage of these technological devices and show how to interpret the information they give you.

The graphics capabilities of computers and calculators vary widely. Examples in this section, and other graphics examples in this text, were run on either a Casio *fx-7700G* or a Texas Instruments *TI-81* graphing calculator or on a Macintosh® computer.

Rather than teach you how to write a graphing program for a computer, we will assume you have access to a program that can be used to graph functions. Some, such as *Master Grapher,*® are specialized graphing programs; some, such as *Lotus 1–2–3,*® are computer "spreadsheets" that have graphing capabilities. Other more specialized programs include *Mathematica,*® *Derive,*® *Microcalc,*® and *Maple,*® all of which might be classified as computer algebra systems.

The purpose of this book is not to explain how to use the graphing capabilities of these programs, but to help you use these computer programs to graph curves and to interpret their graphs.

Using a Graphing Calculator

We begin with a discussion on using graphing calculators. The examples and the instructions in this book are meant to supplement the user's guide for your calculator, not to replace it. Additional details on using a graphing calculator are in Appendix A.

The first equation will be of a straight line, $y = 2x + 7$. Before we begin, we must determine the domain of this function. The range of this function is determined by the allowable values of y that can appear on the calculator's screen.

EXAMPLE 4.38

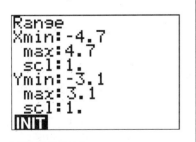

FIGURE 4.24a

Use a graphing calculator to sketch the graph of $f(x) = 2x + 7$.

Solution 1   Casio *fx-7700G*

First, enter $\boxed{\text{SHIFT}}$ $\boxed{\text{F5}}$ $\boxed{\text{EXE}}$ to clear the graphics screen of any previous graphs that might be on the screen or in the calculator's memory.

We begin by setting the domain and range we want displayed on the screen. Press the $\boxed{\text{Range}}$ key. The screen should be similar to Figure 4.24a. Now, press the $\boxed{\text{SHIFT}}$ $\boxed{\text{F1}}$ $\boxed{\text{Range}}$ $\boxed{\text{Range}}$ keys. This clears the memory and sets the domain and range to the default values in the machine.

The domain that will be shown on the screen is indicated by Xmin (the smallest value of x displayed on the screen), Xmax (the corresponding largest x-value), and Xscl (the x-axis' scale or, more precisely, the distance between "tick marks" on the x-axis). The range is similarly restricted by Ymin, Ymax, and Yscl. As explained in Appendix A, these values can be changed and we will do that in later graphing calculator activities. However, for this example, let's use these preset values.

You are now ready to input a formula describing the function you want graphed, in this example $f(x) = 2x + 7$. To input $f(x) = 2x + 7$, press the following sequence of keys:

$$\boxed{\text{Graph}}\ 2\ \boxed{\text{X}, \theta, \text{T}}\ \boxed{+}\ 7\ \boxed{\text{EXE}}$$

Be careful! To input the variable x, you press the $\boxed{\text{X}, \theta, \text{T}}$ key or the $\boxed{\text{ALPHA}}$ $\boxed{\text{X}}$. Do not confuse the $\boxed{\text{X}}$ with the multiplication $\boxed{\times}$ key.

EXAMPLE 4.38 (Cont.)

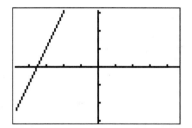

FIGURE 4.24b

The result is the graph shown in Figure 4.24b. Compare this to the earlier graph of this same function in Figure 4.11.

This graph is fine, but it does not show us where the graph crosses the y-axis. Let's see one way that we can change the domain and range of the display screen. Press the $\boxed{\text{SHIFT}}$ $\boxed{\text{F2}}$ $\boxed{\text{F4}}$ keys. This causes the calculator to "zoom out" from the middle of the screen. The result is the graph in Figure 4.24c. Press $\boxed{\text{Range}}$ to see how the domain and range values have changed. As you can see, they have doubled.

You still cannot see where the graph crosses the y-axis, so press the $\boxed{\text{F2}}$ $\boxed{\text{F4}}$ keys again. The calculator "zooms out" once again and produces the graph in Figure 4.24d.

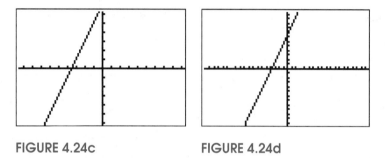

FIGURE 4.24c   **FIGURE 4.24d**

You can now see where the line crosses the x- and y-axes. It is hard to determine the coordinates of these intercepts by looking at the calculator screen. In the next section, we will see how to determine the approximate coordinates of the points.

```
RANGE
Xmin=-10
Xmax=10
Xscl=1
Ymin=-10
Ymax=10
Yscl=1
Xres=1
```

FIGURE 4.24e

Solution 2 *TI-81*

First press $\boxed{\text{ZOOM}}$ $\boxed{6}$. You may have to wait for the screen to clear. This clears the memory and sets the domain and range to the default values in the machine. Wait until the small black rectangle in the upper right-hand corner of the screen disappears, and then press $\boxed{\text{RANGE}}$. The screen should be similar to Figure 4.24e.

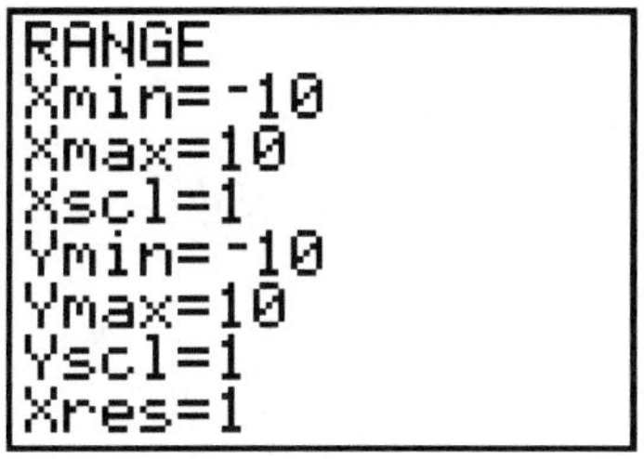

The domain that will be shown on the screen is indicated by Xmin (the smallest value of x displayed on the screen), Xmax (the corresponding largest x-value), and Xscl (the x-axis' scale or, more precisely, the distance between "tick marks" on the x-axis). The range is similarly restricted by Ymin, Ymax, and Yscl. As explained in Appendix A, these values can be changed and we will do that in later graphing calculator activities. However, for this example, let's use these preset values.

You are now ready to input a formula describing the function you want graphed, in this example $f(x) = 2x + 7$. To input $f(x) = 2x + 7$, press the following sequence of keys:

$$\boxed{\text{Y=}}\ \boxed{\text{CLEAR}}\ 2\ \boxed{\text{X|T}}\ \boxed{+}\ 7\ \boxed{\text{GRAPH}}$$

Be careful! To input the variable x, you press the $\boxed{\text{X|T}}$ key or the $\boxed{\text{ALPHA}}$ $\boxed{\text{X}}$ keys. Do not confuse the $\boxed{\text{X}}$ with the multiplication $\boxed{\times}$ key.

The result is the graph shown in Figure 4.24f. Compare this to the earlier graph of this same function in Figure 4.11.

EXAMPLE 4.38 (Cont.)

You will notice that this graph does not show us where the graph crosses the *y*-axis. However, in case we need to zoom out to see more of the graph, let's see one way that we can change the domain and range of the display screen. Press the $\boxed{\text{ZOOM}}$ $\boxed{3}$ $\boxed{\text{ENTER}}$ keys. This causes the calculator to "zoom out" from the middle of the screen. The result is the graph shown in Figure 4.24g. Press $\boxed{\text{RANGE}}$ to see how the domain and range values have changed. As you can see, they are four times as large as they were before.

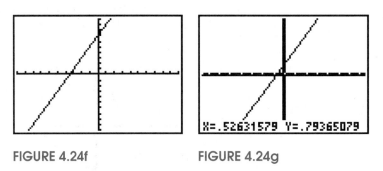

FIGURE 4.24f   FIGURE 4.24g

It is hard to determine the coordinates of these intercepts by looking at the calculator screen. In the next section, we will see how to determine the approximate coordinates of the points.

Let's make the graph of a curve, which is not a straight line. In Section 4.4, we graphed $y = x^2 - 3$ with the result shown in Figure 4.10. Let's see what that looks like on a calculator.

EXAMPLE 4.39

Use a graphing calculator to sketch the graph of $y = x^2 - 3$.

Solution   We will show how to do this on a Casio *fx-7700G*. You will have to make a few changes if you are using a different machine.

First, enter $\boxed{\text{Shift}}$ $\boxed{\text{F5}}$ $\boxed{\text{EXE}}$ to clear the graphics screen of the previous graphs from the screen. As before, press the $\boxed{\text{Range}}$ $\boxed{\text{Shift}}$ $\boxed{\text{F1}}$ $\boxed{\text{Range}}$ $\boxed{\text{Range}}$ keys to clear the memory and set the domain and range to the default values in the machine.

You are now ready to input a formula describing the function you want graphed. To input $y = x^2 - 3$, press the following sequence of keys:

$$\boxed{\text{GRAPH}}\ \boxed{\text{X, }\theta\text{, T}}\ \boxed{\text{Shift}}\ \boxed{x^2}\ \boxed{-}\ 3\ \boxed{\text{EXE}}$$

The result is the graph shown in Figure 4.25. Compare this to the earlier graph of this same function in Figure 4.17.

From this graph we can see that the curve crosses the *y*-axis near $y = -3$ and the *x*-axis around $x = \pm 1.8$. In the next section, we shall see how to obtain better approximations of these values.

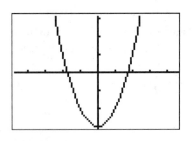

FIGURE 4.25

By now you have probably noticed that in order to tell the calculator what you want it to graph on a Casio *fx-7700G*, you press the GRAPH key and on a *TI-81*, you press the Y= key. In each case you obtain a phrase like Casio's "Graph Y=" and *TI-81*'s "Y1=" on the calculator's screen. This should remind you that you can only graph a function. It may be necessary to solve an equation for *y* before you can graph it.

Exercise Set 4.5

Graph each of these functions with a calculator or on a computer. Compare the machine's graph with the graph you made of the same functions in Section 4.4.

1. $y = x$

2. $y = 2x$

3. $y = -3x$

4. $y = \frac{1}{2}x - 3$

5. $y = -4x + 2$

6. $x + y = 2$

7. $x - y = 2$

8. $2x + y = 1$

9. $3x - 6y = 9$

10. $3x - 6y = -6$

11. $y = x^2$

12. $y = x^2 + 3$

13. $y = x^2 - 2$

14. $y = x^2 + 2x + 1$

15. $y = x^2 - 6x + 9$

16. $y = -x^2 + 2$

17. $y = \frac{1}{x}$

18. $y = \frac{1}{x + 3}$

19. $y = x^3$

20. $y = \sqrt[3]{x}$

21. $y^2 + x^2 = 25$

22. $y = \sqrt{25 - x^2}$

23. $y = |x + 2|$ (HINT: Use the Abs key. Read your manual to see how to access this on your calculator.)

24. $\frac{x^2}{4} + \frac{y^2}{9} = 1$ (HINT: Solve for *y*, then graph two equations.)

25. $\frac{x^2}{4} - \frac{y^2}{9} = 1$

26. $y = x^3 - x^2$

≡ 4.6
USING GRAPHS TO SOLVE EQUATIONS

A graph not only gives you a picture of what an equation looks like, but it can be used to help solve the equation. In Section 4.4, we discussed the points where the graph crosses the axes. The *y*-intercepts were the points where the graph crossed or touched the *y*-axis; the *x*-intercepts were the points where the graph crossed or touched the *x*-axis. The *x*-intercepts are also known as the **roots** or **solutions** of the equation, because these are the values of *x* for which *y* will equal zero. If the equation is a function of *x*, these are called the roots (or **zeros**) of the function because they are the values of *x* when $f(x) = 0$.

EXAMPLE 4.40

Find graphically the approximate roots of $4x^2 + 4x - 15 = 0$.

Solution   First, set $f(x) = 4x^2 + 4x - 15$. Then set up a partial table of values for the function f.

x	-3	-2	-1	0	1	2	3
$y = f(x)$	9	-7	-15	-15	-7	9	33

These points are plotted and connected to form the curve in Figure 4.26. From the graph we can see that the x-intercepts look as if they are at $x = -2\frac{1}{2}$ and $x = 1\frac{1}{2}$. A check of the function confirms that $f\left(-2\frac{1}{2}\right) = 0$ and $f\left(1\frac{1}{2}\right) = 0$. These are the roots of this equation.

EXAMPLE 4.41

Find graphically the approximate roots of $f(x) = x^2 - 3x - 1$.

Solution   Again, if we set up a table of values we get

x	-2	-1	0	1	2	3	4	5
$f(x)$	9	3	-1	-3	-3	-1	3	9

Graphing the curve determined by these points, we get the curve in Figure 4.27. From the graph we can see that there appear to be roots at $x = -0.25$ and $x = 3.25$. If we evaluate the function at these two values, we get $f(-0.25) = -0.1875$ and $f(3.25) = -0.1875$. This shows two things. First, it shows that -0.25 and 3.25 are *not* roots of this function, since $f(x) \neq 0$ at either of these two points. Second, it does show that the roots of this function are "close" to -0.25 and 3.25.

FIGURE 4.26

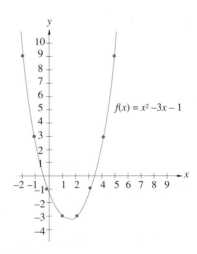

FIGURE 4.27

If we wanted more accurate approximations of the roots of the function in Example 4.41, we could substitute different values for x until we got an approximation that was as accurate as we wanted. In a later chapter, we will find out how to determine the exact roots to this function. In the next example, we will show how to use a graphing calculator to approximate this function's roots.

EXAMPLE 4.42

Use a graphing calculator or computer graphing software with "zoom" capability to find the approximate roots of $f(x) = x^2 - 3x - 1$.

Solution This is the same function we graphed in Example 4.41. We already know that the roots are near $x = -0.25$ and $x = 3.25$. Details on how to use the "zoom" features of a Casio *fx-7700G* and Texas Instrument's *TI-81* graphing calculator are in Appendix A.

When you graph the function using the calculator's default settings for the range, you obtain the graph in Figure 4.28a on an *fx-7700G* or the graph in Figure 4.28b on a *TI-81*.

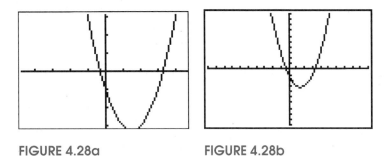

FIGURE 4.28a   FIGURE 4.28b

As you can see from these figures, the graph crosses the x-axis at two points; these are the x-intercepts or roots. Use the "trace" function of your calculator until the cursor is positioned near the x-intercept on the left. Figure 4.28c shows the *fx-7700G* screen and indicates that the cursor is at $(-0.3, -0.01)$, and Figure 4.28d shows that the *TI-81* places the cursor at $(-0.3157895, 0.04709141)$.

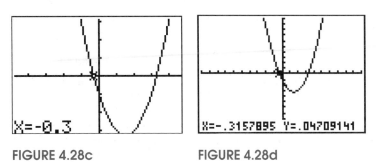

FIGURE 4.28c   FIGURE 4.28d

EXAMPLE 4.42 (Cont.)

For more accurate approximations of this root, you can continue to zoom in around this point on the graph. If you do, you will find that this root is $x \approx -0.302775638$. Later we will show how to find the exact value of this root.

Now, approximate the other root of this function. Return to the original graph in either Figure 4.28a or 4.28b, and zoom in around the right-hand root. If you zoom in enough, you will obtain $x \approx 3.302775638$.

EXAMPLE 4.43

Use a graphing calculator or computer graphing software with "zoom" capability to find the approximate roots of $y = x^4 - 2x^3 - 1$.

Solution   The graph of this function on a *TI-81* is shown in Figure 4.29a. Here Xmin $= -5$, Xmax $= 5$, Ymin $= -3$, and Ymax $= 3$. Zoom in, as in Figure 4.29b, press ⬚Trace⬚ to move the cursor, and see that $x \approx -0.7236842$.

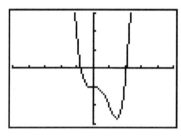
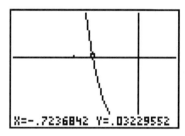

X=-.7236842 Y=.03229552

FIGURE 4.29a   FIGURE 4.29b

Notice that the second root is not on the screen. Returning to the original graph, we move the cursor over near the second root. Zooming in around this point and using the trace function, we see that the other root is approximately $x \approx 2.105$.

EXAMPLE 4.44

Use a graphing calculator or computer graphing software with "zoom" capability to determine any *x*-intercepts of $y = x^2 + 1.5$

Solution   The graph of this function is shown in Figure 4.30. As you can see, the graph does not cross the *x*-axis. Thus, this function does not have any real roots.

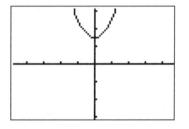

FIGURE 4.30

The graph of an equation can tell the approximate roots or, as in Example 4.44, if the equation has no real roots. There are many times when we will need more accurate values of roots and a quicker procedure than that of graphing the equation. In later chapters, we will learn some alternative methods that can be used to find the roots of an equation.

In Section 4.1, we discussed the inverse of a function. We said that two functions, f and g, are inverses of each other if $f[g(x)] = x$ for every value of x in the domain of g and $g[f(x)] = x$ for every value of x in the domain of f. We also said that if g was the inverse of f, we would write $g(x) = f^{-1}(x)$.

The graph of a function can be used to determine if the function has an inverse and to sketch the graph of the inverse function. You may remember that the vertical line test indicates that if no vertical line intersected a graph more than once, then the graph represents a function. A similar test can be used to determine if a function has an inverse.

> **Horizontal Line Test**
>
> The horizontal line test states that a function has an inverse function if no horizontal line intersects its graph at more than one point.

For example, Figures 4.31a and 4.31b show the graphs of two functions. The function in Figure 4.31a has an inverse function, because it is not possible to draw a horizontal line that will intersect the graph at more than one point. The function in Figure 4.31b does not have an inverse because it is possible to draw a horizontal line that will intersect the graph twice. In fact, the x-axis intersects the graph in Figure 4.31b in at least two places.

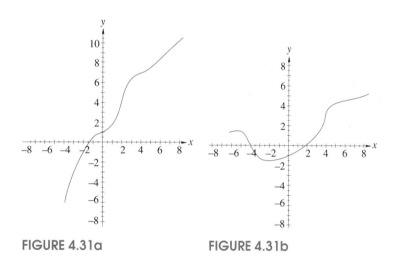

FIGURE 4.31a   FIGURE 4.31b

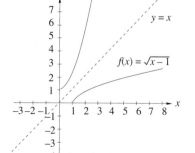

FIGURE 4.32

Now let's use our graphing ability to graph the inverse of a function. The graph of $f(x) = \sqrt{x-1}$ is shown in Figure 4.32. Using the horizontal line test, we can see that this function has an inverse. Now, draw the line $y = x$. (In Figure 4.32 this is the dashed line.) Suppose you placed a mirror on the line $y = x$. The image you would see is indicated by the colored curve in Figure 4.32.

What this essentially does is take any point (a, b) on the function and convert it to its mirror image (b, a). Remember that if (a, b) is a point on a function f, then $f(a) = b$; and if (b, a) is a point on some function g, then $g(b) = a$. But, since $f[f^{-1}(a)] = f(b) = a$, we can see that g must be f^{-1}.

EXAMPLE 4.45

Graph the inverse function of $y = x^3$.

Solution The graph of $y = x^3$ is given in Figure 4.33a and the line $y = x$ is shown by a dashed line. The reflection of $y = x^3$ in the line $y = x$ is shown in Figure 4.33b and is the graph of $y = \sqrt[3]{x}$. If $f(x) = x^3$, then $f^{-1}(x) = \sqrt[3]{x}$.

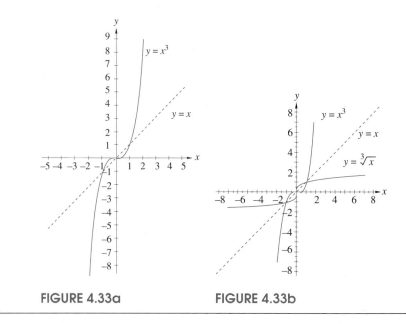

FIGURE 4.33a   FIGURE 4.33b

Application

EXAMPLE 4.46

In Example 4.37, we made the following partial table for the concentration in parts per million (ppm) of a certain pollutant m mi from a certain factory as given by the function

$$C(m) = \frac{50}{m^2 + 4}, m \geq 0$$

m	0	1	2	3	4	5	6
$C(m)$	12.5	10	6.25	3.85	2.5	1.72	1.25

Next, we used our table to sketch the graph of this function.

(a) Use the table to sketch the graph of the inverse of this function.

(b) Use the table or the graph to estimate how far you are from the factory, if the concentration of the pollutant is 3.08 ppm.

(c) Determine the equation that describes the inverse function of C.

(d) Use your equation for C^{-1} to determine $C^{-1}(3.08)$.

Solutions

(a) The graph of the function is shown in Figure 4.34a. The reflection of the graph in the line $y = x$ produces the inverse function shown by the dotted curve in Figure 4.34a.

EXAMPLE 4.46 (Cont.)

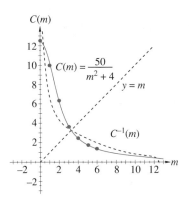

FIGURE 4.34a

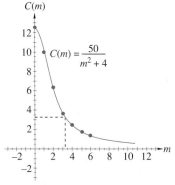

FIGURE 4.34b

(b) We will use the graph of the original function to determine $C^{-1}(3.08)$. Move up the vertical axis for the graph of $C(m)$ until you reach the point that is approximately at 3.08 on the vertical axis. Draw a horizontal line from this point until it reaches the graph of C. Draw a vertical line until it crosses the horizontal axis. This should be near 3.5 as shown in Figure 4.34b. Thus, $C^{-1}(3.08) \approx 3.5$ and we conclude that the pollutant was collected about 3.5 mi from the factory.

(c) As described in bp Section 4.4, we will let $y = \dfrac{50}{m^2 + 4}$. Solving for m, we obtain

$$m = \sqrt{\frac{50}{y} - 4}, \text{ for } y \leq 12.5. \text{ Thus, } C^{-1}(m) = \sqrt{\frac{50}{m} - 4}, \text{ for } m \leq 12.5.$$

(d) Substituting 3.08 in the formula for C^{-1}, we obtain

$$C^{-1}(3.08) = \sqrt{\frac{50}{3.08} - 4} \approx 3.50$$

Thus, the formula and the graph give us approximately the same answers.

Exercise Set 4.6

The equations in Exercises 1–10 all have at least one root between −10 and 10. Write each equation in the form $y = f(x)$. Graph each equation to find the approximate value of the roots. If possible, use a graphing calculator or a computer graphing program to help you.

1. $2x + 5 = 0$

2. $5x - 9 = 0$

3. $x^2 - 9 = 0$

4. $4x^2 - 10 = 0$

5. $x^2 = 5x$

6. $x^2 = 4x - 3$

7. $x^2 + 5x - 3 = 0$

8. $3x^2 + 5x - 3 = 0$

9. $20x^2 + 21x = 54$

10. $x^4 - 4x^3 - 25x^2 + x + 6 = 0$

Use the horizontal line text to determine whether or not the functions graphed in Exercises 11–14 have inverse functions.

11.

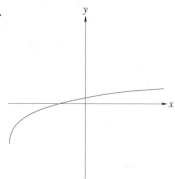

13.

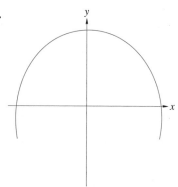

12.

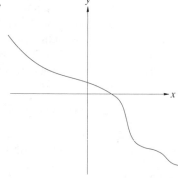

14.

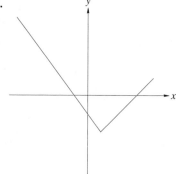

Each of the functions in Exercises 15–18 have inverses. Sketch the graph of the inverse of each function.

15.

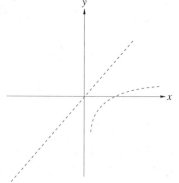

16.

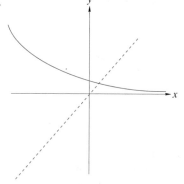

17.

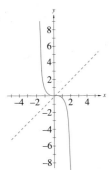

18.

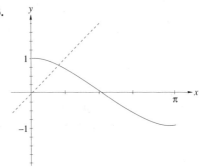

Solve Exercises 19–22.

19. *Automotive technology* Study the following table. It shows that if the antifreeze content of a coolant is increased, the boiling point of the coolant is also increased.

Percent antifreeze in coolant	0	10	20	30	40	50	60	70	80	90	100
Boiling temperature °F	210	212	214	218	222	228	236	246	258	271	330

  (a) What percent of antifreeze will be in the coolant if the boiling point of the system is 218°F?
  (b) What percent of antifreeze will be in the coolant if the boiling point of the system is 236°F?
  (c) What percent of antifreeze will be in the coolant if the boiling point of the system is 250°F?

20. *Environmental science* The population P of a certain species of animal depends on the number n of a smaller animal on which it feeds, with

$$P(n) = 5\sqrt{n} - 10$$

  (a) Determine the inverse function for P.
  (b) If the population of P is 5, how many of the small animals are there?
  (c) Graph the inverse function of P.

21. *Construction* This table contains the results of a series of drillings used to determine the depth of the bedrock at a building site. Drillings were taken along a straight line down the middle of the lot, where the building will be placed. In the table, x is the distance from the front of the parking lot and y is the corresponding depth. Both x and y are given in feet.

x	0	20	40	60	80	100	120	140	160
y	33	35	40	45	42	38	46	40	48

These ordered pairs have been plotted on the following graph. Does this graph have an inverse function?

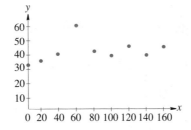

22. *Environmental science* Suppose a cost-benefit model is given by

$$C(x) = \frac{6.4x}{100 - x}$$

where $C(x)$ is the cost in millions of dollars for removing x percent of a given pollutant.
  (a) What is the cost of removing 30% of the pollutant?
  (b) What is the cost of removing 60% of the pollutant?
  (c) What is the inverse function of C?
  (d) If a community can only spend $12,000,000, what percent of the pollutant can be removed?
  (e) Graph the given function and its inverse function.

▤ CHAPTER 4 REVIEW

Important Terms and Concepts

Cartesian coordinate system
Composite function
Dependent variable
Domain
Function
Horizontal line test
Independent variable
Intercepts
Inverse function
Linear equation

Quadrants
Range
Rectangular coordinate system
Relation
Slope
Vertical line test
x-axis
x-intercept
y-axis
y-intercept

Review Exercises

For Exercises 1–6 (a) graph each of the relations, (b) determine the domain, range, x-intercept, and y-intercept of each relation, (c) use the vertical line test to determine if each relation is a function, (d) use the horizontal line test to determine if each function has an inverse function, and (e) graph each inverse function that exists.

1. $y = 8x - 7$

2. $y = 2x^2 - 4$

3. $y = \sqrt{x} - 3$

4. $y = \frac{1}{2}x^3 + 2$

5. $y = x^2 - 2x$

6. $y = x^2 + 4$

In Exercises 7–12, given the function $f(x) = 4x - 12$, determine the following.

7. $f(0)$

8. $f(-2)$

9. $f(3)$

10. $f(a)$

11. $f(a - 2)$

12. $f(x + h)$

In Exercises 13–19, given the function $g(x) = \dfrac{x^2 - 9}{x^2 + 9}$, determine the following.

13. $g(0)$

14. $g(3)$

15. $g(-3)$

16. $g(-2)$

17. $g(4)$

18. $g(-5)$

19. Use your values from Exercises 13–18 to graph $g(x)$. From your graph, determine the zeros of g.

In Exercises 20–33, let $f(x) = 4x - 12$ and $g(x) = \dfrac{x^2 - 9}{x^2 + 9}$. Determine the following.

20. $(f + g)(x)$

21. $(f + g)(3)$

22. $(f - g)(x)$

23. $(f - g)(-2)$

24. $(f \cdot g)(x)$

25. $(f \cdot g)(0)$

26. $(f/g)(x)$

27. $(f/g)(5)$

28. $(g/f)(x)$

29. $(g/f)(2)$

30. $(f \circ g)(x)$

31. $(f \circ g)(4)$

32. $(g \circ f)(x)$

33. $(g \circ f)(3)$

Each of the equations in Exercises 34–37 has a root between -10 and 10. Write each equation in the form $y = f(x)$. Graph each equation to find the approximate value of the roots.

34. $4x + 7y = 0$

35. $x^2 - 20 = 0$

36. $2x^2 + 10x + 4 = 0$

37. $8x^3 - 20x^2 - 34x + 21 = 0$

Solve Exercises 38–41.

38. *Business* The manager of a videotape store has found that n videotapes can be sold if the price is $P(n) = 35 - \dfrac{n}{20}$ dollars.
  (a) What price should be charged in order to sell 101 videotapes? 350 videotapes? 400 videotapes?
  (b) Find an expression for the revenue from the sale of n vidoetapes, where revenue = demand × price.
  (c) How much revenue can be expected if 101 videotapes are sold? if 350 are sold? if 400 are sold?

39. *Business* A videotape store has learned that the function R(n) $= 30n - \dfrac{n^2}{20}$ is a good predictor of its revenue, in dollars, from the sale of n tapes. The cost of operating the store is given by $C(n) = 550 + 10n$.
  (a) If the profit P is given by $P(n) = R(n) - C(n)$, what is the profit function?
  (b) How much profit will the store make if it sells 30 videotapes? 100 videotapes? 150 videotapes? 300 videotapes? 400 videotapes?

40. *Medical technology* A measure of cardiac output can be determined by injecting a dye into a vein near the heart and measuring the concentration of the dye. In a normal heart, the concentration of the dye is given by the function

$$h(t) = -0.02t^4 + 0.2t^3 - 0.3t^2 + 3.2t$$

where t is the number of seconds since the dye was injected. Set up a partial table of values for $0 \le t \le 10$, and sketch the graph of this function.

41. *Automotive technology* The distance s, in feet, needed to stop a car traveling v mph is given by

$$s(v) = 0.04v^2 + v$$

  (a) Set up a partial table of values for $s(v)$ with $0 \le v \le 70$.
  (b) What was the velocity of a car that took 265 ft to stop?
  (c) Sketch the graph of s and s^{-1} on the same set of axes.

≡ CHAPTER 4 TEST

Use the function $f(x) = 7x - 5$ in Exercises 1–2, and determine the indicated value.

1. $f(-2)$ **2.** $f(3 - a)$

In Exercises 3–4, let $g(x) = \dfrac{x^2 - 2x - 15}{x + 3}$ and determine the indicated value.

3. $g(0)$ **4.** $g(5)$

5. (a) Graph $h(x) = \dfrac{1}{2}\sqrt{x + 4} - 3$.
    (b) What is the domain of h?
    (c) What is the range of h?
    (d) What is the x-intercept of h?
    (e) What is the y-intercept of h?

In Exercises 6–11, let $f(x) = 3x - 15$ and $g(x) = \dfrac{x - 5}{x + 5}$, and determine the indicated value.

6. $(f + g)(x)$ **8.** $(f \cdot g)(x)$ **10.** $(f \circ g)(x)$
7. $(f - g)(x)$ **9.** $(f/g)(x)$ **11.** $(g \circ f)(x)$

5

Systems of Linear Equations and Determinants

The ability to schedule the use of trucks is important to gaining the best use of each vehicle and for the company to make a profit. In Section 5.3, you will see how linear equations can be used to determine these factors.

Courtesy of Ruby Gold

Many technical problems require us to consider the effects of several conditions and variables simultaneously. We often need to use more than one equation to show how these variables are related. When this happens, we need to find the solutions that satisfy all of these equations.

For example, in order to determine how many computers can be made using several parts, we have to consider how many of each part are available and the number needed for each computer.

In Chapter 2, we introduced the idea of solving equations. In Chapter 4 we showed how we could use graphs to help find an equation's roots. In this chapter, we will use our algebraic skills to solve linear equations and to solve systems of two or more linear equations.

≡ 5.1
LINEAR EQUATIONS

You may remember that a linear equation is the equation of a straight line. In general, an equation is a **linear equation** if each term contains only one variable, to the first power, or the term is a constant.

EXAMPLE 5.1

The equation $4x + 5 = 25$ is a linear equation in one variable, x. We learned how to solve this equation in Chapter 2. The solution is $x = 5$.

EXAMPLE 5.2

The equation $4x - 5y = 3$ is a linear equation in two variables, x and y. We learned how to solve this equation in Chapter 4. The solution to this equation is all the points that are on the line $y = \frac{4}{5}x - \frac{3}{5}$.

EXAMPLE 5.3

The equation $9x + 2y - 7z = 4$ is a linear equation in three variables x, y, and z. We have not learned how to solve an equation of this type.

A linear equation can have any number of variables. The previous examples show linear equations in 1, 2, and 3 variables. But, a linear equation could just as easily have 4 or 14 variables.

In the last chapter, we found that the slope of a line that went through two points (x_1, y_1) and (x_2, y_2) was given by the formula $m = \dfrac{y_2 - y_1}{x_2 - x_1}$. The slope tells us how steep the line is and whether the line is rising or falling.

Point-Slope Equation of a Line

If we know the slope of a line and we know one of the points on that line, we can then determine an equation for the line. Any point on the line is of the form (x, y). Suppose we know that a specific point (x_1, y_1) is on the line and that the slope of the line is m. (See Figure 5.1.) From the equation for the slope we know that

$$m = \frac{y - y_1}{x - x_1}$$

or $\qquad m(x - x_1) = y - y_1$

Rewriting this equation gives us the point-slope form of a linear equation.

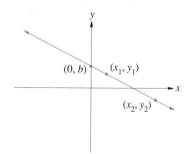

FIGURE 5.1

The point-slope form of a linear equation

If (x_1, y_1) is a point on a line and the slope of the line is m, then

$$y - y_1 = m(x - x_1)$$

is known as the **point-slope equation** of a straight line.

EXAMPLE 5.4

Find the equation of the line through the point (4, 5) with a slope of $\frac{2}{3}$.

Solution   We are told that the slope is $\frac{2}{3}$ and that a point on the line is $(4, 5)$. So we have $m = \frac{2}{3}$, $x_1 = 4$, and $y_1 = 5$. Putting these values in the point-slope form of a linear equation, we get $y - 5 = \frac{2}{3}(x - 4)$.

EXAMPLE 5.5

Find the equation of the line through the points (2, 3) and (5, 9).

Solution   We are not given the slope, but since we have two points we can find it. The slope is

$$m = \frac{9 - 3}{5 - 2} = \frac{6}{3} = 2$$

The point $(2, 3)$ is on the line, so we can let $x_1 = 2, y_1 = 3$, and using the point-slope equation of a line, we get

$$y - 3 = 2(x - 2).$$

We do not have to use the point $(2, 3)$. We can use any known point on the line. If we had used the point $(5, 9)$, then $x_1 = 5, y_1 = 9$, and we would get the equation

$$y - 9 = 2(x - 5)$$

We can show that this is equivalent to the previous equation.

Slope-Intercept Equation of a Line

We learned in Chapter 4 that the y-intercept is the point where the graph crosses the y-axis. The x-coordinate of this point is 0, and if the y-intercept is at b, then the point $(0, b)$ is on the line, as shown in Figure 5.1. If the slope of the line is m, then $y - b = m(x - 0)$. This simplifies to the following equation.

> **The slope-intercept form of a linear equation**
>
> The **slope-intercept form** of the equation for a line is
>
> $$y = mx + b$$
>
> where m is the slope and b is the y-intercept.

Notice that you can easily tell the slope and the y-intercept by looking at a linear equation written in the slope-intercept form.

EXAMPLE 5.6

Find the equation of the straight line with a slope of -7 and a y-intercept of 4.

Solution   Here $m = -7$ and $b = 4$. Using the slope-intercept form, we get $y = -7x + 4$.

EXAMPLE 5.7

What are the slope and y-intercept of the line $2x - 5y - 10 = 0$?

Solution   If we solve this equation for y, it will then be in the slope-intercept form of the line. We can then determine the answers by looking at the equation for the line.

$$2x - 5y - 10 = 0$$

$$-5y = -2x + 10$$

$$y = \frac{2}{5}x - 2$$

This is now in the slope-intercept form for the line, $y = mx + b$. The coefficient of x is the slope and the constant is the y-intercept. So the slope, m, is $\frac{2}{5}$ and the y-intercept, b, is -2.

If we wanted to graph the line in Example 5.7, we would need a second point. Select a value for x and solve for y. For example, if $x = 10$, then $y = \frac{2}{5}(10) - 2 = 4 - 2 = 2$. So, the point $(10, 2)$ is on this line.

We could also let $y = 0$ and find the x-intercept. From the original equation we have $2x - 5(0) - 10 = 0$ or $2x - 10 = 0$ or $x = 5$. This means that the point $(5, 0)$ is also on this line. Plotting these points and drawing the line through them produces the graph in Figure 5.2.

A **horizontal line** is parallel to the x-axis and has a slope of 0. This means that any horizontal line can be written as $y = 0 \cdot x + b = b$ or simply $y = b$. Notice that all points on a horizontal line have the same y-coordinate, b.

A **vertical line** has an undefined slope. This means that a vertical line cannot be written in slope-intercept form. But, all points on a vertical line have the same x-coordinate. If we call this x-coordinate a, we can then write the equation of this vertical line as $x = a$.

Horizontal and Vertical Lines

A horizontal line is parallel to the x-axis, has a slope of 0, and can be written as $y = b$.

A vertical line is perpendicular to the x-axis, has an undefined slope, and can be written as $x = a$.

EXAMPLE 5.8

Graph the lines $x = 5$ and $y = -3$.

Solution   The line $x = 5$ is a vertical line with x-intercept 5. The line $y = -3$ is a horizontal line with y-intercept -3. These two lines are graphed in Figure 5.3.

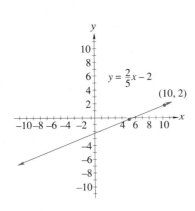

FIGURE 5.2   FIGURE 5.3

Exercise Set 5.1

In Exercises 1–6, determine the slope of the line through the given pair of points.

1. $(2, 5), (3, 8)$

2. $(4, 7), (-2, 1)$

3. $(1, 8), (5, 3)$

4. $(0, 4), (5, 0)$

5. $(9, 3), (2, -7)$

6. $(-6, 1), (2, -5)$

In Exercises 7–14, determine an equation in point-slope form for the line that satisfies the given data.

7. $m = 4$, point: $(5, 3)$

8. $m = -3$, point: $(-6, 1)$

9. $m = \frac{2}{3}$, point: $(1, -5)$

10. $m = \frac{3}{2}$, point: $(0, 5)$

11. $m = -\frac{5}{3}$, point: $(2, 0)$

12. $m = -\frac{3}{4}$, point: $(-3, -1)$

13. points: $(1, 5)$ and $(-3, 2)$

14. points: $(-5, 6)$ and $(1, -6)$

In Exercises 15–18, determine an equation in slope-intercept form for the line that satisfies the given information.

15. $m = 2, b = 4$

16. $m = -3, b = 5$

17. $m = 5, b = -3$

18. $m = -4, b = -2$

In Exercises 19–26, rewrite each equation in slope-intercept form, find the slope m and y-intercept b, and sketch the graph.

19. $y - 3x = 6$

20. $2x - y = 5$

21. $2y - 5x = 8$

22. $2x - 3y = 9$

23. $5x - 2y - 10 = 0$

24. $4y - 3x - 4 = 0$

25. $x = -3y + 7$

26. $3x = 5y - 6$

Solve Exercises 27–30.

27. *Machine technology* A grinding machine operates at 1 780 rev/min. The surface speed s in cm/s is given by the formula $s = \dfrac{1780\pi d}{60}$, where d is the diameter in cm. What is the slope of this equation? (Assume that d is on the horizontal axis.)

28. *Meteorology* The relationship between the Fahrenheit and Celsius temperatures is linear. If the Fahrenheit temperatures are put on the horizontal or x-axis and the Celsius temperatures are put on the vertical or y-axis, then two points are $(-40, -40)$ and $(32, 0)$.
   (a) What is the slope of this line?
   (b) What is the y-intercept?
   (c) Write the equation in point-slope form and in slope-intercept form.
   (d) Graph the line.

29. *Physics* A spring coil has an unstretched or natural length of L_0, and requires a force F of kx to stretch it x units beyond its natural length. The letter k represents a constant known as the **spring constant**. The distance the spring is stretched, x, is equal to $L - L_0$, where L is the length of the spring when it is stretched.
   (a) Write an equation in slope-intercept form for the force F in terms of k, L, and L_0.
   (b) If $k = 4.5$ and $L_0 = 6$ cm, write an equation in slope-intercept form for the force F in terms of L.
   (c) Graph the equation in (b).

30. *Physics* The pressure P at a depth h in a liquid depends on the density of the liquid, ρ. In a certain liquid at 4 ft, the pressure is 250 lb/ft^2. At 9 ft the pressure is 562.5 lb/ft^2. Write an equation in point-slope form for the pressure in terms of the depth.

≡ 5.2

GRAPHICAL AND ALGEBRAIC METHODS FOR SOLVING TWO LINEAR EQUATIONS IN TWO VARIABLES

In this section, we will begin to look at methods for solving a system of simultaneous linear equations. Simultaneous linear equations are equations containing the same variables such as

$$2x + y = 5$$

$$4x - y = 1$$

Our task in this section is to determine all the points, or ordered pairs, that these two equations have in common. We will begin by looking at a way to use graphing to help find these common points. Since we are looking for a common point of these two lines, this point will be where the two lines intersect if we graph each line. A graph of these two lines is shown in Figure 5.4.

As you can see, the lines appear to meet at point $(1, 3)$. A quick check of both equations will show that point $(1, 3)$ is on both lines. To check substitute $x = 1$ and $y = 3$ in the first equation. We obtain

$$2(1) + 3 = 2 + 3 = 5$$

Substituting $x = 1$ and $y = 3$ in the second equation, produces

$$4(1) - 3 = 4 - 3 = 1$$

But, as we saw in Chapter 4, graphical methods are not always accurate ways to determine the roots to an equation. Graphical methods are also not very accurate ways to determine the common solutions of simultaneous equations. What we need are some algebraic methods for solving a system of equations. We will learn two methods in this section. Both methods involve solving for one of the variables by eliminating the other variable. These are called **elimination methods**.

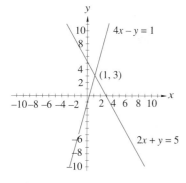

FIGURE 5.4

Substitution Method

The first elimination method involves elimination by substitution. It is generally called the **substitution method**.

Substitution method for solving a system of linear equations

To use the substitution method to solve a system of linear equations:
1. Solve one equation for one of the variables.
2. Substitute the solution from Step 1 into the other equation and solve for the remaining variable.
3. Substitute the value from Step 2 into the equation from Step 1 and solve for the other variable.

In the substitution method, we change two equations in two variables into one equation in one variable. The next two examples show how to use the substitution method to solve systems of two equations in two variables.

EXAMPLE 5.9

Use the substitution method to solve the system of equations

$$\begin{cases} 2x + y = 5 & (1) \\ 4x - y = 1 & (2) \end{cases}$$

Solution   We will solve the first equation for y.

$$y = 5 - 2x \qquad (3)$$

Substitute this solution for y in equation (2). Equation (2) becomes $4x - (5 - 2x) = 1$. Solve this equation for x.

$$4x - 5 + 2x = 1$$

$$6x = 6$$

$$x = 1$$

We then substitute this solution for x in equation (3) and find

$$y = 5 - 2(1)$$

$$y = 3$$

The solution is $(1, 3)$, which was the same answer we got by graphing.

EXAMPLE 5.10

Use the substitution method to solve the system of equations.

$$\begin{cases} -2x + 2y = 5 & (1) \\ x + 6y = 1 & (2) \end{cases}$$

EXAMPLE 5.10 (Cont.)

Solution   This time we will solve equation (2) for x and get

$$x = 1 - 6y \qquad (3)$$

Substituting this value for x in equation (1) we get

$$-2(1 - 6y) + 2y = 5$$
$$-2 + 12y + 2y = 5$$
$$-2 + 14y = 5$$
$$14y = 7$$
$$y = \frac{7}{14} = \frac{1}{2}$$

Replacing the y in equation (3) with $\frac{1}{2}$ we get

$$x = 1 - 6\left(\frac{1}{2}\right)$$
$$x = 1 - 3$$
$$x = -2$$

The solution appears to be $\left(-2, \frac{1}{2}\right)$.

To be certain that we have the correct solution, we should substitute $\left(-2, \frac{1}{2}\right)$ into the original equations—equations (1) and (2)—and see if this solution satisfies both of these equations. It is very important that you always check your work by using the *original* equations. If you made any errors, you might not detect them unless you check your work in the original problem.

Addition Method

The second algebraic method for solving a system of linear equations is normally called the **addition method**. Technically, its name is the **elimination method by addition and subtraction**.

Addition method for solving a system of linear equations

To use the addition method to solve a system of linear equations:

Add or subtract the two equations in order to eliminate one of the variables.

It is sometimes necessary to multiply the original equations by a constant before it is possible to eliminate one of the variables by adding or subtracting the equations.

The next two examples will show how the addition method is used. These are the same examples that we worked with when using the substitution method.

EXAMPLE 5.11

Use the addition method to solve the system

$$\begin{cases} 2x + y = 5 & (1) \\ 4x - y = 1 & (2) \end{cases}$$

Solution   Equation (1) has a $(+y)$ term and equation (2) has a $(-y)$ term. If we add equations (1) and (2), the new equation will not have a y term.

$$\begin{array}{r} 2x + y = 5 \\ 4x - y = 1 \\ \hline \end{array}$$

adding $6x \quad\quad = 6$ $(1) + (2)$

or $x = 1$ (3)

Substituting this value of x into equation (1) we get

$$2(1) + y = 5$$
$$y = 3$$

So, the solution is $x = 1$ and $y = 3$ or the ordered pair $(1, 3)$.

EXAMPLE 5.12

Use the addition method to solve the system

$$\begin{cases} -2x + 2y = 5 & (1) \\ x + 6y = 1 & (2) \end{cases}$$

Solution   Equations (1) and (2) do not have any terms that are equal so we will have to multiply at least one equation by a constant. If we multiply equation (2) by 2, the x-term will become $2x$, which is the additive inverse of the x term in equation (1). After multiplication, equation (2) becomes

$$2x + 12y = 2 \tag{3}$$

and adding equations (1) and (3) we get

$$\begin{array}{r} -2x + 2y = 5 \\ 2x + 12y = 2 \\ \hline \end{array}$$

$$\begin{array}{r} (1) \\ (3) \end{array}$$

adding $14y = 7$ $(1) + (3)$

or $y = \dfrac{1}{2}$ (4)

Substituting this value for y into equation (2) we get

$$x + 6\left(\frac{1}{2}\right) = 1$$
$$x + 3 = 1$$
$$x = -2$$

Thus, we have found $x = -2$, $y = \frac{1}{2}$, and the solution is $\left(-2, \frac{1}{2}\right)$.

You may get very strange looking results when you attempt to solve a system of equations. For example, sometimes all the variables vanish. Consider the next example.

EXAMPLE 5.13

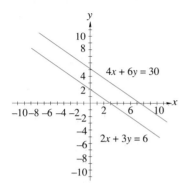

FIGURE 5.5

Use the addition method to solve the system

$$\begin{cases} 2x + 3y = 6 & (1) \\ 4x + 6y = 30 & (2) \end{cases}$$

Solution If we multiply equation (1) by -2 the terms containing the variable x will be additive inverses of each other. This multiplication makes equation (1) into

$$-4x - 6y = -12 \qquad (3)$$

and adding

$$\underline{4x + 6y = 30} \qquad (2)$$

$$0 = 18 \qquad (3) + (2)$$

Now we know that $0 \neq 18$, so something must be wrong. If we check our work, there do not appear to be any errors. Let's graph these equations. The graph in Figure 5.5 indicates the problem. The lines are parallel. They will never intersect, so there is no solution to this system of equations.

When two lines are parallel they will not intersect. Since there is no solution, the equations in the system are said to be **inconsistent.** When you try to solve a system of linear equations that is inconsistent, you will get an untrue equation. In Example 5.13, this equation was $0 = 18$.

In Example 5.14, we examine another kind of system of linear equations in which all the variables vanish.

EXAMPLE 5.14

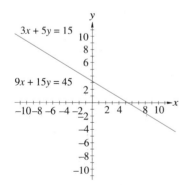

FIGURE 5.6

Solve the system of linear equations

$$\begin{cases} 3x + 5y = 15 & (1) \\ 9x + 15y = 45 & (2) \end{cases}$$

Solution Multiplying equation (1) by -3, we get

$$-9x - 15y = -45 \qquad (3)$$

and adding this to equation (2) we have the following:

$$-9x - 15y = -45 \qquad (3)$$

$$\underline{9x + 15y = 45} \qquad (2)$$

$$0 = 0 \qquad (3) + (2)$$

It is certainly true that $0 = 0$. But, how will that help us solve this system of equations? Once again, we will turn to graphing to help us solve this system. (See Figure 5.6.) The graphs of these two equations are exactly the same. Thus, there are

EXAMPLE 5.14 (Cont.)

an unlimited number of solutions to both equations. In fact, any ordered pairs that satisfy one of the equations will satisfy the other.

If the graphs of two equations coincide, we say that the equations are **dependent**. In the case of dependent equations, every solution to one equation will be a solution to the other equation. When you try to solve a system of dependent equations, you will get an equality of two constants. In Example 5.14, this was $0 = 0$.

If we solve Equation (1) in Example 5.14 for y, we get $y = -\frac{3}{5}x + 3$. (You get the same result if you solve Equation (2) for y.) Hence, every ordered pair (a, b) of the form $(a, -\frac{3}{5}a + 3)$ is a solution of the given system.

Most of the systems of equations we will work with are consistent. A system of linear equations is **consistent** if it has exactly one point as the solution.

In this section, we looked at three methods for solving systems of linear equations—the graphical method, the substitution method, and the elimination method. In the next section, we will learn about a new way of working with numbers that will help us find an easier method for solving systems of linear equations.

Exercise Set 5.2

In Exercises 1–8, graphically solve each system of equations. Estimate each answer to the nearest tenth if necessary. Check your answers, but remember that your estimate may not check exactly.

1. $\begin{cases} x + y = 6 \\ x - y = 2 \end{cases}$

2. $\begin{cases} x + y = 8 \\ 2x - y = 1 \end{cases}$

3. $\begin{cases} 2x + 3y = 15 \\ 4x - 4y = 10 \end{cases}$

4. $\begin{cases} 3x + 2y = 2 \\ 2x - 6y = -39 \end{cases}$

5. $\begin{cases} 3x + 5y = 32 \\ 10x - 5y = -32 \end{cases}$

6. $\begin{cases} -5x + 11y = 11 \\ 5x + 2y = -11 \end{cases}$

7. $\begin{cases} 5x + 10y = 0 \\ 3x - 4y = 11 \end{cases}$

8. $\begin{cases} 4x + y = -8 \\ 3x + 2y = 0 \end{cases}$

In Exercises 9–16, use the substitution method to solve each system of equations.

9. $\begin{cases} y = 3x - 4 \\ x + y = 8 \end{cases}$

10. $\begin{cases} x = -2y + 12 \\ x + y = 5 \end{cases}$

11. $\begin{cases} y = -2x - 2 \\ 3x + 2y = 0 \end{cases}$

12. $\begin{cases} x = 7 + 2y \\ 3x + 4y = 1 \end{cases}$

13. $\begin{cases} 2x + 5y = 6 \\ x - y = 10 \end{cases}$

14. $\begin{cases} 3x - 2y = 5 \\ -7x + 4y = -7 \end{cases}$

15. $\begin{cases} 2x + 3y = 3 \\ 6x + 4y = 15 \end{cases}$

16. $\begin{cases} 2x + 2y = -3 \\ 4x + 9y = 5 \end{cases}$

In Exercises 17–24, use the addition method to solve each system of equations.

17. $\begin{cases} x + y = 9 \\ x - y = 5 \end{cases}$

18. $\begin{cases} 2x + 3y = 5 \\ -2x + 5y = 3 \end{cases}$

19. $\begin{cases} -x + 3y = 5 \\ 2x + 7y = 3 \end{cases}$

20. $\begin{cases} 3x - 2y = 8 \\ 5x + y = 9 \end{cases}$

21. $\begin{cases} 3x - 2y = -15 \\ 5x + 6y = 3 \end{cases}$

22. $\begin{cases} 2x - 3y = 11 \\ 6x - 5y = 13 \end{cases}$

23. $\begin{cases} 3x - 5y = 37 \\ 5x - 3y = 27 \end{cases}$

24. $\begin{cases} x + \dfrac{1}{2}y = 7 \\ 4x - 2y = 5 \end{cases}$

In Exercises 25–32, solve each system of equations by either the substitution method or the addition method. Graph each system of equations.

25. $\begin{cases} 2x + 3y = 5 \\ x - 2y = 6 \end{cases}$

26. $\begin{cases} 2x - 3y = -14 \\ 3x + 2y = 44 \end{cases}$

27. $\begin{cases} 8x + 3y = 13 \\ 3x + 2y = 11 \end{cases}$

28. $\begin{cases} 6x + 12y = 7 \\ 8x - 15y = -1 \end{cases}$

29. $\begin{cases} 10x - 9y = 18 \\ 6x + 2y = 1 \end{cases}$

30. $\begin{cases} 4x - 5y = 7 \\ -8x + 10y = -30 \end{cases}$

31. $\begin{cases} x - 9y = 0 \\ \dfrac{x}{3} = 2y + \dfrac{1}{3} \end{cases}$

32. $\begin{cases} 5x + 3y = 7 \\ \dfrac{3}{2}x - \dfrac{3}{4}y = 9\dfrac{1}{4} \end{cases}$

Solve Exercises 33–44.

33. *Land management*   The perimeter of a rectangular field is 36 km. The length of the field is 8 km longer than the width. What are the length and width of the field? (Hint: To find the length and width of this rectangular field, you need to solve this system of linear equations where L represents the length of the field and w the width.)

$$\begin{cases} 2L + 2w = 36 \\ L = w + 8 \end{cases}$$

34. *Land management*   The perimeter of a rectangular field is 72 mi. The length of the field is 9 mi longer than the width. What are the length and width of the field?

35. *Land management*   The perimeter of a rectangular field is 45 km. The length is 3 times the width. What are the length and width of the field?

36. *Land management*   The perimeter of a field in the shape of an isosceles triangle is 96 yd. The length of each of the two equal sides is $1\frac{1}{2}$ times the length of the third side. What are the lengths of the three sides of this field?

37. *Petroleum technology*   Two different gasohol mixtures are available. One mixture contains 5% alcohol and the other, 13% alcohol. In order to determine how much of each mixture should be used to get 10 000 L of gasohol containing 8% alcohol, you would solve the following equations:

$$\begin{cases} x + y = 10\ 000 \\ 0.05x + 0.13y = (0.08)(10\ 000) \end{cases}$$

where x is the number of liters of 5% gasohol mixture and y is the number of liters of 13% gasohol mixture. Determine the number of liters of each mixture that are needed.

38. *Petroleum technology*   Two different gasohol mixtures are available. One mixture contains 4% alcohol and the other, 12% alcohol. How much of each mixture should be used to get 20 000 L of gasohol containing 9% alcohol?

39. *Physics*   Two forces are applied to the ends of a beam. The force on one end is 8 kg and the force at the other end is not known. The unknown force is 5 m from the centroid; we do not know the distance of the 8-kg force from the centroid. If an additional force of 4 kg is applied to the 8-kg force, the unknown force must be increased by 3.2 kg for equilibrium to be maintained. Find the unknown mass and distance. (See Figure 5.7.)

FIGURE 5.7

40. *Automotive technology*   A 12-L cooling system is filled with 25% antifreeze. How many liters must be replaced with 100% antifreeze to raise the strength to 45% antifreeze?

In Exercises 41–43, use the following information.

The current that flows in each branch of a complex circuit can be found by applying **Kirchhoff's rules** to the circuit. The first rule applies to the junction of three or more wires as in Figure 5.8. The second applies to loops or closed paths in the circuit.

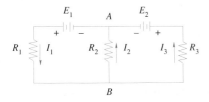

FIGURE 5.8

Rule 1:   The sum of the currents that flow into a junction is equal to the sum of the currents that flow out of the junction. In Figure 5.8, this means that $I_1 = I_2 + I_3$.

Rule 2:   The sum of the voltages around any closed loop equals zero. In Figure 5.8, for the left loop this means that $E_1 = I_1 R_1 + I_2 R_2$ and for the right loop that $E_2 = R_2 I_2 - R_3 I_3$.

41. *Electronics*   Find the currents in the three resistors of the circuit shown in 5.8, given that $E_1 = 8$ V, $E_2 = 5$ V, $R_1 = 3\ \Omega$, $R_2 = 5\ \Omega$, and $R_3 = 6\ \Omega$. [Use $E_1 = I_1 R_1 + I_2 R_2$ and $E_2 = R_2 I_2 - R_3 (I_1 - I_2)$.]

42. *Electronics*   In Figure 5.8, if $E_1 = 10$ V, $E_2 = 15$ V, $R_1 = 2\ \Omega$, $R_2 = 4\ \Omega$, and $R_3 = 8\ \Omega$, find I_1, I_2, and I_3.

43. *Electronics*   In Figure 5.9, we have $I_2 = I_1 + I_3$, $E_1 = R_1 I_1 + R_2 I_2$, and $E_2 = R_3 I_3 + R_2 I_2$. If $E_1 = 6$ V, $E_2 = 10$ V, $R_1 = 8\ \Omega$, $R_2 = 4\ \Omega$, and $R_3 = 7\ \Omega$, find I_1, I_2, and I_3.

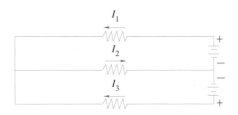

FIGURE 5.9

   44. One computer technique for solving systems of equations is called the **Gauss-Seidel method**. To apply this method, consider a system of two simultaneous linear equations.

$$\begin{cases} ax + by + c = 0 & (1) \\ dx + ey + f = 0 & (2) \end{cases}$$

Solve equation (1) for x and solve equation (2) for y to get

$$\begin{cases} x = -\dfrac{by + c}{a} & (3) \\ y = -\dfrac{dx + f}{e} & (4) \end{cases}$$

Now, guess a value for y. Since this is the first guess for y, we will call it y_1. Substitute the value y_1 in equation (3) to get a first value for x. (We will call this first x-value, x_1.) Substitute x_1 in equation (4) to get y_2. Substitute y_2 in equation (3) to get x_2, and so on. How long do you keep this up? You might continue until two consecutive values of x, say x_n and x_{n+1} are the same and until $y_n = y_{n+1}$.

Write a program to solve two linear equations in two variables using the Gauss-Seidel method. Test your program on the equations in Exercise 15. If, when you run the program, the values of x and y get large, then solve the first equation for y and the second for x.

≡ 5.3
ALGEBRAIC METHODS FOR SOLVING THREE LINEAR EQUATIONS IN THREE VARIABLES

In Section 5.2 we learned to solve a system of two linear equations in two variables by a graphical method and by two algebraic methods of elimination. In this section, we will expand these algebraic techniques to allow us to solve a system of three linear variables. These methods will be used later to allow us to solve n equations with n variables.

The graph of a linear equation in three variables is a plane. Graphing a system of three equations in three variables requires the ability to graph three planes and their intersections. We cannot graph three planes so that we can tell where they intersect. As a result, we will not consider graphical solutions to a system of three equations in three variables.

Substitution Method

We will solve the same system of equations using both elimination methods. The first example uses the substitution method.

EXAMPLE 5.15

Solve this system of equations using the substitution method.

$$\begin{cases} x + 2y - 2z = 3 & (1) \\ 2x - y + 3z = -5 & (2) \\ 4x - 3y + z = 7 & (3) \end{cases}$$

Solution   If we solve equation (3) for z we get

$$z = 3y - 4x + 7 \tag{4}$$

Substituting this value of z in equation (2) changes it to

$$2x - y + 3(3y - 4x + 7) = 2x - y + 9y - 12x + 21 = -5$$

Combining terms results in

$$-10x + 8y = -26 \tag{5}$$

Solving equation (5) for y produces

$$8y = 10x - 26$$

$$\text{or} \qquad y = \frac{5}{4}x - \frac{13}{4} \tag{6}$$

Substituting the value for z from equation (4) and the value for y from equation (6) in equation (1) we get

$$x + 2\left(\frac{5}{4}x - \frac{13}{4}\right) - 2(3y - 4x + 7) = 3$$

$$x + \frac{5}{2}x - \frac{13}{2} - 6y + 8x - 14 = 3$$

$$\frac{23}{2}x - 6y = \frac{47}{2} \tag{7}$$

EXAMPLE 5.15 (Cont.)

Substituting the value of y from equation (6) into equation (7) results in

$$\frac{23}{2}x - 6\left(\frac{5}{4}x - \frac{13}{4}\right) = \frac{47}{2}$$

$$\frac{23}{2}x - \frac{15}{2}x + \frac{39}{2} = \frac{47}{2}$$

$$\frac{8}{2}x = \frac{8}{2}$$

or $1x = 1$

Using this value of $x = 1$ in equation (6) produces

$$y = \frac{5}{4} - \frac{13}{4} = -\frac{8}{4} = -2$$

Finally, using $x = 1$ and $y = -2$ in equation (4) we get

$$z = 3(-2) - 4(1) + 7 = -3$$

So, $x = 1$, $y = -2$, and $z = -3$.

Addition Method

There was no particular reason to begin by solving equation (3) for z in the previous example. We could just as easily have started by solving equation (1) for x or equation (2) for y or even equation (1) for y or z. Let's solve the same system of equations using the addition method.

EXAMPLE 5.16

Solve this system of equations using the addition method.

$$\begin{cases} x + 2y - 2z = 3 & (1) \\ 2x - y + 3z = -5 & (2) \\ 4x - 3y + z = 7 & (3) \end{cases}$$

Solution   We will begin by eliminating one of the variables. When this is done, we will have two equations with two variables. Let's start by eliminating the variable z. To do this, we multiply equation (3) by 2 and add this to equation (1).

$$\begin{array}{ll} 8x - 6y + 2z = 14 & \text{(3) multiplied by 2} \\ \underline{x + 2y - 2z = 3} & \text{(1)} \\ 9x - 4y = 17 & \text{Adding to get equation (4)} \end{array}$$

Next, we multiply equation (3) by 3 and subtract equation (2) from this new equation.

$$\begin{array}{ll} 12x - 9y + 3z = 21 & \text{(3) multiplied by 3} \\ \underline{2x - y + 3z = -5} & \text{(2)} \\ 10x - 8y = 26 & \text{Subtracting to get equation (5)} \end{array}$$

EXAMPLE 5.16 (Cont.)

We now have two equations, (4) and (5), with two variables, x and y.

$$9x - 4y = 17 \tag{4}$$

$$10x - 8y = 26 \tag{5}$$

If we multiply equation (4) by 2 and subtract equation (5) from that equation, we will eliminate the variable y.

$$
\begin{array}{ll}
18x - 8y = 34 & \text{(4) multiplied by 2} \\
\underline{10x - 8y = 26} & \text{(5)} \\
8x \quad\quad = 8 & \text{Subtracting to get (6)}
\end{array}
$$

Solving equation (6) for x, we get $x = 1$. Substituting this value in (4), we get $9 - 4y = 17$ or $-4y = 8$, which simplifies to $y = -2$. Then, if we substitute $x = 1$ and $y = -2$ in equation (3), we get $4 + 6 + z = 7$, or $z = -3$. Again, we get the solution $x = 1$, $y = -2$, and $z = -3$.

As you can see, the elimination method by addition and subtraction is often an easier method to use. We will use this method again in the next example.

Application

EXAMPLE 5.17

By volume, one alloy is 70% copper, 20% zinc, and 10% nickel. A second alloy is 60% copper and 40% nickel. A third alloy is 30% copper, 30% nickel, and 40% zinc. How much of each must be mixed in order to get 1 000 mm³ of a final alloy that is 50% copper, 18% zinc, and 32% nickel?

Solution   We must first determine the equations that are needed to solve this problem. If we let a represent the volume of the first alloy in the final alloy, b the volume of the second, and c the volume of the third, then we know that the total volume of the final alloy, 1 000 mm³, is $a + b + c$.

We know that the final alloy contains 50% or 500 mm³ of copper and that this is $0.7a + 0.6b + 0.3c$. Also, 18% or 180 mm³ of the final solution is zinc, so $0.2a + 0.4c = 180$. This gives you a system of three linear equations in three variables.

$$
\begin{cases}
a + \quad b + \quad c = 1\,000 & (1) \\
0.7a + 0.6b + 0.3c = \quad 500 & (2) \\
0.2a \quad\quad + 0.4c = \quad 180 & (3)
\end{cases}
$$

We can also establish a fourth equation for the amount of nickel in the final solution, 32% or 320 mm³. This is

$$0.1a + 0.4b + 0.3c = 320 \tag{4}$$

We do not need equation (4) to solve the problem, but we can use it to check our answers.

EXAMPLE 5.17 (Cont.)

Since equation (3) does not contain variable b, we will combine equations (1) and (2) to eliminate this variable.

$$
\begin{array}{llr}
0.7a + 0.6b + 0.3c = & 500 & \text{(2)} \\
0.6a + 0.6b + 0.6c = & 600 & \text{(1) multiplied by 0.6} \\
\hline
0.1a \qquad\qquad - 0.3c = -100 & & \text{Subtract to get (5).}
\end{array}
$$

If we now multiply equation (5) by 2 and subtract this from equation (3), we will eliminate variable a.

$$
\begin{array}{llr}
0.2a + 0.4c = & 180 & \text{(3)} \\
0.2a - 0.6c = & -200 & \text{(5) multiplied by 0.2} \\
\hline
\qquad\quad c = & 380 & \text{Subtract to get } c.
\end{array}
$$

So, alloy c is 380 mm^3. Substituting this in (5) we get

$$0.1a - 0.3(380) = -100$$

$$0.1a - 114 = -100$$

$$0.1a = 14$$

$$a = 140$$

Then, substituting these values for a and c in equation (1) we get $140+b+380 = 1\,000$ or $b = 480$. The answer: we need 140 mm^3 of alloy a, 480 mm^3 of alloy b, and 380 mm^3 of alloy c. If you put these values in equation (4) you will see that they check.

Application

EXAMPLE 5.18

A trucking company has three sizes of trucks, large (L), medium (M), and small (S). The trucks are needed to move some packages, which come in three different shapes. We will call these three different shaped packages, A, B, and C. From experience, the company knows that these trucks can hold the combination of packages as shown in this chart.

	Size of Truck		
	Large	Medium	Small
Package A	12	8	0
Package B	10	5	4
Package C	8	7	6

The company has to deliver a total of 64 A packages, 77 B packages, and 99 C packages. How many trucks of each size are needed, if each truck is fully loaded?

Solution From the table we can see that the 64 A packages must be arranged with 12 on each large truck, 8 on each medium truck, and 0 on each small truck. We can

EXAMPLE 5.18 (Cont.)

write this as

$$12L + 8M = 64$$

Similarly, the B packages satisfy $10L + 5M + 4S = 77$ and the C packages satisfy $8L + 7M + 6S = 99$. Thus, we have the system

$$\begin{cases} 12L & + & 8M & & & = & 64 & \quad (1) \\ 10L & + & 5M & + & 4S & = & 77 & \quad (2) \\ 8L & + & 7M & + & 6S & = & 99 & \quad (3) \end{cases}$$

Since equation (1) does not contain variable S, we will combine equations (2) and (3) to eliminate it.

$$\begin{aligned} 30L + 15M + 12S &= 231 & \quad \text{(2) multiplied by 3} \\ 16L + 14M + 12S &= 198 & \quad \text{(3) multiplied by 2} \\ \hline 14L + M &= 33 & \quad \text{Subtracting to get (4)} \end{aligned}$$

Now multiply equation (4) by 8 and subtract equation (1), and variable M is eliminated.

$$\begin{aligned} 112L + 8M &= 264 & \quad \text{(4) multiplied by 8} \\ 12L + 8M &= 64 & \quad \text{(1)} \\ \hline 100L &= 200 & \quad \text{Subtract.} \end{aligned}$$

So, $L = \frac{200}{100} = 2$. Substituting this in (1), we get

$$24 + 8M = 64$$

$$8M = 64 - 24$$

$$8M = 40$$

$$M = 5$$

And finally, substituting these values for L and M in (2), we obtain

$$10(2) + 5(5) + 4S = 77$$

$$20 + 25 + 4S = 77$$

$$4S = 77 - 45$$

$$= 32$$

$$S = 8$$

So, a total of 2 large, 5 medium, and 8 small trucks are needed.

As with a system of two equations in two variables, it is possible to have a system of three equations with three variables that is either inconsistent or dependent. If an elimination method results in an equation of the type $0x + 0y + 0z = c$, or $0 = c$, where $c \neq 0$, then the system is **inconsistent** and has no solutions. Graphically, this would mean that the plane of one equation was parallel to the plane of another equation or the planes intersect in three pairs of parallel lines.

If the elimination method results in an equation of the type $0x + 0y + 0z = 0$, or $0 = 0$, then two or more of the equations graph the same plane and the system is **dependent**.

EXAMPLE 5.19

Solve the system

$$\begin{cases} 3x - y - z = 5 & (1) \\ x - 5y + z = 3 & (2) \\ x + 2y - z = 1 & (3) \end{cases}$$

Solution Adding equations (1) and (2) produces

$$4x - 6y = 8 \tag{4}$$

and adding equations (2) and (3) gives

$$2x - 3y = 4 \tag{5}$$

If we multiply equation (5) by 2 and subtract that result from equation (4), we obtain

$$0 = 0$$

Thus, we see that the given system of equations is dependent; and every ordered triple (a, b, c) of the form $(a, \frac{2}{3}a - \frac{4}{3}, \frac{7}{3}a - \frac{11}{3})$ is a solution of the given equation.

Exercise Set 5.3

In Exercises 1–4, use the substitution method to solve each system of equations.

1. $\begin{cases} 2x + y + z = 7 \\ x - y + 2z = 11 \\ 5x + y - 2z = 1 \end{cases}$

2. $\begin{cases} x + y + 2z = 0 \\ 2x - y + z = 6 \\ 4x + 2y + 2z = 0 \end{cases}$

3. $\begin{cases} 2x - y - z = -8 \\ x + y - z = -9 \\ x - y + 2z = 7 \end{cases}$

4. $\begin{cases} x + y + 5z = -10 \\ x - y - 5z = 11 \\ -x + y - 5z = 13 \end{cases}$

In Exercises 5–14, use the addition method to solve each system of equations. (Exercises 5–8 are the same as Exercises 1–4.)

5. $\begin{cases} 2x + y + z = 7 \\ x - y + 2z = 11 \\ 5x + y - 2z = 1 \end{cases}$

6. $\begin{cases} x + y + 2z = 0 \\ 2x - y + z = 6 \\ 4x + 2y + 2z = 0 \end{cases}$

7. $\begin{cases} 2x - y - z = -8 \\ x + y - z = -9 \\ x - y + 2z = 7 \end{cases}$

8. $\begin{cases} x + y + 5z = -10 \\ x - y - 5z = 11 \\ -x + y - 5z = 13 \end{cases}$

9. $\begin{cases} x + y + z = 2 \\ 8x - 2y + 4z = -3 \\ 6x - 4y - 3z = 3 \end{cases}$

10. $\begin{cases} x + y - z = 7 \\ 8x + 4y + 2z = 21 \\ 4x + 3y + 6z = 2 \end{cases}$

11. $\begin{cases} 3x - y - 2z = 11 \\ -x + 3y + 2z = -1 \\ 2x - 2y - 4z = 17 \end{cases}$

12. $\begin{cases} x - 2y + z = -4 \\ 2x + y + 3z = 5 \\ 6x + 3y + 12z = 6 \end{cases}$

13. $\begin{cases} 2x + 3y + 3z = 9 \\ 5x - 2y + 8z = 6 \\ 4x - y + 5z = -1 \end{cases}$

14. $\begin{cases} x + 2y + 3z = 4 \\ 2x - 3y - 4z = -1 \\ 3x - 4y + 5z = 6 \end{cases}$

Solve Exercises 15–20.

15. *Electronics* Kirchhoff's law for current states that the sum of the currents into any point equals zero. Applying this to junction A in Figure 5.10 produces the equation $I_1 - I_2 + I_3 = 0$. Kirchhoff's voltage law states that the sum of the voltages around any closed loop equals zero. Applying this first to the left loop and then the right loop in Figure 5.10 results in the equations $6I_1 + 6I_2 = 18$ and $6I_2 + I_3 = 14$. What are the values of the currents I_1, I_2, and I_3? (Note that electromotive force E equals current I times resistance or $E = IR$.)

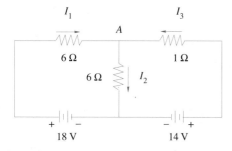

FIGURE 5.10

16. *Electronics* Applying Kirchhoff's laws to Figure 5.11 produces the following equations.

$$\begin{cases} I_1 + I_2 - I_3 = 0 \\ 3I_1 - 5I_2 - 10 = 0 \\ 5I_2 + 6I_3 - 5 = 0 \end{cases}$$

Find the currents associated with I_1, I_2, and I_3.

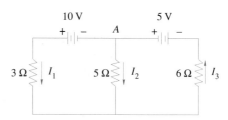

FIGURE 5.11

17. The standard equation for a circle is $x^2 + y^2 + ax + by + c = 0$. A circle passes through the points $P(5, 1)$, $Q(-2, -6)$, and $R(-1, -7)$. When the x- and y-coordinates for point P are put in the standard equation, it becomes $5^2 + 1^2 + a \cdot 5 + b \cdot 1 + c = 0$, or $5a + b + c + 26 = 0$. Use the coordinates of Q and R to obtain two more versions of the standard equation for this circle and then solve your system of equations for a, b, and c.

18. Another circle passes through the points $S(4, 16)$, $T(-6, -8)$, and $U(11, -1)$. Find the values of a, b, and c.

19. *Transportation* A trucking company has three sizes of trucks, large (L), medium (M), and small (S). Experience has shown that the large truck can carry 7 of container A, 6 of container B, and 4 of container C. The medium truck can carry 6 of A, 3 of B, and 2 of C and the small truck can carry 8 of A, 1 of B, and 2 of C. How many trucks of the three sizes are needed to deliver 64 of container A, 33 of B, and 26 of C?

20. *Petroleum technology* Three crude oils are to be mixed and loaded aboard a supertanker that can carry 450 000 tonnes (metric tons, t). The crudes contain the following percentages of light-, medium-, and heavy-weight oils:

	Light	Medium	Heavy
Crude oil A	10%	20%	70%
Crude oil B	30%	40%	30%
Crude oil C	43%	44%	13%

How many tonnes of each crude should be mixed so that the new mixtures contain 24% light-, 32% medium-, and 44% heavy-weight oils?

≡ 5.4
DETERMINANTS

We've learned to solve two equations in two variables and to solve three equations in three variables. Many problems result in having to solve four equations in four variables or five equations in five variables.

Whenever you have to solve more than two equations in two variables, the substitution and the addition methods become very long and difficult. There are other methods that are easier, especially when used with a calculator or a computer. This section gives the background material for one of the easier methods. We begin by giving the definition of a determinant.

If a, b, c, and d are any four real numbers, then the symbol

$$\begin{vmatrix} a & b \\ c & d \end{vmatrix}$$

is called a 2 × 2 **determinant** or a **determinant of the second order**. The numbers a, b, c, and d are called the **elements** or **entries** of the determinant. The value of a determinant is the number $ad - cb$, so we have the following statement.

Evaluating a 2 × 2 Determinant

If $a, b, c,$ and d are any four real numbers, then

$$\begin{vmatrix} a & b \\ c & d \end{vmatrix} = ad - cb$$

As a memory aid, you might want to draw the diagonals of the determinant.

 $= ad - cb$

EXAMPLE 5.20

Evaluate the determinant $\begin{vmatrix} 7 & 5 \\ 2 & 6 \end{vmatrix}$.

Solution $\begin{vmatrix} 7 & 5 \\ 2 & 6 \end{vmatrix} = 7(6) - 2(5) = 42 - 10 = 32$

EXAMPLE 5.21

Evaluate the determinant $\begin{vmatrix} -6 & -2 \\ 3 & 4 \end{vmatrix}$.

Solution $\begin{vmatrix} -6 & -2 \\ 3 & 4 \end{vmatrix} = (-6)(4) - 3(-2) = -24 - (-6) = -18$

Determinants have many useful properties and can be used to solve simultaneous equations. In the next section, we will learn how to use determinants to solve equations. In this section, we will become familiar with their properties. We will use them later.

The determinants that we have used have all been 2×2 (read 2-by-2), or second order determinants. A 3×3 (3-by-3) determinant or a **determinant of the third order** is represented by the symbol

$$\begin{vmatrix} a_1 & a_2 & a_3 \\ b_1 & b_2 & b_3 \\ c_1 & c_2 & c_3 \end{vmatrix}$$

where $a_1, a_2, a_3, b_1, b_2, b_3, c_1, c_2$, and c_3 are any real numbers. We will learn how to evaluate a 3×3 determinant later in this section.

As we mentioned, the numbers $a_1, a_2, \ldots, c_3$, which form the determinant, are called the elements or entries. The **rows** are numbered from top to bottom.

$\begin{vmatrix} a & b \\ c & d \end{vmatrix}$ Row 1 $\quad \begin{vmatrix} a_1 & b_1 & c_1 \\ a_2 & b_2 & c_2 \\ a_3 & b_3 & c_3 \end{vmatrix}$ Row 2, Row 3

The **columns** are numbered from left to right.

$\begin{vmatrix} a & b \\ c & d \end{vmatrix}$

Column 1, Column 2, Column 3

$$\begin{vmatrix} a_1 & b_1 & c_1 \\ a_2 & b_2 & c_2 \\ a_3 & b_3 & c_3 \end{vmatrix}$$

A determinant has the same number of rows and columns.

The **main** or **principal diagonal** of a determinant is the diagonal from upper left to lower right. For example, in the determinant

$$\begin{vmatrix} a & b & c \\ d & e & f \\ g & h & i \end{vmatrix}$$

the entries on the main diagonal are a, e, and i.

A determinant is in **triangular form** if all entries below, or above, the main diagonal are zero. For example, the following determinants are both in triangular form.

$$\begin{vmatrix} 2 & -3 & 5 \\ 0 & 9 & -7 \\ 0 & 0 & \frac{1}{2} \end{vmatrix} \text{ and } \begin{vmatrix} 2 & 0 & 0 \\ -5 & -\frac{3}{5} & 0 \\ 3 & \pi & 17 \end{vmatrix}$$

Property 1 for Determinants

If you interchange (or swap) any two rows or any two columns of a determinant, you change its sign.

EXAMPLE 5.22

If $\begin{vmatrix} 2 & 3 \\ 1 & 9 \end{vmatrix} = 15$ then $\begin{vmatrix} 1 & 9 \\ 2 & 3 \end{vmatrix} = -15$, because the first and second rows were swapped. Thus, $\begin{vmatrix} 2 & 3 \\ 1 & 9 \end{vmatrix} = -\begin{vmatrix} 1 & 9 \\ 2 & 3 \end{vmatrix}$.

EXAMPLE 5.23

If $\begin{vmatrix} 3 & 1 & 5 \\ 2 & 0 & -2 \\ -5 & -1 & 4 \end{vmatrix} = -14$, then $\begin{vmatrix} 5 & 1 & 3 \\ -2 & 0 & 2 \\ 4 & -1 & -5 \end{vmatrix} = 14$, because the first and third columns were swapped.

Property 2 for Determinants

If you multiply every entry in one row or column of a determinant by a constant k, the result is the same as multiplying the value of the determinant by k.

EXAMPLE 5.24

If you multiply each entry in the second column of the determinant $\begin{vmatrix} 2 & 3 \\ 1 & 9 \end{vmatrix} = 15$ by 5 you get $\begin{vmatrix} 2 & 15 \\ 1 & 45 \end{vmatrix} = 75$, and $75 = 5 \times 15$.

EXAMPLE 5.25

If each element in the first row of the determinant $\begin{vmatrix} 3 & 1 & 5 \\ 2 & 0 & -2 \\ -5 & -1 & 4 \end{vmatrix} = -14$ is multiplied by -3, the result is the determinant $\begin{vmatrix} -9 & -3 & -15 \\ 2 & 0 & -2 \\ -5 & -1 & 4 \end{vmatrix}$. It has a value of $(-3)(-14) = 42$.

Property 2 also means that if all the elements of a single row or column have a common factor, then it can be factored out of the entire determinant.

EXAMPLE 5.26

In the determinant $\begin{vmatrix} 2 & 9 & 8 \\ 5 & 7 & 12 \\ 9 & -2 & -16 \end{vmatrix}$, all the entries in column 3 have a common

factor of 4. So this determinant can be rewritten as $4\begin{vmatrix} 2 & 9 & 2 \\ 5 & 7 & 3 \\ 9 & -2 & -4 \end{vmatrix}$.

Property 3 for Determinants

If you add a constant multiple of the entries in any one row (or column) of a determinant to the corresponding entries in any other row (or column), the value of the determinant will not be changed.

EXAMPLE 5.27

In the determinant $\begin{vmatrix} 7 & 3 \\ -5 & 1 \end{vmatrix}$, if you multiply each number in the first row by 4 and add these new numbers to the second row, the value of the determinant will not change.

$$\begin{vmatrix} 7 & 3 \\ -5 & 1 \end{vmatrix} = 7 - (-15) = 22$$

$$\begin{vmatrix} 7 & 3 \\ 28-5 & 12+1 \end{vmatrix} = \begin{vmatrix} 7 & 3 \\ 23 & 13 \end{vmatrix} = 91 - 69 = 22$$

EXAMPLE 5.28

Consider $\begin{vmatrix} 3 & 1 & 5 \\ 2 & 0 & -2 \\ -5 & -1 & 4 \end{vmatrix} = -14$. Multiply the second row by 2 and add this

product to the third row. The result is

$$\begin{vmatrix} 3 & 1 & 5 \\ 2 & 0 & -2 \\ 4-5 & 0-1 & -4+4 \end{vmatrix} = \begin{vmatrix} 3 & 1 & 5 \\ 2 & 0 & -2 \\ -1 & -1 & 0 \end{vmatrix} = -14$$

Property 4 for Determinants

If any two rows (or columns) of a determinant are the same, its value is zero.

If you combine Properties 2 and 4, you can see that if any row (or column) is a multiple of any other row (or column) then the value of the determinant is zero.

EXAMPLE 5.29

(a) $\begin{vmatrix} 2 & 3 & -1 \\ 9 & 8 & 7 \\ 2 & 3 & -1 \end{vmatrix} = 0$, because rows 1 and 3 are the same.

(b) $\begin{vmatrix} 6 & 9 & 9 \\ 1 & 3 & 3 \\ -7 & 4 & 4 \end{vmatrix} = 0$, because columns 2 and 3 are the same.

(c) $\begin{vmatrix} 8 & 4 & -2 \\ -2 & -1 & \frac{1}{2} \\ 7 & 16 & -3 \end{vmatrix} = 0$, because each element in row 1 is -4 times each element in row 2.

Property 5 for Determinants

If a determinant is in triangular form, its value is the product of its main diagonal.

EXAMPLE 5.30

$$\begin{vmatrix} 9 & 3 & -1 & 2 \\ 0 & 4 & -2 & 7 \\ 0 & 0 & -\frac{1}{3} & 17 \\ 0 & 0 & 0 & -1 \end{vmatrix} = 9(4)\left(-\frac{1}{3}\right)(-1) = 12.$$

Minors

Each element in a determinant has a minor associated with it. The **minor** of a given element is the determinant that is formed by deleting the row and column in which the element lies.

EXAMPLE 5.31

Consider the determinant $\begin{vmatrix} 2 & 4 & -4 \\ 5 & 8 & 1 \\ -6 & -3 & 7 \end{vmatrix}$.

(a) The minor of the first element, 2, is found by crossing out the first row and first column.

$$\begin{vmatrix} \cancel{2} & \cancel{4} & \cancel{-4} \\ \cancel{5} & 8 & 1 \\ \cancel{-6} & -3 & 7 \end{vmatrix}$$

The minor of the element 2 is $\begin{vmatrix} 8 & 1 \\ -3 & 7 \end{vmatrix} = 56 + 3 = 59$.

EXAMPLE 5.31 (Cont.)

(b) The minor of the element 8 is found by crossing out the second row and the second column.

$$\begin{vmatrix} 2 & 4 & -4 \\ -5 & 8 & 1 \\ -6 & -3 & 7 \end{vmatrix}$$

The minor of 8 is $\begin{vmatrix} 2 & -4 \\ -6 & 7 \end{vmatrix} = 14 - 24 = -10.$

(c) The minor of 1 is found by crossing out the second row and third column.

$$\begin{vmatrix} 2 & 4 & -4 \\ -5 & 8 & 1 \\ -6 & -3 & 7 \end{vmatrix}$$

The minor of 1 is $\begin{vmatrix} 2 & 4 \\ -6 & -3 \end{vmatrix} = -6 + 24 = 18.$

Cofactors

Each element also has a **cofactor**. The value of the cofactor is determined by first adding the number of the row and the number of the column where the element is located. If this sum is even, the value of the cofactor is equal to the value of the minor for that element. If the sum is odd, the value of the cofactor is then -1 times the value of the minor for that element.

EXAMPLE 5.32

We will consider the same determinant we used in Example 5.31.

(a) The cofactor of 2 is 59. The element 2 is in row 1 and column 1. Since $1 + 1$ is an even number and the minor for 2 had a value of 59 (See Example 5.31a), its cofactor has a value of 59.

(b) The element 8 is in row 2, column 2. Since $2 + 2 = 4$, an even number, the cofactor of 8 is -10, the same as its minor.

(c) The element 1 is in row 2, column 3 and $2 + 3 = 5$, an odd number. The value of the minor of 1 is 18, so the cofactor of 1 is $(-1)18 = -18.$

Evaluating a Determinant

To evaluate any determinant

(1) select any row or column of the determinant,

(2) multiply each element of that row or column by its cofactor, and

(3) add the results.

Notice that this procedure allows you to choose which row or column you want to use to evaluate the determinant.

EXAMPLE 5.33

Evaluate $\begin{vmatrix} 2 & 4 & -4 \\ 5 & 8 & 1 \\ -6 & -3 & 7 \end{vmatrix}$.

Solution   We will evaluate this determinant by using, or expanding, on the second row. (We picked this row because we found the cofactor of the last two elements in Example 5.32.) The cofactor of 5 is $(-1)\begin{vmatrix} 4 & -4 \\ -3 & 7 \end{vmatrix} = (-1)(28 - 12) = -16.$ The cofactor of 8 is -10 and the cofactor of 1 is -18. So, the value of this determinant is

$$\begin{vmatrix} 2 & 4 & -4 \\ 5 & 8 & 1 \\ -6 & -3 & 7 \end{vmatrix} = 5(-16) + 8(-10) + 1(-18)$$

$$= -80 + -80 + -18$$

$$= -178$$

EXAMPLE 5.34

Evaluate $\begin{vmatrix} 2 & 1 & -5 & -2 \\ 1 & 0 & 4 & 5 \\ 7 & 2 & 1 & 0 \\ 5 & 0 & -3 & 2 \end{vmatrix}$.

Solution   Since we can select any row or column, let's select the one that has the most zeros. The second column of this determinant has two zeros, so we shall use column 2 to evaluate it.

$$\begin{vmatrix} 2 & 1 & -5 & -2 \\ 1 & 0 & 4 & 5 \\ 7 & 2 & 1 & 0 \\ 5 & 0 & -3 & 2 \end{vmatrix} = -1\begin{vmatrix} 1 & 4 & 5 \\ 7 & 1 & 0 \\ 5 & -3 & 2 \end{vmatrix} + 0\begin{vmatrix} 2 & -5 & -2 \\ 7 & 1 & 0 \\ 5 & -3 & 2 \end{vmatrix}$$

$$- 2\begin{vmatrix} 2 & -5 & -2 \\ 1 & 4 & 5 \\ 5 & -3 & 2 \end{vmatrix} + 0\begin{vmatrix} 2 & -5 & -2 \\ 1 & 4 & 5 \\ 7 & 1 & 0 \end{vmatrix}$$

$$= -1\begin{vmatrix} 1 & 4 & 5 \\ 7 & 1 & 0 \\ 5 & -3 & 2 \end{vmatrix} - 2\begin{vmatrix} 2 & -5 & -2 \\ 1 & 4 & 5 \\ 5 & -3 & 2 \end{vmatrix}$$

Now we need to evaluate each of these 3×3 determinants. We will evaluate the first determinant along column three.

$$\begin{vmatrix} 1 & 4 & 5 \\ 7 & 1 & 0 \\ 5 & -3 & 2 \end{vmatrix} = 5\begin{vmatrix} 7 & 1 \\ 5 & -3 \end{vmatrix} - 0\begin{vmatrix} 1 & 4 \\ 5 & -3 \end{vmatrix} + 2\begin{vmatrix} 1 & 4 \\ 7 & 1 \end{vmatrix}$$

$$= 5[7(-3) - 5(1)] - 0 + 2[1(1) - 7(4)]$$

$$= 5[-21 - 5] - 0 + 2[1 - 28]$$

EXAMPLE 5.34 (Cont.)

$$= 5(-26) + 2(-27)$$

$$= -130 - 54 = -184$$

The other 3×3 determinant is evaluated along row one:

$$\begin{vmatrix} 2 & -5 & -2 \\ 1 & 4 & 5 \\ 5 & -3 & 2 \end{vmatrix} = 2\begin{vmatrix} 4 & 5 \\ -3 & 2 \end{vmatrix} - (-5)\begin{vmatrix} 1 & 5 \\ 5 & 2 \end{vmatrix} + (-2)\begin{vmatrix} 1 & 4 \\ 5 & -3 \end{vmatrix}$$

$$= 2[4(2) - (-3)5] + 5[1(2) - 5(5)] - 2[1(-3) - 5(4)]$$

$$= 2(8 + 15) + 5(2 - 25) - 2(-3 - 20)$$

$$= 2(23) + 5(-23) - 2(-23)$$

$$= 46 - 115 + 46$$

$$= -23$$

So, the value of the original determinant is

$$-1\begin{vmatrix} 1 & 4 & 5 \\ 7 & 1 & 0 \\ 5 & -3 & 2 \end{vmatrix} - 2\begin{vmatrix} 2 & -5 & -2 \\ 1 & 4 & 5 \\ 5 & -3 & 2 \end{vmatrix} = -1(-184) - 2(-23)$$

$$= 184 + 46$$

$$= 230$$

Needless to say, the last example was quite long. We will next show how to evaluate this same determinant using a calculator.

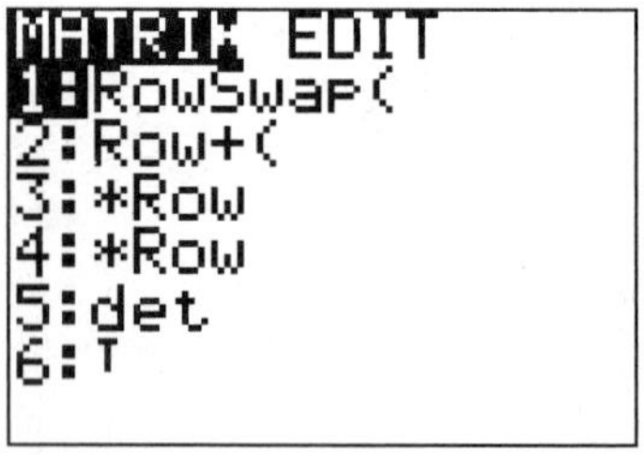

Evaluating Determinants on a Calculator

Many of today's scientific calculators allow you to quickly (and accurately) evaluate a determinant. The following procedure describes how this is done on a Texas Instrument's *TI-81* graphics calculator.

To evaluate a determinant, you need to use the matrix features of this calculator. (We will learn more about a matrix in Chapter 17.)

To begin, turn on your calculator and press the $\boxed{\text{MATRX}}$ key to access the matrix operations. You should see the display shown in Figure 5.12 on your screen. Across the top of the screen are the titles of two menus. The MATRIX menu is highlighted. It deals with several operations on matrices. We will use this menu later. The other menu, EDIT, allows us to define a matrix. In order to evaluate a determinant, we enter it in the calculator as a matrix. To do this, you first select the EDIT menu by pressing the $\boxed{\blacktriangleright}$ key. The screen should now look like Figure 5.13. (What you actually see depends on whether your calculator has previously been used to enter a matrix.)

We are now ready to define the determinant we want to evaluate. The calculator can store three matrices, which it names [A], [B], and [C]. Indicate which matrix you want to define. Press the number shown at the left of each of these matrix names. We

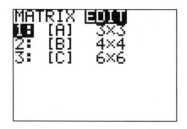

FIGURE 5.12

FIGURE 5.13

will name our matrix [B]. Pressing a $\boxed{2}$ results in the screen display shown in Figure 5.14.

The blinking cursor is on the row dimension of the determinant. We must either accept or change the dimension on the top line. To accept the number, press $\boxed{\text{ENTER}}$. To change the number, enter the number of rows in your determinant, and then press $\boxed{\text{ENTER}}$. We want to evaluate a 4×4 determinant, so press 4 $\boxed{\text{ENTER}}$ 4 $\boxed{\text{ENTER}}$. You should now see the display in Figure 5.15.

We are now ready to enter the elements of our determinant. Start with the element in the upper left-hand corner, 2, and press 2 $\boxed{\text{ENTER}}$. Next, enter the second element in the top row, 1, by pressing 1 $\boxed{\text{ENTER}}$, the third element in the top row, -5, by pressing $\boxed{(-)}$ 5 $\boxed{\text{ENTER}}$, and so on until all 16 elements have been entered. When you complete a row, start with the left-most element in the next row.

When you have finished entering the elements, return to the home screen by pressing $\boxed{\text{2nd}}$ $\boxed{\text{QUIT}}$.

Now, you are ready to evaluate this determinant. If you named your matrix [B], then press $\boxed{\text{MATRX}}$ 5 $\boxed{\text{2nd}}$ $\boxed{\text{[B]}}$ $\boxed{\text{ENTER}}$. (The 5 selects operation #5 on the MATRIX menu. Operation #5, written "det," is an abbreviation for "determinant.") The result, shown in Figure 5.16, shows that the value of the determinant of [B] is 230.

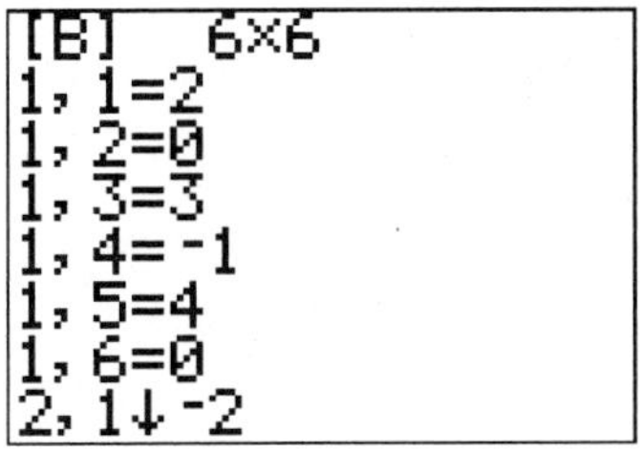

```
[B]      6×6
1,1=2
1,2=0
1,3=3
1,4=-1
1,5=4
1,6=0
2,1↓-2
```

FIGURE 5.14

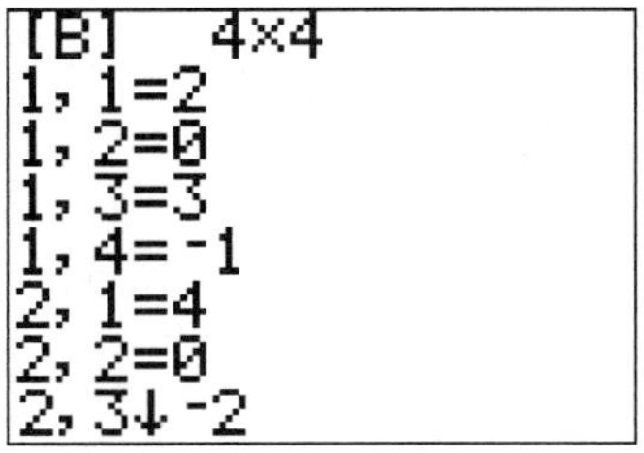

```
[B]      4×4
1,1=2
1,2=0
1,3=3
1,4=-1
2,1=4
2,2=0
2,3↓-2
```

FIGURE 5.15

```
det[B]
              230
```

FIGURE 5.16

Exercise Set 5.4

In Exercises 1–8, evaluate each determinant.

1. $\begin{vmatrix} 2 & 3 \\ 4 & -1 \end{vmatrix}$

2. $\begin{vmatrix} 4 & -5 \\ 6 & 2 \end{vmatrix}$

3. $\begin{vmatrix} 5 & -1 \\ 8 & 1 \end{vmatrix}$

4. $\begin{vmatrix} 9 & -1 \\ -2 & 3 \end{vmatrix}$

5. $\begin{vmatrix} 4 & 7 \\ -3 & 1 \end{vmatrix}$

6. $\begin{vmatrix} -1 & -4 \\ 0 & 1 \end{vmatrix}$

7. $\begin{vmatrix} -9 & \dfrac{1}{2} \\ 2 & 1 \end{vmatrix}$

8. $\begin{vmatrix} 6 & -3 \\ \dfrac{1}{3} & -\dfrac{2}{3} \end{vmatrix}$

In each of Exercises 9–14, use the determinant $\begin{vmatrix} 4 & 2 & -1 \\ 3 & 7 & -4 \\ -2 & 1 & 1 \end{vmatrix}$ and for the indicated position find (a) the element in that position, (b) the minor of that element, and (c) the cofactor of that element.

9. Row 1, Column 2

10. Row 2, Column 1

11. Row 3, Column 2

12. Row 2, Column 3

13. Row 3, Column 1

14. Row 1, Column 3

15. Evaluate the determinant used in Exercises 9–14 by expanding on the second row.

16. Evaluate the determinant used in Exercises 9–14 by expanding on the third column.

In Exercises 17–24, use Properties 1–5 to evaluate each determinant.

17. $\begin{vmatrix} 2 & -5 & 8 \\ 16 & 4 & 3 \\ 2 & -5 & 8 \end{vmatrix}$

18. $\begin{vmatrix} 1 & -2 & 3 \\ 2 & -4 & 6 \\ 0 & 3 & 5 \end{vmatrix}$

19. $\begin{vmatrix} 1 & 2 & 5 \\ -1 & -2 & 3 \\ 3 & 6 & 15 \end{vmatrix}$

20. $\begin{vmatrix} 2 & 4 & 3 \\ 0 & 1 & 19 \\ 0 & 0 & -3 \end{vmatrix}$

21. $\begin{vmatrix} 5 & 2 & 3 \\ 4 & -5 & -6 \\ -2 & 5 & -9 \end{vmatrix}$

22. $\begin{vmatrix} 9 & 3 & 3 \\ -2 & 0 & 6 \\ 6 & 3 & 3 \end{vmatrix}$

23. $\begin{vmatrix} 4 & 3 & 9 \\ -4 & -6 & 16 \\ 2 & 3 & 2 \end{vmatrix}$

24. $\begin{vmatrix} 9 & 18 & -1 \\ 0 & -7 & 5 \\ 0 & 0 & 2 \end{vmatrix}$

Evaluate each of the determinants in Exercises 25–28. You will probably need to use a calculator. (Note: The *TI–81* does not allow computations inside a matrix; the Casio *fx–7700G* does.)

25. $\begin{vmatrix} 2.41 & 3.5 & -5.3 \\ 6.02 & 7.01 & -4.26 \\ 9.1 & -3.2 & 4.5 \end{vmatrix}$

26. $\begin{vmatrix} \sqrt{3} & \dfrac{1}{2} & -5 \\ \dfrac{1}{4} & \sqrt{2} & 4 \\ \dfrac{2}{5} & 7 & -1 \end{vmatrix}$

27. $\begin{vmatrix} 0.071 & 0.069 & -1.095 \\ 0.202 & -0.420 & -0.100 \\ 0.066 & -0.303 & 2.093 \end{vmatrix}$

28. $\begin{vmatrix} \sqrt{7} & -3 & \sqrt{5} \\ -1.4 & \sqrt{6} & 2 \\ \sqrt{8} & 3 & -6 \end{vmatrix}$

In Exercises 29–32, assume that $\begin{vmatrix} a & b & c \\ r & s & t \\ x & y & z \end{vmatrix} = -5$. Find the value of each determinant.

29. $\begin{vmatrix} r & s & t \\ a & b & c \\ x & y & z \end{vmatrix}$

30. $\begin{vmatrix} s & t & r \\ b & c & a \\ y & z & x \end{vmatrix}$

31.
$$\begin{vmatrix} a & b & c \\ 3r & 3s & 3t \\ -2x & -2y & -2z \end{vmatrix}$$

32.
$$\begin{vmatrix} a & b & c \\ r+a & s+b & t+c \\ x-r & y-s & z-t \end{vmatrix}$$

Evaluate each of the determinants in Exercises 33–34.

33.
$$\begin{vmatrix} 1 & 2 & 3 & 4 \\ 5 & 0 & 7 & 0 \\ 9 & 10 & 11 & 12 \\ 13 & 14 & 15 & 16 \end{vmatrix}$$

34.
$$\begin{vmatrix} 2 & 0 & 3 & -1 \\ 4 & 0 & -2 & 5 \\ 9 & -5 & 0 & -2 \\ 0 & 3 & 1 & -6 \end{vmatrix}$$

Solve Exercises 35–37.

 35. Write a computer program to input and evaluate a second-order determinant. Test your program on the determinants in Exercises 1–8.

36. A third-order determinant can be evaluated using minors. Another method is to rewrite the determinant so that it has three rows and five columns. The fourth column is the same as column 1 and the fifth column is the same as column 2. You then multiply the three

numbers in each diagonal. You add the product if the diagonal slopes down and subtract if it slopes up. Try this method on Exercises 21–24.

 37. Write a computer program to input the elements of a third-order determinant and evaluate it using the method described in Exercise 36. Test your program on Exercises 17–28.

≡ 5.5

USING CRAMER'S RULE TO SOLVE SYSTEMS OF LINEAR EQUATIONS

In Section 5.4 we learned how to evaluate determinants. In this section we will learn how to use determinants to solve systems of linear equations. We will first work with second-order determinants.

Consider the system of linear equations

$$\begin{cases} ax + by = h \\ cx + dy = k \end{cases}$$

If we multiply the first equation by d and the second equation by b, we get

$$\begin{cases} adx + bdy = hd \\ bcx + bdy = bk \end{cases}$$

Subtracting the second equation from the first eliminates the y terms and produces the equation

$$adx - bcx = hd - bk$$

or $\qquad (ad - bc)x = hd - bk$

Now, this last equation could have been written using determinants as $\begin{vmatrix} a & b \\ c & d \end{vmatrix} x = \begin{vmatrix} h & b \\ k & d \end{vmatrix}$, and so

$$x = \frac{\begin{vmatrix} h & b \\ k & d \end{vmatrix}}{\begin{vmatrix} a & b \\ c & d \end{vmatrix}} \text{ provided that } \begin{vmatrix} a & b \\ c & d \end{vmatrix} \neq 0$$

Similarly, we can show that $y = \dfrac{\begin{vmatrix} a & h \\ c & k \end{vmatrix}}{\begin{vmatrix} a & b \\ c & d \end{vmatrix}}$.

These last two equations are cumbersome, so we'll abbreviate the determinants by letting $D = \begin{vmatrix} a & b \\ c & d \end{vmatrix}$, $D_x = \begin{vmatrix} h & b \\ k & d \end{vmatrix}$, and $D_y = \begin{vmatrix} a & h \\ c & k \end{vmatrix}$. Then we have $x = \dfrac{D_x}{D}$ and $y = \dfrac{D_y}{D}$. This is **Cramer's rule** for solving two linear equations in two variables.

Cramer's Rule for Solving a 2 × 2 Linear System

The unique solution to the linear system

$$\begin{cases} a_1 x + b_1 y = k_1 \\ a_2 x + b_2 y = k_2 \end{cases}$$

is

$$x = \frac{D_x}{D} = \frac{\begin{vmatrix} k_1 & b_1 \\ k_2 & b_2 \end{vmatrix}}{\begin{vmatrix} a_1 & b_1 \\ a_2 & b_2 \end{vmatrix}} \quad \text{and} \quad y = \frac{D_y}{D} = \frac{\begin{vmatrix} a_1 & k_1 \\ a_2 & k_2 \end{vmatrix}}{\begin{vmatrix} a_1 & b_1 \\ a_2 & b_2 \end{vmatrix}}$$

provided that

$$\begin{vmatrix} a_1 & b_1 \\ a_2 & b_2 \end{vmatrix} \neq 0$$

The determinant D is called the **coefficient determinant**, because its entries are the coefficients of the variables in the system of linear equations:

$$\begin{cases} a_1 x + b_1 y = k_1 \\ a_2 x + b_2 y = k_2 \end{cases} \qquad D = \begin{vmatrix} a_1 & b_1 \\ a_2 & b_2 \end{vmatrix}$$

The determinant D_x is obtained by replacing the first column of D (the coefficients of x) with the constants on the right in the system of equations. Similarly, the determinant D_y is obtained by replacing the second column of D (the coefficients of y) with the constants on the right in the system.

$$\begin{cases} a_1 x + b_1 y = k_1 \\ a_2 x + b_2 y = k_2 \end{cases} \qquad D_x = \begin{vmatrix} k_1 & b_1 \\ k_2 & b_2 \end{vmatrix} \qquad D_y = \begin{vmatrix} a_1 & k_1 \\ a_2 & k_2 \end{vmatrix}$$

≡ **Note**   You cannot use Cramer's rule when $D = 0$. When $D = 0$, the equations in the system are either inconsistent or dependent.

EXAMPLE 5.35

Use Cramer's rule to solve $\begin{cases} 2x - y = 9 \\ x + 5y = 21 \end{cases}$.

Solution   Here we have $D = \begin{vmatrix} 2 & -1 \\ 1 & 5 \end{vmatrix} = 11$, $D_x = \begin{vmatrix} 9 & -1 \\ 21 & 5 \end{vmatrix} = 66$, and

$D_y = \begin{vmatrix} 2 & 9 \\ 1 & 21 \end{vmatrix} = 33$, so $x = \dfrac{D_x}{D} = \dfrac{66}{11} = 6$, and $y = \dfrac{D_y}{D} = \dfrac{33}{11} = 3$.

EXAMPLE 5.36

Use Cramer's rule to solve $\begin{cases} 3x + y = 10 \\ 6x + 2y = 15 \end{cases}$.

Solution   In this system, $D = \begin{vmatrix} 3 & 1 \\ 6 & 2 \end{vmatrix} = 0$, so Cramer's rule does not apply.

Graphing these two lines would show that they are parallel, and so the system is inconsistent.

Application

EXAMPLE 5.37

A contractor mixes some aggregate with cement to make concrete. Two mixtures are on hand. One of the mixtures, which we will call Mixture A, is 40% sand and 60% aggregate. The other mixture, Mixture B, is 70% sand and 30% aggregate. How much of each should be used to get a 500-lb mixture that is 46.2% sand and 53.8% aggregate?

Solution   Let a represent the amount of Mixture A and b the amount of Mixture B used in the final mixture. The final 500-lb mixture will contain 46.2%, or 231 lb, of sand and 53.8%, or 269 lb, of aggregate. Thus, we get two equations. The first

$$0.4a + 0.7b = 231 \tag{1}$$

represents the amounts of sand from each of Mixtures A and B that are needed to make the 231 lb in the final mixture.

The second equation

$$0.6a + 0.3b = 269 \tag{2}$$

represents the amount of aggregate from Mixtures A and B needed to make the aggregate in the final mixture.

Thus, we have the system

$$\begin{cases} 0.4a + 0.7b = 231 \\ 0.6a + 0.3b = 269 \end{cases}$$

Here, $D = \begin{vmatrix} 0.4 & 0.7 \\ 0.6 & 0.3 \end{vmatrix} = -0.3$, $D_a = \begin{vmatrix} 231 & 0.7 \\ 269 & 0.3 \end{vmatrix} = -119$, and $D_b =$

$\begin{vmatrix} 0.4 & 231 \\ 0.6 & 269 \end{vmatrix} = -31$. As a result, we have $a = \dfrac{D_a}{D} = \dfrac{-119}{-0.3} \approx 396.7$ and

EXAMPLE 5.37 (Cont.)

$b = \dfrac{D_b}{D} = \dfrac{-31}{-0.3} \approx 103.3$. The final mixture should contain about 396.7 lb of Mixture A and 103.3 lbof Mixture B.

Using Cramer's Rule on a System of Three Linear Equations

To extend Cramer's rule to a system of three linear equations in three variables, you use a similar procedure, as shown in the following box.

Cramer's Rule for Solving a 3 × 3 Linear System

The unique solution to the linear system

$$\begin{cases} a_1x + b_1y + c_1z = k_1 \\ a_2x + b_2y + c_2z = k_2 \\ a_3x + b_3y + c_3z = k_3 \end{cases}$$

is

$$x = \frac{D_x}{D} \qquad y = \frac{D_y}{D} \qquad \text{and} \qquad z = \frac{D_z}{D}$$

where

$$D_x = \begin{vmatrix} k_1 & b_1 & c_1 \\ k_2 & b_2 & c_2 \\ k_3 & b_3 & c_3 \end{vmatrix}, \quad D_y = \begin{vmatrix} a_1 & k_1 & c_1 \\ a_2 & k_2 & c_2 \\ a_3 & k_3 & c_3 \end{vmatrix}, \quad D_z = \begin{vmatrix} a_1 & b_1 & k_1 \\ a_2 & b_2 & k_2 \\ a_3 & b_3 & k_3 \end{vmatrix}$$

and provided that

$$D = \begin{vmatrix} a_1 & b_1 & c_1 \\ a_2 & b_2 & c_2 \\ a_3 & b_3 & c_3 \end{vmatrix} \neq 0$$

If $D = 0$, the equations are either inconsistent or dependent.

EXAMPLE 5.38

Use Cramer's rule to solve the system of linear equations:

$$\begin{cases} 3x - 2y - z = 2 \\ -x - 3y + z = 10 \\ 2x + 3y - 2z = -14 \end{cases}$$

Solution   For this example, we have

$$D = \begin{vmatrix} 3 & -2 & -1 \\ -1 & -3 & 1 \\ 2 & 3 & -2 \end{vmatrix}$$

$$= 3\begin{vmatrix} -3 & 1 \\ 3 & -2 \end{vmatrix} + 2\begin{vmatrix} -1 & 1 \\ 2 & -2 \end{vmatrix} - 1\begin{vmatrix} -1 & -3 \\ 2 & 3 \end{vmatrix}$$

$$= 3(3) + 2(0) - 1(3)$$

$$= 6$$

EXAMPLE 5.38 (Cont.)

Since $D \neq 0$, the system is consistent, and we can solve it by Cramer's rule. Solving for D_x, D_y, and D_z, we obtain

$$D_x = \begin{vmatrix} 2 & -2 & -1 \\ 10 & -3 & 1 \\ -14 & 3 & -2 \end{vmatrix}$$

$$= 2 \begin{vmatrix} -3 & 1 \\ 3 & -2 \end{vmatrix} + 2 \begin{vmatrix} 10 & 1 \\ -14 & -2 \end{vmatrix} - 1 \begin{vmatrix} 10 & -3 \\ -14 & 3 \end{vmatrix}$$

$$= 2(3) + 2(-6) - 1(-12)$$

$$= 6 - 12 + 12$$

$$= 6,$$

$$D_y = \begin{vmatrix} 3 & 2 & -1 \\ -1 & 10 & 1 \\ 2 & -14 & -2 \end{vmatrix}$$

$$= 3 \begin{vmatrix} 10 & 1 \\ -14 & -2 \end{vmatrix} - 2 \begin{vmatrix} -1 & 1 \\ 2 & -2 \end{vmatrix} - 1 \begin{vmatrix} -1 & 10 \\ 2 & -14 \end{vmatrix}$$

$$= 3(-6) - 2(0) - 1(-6)$$

$$= -18 - 0 + 6$$

$$= -12,$$

and

$$D_z = \begin{vmatrix} 3 & -2 & 2 \\ -1 & -3 & 10 \\ 2 & 3 & -14 \end{vmatrix}$$

$$= 3 \begin{vmatrix} -3 & 10 \\ 3 & -14 \end{vmatrix} + 2 \begin{vmatrix} -1 & 10 \\ 2 & -14 \end{vmatrix} + 2 \begin{vmatrix} -1 & -3 \\ 2 & 3 \end{vmatrix}$$

$$= 3(12) + 2(-6) + 2(3)$$

$$= 36 - 12 + 6$$

$$= 30$$

So, $x = \dfrac{D_x}{D} = \dfrac{6}{6} = 1$, $y = \dfrac{D_y}{D} = \dfrac{-12}{6} = -2$, and $z = \dfrac{D_z}{D} = \dfrac{30}{6} = 5$.

Application

EXAMPLE 5.39

An applied electromotive force (emf) of 200 V produces a current of 40 μA in the simple series circuit shown in Figure 5.17. The voltage drop across resistor R_1 is 60 V more than the combined voltage drop across R_2 and R_3. The voltage drop

EXAMPLE 5.39 (Cont.)

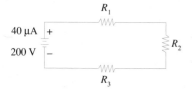

FIGURE 5.17

across R_3 is one-fourth the drop across R_2. What are the values of the three resistors in ohms?

Solution   Since the voltage drop across resistor R_1 is 60 V more than the combined voltage drop across R_2 and R_3, we have

$$V_1 = 60 + V_2 + V_3 \qquad (1)$$

We also are given

$$V_3 = \frac{1}{4} V_2 \qquad (2)$$

The current of $40 \ \mu A = 40 \times 10^{-3} \ A = 0.040 \ A$. We can use Ohm's law to find the total external resistance, R_T in the circuit.

$$E = IR_T$$

$$200 = 0.040R_T$$

$$R_T = \frac{200}{0.040}$$

$$= 5\,000$$

Thus, from Kirchhoff's laws,

$$R_1 + R_2 + R_3 = 5\,000 \qquad (3)$$

If we use Ohm's law, we can use the following equations to express the voltage drop across each resistor:

$$V_1 = 0.040R_1$$

$$V_2 = 0.040R_2$$

$$V_3 = 0.040R_3$$

Substituting these values of V_1, V_2, and V_3 in equations (1) and (2), we obtain

$$0.04R_1 = 60 + 0.04R_2 + 0.04R_3 \qquad (4)$$

$$0.04R_3 = \frac{1}{4} 0.04R_2$$

$$\text{or} \qquad 4R_3 = R_2 \qquad (5)$$

We are now ready to use Cramer's rule to solve the system of equations

$$\begin{cases} R_1 + R_2 + R_3 = 5,000 \\ 0.04R_1 = 60 + 0.04R_2 + 0.04R_3 \\ 4R_3 = R_2 \end{cases}$$

$$\text{or} \qquad \begin{cases} R_1 + R_2 + R_3 = 5,000 \\ 0.04R_1 - 0.04R_2 - 0.04R_3 = 60 \\ R_2 - 4R_3 = 0 \end{cases}$$

EXAMPLE 5.39 (Cont.)

Thus, we have

$$D = \begin{vmatrix} 1 & 1 & 1 \\ 0.04 & -0.04 & -0.04 \\ 0 & 1 & -4 \end{vmatrix} = 0.4$$

$$D_{R_1} = \begin{vmatrix} 5\,000 & 1 & 1 \\ 60 & -0.04 & -0.04 \\ 0 & 1 & -4 \end{vmatrix} = 1\,300$$

$$D_{R_2} = \begin{vmatrix} 1 & 5\,000 & 1 \\ 0.04 & 60 & -0.04 \\ 0 & 0 & -4 \end{vmatrix} = 560$$

and $\quad D_{R_3} = \begin{vmatrix} 1 & 1 & 5\,000 \\ 0.04 & -0.04 & 60 \\ 0 & 1 & 0 \end{vmatrix} = 140$

As a result, we obtain

$$R_1 = \frac{D_{R_1}}{D} = \frac{1\,300}{0.4} = 3\,250$$

$$R_2 = \frac{D_{R_2}}{D} = \frac{560}{0.4} = 1\,400$$

and $\quad R_3 = \frac{D_{R_3}}{D} = \frac{140}{0.4} = 350$

Thus, the three resistors have values of $3\,250\ \Omega$, $1\,400\ \Omega$, and $350\ \Omega$.

Exercise Set 5.5

In Exercises 1–16 use Cramer's rule, when applicable, to solve each system of linear equations.

1. $\begin{cases} 2x + y = 7 \\ 3x - 2y = -7 \end{cases}$

2. $\begin{cases} 3x + y = 3 \\ 2x - 3y = 13 \end{cases}$

3. $\begin{cases} 4x + 3y = 4 \\ 3x - 2y = -14 \end{cases}$

4. $\begin{cases} \dfrac{1}{2}x - \dfrac{2}{3}y = \dfrac{3}{4} \\[2mm] \dfrac{1}{3}x + 2y = \dfrac{5}{6} \end{cases}$

5. $\begin{cases} 1.2x + 3.7y = 9.1 \\ 4.3x - 5.2y = 8.3 \end{cases}$

6. $\begin{cases} 0.02x - 1.22y = 3.74 \\ -0.14x + 0.32y = -1.32 \end{cases}$

7. $\begin{cases} 2.5x + 3.8y = 9.3 \\ 0.5x + 0.76y = -2.44 \end{cases}$

8. $\begin{cases} 4.3x - 2.7y = -4 \\ 3.5x + 4.2y = -1 \end{cases}$

9. $\begin{cases} -5.3x + 2.1y = 4.6 \\ 6.2x - 3.1y = 6 \end{cases}$

10. $\begin{cases} 1.3x - 0.8y = 2.9 \\ 1.7x - 0.7y = 0.4 \end{cases}$

11. $\begin{cases} 6x + 2.5y = 8.2 \\ 13.8x + 5.75y = 18.86 \end{cases}$

12. $\begin{cases} 3.5x - 6.5y = 22.45 \\ 5.5x + 3.3y = -1.21 \end{cases}$

13. $\begin{cases} 4x + 3y = 2 \\ 3x - 2y = -24 \end{cases}$

14. $\begin{cases} x + 3y - z = 7 \\ 5x - 7y + z = 3 \\ 2x - y - 2z = 0 \end{cases}$

15. $\begin{cases} x - 0.5y + 1.5z = 2 \\ x + 0.5y - 3.75z = 0.25 \\ -3x - 2.5y + 4z = -0.75 \end{cases}$

16. $\begin{cases} x + y - z = -4.7 \\ 0.5x - 3.5y + 2.4z = 17.4 \\ 1.5x + 4.5y - 3.6z = -21.6 \end{cases}$

Solve Exercises 17–22.

17. Write a computer program that will allow you to input the numerical values for a system of two linear equations in two variables or a system of three linear equations in three variables, and then use Cramer's rule to solve this system.

18. *Electronics*   Applying Kirchhoff's laws to a certain circuit results in the following system of equations.
$$\begin{cases} I_1 + I_2 + I_3 = 0 \\ 8I_1 - 10I_3 = 8 \\ 6I_1 - 3I_2 = 12 \end{cases}$$
Determine I_1, I_2, and I_3.

19. *Physics*   The displacement s of an object moving in a straight line under constant acceleration from an initial position s_0 is given by the formula
$$s = s_0 + v_0t + \frac{1}{2}at^2$$
where t is elapsed time in seconds, v_0 is the initial velocity, and a the acceleration. The results of three measurements are shown in the table. Find the constants s_0, v_0, and a.

Time, t (s)	2	5	7
Displacement, s (m)	212	156.5	70.5

20. *Metallurgy*   An alloy is composed of three metals A, B, and C. The percentages of each metal are given by the following system of equations
$$\begin{cases} A + B + C = 100 \\ A - 2B = 0 \\ -4A + C = 0 \end{cases}$$
Determine the percentage of each metal in the alloy.

21. *Automotive technology*   A petroleum engineer was testing three different gasoline mixtures A, B, and C in the same car and under the same driving conditions. She noticed that the car traveled 90 km further when it used mixture B than when it used mixture A. Using the fuel C, the car traveled 130 km more than when it used fuel B. The total distance traveled was 1 900 km. Find the distance traveled on the three fuels.

22. *Construction technology*   If three cables are joined at a point and three forces are applied so the system is in equilibrium, the following system of equations results.
$$\begin{cases} \frac{6}{7}F_B - \frac{2}{3}F_C = 2{,}000 \\ -F_A - \frac{3}{7}F_B + \frac{1}{3}F_C = 0 \\ \frac{2}{7}F_B + \frac{2}{3}F_C = 1{,}200 \end{cases}$$
Determine the three forces, F_A, F_B, and F_C, measured in newtons (N).

▤ CHAPTER 5 REVIEW

Important Terms and Concepts

Addition method for solving systems linear equations
Coefficient determinant
Cofactor of determinant
Consistent equations
Cramer's rule for solving linear equations
Dependent equations

Determinant
Elements of a determinant
Elimination methods
   Addition
   Substitution
Entries of a determinant

Horizontal line
  Slope of
Kirchhoff's law
Kirchhoff's rules
Linear equation
  Point-slope form
  Slope-intercept form
Linear system of equations
  Addition method for solving
  Consistent
  Cramer's rule for solving
  Dependent
  Inconsistent

Substitution method for solving
Minor of a determinant
Point-slope form of a linear equation
Rows of a determinant
Slope
  Of a horizontal line
  Of a vertical line
Slope-intercept form of a linear equation
Substitution method for solving systems of linear equations
Vertical line
  Slope of

Review Exercises

Evaluate each of the determinants in Exercises 1–8.

1. $\begin{vmatrix} 2 & 5 \\ -3 & 6 \end{vmatrix}$

2. $\begin{vmatrix} 4 & -7 \\ 3 & -6 \end{vmatrix}$

3. $\begin{vmatrix} 5 & 9 \\ 8 & 1 \end{vmatrix}$

4. $\begin{vmatrix} 3 & -12 \\ 0 & 5 \end{vmatrix}$

5. $\begin{vmatrix} 6 & -2 \\ -4 & 3 \end{vmatrix}$

6. $\begin{vmatrix} 8 & 9 \\ -1 & -2 \end{vmatrix}$

7. $\begin{vmatrix} 9 & 2 & 1 \\ 3 & -4 & 6 \\ 7 & 2 & 1 \end{vmatrix}$

8. $\begin{vmatrix} 9 & -2 & 3 \\ 4 & -5 & 2 \\ 3 & 2 & 1 \end{vmatrix}$

In Exercises 9–12, (a) determine the slope of the line through the given pair of points and then (b) find the equations of the line in point-slope form.

9. $(-2, 3), (5, 1)$

10. $(8, -4), (3, -2)$

11. $(9, -1), (-2, 4)$

12. $(-4, -3), (2, -1)$

In Exercises 13–16, rewrite each of the equations in slope-intercept form.

13. $4x - 2y = 6$

14. $9x + 3y = 5$

15. $4x = 5y - 8$

16. $7x + 3y - 8 = 0$

In Exercises 17–18, solve each system of equations graphically.

17. $\begin{cases} 3x + y = 4 \\ 4x + y = 10 \end{cases}$

18. $\begin{cases} 6x + 2y = 5 \\ 3x - 4y = 7 \end{cases}$

In Exercises 19–22, use the substitution method to solve each system of equations.

19. $\begin{cases} 3x + y = 5 \\ 4x - y = 16 \end{cases}$

20. $\begin{cases} 3x - 2y = 4 \\ x + 8y = 3 \end{cases}$

21. $\begin{cases} 4x - y = 7 \\ 3x + 2y = 5 \end{cases}$

22. $\begin{cases} 6x + y - 5 = 0 \\ 4y - 3x = -7 \end{cases}$

In Exercises 23–28, use the addition method to solve each system of equations.

23. $\begin{cases} x - y = -2 \\ x + y = 8 \end{cases}$

24. $\begin{cases} 6x + 5y = 7 \\ 3x - 7y = 13 \end{cases}$

25. $\begin{cases} x + \dfrac{1}{2}y = 2 \\ 3x - y = 1 \end{cases}$

26. $\begin{cases} \dfrac{1}{6}x + \dfrac{1}{4}y = \dfrac{1}{3} \\ \dfrac{1}{4}x - \dfrac{1}{2}y = 1 \end{cases}$

27. $\begin{cases} 3x + 2y + 3z = -7 \\ 5x - 3y + 2z = -4 \\ 7x + 4y + 5z = 2 \end{cases}$

28. $\begin{cases} x + y + z = 2.7 \\ 6x + 7y - 5z = -8.8 \\ 10x + 16y - 3z = -6.4 \end{cases}$

In Exercises 29–34, use Cramer's rule to solve each system of equations.

29. $\begin{cases} 3x - 2y = 4 \\ 5x + 2y = 12 \end{cases}$

30. $\begin{cases} 4x + 3y = 27 \\ 2x - 5y = -19 \end{cases}$

31. $\begin{cases} 4x - 3y = 9 \\ 11x + 4y = 7 \end{cases}$

32. $\begin{cases} 2x - 5y = -8 \\ 4x + 6y = 17 \end{cases}$

33. $\begin{cases} 5x + 3y - 2z = 5 \\ 3x - 4y + 3z = 13 \\ x + 6y - 4z = -8 \end{cases}$

34. $\begin{cases} 5x - 2y + 3z = 6 \\ 6x - 3y + 4z = 10 \\ -4x + 4y - 9z = 4 \end{cases}$

Solve Exercises 35–37.

35. *Business*   A store owner makes a special blend of coffee from Colombian Supreme costing $4.99/lb and Mocha Java costing $5.99/lb. The mixture sells for $5.39/lb. If this mixture is made in 50-lb batches, how many pounds of each type should be used?

36. *Business*   A computer company make two kinds of computers. One, a personal computer (PC), uses 4 Type A chips and 11 Type B chips. The other, a business computer (BC), uses 9 Type A chips and 6 Type B chips. The company has 670 Type A chips and 1,055 Type B chips in stock. How many PC and BC computers can the company make so that all the chips are used?

37. *Business*   An office building has 146 rooms made into 66 offices. The smallest offices each have 1 room and rent of $300 per month. The middle-sized offices have 2 rooms each and rent for $520 per month. The largest offices have 3 rooms each and rent for $730 each. If the rental income is $37,160 per month, how many of each type of offices are there?

☰ CHAPTER 5 TEST

1. **(a)** Determine the slope of the line through the points $(-4, 2)$ and $(5, 6)$.
   (b) Write the equation of the line through the points in **(a)**.

2. **(a)** Rewrite the equation $6x - 5y = 12$ in slope-intercept form.
   (b) What is the slope of this line?
   (c) What is the y-intercept?
   (d) Sketch the graph of this line.

3. Use the addition method to solve the following system.

$$\begin{cases} 2x + 3y = 1 \\ -3x + 6y = 16 \end{cases}$$

4. Evaluate the determinant.

$$\begin{vmatrix} 4 & -2 \\ 5 & 6 \end{vmatrix}$$

5. Evaluate the determinant.

$$\begin{vmatrix} 4 & 2 & -1 \\ 3 & 0 & 4 \\ -3 & 1 & 1 \end{vmatrix}$$

6. Solve the following system graphically.

$$\begin{cases} 2x + 3y = 5 \\ x - 4y = -14 \end{cases}$$

7. Solve the following system by the substitution method.

$$\begin{cases} 4x + 3y = 9 \\ 2x + y = 2 \end{cases}$$

8. Use Cramer's rule to solve the following system.

$$\begin{cases} 2x - 4y = 2 \\ -6x + 8y = -9 \end{cases}$$

9. Three machine parts cost a total of $60. The first part costs as much as the other two together. The cost of the second part is $3 more than twice the cost of the third part. How much does each part cost?

6

Ratio, Proportion, and Variation

An architect needs to make some scale drawings for a rectangular building. In Example 6.14, we will see how to use proportions to determine the size of the drawing on a blueprint.

Courtesy of Ruby Gold

Ratio and proportion are two ideas that we see very often in mathematics. These ideas are basic to many natural, technical, and scientific relationships. In this chapter, we will examine the concepts of ratio and proportion and look at some of their applications.

☰ 6.1
RATIO AND PROPORTION

The ideas of ratio and proportion include ratio, rate, proportion, and continued proportion. In this section, we will look at all four of these ideas.

Ratio

A **ratio** is a comparison of two quantities. If these quantities are represented by the values a and b, then the ratio comparing a to b is written $\frac{a}{b}$ or $a : b$. A **pure ratio** compares two quantities that have the same units. A pure ratio is written without units.

EXAMPLE 6.1

What is the ratio of **(a)** 4 m to 3 m, **(b)** 5 ft to 4 yd, and **(c)** 2 m to 35 cm?

Solutions   We will write each of these as a pure ratio.

(a) $\dfrac{4 \text{ m}}{3 \text{ m}} = \dfrac{4}{3}$ or $4 : 3$

(b) $\dfrac{5 \text{ ft}}{4 \text{ yd}} = \dfrac{5 \text{ ft}}{12 \text{ ft}} = \dfrac{5}{12}$ or $5 : 12$

(c) $\dfrac{2 \text{ m}}{35 \text{ cm}} = \dfrac{200 \text{ cm}}{35 \text{ cm}} = \dfrac{200}{35} = \dfrac{40}{7}$ or $40 : 7$

Some ratios are always expressed as a quantity compared to 1, as shown in the next example.

Application

EXAMPLE 6.2

A driven gear has 26 teeth and the driving gear has 8 teeth. What is the gear ratio of the driven gear to the driving gear?

Solution $\dfrac{\text{driven gear}}{\text{driving gear}} = \dfrac{26 \text{ teeth}}{8 \text{ teeth}} = \dfrac{26}{8} = \dfrac{13}{4} = \dfrac{3.25}{1}$ or $3.25 : 1$

Application

EXAMPLE 6.3

The **specific gravity** of a substance is the ratio of its density to the density of water. The density of aluminum is 2.7 g/cm^3 and the density of water is 1 g/cm^3. What is the specific gravity of aluminum?

Solution

$$\text{specific gravity of aluminum} = \frac{\text{density of aluminum}}{\text{density of water}}$$

$$= \frac{2.7 \text{ g/cm}^3}{1 \text{ g/cm}^3}$$

$$= 2.7$$

Actually, this is 2.7 : 1, but it is usually written as 2.7.

Rate

A **rate** is a ratio that compares different kinds of units such as miles and gallons.

Application

EXAMPLE 6.4

An automobile travels 243 km in 3 h. What is the rate of speed?

Solution
$$\frac{243 \text{ km}}{3 \text{ h}} = \frac{81 \text{ km}}{1 \text{ h}} = 81 \text{ km/h}$$

The car is traveling at the rate of 81 km/h.

EXAMPLE 6.5

A crankshaft turns 9,500 revolutions (rev) in $2\frac{1}{2}$ min. How many revolutions per minute (rpm) is the crankshaft turning?

Solution
$$\frac{\text{number of revolutions}}{\text{number of minutes}} = \frac{9,500 \text{ rev}}{2\frac{1}{2} \text{ min}}$$

$$= \frac{3,800 \text{ rev}}{1 \text{ min}}$$

$$= 3,800 \text{ rpm}$$

The crankshaft is turning at the rate of 3,800 rpm.

Proportion

A statement that says two ratios are equal is called a **proportion.** If the ratio $a : b$ is equal to the ratio $c : d$, then we have the proportion $a : b = c : d$ or

$$\frac{a}{b} = \frac{c}{d}$$

which is read, "a is to b as c is to d."

The inside terms of a proportion are called the means and the outside terms are the extremes.

$$\overset{\text{means}}{\underset{\text{extremes}}{a : b = c : d}}$$

Two ratios are equal, and hence form a proportion, when the product of the means equals the product of the extremes. This is also referred to as the **cross-product.** This is stated algebraically as shown in the box.

> **Equal Ratios**
>
> Two ratios $a : b$ and $c : d$ are equal, if $ad = bc$.
>
> Stated using the "fractional notation" for ratios, this says that
>
> $$\frac{a}{b} = \frac{c}{d}, \text{ if and only if } ad = bc$$

The following diagram may help you to remember which terms to multiply and also to explain why this is called a cross-product.

$$\frac{a}{b} \bowtie \frac{c}{d}$$

EXAMPLE 6.6

What is the missing number in the proportion $5 : 7 = x : 42$?

Solution   We are given $\dfrac{5}{7} = \dfrac{x}{42}$.

The product of the means is $7x$.

The product of the extremes is $5 \cdot 42 = 210$.

Since this is a proportion, the product of the means equals the product of the extremes. So, $7x = 210$ or $x = 30$, and thus we have determined that $5 : 7 = 30 : 42$.

Application

EXAMPLE 6.7

If 74.5 L of fuel are used to drive 760 km, how many liters are needed to drive 3 754 km?

Solution   We have the rates of 74.5 L to 760 km and x L to 3 754 km. We will form a proportion where the left-hand side equals the ratio $\dfrac{\text{L}}{\text{km}}$. We must then have $\dfrac{\text{L}}{\text{km}}$ on the right-hand side. If we set these two rates equal, we have

$$\frac{74.5 \text{ L}}{760 \text{ km}} = \frac{x \text{ L}}{3\,754 \text{ km}}$$

The product of the extremes is $(74.5)(3\,754) = 279\,673$ and the product of the means is $760x$. Setting these equal to each other we get

$$760x = 279\,673$$

$$x = 367.99079$$

So, it would require about 368 L of fuel to travel 3 754 km.

Continued Proportion

A **continued proportion** is a proportion that involves six or more quantities. If there are six quantities, a continued proportion is of the form $a : b : c = x : y : z$. This is a shorthand notation for writing $\frac{a}{x} = \frac{b}{y} = \frac{c}{z}$. We will use continued proportions when we study similar figures in Section 6.2, and later when we study trigonometry.

EXAMPLE 6.8

Solve the proportion $2 : 5 : 7 = x : 32 : z$.

Solution We will solve this continued proportion in stages. Remember that $2 : 5 : 7 = x : 32 : z$ is a short way of writing $\frac{2}{x} = \frac{5}{32} = \frac{7}{z}$. We will first find x.

Working with the proportion from the two ratios on the left, we obtain the proportion $\frac{2}{x} = \frac{5}{32}$. The product of the extremes is $2 \times 32 = 64$ and is equal to the product of the means, $5x$. If $5x = 64$, then $x = 12.8$.

Next, we work with the proportion from the two ratios on the right of the continued proportion. This proportion is $\frac{5}{32} = \frac{7}{z}$ or $5z = 7(32) = 224$, and so $z = 44.8$.

Thus, the solution to the proportion $2 : 5 : 7 = x : 32 : z$ is $2 : 5 : 7 = 12.8 : 32 : 44.8$.

EXAMPLE 6.9

A 520 mm wire is to be cut into three pieces so that the ratio of the lengths is to be $6 : 4 : 3$. How long should each piece be?

Solution If we represent the length of each piece by $6x$, $4x$, and $3x$, then we have the total length of $6x + 4x + 3x = 13x$. But the length of the wire is 520 mm. Thus,

$$13x = 520$$

$$x = 40$$

The lengths must then be $6(40) = 240$, $4(40) = 160$, and $3(40) = 120$. The proportion would be $6 : 4 : 3 = 240 : 160 : 120$.

Exercise Set 6.1

Express each of the rates in Exercises 1–4 as a ratio.

1. $1.38 for 16 bolts

2. 725 rpm

3. 86 L/km

4. 236 mi in 4 h

Write each of the ratios in Exercises 5–8 as a ratio compared to 1.

5. $9 : 2$

6. $\dfrac{7}{5}$

7. $\dfrac{23}{7}$

8. $37 : 4$

Use cross-products to determine whether the pairs of ratios in Exercises 9–12 are equal.

9. $\dfrac{2}{3}, \dfrac{10}{15}$

10. $\dfrac{1}{3}, \dfrac{4}{12}$

11. $145 : 25, 29 : 5$

12. $3 : 8, 12 : 33$

Solve each of the proportions in Exercises 13–24.

13. $\dfrac{7}{8} = \dfrac{c}{32}$

14. $\dfrac{3}{b} = \dfrac{9}{24}$

15. $\dfrac{124}{62} = \dfrac{158}{d}$

16. $\dfrac{a}{3.5} = \dfrac{8}{20}$

17. $\dfrac{a}{4} = \dfrac{0.16}{0.15}$

18. $\dfrac{7.5}{10.5} = \dfrac{x}{6.3}$

19. $\dfrac{20}{b} = \dfrac{8}{5.6}$

20. $\dfrac{2.4}{10.8} = \dfrac{1.6}{d}$

21. $4 : 9 = x : 4$

22. $3 : x = x : 12$

23. $2 : 5 : 9 = x : 14 : z$

24. $a : 6 : 9 = 20 : y : 45$

Solve Exercises 25–36.

25. *Mechanics*   A driven gear has 54 teeth and a driving gear has 13 teeth. What is the gear ratio of the driven gear to the driving gear?

26. A room is 12 ft wide and 15 ft long. What is the ratio of length to width?

27. *Automotive technology*   The **steering ratio** of an automobile can be determined by dividing the total number of degrees that the steering wheel turns by the number of degrees that the wheels turn. What is the steering ratio if the steering wheel makes $4\frac{2}{3}$ complete turns while the front wheels turn 60°?

28. If the density of gold is 19 g/cm^3, what is its specific gravity?

29. *Electricity*   The **capacitance** of a capacitor is the ratio of the charge on either plate of the capacitor to the potential difference between the plates. The unit of capacitance is the farad (F) where 1 F = 1 coulomb/volt, or 1 C/V. Because the farad is such a large unit the microfarad (μF) is used where 1 μF = 10^{-6} F. If a capacitor has a charge of 5×10^{-4} C when the potential difference across the plates is 300 V, what is the capacitance of this capacitor?

30. *Optics*   The **linear magnification,** *m*, of an optical system is the ratio between the size of the image and the size of the object. What is the linear magnification of a system if an object that is 80 mm long has an image that is 20 mm long?

31. *Automotive technology*   The **compression ratio** compares the volume of a cylinder at Bottom Dead Center (BDC) to the volume at Top Dead Center (TDC). If the compression ratio is 8.7 : 1 and the volume at BDC is 96 cm^3, what is the volume at TDC?

32. *Automotive technology*   A driver added 90 mL of fuel conditioner to 20 L of fuel. How much is needed for 54 L of fuel?

33. *Physics*   It requires 80 calories of heat to change 1 g of ice to water without raising the temperature. How many calories are needed to change 785 kg of ice to water without changing the temperature?

34. *Electricity*   In a transformer, an electric current in the primary wire induces a current in the secondary wire. The ratio of turns in the windings determines the ratio of voltages across them, and is expressed as the proportion $\dfrac{V_1}{V_2} = \dfrac{N_1}{N_2}$, where *V* represents voltage and N represents the number of turns in each coil. A transformer has 100 turns in its primary winding and 750 in its secondary winding. If the primary voltage is 120 V, what is the secondary voltage?

35. The ratio of mm to inches is 25.4 to 1. How many inches are there in 88.9 mm?

36. *Physics*   The acceleration due to gravity is 9.780 39 m/s^2 or 32.087 8 ft/s^2 at the equator. At the north pole, the acceleration due to gravity is 9.832 17 m/s^2. What is the value in ft/s^2 at the north pole?

≡ 6.2
SIMILAR GEOMETRIC SHAPES

In Chapter 3, we studied geometry. As part of that study we examined similar triangles. In this section, we will reexamine similar triangles and expand our work to include other geometric figures.

Similar Triangles

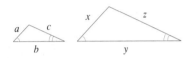

FIGURE 6.1

Two triangles that have the same shape are **similar triangles.** In similar triangles the corresponding angles are congruent and the corresponding sides are proportional. In Figure 6.1, since the triangles are similar, $\dfrac{a}{x} = \dfrac{b}{y} = \dfrac{c}{z}$ or $a : b : c = x : y : z$.

EXAMPLE 6.10

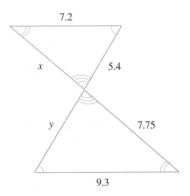

FIGURE 6.2

Find the lengths of the unknown sides of the similar triangles in Figure 6.2.

Solution   The corresponding sides of these triangles are proportional, thus $9.3 : 7.75 : y = 7.2 : x : 5.4$. Solving for x, produces

$$\frac{9.3}{7.2} = \frac{7.75}{x}$$

$$9.3x = (7.75)(7.2)$$

$$= 55.8$$

$$x = 6$$

Now, solving for y, we have the proportion

$$\frac{9.3}{7.2} = \frac{y}{5.4}$$

$$7.2y = (9.3)(5.4)$$

$$= 50.22$$

$$y = 6.975$$

Other Similar Figures

Similar figures other than triangles can be more difficult to work with. If the corresponding angles of a triangle are congruent, we know that the triangles are similar. This is not true for other figures. But it is true that, for two similar figures, the distance between any two points on one figure is proportional to the distance between any two corresponding points on the other figure. This is true for any two similar figures whether they are plane figures or solid ones.

EXAMPLE 6.11

Figure 6.3 shows two similar frustums of cones. The radii of the bases of the smaller frustum are given as is the radius of the smaller base for the larger frustum. What is the radius of the bottom base of the larger frustum?

Solution   Since these are similar figures, we know that the ratios of the radius of the small base to the radius of the large base must be the same for each figure. Thus,

$$\frac{5.6}{8.4} = \frac{9.5}{x}$$

$$5.6x = (8.4)(9.5)$$

$$= 79.8$$

$$x = 14.25 \text{ in.}$$

The radius of the bottom base of the larger frustum is 14.25 in.

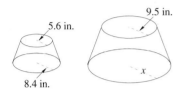

FIGURE 6.3

If we know that two figures are similar, then it is easy to find the area or volume of one if we know the area or volume of the other.

Areas of Similar Figures

Areas of similar figures are related to each other as the *squares* of any two corresponding dimensions. This is true for both plane and solid figures. It is also true for plane areas, lateral surface areas, total surface areas, or cross-sectional areas. For example, suppose two circles are similar. If one circle has a radius of r_1 and the other a radius of r_2, then the ratio of their areas is

$$\frac{A_1}{A_2} = \frac{\pi r_1^2}{\pi r_2^2} = \frac{r_1^2}{r_2^2}$$

Thus, the ratio of the areas of two circles is the same as the ratio of the squares of their radii.

EXAMPLE 6.12

The lateral surface area of the smaller frustum in Figure 6.3 is 220 in.^{2} What is the lateral surface area of the larger figure?

Solution   We will let L represent the lateral surface area that we are to find and use the proportion

$$\frac{220}{L} = \frac{5.6^2}{9.5^2}$$

$$5.6^2 L = (220)(9.5)^2$$

$$31.36L = 220(90.25)$$

$$= 19,855$$

$$L = 633.13138$$

The lateral surface area of the larger figure is about 633.13 in.^{2}

Volumes of Similar Figures

If two solid figures are similar, then their volumes are related to each other as the *cubes* of any two corresponding dimensions. As with the circles, two spheres are similar. If one sphere has a radius of r_1 and the other a radius of r_2, then the ratio of their volumes is

$$\frac{V_1}{V_2} = \frac{\frac{4}{3}\pi r_1^3}{\frac{4}{3}\pi r_2^3} = \frac{r_1^3}{r_2^3}$$

EXAMPLE 6.13

The volume of the smaller frustum in Figure 6.3 is 646.6 in.^{3} What is the volume of the larger figure?

Solution   We will let V represent the volume that we are to find and use the proportion

$$\frac{646.6}{V} = \frac{5.6^3}{9.5^3}$$

$$5.6^3 V = (646.6)(9.5)^3$$

$$175.616V = 554{,}378.67$$

$$V = 3{,}156.7663$$

The volume of the larger frustum is about 3,156.77 in.^{3}

Scale Drawings

Perhaps the most common use of similar figures applies when using scale drawings. Scale drawings are used in maps, blueprints, engineering drawings, and other figures. The ratio of distances on the drawing to corresponding distances on the actual object is called the scale of the drawing.

Application

EXAMPLE 6.14

A rectangular building $200' \times 145'$ is drawn to a scale of $\frac{1}{4}'' = 1'0''$. What is the size of the rectangle that will represent this building on the blueprint?

Solution   This is the continuing proportion $\dfrac{\frac{1}{4}''}{1'0''} = \dfrac{x}{200'} = \dfrac{y}{145'}$. Solving each of these we get $x = 50''$ and $y = 36\frac{1}{4}''$.

The building will be represented by a rectangle that is $50'' \times 36\frac{1}{4}''$.

Application

EXAMPLE 6.15

A map has a scale of 1 : 24 000. What is the actual distance of a map distance of 32 mm?

EXAMPLE 6.15 (Cont.)

Solution We will use the proportion $\dfrac{1}{24\,000} = \dfrac{32\text{ mm}}{x}$.

$$x = (32\text{ mm})(24\,000) = 768\,000\text{ mm}$$

$$= 768\text{ m}$$

So, 32 mm on the map represents an actual distance of 768 m.

Exercise Set 6.2

The pairs of figures in each of Exercises 1–6 are similar. Find the lengths of the unknown sides.

1.

4.
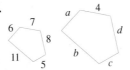

2.
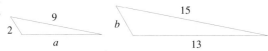

5.
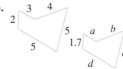

3.
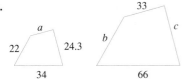

6.
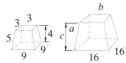

Solve Exercises 7–15.

7. *Architectural technology*   The floor plan of a building has a scale of $\frac{1}{8}$ in. = 1 ft. One room of the floor plan has an area of 20 in.^{2} What is the actual room area in square inches? What is the area in square feet? (Hint: 144 in.^{2} = 1 ft^2)

8. *Wastewater technology*   A pipe at a sewage treatment plant is 75 mm in diameter and discharges 2 000 L of water in a given period of time. If a pipe is to discharge 3 000 L in the same period of time what is its diameter?

9. A square bar of steel, 38 mm on a side, has a mass of 22 kg. What is the mass of another bar of the same length that measures 19 mm on a side?

10. *Agricultural technology*   It cost $982 to fence in a circular field that has an area of 652 ft^2. What will it

cost to enclose another circular field with three times as much area?

11. *Construction*   It requires 700 L to paint a spherical tank that has a radius of 20 m. How much paint will be needed to paint a tank with a radius of 35 m?

12. *Environmental technology*   A water tank that is 12 m high has a volume of 20 kiloliters (kL) or 20 000 L. What is the volume of a similar tank that is 30 m high?

13. *Sheet metal technology*   A cylinder has a capacity of 3 930 mm^3. Its diameter is 15 mm. What is the volume of a similar container with a diameter of 5 mm?

14. *Sheet metal technology* A cylindrical container has a diameter of $4''$ and a height of $5''$. A similar container has a diameter of $2.5''$. What is the height of the second container and the volume of each.

15. *Sheet metal technology* A sphere with a 10-cm radius has a volume of $4\,188.79$ cm^3 and a surface area of $1\,256.64$ cm^2. Find **(a)** the surface area of a sphere with a radius of 2 cm and **(b)** the volume of a sphere with a radius of 2 cm.

≡ 6.3
DIRECT AND INVERSE VARIATION

As we have seen, many scientific laws are given in terms of ratios or proportions. In this section, we will look at some relations between two or more variables.

Direct Variation

When two variables, x and y, are related so that their ratio $\frac{y}{x} = k$, where k is a constant, then y is **directly proportional** to x. Other ways of stating that y **is directly proportional to** x is to say that y is proportional to x, that y **varies directly as** x, or that y varies as x. The constant k is called the **constant of proportionality** or the **constant of variation**. The symbol $\propto$ is sometimes used to indicate that y is directly proportional to x and is written $y \propto x$. Note that $y \propto x$ is not an equation, but that $y \propto x$ means $y = kx$, where k is a constant.

> **Direct Variation**
>
> If x and y are variables and k is a constant, then the formula for direct variation is
> $$y = kx \qquad \text{or} \qquad \frac{y}{x} = k$$

EXAMPLE 6.16

The circumference of a circle is directly proportional to its diameter. Here $C = kd$, where $k = \pi$.

EXAMPLE 6.17

The weight (w) of an object at the earth's surface is directly proportional to the mass (m) of the object. Here $w = mg$, where g is the constant acceleration of gravity. Notice that here g is used instead of k. We saw in Exercise 36 of Exercise Set 6.1 that $g \approx 9.8$ m/s^2 or $g \approx 32$ ft/s^2.

EXAMPLE 6.18

When applied to a spring, **Hooke's law** states that the force F required to stretch a spring is proportional to the distance s the spring is stretched (provided the elastic limit of the spring is not exceeded). Thus $F = ks$, where k is called the **spring constant.**

EXAMPLE 6.19

A force of 10 lb is required to stretch a spring from its natural length of 5 in. to a length of 8 in. Find the spring constant.

EXAMPLE 6.19 (Cont.)

Solution   We know from Hooke's law that $F = ks$. Here, $F = 10$ lb and $s = 8 - 5 = 3$ in. So, $10 = k \cdot 3$ and $k = \frac{10}{3}$.

It is also possible to state that y varies as the square of x, $y = kx^2$, or y varies as the cube of x, $y = kx^3$, and so on.

EXAMPLE 6.20

The surface area of a sphere varies as the square of its radius. Since $A = 4\pi r^2$, $k = 4\pi$.

EXAMPLE 6.21

The volume of a sphere varies directly as the cube of its radius. Since $V = \frac{4}{3}\pi r^3$, the constant of proportionality is $\frac{4}{3}\pi$.

EXAMPLE 6.22

The distance h that an object falls when dropped from rest is directly proportional to the square of the time t that the object has fallen. If a stone is dropped 30.625 m from a bridge and takes 2.5 s to strike the water, what is the constant of proportionality?

Solution   The statement "the distance h that an object falls when dropped from rest is directly proportional to the square of the time t the object has fallen," can be written as $h = kt^2$. We are given $h = 30.625$ m and $t = 2.5$ s.

$$\text{Thus,} \qquad 30.625 = k(2.5)^2$$

$$= k(6.25)$$

$$4.9 = k$$

The constant is 4.9 m/s². Since the acceleration of gravity g is about 9.8 m/s², then $k = \frac{1}{2}g$.

If we have two pairs of values, x_1, y_1, and x_2, y_2, from a direct variation then we know that $\frac{y_1}{x_1} = k$ and $\frac{y_2}{x_2} = k$, so $\frac{y_1}{x_1} = \frac{y_2}{x_2}$, or $y_1 x_2 = x_1 y_2$.

EXAMPLE 6.23

Charles' law states that at a constant pressure the temperature of a gas varies directly as its volume. This is often expressed as $\frac{V_1}{T_1} = \frac{V_2}{T_2}$, where V_1 represents the volume of the gas at an absolute temperature T_1, and similarly for V_2 and T_2. If a gas sample of 5 m³ at 0°C is cooled until its volume is 2 m³, what is the final temperature?

Solution   Charles' law is true only when we are working in degrees expressed in Kelvin (K). The official temperature unit in the SI Metric system is Kelvin but people more often use the Celsius scale. If T_C is the temperature in degrees Celsius and T_K is the temperature in Kelvin, then $T_K = T_C + 273.18$.

EXAMPLE 6.23 (Cont.)

In this problem, $0°C = 273.18$ K, and we have $V_1 = 5 \text{ m}^3$, $V_2 = 2 \text{ m}^3$, and $T_1 = 273.18$ K. According to Charles' law

$$\frac{V_1}{T_1} = \frac{V_2}{T_2}$$

$$\frac{5}{273.18} = \frac{2}{T_2}$$

$$5T_2 = 546.36$$

$$T_2 = 109.272 \text{ K}$$

The temperature is 109.272 K or $-163.908°C$.

EXAMPLE 6.24

The area of a circle is directly proportional to the square of its radius. If the area of a circle with radius 2 cm is $4\pi \text{ cm}^2$, what is the area of a circle with a radius of 6 cm?

Solution   We could use the formula for the area of a circle. However, we will apply our knowledge of direct variation. If A_1, r_1, A_2, and r_2 are the areas and radii of two circles, then we know

$$\frac{A_1}{r_1^2} = \frac{A_2}{r_2^2}$$

Since $A_1 = 4\pi$, $r_1 = 2$, and $r_2 = 6$, then

$$\frac{4\pi}{2^2} = \frac{A_2}{6^2}$$

or $4A_2 = (4\pi)(36)$

$$A_2 = 36\pi \text{ cm}^2$$

Inverse Variation

Another type of variation is **inverse variation,** where the product of two variables, x and y, is a constant. In inverse variation $xy = k$, or $y = \dfrac{k}{x}$, and we say that y is **inversely proportional** to x, or y **varies inversely** as x. Again k is called the constant of variation or the constant of proportionality.

Inverse Variation

If x and y are variables and k is a constant, then the formula for inverse variation is

$$y = \frac{k}{x} \qquad \text{or} \qquad yx = k$$

EXAMPLE 6.25

The electrical resistance of a wire varies inversely as its cross-sectional area. If R is the resistance of a wire and A is the area of its cross-section, then $RA = k$ or $R = \dfrac{k}{A}$.

EXAMPLE 6.26

Boyle's law states that at a constant temperature the volume of a sample of gas is inversely proportional to the absolute pressure applied to the gas. Thus, if p is the gas pressure and V its volume, then $pV = k$, or $V = \dfrac{k}{p}$.

EXAMPLE 6.27

If 250 L of air in a cylinder exert a gauge pressure of 1 380 kilopascals (kPa), what is the constant of proportionality?

Solution In Example 6.26, we learned that $pV = k$. In this example $p = 1\,380$ kPa and $V = 250$ L, so $k = (1\,380)(250) = 345\,000$.

As was true with direct variation, it is also possible to state that y varies inversely as the square of x, thus $yx^2 = k$ $\left(\text{or } y = \dfrac{k}{x^2}\right)$, or that y varies inversely as the cube of x, thus $yx^3 = k$ $\left(\text{or } y = \dfrac{k}{x^3}\right)$, and so on.

EXAMPLE 6.28

The resistance R of a wire varies inversely as its cross-sectional area A, and so $RA = k$. The area can be determined from the diameter, d. In fact, $A = \dfrac{\pi}{4}d^2$, thus, because $\dfrac{\pi}{4}$ is a constant, we could say that the resistance of a wire varies inversely as the square of its diameter. Thus, $Rd^2 = k$, or $R = \dfrac{k}{d^2}$. Note that this constant of variation is not the same constant as in Example 6.25.

If we have two pairs of values, x_1, y_1 and x_2, y_2, from the same inverse variation we then know that, since $x_1y_1 = k$ and $x_2y_2 = k$, then $x_1y_1 = x_2y_2$ or $\dfrac{x_1}{x_2} = \dfrac{y_2}{y_1}$.

EXAMPLE 6.29

Boyle's law (see Example 6.26) is often given as $p_1V_1 = p_2V_2$, where p_1 is the gas pressure when the volume is V_1 and p_2 is its pressure when the volume is V_2. In Example 6.27, we had 250 L of air in a cylinder with a pressure of 1 380 kPa. If the volume of the cylinder is increased to 1 m³, what is the pressure?

Solution We must first change the units for one of the volumes so that both volumes are in the same unit. Since 1 m³ = 1 000 L, we can let $V_2 = 1\,000$ L, and

EXAMPLE 6.29 (Cont.)

$V_1 = 250$ L, while $p_1 = 1\,380$ kPa. So,

$$p_1V_1 = p_2V_2$$

$$(1\,380)(250) = p_2(1\,000)$$

$$345 = p_2$$

The pressure is 345 kPa when the volume is 1 m^3.

Exercise Set 6.3

In Exercises 1–6, express each of the given quantities as variations.

1. The resistance R of a wire is directly proportional to its length l.

2. The pressure P of a column of liquid varies directly as the height h of the liquid.

3. The area A of a geometric figure varies directly as the square of any dimension, d.

4. The energy E of a photon of light is directly proportional to its frequency, f.

5. The current I in a transformer is inversely proportional to the number of turns N in the winding.

6. The intensity of illumination E produced by a source of light varies inversely as the square of the distance d from the light source.

In Exercises 7–10, give the equation relating the variables and find the constant of proportionality for the given set of values.

7. r varies directly as t and $r = 4$ when $t = 6$.

8. a varies inversely as b and $a = 8$ when $b = 3$.

9. d varies directly as the square of r and $d = 8$ when $r = 4$.

10. p varies inversely as the cube of s and $p = 8$ when $s = 2$.

In Exercises 11–16, find the required value by setting up the equation and solving.

11. Find a when $d = 18$, if a varies directly as d and $a = 4$ when $d = 8$.

12. If r varies inversely with t, and $r = 6$ when $t = 3$, find the value of r when $t = 9$.

13. If v varies directly as t and $v = 60$ m/s when $t = 20$ s, what is v when $t = 1$ min?

14. If p varies inversely with q and $p = 30$ mm^3 when $q = 10$ kg/mm^2, what is p when $q = 40$ kg/mm^2?

15. If n varies directly as d^2 and $n = 10\,890$ neutrons when $d = 10$ μm, what is n when $d = 2$ μm?

16. If m varies inversely as r^3 and $m = 25$ kPa when $r = 30$ mm, what is m when $r = 5$ mm?

Solve Exercises 17–25.

17. *Chemistry*   A balloon filled with 25 m^3 of air at 20°C is heated under pressure to 85°C. According to Charles' law, $V = kT$. Find **(a)** the value of the constant of proportionality and **(b)** the increase in volume. (Remember to change the temperatures from °C to K.)

18. *Navigation* If an airplane is traveling at a uniform rate, the distance traveled will vary directly as the time it has traveled. If a plane covers 475 mi in 1 h, how long will it take to travel 1,235 mi?

19. *Electricity* The resistance of an electrical wire varies directly with its length. If a wire 1,500 ft long has a resistance of 32 ohms what is the resistance in 5,000 ft of the same wire?

20. *Physics* The distance d that an object falling from rest will travel varies directly as the square of the time t that the object has been falling. (a) If $d = 29.4$ m when $t = 3$ s, find the formula for d in terms of t. (b) What is d when $t = 8$ s? (c) What is t when $d = 150$ m?

21. *Physics* Using Boyle's law (see Example 6.26) and the fact that you know that the volume is 200 in.^{3} when the pressure is 40 psi, find the pressure when the volume is 50 in.^{3}

22. *Physics* Use Exercise 6 and the fact that a given light source produces an illumination of 600 lux (1 lux = 1 lumen per m^2) on a surface 2 m away

to determine how much illumination the same light source produces on a surface 6 m away?

23. *Electricity* The current in the winding of a transformer is inversely proportional to the number of turns. A transformer has 200 turns in its primary winding and 50 turns in its secondary winding. If the current in the secondary circuit is 0.3 A what is the current in the primary circuit?

24. *Electricity* The voltage in a winding of a transformer is directly proportional to the number of turns in the winding. A 120-V ac power line has 300 turns in its primary winding. If there are 40 turns in its secondary winding, what is the voltage across its secondary winding?

25. *Space technology* The weight of a body in space varies inversely as the square of its distance from the center of earth. If a person weighs 180 lb on the surface of the earth, how much does he or she weigh 500 mi above the surface if the radius of the earth is 4,000 mi?

≣ 6.4
JOINT AND COMBINED VARIATION

In Section 6.3 we studied about direct and inverse variation. In this section, we will study two more types of variation—joint and combined.

Both direct and inverse variation are functions of one variable. In each case we had $y = f(x)$. For direct variation, $y = kx$, and for inverse variation, $y = \frac{k}{x}$. The types of variation we will study in this section, joint and combined variation, are functions of two or more variables. Thus joint and combined variation are of the forms $y = f(x, z)$, $y = f(x, w, z)$, or $y = f(x, w, z, p)$.

Joint Variation

Joint variation is when one quantity, y, varies directly as the product of two or more quantities. Thus, if y is jointly proportional to x and z, then $y = kxz$, or if y is jointly proportional to x, w, and z, then $y = kxwz$. The term "jointly" is sometimes omitted and we simply say that y varies as w and z or that y is proportional to w and z when $y = kwz$. Again, k is the constant of variation or the constant of proportionality.

Joint Variation

If x, y, and z are variables and k is a constant, then the formula for joint variation of y with x and z is

$$y = kxz$$

| EXAMPLE 6.30 | The absolute temperature T of a perfect gas varies jointly as its pressure P and volume V. So, $T = kPV$. |

| EXAMPLE 6.31 | The volume of a prism V varies jointly as the area of its base B and its height h. So, $V = kBh$. In this case, we know from our study of geometry in Chapter 3 that $k = 1$. |

Combined Variation

If a quantity y varies with two or more variables in ways more complicated than that described in joint variation, it is then referred to as **combined variation.** Unlike direct, inverse, and joint variation, the term combined variation is seldom used in the description of the problem. In fact, a combined variation relationship will often use both "directly" and "inversely" in the description of the relationship.

Guidelines for Solving Variations

1. Write the appropriate variation formula.
2. Use the given information to find k.
3. Rewrite the variation formula using the value for k found in Step 2.

If asked to find a particular value from other given information, then use Step 4.

4. Substitute this other information into the equation. Solve for the required value.

| EXAMPLE 6.32 | The resistance R of a conductor is directly proportional to its length l, and inversely proportional to its cross-sectional area A. Thus, $R = \dfrac{kl}{A}$. |

| EXAMPLE 6.33 | The potential energy W of a capacitor is directly related to the square of its charge Q, and inversely related to its capacitance C. This relation is given by the formula $W = k\dfrac{Q^2}{C}$. |

| EXAMPLE 6.34 | The volume of a cone V is directly related to the square of the radius of its base r and its height h. Here $V = kr^2h$ and we know that the constant of proportionality $k = \frac{1}{3}\pi$. |

Just as was possible with direct and inverse variation, it is possible to use a given set of values from a joint or combined variation to find an unknown value from another set.

Application

EXAMPLE 6.35

The resistance R of an electrical wire varies directly as its length l, and inversely as the square of its diameter d. If a wire 600 m long with a diameter of 5 mm has a resistance of 32 ohms (Ω), determine the resistance in a 1 500-m wire made of the same material with a diameter of 10 mm.

Solution We know that $R = \dfrac{kl}{d^2}$, where k is the constant of proportionality. From the given values of $R = 32$ Ω, $d = 5$ mm, and $l = 600$ m, we have $32 = \dfrac{k(600)}{5^2}$, or $k = \dfrac{4}{3}$. So,

$$R = \frac{\frac{4}{3}\,(1\,500)}{10^2}$$

$$= \frac{2\,000}{100} = 20$$

The resistance is 20 Ω.

Application

EXAMPLE 6.36

Newton's law of universal gravitation states that the gravitational force of attraction F (in newtons, N) between two bodies is directly proportional to each of their masses m_1 and m_2 and inversely proportional to the square of the distance between them, d. Thus $F = \dfrac{km_1m_2}{d^2}$, where k is the **gravitational constant.** Two objects that each have a mass of 1 kg and are 1 m apart exert a mutual gravitational attraction of $F = \dfrac{k(1)(1)}{1^2} = 6.67 \times 10^{-11}$ N. So, $k \approx 6.67 \times 10^{-11}$ N.

The earth's mass is approximately 5.98×10^{24} kg. What is the earth's gravitational force on a space shuttle that has a mass of 29 500 kg and is in orbit 10^8 m from the center of the earth?

Solution Since $k = 6.67 \times 10^{-11}$ N, we have $F = 6.67 \times 10^{-11} \dfrac{m_1m_2}{d^2}$, where $m_1 = 5.98 \times 10^{24}$ kg, $m_2 = 29\,500$ kg $= 2.95 \times 10^4$ kg, and $d = 10^8$ m.

$$F = 6.67 \times 10^{-11} \frac{\left(5.98 \times 10^{24}\right)\left(2.95 \times 10^4\right)}{\left(10^8\right)^2}$$

$$= (6.67)(5.98)(2.95) \times 10^{-11+24+4-16}$$

$$= 117.67 \times 10^1 = 1\,176.7 \text{ N}$$

The earth's gravitational force on this space shuttle is about 1 176.7 N.

The previous examples have all used the guidelines for solving variation problems.

Exercise Set 6.4

In Exercises 1–6, express each of the given quantities as variations.

1. The kinetic energy K of a moving body is jointly proportional to its mass m and the square of its speed v.

2. The volume V of a parallelepiped varies jointly as its length l, width w, and height h.

3. The frequency f in hertz of a musical tone is directly proportional to the speed v of sound in air, and inversely proportional to the wavelength l of the sound wave.

4. The revolutions per minute r of a shaft being turned in a lathe varies directly as the cutting speed s of the shaft, and inversely as the diameter d of the shaft.

5. The voltage drop E in a wire varies directly as its length l and the current I, and inversely as the square of its diameter d.

6. The inductance of a solenoid L varies directly as the square of the number of turns of the coil N and the cross-sectional area of the solenoid, A, and inversely as its length l.

In Exercises 7–10, give the equation relating the variables and find the constant of proportionality for the given set of values.

7. a varies jointly as b and c, and $a = 20$, when $b = 4$ and $c = 2$.

8. p varies jointly as r, s, and t, and $p = 15$, when $r = 3$, $s = 10$, and $t = 25$.

9. u varies directly as v and inversely as w, and $u = 20$, when $v = 5$ and $w = 2$.

10. x varies directly as y and the cube of w, and inversely as the square of z, and $x = 175$, when $y = 3$, $w = 5$, and $z = 4$.

In Exercises 11–16, find the required value by setting up the equation and solving.

11. Find r when $s = 3$ and $t = 4$, if r varies jointly as s and t, and $r = 10$, when $s = 6$ and $t = 5$.

12. If f varies directly as r and inversely as l, and $f = 125$ when $r = 10$ and $l = 5$, then find the value of f when $r = 20$ and $l = 10$.

13. If a varies jointly as x and the square of y, and $a = 9$ when $x = 3$ and $y = 5$, then find a when $x = 6$ and $y = 15$.

14. Using the relationship from Exercise 13, find y when $x = 9$ and $a = 10$.

15. If p varies directly as r and the square root of t and inversely as w, and $p = 9$ when $r = 18$, $t = 9$, and $w = 6$, then find p when $r = 36$, $t = 4$, and $w = 2$.

16. Using the relationships from Exercise 15, find t when $p = 18$, $r = 9$, and $w = 4$.

Solve Exercises 17–25.

17. The area of a triangle varies jointly as the length of the base b and the height h of the triangle. If the area of a triangle is 12 when the base is 3 and the height is 8, what is (a) the constant of proportionality and (b) the area, when the base is 10 and the height is 5?

18. *Electricity* The energy W used by a device is directly related to the current I, the voltage V, and the length of time t, for which the energy is used. If a 240-V clothes drier draws a current of 15 A and in 45 min used 2.4 kWh (kilowatt hours) of energy, how much energy will a 120-V light bulb use, which draws a current of 1.25 A in 8 h of use?

19. *Electricity* The electric power consumed by a device is directly related to the square of the voltage and inversely related to the resistance. The 18 Ω filament of a tube is rated at 2 W and has an operating voltage of 6 V. What is the maximum voltage in a 50 Ω, 10 W resister?

20. *Electricity* The capacitance of a parallel-plate capacitor varies directly as the area of either of its plates and inversely as the distance between them. If the plates are 0.5 mm apart and each is a square that measures 4 m on a side, the capacitance is 0.3 μF. What is the capacitance if the plates are moved so they are 0.1 mm apart?

21. *Electricity* The resistance of a copper wire varies directly as its length and inversely as its cross-sectional area. If a copper wire 80 m long with a cross-sectional area of 2.5 mm^2 has a resistance of 0.56 Ω, what length of 0.1-mm^2 copper wire is needed for a resistance of 3 Ω?

22. *Electricity* The specific resistances of most metals vary with temperature. The change in the resistance of a metal is jointly proportional to the resistance of the metal and the change in temperature. If a copper wire has a resistance of 2.25 Ω at 25°C and 2.09 Ω at 10°C, what is its resistance at 50°C?

23. *Energy* The rate of heat conduction through a flat plate varies jointly as the area of one face of the plate and the difference between the temperature of the opposite faces, and inversely as the thickness of the plate. A 3-in. thick brick with a 12-in.^{2} area on one face loses heat at the rate of 50 Btu/h when the temperature is 72°F on one side and 32°F on the other side of the brick. What is the rate of conduction if the outside temperature of 32°F drops to 10°F?

24. *Sound* The frequency in hertz (Hz) of a guitar wire varies directly as the square root of the tension and inversely as the length. If a wire 500 mm long under 25 kg tension vibrates at 256 Hz, what is the frequency of a wire 450 mm long under a tension of 30 kg?

25. *Physics* The **ideal gas law** states that the pressure varies directly as the absolute temperature and inversely as the volume. A gas sample has a volume of 5 ft^3 at 70°F at a pressure of 15 lb/in.^{2} (psi). Find the volume at 200°F at a pressure of 75 lb/in.^{2} (The **Rankine scale** is the absolute temperature based on the Fahrenheit scale. If T_R represents a temperature on the Rankine scale and T_F represents the equivalent temperature on the Fahrenheit scale, then $T_R \approx T_F + 460°$. The freezing point of water is 32°F or about 492°R and the boiling point of water is 212°F or about 672°R.)

☰ CHAPTER 6 REVIEW

Important Terms and Concepts

Combined variation

Constant of proportionality

Continued proportion

Direct variation

Directly proportional

Inverse variation

Joint variation	Similar figures
Proportion	Area
Rate	Volume
Ratio	Similar triangles
Scale drawings	

Review Exercises

Solve each of the proportions in Exercises 1–8.

1. $3 : x = 4 : 6$

2. $x : 5 = 3 : 15$

3. $\dfrac{7}{9} = \dfrac{21}{d}$

4. $\dfrac{14}{6} = \dfrac{c}{27}$

5. $4 : 8 = 19 : x$

6. $x : 12 = 15 : 32$

7. $7 : 24.5 : x = 8 : y : 42$

8. $\dfrac{12.5}{x} = \dfrac{y}{47} = \dfrac{8}{5}$

Solve Exercises 9–19.

9. *Electricity*   The **turn ratio** in a transformer is the number of turns in the primary winding to the number of turns in the secondary winding. If the turn ratio for a transformer is 25, and there are 4,000 turns in the primary winding, how many turns are in the secondary winding?

10. *Economics*   A worker's salary increase is proportional to the cost-of-living index. If the worker received a raise of $12.50/wk when the index was 6.5, how much should the raise be when the index is 9.6?

11. The longest side of triangle A is 180 mm. Triangle B has sides of 4, 5, and 8 mm. Triangles A and B are similar. What are the lengths of the other two sides of triangle A?

12. *Physics*   The **theoretical mechanical advantage (TMA)** of an inclined plane or ramp is equal to the ratio between its length and height. A ramp 90 ft long slopes down 5 ft to the edge of a lake. What is the TMA of the ramp?

13. *Machine technology*   The efficiency of a machine is the ratio between its actual mechanical advantage (AMA) and its TMA. The AMA of the boat ramp in Exercise 12 is 16 because of the friction in a boat trailer's wheels. What is the efficiency of the ramp?

14. *Physics*   The rate of flow of a liquid through a pipe is jointly proportional to its velocity and the cross-sectional area of the pipe. A hose with a 1-cm diameter has a liquid flow through it at 3 m/s. **(a)** If a hose with a diameter of 5 mm is attached to one end of this hose, what is the velocity of the liquid through this smaller hose? **(b)** What size hose should have been attached if the velocity was to be 30 m/s?

15. *Petroleum technology*   A horizontal pipe 0.5 m in diameter is joined to an oil pipeline 2.5 m in diameter. If the velocity of the oil in the smaller pipe is 5 m/s, what is the velocity in the large pipe? (See Exercise 14.)

16. *Energy*   The rate at which an object radiates energy, as given by the **Stefan-Boltzmann law,** varies directly as the fourth power of its absolute temperature. The sun radiates energy at the rate of 6.5×10^7 W/m^2 from its surface. If the surface temperature of the sun is 5 800 K, what is the constant of proportionality? (This is the **Stefan-Boltzmann constant.**)

17. *Electricity*   The force that one charge exerts on another is given by **Coulomb's law** and is jointly proportional to the magnitudes of the charges and inversely proportional to the square of the distance between them. If a charge of 4×10^{-9} C is 5 cm from a charge of 5×10^{-8} C, and has a force of 7.192×10^{-4} N, what is the constant of proportionality (**Coulomb's constant**)? (Hint: Change 5 cm to m.)

18. *Electricity*   The magnetic field of a straight current varies directly as the magnitude of the current and inversely as the distance from the current. A cable

5 m above the ground carrying a 100-A current has a magnetic field of 4×10^{-6} T. What is the magnetic field of a cable that carries a 150-A current and is 15 m above the ground?

19. *Light*   The **index of refraction** of a transparent medium is the ratio between the velocity of light in

free space and its velocity in the medium. If the velocity of light is 3×10^8 m/s and the index of refraction in a diamond is 2.42, what is the speed of light in a diamond?

≡ CHAPTER 6 TEST

In Exercises 1–6, solve the proportions.

1. $4 : 9 = x : 51$

2. $8 : x = 68 : 93.5$

3. $\dfrac{a}{28} = \dfrac{12}{20}$

4. $\dfrac{24}{42} = \dfrac{78}{d}$

5. $6 : 8 : x = y : 14 : 24.5$

6. $\dfrac{a}{9} = \dfrac{b}{30} = \dfrac{75}{45}$

In Exercises 7–8, express in writing each of the given quantities as a variation.

7. E varies directly as the square of d.

8. h varies directly as r and inversely as the cube of t.

Solve Exercises 9–12.

9. A right triangle with sides of 5, 12, and 13 is proportional to a triangle with hypotenuse 71.5. What are the lengths of the other two sides?

10. The perimeter of a rectangular solar panel is 540 cm. The ratio of the length to the width is 3 : 2. What are the length and width?

11. The area of a geometric figure varies directly as the square of any dimension. Two similar triangles have corresponding sides of length 12 m and 18 m. The smaller triangle has an area of 72 m². What is the area of the other triangle?

12. The number of days d needed to erect a certain building varies inversely with the number of people working on the building p and the number of hours h they work each day. If a building requires 45 days to complete when 80 people work 8 h/day, how much time is needed if 60 people work 10 h/day?

Factoring and Algebraic Fractions

We often want to make a container that will have the largest possible volume while using the least possible amount of material. In Sections 7.1 and 7.2, we will learn some of the basic algebra needed to solve this type of problem.

Courtesy of Ruby Gold

This chapter uses the algebraic foundations established in Chapter 2. In Chapter 2, we learned how to multiply algebraic expressions. In this chapter we will learn how to do the reverse—to find the factors that, when multiplied together, give the original expression. Once we have the ability to factor algebraic expressions, we will use it to help solve second degree or quadratic equations.

We will also use factoring to help add, subtract, multiply, divide, and simplify algebraic fractions. In spite of all the scientific and technical examples we have used that involve linear equations, many scientific and technical situations must be described with quadratic equations or with algebraic fractions.

☰ 7.1
SPECIAL PRODUCTS

In mathematics we use certain products so often that we need to take the time to learn them. Success in factoring depends on your ability to recognize patterns in any form that they appear. All of these special products were developed in Section 2.2, but most were not stated in the general form. We will present each special product, give some examples of how it is used, and then summarize all of the special products at the end of this section.

The first special product is the distributive law of multiplication over addition, which we saw in Chapter 1.

Distributive Law $a(x + y) = ax + ay$ #1

EXAMPLE 7.1

$$8(4m + 3n) = 8(4m) + 8(3n)$$
$$= 32m + 24n$$

Here $a = 8$, $x = 4m$, and $y = 3n$. As you can see, the variable x can represent a product of a constant and a variable.

EXAMPLE 7.2

$$3x(2y + 5ax) = (3x)(2y) + (3x)(5ax)$$
$$= 6xy + 15ax^2$$

The term a in the distributive law can represent a constant, a variable, or the product of a constant and a variable.

The next special product is formed by multiplying the sum and difference of two terms. It is equal to the square of the first term minus the square of the second term.

Difference of Two Squares $(x + y)(x - y) = x^2 - y^2$ #2

EXAMPLE 7.3

$$(x + 6)(x - 6) = x^2 - 6^2$$
$$= x^2 - 36$$

Notice that each term to the right of the equal sign is a perfect square.

EXAMPLE 7.4

$$(2a + 3\sqrt{y})(2a - 3\sqrt{y}) = (2a)^2 - (3\sqrt{y})^2$$
$$= 4a^2 - 9y$$

EXAMPLE 7.4 (Cont.)

Again, once you realize that $y = (\sqrt{y})^2$, you can see that each term to the right of the equal sign is a perfect square. As you gain practice, you will be able to omit the middle step.

The third and fourth special products will be considered together.

Perfect Square Trinomials $(x + y)^2 = x^2 + 2xy + y^2$	#3
$(x - y)^2 = x^2 - 2xy + y^2$	#4

Note that both of these demonstrate that the square of a binomial is a trinomial.

 Caution

Be careful, many students make the mistake of thinking that $(x + y)^2 = x^2 + y^2$. This is not true, as you can easily verify by using the FOIL method from Chapter 2.

EXAMPLE 7.5

$$(p + 7)^2 = p^2 + 2(7)p + 7^2$$
$$= p^2 + 14p + 49$$

Here $x = p$ and $y = 7$.

EXAMPLE 7.6

$$(2a - 5t^2)^2 = (2a)^2 - 2(2a)(5t^2) + (5t^2)^2$$
$$= 4a^2 - 20at^2 + 25t^4$$

Here $x = 2a$ and $y = 5t^2$.

The next special product is the product of two binomials that have the same first term. It can be verified by using the FOIL method.

Two Binomials, Same First Term $(x + a)(x + b) = x^2 + (a + b)x + ab$	#5

EXAMPLE 7.7

$$(p + 3)(p + 7) = p^2 + (3 + 7)p + (3)(7)$$
$$= p^2 + 10p + 21$$

EXAMPLE 7.8

$$(r + 2)(r - 8) = r^2 + (2 - 8)r + (2)(-8)$$
$$= r^2 - 6r - 16$$

In this example, $a = 2$ and $b = -8$. Remember that $r - 8$ can be written as $r + (-8)$.

The next special product is a general version of the last one. Again, this special product can be checked by using the FOIL method.

General Quadratic Trinomial $(ax + b)(cx + d) = acx^2 + (ad + bc)x + bd$ #6

EXAMPLE 7.9

Use Special Product 6 to multiply $(4x + 2)(3x + 5)$.

Solution   Here $a = 4$, $b = 2$, $c = 3$, and $d = 5$, so

$$(4x + 2)(3x + 5) = 4 \cdot 3x^2 + (4 \cdot 5 + 2 \cdot 3)x + 2 \cdot 5$$
$$= 12x^2 + (20 + 6)x + 10$$
$$= 12x^2 + 26x + 10$$

EXAMPLE 7.10

Multiply $(7x + 3y)(5x - 2y)$.

Solution   Here $a = 7$, $b = 3y$, $c = 5$, and $d = -2y$. Using Special Product 6 produces

$$(7x + 3y)(5x - 2y) = (7 \cdot 5x^2) + [7(-2y) + (3y)5]x + (3y)(-2y)$$
$$= 35x^2 + (-14y + 15y)x + (-6y^2)$$
$$= 35x^2 + xy - 6y^2$$

Of course, there is nothing to prevent the combination of two or more of these special products.

≡ **Note**   Remember the order of operations from Chapter 1. Powers are executed before multiplication or division. So, in the following example, we first square $(3x - 5)$ and then multiply that result by $4x$.

EXAMPLE 7.11

$$4x(3x - 5)^2 = 4x(9x^2 - 30x + 25)$$
$$= 36x^3 - 120x^2 + 100x$$

EXAMPLE 7.12

$$(x^2 + 9)(x + 3)(x - 3) = (x^2 + 9)(x^2 - 9)$$
$$= x^4 - 81$$

In this example, notice that the left-hand side has the special product $(x + 3)(x - 3)$. When these are multiplied, we get another special product: $(x^2 + 9)(x^2 - 9)$. The

EXAMPLE 7.12 (Cont.)

time spent looking at a problem for special products can save some computation and time. It can also reduce errors.

There are four other special products that you will use less often than the ones already listed. They will be presented in pairs. The first two are the cubes of a sum or difference.

Perfect Cubes $(x+y)^3 = x^3 + 3x^2y + 3xy^2 + y^3$ #7

$(x-y)^3 = x^3 - 3x^2y + 3xy^2 - y^3$ #8

EXAMPLE 7.13

$$(x+5)^3 = x^3 + 3x^2(5) + 3x(5^2) + 5^3$$
$$= x^3 + 15x^2 + 75x + 125$$

EXAMPLE 7.14

$$(2a-4)^3 = (2a)^3 - 3(2a)^2(4) + 3(2a)(4^2) - (4^3)$$
$$= 8a^3 - 48a^2 + 96a - 64$$

The last two special products also deal with cubes.

Sum and Difference of Two Cubes $(x+y)(x^2 - xy + y^2) = x^3 + y^3$ #9

$(x-y)(x^2 + xy + y^2) = x^3 - y^3$ #10

EXAMPLE 7.15

$$(x-3)(x^2 + 3x + 9) = x^3 - (3)^3$$
$$= x^3 - 27$$

EXAMPLE 7.16

$$(2d+5)(4d^2 - 10d + 25) = (2d)^3 + (5)^3$$
$$= 8d^3 + 125$$

Application

EXAMPLE 7.17

An open box is going to be formed by cutting a square out of each corner of a rectangular piece of cardboard and folding up the sides as shown in Figure 7.1a. Suppose the piece of cardboard measures 10 cm × 12 cm. **(a)** Determine a function for the volume of the box. **(b)** What is the domain of this function? **(c)** Set up a partial table of values and sketch the graph of this function. **(d)** What size square seems to produce a box with the largest volume?

EXAMPLE 7.17 (Cont.)

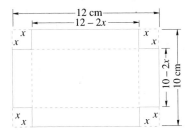

FIGURE 7.1a

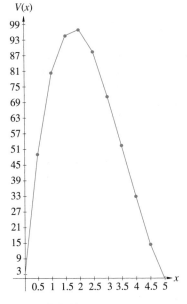

FIGURE 7.1b

Solutions

(a) If the square measures x cm $\times$ x cm, then the lengths of the sides that will be folded up are $10 - 2x$ cm and $12 - 2x$ cm. So, the volume of the box is given by

$$V(x) = x(10 - 2x)(12 - 2x)$$

$$= 120x - 44x^2 + 4x^3$$

(b) The domain of the function is all real numbers. But, we cannot cut out a square that measures 0 cm or less and we cannot cut out a square that measures 5 cm or more. So, the "realistic" domain is represented by $\{x : 0 < x < 5\}$. Notice how the factored form helped us determine the domain. The expanded form will be more useful later in this book when we get to calculus.

(c) We have the following partial table of values.

x	0.5	1.0	1.5	2.0	2.5	3.0	3.5	4.0	4.5
$V(x)$	49.5	80.0	94.5	96.0	87.5	72.0	52.5	32.0	13.5

A graph formed from this table is shown in Figure 7.1b.

(d) From looking at the table, or the graph, it seems as if a 2 cm $\times$ 2 cm square will produce a box with the largest volume. Calculus can be used to show that $x \approx 1.8$ cm will produce the largest volume: 96.768 cm^3.

Thus, we have a total of 10 special products. Study them. Become familiar with them. Learn to recognize them and to use them to help simplify your work. The ten special products are listed here together.

The Special Products		
$a(x + y) = ax + ay$		#1
$(x + y)(x - y) = x^2 - y^2$		#2
$(x + y)^2 = x^2 + 2xy + y^2$		#3
$(x - y)^2 = x^2 - 2xy + y^2$		#4
$(x + a)(x + b) = x^2 + (a + b)x + ab$		#5
$(ax + b)(cx + d) = acx^2 + (ad + bc)x + bd$		#6
$(x + y)^3 = x^3 + 3x^2y + 3xy^2 + y^3$		#7
$(x - y)^3 = x^3 - 3x^2y + 3xy^2 - y^3$		#8
$(x + y)(x^2 - xy + y^2) = x^3 + y^3$		#9
$(x - y)(x^2 + xy + y^2) = x^3 - y^3$		#10

Exercise Set 7.1

In Exercises 1–44, find the indicated products by direct use of one of the 10 special products. It should not be necessary to write intermediate steps.

1. $3(p + q)$

2. $7(2 + x)$

3. $3x(5 - y)$

4. $2a(4 + a^2)$

5. $(p + q)(p - q)$

6. $(3a + b)(3a - b)$

7. $(2x - 6p)(2x + 6p)$

8. $\left(\dfrac{y}{2} + \dfrac{2p}{3} \right)\left(\dfrac{y}{2} - \dfrac{2p}{3} \right)$

9. $(r + w)^2$

10. $(q - f)^2$

11. $(2x + y)^2$

12. $\left(\dfrac{1}{2}a + b \right)^2$

13. $\left(\dfrac{2}{3}x + 4b \right)^2$

14. $\left(a - \dfrac{1}{2}y \right)^2$

15. $\left(2p - \dfrac{3}{4}r \right)^2$

16. $\left(\dfrac{2}{3}r - 5t \right)^2$

17. $(a + 2)(a + 3)$

18. $(x + 5)(x + 7)$

19. $(x - 5)(x + 2)$

20. $(a + 9)(x - 12)$

21. $(2a + b)(3a + b)$

22. $(4x + 2)(3x + 5)$

23. $(3x + 4)(2x - 5)$

24. $(3m - n)(4m + 2n)$

25. $(a + b)^3$

26. $(r - s)^3$

27. $(x + 4)^3$

28. $(2a + b)^3$

29. $(2a - b)^3$

30. $(3x + 2y)^3$

31. $(3x - 2y)^3$

32. $(4r - s)^3$

33. $(m + n)(m^2 - mn + n^2)$

34. $(a + 2)(a^2 - 2a + 4)$

35. $(r - t)(r^2 + rt + t^2)$

36. $(h - 3)(h^2 + 3h + 9)$

37. $(2x + b)(4x^2 - 2xb + b^2)$

38. $\left(3a + \dfrac{2}{3}c \right)\left(9a^2 - 2ac + \dfrac{4}{9}c^2 \right)$

39. $(3a - d)(9a^2 + 3da + d^2)$

40. $\left(\dfrac{2e}{5} - \dfrac{5r}{4} \right)\left(\dfrac{4e^2}{25} + \dfrac{er}{2} + \dfrac{25r^2}{16} \right)$

41. $3(a + 2)^2$

42. $5(2a - 4)^2$

43. $\dfrac{5r}{t}\left(t + \dfrac{r}{5} \right)^2$

44. $(x^2 + 4)(x + 2)(x - 2)$

In Exercises 45–60, perform the indicated operations.

45. $(x^2 - 6)(x^2 + 6)$

46. $\left(\dfrac{2a}{b} + \dfrac{b}{2} \right)^2$

47. $\left(\dfrac{3x}{y} - \dfrac{y}{x} \right)\left(\dfrac{3x}{y} + \dfrac{y}{x} \right)x^2y^2$

48. $(x + 1)(x + 2)(x + 3)$

49. $(x - y)^2 - (y - x)^2$

50. $(x + y)^2 - (y - x)^2$

51. $(x + 3)(x - 3)^2$

52. $(y - 4)(y + 4)^2$

53. $r(r - t)^2 - t(t - r)^2$

54. $5(x + 4)(x - 4)(x^2 + 16)$

55. $(5 + 3x)(25 - 15x + 9x^2)$

56. $(2 - \sqrt{x})(2 + \sqrt{x})(4 + x)$

57. $[(x + y) - (w + z)]^2$

58. $[(r + s) + (t + u)]^2$

59. $(x + y - z)(x + y + z)$

60. $(a + b + 2)(a - b + 2)$

Solve Exercises 61–64.

61. *Electricity* The impedance z of an ac circuit at frequency f is given by the formula $z^2 = R^2 + (x_L - x_C)^2$, where R is the resistance, x_L is the inductive resistance and x_C is the capacitive resistance at the frequency f. Use a special product to expand the equation.

62. *Physics* The kinetic energy KE of an object is given by the formula $KE = \frac{1}{2}mv^2$, where m is the mass of the object and v its velocity. If the velocity of an object at any time t is given by the equation $v = 3t + 1$, find an equation for the kinetic energy in terms of m and t and expand your result using a special formula.

63. *Physics* The magnitude of the centripetal acceleration of a body in uniform circular motion is given by

the formula $a_c = \dfrac{v^2}{r}$, where v is the velocity of the body and r is the radius of the circular path. If the velocity at any given time t is expressed as $v = 2t^2 - t$, find an equation for the centripetal acceleration in terms of r and t, and expand your result.

64. *Energy* The work W done by a steam turbine in a certain period of time is given by the formula $W = \frac{1}{2}m(v_1 - v_2)(v_1 + v_2)$, where m is the mass of the steam that passes through the turbine during that period, v_1 is the velocity of the steam when it enters, and v_2 the velocity when it leaves. Simplify this formula by multiplying the factors together.

≡ 7.2
FACTORING

Sometimes it is important to determine what expressions were multiplied together to form a product, instead of multiplying quantities together as we did in the special products. Recall from Section 2.1 that an algebraic term consists of several factors. For example, the term $5xy$ has factors of $1, 5, x, y, 5x, 5y, xy$, and $5xy$. In Section 7.1, we learned to multiply several factors together to form the special products. In this section, we will begin to learn how to factor an algebraic expression. Once we know how to factor we will be able to solve more complicated problems. Determining the factors of an algebraic expression is called **factoring**.

When factoring a polynomial, we will use only those factors that are also polynomials. Further, we will continue factoring a polynomial until the only remaining factors are 1 and -1. When this has been completed, we will be able to say that the polynomial has been factored completely and all of the factors will be **prime factors**.

Common Factors

The simplest type of factoring is the reverse of the distributive law of multiplication. This type of factoring is called **removing a common factor**. If each term in an expression contains the same factor, then this is a common factor and it can be factored out using the distributive law.

Common factor $ax + ay = a(x + y)$

EXAMPLE 7.18

In the expression $6a + 3a^2$, each term contains a factor of $3a$.

$$6a + 3a^2 = 2(3a) + a(3a)$$
$$= (2 + a)3a$$
$$= 3a(2 + a)$$

EXAMPLE 7.19

$8y^2 + 2y = 2y(4y + 1)$

Remember that $2y = (2y)(1)$. When you factor $2y$, you are left with a factor of 1.

EXAMPLE 7.20

$12x^5y + 8x^3y^2 = 4x^3y(3x^2 + 2y)$

You could have factored this in several other ways, such as $2xy(6x^4 + 4x^2y)$. This would not have been considered completely factored, since $6x^4 + 4x^2y$ can be factored further.

You may not see how to completely factor out all of the common factors in one step. For instance, in the last example you might have first written

$$12x^5y + 8x^3y^2 = 2xy(6x^4 + 4x^2y)$$

and then noticed that you could factor $6x^4 + 4x^2y$ as $2x^2(3x^2 + 2y)$. Then you would have had

$$12x^5y + 8x^3y^2 = 2xy(6x^4 + 4x^2y)$$
$$= 2xy(2x^2)(3x^2 + 2y)$$
$$= 4x^3y(3x^2 + 2y)$$

Notice that in the last step the two monomials were combined.

EXAMPLE 7.21

Factor $6x^2y + 9xy^2z - 3xyz$.

Solution We can see that each term contains a multiple of 3, as well as an x and y factor. If we factor $3xy$ out of each term, we have $(3xy)(2x) + (3xy)(3yz) + (3xy)(-z) = 3xy(2x + 3yz - z)$. So, $6x^2y + 9xy^2z - 3xyz = 3xy(2x + 3yz - z)$.

The easiest factors to locate (and sometimes the easiest to overlook) are the monomials or common factors.

Hint

You should always begin factoring an algebraic expression by looking for common factors. Once you have factored out all common factors, the remaining expression is easier to factor.

Using the Special Products

Special product #2 gives us our second important form of factoring. Since $(x + y)(x - y) = x^2 - y^2$, we can reverse this to get

| **Difference of two squares** | $x^2 - y^2 = (x + y)(x - y)$ |

As you can see, to factor the difference between two squares you get factors that are the sum and difference of the quantities.

EXAMPLE 7.22

Factor $x^2 - 25$.

Solution $\quad x^2 - 25 = x^2 - 5^2$

$$= (x + 5)(x - 5)$$

You would have had to notice that 25 was 5^2 before you could recognize that this was the difference of two squares.

EXAMPLE 7.23

Factor $9a^2 - 49b^4$.

Solution $\quad 9a^2 - 49b^4 = (3a)^2 - (7b^2)^2$

$$= (3a + 7b^2)(3a - 7b^2)$$

Again, you have to recognize the perfect squares: $9a^2 = (3a)^2$ and $49b^4 = (7b^2)^2$.

EXAMPLE 7.24

Factor $27x^3 - 75xy^2$.

Solution $\quad 27x^3 - 75xy^2 = 3x(9x^2 - 25y^2)$

$$= 3x(3x + 5y)(3x - 5y)$$

This example demonstrates the value of first looking for common factors. There is a common factor of $3x$. When it is factored out, it is easier to see that the remaining factor is the difference of two squares.

EXAMPLE 7.25

Factor $5x^4 - 80y^4$.

Solution $\quad 5x^4 - 80y^4 = 5(x^4 - 16y^4)$

$$= 5[(x^2)^2 - (4y^2)^2]$$

$$= 5(x^2 + 4y^2)(x^2 - 4y^2)$$

$$= 5(x^2 + 4y^2)(x + 2y)(x - 2y)$$

Again, you should have noticed that there was a common factor of 5 and that each of the remaining terms was a perfect square. After factoring out $x^4 - 16y^4$ you were not finished, because $x^2 - 4y^2$ is also the difference of two squares. The other factor $x^2 + 4y^2$ cannot be factored further.

Application

EXAMPLE 7.26

In order to shield it from stray electromagnetic radiation, an electronic device is housed in a metal canister that is shaped like a right circular cylinder. The surface

EXAMPLE 7.26 (Cont.)

area of the cylinder is given by $A = 2\pi r^2 + 2\pi rh$, where r is the radius of the base and h is the height of the cylinder. Factor the right-hand side of this formula.

Solution This formula has common factors of 2π and r. Factoring, we obtain

$$A = 2\pi r^2 + 2\pi rh$$

$$= 2\pi r(r + h)$$

Exercise Set 7.2

Completely factor each of the expressions in Exercises 1–40.

1. $6x + 6$

2. $12x + 12$

3. $12a - 6$

4. $15d - 5$

5. $4x - 2y + 8$

6. $6a + 9b - 3c$

7. $5x^2 + 10x + 15$

8. $16a^2 + 8b - 24$

9. $10x^2 - 15$

10. $14x^4 + 21$

11. $4x^2 + 6x$

12. $8a - 4a^2$

13. $7b^2y + 28b$

14. $9ax^2 + 27bx$

15. $3ax + 6ax^2 - 2ax$

16. $6by - 12b^2y + 7by^2$

17. $4ap^2 + 6a^2pq + 8apq^2$

18. $12p^2r^2 - 8p^3r + 24pr^2$

19. $a^2 - b^2$

20. $p^2 - r^2$

21. $x^2 - 4$

22. $a^2 - 16$

23. $y^2 - 81$

24. $m^2 - 49$

25. $4x^2 - 9$

26. $49y^2 - 64$

27. $9a^4 - b^2$

28. $16t^6 - a^2$

29. $25a^2 - 49b^2$

30. $121r^2 - 81t^2$

31. $144 - 25b^4$

32. $81 - 49r^4$

33. $5a^2 - 125$

34. $7x^2 - 63$

35. $28a^2 - 63b^4$

36. $81x^2 - 36t^6$

37. $a^4 - 81$

38. $b^4 - 256$

39. $16x^4 - 256y^4$

40. $25a^5 - 400ab^8$

Solve Exercises 41–46.

41. The total surface area of a cone is given by the formula $\pi r^2 + \pi r\sqrt{h^2 + r^2}$, where r is the radius of the base and h is the height of the cone. Factor this expression.

42. *Thermodynamics* The amount of heat that must be added to a metal object in order for it to melt is given by the formula $Q = mc\Delta t + mL_f$. Factor the right side of this equation.

43. *Wastewater technology* According to Bernoulli's equation, if a fluid of density d is flowing horizontally in a pipe and its pressure and velocity at one location are p_1 and v_1, respectively, and at a second location, they are p_2 and v_2, then the difference in their pres-

sures is given by $p_1 - p_2 = \frac{1}{2}dv_2^2 - \frac{1}{2}dv_1^2$. Factor the right-hand side of this equation.

44. The cross-sectional area A of a tube can be determined from the formula $A = \pi R^2 - \pi r^2$, where R is the outside radius and r is the inside radius of the tube. Factor the right-hand side of this equation.

45. *Acoustical engineering* The angular acceleration α of a stereo turntable during a time period can be determined using the formula

$$\alpha = \frac{\omega_f^2 - \omega_0^2}{2\theta}.$$

where ω_0 is the angular velocity at the beginning of the time interval, ω_f is the angular velocity at the end of the time interval, and θ is the number of radians the turntable moved during the interval. Factor the right-hand side of this equation.

46. *Thermodynamics* A black body is a hypothetical body that absorbs, without reflection, all of the electromagnetic radiation that strikes its surface. The

energy E radiated by a black body is given by $E = e\sigma T^4 - e\sigma T_0^4$, where T and T_0 are the absolute temperatures of the body and the surroundings, respectively, σ is the Stefan-Boltzmann constant, and e is the emissivity of the body. Factor the right-hand side of this equation.

7.3
FACTORING
TRINOMIALS

In Section 7.2, we introduced the idea of factoring and learned how to factor two types of problems. These problems were based on the first two special products that we learned in Section 7.1.

An algebraic expression that has three terms is called a **trinomial**. Special products #3 through #6 all resulted in quadratic trinomials. In this section we will focus on factoring quadratic trinomials with the purpose of reversing one of these four special products.

Not all quadratic trinomials can be factored using real numbers. The general quadratic trinomial is of the form $ax^2 + bx + c$, where a, b, and c represent constants.

 Note   You can determine if a quadratic trinomial can be factored by examining the discriminant $b^2 - 4ac$. If the discriminant is a perfect square, then it is possible to factor the quadratic using rational numbers.

Discriminant

The **discriminant** of the trinomial $ax^2 + bx + c$ is $b^2 - 4ac$.
If $b^2 - 4ac > 0$, then $ax^2 + bx + c$ can be factored using real numbers.
If $b^2 - 4ac$ is a perfect square, then $ax^2 + bx + c$ can be factored using rational numbers.

EXAMPLE 7.27

Can the trinomial $4x^2 + 3x - 7$ be factored?

Solution   In this quadratic trinomial $a = 4$, $b = 3$, and $c = -7$, so

$$b^2 - 4ac = (3)^2 - 4(4)(-7)$$
$$= 9 - (-112)$$
$$= 121$$

Using our knowledge (or a calculator) we see that $\sqrt{121} = 11$, so this trinomial can be factored. Later in this section, our job will be to find its factors.

EXAMPLE 7.28

Can the quadratic trinomial $5x^2 - 3x - 7$ be factored?

Solution   Here $a = 5$, $b = -3$, and $c = -7$, so

$$b^2 - 4ac = (-3)^2 - 4(5)(-7)$$

EXAMPLE 7.28 (Cont.)

$$= 9 - (-140)$$

$$= 149$$

Using a calculator, we find that $\sqrt{149} \approx 12.207$ and is not a perfect square. Thus, it is possible to factor this quadratic equation with real numbers.

We will skip special products #3 and #4 and consider special product #5: $(x + a)(x + b) = x^2 + (a + b)x + ab$. If you examine special products #3 and #4 you can see that they are just special cases of #5. In the special product where $(x + a)(x + b) = x^2 + (a + b)x + ab$, the leading coefficient (the coefficient of the x^2 term) is 1. This makes the job of factoring somewhat easier, because all we need to do is determine a and b. Notice that $a + b$ is the coefficient of the x-term and ab is the constant. Thus, we have the reverse of special product #5.

$$x^2 + (a + b)x + ab = (x + a)(x + b) \qquad \text{#5}$$

Of course, the difficulty is determining the values of a and b.

EXAMPLE 7.29

Factor $x^2 - 3x - 10$.

Solution   Since the value of the discriminant is $49 = 7^2$, we know that this can be factored, and from the formula in special product #5 we know that it factors into $(x + a)(x + b)$ where $a + b = -3$ and $ab = -10$.

What possible choices are there for a and b? If we look at the factors of -10 we get these pairs.

$$-10 \text{ and } 1 \qquad 10 \text{ and } -1$$

$$-5 \text{ and } 2 \qquad 5 \text{ and } -2$$

Now, which of these pairs add to -3? There is only one: -5 and 2. So, if we let $a = -5$ and $b = 2$ then we have

$$x^2 - 3x - 10 = (x - 5)(x + 2)$$

EXAMPLE 7.30

Factor $x^2 + 10x + 16$.

Solution   The discriminant is 36, a perfect square, so this can be factored into the form $(x + a)(x + b)$. We want to find a and b so that $a + b = 10$ and $ab = 16$. Again, we will begin with the factors of the product $ab = 16$.

$$16 \text{ and } 1 \qquad -16 \text{ and } -1$$

$$8 \text{ and } 2 \qquad -8 \text{ and } -2$$

$$4 \text{ and } 4 \qquad -4 \text{ and } -4$$

EXAMPLE 7.30 (Cont.)

Do any of these add to 10? Yes, 8 and 2 have a sum of 10, so we can let $a = 8$ and $b = 2$. Then we have

$$x^2 + 10x + 16 = (x + 8)(x + 2)$$

A check to see if the constant is positive or negative will give you some help in factoring. If the constant is positive, then the factors will have the same sign—both will be positive or both negative. If the constant term is negative, then the two factors will have different or unlike signs—one will be positive and the other negative.

EXAMPLE 7.31

Factor each trinomial completely.
Here the constant term is positive.

a. $x^2 + 7x + 12 = (x + 4)(x + 3)$ Both factors have positive signs.

b. $x^2 - 7x + 12 = (x - 4)(x - 3)$ Both factors have negative signs.

Here the constant term is negative.

c. $x^2 + x - 12 = (x + 4)(x - 3)$

d. $x^2 - x - 12 = (x - 4)(x + 3)$

Factors have unlike signs—one is positive and one is negative.

Factoring a quadratic trinomial with a leading coefficient that is not 1 is not as easy. Here we are looking at the reverse of special product #6 or

$$acx^2 + (ad + bc)x + bd = (ax + b)(cx + d) \qquad \text{#6}$$

There are three techniques used to factor these quadratic trinomials. We will look at two of them in this section and examine the third in Section 7.4.

The first technique is called trial and error. In using the trial and error method, we make use of the fact that ac is the leading coefficient, bd is the constant term, and $ad + bc$ is the middle coefficient. We can also use our knowledge of signs to help find the signs of the factors.

EXAMPLE 7.32

Factor $3x^2 - 8x + 4$.

Solution The discriminant is $16 = 4^2$, so this equation will factor. Since 3 is a prime number, we know that its only factors are 3 and 1. Also, since the constant

EXAMPLE 7.32 (Cont.)

is positive, both factors have the same sign. Thus, we know that the factors are $(3x + b)(x + d)$. Now, all we need to find are b and d.

Since $bd = 4$, the possible pairs of factors are $-4, -1$; $4, 1$; $2, 2$; and $-2, -2$. Next, we know that $3d + b = -8$ and the only choices from the pairs of factors that satisfy this are $b = -2$ and $d = -2$. Thus, the factors of $3x^2 - 8x + 4$ are $(3x - 2)(x - 2)$. You should multiply these factors together to check that their product is the original trinomial.

EXAMPLE 7.33

Factor $6x^2 + 7x - 20$.

Solution The discriminant is $529 = 23^2$, so this equation will factor. The leading coefficient, 6, has factors of 6 and 1, and 2 and 3. We will try 6 and 1.

Since the constant term is negative, the factors will have unlike signs. Thus, we have $(6x + b)(x + d)$, where either b or d is negative, $bd = -20$, and $6d + b = 7$. The possible choices for the pairs b and d are $1, 20$; $2, 10$; and $4, 5$, where one is positive and the other is negative.

If we try $b = 4$ and $d = -5$, we get $6(-5) + 4 = -26$. Since this is not 7, this is not the correct solution. All other possible combinations for b and d also fail. Thus, we must make another choice for a and c.

If $a = 2$ and $c = 3$, we have $(2x + b)(3x + d)$. If $b = 5$ and $d = -4$, we get $ad + bc = 2(-4) + (5)(3) = -8 + 15 = 7$. This is the correct coefficient for the x-term, so the factors of $6x^2 + 7x - 20$ are $(2x + 5)$ and $(3x - 4)$.

As you can see, the trial and error method can be very long and frustrating, but some people learn to factor a quadratic trinomial quickly by using this method.

The other method we will look at in this section is called either the "grouping" or "split the middle" method. It is longer than the trial and error method, but it is a "sure-fire" technique. We will use the grouping method on the same problem that we just worked.

EXAMPLE 7.34

Factor $6x^2 + 7x - 20$ using the grouping method.

Solution

Step 1. Multiply the leading coefficient and the constant term: $(6)(-20) = -120$.

Step 2. Find two factors of this product whose sum is the coefficient of the middle term of the trinomial. For this problem we want to find factors p and q, where $pq = -120$ and $p + q = 7$. We get $p = 15$ and $q = -8$.

Step 3. Rewrite the trinomial by splitting the middle term into $px + qx$ and grouping the first two terms and the last two terms. In this problem, $7x = 15x - 8x$ and the trinomial becomes $(6x^2 + 15x) + (-8x - 20)$.

EXAMPLE 7.34 (Cont.)

Step 4. Distribute the common factors from each grouping.

$$(6x^2 + 15x) + (-8x - 20) = 3x(2x + 5) - 4(2x + 5)$$

Step 5. Distribute the common factor $2x + 5$ from the entire expression.

$$3x(2x + 5) - 4(2x + 5) = (3x - 4)(2x + 5)$$

These are the required factors. Thus, we have factored $6x^2 + 7x - 20$ as $(3x - 4)(2x + 5)$.

EXAMPLE 7.35

Factor $21x^2 - 41x + 10$ using the grouping method.

Solution

Step 1: $(21)(10) = 210$

Step 2: $210 = (-35)(-6)$ and $-35 - 6 = -41$

Step 3: $(21x^2 - 35x) + (-6x + 10)$

Step 4: $7x(3x - 5) - 2(3x - 5)$

Step 5: $(7x - 2)(3x - 5)$

≡ Note   Remember to look for any common factors before you start to use either the trial and error or the grouping method.

EXAMPLE 7.36

Factor $20x^3 + 22x^2 - 12x$.

Solution   There is a common factor of $2x$, so

$$20x^3 + 22x^2 - 12x = 2x(10x^2 + 11x - 6)$$

$$= 2x(5x - 2)(2x + 3)$$

Not all trinomials have just one variable. The grouping method is the best method to use when factoring trinomials that have more than one variable. For example, as shown in the next example, $2x^2 - 17xy + 36y^2$ can be factored using the grouping method.

EXAMPLE 7.37

Factor $2x^2 - 17xy + 36y^2$ completely.

Solution

Step 1: $2x^2(36y^2) = 72x^2y^2$

Step 2: $72x^2y^2 = (-9xy)(-8xy)$ and $-9xy - 8xy = -17xy$

Step 3: $(2x^2 - 9xy) + (-8xy + 36y^2)$

Step 4: $x(2x - 9y) - 4y(2x - 9y)$

EXAMPLE 7.37 (Cont.)

Step 5: $(x - 4y)(2x - 9y)$

Multiplying the two factors in Step 5 gives the original expression, and so we can say that $2x^2 - 17xy + 36y^2 = (x - 4y)(2x - 9y)$.

It is also possible to factor trinomials with powers greater than 2 if one exponent is twice the other.

EXAMPLE 7.38

Factor $x^8 + 5x^4 + 6$.

Solution
$$x^8 + 5x^4 + 6 = (x^4)^2 + 5x^4 + 6$$
$$= (x^4 + 3)(x^4 + 2)$$

Two other useful methods are based on special products #9 and #10. These are the sum and difference of two cubes.

Sum of Two Cubes $\quad (x + y)(x^2 - xy + y^2) = x^3 + y^3$	#9

Difference of Two Cubes $\quad (x - y)(x^2 + xy + y^2) = x^3 - y^3$	#10

EXAMPLE 7.39

Factor $8x^3 + 125$.

Solution $8x^3 = (2x)^3$ and $125 = 5^3$, so this is the sum of two cubes, and $8x^3 + 125 = (2x)^3 + 5^3 = (2x + 5)(4x^2 - 10x + 25)$.

Application

EXAMPLE 7.40

The volume V of a box of height x can be given by $V(x) = 4x^3 - 64x^2 + 252x$. Factor the right-hand side of this equation to determine the "realistic" domain for the height x.

Solution First, we notice that each term on the right-hand side of $V(x) = 4x^3 - 64x^2 + 252x$ has a common factor of $4x$. So, $V(x) = 4x(x^2 - 16x + 63)$. The quadratic expression factors as $x^2 - 16x + 63 = (x - 9)(x - 7)$. So, the completely factored form is

$$V(x) = 4x(x - 9)(x - 7)$$

From this, we see that the domain is $0 < x < 7$ or $x > 9$. We will see later that $x > 9$ will not satisfy many methods used to make such a box.

The six special factors are listed here. Study them, and learn to recognize them either as factors or as one of the special factors.

The Special Factors	$ax + ay = a(x + y)$	#1
	$x^2 - y^2 = (x + y)(x - y)$	#2
	$x^2 + (a + b)x + ab = (x + a)(x + b)$	#5
	$acx^2 + (ad + bc)x + bd = (ax + b)(cx + d)$	#6
	$x^3 + y^3 = (x + y)(x^2 - xy + y^2)$	#9
	$x^3 - y^3 = (x - y)(x^2 + xy + y^2)$	#10

Exercise Set 7.3

Determine if each of the trinomials in Exercises 1–6 can be factored.

1. $x^2 + 9x - 8$

2. $x^2 + 7x - 8$

3. $3x^2 - 10x - 8$

4. $2x^2 + 16x + 14$

5. $5x^2 + 23x + 18$

6. $7x^2 - 5x + 16$

Factor each of the following trinomials completely.

7. $x^2 + 7x + 10$

8. $x^2 + 8x + 15$

9. $x^2 - 12x + 27$

10. $x^2 - 14x + 33$

11. $x^2 - 27x + 50$

12. $x^2 + 19x + 48$

13. $x^2 - x - 2$

14. $x^2 - 4x - 5$

15. $x^2 - 3x - 10$

16. $p^2 - 16p + 64$

17. $r^2 + 10r + 25$

18. $v^2 - 14v + 49$

19. $a^2 + 22a + 121$

20. $e^2 + 26e + 169$

21. $f^2 - 30f + 225$

22. $3x^2 + 4x + 1$

23. $6y^2 - 7y + 1$

24. $3p^2 + 5p + 2$

25. $7t^2 + 9t + 2$

26. $5a^2 + 14a - 3$

27. $7b^2 - 34b - 5$

28. $2y^2 + y - 6$

29. $4e^2 + 19e - 5$

30. $6m^2 - 19m + 3$

31. $3u^2 + 10u + 8$

32. $7r^2 + 13r - 2$

33. $9t^2 - 25t - 6$

34. $4x^2 + 8x + 3$

35. $6x^2 + 13x - 5$

36. $8y^2 - 8y - 6$

37. $15a^2 - 16a - 15$

38. $15d^2 + 16d - 15$

39. $15e^2 + 34e + 15$

40. $14a^2 - 39a + 10$

41. $10x^2 - 19x + 6$

42. $3x^2 + 18x + 27$

43. $3r^2 - 18r - 21$

44. $15x^2 + 50x + 35$

45. $49t^4 - 105t^3 + 14t^2$

46. $2y^4 - 9y^2 + 7$

47. $6x^2 - 11xy - 10y^2$

48. $4p^2 + 20pq + 25q^2$

49. $8a^2 - 14ab - 9b^2$

50. $6d^9 + 15d^5e^2 + 6de^4$

51. $a^3 - b^3$

52. $y^3 - 8$

53. $8x^3 - 27$

54. $64p^3 + 125t^6$

Solve Exercises 55–59.

55. *Electronics* The current i, in amperes, in a certain circuit varies with time, in seconds, according to the equation

$$i = 0.7t^2 - 2.1t - 2.8$$

Factor the right-hand side of this equation.

56. *Sheet metal technology* A box with an open top is made from a rectangular sheet of metal by cutting equal-sized squares from the corners and folding up the sides. If the length of the side of a square that is removed is x, then the volume of this box is given by $V = 180x - 58x^2 + 4x^3$. Factor this expression.

57. *Business* The cost C for a certain company to produce n items is given by the equation $C(n) = 0.0001n^3 - 0.2n^2 - 3n + 6{,}000$. Factor the right-hand side of this equation.

58. *Dynamics* A ball is thrown upward with a speed of 48 ft/s from the edge of a cliff 448 ft above the ground. The distance s of this ball above the ground

at any time t is given by $s(t) = -16t^2 + 48t + 448$. Factor the right-hand side of this equation.

59. *Ecology* An ecology center wants to make an experimental garden. A gravel border of uniform width will be placed around the rectangular garden. The garden is 10 m long and 6 m wide. The builder has only enough gravel to cover 36 m^2 to the desired depth. In order to determine the width of the border, the equation

$$(6 + 2x)(10 + 2x) - 60 = 36$$

must be solved. **(a)** Simplify this equation and **(b)** factor your answer.

≡ 7.4
FRACTIONS

Working with algebraic expressions in technical situations often requires work with algebraic fractions. Working with algebraic fractions is very similar to working with fractions based on real numbers. In this section, we will work with some fundamental properties of fractions, and in Sections 7.5 and 7.6 we will use these properties on the basic operations of addition, subtraction, multiplication, and division with fractions. We begin by stating the Fundamental Principle of Fractions.

> **Fundamental Principle of Fractions**
>
> Multiplying or dividing both the numerator and denominator of a fraction by the same number, except zero, results in a fraction that is equivalent to the original fraction. Two fractions are equivalent if their cross-products are equal.

EXAMPLE 7.41

Since $\dfrac{3}{4} = \dfrac{3 \cdot 5}{4 \cdot 5} = \dfrac{15}{20}$ then the fractions $\dfrac{15}{20}$ and $\dfrac{3}{4}$ are equivalent by the Fundamental Principle of Fractions.

You can check their cross-products to verify that $\dfrac{3}{4}$ and $\dfrac{15}{20}$ are equivalent. Since $(3)(20) = 60$ and $(4)(15) = 60$, their cross-products are equal and the fractions are equivalent.

EXAMPLE 7.42

$$\frac{a}{x} = \frac{a(xy)}{x(xy)} = \frac{axy}{x^2y}$$

Solution We can verify that $\dfrac{a}{x} = \dfrac{axy}{x^2y}$ by comparing their cross-products:

$$a(x^2y) = ax^2y \qquad \text{and} \qquad x(axy) = ax^2y$$

Since the cross-products are equal, the fractions are equivalent as well.

EXAMPLE 7.43

$$\frac{x+3}{x-2} = \frac{(x+3)(x-3)}{(x-2)(x-3)} = \frac{x^2-9}{x^2-5x+6}$$

As long as $x \neq 3$, we can say that $\dfrac{x+3}{x-2} = \dfrac{x^2-9}{x^2-5x+6}$.

Saying that two fractions are equivalent is another way of saying that they represent the same number. Remembering the rules for signed numbers in Section 1.3, we stated that the number of negative signs can be reduced (or increased) by twos without changing the value of the expression. The Fundamental Principle of Fractions says essentially the same thing.

EXAMPLE 7.44

$$\frac{-a}{b} = \frac{-a(-1)}{b(-1)} = \frac{a}{-b}$$

One of the most important applications of the Fundamental Principle of Fractions is in reducing a fraction to **lowest terms** or **simplest form**. A fraction is in lowest terms when the numerator and denominator have no factors in common other than +1.

EXAMPLE 7.45

Reduce $\dfrac{14ab^2x}{7ax^3}$ to simplest form.

Solution The numerator and denominator have a common factor of $7ax$. We can then write $\dfrac{14ab^2x}{7ax^3} = \dfrac{7ax(2b^2)}{7ax(x^2)} = \dfrac{2b^2}{x^2}$. Notice that we used a part of the Fundamental Principle of Fractions that we had not used before. We *divided* both the numerator and the denominator by $7ax$, the common factor.

EXAMPLE 7.46

Reduce $\dfrac{x^2(x-3)}{x^2-9}$ to lowest terms.

Solution We need to find a common factor of both the numerator and the denominator; x^2 is *not* that common factor. While x^2 is a factor of the numerator, it is not a factor of the denominator. However, the denominator will factor into $(x-3)(x+3)$, so

$$\frac{x^2(x-3)}{x^2-9} = \frac{x^2(x-3)}{(x-3)(x+3)} = \frac{x^2}{x+3}$$

Again, we used the Fundamental Principle of Fractions and divided both the numerator and denominator by $x-3$.

≡ **Note**   Some factors differ only in sign. While this difference may not seem like much, it is often overlooked. In particular, you should note that

$$x - y = (-1)(-x + y) = -(-x + y) = -(y - x)$$

Of these four, the first and last are the ones you will use the most. Remember, $x - y$ and $y - x$ differ only in sign.

EXAMPLE 7.47

Reduce $\dfrac{5x - xy}{3y - 15}$ to lowest terms.

Solution   First, factor the numerator and the denominator.

$$\frac{5x - xy}{3y - 15} = \frac{x(5 - y)}{3(y - 5)}$$

Since $5 - y = -(y - 5)$, replace $5 - y$ with $-(y - 5)$.

$$\frac{x(5 - y)}{3(y - 5)} = \frac{-x(y - 5)}{3(y - 5)} = \frac{-x}{3}$$

Is $\dfrac{-x}{3}$ equal to $\dfrac{5x - xy}{3y - 15}$? You can always verify your answer by checking the cross-products. In this case, the cross-products are equal.

EXAMPLE 7.48

Reduce $\dfrac{2x^3 + 4x^2 - 30x}{15x^2 + x^3 - 2x^4}$ to simplest form.

Solution   First remove the common factors.

$$\frac{2x^3 + 4x^2 - 30x}{15x^2 + x^3 - 2x^4} = \frac{2x(x^2 + 2x - 15)}{x^2(15 + x - 2x^2)}$$

Factor each of the expressions in parentheses.

$$\frac{2x(x^2 + 2x - 15)}{x^2(15 + x - 2x^2)} = \frac{2x(x + 5)(x - 3)}{x^2(5 + 2x)(3 - x)}$$

Notice that $x - 3 = -(3 - x)$ and rewrite the numerator.

$$\frac{-2x(x + 5)(3 - x)}{x^2(5 + 2x)(3 - x)}$$

Divide both numerator and denominator by $x(3 - x)$, obtaining

$$\frac{-2x(x + 5)(3 - x)}{x^2(5 + 2x)(3 - x)} = \frac{-2(x + 5)}{x(5 + 2x)}$$

and we have the final result.

$$\frac{2x^3 + 4x^2 - 30x}{15x^2 + x^3 - 2x^4} = \frac{-2(x + 5)}{x(5 + 2x)}$$

Exercise Set 7.4

In Exercises 1–10, multiply the numerator and denominator of each fraction by the given factor. Check the cross-products to verify that your answer is equivalent to the given fraction.

1. $\dfrac{7}{8}$ (by 5)

2. $\dfrac{-5}{9}$ (by −4)

3. $\dfrac{x}{y}$ (by a)

4. $\dfrac{r}{t}$ (by z)

5. $\dfrac{x^2 y}{a}$ (by $3ax$)

6. $\dfrac{a^3 b}{ca}$ (by $3ab$)

7. $\dfrac{4}{x-y}$ (by $x+y$)

8. $\dfrac{a+b}{4}$ (by $a-4$)

9. $\dfrac{a+b}{a-b}$ (by $a+b$)

10. $\dfrac{x-2}{x+3}$ (by $x-3$)

In Exercises 11–18, divide the numerator and denominator of each of the fractions by the given factor. Check the cross-products to verify that your answer is equivalent to the given fraction.

11. $\dfrac{38}{24}$ (by 2)

12. $\dfrac{51}{119}$ (by 17)

13. $\dfrac{3x^2}{12x}$ (by $3x$)

14. $\dfrac{15a^3 x^2}{3a^4 x}$ (by $3a^3 x$)

15. $\dfrac{4(x+2)}{(x+2)(x-3)}$ (by $x+2$)

16. $\dfrac{7(x-3)(x+5)}{14(x+5)(x-1)}$ [by $7(x+5)$]

17. $\dfrac{x^2-16}{x^2+8x+16}$ (by $x+4$)

18. $\dfrac{(x-a)(x-b)(x-c)}{(x-c)(x-b)(x-d)}$ [by $(x-c)(x-b)$]

In Exercises 19–40, reduce each fraction to lowest terms.

19. $\dfrac{4x^2}{12x}$

20. $\dfrac{9y}{3y^2}$

21. $\dfrac{x^2+3x}{x^3+5x}$

22. $\dfrac{y^2-4y}{2y+y^3}$

23. $\dfrac{6m^2-3m^3}{9m+18m^3}$

24. $\dfrac{4r^2+12r^3}{8r+12r^2}$

25. $\dfrac{x^2+3x}{x^2-9}$

26. $\dfrac{a^2-9a}{a^2-81}$

27. $\dfrac{2b^2-10b}{3b^2-75}$

28. $\dfrac{4e^2-196}{14e-2e^2}$

29. $\dfrac{z^2-9}{z^2-6z+9}$

30. $\dfrac{x^2-16}{x^2+8x+16}$

31. $\dfrac{x^2+4x+3}{x^2+7x+12}$

32. $\dfrac{a^2-5a+6}{a^2+5a-14}$

33. $\dfrac{2x^2+9x+4}{x^2+9x+20}$

34. $\dfrac{15m^2-22m-5}{3m^2+4m-15}$

35. $\dfrac{12y^3+12y^2+3y}{6y^2-3y-3}$

36. $\dfrac{45x^2-60x+20}{6x^2+5x-6}$

37. $\dfrac{x^3 y^6-y^3 x^6}{2x^3 y^4-2x^4 y^3}$

38. $\dfrac{x^3-y^3}{y^2-x^2}$

39. $\dfrac{x^2-y^2}{x+y}$

40. $\dfrac{y-x}{x^2-y^2}$

≡ 7.5
MULTIPLICATION AND DIVISION OF FRACTIONS

The ability to simplify fractions is a skill that will be helpful in this section and the next, as we learn to operate with fractions. After that, it is a skill that will be required throughout this book.

In Section 1.3, we learned the basic operations with real numbers. Among those were Rules 7 and 8, which dealt with multiplying and dividing rational numbers. To refresh your memory, they are repeated here.

Rule 7

To multiply two rational numbers, multiply the numerators and multiply the denominators.

EXAMPLE 7.49

$$\frac{3}{4} \times \frac{-5}{8} = \frac{3 \times (-5)}{4 \times 8} = \frac{-15}{32}$$

Rule 8

To divide one rational number by another, multiply the first by the reciprocal of the second.

EXAMPLE 7.50

Compute $\frac{-5}{8} \div \frac{2}{3}$.

Solution   The reciprocal of $\frac{2}{3}$ is $\frac{3}{2}$, so

$$\frac{-5}{8} \div \frac{2}{3} = \frac{-5}{8} \times \frac{3}{2}$$
$$= \frac{(-5)(3)}{(8)(2)} = \frac{-15}{16}$$

If we express these two rules symbolically, we will have the rules for multiplying or dividing any two fractions, whether they are rational numbers or algebraic fractions. The rule for multiplying two rational numbers can be restated as the following.

Multiplying Fractions

If $\frac{a}{b}$ and $\frac{c}{d}$ are fractions, then their product is

$$\frac{a}{b} \cdot \frac{c}{d} = \frac{ac}{bd}$$

Similarly, we can state the following rule for dividing one rational number by another.

> **Dividing Fractions**
>
> If $\dfrac{a}{b}$ and $\dfrac{c}{d}$ are fractions, then their quotient is
>
> $$\frac{a}{b} \div \frac{c}{d} = \frac{a}{b} \times \frac{d}{c} = \frac{ad}{bc}$$

EXAMPLE 7.51

Find the product of $\dfrac{3a^2}{x}$ and $\dfrac{7y}{5p^3}$.

Solution $\quad \dfrac{3a^2}{x} \cdot \dfrac{7y}{5p^3} = \dfrac{(3a^2)(7y)}{(x)(5p^3)} = \dfrac{21a^2y}{5xp^3}$

EXAMPLE 7.52

Find the product of $\dfrac{x-2}{x+3}$ and $\dfrac{x+2}{x-5}$.

Solution $\quad \dfrac{x-2}{x+3} \cdot \dfrac{x+2}{x-5} = \dfrac{(x-2)(x+2)}{(x+3)(x-5)}$

$$= \dfrac{x^2 - 4}{x^2 - 2x - 15}$$

EXAMPLE 7.53

Find the quotient when $\dfrac{7x^2}{3a}$ is divided by $\dfrac{5y}{4x}$.

Solution $\quad \dfrac{7x^2}{3a} \div \dfrac{5y}{4x} = \dfrac{7x^2}{3a} \cdot \dfrac{4x}{5y}$

$$= \dfrac{(7x^2)(4x)}{(3a)(5y)}$$

$$= \dfrac{28x^3}{15ay}$$

EXAMPLE 7.54

Find $\dfrac{x-2}{x+3} \div \dfrac{x-4}{x-3}$.

Solution $\quad \dfrac{x-2}{x+3} \div \dfrac{x-4}{x-3} = \dfrac{x-2}{x+3} \cdot \dfrac{x-3}{x-4} = \dfrac{(x-2)(x-3)}{(x+3)(x-4)}$

$$= \dfrac{x^2 - 5x + 6}{x^2 - x - 12}$$

≡ **Note** It is often beneficial to leave the answer in factored form. For example, in Example 7.54, you might have wanted to leave the answer as $\dfrac{(x-2)(x-3)}{(x+3)(x-4)}$. This version has 2 multiplication operations, 1 division operation, 3 subtraction operations, and

1 addition operation, for a total of 7 operations. The answer $\dfrac{x^2 - 5x + 6}{x^2 - x - 12}$ has 3 multiplication operations, 1 division operation, 3 subtractions, and 1 addition, for a total of 8 operations. In computer programming, more operations require more computer time, which costs more money.

Caution When you divide, make sure that you multiply by the reciprocal of the *divisor*. Do not invert the dividend.

The following examples use all of the skills we have learned for simplifying fractions.

EXAMPLE 7.55

Multiply $\dfrac{x^2 - 9}{4x - 8}$ and $\dfrac{2x + 8}{x + 3}$.

Solution If we proceed as before, we get

$$\frac{x^2 - 9}{4x - 8} \cdot \frac{2x + 8}{x + 3} = \frac{(x^2 - 9)(2x + 8)}{(4x - 8)(x + 3)} = \frac{2x^3 + 8x^2 - 18x - 72}{4x^2 + 4x - 24}$$

While this is correct, it is not the easiest approach. It is a good idea to study a problem for a few seconds before you start to work it. If we had stopped to factor these fractions we could have saved some work.

$$\frac{x^2 - 9}{4x - 8} \cdot \frac{2x + 8}{x + 3} = \frac{(x + 3)(x - 3)}{4(x - 2)} \cdot \frac{2(x + 4)}{(x + 3)}$$

$$= \frac{2(x + 3)(x - 3)(x + 4)}{4(x - 2)(x + 3)}$$

$$= \frac{(x - 3)(x + 4)}{2(x - 2)}$$

or $\qquad\qquad = \dfrac{x^2 + x - 12}{2x - 4}$

EXAMPLE 7.56

Compute $\dfrac{2x^2 + 9x - 5}{3x^2 - 3x - 60} \cdot \dfrac{3x + 12}{2x + 10}$.

Solution $\dfrac{2x^2 + 9x - 5}{3x^2 - 3x - 60} \cdot \dfrac{3x + 12}{2x + 10} = \dfrac{(2x - 1)(x + 5)(3)(x + 4)}{3(x - 5)(x + 4)(2)(x + 5)}$

$$= \frac{2x - 1}{2(x - 5)}$$

The common factor $3(x + 5)(x + 4)$ is easily seen using this procedure.

Caution Be sure to factor the numerator and denominator first. Only common factors can be "canceled."

 Hint When a polynomial is factored, all the + and − signs are inside parentheses.

EXAMPLE 7.57

Compute $\dfrac{x^2 - 9}{6x^2 - 21x} \div \dfrac{(x+3)^2}{2x - 7}$.

Solution $\dfrac{x^2 - 9}{6x^2 - 21x} \div \dfrac{(x+3)^2}{2x - 7} = \dfrac{x^2 - 9}{6x^2 - 21x} \cdot \dfrac{2x - 7}{(x+3)^2}$

$$= \dfrac{(x-3)(x+3)(2x-7)}{3x(2x-7)(x+3)(x+3)}$$

$$= \dfrac{x-3}{3x(x+3)} = \dfrac{x-3}{3x^2 + 9x}$$

There are times when it is just as useful to leave the final answer in the factored form rather than multiplying the factors together. Thus, we could have left the last answer in the form $\dfrac{x-3}{3x(x+3)}$.

EXAMPLE 7.58

Simplify $\dfrac{\dfrac{6x}{x^2 - 4}}{\dfrac{2x^2 + 10x}{x + 2}}$.

Solution   We have to remember that $\dfrac{a}{b}$ means $a \div b$. So, this is the division problem:

$$\dfrac{6x}{x^2 - 4} \div \dfrac{2x^2 + 10x}{x + 2} = \dfrac{6x}{x^2 - 4} \cdot \dfrac{x + 2}{2x^2 + 10x}$$

$$= \dfrac{6x}{(x-2)(x+2)} \cdot \dfrac{x + 2}{2x(x + 5)}$$

$$= \dfrac{6x(x + 2)}{(x-2)(x+2)2x(x + 5)}$$

$$= \dfrac{3}{(x-2)(x + 5)}$$

Exercise Set 7.5

In Exercises 1–40, perform the indicated operation and simplify.

1. $\dfrac{2}{x} \cdot \dfrac{5}{y}$

2. $\dfrac{4}{y} \cdot \dfrac{x}{3}$

3. $\dfrac{4x^2}{5} \cdot \dfrac{3}{y^3}$

4. $\dfrac{7x}{6} \cdot \dfrac{5y}{2t}$

5. $\dfrac{3}{x} \div \dfrac{7}{y}$

6. $\dfrac{a}{3} \div \dfrac{b}{4}$

7. $\dfrac{2x^2}{3} \div \dfrac{7y}{4x}$

8. $\dfrac{9x}{2y} \div \dfrac{4y}{3a}$

9. $\dfrac{2x}{3y} \cdot \dfrac{5}{4x^2}$

10. $\dfrac{3xy}{7} \cdot \dfrac{14x}{5y^2}$

11. $\dfrac{3a^2b}{5d} \cdot \dfrac{25ad^2}{6b^2}$

12. $\dfrac{x^2y^2t}{abc} \cdot \dfrac{b^2c}{y^3t}$

13. $\dfrac{3y}{5x} \div \dfrac{15x^2}{8xy}$

14. $\dfrac{4y^2}{7x} \div \dfrac{8y^3}{21x}$

15. $\dfrac{3x^2y}{7p} \div \dfrac{15x^2p}{7y^2}$

16. $\dfrac{9xyz}{7a} \div \dfrac{3ayz}{14z}$

17. $\dfrac{4y+16}{5} \cdot \dfrac{15y}{3y+12}$

18. $\dfrac{x^2+3x}{6a} \cdot \dfrac{a^2}{x^2-9}$

19. $\dfrac{a^2-b^2}{a+3b} \cdot \dfrac{5a+15b}{a+b}$

20. $(x+y)\dfrac{x^2+2x}{x^2-y^2}$

21. $\dfrac{x^2-100}{10} \div \dfrac{2x+10}{15}$

22. $\dfrac{5a^2}{x^2-49} \div \dfrac{25ax-25a}{x^2+7x}$

23. $\dfrac{4x^2-1}{9x-3x^2} \div \dfrac{2x+1}{x^2-9}$

24. $\dfrac{x+y}{3x-3y} \div \dfrac{(x+y)^2}{x^2-y^2}$

25. $\dfrac{a^2-8a}{a-8} \cdot \dfrac{a+2}{a}$

26. $\dfrac{49-x^2}{x+y} \cdot \dfrac{x}{7-x}$

27. $\dfrac{2a-b}{4a} \cdot \dfrac{2a-b}{4a^2-4ab+b^2}$

28. $\dfrac{x^4-81}{(x-3)^2} \cdot \dfrac{x-3}{4-x^2}$

29. $\dfrac{y^2}{x^2-1} \div \dfrac{y^2}{x-1}$

30. $\dfrac{m^2-49}{m^2-5m-14} \div \dfrac{m+7}{2m^2-13m-7}$

31. $\dfrac{2y^2-y}{4y^2-4y+1} \div \dfrac{y^2}{8y-4}$

32. $\dfrac{a-1}{a^2-1} \div \dfrac{(a-1)^2}{a^2-1}$

33. $\dfrac{x^2-3x+2}{x^2+5x+6} \cdot \dfrac{x+3}{3x-6}$

34. $\dfrac{2x+2}{x^2+2x-8} \cdot \dfrac{x^2-4}{x^2+4x+4}$

35. $\dfrac{x^2+xy-6y^2}{x^2+6xy+8y^2} \cdot \dfrac{x^2-9xy+20y^2}{x^2-4xy-21y^2}$

36. $\dfrac{y^2+14xy+49x^2}{y^2-7xy-30x^2} \cdot \dfrac{y^2-100x^2}{y^3+7xy^2}$

37. $\dfrac{9x^2-25}{x^2+6x+9} \div \dfrac{3x+5}{x+3}$

38. $\dfrac{x^2-16}{x^2-6x+8} \div \dfrac{x^3+4x^2}{x^2-9x+14}$

39. $\dfrac{x^2+4xy+4y^2}{x^2-4y^2} \div \dfrac{x^2+xy-2y^2}{x^2-xy-2y^2}$

40. $\dfrac{p^3-27q^3}{3p^2+9pq+27q^2} \div \dfrac{9q^2-p^2}{6p+18q}$

≡ 7.6
ADDITION AND SUBTRACTION OF FRACTIONS

In Section 7.5, we learned how to multiply and divide two fractions. In this section, we will look at two other operations with fractions—addition and subtraction.

As we mentioned earlier, much of this work with algebraic fractions is patterned after our work with rational numbers from Section 1.3 . Rule 6 dealt with the addition and subtraction of rational numbers. This rule is repeated here.

Rule 6

To add (or subtract) two rational numbers, change both denominators to the same positive integer (the common denominator), add (or subtract) the numerators, and place the result over the common denominator.

EXAMPLE 7.59

Perform the indicated operations and simplify **(a)** $\frac{2}{3} + \frac{-5}{6}$ and **(b)** $\frac{-5}{7} - \frac{-8}{3}$.

Solutions

(a) A common denominator of 3 and 6 is 6, so

$$\frac{2}{3} + \frac{-5}{6} = \frac{4}{6} + \frac{-5}{6} = \frac{4+(-5)}{6} = \frac{-1}{6}$$

(b) A common denominator of 7 and 3 is 21, so

$$\frac{-5}{7} = \frac{-15}{21} \qquad \text{and} \qquad \frac{-8}{3} = \frac{-56}{21}$$

EXAMPLE 7.59 (Cont.)

As a result, we obtain

$$\frac{-5}{7} - \frac{-8}{3} = \frac{-15}{21} - \frac{-56}{21} = \frac{-15 - (-56)}{21} = \frac{41}{21}$$

Before we restate Rule 6 in symbols, we will consider a special case of adding or subtracting fractions. If two fractions have the same denominator, then you need only add or subtract the numerators. Symbolically, this is represented as:

$$\frac{a}{c} + \frac{b}{c} = \frac{a+b}{c} \qquad \text{and} \qquad \frac{a}{c} - \frac{b}{c} = \frac{a-b}{c}$$

EXAMPLE 7.60

Simplify (a) $\dfrac{7}{2x} + \dfrac{9y}{2x}$ and (b) $\dfrac{x}{x+5} - \dfrac{2x-y}{x+5}$.

Solutions

(a) $\dfrac{7}{2x} + \dfrac{9y}{2x} = \dfrac{7+9y}{2x}$

(b) $\dfrac{x}{x+5} - \dfrac{2x-y}{x+5} = \dfrac{x-(2x-y)}{x+5}$

$$= \frac{x-2x+y}{x+5}$$

$$= \frac{-x+y}{x+5}$$

If the denominators are not the same, then addition and subtraction become somewhat more complicated. As Rule 6 indicates, you need to find a common denominator and rewrite each fraction as an equivalent fraction with this common denominator. The quickest way to find a common denominator is to multiply the denominators together. This method is demonstrated here.

Adding and Subtracting Fractions

The sum of two fractions $\dfrac{a}{b}$ and $\dfrac{c}{d}$ is

$$\frac{a}{b} + \frac{c}{d} = \frac{ad}{bd} + \frac{bc}{bd} = \frac{ad+bc}{bd}$$

The difference of two fractions $\dfrac{a}{b}$ and $\dfrac{c}{d}$ is

$$\frac{a}{b} - \frac{c}{d} = \frac{ad}{bd} - \frac{bc}{bd} = \frac{ad-bc}{bd}$$

As you will see in Example 7.61(b), this may not be the lowest common denominator. You must also remember that both the numerator and denominator must by multiplied by the same quantity.

Caution
When adding or subtracting fractions, remember to multiply both the numerator and denominator of a fraction by the same quantity.

EXAMPLE 7.61

Simplify **(a)** $\dfrac{3}{x+5} + \dfrac{x}{x-5}$ and **(b)** $\dfrac{2x}{x+3} - \dfrac{x-4}{x^2-9}$.

Solutions

(a) $\dfrac{3}{x+5} + \dfrac{x}{x-5} = \dfrac{3}{x+5} \cdot \dfrac{x-5}{x-5} + \dfrac{x}{x-5} \cdot \dfrac{x+5}{x+5}$

$= \dfrac{3(x-5)}{(x+5)(x-5)} + \dfrac{x(x+5)}{(x+5)(x-5)}$

$= \dfrac{3(x-5) + (x^2+5x)}{(x+5)(x-5)}$

$= \dfrac{3x - 15 + x^2 + 5x}{(x+5)(x-5)}$

$= \dfrac{x^2 + 8x - 15}{x^2 - 25}$

(b) $\dfrac{2x}{x+3} - \dfrac{x-4}{x^2-9} = \dfrac{2x}{x+3} \cdot \dfrac{x^2-9}{x^2-9} - \dfrac{x-4}{x^2-9} \cdot \dfrac{x+3}{x+3}$

$= \dfrac{2x(x^2-9)}{(x+3)(x^2-9)} - \dfrac{(x-4)(x+3)}{(x+3)(x^2-9)}$

$= \dfrac{(2x^3 - 18x) - (x^2 - x - 12)}{(x+3)(x^2-9)}$

$= \dfrac{2x^3 - x^2 - 17x + 12}{x^3 + 3x^2 - 9x - 27}$

Least Common Denominator

Perhaps you wondered if the last answer was in simplest form. It is not, since

$$\frac{2x^3 - x^2 - 17x + 12}{x^3 + 3x^2 - 9x - 27} = \frac{(x+3)(2x^2 - 7x + 4)}{(x+3)(x^2 - 9)} = \frac{2x^2 - 7x + 4}{x^2 - 9}$$

If we want our answers in the simplest form, then simply multiplying the denominators together is not the best method to use. What we need to do is determine the least common denominator, or LCD, of the fractions to be added or subtracted.

There are three steps to determining the LCD:

How to Find the Least Common Denominator

(1) Factor the denominator of each of the fractions in the problem,

(2) Determine the different factors and the highest power of each factor that occurs in any denominator, and

(3) Multiply the distinct factors from Step 2 after each has been raised to its highest power.

EXAMPLE 7.62

Find the LCD of the fractions $\dfrac{7x+1}{x^4+x^3}$, $\dfrac{14}{x^3-4x^2+4x}$, and $\dfrac{9}{2x^2-2x-4}$.

Solution

Step 1: Factor the denominator of each of the fractions: $\dfrac{7x+1}{x^3(x+1)}$, $\dfrac{14}{x(x-2)^2}$, and

$\dfrac{9}{2(x+1)(x-2)}$.

Step 2: List each factor and the highest exponent of each.

factor	highest exponent	final factors
2	1	2^1
x	3	x^3
$x+1$	1	$(x+1)^1$
$x-2$	2	$(x-2)^2$

Step 3: The LCD is $2x^3(x+1)(x-2)^2$.

EXAMPLE 7.63

Find the least common denominator of $\dfrac{2x}{x^2+5x+6}$, $\dfrac{x-3}{x^3+2x^2}$, and $\dfrac{x^2+x}{x^3+6x^2+9x}$.

Solution

Step 1: Factor each denominator: $\dfrac{2x}{(x+2)(x+3)}$, $\dfrac{x-3}{x^2(x+2)}$, and $\dfrac{x^2+x}{x(x+3)^2}$.

Step 2: List each factor and the highest exponent of each.

factor	highest exponent	final factors
$x+2$	1	$x+2$
$x+3$	2	$(x+3)^2$
x	2	x^2

Step 3: The LCD is $x^2(x+2)(x+3)^2$.

Now we have the foundation for a much better way to add or subtract algebraic fractions, or any fractions. For each fraction, multiply both the numerator and denominator by a quantity that makes the denominator equal to the LCD of the fractions being added or subtracted. Then, add or subtract the numerators; place the result over the common denominator; and, if possible, simplify.

EXAMPLE 7.64

Calculate $\dfrac{2x}{x+3} - \dfrac{x-4}{x^2-9}$.

Solution This is the same difference we were asked to compute in Example 7.61(b). First we find the LCD, which is $(x+3)(x-3)$.

EXAMPLE 7.64 (Cont.)

We rewrite the first fraction as $\dfrac{2x}{x+3} = \dfrac{2x(x-3)}{(x+3)(x-3)}$. The second fraction,

$\dfrac{x-4}{x^2-9}$, is already written with the common denominator. So,

$$\frac{2x}{x+3} - \frac{x-4}{x^2-9} = \frac{2x(x-3)}{(x+3)(x-3)} - \frac{x-4}{x^2-9}$$

$$= \frac{2x(x-3) - (x-4)}{x^2-9}$$

$$= \frac{(2x^2-6x) - (x-4)}{x^2-9}$$

$$= \frac{2x^2-7x+4}{x^2-9}$$

This was the same problem we worked in Example 7.61(b). Notice how much simpler this answer looks compared to the answer we found before.

EXAMPLE 7.65

Calculate $\dfrac{2x}{x+2} + \dfrac{x}{x-2} - \dfrac{1}{x^2-4}$.

Solution   The LCD of these fractions is $(x+2)(x-2)$. So,

$$\frac{2x}{x+2} = \frac{2x(x-2)}{(x+2)(x-2)},$$

$$\frac{x}{x-2} = \frac{x(x+2)}{(x-2)(x+2)},$$

and $$\frac{1}{x^2-4} = \frac{1}{(x-2)(x+2)}.$$

Thus, we get

$$\frac{2x}{x+2} + \frac{x}{x-2} - \frac{1}{x^2-4} = \frac{2x(x-2)}{(x+2)(x-2)} + \frac{x(x+2)}{(x-2)(x+2)}$$

$$- \frac{1}{x^2-4}$$

$$= \frac{(2x^2-4x) + (x^2+2x) - 1}{x^2-4}$$

$$= \frac{3x^2-2x-1}{x^2-4}$$

Complex Fractions

A **complex fraction** is a fraction in which the numerator, the denominator, or both, contain a fraction. There are two methods that are commonly used to simplify complex fractions.

Method 1: Find the LCD of all the fractions that appear in the numerator and denominator. Multiply both the numerator and denominator by the LCD.

Method 2: Combine the terms in the numerator into a single fraction. Combine the terms in the denominator into a single fraction. Divide the numerator by the denominator.

    We will work each of the next two examples using both methods. Then you will be better able to select the method you prefer.

EXAMPLE 7.66

Simplify $\dfrac{2 + \dfrac{1}{x}}{x - \dfrac{2}{x^2}}$.

Solution

Method 1: The LCD of 2, $\dfrac{1}{x}$, x, and $\dfrac{2}{x^2}$ is x^2, so

$$\frac{2 + \dfrac{1}{x}}{x - \dfrac{2}{x^2}} = \frac{2 + \dfrac{1}{x}}{x - \dfrac{2}{x^2}} \cdot \frac{x^2}{x^2}$$

$$= \frac{2x^2 + x}{x^3 - 2}$$

$$= \frac{x(2x + 1)}{x^3 - 2}$$

Method 2:

$$2 + \frac{1}{x} = \frac{2x}{x} + \frac{1}{x} = \frac{2x + 1}{x}$$

$$x - \frac{2}{x^2} = \frac{x^3}{x^2} - \frac{2}{x^2} = \frac{x^3 - 2}{x^2}$$

$$\frac{2 + \dfrac{1}{x}}{x - \dfrac{2}{x^2}} = \frac{\dfrac{2x + 1}{x}}{\dfrac{x^3 - 2}{x^2}} = \frac{2x + 1}{x} \div \frac{x^3 - 2}{x^2}$$

$$= \frac{2x + 1}{x} \cdot \frac{x^2}{x^3 - 2} = \frac{x(2x + 1)}{x^3 - 2}$$

💡 **Hint**   Method 1 is often simpler and you are less likely to make a mistake with this method.

EXAMPLE 7.67

Simplify $\dfrac{\dfrac{1}{2x} - \dfrac{6}{y}}{\dfrac{1}{x} + \dfrac{2}{3y}}$.

Solution

Method 1: The LCD of $\dfrac{1}{2x}$, $\dfrac{6}{y}$, $\dfrac{1}{x}$, and $\dfrac{2}{3y}$ is $6xy$, so

$$\frac{\dfrac{1}{2x} - \dfrac{6}{y}}{\dfrac{1}{x} + \dfrac{2}{3y}} = \frac{\left(\dfrac{1}{2x} - \dfrac{6}{y}\right)}{\left(\dfrac{1}{x} + \dfrac{2}{3y}\right)} \cdot \frac{6xy}{6xy}$$

$$= \frac{3y - 36x}{6y + 4x} = \frac{3(y - 12x)}{2(3y + 2x)}$$

Method 2:

$$\frac{1}{2x} - \frac{6}{y} = \frac{y}{2xy} - \frac{12x}{2xy} = \frac{y - 12x}{2xy}$$

$$\frac{1}{x} + \frac{2}{3y} = \frac{3y}{3xy} + \frac{2x}{3xy} = \frac{3y + 2x}{3xy}$$

$$\frac{\dfrac{1}{2x} - \dfrac{6}{y}}{\dfrac{1}{x} + \dfrac{2}{3y}} = \frac{\dfrac{y - 12x}{2xy}}{\dfrac{3y + 2x}{3xy}}$$

$$= \frac{y - 12x}{2xy} \div \frac{3y + 2x}{3xy}$$

$$= \frac{y - 12x}{2xy} \cdot \frac{3xy}{3y + 2x}$$

$$= \frac{3(y - 12x)}{2(3y + 2x)}$$

Application

EXAMPLE 7.68

If four resistances are connected in a series-parallel circuit, the total resistance R is given by the equation

$$R = \frac{1}{R_1 + R_2} + \frac{1}{R_3 + R_4}$$

where R_1, R_2, R_3, and R_4 represent four resistances. Simplify this equation by adding the right-hand side.

Solution We find that the LCD of the right-hand side of the given equation is $(R_1 + R_2)(R_3 + R_4)$. So,

$$R = \frac{1}{R_1 + R_2} \cdot \frac{R_3 + R_4}{R_3 + R_4} + \frac{1}{R_3 + R_4} \cdot \frac{R_1 + R_2}{R_1 + R_2}$$

EXAMPLE 7.68 (Cont.)

$$= \frac{R_3 + R_4}{(R_1 + R_2)(R_3 + R_4)} + \frac{R_1 + R_2}{(R_1 + R_2)(R_3 + R_4)}$$

$$= \frac{R_1 + R_2 + R_3 + R_4}{(R_1 + R_2)(R_3 + R_4)}$$

Exercise Set 7.6

In Exercises 1–44, perform the indicated operations and simplify.

1. $\dfrac{2}{7} + \dfrac{5}{7}$

2. $\dfrac{4}{5} + \dfrac{-11}{5}$

3. $\dfrac{7}{3} - \dfrac{5}{3}$

4. $\dfrac{-2}{9} - \dfrac{8}{9}$

5. $\dfrac{1}{2} + \dfrac{1}{3}$

6. $\dfrac{3}{4} + \dfrac{-2}{3}$

7. $\dfrac{4}{5} - \dfrac{2}{3}$

8. $-\dfrac{5}{7} - \dfrac{3}{5}$

9. $\dfrac{1}{x} + \dfrac{5}{x}$

10. $\dfrac{2}{y} + \dfrac{-5}{y}$

11. $\dfrac{4}{a} - \dfrac{3}{a}$

12. $\dfrac{-5}{p} - \dfrac{-7}{p}$

13. $\dfrac{2x}{y} + \dfrac{3x}{y}$

14. $\dfrac{4p}{q} - \dfrac{6p}{q}$

15. $\dfrac{3r}{2t} + \dfrac{-r}{2t} - \dfrac{5r}{2t}$

16. $\dfrac{3x}{2y} - \dfrac{5x}{2y} + \dfrac{x}{2y}$

17. $\dfrac{3}{x+2} + \dfrac{x}{x+2}$

18. $\dfrac{5}{y-3} + \dfrac{y}{y-3}$

19. $\dfrac{t}{t+1} - \dfrac{2}{t+1}$

20. $\dfrac{a}{b-3} - \dfrac{4}{3-b}$

21. $\dfrac{y-3}{x+2} + \dfrac{3+y}{x+2}$

22. $\dfrac{x+4}{x-2} + \dfrac{x-5}{x-2}$

23. $\dfrac{x+2}{a+b} - \dfrac{x-5}{a+b}$

24. $\dfrac{x+4}{y-5} - \dfrac{2-x}{y-5}$

25. $\dfrac{2}{x} + \dfrac{3}{y}$

26. $\dfrac{x}{y} + \dfrac{5}{x}$

27. $\dfrac{a}{b} - \dfrac{4}{d}$

28. $\dfrac{2x}{y} - \dfrac{3y}{x}$

29. $\dfrac{3}{x(x+1)} + \dfrac{4}{x^2 - 1}$

30. $\dfrac{5}{y(x+1)} + \dfrac{x}{y(x+2)}$

31. $\dfrac{2}{x^2 - 1} - \dfrac{4}{(x+1)^2}$

32. $\dfrac{6}{y-2} - \dfrac{3}{y+2}$

33. $\dfrac{x}{x^2 - 11x + 30} + \dfrac{2}{x^2 - 36}$

34. $\dfrac{a}{a^2 - 9a + 18} + \dfrac{a}{a^2 - 9}$

35. $\dfrac{2}{x^2 - x - 6} - \dfrac{5}{x^2 - 4}$

36. $\dfrac{b}{b^2 - 10b + 21} - \dfrac{b}{b^2 - 9}$

37. $\dfrac{x - 1}{3x^2 - 13x + 4} + \dfrac{3x + 1}{4x - x^2}$

38. $\dfrac{x - 3}{x^2 + 3x + 2} + \dfrac{2x - 5}{x^2 + x - 2}$

39. $\dfrac{x - 3}{x^2 - 1} + \dfrac{2x - 7}{x^2 + 5x + 4}$

40. $\dfrac{x + 4}{x^2 - 9} - \dfrac{x - 3}{x^2 + 6x + 9}$

41. $\dfrac{y + 3}{y^2 - y - 2} - \dfrac{2y - 1}{y^2 + 2y - 8}$

42. $\dfrac{1}{(a - b)(a - c)} + \dfrac{1}{(b - a)(b - c)} - \dfrac{1}{(b - c)(a - c)}$

43. $\dfrac{x}{(x^2 + 3)(x - 1)} + \dfrac{3x^2}{(x - 1)^2(x + 2)} - \dfrac{x + 2}{x^2 + 3}$

44. $\dfrac{2x - 1}{x^2 + 5x + 6} - \dfrac{x - 2}{x^2 + 4x + 3} + \dfrac{x - 4}{x^2 + 3x + 2}$

Use Method 1 to simplify each of the complex fractions in Exercises 45–50.

45. $\dfrac{1 + \frac{2}{x}}{1 - \frac{3}{x}}$

46. $\dfrac{x + \frac{1}{x}}{2 - \frac{1}{x}}$

47. $\dfrac{x - 1}{1 + \frac{1}{x}}$

48. $\dfrac{x^2 - 25}{\frac{1}{x} - \frac{1}{5}}$

49. $\dfrac{\frac{x}{x + y} - \frac{y}{x - y}}{\frac{x}{x + y} + \frac{y}{x - y}}$

50. $\dfrac{x + 3 - \frac{16}{x + 3}}{x - 6 + \frac{20}{x + 6}}$

Use Method 2 to simplify each of the complex fractions in Exercises 51–56.

51. $\dfrac{1 + \frac{3}{x}}{1 + \frac{2}{x}}$

52. $\dfrac{y + \frac{1}{y}}{3 + \frac{2}{y}}$

53. $\dfrac{t - 1}{t + \frac{1}{t}}$

54. $\dfrac{x^2 - 36}{\frac{1}{6} - \frac{1}{x}}$

55. $\dfrac{\frac{x}{x - y} - \frac{y}{x + y}}{\frac{1}{x - y} + \frac{1}{x + y}}$

56. $\dfrac{t - 5 + \frac{25}{t - 5}}{t + 3 + \frac{10}{t - 3}}$

Solve Exercises 57–60.

57. *Electricity* If two resistors, R_1 and R_2, are connected in parallel, the equivalent resistance of the combination can be found using $\dfrac{1}{R} = \dfrac{1}{R_1} + \dfrac{1}{R_2}$. Add the right-hand side of the equation.

58. *Optics* The lensmakers equation states that if p is the object distance, q the image distance, and f the focal length of a lens, then $\dfrac{1}{p} + \dfrac{1}{q} = \dfrac{1}{f}$. Simplify the left-hand side of this equation.

59. *Electricity* If three capacitors with capacitance C_1, C_2, and C_3 are connected together in series, then they can be replaced by a single capacitor of capacitance C. The value of C can be determined from the equation $\dfrac{1}{C} = \dfrac{1}{C_1} + \dfrac{1}{C_2} + \dfrac{1}{C_3}$. Simplify the right-hand side of the equation.

60. *Transportation* A car travels the first part of a trip for a distance d_1 at velocity v_1, and the second part of the trip it travels d_2 at the velocity v_2. The average speed for these two parts of the trip is given by

$\dfrac{d_1 + d_2}{\dfrac{d_1}{v_1} + \dfrac{d_2}{v_2}}$. Simplify this fraction.

☰ CHAPTER 7 REVIEW

Important Terms and Concepts

Common factor

Denominator

Discriminant

Factor

Fractions
   Addition
   Complex
   Division

Equivalent

Multiplication

Reducing

Subtraction

Least common denominator (LCD)

Numerator

Trinomial

Review Exercises

In Exercises 1–10, find the indicated products by direct use of one of the special products.

1. $5x(x - y)$

2. $(3 + x)^2$

3. $(x - 2y)^3$

4. $(x + y)(x - 6)$

5. $(2x + 3)(x - 6)$

6. $(x + 7)(x - 7)$

7. $(x^2 - 5)(x^2 + 5)$

8. $(7x - 1)(x + 5)$

9. $(2 + x)^3$

10. $(x - 7)^2$

Completely factor each of the expressions in Exercises 11–20.

11. $9 + 9y$

12. $x^2 - 4$

13. $7x^2 - 63$

14. $x^2 - 12x + 36$

15. $x^2 - 11x + 30$

16. $x^2 + 15x + 36$

17. $x^2 + 6x - 16$

18. $x^2 - 4x - 45$

19. $2x^2 - 3x - 9$

20. $8x^3 + 6x^2 - 20x$

In Exercises 21–26, reduce each fraction to lowest terms.

21. $\dfrac{2x}{6y}$

22. $\dfrac{7x^2 y}{9xy^2}$

23. $\dfrac{x^2 - 9}{(x + 3)^2}$

24. $\dfrac{x^2 - 4x - 45}{x^2 - 81}$

25. $\dfrac{x^3 + y^3}{x^2 + 2xy + y^2}$

26. $\dfrac{x^3 - 16x}{x^2 + 2x - 8}$

In Exercises 27–40, perform the indicated operations and simplify.

27. $\dfrac{x^2}{y} \cdot \dfrac{3y^2}{7x}$

28. $\dfrac{x^2 - 9}{x + 4} \cdot \dfrac{x^3 - 16x}{x - 3}$

29. $\dfrac{4x}{3y} \div \dfrac{2x^2}{6y}$

30. $\dfrac{x^2 - 25}{x^2 - 4x} \div \dfrac{2x^2 + 2x - 40}{x^3 - x}$

31. $\dfrac{4x}{y} + \dfrac{3x}{y}$

32. $\dfrac{4}{x - y} + \dfrac{6}{x + y}$

33. $\dfrac{3(x - 3)}{(x + 2)(x - 5)^2} + \dfrac{4(x - 1)}{(x + 2)^2(x - 5)}$

34. $\dfrac{8a}{b} - \dfrac{3}{b}$

35. $\dfrac{x}{y + x} - \dfrac{x}{y - x}$

36. $\dfrac{2(x + 3)}{(x + 1)^2(x + 2)} - \dfrac{3(x - 1)}{(x + 1)(x + 2)^2}$

37. $\dfrac{x^2 - 5x - 6}{x^2 + 8x + 12} + \dfrac{x^2 + 7x + 6}{x^2 - 4x - 12}$

38. $\dfrac{2x - 1}{4x^2 - 12x + 5} - \dfrac{x + 1}{4x^2 - 4x - 15}$

39. $\dfrac{x^2 - 5x - 6}{x^2 + 8x + 12} \div \dfrac{x^2 + 7x + 6}{x^2 - 4x - 12}$

40. $\dfrac{x^2 + x - 2}{7a^2x^2 - 14a^2x + 7a^2} \cdot \dfrac{14ax - 28a}{1 - 2x + x^2}$

Simplify each of the complex fractions in Exercises 41–46.

41. $\dfrac{\dfrac{1}{x} - \dfrac{1}{y}}{\dfrac{1}{x} + \dfrac{1}{y}}$

42. $\dfrac{\dfrac{1}{x} + \dfrac{1}{y}}{x + y}$

43. $\dfrac{\dfrac{1}{x} - \dfrac{1}{y}}{\dfrac{x - y}{xy}}$

44. $\dfrac{1 - \dfrac{1}{x}}{x - 2 + \dfrac{1}{x}}$

45. $\dfrac{\dfrac{x}{1 + x} - \dfrac{1 - x}{x}}{\dfrac{x}{1 + x} + \dfrac{1 - x}{x}}$

46. $\dfrac{x - \dfrac{xy}{x - y}}{\dfrac{x^2}{x^2 - y^2} - 1}$

☰ CHAPTER 7 TEST

1. Multiply $(x + 5)(x - 3)$.

2. Multiply $(2x - 3)(2x + 3)$.

3. Completely factor $2x^2 - 128$.

4. Completely factor $x^2 - 12x + 32$.

5. Reduce $\dfrac{x^2 - 25}{x^2 + 6x + 5}$ to lowest terms.

6. Calculate $\dfrac{3x}{x + 2} \cdot \dfrac{x - 1}{x + 2}$.

7. Calculate $\dfrac{2x + 6}{x - 2} \div \dfrac{3x + 9}{x^2 - 4}$.

8. Calculate $\dfrac{6}{x - 5} + \dfrac{x^2 - 2x}{x - 5}$.

9. Calculate $\dfrac{2x}{x + 3} - \dfrac{x + 4}{x - 2}$.

10. Simplify $\dfrac{x - \dfrac{1}{x}}{x - \dfrac{2}{x + 1}}$.

11. The total resistance, R, in a parallel electrical circuit with three resistances is given by

$$\frac{1}{R} = \frac{1}{R_1} + \frac{1}{R_2} + \frac{1}{R_3}$$

Express the sum on the right-hand side as a single fraction.

8

Fractional and Quadratic Equations

Quadratic equations are needed to determine lengths or other dimensions. In Sections 8.3 and 8.4, you will learn two ways to determine the lengths of the rafters in this solar collector.

Courtesy of New York State Energy Research and Development Authority

In Chapter 7, we learned the special products, how to factor some algebraic expressions, and how to simplify and operate with fractions. In this chapter, we will use that knowledge to help solve problems.

You already know something about solving equations; but your knowledge has been limited to working with linear equations. In this chapter, we will venture into learning techniques for solving two new types of equations—fractional equations and quadratic equations.

☰ 8.1
FRACTIONAL EQUATIONS

An equation in which one or more terms is a fraction is called a **fractional equation**. Solving a fractional equation requires a technique that we used in solving systems of linear equations. In order to add or subtract two linear equations in Section 5.2, you often had to multiply one or both equations by a nonzero number. To solve fractional equations, we will use that same technique—we will multiply the equation by a nonzero quantity. In particular, we will multiply the equation by the LCD. This is often referred to as **clearing the equation**.

The easiest type of fractional equations to solve are those in which the variables occur only in the numerator.

EXAMPLE 8.1

Solve $\frac{2x}{3} - \frac{3x}{5} = \frac{1}{10}$ for x.

Solution The LCD of $\frac{2x}{3}$, $\frac{3x}{5}$, and $\frac{1}{10}$ is 30, so we will multiply both sides of the equation by 30.

$$30\left(\frac{2x}{3} - \frac{3x}{5}\right) = 30\left(\frac{1}{10}\right)$$

$$30\left(\frac{2x}{3}\right) - 30\left(\frac{3x}{5}\right) = 30\left(\frac{1}{10}\right)$$

$$20x - 18x = 3$$

$$2x = 3$$

$$x = \frac{3}{2}$$

If we check our answer in the original problem, we see that $\dfrac{2\left(\frac{3}{2}\right)}{3} - \dfrac{3\left(\frac{3}{2}\right)}{5} = 1 - \frac{9}{10} = \frac{1}{10}$. The answer checks.

If the variables are in the denominator, we then need to use more caution.

Caution The original equation will not be defined for any values of the variable that give any of the denominators a value of 0. If you forget this, you may get an answer that does not satisfy the original problem. This type of answer is called an **extraneous solution** because it seems to be a solution but is not a valid one. For this reason, it is a good idea to study the equation first and to note any values that make the denominator 0.

In these next examples, we multiply the equation by the least common denominator for the fractions. Notice that we begin each solution by finding out which values make a denominator 0.

EXAMPLE 8.2

Solve $\dfrac{2}{x-5} = \dfrac{1}{4x-12}$ for x.

Solution   The LCD of $\dfrac{2}{x-5}$ and $\dfrac{1}{4x-12}$ is $4(x-5)(x-3)$. Since the LCD has a value of 0 when $x = 5$ or $x = 3$, neither of these values is a possible solution for this equation.

If we multiply both sides of the equation by the LCD, we obtain

$$4(x-5)(x-3)\left(\frac{2}{x-5}\right) = 4(x-5)(x-3)\left(\frac{1}{4x-12}\right)$$

$$4(x-3)(2) = x-5$$

$$8x - 24 = x - 5$$

$$7x = 19$$

$$x = \frac{19}{7}$$

Thus, $x = \frac{19}{7}$ appears to be the solution. But, we should check our work to ensure that we have made no errors.

Check: The left-hand side of the equation becomes

$$\frac{2}{\frac{19}{7}-5} = \frac{2}{\frac{19}{7}-\frac{35}{7}} = \frac{2}{\frac{-16}{7}} = -\frac{7}{8}$$

The value of the right-hand side is

$$\frac{1}{4\left(\frac{19}{7}\right)-12} = \frac{1}{\frac{76}{7}-12} = \frac{1}{\frac{76}{7}-\frac{84}{7}} = \frac{1}{\frac{-8}{7}} = -\frac{7}{8}$$

Both sides of the equation have a value of $-\frac{7}{8}$ when $x = \frac{19}{7}$, so $x = \frac{19}{7}$ must be the correct solution.

EXAMPLE 8.3

Solve $\dfrac{4}{x^2-1} = \dfrac{2}{x-1} - \dfrac{3}{x+1}$.

Solution   The LCD of $\dfrac{4}{x^2-1}$, $\dfrac{2}{x-1}$, and $\dfrac{3}{x+1}$ is $(x+1)(x-1) = x^2 - 1$. Notice that $x \neq 1$ and $x \neq -1$, because each of these values makes two of the denominators 0. Multiplying both sides of the given equation by $x^2 - 1$, we obtain

$$(x^2-1)\left(\frac{4}{x^2-1}\right) = (x^2-1)\left(\frac{2}{x-1}\right) - (x^2-1)\left(\frac{3}{x+1}\right)$$

$$4 = (x+1)2 - (x-1)3$$

$$4 = 2x + 2 - (3x - 3)$$

$$4 = 2x + 2 - 3x + 3$$

$$4 = -x + 5$$

EXAMPLE 8.3 (Cont.)

$$-1 = -x$$

or $$x = 1$$

Since $x = 1$ is not an allowable solution, the "solution" $x = 1$ is extraneous. *This equation has no solution.*

EXAMPLE 8.4

Solve $\dfrac{3x}{x - 2} + 5 = \dfrac{7x}{x - 2}$.

Solution   The LCD is $x - 2$, so $x \neq 2$. Multiplying both sides by $x - 2$, we get

$$(x - 2)\left(\frac{3x}{x - 2} + 5\right) = (x - 2)\left(\frac{7x}{x - 2}\right)$$

$$(x - 2)\left(\frac{3x}{x - 2}\right) + (x - 2)5 = (x - 2)\left(\frac{7x}{x - 2}\right)$$

$$3x + 5x - 10 = 7x$$

$$8x - 10 = 7x$$

$$x = 10$$

Substituting $x = 10$ into the original equation shows that it satisfies the equation.

Application

EXAMPLE 8.5

In a lens, if the object distance is p, the image distance is q, and the focal length is f, then the relation exists where $\dfrac{1}{f} = \dfrac{1}{p} + \dfrac{1}{q}$. Solve this equation for q.

Solution   The LCD of $\dfrac{1}{f}$, $\dfrac{1}{p}$, and $\dfrac{1}{q}$ is fpq. Multiplying both sides of the equation by fpq, we obtain

$$fpq\left(\frac{1}{f}\right) = fpq\left(\frac{1}{p} + \frac{1}{q}\right)$$

$$= fpq\left(\frac{1}{p}\right) + fpq\left(\frac{1}{q}\right)$$

Multiplying further, we obtain

$$pq = fq + fp$$

To solve for q, we put the terms containing q on the left-hand side with all other terms on the right-hand side of the equation.

$$pq - fq = fp$$

At this point, we need to determine the coefficient of q. We do this by factoring.

$$q(p - f) = fp$$

EXAMPLE 8.5 (Cont.)

We see that $p - f$ acts as the coefficient of q. Dividing by this coefficient, we obtain the desired solution.

$$q = \frac{fp}{p - f}$$

Application

EXAMPLE 8.6

A technician can assemble an instrument in 12.5 hours. After working for 3 hours on a job, the technician is joined by another technician, who is able to assemble the instrument alone in 9.5 hours. How long does it take to assemble this instrument?

Solution The problem is similar to the work problem we solved in Chapter 2. To solve this example, let h represent the number of hours that the technicians worked together on the instrument. The time to assemble the instrument will be $h + 3$ hours because the first technician worked alone for 3 hours.

The first technician, working alone, can complete the job in 12.5 hours. So, each hour this technician works, $\frac{1}{12.5}$ of the instrument is assembled. Similarly, the second technician will assemble $\frac{1}{9.5}$ of the instrument for each hour worked. The first technician works $h + 3$ hours and is able to complete $\frac{1}{12.5}(h + 3)$ of the work. The second technician works h hours and completes $\frac{1}{9.5}(h)$ of the work. Together, they assemble the entire instrument, so we get the equation

$$\frac{1}{12.5}(h + 3) + \frac{1}{9.5}(h) = 1$$

Multiplying by the common denominator $(12.5)(9.5)$, we obtain

$$9.5(h + 3) + 12.5(h) = (9.5)(12.5)$$

$$9.5h + 28.5 + 12.5h = 118.75$$

$$22h = 90.25$$

$$h \approx 4.1$$

Thus, the two technicians will be able to completely assemble the instrument in about 7.1 hours or 7 hours 6 minutes. (Remember, the total time of 7.1 is $h + 3$ hours.)

Exercise Set 8.1

In Exercises 1–30, solve the given equations and check the results.

1. $\frac{x}{2} + \frac{x}{3} = \frac{1}{4}$

2. $\frac{x}{3} - \frac{x}{4} = \frac{1}{2}$

3. $\frac{y}{2} + 3 = \frac{4y}{5}$

4. $\frac{y}{5} - 5\frac{1}{2} = \frac{3y}{4}$

5. $\frac{x - 1}{2} + \frac{x + 1}{3} = \frac{x - 1}{4}$

6. $\frac{x + 2}{3} - \frac{x + 4}{2} = \frac{x - 1}{6}$

7. $\dfrac{1}{x} + \dfrac{2}{x} = \dfrac{1}{3}$

8. $\dfrac{3}{x} - \dfrac{4}{x} = \dfrac{2}{5}$

9. $\dfrac{7}{w - 4} = \dfrac{1}{2w + 5}$

10. $\dfrac{5}{y + 1} = \dfrac{3}{y - 3}$

11. $\dfrac{2}{2x - 1} = \dfrac{5}{x + 5}$

12. $\dfrac{3}{4x + 2} = \dfrac{1}{x + 2}$

13. $\dfrac{4x}{x - 3} - 1 = \dfrac{3x}{x - 3}$

14. $7 - \dfrac{3x}{x + 2} = \dfrac{4x}{x + 2}$

15. $\dfrac{4}{x + 2} - \dfrac{3}{x - 1} = \dfrac{5}{(x - 1)(x + 2)}$

16. $\dfrac{3}{x - 3} + \dfrac{2}{2 - x} = \dfrac{5}{(x - 3)(x - 2)}$

17. $\dfrac{x + 1}{x + 2} + \dfrac{x + 3}{x - 2} = \dfrac{2x^2 + 3x - 5}{x^2 - 4}$

18. $\dfrac{x + 2}{x + 3} - \dfrac{x + 5}{x - 3} = \dfrac{2x - 1}{x^2 - 9}$

19. $\dfrac{3}{a + 1} + \dfrac{a + 1}{a - 1} = \dfrac{a^2}{a^2 - 1}$

20. $\dfrac{5}{x - 4} - \dfrac{x + 2}{x + 4} = \dfrac{x^2}{16 - x^2}$

21. $\dfrac{2}{x - 1} + \dfrac{5}{x + 1} = \dfrac{4}{x^2 - 1}$

22. $\dfrac{3x + 4}{x + 2} - \dfrac{3x - 5}{x - 4} = \dfrac{12}{x^2 - 2x - 8}$

23. $\dfrac{5x - 2}{x - 3} + \dfrac{4 - 5x}{x + 4} = \dfrac{10}{x^2 + x - 12}$

24. $\dfrac{2x}{x - 1} - \dfrac{3}{x + 2} = \dfrac{4x}{x^2 + x - 2} + 2$

25. $\dfrac{5}{x} + \dfrac{3}{x + 1} = \dfrac{x}{x + 1} - \dfrac{x + 1}{x}$

26. $\dfrac{y}{y + 2} + \dfrac{5}{y - 1} = \dfrac{3}{y + 2} + \dfrac{y}{y - 1}$

27. $\dfrac{2t - 4}{2t + 4} = \dfrac{t + 2}{t + 4}$

28. $\dfrac{3x + 5}{x - 5} = \dfrac{3x - 1}{x + 3}$

29. $\dfrac{3x + 1}{x - 1} - \dfrac{x - 2}{x + 3} = \dfrac{2x - 3}{x + 3} + \dfrac{4}{x - 1}$

30. $\dfrac{7x + 2}{x + 2} + \dfrac{3x - 1}{x + 3} = \dfrac{6x + 1}{x + 3} + \dfrac{4x - 3}{x + 2}$

Solve each of Exercises 31–38 for the indicated variable.

31. $\dfrac{1}{r} + \dfrac{1}{s} = \dfrac{1}{t}$ for s

32. $\dfrac{P_1 V_1}{T_1} = \dfrac{P_2 V_2}{T_2}$ for T_1

33. $\dfrac{1}{R} = \dfrac{1}{R_1} + \dfrac{1}{R_2} + \dfrac{1}{R_3}$ for R

34. $P = \dfrac{E^2}{R + r} - \dfrac{E^2}{(R + r)^2}$ for E^2

35. $V = 2\pi rh + 2\pi r^2$ for h

36. $\dfrac{5}{9}(F - 32) = C$ for F

37. $\dfrac{1}{f} = (n - 1)\left(\dfrac{1}{R_1} + \dfrac{1}{R_2}\right)$ for R_2

38. $\dfrac{P_1}{g} + \dfrac{V_1{}^2}{2g} + h_1 \doteq \dfrac{P_2}{dg} + \dfrac{V_2{}^2}{2g} + h_2$ for g

Solve Exercises 39–42.

39. Pipe A can fill a tank in 6 h and Pipe B can fill it in 4 h. If Pipe A is opened 1 h before Pipe B is opened, how long does it take to fill the tank?

40. *Computer technology*   One microprocessor can process a set of data in 5 μs (microseconds) and a second microprocessor can process the same amount of data in 8 μs. If they process the data together, how many microseconds should it take?

41. *Electricity*   A generator can charge a group of batteries in 18 h. It begins charging the batteries, and

4 h later a second generator starts charging the same set of batteries. If the second generator alone could charge the batteries in 12 h, how long will it take both to charge the batteries?

42. *Transportation*   An airplane traveling against the wind travels 500 km in the same time it takes it to travel 650 km with the wind. If the wind speed is 20 km/h, find the speed of the airplane in still air.

☰ **8.2**
QUADRATIC EQUATIONS AND FACTORING

Until now, all the equations we have solved have been first degree, or linear, equations and systems of linear equations. Many technical problems require the ability to solve more complicated equations. In the remainder of this chapter, we will focus on second-degree, or quadratic, equations. As we continue through the book, we will learn how to solve more types of equations.

Quadratic Equations ▬▬▬▬▬

We worked with quadratic trinomials and binomials in Chapter 7. A polynomial equation of the second degree is a **quadratic equation**.

> **Quadratic Equation**
>
> If a, b, and c are constants and $a \neq 0$, then
> $$ax^2 + bx + c = 0$$
> is the **standard quadratic equation**.

EXAMPLE 8.7

The following are all quadratic equations written in the standard form:

(a) $2x^2 - 3x + 5 = 0$ $\qquad$ $a = 2,\ b = -3,\ c = 5$

(b) $4x^2 + 7x = 0$ $\qquad$ $a = 4,\ b = 7,\ c = 0$

(c) $5x^2 - 125 = 0$ $\qquad$ $a = 5,\ b = 0,\ c = -125$

(d) $(p + 3)x^2 + px - p + 2 = 0$ $\qquad$ $a = p + 3,\ b = p,\ c = 2 - p$

EXAMPLE 8.8

The following are also quadratic equations, but are not in the standard form.

(a) $x^2 = 49$ $\qquad$ $a = 1,\ b = 0,\ c = -49$

(b) $8 + 2x = \dfrac{7x^2}{2}$ $\qquad$ $a = \dfrac{7}{2},\ b = -2,\ c = -8$

☰ **Note** Quadratic equations that contain fractions are often simplified by writing them without fractions and with $a > 0$. This simplification is achieved by multiplying the equation by the LCD of the coefficients. (We could write the equation in Example 8.8(b) as $-\frac{7}{2}x^2 + 2x + 8 = 0$, but it would be better to multiply the equation by -2 and obtain the equivalent equation $7x^2 - 4x - 16 = 0$, since integer coefficients are usually easier to use.)

EXAMPLE 8.9

The following are not quadratic equations.

(a) $2x^3 - x^2 + 5 = 0$ This equation has a term of degree 3. The highest degree of any term in a quadratic equation is 2.

(b) $4x + 5 = 0$ This does not have a term of degree 2. This means that $a = 0$, which contradicts part of the definition of a quadratic equation.

In order to solve a quadratic equation, we need another property for the real numbers. We have not needed the **zero-product rule** until now.

Zero-Product Rule for Real Numbers

If a and b are numbers and $ab = 0$, then $a = 0$, $b = 0$, or both a and b are 0.

This is a very simple but powerful statement, as shown by the next example.

EXAMPLE 8.10

Use the zero-product rule to solve $(x - 1)(x + 5) = 0$.

Solution   Here a has the value $x - 1$ and b has the value $x + 5$. According to the zero-product rule, either $x - 1 = 0$ or $x + 5 = 0$ (or both). If $x - 1 = 0$, then $x = 1$. If $x + 5 = 0$, then $x = -5$. These are the roots (also called solutions or zeros) of the equation.

Substitute 1 into the original equation. Did you get 0? Now substitute -5 and you will get 0 again. So, both of these answers check. The solutions are $x = -5$ and $x = 1$.

Roots of Quadratic Equations

The zero-product rule indicates that if we can factor a quadratic equation, then we can find its roots or solutions. A quadratic equation will never have more than two roots. Normally, all quadratic equations are considered to have two roots. But, there are times when both of these roots are the same number. In this case, the roots are referred to as **double roots**. There are also times when there will be no real numbers that are roots of a quadratic equation. In this case, the roots will be imaginary numbers. (We will talk more about this later in this chapter and again in Chapter 14.)

Finding Roots by Factoring

Let's begin by looking at the general idea behind the method of factoring to find roots of a quadratic equation. Suppose we have a general quadratic equation $ax^2 + bx + c = 0$

and that this equation can be factored as follows.

$$ax^2 + bx + c = (rx + t)(sx + v) = 0$$

From the zero-product rule, we know that $rx + t = 0$ or $sx + v = 0$. Solving each of these linear equations, we get $x = \dfrac{-t}{r}$ and $x = \dfrac{-v}{s}$, which are the roots of the quadratic equation. Now, let's look at some examples showing how to use this method.

EXAMPLE 8.11

Find the roots of $x^2 - x - 6 = 0$.

Solution   This quadratic equation factors to $(x - 3)(x + 2)$. So, from the zero-product rule, we have

$$x^2 - x - 6 = 0$$

$$(x - 3)(x + 2) = 0$$

$$x - 3 = 0 \qquad \text{and} \qquad x = 3$$

$$\text{or} \qquad x + 2 = 0 \qquad \text{and} \qquad x = -2$$

EXAMPLE 8.12

Find the roots of $x^2 + 6x + 9 = 0$.

Solution   Factoring $x^2 + 6x + 9$, we get $(x + 3)^2$, so

$$x^2 + 6x + 9 = 0$$

$$(x + 3)(x + 3) = 0$$

$$x + 3 = 0 \qquad \text{and} \qquad x = -3$$

This is a double root. Both roots are the same: -3.

EXAMPLE 8.13

Find the roots of $6x^2 - 11x - 35 = 0$.

Solution

$$6x^2 - 11x - 35 = 0$$

$$(3x + 5)(2x - 7) = 0$$

$$3x + 5 = 0 \qquad \text{and} \qquad x = -\frac{5}{3}$$

$$\text{or} \qquad 2x - 7 = 0 \qquad \text{and} \qquad x = \frac{7}{2}$$

It is very important to be sure that the right-hand side of the equation is equal to 0.

 Caution   The property that we are using, the zero-product rule, says that if $ab = 0$, then $a = 0$ or $b = 0$ (or both). It is guaranteed to work *only* if the right-hand side of the equation is 0.

Suppose you had a problem such as $(x - 2)(x + 4) = 6$, where the right-hand side of the equation was not 0, and you tried to use the zero-product rule to solve the equation. If $(x - 2)(x + 4) = 6$, you would say $x - 2 = 6$ or $x + 4 = 6$. In the first case, $x - 2 = 6$, we get a possible solution of $x = 8$. In the second case, $x + 4 = 6$, we obtain an answer of $x = 2$.

Now check these answers.

If $x = 8$, then $(x - 2)(x + 4) = (8 - 2)(8 + 4) = (6)(12) = 72$, which is certainly not 6. This answer does not check. Let's try the other solution.

If $x = 2$, then $(x - 2)(x + 4) = (2 - 2)(2 + 4) = (0)(6) = 0$. Again, we do not get an answer of 6.

This example was intended to show you that it is important to make sure that the right-hand side of the equation is 0 before applying the zero-product rule. The next example will show you what to do when the right-hand side of the equation is not 0.

EXAMPLE 8.14

Find the roots of $4x^2 - 10 = 3x$.

Solution   Before we can factor this problem, we have to get all the terms on the left-hand side of the equation; then the right-hand side will be 0.

$$4x^2 - 10 = 3x$$

$$4x^2 - 3x - 10 = 0$$

$$(4x + 5)(x - 2) = 0$$

$$4x + 5 = 0 \quad \text{and} \quad x = -\frac{5}{4}$$

$$\text{or} \qquad x - 2 = 0 \quad \text{and} \quad x = 2$$

So, the two roots of this equation are $x = -\frac{5}{4}$ and $x = 2$.

It is also possible to solve some fractional equations by factoring. You will need to first multiply the equation by the LCD, and then put all the nonzero terms on the left-hand side of the equation before you begin to factor the equation. The next example shows how to do this.

EXAMPLE 8.15

Find the roots of $\dfrac{6}{x(3 - 2x)} - \dfrac{4}{3 - 2x} = 1$.

Solution   The LCD is $x(3 - 2x)$, so $x \neq 0$ and $x \neq \frac{3}{2}$. Multiplying both sides of the equation by $x(3 - 2x)$ provides

$$6 - 4x = x(3 - 2x)$$

$$6 - 4x - x(3 - 2x) = 0$$

$$6 - 4x - 3x + 2x^2 = 0$$

$$2x^2 - 7x + 6 = 0$$

$$(2x - 3)(x - 2) = 0$$

EXAMPLE 8.15 (Cont.)

$$2x - 3 = 0 \quad \text{and} \quad x = \frac{3}{2}$$

or $\qquad x - 2 = 0 \quad \text{and} \quad x = 2$

Since $x \neq \frac{3}{2}$, the only root is $x = 2$.

This section will not place much emphasis on graphing. But, you should realize that a quadratic function and its roots can be graphically represented. Figures 8.1 and 8.2 show the graphs for the quadratic functions $f(x) = x^2 - x - 6$ and $f(x) = x^2 + 6x + 9$, formed by the quadratic equations in Examples 8.11 and 8.12.

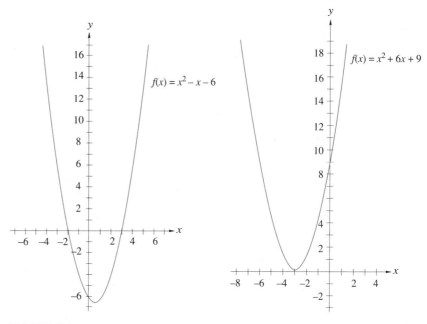

FIGURE 8.1   FIGURE 8.2

In Example 8.11, we found that the roots were -2 and 3 and that these are the points where the graph crosses the x-axis. In Example 8.12, the quadratic equation $x^2 + 6x + 9 = 0$ had a double root, $x = -3$. The graph of the quadratic function for Example 8.12 intersects the x-axis at exactly one point, and that one point is when $x = -3$.

EXAMPLE 8.16

Find the roots of $x^2 = 16$.

Solution   This equation is equivalent to $x^2 - 16 = 0$. Factoring, we obtain $(x - 4)(x + 4) = 0$ and $x = 4$ or $x = -4$. We could have solved this problem with less

EXAMPLE 8.16 (Cont.)

work if we had taken the square root of both sides.

$$x^2 = 16$$

$$x = \pm\sqrt{16}$$

$$x = \pm 4$$

In general, if $c \geq 0$ and $x^2 = c$, then $x = \pm\sqrt{c}$.

EXAMPLE 8.17

Find the roots of $9x^2 = 25$.

Solution
$$9x^2 = 25$$

$$x^2 = \frac{25}{9}$$

$$x = \pm\sqrt{\frac{25}{9}} = \pm\frac{5}{3}$$

EXAMPLE 8.18

Find the roots of $(x - 4)^2 = 25$.

Solution   There are two ways to work this problem.

Method 1: First, we will square the left-hand side of the equation and collect like terms.

$$(x - 4)^2 = 25$$

$$x^2 - 8x + 16 = 25$$

$$x^2 - 8x + 16 - 25 = 0$$

$$x^2 - 8x - 9 = 0$$

$$(x - 9)(x + 1) = 0$$

$$x = 9 \qquad \text{or} \qquad x = -1$$

Method 2: This is a short cut. It will save time, but you have to be careful that you do not make errors. We will first take the square root of both sides. Once this is done, we will solve the linear equation for both values of x.

$$(x - 4)^2 = 25$$

$$\sqrt{(x - 4)^2} = \pm\sqrt{25}$$

$$x - 4 = \pm 5$$

$$x = 4 + 5 = 9$$

$$\text{or} \qquad\qquad x = 4 - 5 = -1$$

This is the same answer we got using the first method.

Application

EXAMPLE 8.19

A rectangular piece of aluminum is to be used to form a box. (See Figure 8.3a.) A 2-in. square is to be cut from each corner and the ends are to be folded up to form an open box. (See Figure 8.3b.) If the original piece of aluminum was twice as long as it was wide, and the volume of the box is 672 in.^{3}, what were the dimensions of the original rectangle?

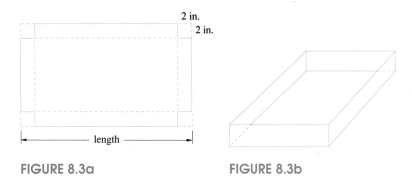

FIGURE 8.3a   **FIGURE 8.3b**

Solution   If the width of the original rectangle is w, then the length is $2w$. The width of the box is $w - 4$ and the length of the box is $2w - 4$. The volume of the box is the product of the width, length, and height or

$$wlh = V$$

$$(w - 4)(2w - 4)2 = 672$$

$$4w^2 - 24w + 32 = 672 \qquad \text{Multiply.}$$

$$4w^2 - 24w - 640 = 0 \qquad \text{Put equation in standard form.}$$

$$w^2 - 6w - 160 = 0 \qquad \text{Divide both sides by 4.}$$

$$(w - 16)(w + 10) = 0 \qquad \text{Factor.}$$

If $w - 16 = 0$, then $w = 16$, and if $w + 10 = 0$, then $w = -10$. The last answer does not make sense. We cannot have a rectangle with a width of -10 in. So, the width of the original rectangle is 16 in. and the length is 32 in.

Exercise Set 8.2

In Exercises 1–34, solve the quadratic equations by factoring.

1. $x^2 - 9 = 0$

2. $x^2 - 100 = 0$

3. $x^2 + x - 6 = 0$

4. $x^2 - 6x - 7 = 0$

5. $x^2 - 11x - 12 = 0$

6. $x^2 - 5x + 4 = 0$

7. $x^2 + 2x - 8 = 0$

8. $x^2 + 2x - 15 = 0$

9. $x^2 - 5x = 0$

10. $x^2 + 10x = 0$

11. $x^2 + 12 = 7x$

12. $x^2 = 7x - 10$

13. $2x^2 - 3x - 14 = 0$

14. $2x^2 + x - 15 = 0$

15. $2x^2 + 12 = 11x$

16. $2x^2 + 18 = 15x$

17. $3x^2 - 8x - 3 = 0$

18. $3x^2 - 4x - 4 = 0$

19. $4x^2 - 24x + 35 = 0$

20. $6x^2 - 13x + 6 = 0$

21. $6x^2 + 11x - 35 = 0$

22. $10x^2 + 9x - 9 = 0$

23. $10x^2 - 17x + 3 = 0$

24. $14x^2 - 29x - 15 = 0$

25. $6x^2 = 31x + 60$

26. $15x^2 = 23x - 4$

27. $(x - 1)^2 = 4$

28. $(x + 2)^2 = 9$

29. $(5x - 2)^2 = 16$

30. $(3x + 2)^2 = 64$

31. $\dfrac{x}{x + 1} = \dfrac{x + 2}{3x}$

32. $(x + 2)^3 - x^3 = 56$

33. $\dfrac{1}{x - 3} + \dfrac{1}{x + 4} = \dfrac{1}{12}$

34. $\dfrac{1}{x - 5} + \dfrac{1}{x + 3} = \dfrac{1}{3}$

Solve Exercises 35–40.

35. *Dynamics* A ball is thrown vertically upward into the air from the roof of a building 192 ft high. The height of the ball above the ground is a function of the time in seconds and the initial velocity of the ball. If the initial velocity is 64 ft/s, then the height is given by $h(t) = -16t^2 + 64t + 192$. How many seconds will it take for the ball to return to the roof? (That is, when will $h(t) = 192$?)

36. *Dynamics* If the ball in Exercise 35 misses the building when it comes down, how long will it take for the ball to hit the ground? (When will $h(t) = 0$?)

37. The length of a rectangle is 5 cm more than its width. Find the length and width, if the area is 104 cm^2.

38. *Dynamics* The World Trade Center is 411 m high. A ball dropped from the top will fall according to $4.9t^2$ m. How long will it take for this ball to strike the ground? (Hint: Solve $4.9t^2 = 411$.)

39. *Solar Energy* In order to support a solar collector at the correct angle, the roof trusses for a building are designed as right triangles, as shown in Figure 8.4. The rafter on the same side of the solar collector is 7 m shorter than the other rafter and the base of each truss is 13 m long. What are the lengths of the rafters?

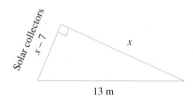

13 m

FIGURE 8.4

40. *Computer technology* Working alone, computer A can complete a data-processing job in 6 h less than computer B working alone. Together the two computers can complete the job in 4 h. How long would it take each computer by itself?

☰ **8.3**

COMPLETING THE SQUARE

Factoring is one method that can be used to solve quadratic equations. However, it is very difficult to factor some equations. In fact, most equations cannot be factored using integers or with rational numbers. We are going to introduce a technique called **completing the square**, which we can use to solve these quadratic equations.

In Section 8.2, we worked Example 8.18 using two different methods. We will begin our work in this section by reviewing the second method that we used in that example.

$$(x - 4)^2 = 25$$
$$\sqrt{(x-4)^2} = \pm\sqrt{25}$$
$$x - 4 = \pm 5$$
$$x = 4 + 5 = 9$$

or $\qquad x = 4 - 5 = -1$

Suppose that this was a slightly different problem: $(x - 4)^2 = 26$. Let's solve it the same way.

$$(x - 4)^2 = 26$$
$$\sqrt{(x-4)^2} = \pm\sqrt{26}$$
$$x - 4 = \pm\sqrt{26}$$
$$x = 4 \pm \sqrt{26}$$
$$x = 4 + \sqrt{26}$$

or $\qquad x = 4 - \sqrt{26}$

Now, let's look at some special products—those that are perfect squares. The general form is $(x + k)^2 = x^2 + 2kx + k^2$. Notice that the constant, k^2, is the square of one-half of $2k$, the coefficient of x. If we combine the method for solving quadratic equations and the special product for perfect squares, we can develop the completing-the-square method.

Suppose you had the equation $x^2 + 6x - 10 = 0$. A quick check of the discriminant ($b^2 - 4ac$) shows that it is $6^2 - 4(1)(-10) = 36 + 40 = 76$. Since 76 is not a perfect square, we cannot factor this equation. Rewrite the equation so that the variables are on the left-hand side and the constant term is on the right-hand side. (We have placed an empty box on each side of the equation to show that we will add something to both sides when we complete the square.)

$$x^2 + 6x + \boxed{} = 10 + \boxed{}$$

Complete the square on the left-hand side by taking one-half of the coefficient of the x-term, squaring it, and adding this number to both sides of the equation. The coefficient of x is 6, half of that is 3, and $3^2 = 9$. This is added to both sides of the equation and placed inside the empty boxes.

$$x^2 + 6x + \boxed{9} = 10 + \boxed{9}$$

or $\qquad x^2 + 6x + 9 = 19$

The left-hand side is now a perfect square.

$$(x+3)^2 = 19$$

We can solve this equation by using the second method from Example 8.18.

$$\sqrt{(x+3)^2} = \pm\sqrt{19}$$

$$x + 3 = \pm\sqrt{19}$$

$$x = -3 \pm \sqrt{19}$$

$$x = -3 + \sqrt{19} \qquad \text{or} \qquad x = -3 - \sqrt{19}$$

Check these answers. (The easiest way is to use your calculator.) You should see that they check when they are substituted in the original equation $x^2 + 6x - 10 = 0$.

EXAMPLE 8.20

Find the roots of $x^2 - 9x - 5 = 0$.

Solution
$$x^2 - 9x - 5 = 0$$

$$x^2 - 9x + \boxed{} = 5 + \boxed{}$$

$$x^2 - 9x + \boxed{\left(\frac{9}{2}\right)^2} = 5 + \boxed{\left(\frac{9}{2}\right)^2}$$

$$= 5 + \frac{81}{4}$$

$$= \frac{101}{4}$$

$$\left(x - \frac{9}{2}\right)^2 = \frac{101}{4}$$

$$\sqrt{\left(x - \frac{9}{2}\right)^2} = \pm\sqrt{\frac{101}{4}} = \pm\frac{\sqrt{101}}{2}$$

$$x - \frac{9}{2} = \pm\frac{\sqrt{101}}{2}$$

$$x = \frac{9}{2} \pm \frac{\sqrt{101}}{2}$$

$$x = \frac{9 + \sqrt{101}}{2} \qquad \text{or} \qquad x = \frac{9 - \sqrt{101}}{2}$$

Before you complete the square, the coefficient on the x^2-term must be 1. One way to do this is shown in the next example.

EXAMPLE 8.21

Find the roots of $2x^2 - 8x + 3 = 0$.

Solution This is slightly complicated by the fact the coefficient of $2x^2$ is not 1. Our first step will be to divide the equation by 2 and then proceed as we have before.

$$2x^2 - 8x + 3 = 0$$

$$x^2 - 4x + \frac{3}{2} = 0$$

$$x^2 - 4x + \boxed{} = -\frac{3}{2} + \boxed{}$$

$$x^2 - 4x + \boxed{2^2} = -\frac{3}{2} + \boxed{2^2}$$

$$(x - 2)^2 = -\frac{3}{2} + 4 = \frac{5}{2}$$

$$x - 2 = \pm\sqrt{\frac{5}{2}}$$

$$x = 2 \pm \sqrt{\frac{5}{2}}$$

Application

EXAMPLE 8.22

The photograph in Figure 8.5a shows a solar collector. The roof trusses for a solar collector are often designed as right triangles, so the solar collector will be supported at the correct angle. Rafters form the legs of the right triangle and the base of the truss forms the hypotenuse. Suppose the rafter along the back of the solar collector is 3.5 m shorter than the other rafter and that the base of each truss is 6.5 m long. What is the length of each rafter?

Solution A triangle has been drawn over the photograph of the solar collector in Figure 8.5a. The triangle is labeled with the given information and shown by itself in Figure 8.5b. Since this is a right triangle, we can use the Pythagorean theorem. Thus, we have

$$x^2 + (x - 3.5)^2 = 6.5^2$$

Squaring both sides produces

$$x^2 + (x^2 - 7x + 12.25) = 42.25$$

or $\qquad\qquad 2x^2 - 7x = 30$

Dividing by 2, we get

$$x^2 - 3.5x = 15$$

Next, we complete the square.

$$x^2 - 3.5x + \boxed{} = 15 + \boxed{}$$

$$x^2 - 3.5x + \boxed{\left(\frac{3.5}{2}\right)^2} = 15 + \boxed{\left(\frac{3.5}{2}\right)^2}$$

FIGURE 8.5

(a)

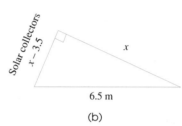

(b)

EXAMPLE 8.22 (Cont.)

$$= 15 + \frac{12.25}{4}$$

$$= \frac{72.25}{4}$$

$$\left(x - \frac{3.5}{2} \right)^2 = \frac{72.25}{4}$$

$$\sqrt{\left(x - \frac{3.5}{2} \right)^2} = \pm \sqrt{\frac{72.25}{4}} = \pm \frac{8.5}{2}$$

$$x - \frac{3.5}{2} = \pm \frac{8.5}{2}$$

$$x = \frac{3.5}{2} \pm \frac{8.5}{2}$$

$$= 6 \quad \text{or} \quad x = -2.5$$

It would not make sense for a rafter to be -2.5 m long. The lengths of the rafters must be 6 m and $6 - 3.5 = 2.5$ m long.

Exercise Set 8.3

Find the roots of each of the quadratic equations in Exercises 1–14 by completing the square.

1. $x^2 + 6x + 8 = 0$

2. $x^2 - 7x - 8 = 0$

3. $x^2 - 10x = 11$

4. $2x^2 - 3x = 14$

5. $x^2 + 6x + 3 = 0$

6. $x^2 + 8x + 10 = 0$

7. $x^2 - 5x + 5 = 0$

8. $x^2 - 7x + 11 = 0$

9. $2x^2 - 6x - 10 = 0$

10. $3x^2 + 12x - 18 = 0$

11. $4x^2 - 12x - 18 = 0$

12. $3x^2 - 9x = 33$

13. $x^2 + 2kx + c = 0$

14. $px^2 + 2qx + r = 0$

Solve Exercises 15–19.

15. *Physics* When an object is dropped from a building that is 196 ft tall, its height h at any time t after it was dropped is given by the function $h(t) = 196 - 16t^2$. Find the time it takes for the object to strike the ground.

16. *Business* Total revenue from selling a certain object is the product of the price and the quantity of the object sold. If q represents the quantity sold and the price is $2{,}520 - 3q$, find the quantity that produces a total revenue of $140,400.

17. *Business* Total revenue from selling a certain object is the product of the price and the quantity of the object sold. If q represents the quantity sold and the price of the object is $1{,}560 - 4q$, find the quantity that produces a total revenue of $29,600.

18. *Agriculture* A farmer has 1,500 ft of fencing to enclose a rectangular field. Find the dimensions of the field so that the enclosed area will be 137,600 ft^2.

19. *Agriculture* A farmer has 2,700 ft of fencing to enclose two adjacent rectangular fields. Find the dimensions of the fields, so that the enclosed area will be 270,000 ft^2.

THE QUADRATIC FORMULA

In Section 8.3, we learned how to solve a quadratic equation by completing the square. In this section, we will develop a general formula that can be used to find the roots of any quadratic equation.

If there is a formula that will allow us to solve a quadratic equation with very little difficulty, why did we take the time to learn how to complete the square? There are two reasons. The first reason is that we will use completing the square to develop the quadratic formula. The second reason is that it is a tool that we will need when working with conic sections.

Suppose we have a standard quadratic equation

$$ax^2 + bx + c = 0, \text{ with } a \neq 0$$

What are the roots of this quadratic equation? If we complete the square, we can find out.

$$ax^2 + bx + c = 0$$

$$x^2 + \frac{b}{a}x + \frac{c}{a} = 0 \qquad \text{Divide both sides by } a.$$

$$x^2 + \frac{b}{a}x = -\frac{c}{a} \qquad \text{Add } -\frac{c}{a} \text{ to both sides.}$$

$$x^2 + \frac{b}{a}x + \left(\frac{b}{2a}\right)^2 = -\frac{c}{a} + \left(\frac{b}{2a}\right)^2$$

Complete the square by adding $\left(\dfrac{b}{2a}\right)^2$ to both sides.

$$\left(x + \frac{b}{2a}\right)^2 = \frac{b^2}{4a^2} - \frac{4ac}{4a^2}$$

Factor the left-hand side; reverse terms on the right-hand side.

$$= \frac{b^2 - 4ac}{4a^2}$$

Collect terms on the right-hand side.

$$x + \frac{b}{2a} = \pm\sqrt{\frac{b^2 - 4ac}{4a^2}}$$

Take the square root of both sides.

$$= \pm\frac{\sqrt{b^2 - 4ac}}{2a}$$

$$x = \frac{-b}{2a} \pm \frac{\sqrt{b^2 - 4ac}}{2a}$$

Solve for x.

$$= \frac{-b \pm \sqrt{b^2 - 4ac}}{2a}$$

This is the **quadratic formula**.

Quadratic Formula

The solutions of the equation $ax^2 + bx + c = 0$, $(a \neq 0)$, are given by

$$x = \frac{-b \pm \sqrt{b^2 - 4ac}}{2a}$$

The expression $b^2 - 4ac$ is called the **discriminant**.

To solve a quadratic equation by using the quadratic formula, write the equation in the standard form, identify a, b, and c, and substitute these numbers into the equation. The quadratic formula is a very useful tool, but many equations are easier to solve by factoring.

You may recognize something we have used before in part of the quadratic formula. The quantity $b^2 - 4ac$ is the discriminant. We used it to help tell us when a quadratic equation can be factored. Now we can use it to tell us something else. Since the quadratic formula takes the square root of the discriminant, a quadratic equation will have only real numbers as roots when $b^2 - 4ac \geq 0$. In Chapter 14, when we study complex numbers, we will consider quadratic equations in which the discriminant is negative.

EXAMPLE 8.23

Solve $x^2 + 7x - 8 = 0$.

Solution In this equation, $a = 1$, $b = 7$, and $c = -8$. Putting these values in the quadratic formula we get

$$x = \frac{-7 \pm \sqrt{7^2 - 4(1)(-8)}}{2(1)}$$

EXAMPLE 8.23 (Cont.)

$$= \frac{-7 \pm \sqrt{49 + 32}}{2}$$

$$= \frac{-7 \pm \sqrt{81}}{2}$$

$$= \frac{-7 \pm 9}{2}$$

$$x = \frac{-7 + 9}{2} = \frac{2}{2} = 1$$

or $\quad x = \frac{-7 - 9}{2} = \frac{-16}{2} = -8$

The roots are 1 and -8. This is an equation that we could have solved by factoring.

EXAMPLE 8.24

Solve $2x^2 + 5x - 3 = 0$.

Solution In this equation $a = 2$, $b = 5$, and $c = -3$. Putting these values in the quadratic formula we get

$$x = \frac{-5 \pm \sqrt{5^2 - 4(2)(-3)}}{2(2)}$$

$$= \frac{-5 \pm \sqrt{25 + 24}}{4}$$

$$= \frac{-5 \pm \sqrt{49}}{4}$$

$$= \frac{-5 \pm 7}{4}$$

So,

$$x = \frac{-5 + 7}{4} = \frac{2}{4} = \frac{1}{2} \quad \text{or} \quad x = \frac{-5 - 7}{4} = \frac{-12}{4} = -3$$

The roots are $\frac{1}{2}$ and -3.

EXAMPLE 8.25

Solve $9x^2 + 49 = 42x$.

Solution This equation is not in the standard form. If we subtract $42x$ from both sides, we get $9x^2 - 42x + 49 = 0$, with $a = 9$, $b = -42$, and $c = 49$. Substituting

EXAMPLE 8.25 (Cont.)

these values for a, b, and c in the quadratic formula, we obtain

$$x = \frac{42 \pm \sqrt{(-42)^2 - 4(9)(49)}}{2(9)}$$

$$= \frac{42 \pm \sqrt{1764 - 1764}}{18}$$

$$= \frac{42 \pm 0}{18}$$

$$= \frac{7}{3}$$

This is a double root; in this case both roots are $\frac{7}{3}$.

EXAMPLE 8.26

Find the roots of $3x^2 + 7x + 3 = 0$.

Solution   In this equation, $a = 3$, $b = 7$, and $c = 3$, so

$$x = \frac{-7 \pm \sqrt{7^2 - 4(3)(3)}}{2(3)}$$

$$= \frac{-7 \pm \sqrt{49 - 36}}{6}$$

$$= \frac{-7 \pm \sqrt{13}}{6}$$

So, $x = \frac{-7 + \sqrt{13}}{6}$ or $x = \frac{-7 + \sqrt{13}}{6}$

Notice that we really needed the quadratic formula to find the factors.

Application

EXAMPLE 8.27

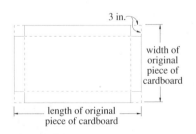

FIGURE 8.6

The length of a rectangular piece of cardboard is 4 in. more than its width. A 3-in. square is removed from each corner as shown in Figure 8.6. The remaining cardboard is bent to form an open box. If the volume of the box is 420 in.³, what are the dimensions of the original piece of cardboard?

Solution   We will begin by letting x be the width, in inches, of the original piece of cardboard. The length of this cardboard is $x + 4$ in. After the squares are removed and the sides have been folded up, the dimensions of the box are

$$\text{length} = (x + 4) - 2 \cdot 3 = x - 2$$

$$\text{width} = x - 6$$

$$\text{height} = 3$$

EXAMPLE 8.27 (Cont.)

The volume, 420 in.^{3}, is: length × width × height, or $V = lwh$, and so we obtain

$$(x - 2)(x - 6)3 = 420$$

$$(x - 2)(x - 6) = 140$$

$$x^2 - 8x + 12 = 140$$

$$x^2 - 8x - 128 = 0$$

$$(x - 16)(x + 8) = 0$$

$$x = 16 \quad \text{or} \quad x = -8$$

It would make no sense for the width of a piece of cardboard to be -8 in., so we reject $x = -8$ as an answer.

Thus, the piece of cardboard must have a width of 16 in. and a length of $16 + 4 = 20$ in.

Application

EXAMPLE 8.28

A ball is thrown upward from the top of a building that is 555 ft high with an initial velocity of 64 ft/s. The height of the ball in feet above the ground at any time t is given the formula $s(t) = 555 + 64t - 16t^2$. When will the ball hit the ground?

Solution   When the ball hits the ground, its height will be 0. So, we want to solve the equation

$$0 = 555 + 64t - 16t^2$$

for t. First write the equation in standard form and then use the quadratic formula to solve it. Multiplying the equation by -1 produces

$$16t^2 - 64t - 555 = 0$$

Here $a = 16$, $b = -64$, and $c = -555$. Substituting these into the quadratic formula we get

$$t = \frac{-b \pm \sqrt{b^2 - 4ac}}{2a}$$

$$= \frac{-(-64) \pm \sqrt{(-64)^2 - 4(16)(-555)}}{2(16)}$$

$$= \frac{64 \pm \sqrt{4096 + 35{,}520}}{32}$$

$$= \frac{64 \pm \sqrt{39{,}616}}{32}$$

$$\approx \frac{64 \pm 199.03768}{32}$$

$$\approx 2 \pm 6.22$$

Thus, $t \approx 8.22$ s or $t \approx -4.22$ s. The second answer makes no sense, because this would mean that the ball struck the ground before it was thrown. The first answer,

EXAMPLE 8.28 (Cont.) about 8.22 s, checks when it is substituted into the original equation, and so it is the correct answer.

Exercise Set 8.4

In Exercises 1–38, use the quadratic formula to find the roots of each equation.

1. $x^2 + 3x - 4 = 0$

2. $x^2 - 8x - 33 = 0$

3. $3x^2 - 5x - 2 = 0$

4. $7x^2 + 5x - 2 = 0$

5. $7x^2 + 6x - 1 = 0$

6. $2x^2 - 3x - 20 = 0$

7. $2x^2 - 5x - 7 = 0$

8. $3x^2 + 4x - 7 = 0$

9. $3x^2 + 2x - 8 = 0$

10. $9x^2 - 6x + 1 = 0$

11. $9x^2 + 12x + 4 = 0$

12. $3x^2 + 3x - 7 = 0$

13. $2x^2 - 3x - 1 = 0$

14. $2x^2 - 5x + 1 = 0$

15. $x^2 + 5x + 2 = 0$

16. $3x^2 - 6x - 2 = 0$

17. $2x^2 + 6x - 3 = 0$

18. $5x^2 + 2x - 1 = 0$

19. $x^2 - 2x - 7 = 0$

20. $x^2 + 3 = 0$

21. $2x^2 - 3 = 0$

22. $2x^2 = 5$

23. $3x^2 + 4 = 0$

24. $3x^2 + 1 = 5x$

25. $\frac{2}{3}x^2 - \frac{1}{9}x + 3 = 0$

26. $\frac{1}{2}x^2 - 2x + \frac{1}{3} = 0$

27. $0.01x^2 + 0.2x = 0.6$

28. $0.16x^2 = 0.8x - 1$

29. $\frac{1}{4}x^2 + 3 = \frac{5}{2}x$

30. $\frac{3}{2}x^2 + 2x = \frac{7}{2}$

31. $1.2x^2 = 2x - 0.5$

32. $1.4x^2 + 0.2x = 2.3$

33. $3x^2 + \sqrt{3}x - 7 = 0$

34. $2x^2 - \sqrt{89}x + 5 = 0$

35. $\frac{x-3}{7} = 2x^2$

36. $\frac{x-5}{3} = 5x^2$

37. $\frac{2}{x-1} + 3 = \frac{-2}{x+1}$

38. $\frac{3x}{x+2} + 2x = \frac{2x^2 - 1}{x+1}$

Solve Exercises 39–51.

39. *Dynamics*   The World Trade Center is 411 m high. If a ball is dropped from the top of the World Trade Center, its height at time t is given by the formula $h(t) = -4.9t^2 + 411$. The ball will hit the ground when $h(t) = 0$. How long does it take for the ball to fall to the ground?

40. *Dynamics*   If the ball in the previous problem is thrown downward with an initial velocity of 20 m/s, its height at time t is given by $h(t) = -4.9t^2 - 20t + 411$. When does it hit the ground?

41. *Dynamics*   If the ball in the previous problem is thrown upward with an initial velocity of 20 m/s, its height at time t is given by $h(t) = -4.9t^2 + 20t + 411$. How long will it take to hit the ground?

42. *Sheet metal technology*   An open box is to be made from a square piece of aluminum by cutting out a 4-cm square from each corner and folding up the sides. If the box is to have a volume of 100 cm^3, find the dimensions of the piece of aluminum that is needed.

43. *Sheet metal technology*   An open box is to be made from a rectangular piece of aluminum. A 3-cm square is to be cut from each corner and the sides will be folded up. If the original piece of aluminum was 1.5 times as long as its width and the volume of the box is 578 cm^3, what were the dimensions of the original rectangle?

44. *Construction*   An angle beam is to be constructed as shown in Figure 8.7. If the cross-sectional area of the angle beam is 81.25 cm^2, what is the thickness of the beam?

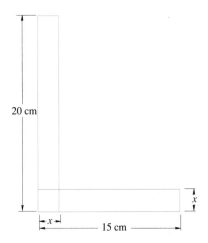

FIGURE 8.7

45. *Sheet metal technology*   A gutter is to be made by folding up the edges of a strip of metal as shown in Figure 8.8. If the metal is 12 in. wide and the cross-sectional area of the gutter is to be $16\frac{7}{8}$ in.^{2}, what are the width and depth of the gutter?

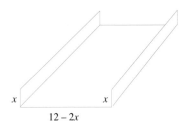

FIGURE 8.8

46. *Package design*   A cylindrical container is to be made with a height of 10 cm. If the total surface area of the container is to be 245 cm^2, what is the radius of the base?

47. *Business*   An oil distributor has 1,000 commercial customers who each pay a base rate of $30 per month for oil. The distributor figures that for each $1-per-month increase in the base rate, 5 customers will convert to coal. The distributor needs to increase its monthly income to $45,000 and lose as few customers as possible. How much should the base rate be increased?

48. *Electricity*   In an ac circuit that contains resistance, inductance, and capacitance in series, the applied voltage V can be found by solving $V^2 = V_R{}^2 + (V_L - V_C)^2$. If $V = 5.8$ V, $V_R = 5$ V, and $V_C = 10$ V, find V_L.

49. *Electricity*   In an ac circuit that contains resistance, inductance, and capacitance in series, the impedance of the circuit Z is related to the resistance R, the inductive reactance X_L, and the capacitive reactance X_C, by the formula, $Z^2 = R^2 + (X_L - X_C)^2$. If a circuit has $Z = 610$ Ω, $R = 300$ Ω, and $X_C = 531$ Ω, find X_L.

50. *Construction*   For a simply supported beam of length l having a distributed load of w kg/m, the binding moment M at any distance x from one end is given by

$$M = \frac{1}{2}wlx - \frac{1}{2}wx^2$$

At which locations is the binding moment zero?

 51. Write a program to solve a quadratic equation by using the quadratic formula.

☰ CHAPTER 8 REVIEW
_____

Important Terms and Concepts

Completing the square

Discriminant

Fractional equation

General quadratic equation

Quadratic equation

Quadratic formula

Zero-product rule

Review Exercises

Solve each of Exercises 1–6 for x.

1. $\dfrac{x}{3} + \dfrac{x}{2} = 5$

2. $\dfrac{2}{x} - \dfrac{3}{x} = \dfrac{1}{5}$

3. $\dfrac{x-2}{4} - \dfrac{x+2}{5} = \dfrac{x}{2}$

4. $\dfrac{2}{x-1} + \dfrac{3}{x-2} = \dfrac{4}{x^2 - 3x + 2}$

5. $\dfrac{3x}{x-1} - \dfrac{3x+2}{x} = 3$

6. $\dfrac{4x}{2x-3} + \dfrac{1}{x-1} = \dfrac{2x+1}{x-1}$

Solve each of Exercises 7–10 for the indicated variable.

7. $A = 2lw + 2(l + w)h$, for h

8. $A = 2lw + 2(l + w)h$, for l

9. $\dfrac{1}{f} = \dfrac{1}{p} + \dfrac{1}{q}$, for f

10. $X = wL - \dfrac{1}{wc}$, for c

Use factoring to find the roots of each of the quadratic equations in Exercises 11–22.

11. $x^2 - 8x + 7 = 0$

12. $x^2 + 4x + 3 = 0$

13. $x^2 - 11x + 10 = 0$

14. $x^2 + 15x + 14 = 0$

15. $x^2 + 15x + 56 = 0$

16. $x^2 - 8x + 12 = 0$

17. $2x^2 - 5x + 3 = 0$

18. $3x^2 + 10x + 7 = 0$

19. $6x^2 + 7x - 10 = 0$

20. $8x^2 + 22x + 9 = 0$

21. $4x^2 - 9 = 0$

22. $9x^2 - 16 = 0$

Find the roots of each of the quadratic equations in Exercises 23–28 by completing the square.

23. $x^2 + 4x - 5 = 0$

24. $x^2 - 7x + 6 = 0$

25. $x^2 + 19x = 11$

26. $2x^2 + 7x = 15$

27. $3x^2 + 5x - 14 = 0$

28. $4x^2 + 2x - 5 = 0$

Use the quadratic formula to find the roots of each of the equations in Exercises 29–40.

29. $x^2 - 8x + 5 = 0$

30. $x^2 + 7x - 6 = 0$

31. $2x^2 + 3x - 5 = 0$

32. $2x^2 - 7x + 4 = 0$

33. $3x^2 + 2x - 4 = 0$

34. $4x^2 - 3x = 1$

35. $5x^2 + 2 = 8x$

36. $6x^2 + 2x = 3$

37. $3x^2 - 8x + 10 = 0$

38. $8x^2 = 4x + 3$

39. $\dfrac{x}{x-1} + \dfrac{2}{x+1} = 3$

40. $\dfrac{2}{x} - \dfrac{3}{x+2} = 4$

Solve Exercises 41–44.

41. *Electricity*   Figure 8.9 shows two currents flowing in a single resistor R. The total current in the resistor is $i_1 + i_2$ and the power dissipated P is

$$P = (i_1 + i_2)^2 R$$

If $R = 50 \, \Omega$ and $i_2 = 0.4 \, A$, find the current i_1 needed to produce a power of 12 W.

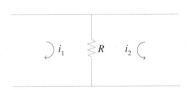

FIGURE 8.9

42. John can mow a field in 7 h and Matt can mow the same field in 8 h. If they work together, how long will it take to mow the field?

43. *Dynamics*   A stone is dropped from the edge of a cliff that is 144 ft high. If the stone's height at time t is given by $L(t) = 144 - 16t^2$, how long does it take for the stone to reach the bottom of the cliff?

44. *Industrial design*   A wedge is being designed in the shape of a right triangle. The hypotenuse will be 27 mm long and one leg will be 6 mm shorter than the other leg. What are the lengths of the sides of the triangle?

≡ CHAPTER 8 TEST

In Exercises 1–8, solve each equation for x.

1. $\dfrac{x}{8} + \dfrac{3x}{4} = \dfrac{7}{2}$

2. $\dfrac{12}{x^2 - 9} = \dfrac{x}{x - 3} - \dfrac{x}{x + 3}$

3. $x^2 - 4x - 5 = 0$

4. $3x^2 + x - 2 = 0$

5. $16x^2 + 9 = 24x$

6. $2x^2 + 3x + 3 = 0$

7. $x^2 + 3x + 1 = 0$

8. Solve $x^2 - 8x + 5 = 0$ by completing the square.

Solve Exercises 9–10.

9. A rectangular work area is to be 3 m longer than it is wide and will have an area of 46.75 m². What are the dimensions of the work area?

10. The position s at time t of an object moving rectilinearly with an initial velocity of v and an initial acceleration of a is

$$s = vt + \frac{at^2}{2}$$

Solve this equation for v in terms of a, t, and s.

Trigonometric Functions

Trigonometry allows us to measure distances when we cannot measure them directly. In Section 9.3, we will determine the height that this balloon is above the ground.

Courtesy of Michael A. Gallitelli, Metroland Photo Inc.

Until now, most of the functions we have used involved polynomials. Even our work with systems of linear equations and with quadratic equations dealt with polynomials. But much of the mathematics used by people working in technical areas cannot be solved by using polynomial functions.

In this chapter, we will learn about a new type of function—the trigonometric function. Trigonometric functions were originally developed to describe the relationship between the sides and angles of triangles. But, as so often happens in mathematics, other uses were discovered for these functions. Technical and scientific areas rely a great deal on trigonometric functions.

We will begin our study of trigonometry with the study of angles and angular measurements. We will then define the six trigonometric functions and their inverse functions, and study how we can use trigonometry to solve problems involving right triangles. (Chapter 10 will expand the applications to all types of triangles.) This chapter also provides some good opportunities to use our calculators and computers in new ways.

≡ 9.1
ANGLES, ANGLE MEASURE, AND TRIGONOMETRIC FUNCTIONS

At the beginning of this section, we will review our knowledge of angles and how they are measured. From this introduction we will quickly move into a study of the trigonometric functions.

Positive and Negative Angles

In Chapter 3, we gave two definitions of an angle. One of these was to think of generating a ray from an initial position to a terminal position. One revolution is the amount a ray would turn to return to its original position. As can be seen in Figure 9.1, if the rotation of the terminal side from the initial side is counterclockwise, the angle is a positive angle, but if the rotation is clockwise, the angle is a negative angle.

Degrees and Radians

Angles are measured using several different systems. The two most common are degrees and radians. In Chapter 3, we discussed how to change from degrees to radians and from radians to degrees. This is an important skill. We will briefly review this technique, but for more details you should refer to Section 3.1.

A degree is $\frac{1}{360}$ of a circle and the symbol $°$ is used to indicate degree(s). A radian is $\frac{1}{2\pi}$ of a circle. An entire circle contains $360°$ or 2π rad.

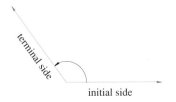

Positive Angle
(counterclockwise rotation)

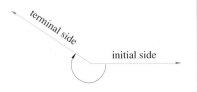

Negative Angle
(clockwise rotation)

FIGURE 9.1

Degree-Radian Conversions

To convert from degrees to radians, multiply the number of degrees by $\frac{\pi}{180°} \approx$ 0.01745, or divide by its reciprocal, 57.296.

To convert from radians to degrees, multiply the number of radians by $\frac{180°}{\pi} \approx 57.296$ or divide by its reciprocal, 0.01745.

Some scientific calculators have a key that will allow you to convert between degrees and radians. (Consult your owner's manual to see if your calculator can do this.)

EXAMPLE 9.1

Convert $72°$ to radians.

Solution

$$72° \approx 72 \times 0.01745 \text{ rad}$$

$$72° = 1.2564 \text{ rad}$$

or $\quad 72° = 72° \left(\frac{\pi}{180°} \right)$

$$= \frac{72\pi}{180}$$

$$= \frac{2\pi}{5} \text{ rad}$$

EXAMPLE 9.2

Convert 0.62 rad to degrees

Solution $0.62 \text{ rad} = 0.62 \times 57.296°$

$$= 35.52352°$$

EXAMPLE 9.3

Convert $\dfrac{3\pi}{4}$ rad to degrees.

Solution   Because this angle is given in terms of π, we will multiply by $\dfrac{180°}{\pi}$ rather than 57.296°.

$$\frac{3\pi}{4} = \frac{3\pi}{4} \times \frac{180°}{\pi}$$

$$= \frac{3}{4} \times 180°$$

$$= 135°$$

EXAMPLE 9.4

Use a calculator to convert 110° to radians.

Solution

	PRESS	DISPLAY
Algebraic calculator:	110 [2nd] [D·R]	1.9198622
RPN calculator:	110 [2nd] [RAD]	1.91862177

EXAMPLE 9.5

Use a calculator to convert 1.6 rad to degrees.

Solution

	PRESS	DISPLAY
Algebraic calculator:	1.6 [INV] [2nd] [D·R]	91.673247
RPN calculator	1.6 [2nd] [→DEG]	91.67324722

If you cannot remember these conversion values and if your calculator is not handy, use your knowledge of proportions. Since $180° = \pi$ rad, we can use the proportion given in the box.

Converting Between Degrees and Radians

To convert d degrees to r radians (or vice versa), use the proportion

$$\frac{d}{180°} = \frac{r}{\pi}$$

When you use this ratio, you will know either d or r and want to find the other.

EXAMPLE 9.6

Use this ratio to convert $72°$ to radians.

Solution　Here $d = 72°$, so

$$\frac{72°}{180°} = \frac{r}{\pi}$$

and $\frac{72°}{180°} \pi = \frac{2}{5} \pi \approx 1.25664$ gives you the same answer we got in Example 9.1. Notice that $72°$ is exactly $\frac{2}{5}\pi$, while 1.25664 is an approximation.

Notice that the degree-radian method used in Example 9.1 gave an answer of $72° \approx 1.2564$, while the proportion method used in Example 9.6 produced $72° = \frac{2}{5}\pi \approx 1.25664$. The difference in these two results is caused by the multiplication of an approximate number in the first method. As you can see, the proportion method used in Example 9.6, and used next in Example 9.7, produces more accurate results.

EXAMPLE 9.7

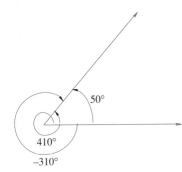

FIGURE 9.2

Use the ratio to convert $\dfrac{3\pi}{4}$ to degrees.

Solution　This is the same problem we worked in Example 9.3. Here $r = \dfrac{3\pi}{4}$, so

$$\frac{d}{180°} = \frac{3\pi/4}{\pi}$$

and $d = \frac{3}{4} \cdot 180° = 135°$.

Coterminal Angles

If two angles have the same initial side and the same terminal side, they are **coterminal angles**. An example of coterminal angles is given in Figure 9.2. One way to find a coterminal angle of a given angle is to add $360°$ to the original angle. In Figure 9.2, the original angle is $50°$ and one coterminal angle is $50° + 360° = 410°$. In fact, you could add any integer multiple of $360°$ to the original angle to find a coterminal angle. So, $50° + 4(360°) = 50° + 1,440° = 1,490°$ is another coterminal angle of a $50°$ angle.

In the same way, you could subtract an integer multiple of $360°$ from the original angle to get a coterminal angle. Thus, $50° - 360° = -310°$ is a coterminal angle of a $50°$ angle, and so is $50° - 2(360°) = 50° - 720° = -670°$.

An angle is in **standard position** if its vertex is at the origin of a rectangular coordinate system and its initial side coincides with the positive x-axis. The angle is determined by the position of the terminal side. The angle is said to be in a certain quadrant if its terminal side lies in that quadrant. If the terminal side coincides with one of the coordinate axes, the angle is a **quadrantal angle**.

Consider an angle θ in a standard position and let $P(x, y)$ be a fixed point on the terminal side of θ, as in Figure 9.3. We will call r the distance from O to P. From the Pythagorean theorem we know $r = \sqrt{x^2 + y^2}$. The length r is also called the **radius vector**. Suppose $Q(x_1, y_1)$ is any other point on the terminal side θ. If PR and QS are both perpendicular to the x-axis, then $\triangle POR$ and $\triangle QOS$ are similar. As you

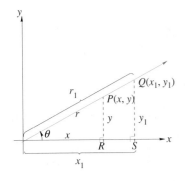

FIGURE 9.3

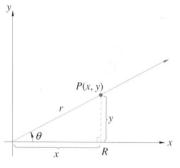

FIGURE 9.4

remember from our chapter on proportions (Chapter 6), the corresponding sides of similar triangles are proportional. So, the ratios of the corresponding sides are equal. For example, $\frac{y}{r} = \frac{y_1}{r_1}$ and $\frac{x}{y} = \frac{x_1}{y_1}$. As long as θ does not change, these ratios will not change.

The Trigonometric Functions

There are six possible ratios of two sides of a triangle. Each of these ratios has been given a name. These names (and their abbreviations) are sine (sin), cosine (cos), tangent (tan), cosecant (csc), secant (sec), and cotangent (cot). Because these ratios depend on the size of angle θ, they are written $\sin \theta$, $\cos \theta$, and so on. So, what you have are ratios that are functions of θ—the trigonometric functions. The six trigonometric, or trig, functions are defined using the triangle in Figure 9.4.

Trigonometric Functions

$$\sin \theta = \frac{y}{r} \qquad \csc \theta = \frac{r}{y}$$

$$\cos \theta = \frac{x}{r} \qquad \sec \theta = \frac{r}{x}$$

$$\tan \theta = \frac{y}{x} \qquad \cot \theta = \frac{x}{y}$$

EXAMPLE 9.8

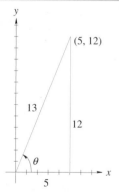

FIGURE 9.5

Given the point $(5, 12)$ on the terminal side of an angle θ, find the six trigonometric functions of θ.

Solution   A sketch of the angle is in Figure 9.5. Since $x = 5$, and $y = 12$, we can find the radius vector r by using the Pythagorean theorem.

$$r = \sqrt{x^2 + y^2} = \sqrt{5^2 + 12^2} = \sqrt{25 + 144} = \sqrt{169} = 13$$

$$\sin \theta = \frac{y}{r} = \frac{12}{13} \qquad \csc \theta = \frac{r}{y} = \frac{13}{12}$$

$$\cos \theta = \frac{x}{r} = \frac{5}{13} \qquad \sec \theta = \frac{r}{x} = \frac{13}{5}$$

$$\tan \theta = \frac{y}{x} = \frac{12}{5} \qquad \cot \theta = \frac{x}{y} = \frac{5}{12}$$

≡ Note   The values of x and y may be positive, negative, or zero; but r is always positive or zero. Remember, r is never negative because $r = \sqrt{x^2 + y^2}$ and the expression $\sqrt{x^2 + y^2}$ is never negative.

Of course, there is nothing to restrict the terminal side to the first quadrant. In the next two examples, the terminal side is in Quadrants II and IV, respectively.

EXAMPLE 9.9

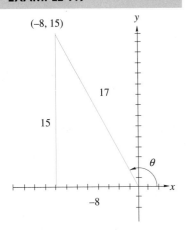

FIGURE 9.6

Given the point $(-8,\ 15)$ on the terminal side of an angle θ, find the six trigonometric functions of θ.

Solution   Here we have $x = -8$ and $y = 15$, so the radius vector r is given by $\sqrt{(-8)^2 + (15)^2} = \sqrt{64 + 225} = \sqrt{289} = 17$ as shown in Figure 9.6. The six trigonometric functions are

$$\sin \theta = \frac{15}{17} \qquad \csc \theta = \frac{17}{15}$$

$$\cos \theta = \frac{-8}{17} \qquad \sec \theta = \frac{-17}{8}$$

$$\tan \theta = -\frac{15}{8} \qquad \cot \theta = -\frac{8}{15}$$

As a final example, we will consider a case where the radius vector is not an integer. Note that none of the sides of the triangle have to be integers.

EXAMPLE 9.10

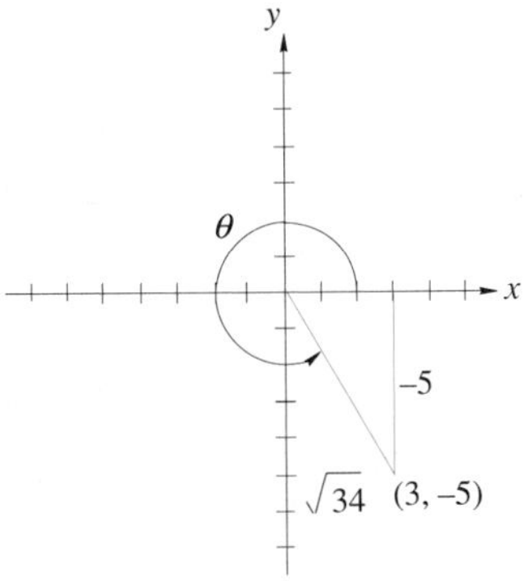

Given the point $(3,\ -5)$ on the terminal side of an angle θ, find the six trigonometric functions of θ.

Solution   Since $x = 3$ and $y = -5$ the radius vector, r, is $\sqrt{9 + 25} = \sqrt{34}$, as shown in Figure 9.7. The trigonometric functions of θ are

$$\sin \theta = \frac{-5}{\sqrt{34}} \approx -0.8575 \qquad \csc \theta = -\frac{\sqrt{34}}{5} \approx -1.1662$$

$$\cos \theta = \frac{3}{\sqrt{34}} \approx 0.5145 \qquad \sec \theta = \frac{\sqrt{34}}{3} \approx 1.9437$$

$$\tan \theta = -\frac{5}{3} \approx -1.6667 \qquad \cot \theta = -\frac{3}{5} = -0.6$$

FIGURE 9.7

Exercise Set 9.1

Convert each of the angle measures in Exercises 1–8 from degrees to radians.

1. $90°$

2. $45°$

3. $80°$

4. $-15°$

5. $155°$

6. $-235°$

7. $215°$

8. $180°$

Convert each of the angle measures in Exercises 9–16 from radians to degrees.

9. 2

10. 3

11. 1.5

12. π

13. $\dfrac{\pi}{3}$

14. $\dfrac{5\pi}{6}$

15. $-\dfrac{\pi}{4}$

16. -1.3

In Exercises 17–20, (a) draw each of the angles in standard position; (b) draw an arrow to indicate the rotation; (c) for each angle, find two other angles, one positive and one negative, which are coterminal with the given angle. (Note: There are many possible correct answers.)

17. $150°$

18. $315°$

19. $-135°$

20. $-30°$

Each point in Exercises 21–30 is on the terminal side of angle θ in standard position. Find the six trigonometric functions of the angle θ associated with each of these points.

21. $(4, 3)$

22. $(-6, -8)$

23. $(8, -15)$

24. $(-20, 21)$

25. $(1, 2)$

26. $(-2, 4)$

27. $(10, -8)$

28. $(-5, -2)$

29. $(\sqrt{11}, 5)$

30. $\left(-6, \sqrt{13}\right)$

Find the trigonometric functions that exist for each of the quadrantal angles θ when drawn in standard position for each of the points in Exercises 31–34.

31. $(3, 0)$

32. $(0, -4)$

33. $(-5, 0)$

34. $(0, 6)$

For each of Exercises 35–40, there is a point P on the terminal side of an angle θ in standard position. From the information given, determine the six trigonometric functions of θ.

35. $x = 6, r = 10, y > 0$

36. $y = -9, r = 15, x > 0$

37. $x = -20, r = 29, y > 0$

38. $y = 5, r = 13, x < 0$

39. $x = -7, r = 8, y < 0$

40. $y = 5, r = 30, x < 0$

▤ 9.2
VALUES OF THE TRIGONOMETRIC FUNCTIONS

We have learned how to calculate the values of the trigonometric functions for an angle θ in standard position when we are given a point on the terminal side of θ. This is not always the most convenient way to find the values of the trigonometric functions for θ.

    If we have an angle θ in the standard position and draw the triangle as we did in Figure 9.4, we get a picture that helps us to determine the values of the trigonometric functions for θ. If we look at just $\triangle POR$, we get a figure much like the one in

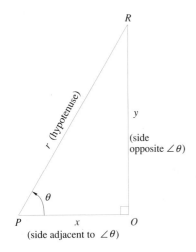

FIGURE 9.8

Figure 9.8. The length of the hypotenuse is r, the length y is for the side opposite angle θ, and the other side x is the side adjacent to angle θ. We can use these descriptions of the sides to rephrase our definitions for the trigonometric function.

Trigonometric Functions

$$\sin \theta = \frac{y}{r} = \frac{\text{side opposite } \theta}{\text{hypotenuse}} \qquad \csc \theta = \frac{r}{y} = \frac{\text{hypotenuse}}{\text{side opposite } \theta}$$

$$\cos \theta = \frac{x}{r} = \frac{\text{side adjacent to } \theta}{\text{hypotenuse}} \qquad \sec \theta = \frac{r}{x} = \frac{\text{hypotenuse}}{\text{side adjacent to } \theta}$$

$$\tan \theta = \frac{y}{x} = \frac{\text{side opposite } \theta}{\text{side adjacent to } \theta} \qquad \cot \theta = \frac{x}{y} = \frac{\text{side adjacent to } \theta}{\text{side opposite } \theta}$$

With these relationships, we can find the trigonometric functions of any angle of a right triangle.

EXAMPLE 9.11

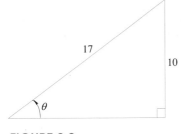

FIGURE 9.9

Determine the values of the trigonometric functions for an angle of a triangle with a hypotenuse of 17 and the opposite side of length 10, as shown in Figure 9.9.

Solution   The length of the adjacent side x is missing. Use the Pythagorean theorem, $x = \sqrt{r^2 - y^2}$. Since $r = 17$ and $y = 10$, $x = \sqrt{17^2 - 10^2} = \sqrt{189} = 3\sqrt{21}$. The trigonometric functions are

$$\sin \theta = \frac{\text{side opposite}}{\text{hypotenuse}} = \frac{10}{17} \qquad \csc \theta = \frac{\text{hypotenuse}}{\text{side opposite}} = \frac{17}{10}$$

$$\cos \theta = \frac{\text{side adjacent}}{\text{hypotenuse}} = \frac{3\sqrt{21}}{17} \qquad \sec \theta = \frac{\text{hypotenuse}}{\text{side adjacent}} = \frac{17}{3\sqrt{21}}$$

$$\tan \theta = \frac{\text{side opposite}}{\text{side adjacent}} = \frac{10}{3\sqrt{21}} \qquad \cot \theta = \frac{\text{side adjacent}}{\text{side opposite}} = \frac{3\sqrt{21}}{10}$$

We could have worked the problem had we been given the value of one of the trigonometric functions rather than the lengths of two of the sides.

EXAMPLE 9.12

In Figure 9.10, if θ is an angle of a triangle and $\cos \theta = \frac{2}{3}$, what are the values of the other trigonometric functions?

EXAMPLE 9.12 (Cont.)

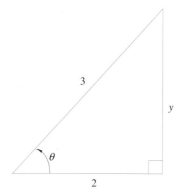

FIGURE 9.10

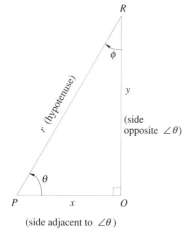

FIGURE 9.11

Solution The length y of the side opposite θ is $\sqrt{3^2 - 2^2} = \sqrt{5}$. Thus, we have the values for the other functions

$$\sin\theta = \frac{\sqrt{5}}{3} \qquad \csc\theta = \frac{3}{\sqrt{5}} \qquad \sec\theta = \frac{3}{2}$$

$$\tan\theta = \frac{\sqrt{5}}{2} \qquad \cot\theta = \frac{2}{\sqrt{5}}$$

Look at Figure 9.11. This is the same triangle that was in Figure 9.8. There is another acute angle in that triangle, labeled ϕ. Angles θ and ϕ are complementary. Remember, complementary angles are two angles that measure 90° when added. So, $\theta + \phi = 90°$. What are the trigonometric functions for ϕ? The side opposite ϕ is x and the side adjacent to ϕ is y, so

$$\sin\phi = \frac{x}{r} \qquad \csc\phi = \frac{r}{x}$$

$$\cos\phi = \frac{y}{r} \qquad \sec\phi = \frac{r}{y}$$

$$\tan\phi = \frac{x}{y} \qquad \cot\phi = \frac{y}{x}$$

From this, we get the principle of cofunctions of complementary angles.

Trigonometric Functions of Complementary Angles

If θ and ϕ are complementary angles, then

$$\sin\theta = \cos\phi \qquad \tan\theta = \cot\phi \qquad \csc\theta = \sec\phi$$

$$\cos\theta = \sin\phi \qquad \cot\theta = \tan\phi \qquad \sec\theta = \csc\phi$$

You may have begun to notice a pattern with the trigonometric functions. The values of some of the trig functions are the reciprocals of other functions. These are known as the **reciprocal identities**. There are three reciprocal identities. (An identity is an equation that is true for every value in the domain of its variables.)

Reciprocal Identities

$$\csc\theta = \frac{1}{\sin\theta} \qquad \sec\theta = \frac{1}{\cos\theta} \qquad \cot\theta = \frac{1}{\tan\theta}$$

There is also a relationship between the $\sin\theta$, $\cos\theta$, and the $\tan\theta$. This relationship, and its reciprocal, are known as the **quotient identities**.

Quotient Identities $\qquad\qquad \tan\theta = \dfrac{\sin\theta}{\cos\theta} \qquad \cot\theta = \dfrac{\cos\theta}{\sin\theta}$

None of these identities is true when the denominator is zero.

EXAMPLE 9.13

If $\sin \theta = \frac{1}{2}$ and $\cos \theta = \frac{\sqrt{3}}{2}$, find the values of the other four trigonometric functions.

Solution Using the reciprocal and quotient identities we have

$$\tan \theta = \frac{\sin \theta}{\cos \theta} = \frac{\frac{1}{2}}{\frac{\sqrt{3}}{2}} = \frac{1}{\sqrt{3}}$$

$$\cot \theta = \frac{1}{\tan \theta} = \frac{1}{\frac{1}{\sqrt{3}}} = \frac{\sqrt{3}}{1} = \sqrt{3}$$

$$\csc \theta = \frac{1}{\sin \theta} = \frac{1}{\frac{1}{2}} = \frac{2}{1} = 2$$

$$\sec \theta = \frac{1}{\cos \theta} = \frac{1}{\frac{\sqrt{3}}{2}} = \frac{2}{\sqrt{3}}$$

FIGURE 9.12

FIGURE 9.13

FIGURE 9.14

Trigonometric Values

Until now, we have been finding the values of trigonometric functions without knowing the size of the angle θ. But, many times we are given the value of θ and are asked to determine the values of the trigonometric functions for that angle.

There are two basic ways that are used to find values of trigonometric functions. By far, the easiest way is with a calculator or a computer. The other basic method is with a table of trigonometric values. There are some trigonometric angles that seem to be used quite frequently. Some people learn the values of these basic angles so they can readily use these values when they are needed. We will concentrate on calculators and computers for determining the value of a trigonometric function.

Before you use a calculator or a computer, you need to decide if the angle is measured in degrees or radians. Most calculators can compute the trigonometric functions of an angle whether it is in degrees or radians, as long as the calculator is set to work in that mode. Some calculators have a $\boxed{\text{DRG}}$ button. When you turn on the calculator, a 0 is the only figure displayed, as shown in Figure 9.12. The calculator is in degree mode when it is turned on.

Press the $\boxed{\text{DRG}}$ button. Now the screen displays a 0 and the abbreviation RAD, as in Figure 9.13. The calculator is now ready to work with angles in radians.

Press the $\boxed{\text{DRG}}$ again. The RAD disappears and the word GRAD appears, as shown in Figure 9.14. GRAD stands for another type of angle measure in which the circle is divided into 400 equal parts. One grad is $\frac{1}{400}$ of a circle. We will not use grads in this book.

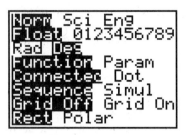

FIGURE 9.15

Press the ⬚DRG⬚ again. The calculator returns to the degree mode shown in Figure 9.12. You can reverse the order by using the ⬚INV⬚ key. Thus, to change from the radian mode to the degree mode press ⬚INV⬚ ⬚DRG⬚.

Some calculators have separate keys for each function. Again, the calculator will be in degree mode when it is turned on. To get the calculator to the radian mode, press ⬚2nd⬚ ⬚RAD⬚. Unfortunately, you may not be able to see any difference in the display.

Other calculators have different symbols or different locations for the symbols that let you know which mode the calculator is in. For example, on a graphing calculator, such as the Casio *fx-7700G* or Texas Instruments *TI-81*, press ⬚MODE⬚. The *TI-81* will display a screen like that in Figure 9.15. (The calculator displayed in Figure 9.15 is in degree mode.)

Use the cursor keys ⬚▲⬚, ⬚▼⬚, ⬚◀⬚, and ⬚▶⬚ to darken the appropriate degree or radian mode. Press ⬚ENTER⬚ or ⬚EXE⬚ to set the calculator in that mode. Make sure that you consult the owner's manual for your calculator.

Trigonometric Values on a Calculator

Some calculators have three trigonometric function keys— ⬚sin⬚, ⬚cos⬚, and ⬚tan⬚. If you want one of the other trigonometric functions, you will have to use one of these keys and the ⬚1/x⬚ key. Here you need to know the reciprocal identities discussed earlier in this section. You must press the ⬚1/x⬚ or ⬚x⁻¹⬚ key *after* you have pressed the trigonometric function key.

If you are using a nongraphing calculator to find the trigonometric function for a particular angle in degrees, enter the measure of the angle and press the appropriate function key to obtain sin, cos, or tan. To obtain csc, sec, or cot, press the ⬚1/x⬚ or ⬚x⁻¹⬚ key *after* you have pressed the appropriate key for its reciprocal function. On graphing calculators, you press the function key first.

Make sure that your calculator is set in the correct degree or radian mode before you begin to work the problem.

EXAMPLE 9.14

Determine the trigonometric functions of 48.9°.

Solution Put the calculator in degree mode, and then proceed as follows.

Function	ENTER (Algebraic)	ENTER (Graphing)	DISPLAY
$\sin 48.9°$	48.9 ⬚sin⬚	⬚sin⬚ 48.9 ⬚ENTER⬚	0.753563
$\cos 48.9°$	48.9 ⬚cos⬚	⬚cos⬚ 48.9 ⬚ENTER⬚	0.6573752
$\tan 48.9°$	48.9 ⬚tan⬚	⬚tan⬚ 48.9 ⬚ENTER⬚	1.1463215
$\csc 48.9°$	48.9 ⬚sin⬚ ⬚1/x⬚	⬚(⬚ ⬚sin⬚ 48.9 ⬚)⬚ ⬚x⁻¹⬚ ⬚ENTER⬚	1.3270284
$\sec 48.9°$	48.9 ⬚cos⬚ ⬚1/x⬚	⬚(⬚ ⬚cos⬚ 48.9 ⬚)⬚ ⬚x⁻¹⬚ ⬚ENTER⬚	1.5212012
$\cot 48.9°$	48.9 ⬚tan⬚ ⬚1/x⬚	⬚(⬚ ⬚tan⬚ 48.9 ⬚)⬚ ⬚x⁻¹⬚ ⬚ENTER⬚	0.8723556

To find the values of the trigonometric functions for an angle in radians, put the calculator in the radian mode. Then proceed as in Example 9.15.

EXAMPLE 9.15

Use a calculator to determine the trigonometric functions of 0.65 rad.

Solution Put the calculator in radian mode and then proceed as follows.

Function	ENTER (Algebraic)	ENTER (Graphing)	DISPLAY
sin 0.65	0.65 $\boxed{\sin}$	$\boxed{\sin}$ 0.65 $\boxed{\text{ENTER}}$	0.6051864
cos 0.65	0.65 $\boxed{\cos}$	$\boxed{\cos}$ 0.65 $\boxed{\text{ENTER}}$	0.7960838
tan 0.65	0.65 $\boxed{\tan}$	$\boxed{\tan}$ 0.65 $\boxed{\text{ENTER}}$	0.7602044
csc 0.65	0.65 $\boxed{\sin}$ $\boxed{1/x}$	$\boxed{(}$ $\boxed{\sin}$ 0.65 $\boxed{)}$ $\boxed{x^{-1}}$ $\boxed{\text{ENTER}}$	1.6523834
sec 0.65	0.65 $\boxed{\cos}$ $\boxed{1/x}$	$\boxed{(}$ $\boxed{\cos}$ 0.65 $\boxed{)}$ $\boxed{x^{-1}}$ $\boxed{\text{ENTER}}$	1.2561492
cot 0.65	0.65 $\boxed{\tan}$ $\boxed{1/x}$	$\boxed{(}$ $\boxed{\tan}$ 0.65 $\boxed{)}$ $\boxed{x^{-1}}$ $\boxed{\text{ENTER}}$	1.3154357

Trigonometric Values on a Computer

Computers operate in much the same way as do calculators. There are two main differences. Computers normally operate in radians. If you want to work in degrees, you will need to write a program that will convert radians to degrees. The second difference is that you must put parentheses around the size of the angle. Like calculators, computers do not have the secant, cosecant, and cotangent built into them. In Example 9.16, we will use the same angle that we used in Example 9.15 and we will only find the values of the sin, tan, and sec.

EXAMPLE 9.16

Use a computer to determine the sin, tan, and sec of 0.65 rad.

Solution	Function	ENTER	DISPLAY
	sin .65	PRINT SIN(.65)	.605186406
	tan .65	PRINT TAN(.65)	.760204399
	sec .65	PRINT 1/COS(.65)	1.25614917

Trigonometric Tables

The other principal way to determine the values of trigonometric functions is with a table of trigonometric functions. Until the advent of electronic calculators, this was the main way people determined the value of a trigonometric function. Trigonometric tables can be found in many mathematics books or in special books of mathematical tables. We have not included the trigonometric tables in the back of this book, because we assume that you will use a calculator or a computer to obtain values of trigonometric functions.

Exercise Set 9.2

In Exercises 1–6, find the six trigonometric functions for angle θ for the indicated sides of $\triangle ABC$. (See Figure 9.16.)

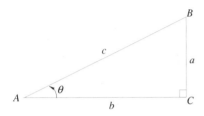

FIGURE 9.16

1. $a = 5, c = 13$

2. $a = 7, b = 8$

3. $a = 1.2, c = 2$

4. $a = 2.1, b = 2.8$

5. $a = 1.4, b = 2.3$

6. $c = 3.5, b = 2.1$

In Exercises 7–12, let θ be an angle of a right triangle with the given trigonometric function. Find the values of the other trigonometric functions.

7. $\cos \theta = \dfrac{8}{17}$

8. $\tan \theta = \dfrac{21}{20}$

9. $\sin \theta = \dfrac{3}{5}$

10. $\sec \theta = \dfrac{10}{6}$

11. $\csc \theta = \dfrac{29}{20}$

12. $\csc \theta = \dfrac{15}{12}$

In Exercises 13–16, the values of two of the trigonometric functions of angle θ are given. Use this information to determine the values of the other four trigonometric functions for θ.

13. $\sin \theta = 0.866, \cos \theta = 0.5$

14. $\sin \theta = 0.975, \cos \theta = 0.222$

15. $\sin \theta = 0.085, \sec \theta = 1.004$

16. $\csc \theta = 4.872, \cos \theta = 0.979$

Use a calculator to determine the values of each of the indicated trigonometric functions in Exercises 17–32.

17. $\sin 18.6°$

18. $\cos 38.4°$

19. $\tan 18.3°$

20. $\sin 20°15'$

21. $\tan 76°32'$

22. $\sec 24° 14'$

23. $\cot 82.6°$

24. $\csc 19° 50'$

25. $\sin 0.25$ rad

26. $\cos 0.4$ rad

27. $\tan 0.63$ rad

28. $\sec 1.35$ rad

29. $\cot 1.43$ rad

30. $\sin 1.21$ rad

31. $\csc 0.21$ rad

32. $\tan 1.555$ rad

Solve Exercises 33–37.

33. *Dynamics* If there is no air resistance, a projectile fired at an angle θ above the horizontal with an initial velocity of v_0 has a range R of

$$R = \frac{v_0^2}{g} \sin 2\theta$$

A football is thrown with initial velocity of 15 m/s at an angle of $40°$ above the horizontal. If $g = 9.8$ m/s^2, how far will the football travel?

34. *Dynamics* If the football in Exercise 33 had been thrown at an angle of $45°$ above the horizontal, how

much further, or shorter, would the ball have traveled? If the angle had been $48°$, how would the results have differed?

35. *Electricity* In an ac circuit that contains resistance, inductance, and capacitance in series, the angle of the applied voltage v and the voltage drop across the resistance V_R is the **phase angle** ϕ, and $V_R = V \cos \phi$. If the phase angle is $32°$ and the applied voltage is 5.8 V, what is the effective voltage across the resistor V_R?

You will need the following information in order to work Exercises 36–37.

Tables of trigonometric functions are usually computed using the Maclaurin series. For the sin x and cos x the series are

$$\sin x = x - \frac{x^3}{3!} + \frac{x^5}{5!} - \frac{x^7}{7!} + \cdots$$

and

$$\cos x = 1 - \frac{x^2}{2!} + \frac{x^4}{4!} - \frac{x^6}{6!} + \frac{x^8}{8!} - \cdots$$

where x is the angle in radians, and $3! = 3 \cdot 2 \cdot 1 = 6$, $4! = 4 \cdot 3 \cdot 2 \cdot 1 = 24$, $5! = 5 \cdot 4 \cdot 3 \cdot 2 \cdot 1 = 120$, and so on. (Some calculators have an $\boxed{x!}$ key.) The symbols $x!$, $5!$, $2!$, and so on, are read x factorial, 5 factorial, 2 factorial, and so on.

36. Write a program to compute and print the sines and cosines of an angle from 0 rad to 1.50 rad every 0.1 rad, using the first four terms of the series.

37. Convert your program from Exercise 36 so that it will compute the sines and cosines from $0°$ to $90°$ every $1°$.

≡ 9.3
THE RIGHT TRIANGLE

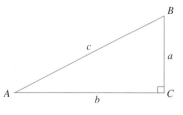

FIGURE 9.17

In Section 9.2, we learned how to determine trigonometric functions given the lengths of two sides of a right triangle. We are now ready to begin finding the unknown parts of a triangle if we know the length of one side and the size of one of the acute angles. Once we can do that, we will begin to apply our knowledge to solving some problems that involve triangles.

Look at the right triangle in Figure 9.17. The right angle is labeled C and the other two vertices A and B. The side opposite each of these vertices is labeled with the lowercase version of the same letter. Thus, side a is opposite $\angle A$, side b is opposite $\angle B$, and side c is opposite $\angle C$. Since this is a right triangle, angles A and B are complementary.

We now have the tools to solve right triangles. To solve a triangle means to find the sizes of all unknown sides and angles.

EXAMPLE 9.17

Given $A = 55°$ and $b = 7.92$, solve the right triangle ABC.

Solution This data chart contains the information that we know and shows what we have left to find. Notice that, since this is a right triangle, we wrote that $C = 90°$.

sides	angles
$a = \underline{}$	$A = 55°$
$b = 7.92$	$B = \underline{}$
$c = \underline{}$	$C = 90°$

Since $A = 55°$, then $B = 90° - 55° = 35°$. If you look at Figure 9.18 you see that $\tan A = \dfrac{a}{b}$ or $a = b \tan A$. We know $A = 55°$ and $b = 7.92$, so $a = 7.92 \tan 55° = (7.92)(1.428148) = 11.310932$ or about 11.31. Now $\sec A = \dfrac{c}{b}$, so $c = b \sec A = (7.92)(1.7434468) = 13.808099$ or 13.81. Since your calculator does not have a sec key, it may be better to find c using cos. Now,

$$\cos A = \frac{b}{c}$$

so

$$c = \frac{b}{\cos A}$$

$$= \frac{7.92}{\cos 55°} \approx \frac{7.92}{0.5736}$$

$$\approx 13.81$$

Look at the data chart again. This time all of the parts of the chart have been filled in, so we know that we have solved for all the missing parts of this triangle. From the completed chart, we can easily see the measures of any part of the triangle.

sides	angles
$a \approx 11.31$	$A = 55°$
$b = 7.92$	$B = 35°$
$c \approx 13.81$	$C = 90°$

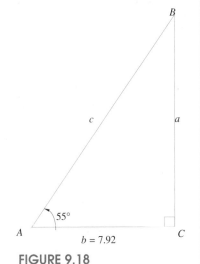

FIGURE 9.18

We could have used our derived value of a and the Pythagorean theorem to find c, or we could have used a and one of the other trigonometric functions.

 Hint

It is usually best not to use the derived values because of errors that can be introduced by rounding.

By the way, if you use the Pythagorean theorem to check the results in Example 9.17, you will see that $(7.92)^2 + (11.31)^2 \neq (13.81)^2$. The Pythagorean theorem says that these should be equal, but because we are using rounded-off values, they are not. However, they are equal to three significant digits.

Using the INV Key

Until now, we have calculated the values of a trigonometric function from the size of the angle. But, what if we know the value of a trigonometric function and want to know the size of the angle? In this case, press the INV or 2nd key on the calculator before entering the function key.

EXAMPLE 9.18

Use a calculator to determine the angles that have each of the following values of trigonometric functions: $\sin A = 0.785$, $\cos B = 0.437$, $\tan C = 4.213$, $\csc D = 8.915$.

Function	ENTER	DISPLAY
$\sin A = 0.785$	0.785 INV sin	51.720678
$\cos B = 0.437$	0.437 INV cos	64.087375
$\tan C = 4.213$	4.213 INV tan	76.647345
$\csc D = 8.915$	8.915 1/x INV sin	6.4404506

So, $A = 51.72°$, $B = 64.09°$, $C = 76.65°$, and $D = 6.44°$. Notice that to find angle D, we had to first take the reciprocal of $\csc D$, to get $\sin D$.

Some calculators have second keys with names $\boxed{\sin^{-1}}$, $\boxed{\cos^{-1}}$, or $\boxed{\tan^{-1}}$. Pressing 2nd $\boxed{\sin^{-1}}$ on these calculators has the same effect as pressing the INV sin key combination just described. Thus, if $\sin A = 0.785$, then the size of angle A, in degrees, is determined by pressing 0.785 2nd $\boxed{\sin^{-1}}$ to obtain 51.72067826. On a graphics calculator you would press 2nd $\boxed{\sin^{-1}}$ 0.785 ENTER.

In the BASIC computer language, you are able to find an angle, in radian measure, if you know its tangent. Thus, if $\tan B = 4.213$, to find the size of B, you would enter PRINT ATN (4.213) and get the display 1.33774853. This is in radians and converts to 76.64°. ATN is the BASIC function for arctan or INV tan. If you want to find an angle in BASIC when you know $\sin x$ or $\cos x$, you need to define special functions. We will discuss this later in the chapter.

EXAMPLE 9.19

Solve right triangle ABC ($\triangle ABC$), if $a = 23.5$ and $c = 42.7$, as shown in Figure 9.19.

Solution   The following data chart shows the information we are starting with and what we have to find.

sides	angles
$a = 23.5$	$A = \underline{}$
$b = \underline{}$	$B = \underline{}$
$c = 42.7$	$C = 90°$

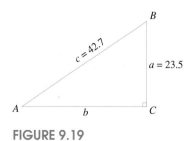

FIGURE 9.19

From the Pythagorean theorem we know that $b^2 = c^2 - a^2 = (42.7)^2 - (23.5)^2 = 1271.04$, so $b = 35.7$. Now, $\sin A = \frac{23.5}{42.7} \approx 0.5503513$. Pressing INV sin, we get

EXAMPLE 9.19 (Cont.)

33.391116. Thus, $A \approx 33.4°$ and $B \approx 90° - 33.4° = 56.6°$. The completed data chart follows.

sides	angles
$a = 23.5$	$A \approx 33.4°$
$b \approx 35.7$	$B \approx 56.6°$
$c = 42.7$	$C = 90°$

EXAMPLE 9.20

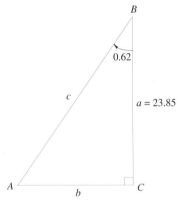

FIGURE 9.20

Solve for the right triangle ABC if $B = 0.62$ rad and $a = 23.85$, as shown in Figure 9.20.

Solution   The data chart for this triangle is

sides	angles
$a = 23.85$	$A = \underline{}$
$b = \underline{}$	$B = 0.62$
$c = \underline{}$	$C = \dfrac{\pi}{2} \approx 1.57$

In this example, we can use angle B as our reference angle. So, $\tan B = \dfrac{b}{a}$ and $b = a \tan B = (23.85) \tan 0.62 = (23.85)(0.713909) = 17.02673$, or $b \approx 17.03$. (Did you remember to put your calculator in radian mode?)

Also, $\cos B = \dfrac{a}{c}$, so

$$
\begin{aligned}
c &= \frac{a}{\cos B} \\
&= \frac{23.85}{\cos 0.62} \\
&= \frac{23.85}{0.81387846} \\
&= 29.30413
\end{aligned}
$$

or ≈ 29.30

Finally, we have $\tan A = \frac{23.85}{17.03} = 1.4004698$, and pressing $\boxed{\text{INV}}$ $\boxed{\text{tan}}$, we get 0.9507055, or $A \approx 0.95$ rad.

Remember from Exercise 1 in Exercise Set 9.1 that $90° \approx 1.57$ rad. If you used your calculator to convert $90°$ to radians, you could then have found A as $1.57 - 0.62 = 0.95$.

The completed data chart for this example is

sides	angles
$a = 23.85$	$A \approx 0.95$
$b \approx 17.03$	$B = 0.62$
$c \approx 29.30$	$C = \dfrac{\pi}{2} \approx 1.57$

Angles of Elevation or Depression

Frequently when solving problems involving trigonometry, we have to use the **angle of elevation** (see Figure 9.21), which is the angle, measured from the horizontal, through which an observer would have to elevate his or her line of sight in order to see an object. Similarly, the **angle of depression** is the angle, measured from the horizontal, through which an observer has to lower his or her line of sight in order to see an object. (See Figure 9.22.)

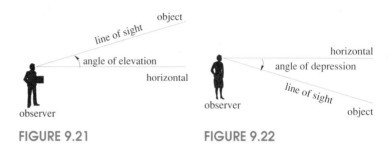

FIGURE 9.21   **FIGURE 9.22**

Application

EXAMPLE 9.21

A person is standing 50 m from the base of a tower. (See Figure 9.23.) The angle of elevation to the top of the tower is 76°. How high is the tower?

Solution We have a right triangle as sketched in Figure 9.23. The height of the tower is labeled x, so we have $\tan 76° = \dfrac{x}{50}$, or

$$x = 50 \tan 76°$$
$$= 50(4.0107809)$$
$$= 200.53905$$

The tower is about 200.5 m high.

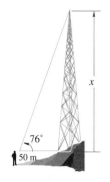

FIGURE 9.23

Application

EXAMPLE 9.22

Two people are in a hot air balloon. One of them is able to get a sighting from the gondola of the balloon as it passes over one end of a football field, as shown in Figure 9.24. The angle of depression to the other end of the football field is 53.8°. This person knows that the length of the football field, including the end zones, is 120 yd. How high was the balloon when it went over the football field?

EXAMPLE 9.22 (Cont.)

FIGURE 9.24

Solution We want to determine the height of the right triangle. The height has been labeled h in Figure 9.24. So, we have $\tan 53.8° = \dfrac{h}{120}$, or

$$h = 120 \tan 53.8°$$

$$= 120(1.366326733)$$

$$= 163.9592079$$

The balloon was about 164 yd high.

Application

EXAMPLE 9.23

A surveyor marks off a right-angle corner of a rectangular house foundation. In sighting on the diagonally opposite corner of the foundation, the line of sight has to move through an angle of 35° as shown in Figure 9.25a. If the length of the short side of the foundation is 46.3 ft, what is the length of the long side?

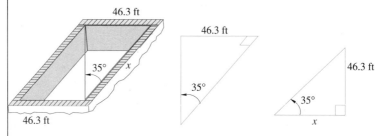

FIGURE 9.25a   **FIGURE 9.25b**   **FIGURE 9.25c**

Solution We have the situation sketched in Figure 9.25b. (Notice that in Figure 9.25b, the right angle does not look like a right angle.) But, this is confusing; so in Figure 9.25c the triangle has been rotated to the more "typical" position with the right angle along the bottom. Since $\tan 35° = \dfrac{46.3}{x}$, we have

$$x = \frac{46.3}{\tan 35°}$$

$$= \frac{46.3}{0.70020754}$$

$$= 66.123253$$

So the long side is about 66.1 ft.

Exercise Set 9.3

Find the acute angles for the trigonometric functions given in Exercises 1–12. (Give answers to the nearest 0.01.)

1. $\sin A = 0.732$
2. $\cos B = 0.285$
3. $\tan C = 4.671$
4. $\sin D = 0.049$

5. $\cos E = 0.839$
6. $\tan F = 0.539$
7. $\sec G = 3.421$
8. $\csc H = 1.924$

9. $\cot I = 0.539$
10. $\sec J = 1.734$
11. $\csc K = 4.761$
12. $\cot L = 4.021$

In Exercises 13–30, sketch each right triangle and solve it. Use your knowledge of significant figures to round off appropriately.

13. $A = 16.5°, a = 7.3$
14. $B = 53°, b = 9.1$
15. $A = 72.6°, c = 20$
16. $B = 12.7°, a = 19.4$
17. $A = 43°, b = 34.6$
18. $B = 67°, c = 32.4$

19. $A = 0.92$ rad, $a = 6.5$
20. $B = 1.13$ rad, $b = 24$
21. $A = 0.15$ rad, $c = 18$
22. $B = 0.23$ rad, $a = 9.7$
23. $A = 1.41$ rad, $b = 40$
24. $A = 1.15$ rad, $c = 18$

25. $a = 9, b = 15$
26. $a = 19.3, c = 24.4$
27. $b = 9.3, c = 18$
28. $a = 14, b = 9.3$
29. $a = 20, c = 30$
30. $b = 15, c = 25$

Solve Exercises 31–39.

31. *Electricity*   In an ac circuit that has inductance x_L and resistance R, the phase angle can be determined from the equation $\tan \phi = \frac{x_L}{R}$. If $x_L = 12.3 \ \Omega$ and $R = 19.7 \ \Omega$, what is the phase angle?

32. From the top of a lighthouse, the angle of depression to the waterline of a boat is $23.2°$. If the lighthouse is 222 ft high, how far away is the ship from the bottom of the lighthouse?

33. *Navigation*   An airplane is flying at an altitude of 700 m when the copilot spots a ship in distress at an angle of depression of $37.6°$. How far is it from the plane to the ship?

34. *Construction*   A bridge is to be constructed across a river. As shown in Figure 9.26, a piling is to be placed at point A and another at C. To find the distance

between A and C a surveyor locates point B exactly 95 m from C so that $\angle C$ is a right angle. If $\angle B$ is $57.62°$, how far apart are the pilings?

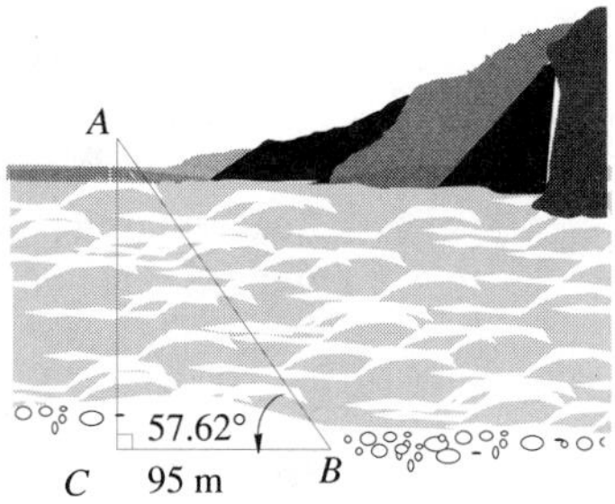

FIGURE 9.26

35. A vector is usually resolved into component vectors that are perpendicular to each other. Thus, you get a rectangle with one side F_x, the horizontal component and the other F_y, the vertical component, as shown in Figure 9.27. The original vector F is the diagonal of

the rectangle. If $F = 10$ N and $\theta = 30°$, what are the horizontal and vertical components of this force?

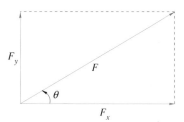

FIGURE 9.27

36. Two forces, one of 20 lb and the other of 12 lb, act on a body in directions perpendicular to each other. What is the magnitude of the resultant of these forces? What is the size of the angle the resultant of these forces makes with the smaller force?

37. *Transportation engineering*   Highway curves are usually banked at an angle ϕ. The proper banking angle for a car making a turn of radius r at velocity v is $\tan \phi = \dfrac{v^2}{gr}$, where $g = 32$ ft/s^2 or 9.8 m/s^2. Find the proper banking angle for a car moving at 55 mph to go around a curve 1,200 ft in radius. (First, convert 55 mph to 80.67 ft/s.)

38. *Transportation engineering*   Find the proper banking angle for a car moving 88 km/h to go around a curve 500 m in radius. (First, convert 88 km/h to 24.44 m/s.)

39. *Transportation engineering*   Two straight highways A and B intersect at an angle 67°. A service station is located on highway A 300 m from the intersection. What is the location of the point on highway B that is closest to the service station? How near is it?

≣ 9.4
TRIGONOMETRIC FUNCTIONS OF ANY ANGLE

In Sections 9.2 and 9.3, we have concentrated on the trigonometric functions for angles between 0° and 90° or 0 and $\dfrac{\pi}{2}$ rad. We will now examine other angles.

If you remember, we originally defined the trigonometric functions in terms of a point on the terminal side of an angle in standard position. We want to return to that definition as we look at angles that are not acute.

Reference Angles

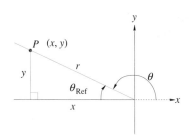

FIGURE 9.28

Consider angle θ in Figure 9.28. Point P is on the terminal side and is in the second quadrant. If a perpendicular line is dropped from P to the x-axis, a triangle is formed with an acute angle θ_{Ref}. If we use the definition of the trigonometric functions from Section 9.1, we seem to get the same values we got for angle θ of the triangle.

Reference Angle
The acute angle, θ_{Ref}, between the terminal side of θ and the x-axis is called the **reference angle** for θ. Thus, we can see that the trigonometric functions of θ are numerically the same as those of its reference angle, θ_{Ref}. The reference angle θ_{Ref} for any angle θ in each of four quadrants is shown in Figures 9.29a–d.

EXAMPLE 9.24

Find the reference angle θ_{Ref} for each angle θ: **(a)** $\theta = 75°$, **(b)** $\theta = 218°$, **(c)** $\theta = 320°$, **(d)** $\theta = \dfrac{3\pi}{4}$, **(e)** $\theta = \dfrac{7\pi}{6}$.

EXAMPLE 9.24 (Cont.)

Solutions The solutions will use the guidelines demonstrated in Figure 9.29.

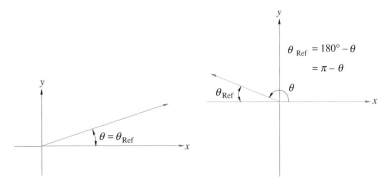

FIGURE 9.29a   **FIGURE 9.29b**

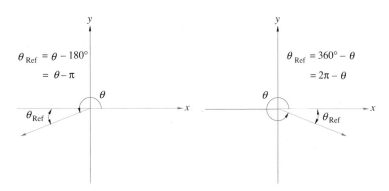

FIGURE 9.29c   **FIGURE 9.29d**

(a) $75°$ is in Quadrant I, so $\theta_{Ref} = 75°$.

(b) $218°$ is Quadrant III, so $\theta_{Ref} = 218° - 180° = 38°$.

(c) $320°$ is in Quadrant IV, so $\theta_{Ref} = 360° - 320° = 40°$.

(d) $\dfrac{3\pi}{4}$ is in Quadrant II, so $\theta_{Ref} = \pi - \dfrac{3\pi}{4} = \dfrac{4\pi}{4} - \dfrac{3\pi}{4} = \dfrac{\pi}{4}$.

(e) $\dfrac{7\pi}{6}$ is in Quadrant III, so $\theta_{Ref} = \dfrac{7\pi}{6} - \pi = \dfrac{7\pi}{6} - \dfrac{6\pi}{6} = \dfrac{\pi}{6}$.

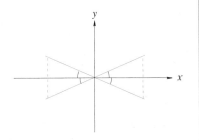

FIGURE 9.30

 Hint

The reference angle is always measured from the *x*-axis and never from the *y*-axis. The "bowtie" in Figure 9.30 may help you remember how to determine the reference angle.

EXAMPLE 9.25

A point on the terminal side of an angle θ has the coordinates $(-5, -12)$. Write the six trigonometric functions of θ to three decimal places.

EXAMPLE 9.25 (Cont.)

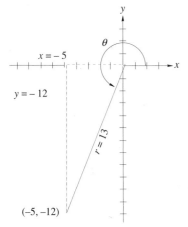

FIGURE 9.31

Solution   A sketch of the angle is in Figure 9.31. This angle is in the third quadrant. From the Pythagorean theorem we get

$$r = \sqrt{(-5)^2 + (-12)^2} = \sqrt{25 + 144} = \sqrt{169} = 13$$

So, the trigonometric functions of θ are

$$\sin\theta = \frac{y}{r} = \frac{-12}{13} \approx -0.923 \qquad \csc\theta = \frac{r}{y} = \frac{13}{-12} \approx -1.083$$

$$\cos\theta = \frac{x}{r} = \frac{-5}{13} \approx -0.385 \qquad \sec\theta = \frac{r}{x} = \frac{13}{-5} = -2.6$$

$$\tan\theta = \frac{y}{x} = \frac{-12}{-5} = 2.4 \qquad \cot\theta = \frac{x}{y} = \frac{-5}{-12} \approx 0.417$$

As you can see, the sine and cosine and their reciprocals are negative. The tangent, and its reciprocal are positive.

In our previous work with right triangles, we found that the values of the trigonometric functions were always positive. You can see here and from our work in Section 9.1 that they can sometimes be negative. Whenever the angle is in the second, third, or fourth quadrant, some of the trigonometric functions will be negative.

EXAMPLE 9.26

Use a calculator to find the values of the following trigonometric functions: sin 215°, cos 110°, tan 332°, csc 163°, sec 493°, and cot(−87°).

Solution   We find these the same way we used the calculator to find the values of angles between 0° and 90°:

Function	ENTER (Algebraic)	ENTER (Graphing)	DISPLAY
sin 215°	215 [sin]	[sin] 215 [ENTER]	−0.5735764
cos 110°	110 [cos]	[cos] 110 [ENTER]	−0.3420201
tan 332°	332 [tan]	[tan] 332 [ENTER]	−0.5317094
csc 163°	163 [sin] [1/x]	([[sin] 163 [)] [x^{-1}] [ENTER]	3.4203036
sec 493°	493 [cos] [1/x]	([[cos] 493 [)] [x^{-1}] [ENTER]	−1.4662792
cot(−87°)	87 [+/−] [tan] [1/x]	([[tan] [(−)] 87 [)] [x^{-1}] [ENTER]	−0.0524078

There are times when we need to know the sign of an angle and we may not have our calculator handy. One case might be if you are using the table of trigonometric functions. But you will need it more often when you have to find the size of an angle and you know the value of a trigonometric function. You will see in Section 9.5 that the calculator will give you an answer, but it may not be the correct answer to the problem. You will have to determine the correct answer from the calculator's answer and from your knowledge of the signs of the trigonometric functions.

To find the sign of a trigonometric function, make a sketch of the angle. You do not need a very accurate sketch, but make sure you get the angle in the correct quadrant. Draw the triangle for the reference angle and note the signs of x and y. Remember that the radius vector r is always positive. Then set up the ratio using the two values from x, y, and r that are appropriate for this angle.

EXAMPLE 9.27

What is the sign of $\sec 215°$?

Solution   Examine the sketch in Figure 9.32. We have drawn a right triangle for the reference triangle and labeled the sides with the sign in parentheses after the letter x, y, or r. Since $\sec 215° = \dfrac{r}{x}$ where r is positive and x is negative, the sign of the $\sec 215°$ is negative.

Figure 9.33 summarizes the signs of the trigonometric functions in each quadrant. If a function is not mentioned in a quadrant, then the function is negative. Thus in Quadrant II, since they are not mentioned, the cos, tan, cot, and sec are all negative.

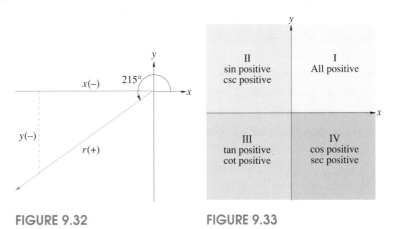

FIGURE 9.32   FIGURE 9.33

We will finish this section with a discussion of how to find the values of trigonometric functions from a table. If an angle is not an acute angle, you need to use the reference angle and the coterminal angle.

If an angle is negative, or if it is a positive angle with measure more than $360°$ or 2π rad, replace it with a coterminal angle that is between $0°$ and $360°$ or 0 and 2π rad. To do this, you will add or subtract multiples of $360°$ if the angle is in degrees, or 2π if it is in radians, until you have the desired coterminal angle.

EXAMPLE 9.28

For each angle find a coterminal angle between $0°$ and $360°$ or 0 and 2π rad:
(a) $1,292°$, **(b)** $-683°$, **(c)** $\frac{17}{4}\pi$, **(d)** $-\frac{5\pi}{3}$, **(e)** -8.7 rad.

Solutions
(a) $1,292° - 3(360°) = 1,292° - 1,080° = 212°$
(b) $-683° + 2(360°) = -683° + 720° = 37°$

EXAMPLE 9.28 (Cont.)

(c) $\dfrac{17}{4}\pi - 2(2\pi) = \dfrac{17}{4}\pi - 2\left(\dfrac{8\pi}{4}\right) = \dfrac{17\pi}{4} - \dfrac{16\pi}{4} = \dfrac{\pi}{4}$

(d) $-\dfrac{5\pi}{3} + 1(2\pi) = \dfrac{-5\pi}{3} + \dfrac{6\pi}{3} = \dfrac{\pi}{3}$

(e) $-8.7 + 2(2\pi) = -8.7 + 12.566371 = 3.8663706$ rad

The value of any trigonometric function of an angle θ is the same as the value of the function for the reference angle θ_{Ref}, except for an occasional change in the algebraic sign. If the angle is not between $0°$ and $360°$ or 0 and 2π rad, first find the coterminal angle in that interval and then find the reference angle of the coterminal angle.

Exercise Set 9.4

Give the reference angle for each of the angles in Exercises 1–8.

1. $87°$

2. $200°$

3. $137°$

4. $298°$

5. $\dfrac{9\pi}{8}$ rad

6. 2.1 rad

7. 4.5 rad

8. 5.85 rad

Give the coterminal angle for each of the angles in Exercises 9–16.

9. $518°$

10. $1{,}673°$

11. $-871°$

12. $-137°$

13. 7.3 rad

14. $\dfrac{16\pi}{3}$ rad

15. -2.17 rad

16. -8.43 rad

State which quadrant or quadrants the terminal side of θ is in for each of the angles or expressions in Exercises 17–30.

17. $165°$

18. $285°$

19. $-47°$

20. $312°$

21. $250°$

22. $197°$

23. $98°$

24. $\sin\theta$ is positive.

25. $\tan\theta$ is negative.

26. $\cos\theta$ is positive.

27. $\sec\theta$ is negative.

28. $\cot\theta$ is negative.

29. $\csc\theta$ is negative and $\cos\theta$ is positive.

30. $\tan\theta$ and $\sec\theta$ are negative.

State whether each of the expressions in Exercises 31–38 is positive or negative.

31. $\sin 105°$

32. $\sec 237°$

33. $\tan 372°$

34. $\cos(-53°)$

35. $\cos 1.93$ rad

36. $\cot 4.63$ rad

37. $\sin 215°$

38. $\csc 5.42$ rad

Find the values of each of the trigonometric functions in Exercises 39–62.

39. $\sin 137°$

40. $\cos 263°$

41. $\tan 293°$

42. $\sin 312°$

43. $\tan 164.2°$

44. $\csc 197.3°$

45. $\sin 2.4$ rad

46. $\cos 1.93$ rad

47. $\tan 6.1$ rad

48. $\sec 4.32$ rad

49. $\tan 1.37$ rad

50. $\sin 3.2$ rad

51. $\sin 415.5°$

52. $\tan 512.1°$

53. $\cot -87.4°$

54. $\cos 372.1°$

55. $\csc -432.4°$

56. $\cos 357.3°$

57. $\sin 6.5$ rad

58. $\sec(-4.3$ rad$)$

59. $\tan 8.35$ rad

60. $\sin 9.42$ rad

61. $\cos -0.43$ rad

62. $\sec 9.34$ rad

≡ 9.5
INVERSE TRIGONOMETRIC FUNCTIONS

In Section 9.3, we were introduced to the inverse trigonometric functions. For example, when we knew that $\sin \theta = 0.2437$ we could enter 0.2437 $\boxed{\text{INV}}$ $\boxed{\sin}$ or $\boxed{\text{2nd}}$ $\boxed{\sin}$ 0.2437 into a calculator and find that $\theta = 14.105021°$. In this section, we look more closely at these functions.

Inverse Trigonometric Functions

In our earlier work with inverse trigonometric functions, we were concerned with right triangles. Thus, all the angles we had to find were acute angles. In Section 9.4, we learned that, simply knowing the value of a trigonometric function for an angle did not allow us to determine the size of the angle. For example, if $\sin \theta = 0.2437$, we know that θ must be in Quadrant I or II. These are the only quadrants where sine is positive. This means that $\theta = 14.105021°$ or $\theta = 180° - 14.105021° = 165.894979°$. It is possible that θ could be any angle coterminal with either of these two angles.

In a similar manner, if $\tan \theta = -1.5$, then using a calculator you obtain, $\theta = -56.309932°$. This is coterminal with $\theta = 303.690068°$. This angle is in Quadrant IV. The tangent is also negative for values in Quadrant II. In Quadrant II, the angle would be $123.69007°$.

Using Calculators for Inverse Trigonometric Functions

Using a calculator as we did in the two previous examples, is certainly fast and accurate. But, it may not always give us the angle that we want. For example, it will never give an angle in Quadrant III. Why is this? It goes back to the basic idea of a function—the idea that for each different value of x there is exactly one value of y.

Because people wanted inverse trigonometric *functions* they had to restrict the answers to exactly one value of y for each value of x. That is why, when you use your calculator to determine inverse sin 0.5, you obtain $30°$ by pressing 0.5 $\boxed{\text{INV}}$ $\boxed{\sin}$ or $\boxed{\text{2nd}}$ $\boxed{\sin}$ 0.5.

This is nothing new. At one time we studied a function $f(x) = x^2$ and its inverse $f^{-1}(x) = \sqrt{x}$. We knew that $f(-5) = f(5) = 5^2 = 25$, but that $f^{-1}(25) = \sqrt{25} = 5$. The inverse function of x^2, $\sqrt{x}$, only gave the principal value of the square root. In the same way, the inverse trigonometric functions have only the *principal value* of the angle defined for each value of x. Thus, we have the following inverse trigonometric functions.

Inverse Trigonometric Functions

If $\sin \theta = x$, then $\arcsin x = \theta$, where $-90° \leq \theta \leq 90°$

or $\quad -\dfrac{\pi}{2} \leq \theta \leq \dfrac{\pi}{2}$

If $\cos \theta = x$, then $\arccos x = \theta$, where $0° \leq \theta \leq 180°$

or $\quad 0 \leq \theta \leq \pi$

If $\tan \theta = x$, then $\arctan x = \theta$, where $-90° \leq \theta \leq 90°$

or $\quad \dfrac{-\pi}{2} < \theta < \dfrac{\pi}{2}$

If $\csc \theta = x$, then $\text{arccsc}\, x = \theta$, where $-90° \leq \theta \leq 90°, \theta \neq 0°$

or $\quad -\dfrac{\pi}{2} < \theta < \dfrac{\pi}{2}, \theta \neq 0$

If $\sec \theta = x$, then $\text{arcsec}\, x = \theta$, where $0° \leq \theta \leq 180°, \theta \neq 90°$

or $\quad 0 \leq \theta \leq \pi, \theta \neq \dfrac{\pi}{2}$

If $\cot \theta = x$, then $\text{arccot}\, x = \theta$, where $0° < \theta < 180°$

or $\quad 0 < \theta < \pi$

As you can see, for positive values you always get an angle in the first quadrant. If you want the inverse value of a negative number, then you will get an angle in the fourth quadrant for INV sin and INV tan and an angle in the second quadrant for INV cos. Working with a calculator will also show you that the angles in the fourth quadrant are given as negative angles. Thus on a calculator, $\arcsin(-0.5) = -30°$ and not $330°$.

In the previous list, we used the symbol $\arcsin x$ to represent the inverse of the $\sin \theta$. Arcsin is an accepted symbol for the inverse of the sine function. Another symbol that is often used is $\sin^{-1} x$. Just as we used $f^{-1}(x) = \sqrt{x}$; here we use $f^{-1}(x) = \sin^{-1} x$. Similarly, $\arccos x = \cos^{-1} x$, $\arctan x = \tan^{-1} x$, and so on.

Finding All Angles for Inverse Trigonometric Functions ▬▬▬

Since inverse trigonometric functions only give one answer to problems like $\sin \theta = \frac{1}{2}$ where $\theta = \sin^{-1} \frac{1}{2}$, we must develop a method for finding other angles that satisfy this equation. We will use our knowledge of reference angles and the sign of the functions in the four quadrants to do this.

To find the reference angle of an equation of the form $\sin \theta = n$, where n is a number in the range of sine, we will use the absolute value of n, $|n|$. Then the desired answers can be found by adding or subtracting the reference angle from $180°$ (or π) or subtracting from $360°$ (or 2π).

EXAMPLE 9.29

Find all angles θ in the interval $[0°, 360°)$ where $\cos\theta = -0.42$.

Solution Here, we have $n = -0.42$, so $|n| = |-0.42| = 0.42$ and $\theta_{Ref} = \cos^{-1} 0.42 = 65.2°$. Since n was negative, we have answers in Quadrants II and III, the quadrants where cosine is negative.

$$\theta_{II} = 180° - \theta_{Ref} = 180° - 65.2° = 114.8°$$

$$\theta_{III} = 180° + \theta_{Ref} = 180° + 65.2° = 245.2°$$

The answers are $114.8°$ and $245.2°$.

EXAMPLE 9.30

Find all angles θ in the interval $[0, 2\pi)$ where $\cot\theta = -1.73$.

Solution Here, we have $n = -1.73$, so $|n| = |-1.73| = 1.73$ and $\theta_{Ref} = \cot^{-1} 1.73 = \tan^{-1} \frac{1}{1.73} = 0.5241$. Since n was negative, we have answers in Quadrants II and IV, the quadrants where cotangent is negative.

$$\theta_{II} = \pi - \theta_{Ref} = \pi - 0.5241 \approx 2.6175$$

$$\theta_{IV} = 2\pi - \theta_{Ref} = 2\pi - 0.5241 \approx 5.7591$$

The answers are 2.6175 rad and 5.7591 rad.

In the previous two examples, we only found two answers. There are, however, an infinite number of angles coterminal with these answers. If we want to represent all these answers, we proceed as follows.

For Example 9.29:

$$114.8° + 360°k \text{ and } 245.2° + 360°k, \text{ where } k \text{ is an integer}$$

For Example 9.30:

$$2.6175 + 2\pi k \text{ rad and } 5.7591 + 2\pi k \text{ rad, where } k \text{ is an integer}$$

 ## Using Computers for Inverse Trigonometric Functions

A computer operates much like a calculator. If you want to find the arctangent of a value, you use the BASIC function ATN. The answer will be in radians. For example, if you would enter PRINT ATN (4.213) and press ⟨RETURN⟩, the computer would display 1.33774853.

Some BASIC languages do not have BASIC functions equivalent to the inverses of the other trigonometric functions. (Some languages, such as True BASIC™, have libraries that do contain these functions.) You can define the inverses of the other trigonometric functions in terms of ATN(x), which follows. In a program, you must check to see that the value of x, is in the domains of the functions previously listed. For example, for the $\sin^{-1}(x)$ you have to make sure that x is not larger than

$\frac{\pi}{2} = 1.57079633$ or that x is not less than $-\frac{\pi}{2}$. For $\csc^{-1} x$ you must also check to make sure that $x \neq 0$.

```
ARCSIN(X)=ATN(X/SQR(-X*X+1))
ARCCOS(X)=-ATN(X/SQR(-X*X+1))+1.57079633
ARCCSC(X)=ATN(1/SQR(X*X-1))+(SGN(X)-1)
*1.57079633
```

Note $\lvert x \rvert > 1$

```
ARCSEC(X)=ATN(SQR(X*X-1))+(SGN(X)-1)
*1.57079633
ARCCOT(X)=-ATN(X)+1.57079633
```

A new function was used. The SGN function determines if a value is positive, negative, or zero. SGN(X) = +1 if x is positive, SGN(X) = −1 if x is negative, and SGN(X) = 0 if x is zero.

Exercise Set 9.5

State in which quadrants the angles θ lie for the given trigonometric functions in Exercises 1–8.

1. $\sin \theta = \dfrac{1}{2}$

2. $\tan \theta = \dfrac{3}{4}$

3. $\cot \theta = -2$

4. $\cos \theta = -0.25$

5. $\sec \theta = 4.3$

6. $\cos \theta = 0.8$

7. $\csc \theta = -6.1$

8. $\sin \theta = -\dfrac{\sqrt{3}}{2}$

In Exercises 9–16 find, to the nearest tenth of a degree, all angles θ, where $0° \leq \theta < 360°$, with the given trigonometric function.

9. $\sin \theta = \dfrac{1}{2}$

10. $\tan \theta = \dfrac{3}{4}$

11. $\cot \theta = -2$

12. $\cos \theta = -0.25$

13. $\sec \theta = 4.3$

14. $\cos \theta = 0.8$

15. $\csc \theta = -6.1$

16. $\sin \theta = -\dfrac{\sqrt{3}}{2}$

In Exercises 17–24 find, to the nearest hundredth of a radian, all angles θ, where $0 \leq \theta < 2\pi$, with the given trigonometric function.

17. $\sin \theta = 0.75$

18. $\tan \theta = 1.6$

19. $\cot \theta = -0.4$

20. $\cos \theta = 0.25$

21. $\csc \theta = 4.3$

22. $\sin \theta = -0.08$

23. $\sec \theta = 2.7$

24. $\cos \theta = -0.95$

Evaluate each of the functions in Exercises 25–32. Write each answer to the nearest tenth of a degree.

25. $\arcsin 0.84$

26. $\arccos(-0.21)$

27. $\arctan 4.21$

28. $\text{arccot}(-0.25)$

29. $\sin^{-1} 0.32$

30. $\cos^{-1} 0.47$

31. $\tan^{-1}(-0.64)$

32. $\csc^{-1}(-3.61)$

Evaluate each of the functions in Exercises 33–40. Write each answer to the nearest hundredth of a radian.

33. $\arccos(-0.33)$

34. $\arctan 1.55$

35. $\arccos 0.29$

36. $\text{arcsec}(-3.15)$

37. $\sin^{-1} 0.95$

38. $\cos^{-1}(-0.67)$

39. $\tan^{-1} 0.25$

40. $\cot^{-1}(-0.75)$

Solve Exercises 41–44.

 41. Write a computer program that will allow you to input a value of x and that will print x and $\arctan x$.

 42. Write a computer program that will allow you to input a value of x. Check to see if it is an acceptable value. Then print x and $\arccos x$.

 43. Write a computer program that will allow you to input a value of x and that will print x and all possible inverse trigonometric functions for that value of x.

 44. Adapt each of the programs that you just wrote to print the answers in degrees instead of radians.

≡ **9.6**
APPLICATIONS OF TRIGONOMETRY

In earlier sections, we looked at some applications of trigonometry, particularly applications dealing with right triangles. In this section, we will look at some additional applications of trigonometry.

We defined a radian as $\frac{1}{2\pi}$ of a circle. Another way to look at a radian is to examine the circle in Figure 9.34. If r is the radius of the circle and s is the length of the arc opposite θ, then $\theta = \frac{s}{r}$ or $\theta r = s$, where θ is in radians.

We can use this relationship to find the area of the sector of a circle. For the shaded sector in Figure 9.34, the area $A = \frac{rs}{2} = \frac{r^2\theta}{2} = \frac{1}{2}r^2\theta$. Both of these formulas work if the angles are in radians.

EXAMPLE 9.31

Find the arc length and area of a sector of a circle, with radius 2.7 cm and central angle 1.3 rad.

Solution To find the arc length we use $s = \theta r$ with $\theta = 1.3$ rad and $r = 2.7$ cm.

$$s = \theta r$$

$$= (1.3)(2.7)$$

$$= 3.51 \text{ cm}$$

EXAMPLE 9.31 (Cont.)

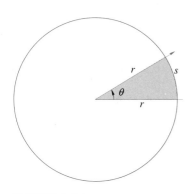

FIGURE 9.34

The area is found using $A = \frac{1}{2} r^2 \theta$.

$$A = \frac{1}{2} r^2 \theta$$

$$= \frac{1}{2} (2.7)^2 (1.3)$$

$$= 4.7385 \qquad \text{or} \qquad 4.74 \text{ cm}^2$$

Radian measure was developed with the aid of a circle and is used to measure angles. One application of radian measure involves rotational motion. When we work with motion in a straight line, we use the formula $d = vt$ or distance = velocity (or rate) × time. We assume that the rate is constant.

Suppose instead that we had an object moving around a circle at a constant rate of speed. If we let s represent the distance around the circle and v the velocity (or rate), then $s = vt$ or $v = \frac{s}{t}$. Since $s = \theta r$, where r is the radius of the circle, we have $v = \frac{\theta r}{t}$. The centripetal acceleration is $a_c = \frac{v^2}{r} = \frac{\theta^2 r}{t^2}$.

Application

EXAMPLE 9.32

A ball is spun in a horizontal circle 60 cm in radius at the rate of one revolution every 3 s. What is the ball's velocity and centripetal acceleration?

Solution   The velocity $v = \frac{s}{t}$. We know that $t = 3$ s, but we have to find s. In one revolution, the ball covers one complete circle, so $\theta = 2\pi$, and since $r = 60$ cm, we have $s = \theta r = 120\pi$. Thus,

$$v = \frac{s}{t}$$

$$= \frac{120\pi \text{ cm}}{3 \text{ s}}$$

$$= 40\pi \text{ cm/s}$$

$$= 125.66 \text{ cm/s}$$

The centripetal acceleration is

$$a_c = \frac{v^2}{r} = \frac{(40\pi \text{ cm/s})^2}{60 \text{ cm}}$$

$$= 263.19 \text{ cm/s}^2$$

The angular distance D (measured as the earth's center) between two points P_1 and P_2 on the earth's surface is determined by

$$\cos D = \sin L_1 \sin L_2 + \cos L_1 \cos L_2 \cos(M_1 - M_2)$$

where L_1, L_2 and M_1, M_2 are the respective latitudes and longitudes of two points. The latitude is the angle measured at the earth's surface between a point on the earth and the equator on the same meridian. Meridians are imaginary lines on the surface of the earth, which pass through both the north and south poles. (See Figure 9.35.) The longitude is the angle between the meridian passing through a point on the earth and the $0°$ meridian passing through Greenwich, England.

Application

EXAMPLE 9.33

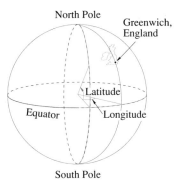

North Pole

Greenwich, England

Latitude

Equator

Longitude

South Pole

FIGURE 9.35

If Knoxville, Tennessee, is at latitude $36°$ N and longitude $83°55'$ W, and Denver, Colorado, is at latitude $39°44'$ N and longitude $104°59'$ W, and the radius of the earth is 3,960 mi, how far is it from Knoxville to Denver?

Solution We have $L_1 = 36°, L_2 = 39°44', M_1 = 83°55'$, and $M_2 = 104°59'$, so $M_1 - M_2 = -21°04'$.

$$-\cos D = (\sin 36°)[\sin(39°44')]$$
$$+ (\cos 36°)(\cos 39°44')[\cos(-21°04')]$$
$$= (0.5878)(0.6392) + (0.8090)(0.7690)(0.9332)$$
$$= 0.3757 + 0.5806$$
$$= 0.9563$$
$$D = \cos^{-1}(0.9563)$$
$$= 17.00°$$
$$= 0.2967 \text{ rad}$$

Notice that we had to convert our answer for D from degrees to radians.

Since $r = 3,960$ mi, the distance from Knoxville to Denver is

$$s = \theta r = (0.2967)(3,960) = 1,174.93 \text{ mi}$$

This would, of course, be air miles and not the distance you would travel by automobile.

One use of radian measure involves rotational motion. Remember that for motion in a straight line, distance = velocity (or rate) × time or $d = vt$, where we assumed that the rate remains constant. Now, if the motion is in a circular direction instead of a straight line, we would have the formula $s = vt$. Since $s = \theta r$ and $v = \frac{s}{t}$, then $v = \frac{\theta r}{t} = r\left(\frac{\theta}{t}\right)$. This ratio, $\frac{\theta}{t}$, is called the **angular velocity** and is usually denoted by the Greek letter ω (omega). Thus it is known that angular velocity ω of an object that turns through the angle θ in time t is given by $\omega = \frac{\theta}{t}$. The linear velocity v of a point that moves in a circle of radius r with uniform angular velocity ω is $v = \omega r$. A rotating body with an angular velocity that changes from ω_0 to ω_f in the time interval t has an angular acceleration of $\alpha = \frac{\omega_f - \omega_0}{t}$.

Application

EXAMPLE 9.34

A steel cylinder 8 cm in diameter is to be machined in a lathe. If the desired linear velocity of the cylinder's surface is to be 80.5 cm/s, at how many rpm should it rotate?

Solution   We have $v = \omega r$. We want to find ω and $\omega = \dfrac{v}{r}$, where $v = 80.5$ cm/s and $r = 4$ cm.

$$\omega = \frac{80.5 \text{ cm/s}}{4 \text{ cm}}$$

$$= 20.125 \text{ rad/s}$$

Now, 1 rpm is 2π rad/60 s, so using the proportion

$$\frac{2\pi \text{ rad}}{60 \text{ s}} = \frac{x \text{ rad}}{1 \text{ s}}$$

we get 1 rpm = 0.1047 rad/s. Thus

$$\omega = \frac{20.125 \text{ rad/s}}{(0.1047 \text{ rad/s})/\text{rpm}}$$

$$= 192.22 \text{ rpm}$$

Application

EXAMPLE 9.35

An engine requires 6 s to go from its idling speed of 600 rpm to 1 500 rpm. What is its angular acceleration?

Solution   The initial velocity of the engine is

$$\omega_0 = 600 \text{ rpm}$$

$$= 600(0.1047)$$

$$= 62.82 \text{ rad/s}$$

and the final velocity is

$$\omega_f = 1\,500 \text{ rpm}$$

$$= 1\,500(0.1047)$$

$$= 157.05 \text{ rad/s}$$

The angular acceleration is

$$\alpha = \frac{\omega_f - \omega_0}{t}$$

$$= \frac{157.05 - 62.82}{6}$$

$$= \frac{94.23 \text{ rad/s}}{6 \text{ s}}$$

$$= 15.705 \text{ rad/s}^2$$

Exercise Set 9.6

Solve Exercises 1–21.

1. *Transportation engineering*   A circular highway curve has a radius of 1 050.250 m and a central angle of 47° measured to the center line of the road. What is the length of the curve?

2. *Physics*   An object is moving around a circle of radius 8 in. with an angular velocity of 5 rad/s. What is the linear velocity?

3. *Physics*   A flywheel rotates with an angular velocity of 30 rpm. If its radius is 12 in., what is the linear velocity of the rim?

4. *Industrial technology*   A pulley belt 4 m long takes 2 s to complete one revolution. If the radius of the pulley is 180 mm, what is the angular velocity of a point on the rim of the pulley?

5. In a circle of radius 15 cm, what is the length of the arc intercepted by a central angle of $\frac{5\pi}{8}$.

6. In a circle of radius 235 mm, what is the length of the arc intercepted by a central angle of 85°?

7. What is the area of the sector in Exercise 5?

8. What is the area of the sector in Exercise 6?

9. *Physics*   A clock pendulum 1.2 m long oscillates through an angle of 0.07 rad on each side of the vertical. What is the distance the end of the pendulum travels from one end to the other?

10. *Space technology*   A communication satellite is in orbit at 22,300 mi above the surface of the earth. The satellite is in permanent orbit above a certain point on the equator. Thus, the satellite makes one revolution every 24 h. What is the angular velocity of the satellite? What is its linear velocity? (The radius of the earth at the equator is 3,963 mi.)

11. *Acoustical engineering*   A phonograph record is 175.26 mm in diameter and rotates at 45 rpm. **(a)** What is the linear velocity of a point on the rim? **(b)** How far does this point travel in 1 min?

12. *Acoustical engineering*   A phonograph turntable rotating at 4.2 rad/s makes 4 complete turns before it stops after 11.92 s. What is its angular acceleration?

13. *Navigation*   City A is at 35°30′ N, 78°40′ W and City B is at 40°40′ N, 88°50′ W. What is the angular distance between the two cities? How far is it between the cities?

14. *Mechanical engineering*   A 1,200-rpm motor is directly connected to a 12-in.-diameter circular saw blade. What is the linear velocity of the saw's teeth in ft/min?

15. *Computer science*   The outer track on a $5\frac{1}{4}''$ diameter diskette for a microcomputer is $2\frac{1}{2}''$ from the center of the diskette. If 8 bytes of a data can be stored on $\frac{1}{4}''$ of track, how many bytes can be stored in the length of the track subtended by $\frac{\pi}{3}$ rad?

16. *Computer science*   Some diskettes are divided into 16 sectors per track. What is the length of one sector of the track in Exercise 15?

17. *Mechanical engineering*   A flywheel makes 850 rev in 1 min. How many degrees does it rotate in 1 s?

18. *Mechanical engineering*   If the flywheel in Exercise 17 has a 15 in. diameter, what is the linear velocity of a point on its outer edge?

19. *Energy technology*   A wind generator has blades 4.2 m long. What is the speed of a blade tip when the blades are rotating in 20 rpm?

20. *Physics*   A 2 500 newton weight is resting on an inclined plane that makes an angle of 23° with the horizontal. Find the component F_x and F_y of the weight parallel to and perpendicular to the surface of the plane, as shown in Figure 9.36.

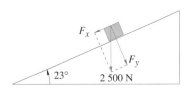

FIGURE 9.36

21. The arctic circle is approximately at latitude 66.5° N. If the radius of the earth is 6 370 km, how far is the arctic circle from the equator?

☰ CHAPTER 9 REVIEW

Important Terms and Concepts

Angle(s)
  Complementary
  Coterminal
  Negative
  of Depression
  of Elevation
  Positive
  Quadrantal
  Reference
  Standard position
Degrees
Inverse trigonometric functions
  Arccos or $\cos^{-1}$
  Arccot or $\cot^{-1}$

Arccsc or $\csc^{-1}$
Arcsec or $\sec^{-1}$
Arcsin or $\sin^{-1}$
Arctan or $\tan^{-1}$
Quotient identities
Radians
Reciprocal identities
Trigonometric functions
  Cosecant (csc)
  Cosine (cos)
  Cotangent (cot)
  Secant (sec)
  Sine (sin)
  Tangent (tan)

Review Exercises

Convert each of the angle measures in Exercises 1–6 from degrees to radians.

1. $60°$

2. $198°$

3. $325°$

4. $180°$

5. $-115°$

6. $435°$

Convert each of the angle measures to Exercises 7–12 from radians to degrees.

7. $\dfrac{3\pi}{4}$ rad

8. 1.10 rad

9. 2.15 rad

10. $\dfrac{7\pi}{3}$ rad

11. -4.31 rad

12. 5.92 rad

For Exercises 13–24, (a) tell which quadrant the angle is in, (b) give the reference angle for the given angle, and (c) give two coterminal angles, one positive and one negative.

13. $60°$

14. $198°$

15. $325°$

16. $180°$

17. $-115°$

18. $435°$

19. $\dfrac{3\pi}{4}$ rad

20. 1.10 rad

21. 2.15 rad

22. $\dfrac{7\pi}{3}$ rad

23. -4.31 rad

24. 5.92 rad

Each point in Exercises 25–30 is on the terminal side of an angle θ in standard position. Find the six trigonometric functions of the angle θ associated with each of these points.

25. $(3, -4)$

26. $(5, 12)$

27. $(-20, 21)$

28. $(-4, -7)$

29. $(7, 1)$

30. $(-12, 8)$

In Exercises 31–36, find the trigonometric functions of angle θ at vertex A of triangle ABC for the indicated sides.

31. $a = 8, c = 17$

32. $a = 5, b = 12$

33. $b = 8, c = \dfrac{40}{3}$

34. $a = 6, c = 7.5$

35. $b = 7, c = 18.2$

36. $a = 42, b = 44.1$

In Exercises 37–39, let θ be an angle of a triangle with the given trigonometric function. Find the values of the other trigonometric functions.

37. $\sin \theta = \dfrac{12.8}{27.2}$

38. $\tan \theta = \dfrac{16}{16.8}$

39. $\sec \theta = \dfrac{4}{2.5}$

In Exercises 40–42, the values of two trigonometric functions of an angle θ are given. Use this information to determine the values of the other four trigonometric functions for θ.

40. $\sin \theta = 0.532, \tan \theta = 0.628$

41. $\sin \theta = 0.5, \cos \theta = 0.866$

42. $\cos \theta = 0.680, \csc \theta = 1.364$

Use a calculator to determine the values of each of the indicated trigonometric functions in Exercises 43–50.

43. $\sin 45°$

44. $\cos 82.5°$

45. $\tan 213.5°$

46. $\sec(-81°)$

47. $\cos 2.3$ rad

48. $\sin 4.75$ rad

49. $\tan(-3.2)$ rad

50. $\csc 0.21$ rad

In Exercises 51–54, find to the nearest tenth of a degree, all angles θ, where $0° \leq \theta < 360°$, with the given trigonometric function.

51. $\sin \theta = 0.5$

52. $\tan \theta = 2.5$

53. $\cos \theta = -0.75$

54. $\csc \theta = 3.0$

In Exercises 55–58, find to the nearest hundredth of a radian, all angles θ, where $0 \leq \theta < 2\pi$, with the given trigonometric function.

55. $\cos \theta = -0.5$

56. $\sin \theta = 0.717$

57. $\tan \theta = -0.95$

58. $\sec \theta = 2.25$

Evaluate each of the functions in Exercises 59–64. Write each answer to the nearest tenth of a degree.

59. $\arcsin 0.866$

60. $\arccos 0.5$

61. $\arctan(-1)$

62. $\cos^{-1}(-0.707)$

63. $\sin^{-1} 0.385$

64. $\cot^{-1}(3.5)$

Solve Exercises 65–80.

65. *Optics*   The index of refraction n of a medium is the ratio of the speed of light c in air to the speed of light in the medium c_m. According to Snell's law, the angles of incidence i and refraction r (see Figure 9.37) of a light ray are related by the formula

$$n = \frac{\sin i}{\sin r}$$

The index of refraction for water is 1.33. If a light beam enters a lake at an angle of incidence of 30°, what is the angle of refraction?

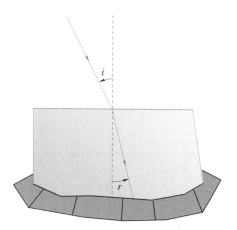

FIGURE 9.37

66. *Optics*   A beam of light enters a pane of glass at an angle of incidence of 0.90 rad. If the angle of refraction is 0.55 rad, what is the index of refraction of the glass?

67. *Electricity*   The current in an ac circuit varies with time t and the maximum value of the current I_m according to the formula

$$I = I_m \sin \omega t$$

where ω is the angular frequency of the alternating current. Find I, if $I_m = 12.6$ A and ωt is $\frac{\pi}{5}$ rad.

68. *Machine technology*   An engine valve is shown in Figure 9.38. What is the angle θ of the valve face, in degrees?

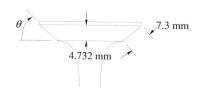

FIGURE 9.38

69. *Electricity*   If resistor and capacitor are connected in series to an ac power source, the phase angle θ is given by the formula

$$\tan \theta = \frac{V_L}{V_R}$$

where θ is the phase angle between the total voltage V and the resistive voltage V_R. If V_L, the voltage across the inductor, is 44 V and $V_R = 50$ V, what is the phase angle θ? What is the total voltage V? (See Figure 9.39.)

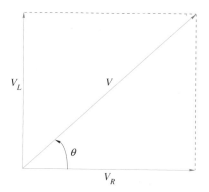

FIGURE 9.39

70. *Machine technology*   A tapered shaft in the shape of an equilateral trapezoid is shown in Figure 9.40. What is the height of the shaft?

FIGURE 9.40

71. *Transportation engineering*   A guard rail is going to be constructed around the outer edge of a highway curve. If the curve has a radius of 900 ft and an angle of 37°, how many feet of railing are needed?

72. *Computer technology*   The drive speed of a disk drive for a microcomputer is 300 rpm. How many radians does it rotate in 1 s? What is the linear velocity of a point on the rim of a $5\frac{1}{4}''$ diskette?

73. *Space technology*   The planet Earth is approximately 93,000,000 mi from the sun and revolves around the sun in an almost circular orbit every 365 days. What is the approximate linear speed in miles per hour of Earth in its orbit?

74. *Construction*   A guy wire for an electric pole is anchored to the ground at a point 15 ft from the base of the pole. The wire makes an angle of 68° with the level ground. How high up the pole is the wire attached?

75. Find the size of angle x in Figure 9.41 to the nearest tenth of a degree.

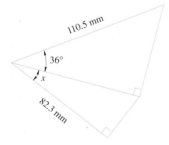

FIGURE 9.41

76. *Forestry*   A tree 56 ft tall casts a shadow 43 ft long. What is the angle of the elevation of the sun at that moment?

77. *Construction*   The light on the top of a tower is at an angle of elevation of 53.7°, when a person is 150 ft from the base of the tower. How high is the light?

78. Determine the size of angle θ in Figure 9.42 to the nearest tenth of a degree?

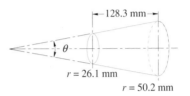

FIGURE 9.42

79. *Automotive technology*   A 3,500 lb automobile rests on a ramp that makes an angle of 34.3° with the horizontal. Determine the components F_x and F_y, as shown in Figure 9.43.

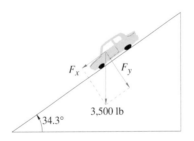

FIGURE 9.43

80. *Construction*   Two guy wires from the top of a tower are anchored in the ground 35 m apart and in direct line with the tower. If the wires make angles of 36.4° and 23.7° with the top of the tower as shown in Figure 9.44, what is the height of the tower?

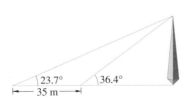

FIGURE 9.44

☰ CHAPTER 9 TEST

1. Convert $50°$ to radians.

2. Convert $\dfrac{4\pi}{3}$ to degrees.

3. (a) In what quadrant is a $237°$ angle?
   (b) Give the reference angle for a $237°$ angle.

4. The point $(-5,\ 12)$ is on the terminal side of an angle θ in standard position. Give the six trigonometric functions of this angle.

5. If angle θ is at vertex A of right $\triangle ABC$ with sides $a = 9$ and $c = 23.4$, then determine
   (a) $\sin\theta$
   (b) $\tan\theta$

6. If $\cot\theta = \dfrac{6.75}{30}$, then determine
   (a) $\tan\theta$
   (b) $\sin\theta$

7. Determine each of the indicated values:
   (a) $\sin 53°$
   (b) $\tan(-112°)$
   (c) $\sec 127°$
   (d) $\cos 4.2$

8. Find, to the nearest tenth of a degree, all angles θ, where $0° \le \theta < 360°$, with the given function.
   (a) $\tan\theta = 1.2$
   (b) $\sin\theta = -0.72$

9. A guy wire for a radio tower is anchored to the ground 135 ft from the base of the tower. The wire makes an angle of $53°$ with the level ground. How high up the tower is the wire attached?

10. The drive speed of a computer hard disk drive is 3,600 rpm. How many radians does it rotate in 1 s?

10

Vectors and Trigonometric Functions

A police officer uses a wheel tape to measure the length of skid marks at an accident. In Section 10.1, we will learn how to determine the horizontal and vertical forces the officer exerts.

Courtesy of Michael A. Gallitelli, Metroland Photo Inc.

Many applications of mathematics involve quantities that have both a magnitude (or size) and a direction. The quantities include force, displacement, velocity, acceleration, torque, and magnetic flux density. In this chapter, we will focus on ways to represent these quantities and how they can be used in conjunction with trigonometry to solve problems. These ways will require that you learn about vectors and how to use them with trigonometry.

☰ 10.1

INTRODUCTION TO VECTORS

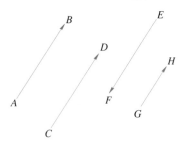

FIGURE 10.1

FIGURE 10.2a

FIGURE 10.2b

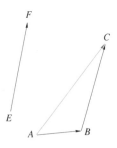

FIGURE 10.2c

Our study will begin with the introduction of two quantities: scalars and vectors. After we finish this introductory material, you will be ready to learn some of the basic operations with scalars and vectors. We will use these operations to help solve some applied problems.

Scalars

Quantities that have size, or magnitude, but no direction are called **scalars**. Time, volume, mass, speed, distance, and temperature are some examples of scalars.

Vectors

A **vector** is a quantity that has both magnitude and direction. The magnitude of the vector is indicated by its length. The direction is often given by an angle. A vector is usually pictured as an arrow with the arrowhead pointing in the direction of the vector. Force, velocity, acceleration, torque, and electric and magnetic fields are all examples of vectors.

Vectors are usually represented with boldface letters such as **a** or **A** or by letters with arrows over them, $\vec{a}$ or $\vec{A}$. If a vector extends from a point A, called the **initial point**, to a point B, called the **terminal point**, then the vector can be represented by $\overrightarrow{AB}$ with the arrowhead over the terminal point B. In written work, it is hard to write in boldface, so you should use $\vec{a}$ or $\vec{b}$ or $\overrightarrow{AB}$.

The magnitude of a vector **A** is usually denoted by $|\mathbf{A}|$ or A. The magnitude of a vector is never negative. Two vectors, $\overrightarrow{AB}$ and $\overrightarrow{CD}$, are **equal**, or **equivalent**, if they have the same magnitude and direction, and we write $\overrightarrow{AB} = \overrightarrow{CD}$. In Figure 10.1, $\overrightarrow{AB} = \overrightarrow{CD}$ but $\overrightarrow{AB} \neq \overrightarrow{EF}$ because, while they have the same magnitude, they are not in the same direction. Also, $\overrightarrow{AB} \neq \overrightarrow{GH}$, because they do not have the same magnitude, even though they are in the same direction.

Resultant Vectors

If $\overrightarrow{AB}$ is the vector from A to B and $\overrightarrow{BC}$ is the vector from B to C, then the vector $\overrightarrow{AC}$ is the **resultant vector** and represents the sum of $\overrightarrow{AB}$ and $\overrightarrow{BC}$.

$$\overrightarrow{AC} = \overrightarrow{AB} + \overrightarrow{BC}$$

The sum of two vectors is shown geometrically in Figure 10.2a. It is also possible to add vectors such as $\overrightarrow{AB} + \overrightarrow{AD}$, as shown in Figure 10.2b. Here $\overrightarrow{AC} = \overrightarrow{AB} + \overrightarrow{AD}$. As you can probably guess from the shape of Figure 10.2b, this method of adding vectors is called the **parallelogram method**.

The vectors that are being added do not need an endpoint in common. For example, in Figure 10.2c to find $\overrightarrow{AB} + \overrightarrow{EF}$ we would find the vector $\overrightarrow{BC}$, which is equivalent to $\overrightarrow{EF}$. That is, $\overrightarrow{BC} = \overrightarrow{EF}$. We are then back to the first method and $\overrightarrow{AB} + \overrightarrow{EF} = \overrightarrow{AB} + \overrightarrow{BC} = \overrightarrow{AC}$.

Vectors can also be subtracted. Here you need to realize that the vector $-\mathbf{V}$ has the opposite direction of the vector $\mathbf{V}$. An example of vector subtraction, $\overrightarrow{AB} - \overrightarrow{CD} = \vec{R}$ is shown in Figure 10.3.

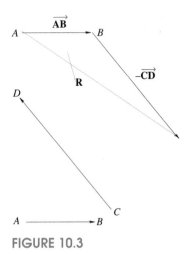

FIGURE 10.3

Another way to represent subtraction of vectors is shown in Figure 10.4. If we want to subtract $\overrightarrow{AB} - \overrightarrow{CD}$ as in Figure 10.3, we could draw $\overrightarrow{AE}$, where $\overrightarrow{AE} = \overrightarrow{CD}$. Then $\overrightarrow{EB}$ would be the vector that represents $\overrightarrow{AB} - \overrightarrow{CD} = \overrightarrow{AB} - \overrightarrow{AE}$. Notice that in Figure 10.4, $\overrightarrow{EB}$ is drawn from the terminal point of the second vector in the difference $(\overrightarrow{AE})$ to the terminal point of the first vector $(\overrightarrow{AB})$.

A vector can be multiplied by a scalar. Thus, 2**a** has twice the magnitude but the same direction as **a** and 5**a** has five times the magnitude and the same direction as **a**. (See Figure 10.5.)

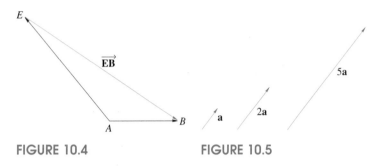

FIGURE 10.4   FIGURE 10.5

Component Vectors

In addition to adding two vectors together to find a resultant vector, we often need to reverse the process and think of a vector as the sum of two other vectors. Any two vectors that can be added together to give the original vector are called **component vectors**. To resolve a vector means to replace it by its component vectors. Usually a vector is resolved into component vectors that are perpendicular to each other.

Position Vector

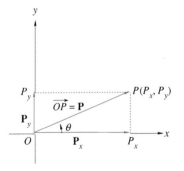

FIGURE 10.6

If P is any point in a coordinate plane and O is the origin, then $\overrightarrow{OP}$ is called the **position vector** of P. It is relatively easy to resolve a position vector into two component vectors by using the x- and y-axes. These are called the **horizontal** (or **x-**)**component** and the **vertical** (or **y-**)**component**. In Figure 10.6, $\mathbf{P}_x$ is the horizontal component of $\overrightarrow{OP}$ and $\mathbf{P}_y$ is the vertical component.

Finding Component Vectors

If you study Figure 10.6, you can see that the coordinates of P are (P_x, P_y). (Remember that P_x is the magnitude of vector $\mathbf{P}_x$.) Since every position vector has O as its initial point, it is easier to refer to a position vector by its terminal point. From now on we will refer to $\overrightarrow{OP}$ as **P**.

If the angle that a position vector **P** makes with the positive x-axis is θ, then the components of **P** are found as follows.

Components of a Vector

A position vector **P** that makes an angle θ with the positive x-axis can be resolved into component vectors $\mathbf{P}_x$ and $\mathbf{P}_y$ along the x- and y-axis respectively, with magnitudes P_x and P_y, where

$$\mathbf{P}_x = P \cos \theta$$

$$\text{and} \quad \mathbf{P}_y = P \sin \theta$$

EXAMPLE 10.1

Resolve a vector 12.0 units long and at an angle of $150°$ into its horizontal and vertical components.

Solution Consider this to be a position vector and put the initial point at the origin. The vector will look like vector **P** in Figure 10.7. We are told that $P = 12$ and $\theta = 150°$, so

$$\mathbf{P}_x = 12 \cos 150°$$

$$\approx 12(-0.866)$$

$$= -10.392$$

$$\text{and} \quad \mathbf{P}_y = 12 \sin 150°$$

$$= 12(0.5)$$

$$= 6$$

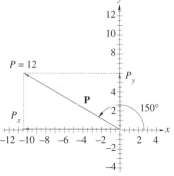

FIGURE 10.7

We have resolved **P** into two component vectors. One component is along the negative x-axis and has approximate magnitude 10.4 units. The other component is along the positive y-axis and has magnitude 6.00 units.

EXAMPLE 10.2

Vector **P** is shown in Figure 10.8a. Resolve **P** into its horizontal and vertical components.

Solution This vector is in Quadrant III, so both components will be negative. The reference angle is $67°$. We want to know the angle that this vector makes with the positive x-axis. Since the vector is in Quadrant III, $\theta = 180° + 67° = 247°$. We see

EXAMPLE 10.2 (Cont.)

that $P = 130$, so we determine

$$\mathbf{P}_x = 130\cos 247°$$

$$\approx 130(-0.3907)$$

$$\approx -50.7950$$

$$\approx -50.8$$

$$\mathbf{P}_y = 130\sin 247°$$

$$\approx 130(-0.9205)$$

$$\approx -119.6656$$

$$\approx -120$$

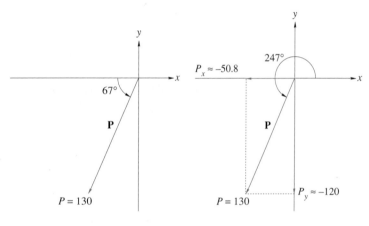

FIGURE 10.8a   FIGURE 10.8b

Thus, $\mathbf{P}$ has been resolved into component vectors $\mathbf{P}_x$ of magnitude 50.8 units along the negative x-axis and $\mathbf{P}_y$ of magnitude 120 units along the negative y-axis, as shown in Figure 10.8b.

Application

EXAMPLE 10.3

The police officer in Figure 10.9a is measuring skid marks at an accident scene by pushing a wheel tape with a force of 10 lb and holding the handle at an angle of 46° with the ground. Resolve this into its horizontal and vertical component vectors.

Solution   A vector diagram has been drawn over the photograph (Figure 10.9a) of the police officer operating the wheel tape. The vector diagram is shown alone in Figure 10.9b. The initial point of the vector is at the officer's hand and the terminal point is at the hub, or axle, of the wheel tape. We have placed the origin of our

EXAMPLE 10.3 (Cont.)

FIGURE 10.9

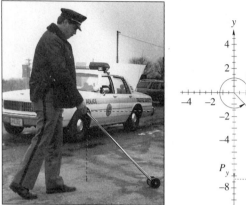

Courtesy of Michael A. Gallitelli, Metroland Photo Inc.
(a)   (b)

coordinate system at the initial point of the vector, and the horizontal, or x-axis, is parallel to the ground.

The vector makes an angle of $360° - 46° = 314°$ with the positive x-axis. Thus, we have $P = 10$ and $\theta = 314°$, so

$$\mathbf{P}_x = 10 \cos 314°$$

$$\approx 6.9466$$

and $\mathbf{P}_y = 10 \sin 314°$

$$\approx -7.1934$$

The police officer exerts a horizontal force of about 6.9 lb and a vertical force of approximately 7.2 lb.

Application

EXAMPLE 10.4

A cable supporting a television tower exerts a force of 723 N at an angle of 52.7° with the horizontal, as shown in Figure 10.10. Resolve this force into its vertical and horizontal components.

Solution   A vector diagram has been drawn in Figure 10.10. If the cable is represented by vector $\mathbf{V}$, then the horizontal component is $\mathbf{V}_x$ and the vertical component is $\mathbf{V}_y$. In this example, $\theta = 52.7°$, and so,

$$V_x = 723 \cos 52.7°$$

$$\approx 438.1296$$

and $V_y = 723 \sin 52.7°$

$$\approx 575.1273$$

EXAMPLE 10.4 (Cont.)

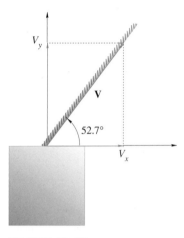

FIGURE 10.10

We see that this cable exerts a horizontal force of approximately 438 N and a vertical force of about 575 N.

Finding the Magnitude and Direction of Vectors

If we have the horizontal and vertical components of a vector **P**, then we can use the components to determine the magnitude and direction of the resultant vector.

Magnitude and Direction of a Vector

If $\mathbf{P}_x$ is the horizontal component of vector **P** and $\mathbf{P}_y$ is its vertical component, then

$$|\mathbf{P}| = P = \sqrt{P_x^2 + P_y^2}$$

and

$$\theta_{\text{Ref}} = \tan^{-1}\left|\frac{P_y}{P_x}\right|$$

EXAMPLE 10.5

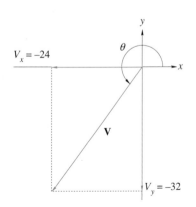

FIGURE 10.11

A position vector **V** has its horizontal component $\mathbf{V}_x = -24$ and its vertical component $\mathbf{V}_y = -32$. What are the direction and magnitude of **V**?

Solution A sketch of this problem in Figure 10.11 shows that **V** is in the third quadrant. Since $\theta_{\text{Ref}} = \tan^{-1}\left|\frac{-32}{-24}\right| \approx 53.13°$, we see that $\theta \approx 180° + 53.13° = 233.13°$.

We know that

$$V = \sqrt{V_x^2 + V_y^2}$$
$$= \sqrt{(-24)^2 + (-32)^2} = \sqrt{1,600}$$
$$= 40$$

So, the magnitude of **V** is 40 and **V** is at an angle of 233.13°.

Application

EXAMPLE 10.6

A pilot heads a jet plane due east at a ground speed of 425.0 mph. If the wind is blowing due north at 47 mph, find the true speed and direction of the jet.

Solution A sketch of this situation is shown in Figure 10.12. If **V** represents the vector with components $\mathbf{V}_x$ and $\mathbf{V}_y$, then we are given $V_x = 425.0$ and $V_y =$

EXAMPLE 10.6 (Cont.)

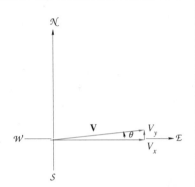

FIGURE 10.12

47. Thus,

$$V = \sqrt{V_x{}^2 + V_y{}^2}$$

$$= \sqrt{425.0^2 + 47^2}$$

$$\approx 427.59$$

and $\theta = \tan^{-1}\left(\dfrac{47}{425}\right)$

$$= \tan^{-1} 0.110588$$

and so, $\theta = 6.31°$

The jet is flying at a speed of approximately 427.6 mph in a direction 6.31° north of due east.

Exercise Set 10.1

In Exercises 1–4, add the given vectors by drawing the resultant vector.

1.

2.

3.

4.

For Exercises 5–20, trace each of the vectors A–D in Figure 10.13. Use these vectors to find each of the indicated sums or differences.

FIGURE 10.13

5. A + B

6. B + D

7. A + B + C

8. B + C + D

9. A − B

10. C − D

11. B − A

12. D − C

13. A + 3B

14. C + 2D

15. 2C − D

16. 3B − A

17. A + 2B − C

18. A + 3C − D

19. 2B − C + 2.5D

20. 4C − 3B − 2D

In Exercises 21–26, use trigonometric functions to find the horizontal and vertical components of the given vectors.

21. Magnitude 20, $\theta = 75°$

22. Magnitude 16, $\theta = 212°$

23. Magnitude 18.4, $\theta = 4.97$ rad

24. Magnitude 23.7, $\theta = 2.22$ rad

25. $V = 9.75, \theta = 16°$

26. $P = 24.6, \theta = 317°$

In Exercises 27–30, the horizontal and vertical components are given for a vector. Find the magnitude and direction of each resultant vector.

27. $A_x = -9;\ A_y = 12$

28. $B_x = 10;\ B_y = -24$

29. $C_x = 8;\ C_y = 15$

30. $D_x = -14;\ D_y = -20$

Solve Exercises 31–37.

31. *Navigation* A ship heads into port at 12.0 km/h. The current is perpendicular to the ship at 5 km/h. What is the resultant velocity of the ship?

32. *Navigation* A pilot heads a jet plane due east at a ground speed of 756.0 km/h. If the wind is blowing due north at 73 km/h, find the true speed and direction of the jet.

33. *Construction* A cable supporting a tower exerts a force of 976 N at an angle of 72.4° with the horizontal. Resolve this force into its vertical and horizontal components.

34. *Construction* A sign of mass 125.0 lb hangs from a cable. A worker is pulling the sign horizontally by a force of 26.50 lb. Find the force and the angle of the resultant force on the sign.

35. *Medical technology* A ramp for the physically challenged makes an angle of 12° with the horizontal. A woman and her wheelchair weigh 153 lb. What are the components of this weight parallel and perpendicular to the ramp?

36. *Electricity* If a resistor, a capacitor, and an inductor are connected in series to an ac power source, then the effective voltage of the source is given by the vector **V**, where **V** is the sum of the vector quantities **V**$_R$ and **V**$_L$ − **V**$_C$, as shown in Figure 10.14. The phase angle, ϕ, is the angle between **V** and **V**$_R$, where $\tan \phi = \dfrac{\mathbf{V}_L - \mathbf{V}_C}{\mathbf{V}_R}$. If the effective voltages across the circuit components are $V_R = 12$ V, $V_C = 10$ V, and $V_L = 5$ V, determine the effective voltage and the phase angle.

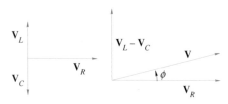

FIGURE 10.14a   FIGURE 10.14b

37. *Electricity* A resistor, capacitor, and inductor are connected in series to an ac power source. If the effective voltages across the circuit components are $V_R = 15.0$ V, $V_C = 17.0$ V, and $V_L = 8.0$ V, determine the effective voltage and the phase angle.

≡ 10.2
ADDING AND SUBTRACTING VECTORS

In Section 10.1, we learned how to use diagrams to add and subtract vectors. We also learned how to use trigonometry to determine the horizontal and vertical components of a vector. In this section, we will learn how to use trigonometry and the Pythagorean theorem to add and subtract vectors.

We will begin by looking at two special cases. In the first case, both vectors are on the same axis. In the second case, the vectors are on different axes.

EXAMPLE 10.7

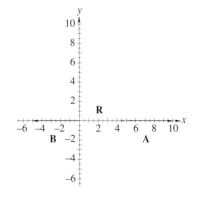

FIGURE 10.15

Add vectors **A** and **B**, where $A = 9.6$, $\theta_A = 0°$ and $B = 4.3$, $\theta_B = 180°$.

Solution If we find the horizontal and vertical components of these two vectors we see that

$$\mathbf{A}_x = 9.6 \cos 0° = 9.6$$

$$\mathbf{A}_y = 9.6 \sin 0° = 0$$

$$\mathbf{B}_x = 4.3 \cos 180° = -4.3$$

$$\mathbf{B}_y = 4.3 \sin 180° = 0$$

The resultant vector **R** has horizontal component $\mathbf{R}_x = \mathbf{A}_x + \mathbf{B}_x = 9.6 + -4.3 = 5.3$ and a vertical component $\mathbf{R}_y = 0 + 0 = 0$. The angle of the resultant vector θ_R is found using $\tan \theta_R = \dfrac{R_y}{R_x} = \dfrac{0}{5.3} = 0$. Since $R_x > 0$, we know that θ_R is 0° rather than 180°. (See Figure 10.15.)

In this first case, the component method required a little extra work. But, the example gave us a foundation for the next example.

EXAMPLE 10.8

Find the resultant of **C** and **D**, when $C = 6.2, \theta_C = 270°, D = 12.4$, and $\theta_D = 180°$, as shown in Figure 10.16.

Solution Since **C** and **D** are perpendicular, the length of the resultant vector can be found by using the Pythagorean theorem.

$$R = \sqrt{C^2 + D^2} = \sqrt{(6.2)^2 + (12.4)^2} \approx 13.9$$

We also know that $\tan \theta = \dfrac{C}{D} = \dfrac{6.2}{12.4} = \dfrac{1}{2}$ and that $\tan^{-1} \dfrac{1}{2} = 26.57°$. From Figure 10.16, we can see that **R** is in the third quadrant. So, $\theta_R = 206.57° \approx 207°$.

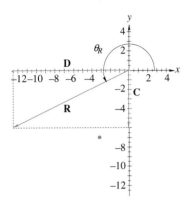

FIGURE 10.16

If we want to find the resultant of two vectors that are not at right angles, the method takes a bit longer. We first resolve each vector into its horizontal and vertical components and then add the horizontal and vertical components, as outlined in the following box.

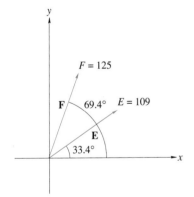

FIGURE 10.17a

Adding Vectors Using Components

1. Resolve each vector into its horizontal and vertical components.
2. Add the horizontal components. This sum is $\mathbf{R}_x$, the horizontal component of the resultant vector.
3. Add the vertical components. This sum is $\mathbf{R}_y$, the vertical component of the resultant vector.
4. Use R_x and R_y to determine the magnitude R and direction θ of the resultant vector, where

$$R = \sqrt{R_x^2 + R_y^2} \qquad \text{and} \qquad \tan \theta = \frac{R_y}{R_x}$$

EXAMPLE 10.9

Find the resultant of two vectors **E** and **F**, where $E = 109, \theta_E = 33.4°, F = 125$ and $\theta_F = 69.4°$. (See Figure 10.17a.)

Solution Resolving **E** into its horizontal and vertical components we get

$$\mathbf{E}_x = 109 \cos 33.4° = 91$$

and $$\mathbf{E}_y = 109 \sin 33.4° = 60$$

The components for vector **F** are

$$\mathbf{F}_x = 125 \cos 69.4° = 44$$

$$\mathbf{F}_y = 125 \sin 69.4° = 117$$

EXAMPLE 10.9 (Cont.)

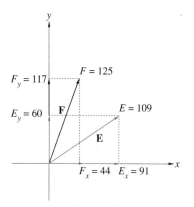

FIGURE 10.17b

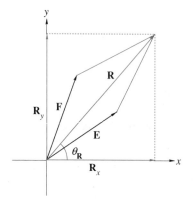

FIGURE 10.17c

These components are shown in Figure 10.17b. The components of the resultants vector are found by adding the horizontal and vertical components of **E** and **F**.

$$\mathbf{R}_x = \mathbf{E}_x + \mathbf{F}_x$$
$$= 91 + 44$$
$$= 135$$
$$\mathbf{R}_y = \mathbf{E}_y + \mathbf{F}_y$$
$$= 60 + 117$$
$$= 177$$

These two component vectors and their resultant vector are shown in Figure 10.17c.

$$R = \sqrt{R_x^2 + R_y^2}$$
$$= \sqrt{135^2 + 177^2}$$
$$\approx 222.61$$

and $\quad \tan\theta_R = \dfrac{R_y}{R_x}$

$$\tan\theta_R = \frac{177}{135}$$

$$\theta_R = \tan^{-1}\frac{177}{135}$$

so $\quad \theta_R = 52.7°$

It is often helpful to use a table to help keep track of vectors and their components. It is an effective way to organize this information. A table for this example follows. It lists the horizontal and vertical components of each vector and the resultant vector.

Vector	Horizontal component		Vertical component	
E	$\mathbf{E}_x = 109\cos 33.4° =$	91	$\mathbf{E}_y = 109\sin 33.4° =$	60
F	$\mathbf{F}_x = 125\cos 69.4° =$	44	$\mathbf{F}_y = 125\sin 69.4° =$	117
R	$\mathbf{R}_x$	135	$\mathbf{R}_y$	177

From these values for R_x and R_y, we can determine $R \approx 223$ and $\theta_R = 52.7°$.

EXAMPLE 10.10

Find the resultant of two vectors **G** and **H**, where $G = 449$, $\theta_G = 128.6°$, $H = 521$, and $\theta_H = 327.6°$. (See Figure 10.18a.)

Solution   Again, we will resolve each vector into its horizontal and vertical components.

$$\mathbf{G}_x = 449\cos 128.6°$$
$$= -280.1$$
$$\mathbf{G}_y = 449\sin 128.6°$$
$$= 350.9$$
$$\mathbf{H}_x = 521\cos 327.6°$$
$$= 439.9$$
$$\mathbf{H}_y = 521\sin 327.6°$$
$$= -279.2$$

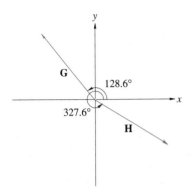

FIGURE 10.18a

Figure 10.18b shows each vector and its horizontal and vertical components.

Adding the horizontal components gives the horizontal component of the resultant vector.

$$\mathbf{R}_x = \mathbf{G}_x + \mathbf{H}_x$$
$$= -280.1 + 439.9$$
$$= 159.8$$

Similarly, we can determine the vertical component of the resultant vector.

$$\mathbf{R}_y = \mathbf{G}_y + \mathbf{H}_y$$
$$= 350.9 - 279.2$$
$$= 71.7$$

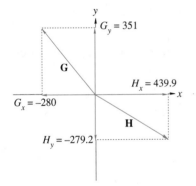

FIGURE 10.18b

The resultant vector is shown in Figure 10.18c.

The magnitude of the resultant vector is

$$R = \sqrt{R_x{}^2 + R_y{}^2}$$
$$= \sqrt{159.8^2 + 71.7^2}$$
$$\approx 175.1$$

and the direction of the resultant vector is found from

$$\tan \theta_R = \frac{R_y}{R_x}$$
$$= \frac{71.7}{159.8}$$
$$= 0.4487$$

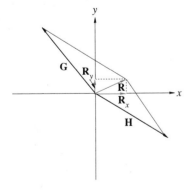

FIGURE 10.18c

EXAMPLE 10.10 (Cont.)

so

$$\theta_R = \tan^{-1} 0.4487$$

$$= 24.2°$$

The table method for this example would appear as follows.

Vector	Horizontal component		Vertical component	
G	$G_x = 449 \cos 128.6° =$	-280.1	$G_y = 449 \sin 128.6° =$	350.9
H	$H_x = 521 \cos 327.6° =$	439.9	$H_y = 521 \sin 327.6° =$	-279.2
R	R_x	159.8	R_y	71.7

These values for R_x and R_y can be used as before to find $R = 175.1$ and $\theta_R = 24.2°$.

EXAMPLE 10.11

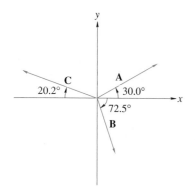

FIGURE 10.19

Find the resultant of the three vectors shown in Figure 10.19, if $A = 137, B = 89.4$, and $C = 164.6$.

Solution   The table that follows lists the horizontal and vertical components of each of the vectors and the resultant vector **R**.

Vector	Horizontal component		Vertical component	
A	$A_x = 137 \cos 30° =$	118.6	$A_y = 137 \sin 30° =$	68.5
B	$B_x = 89.4 \cos 287.5° =$	26.9	$B_y = 89.4 \sin 287.5° =$	-85.3
C	$C_x = 164.6 \cos 159.8° =$	-154.5	$C_y = 164.6 \sin 159.8° =$	56.8
R	R_x	-9.0	R_y	40.0

Once again, the values for R_x and R_y can be used to find

$$R = \sqrt{R_x{}^2 + R_y{}^2}$$

$$= \sqrt{(-9)^2 + 40^2}$$

$$= 41$$

and

$$\theta_R = \tan^{-1} \frac{40}{-9}$$

$$\theta_R = 102.7°$$

Application

EXAMPLE 10.12

A sign has been lifted into position by two cranes, as shown in Figure 10.20a. If the sign weighs 420 lb, what is the tension in each of the three cables?

EXAMPLE 10.12 (Cont.)

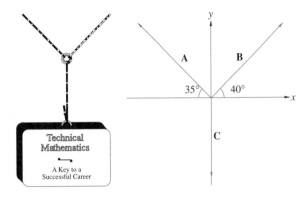

FIGURE 10.20a   FIGURE 10.20b

Solution   We will draw the coordinate axes and label the angles as shown in Figure 10.20b. The origin is placed at the ring where the three cables meet, because the tension on all three cables acts on this ring.

As usual, we make a table listing the horizontal and vertical components of each vector.

Vector	Horizontal component	Vertical component
A	$\mathbf{A}_x = A \cos 145°$	$\mathbf{A}_y = A \sin 145°$
B	$\mathbf{B}_x = B \cos 40°$	$\mathbf{B}_y = B \sin 40°$
C	$\mathbf{C}_x = C \cos 270° = \underline{0}$	$\mathbf{C}_y = C \sin 270° = -C = \underline{-420}$
R	$\mathbf{R}_x$	$\mathbf{R}_y$

The ring is at rest as a result of these three forces, so $R_x = 0$ and $R_y = 0$. Thus, we have

$$R_x = A \cos 145° + B \cos 40° = 0$$

$$R_y = A \sin 145° + B \sin 40° - 420 = 0$$

or $$-0.81915A + 0.76604B = 0$$

$$0.57358A + 0.64279B = 420$$

Solving this system of two equations in two variables (by Cramer's rule), we get

$$A \approx 333.0862 \text{ lb}$$

and $$B \approx 356.1792 \text{ lb}$$

The tension in the cable on the left is about 333 lb and the tension in the cable on the right side is about 356 lb.

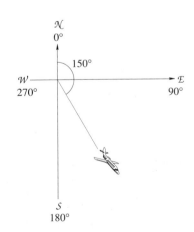

FIGURE 10.21

≡ **Note**   In flight terminology, the **heading** of an aircraft is the direction in which the aircraft is pointed. Usually the wind is pushing the aircraft so that it is actually moving in a different direction, called the **track** or **course**. The angle between the heading and the course is the **drift angle**. The **air speed** is the speed of the plane relative to the air. The **ground speed** is the speed of the aircraft relative to the ground.

≡ **Note**   In navigation, directions are usually given in terms of the size of the angle measured clockwise from true north. For example, the airplane in Figure 10.21 has a heading of 150°.

Application

EXAMPLE 10.13

An airplane is flying at 340.0 mph with a heading of 210°. If a 50 mph wind is blowing from 165°, find the ground speed, drift angle, and course of the airplane.

Solution   In Figure 10.22a, $\overrightarrow{OA}$ represents the airspeed of 340 mph with a heading of 210°, and $\overrightarrow{OW}$ represents a wind of 50 mph from 165°. In Figure 10.22b, we have completed the parallelogram to obtain the vector $\overrightarrow{OR} = \overrightarrow{OA} + \overrightarrow{OW}$. Notice that, while the wind is from 165°, it has a heading of 180° + 165° = 345°.

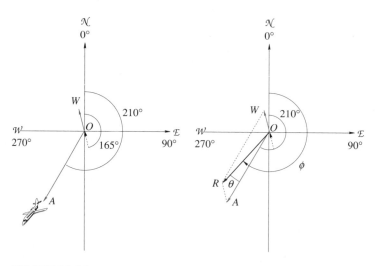

FIGURE 10.22a   FIGURE 10.22b

The length $\overrightarrow{OR}$ represents the ground speed, θ is the drift angle, and ϕ is the track or course. The following table lists the horizontal and vertical components of each vector $\overrightarrow{OA}$ and $\overrightarrow{OW}$ and the resultant vector $\overrightarrow{OR}$.

Vector	Horizontal component	Vertical component
$\overrightarrow{OA} = \mathbf{A}$	$\mathbf{A}_x = 340 \cos 210° \approx -294.45$	$\mathbf{A}_y = 340 \sin 210° = -170.00$
$\overrightarrow{OW} = \mathbf{W}$	$\mathbf{W}_x = 50 \cos 345° \approx \underline{48.30}$	$\mathbf{W}_y = 50 \sin 345° \approx \underline{-12.94}$
$\overrightarrow{OR} = \mathbf{R}$	$\mathbf{R}_x -246.15$	$\mathbf{R}_y -182.94$

EXAMPLE 10.13 (Cont.)

Using these values for R_x and R_y, we see that the length of the resultant vector is
$\overrightarrow{OR} = \sqrt{(-246.15)^2 + (-182.94)^2} \approx 306.69$ and $\phi \approx \tan^{-1}\left(\frac{-182.94}{-246.15}\right) = 36.62°$.
Since ϕ is in the third quadrant, $\phi = 180° + 36.62° = 216.62°$.

So, the plane has a ground speed of 306.69 mph, a drift angle of 6.62°, and a course of 216.62°.

Exercise Set 10.2

In Exercises 1–4, vectors A and B are both on the same axis. Find the magnitude and direction of the resultant vector.

1. $A = 20.0, \theta_A = 0°, B = 32.5, \theta_B = 180°$

2. $A = 14.3, \theta_A = 90°, B = 7.2, \theta_B = 90°$

3. $A = 121.7, \theta_A = 270°, B = 86.9, \theta_B = 90°$

4. $A = 63.1, \theta_A = 180°, B = 43.5, \theta_B = 180°$

In Exercises 5–8, vectors C and D are perpendicular. Find the magnitude and direction of the resultant vectors. It may help if you draw vectors C, D, and the resultant vector.

5. $C = 55, \theta_C = 90°, D = 48, \theta_D = 180°$

6. $C = 65, \theta_C = 270°, D = 72, \theta_C = 180°$

7. $C = 81.4, \theta_C = 0°, D = 37.6, \theta_D = 90°$

8. $C = 63.4, \theta_C = 270°, D = 9.4, \theta_D = 0°$

In Exercises 9–16, find the magnitude and direction of the vector with the given components.

9. $\mathbf{A}_x = 33, \mathbf{A}_y = 56$

10. $\mathbf{B}_x = 231, \mathbf{B}_y = 520$

11. $\mathbf{C}_x = 11.7, \mathbf{C}_y = 4.4$

12. $\mathbf{D}_x = 31.9, \mathbf{D}_y = 36.0$

13. $\mathbf{E}_x = 6.3, \mathbf{E}_y = 1.6$

14. $\mathbf{F}_x = 5.1, \mathbf{F}_y = 14.0$

15. $\mathbf{G}_x = 8.4, \mathbf{G}_y = 12.6$

16. $\mathbf{H}_x = 15.3, \mathbf{H}_y = 9.2$

In Exercises 17–30, add the given vectors by using the trigonometric functions and the Pythagorean theorem.

17. $A = 4, \theta_A = 60°, B = 9, \theta_B = 20°$

18. $C = 12, \theta_C = 75°, D = 15, \theta_D = 37°$

19. $C = 28, \theta_C = 120°, D = 45, \theta_D = 210°$

20. $E = 72, \theta_E = 287°, F = 65, \theta_F = 17°$

21. $A = 31.2, \theta_A = 197.5°, B = 62.1, \theta_B = 236.7°$

22. $C = 53.1, \theta_C = 324.3°, D = 68.9, \theta_D = 198.6°$

23. $E = 12.52, \theta_E = 46.4°, F = 18.93, \theta_F = 315°$

24. $G = 76.2, \theta_G = 15.7°, H = 89.4, \theta_H = 106.3°$

25. $A = 9.84, \theta_A = 215°30', B = 12.62, \theta_B = 105°15'$

26. $C = 79.63, \theta_C = 262°45', D = 43.72, \theta_D = 196°12'$

27. $E = 42.0, \theta_E = 3.4 \text{ rad}, F = 63.2, \theta_F = 5.3 \text{ rad}$

28. $G = 37.5, \theta_G = 0.25 \text{ rad}, H = 49.3, \theta_H = 1.92 \text{ rad}$

29.

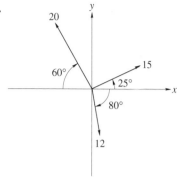

30.

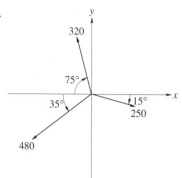

Solve Exercises 31–34.

31. *Navigation* An airplane is flying at 480 mph with a heading of 63°. A 45.0 mph wind is blowing from 325°. Find the ground speed, course, and drift angle of the airplane.

32. *Navigation* An airplane is flying at 320.0 mph with a heading of 172°. A 72.0 mph wind is blowing from 137°. Find the ground speed, course, and drift angle of the airplane.

 33. Write a computer program to determine the component vectors for a position vector. Remember, compute the sines and cosines of the given angles in radians.

 34. Write a computer program to determine the resultant vector when two or more vectors are added.

≡ 10.3
APPLICATIONS OF VECTORS

Vectors are used in science and technology, as well as in mathematics. In this section, we will look at some of those applications.

Application

EXAMPLE 10.14

Two forces, F_1 and F_2, act on an object. If F_1 is 40 lb, F_2 is 75 lb, and the angle θ between them is 50°, find the magnitude and direction of the resultant force.

Solution We sketch the two forces as vectors and place the object at the origin with $\mathbf{F_1}$ along the positive x-axis as in Figure 10.23. We will use a table similar to the one in the last example to find the components of $\mathbf{F_1}, \mathbf{F_2}$, and the resultant vector.

Vector	Horizontal component		Vertical component	
$\mathbf{F_1}$		40.0		0.0
$\mathbf{F_2}$	$75 \cos 50° =$	48.2	$75 \sin 50° =$	57.5
$\mathbf{R}$	$\mathbf{R}_x$	88.2	$\mathbf{R}_y$	57.5

So, $R_x = 88.2$ and $R_y = 57.5$. The magnitude of $\mathbf{R}$ is $R = \sqrt{88.2^2 + 57.5^2} = 105.3$ lb and $\theta_R = \tan^{-1} \frac{57.5}{88.2} = 33.1°$.

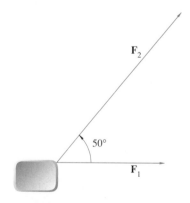

FIGURE 10.23

Application

EXAMPLE 10.15

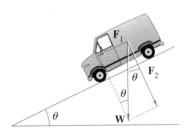

FIGURE 10.24

A truck weighing 22,500 lb is on a 25° hill. Find the components of the truck's weight parallel and perpendicular to the road.

Solution The weight of an object (truck, car, building, etc.) is the gravitational force with which earth attracts it. This force always acts vertically downward and is indicated by the vector $\mathbf{W}$ in Figure 10.24. The components of $\mathbf{W}$ have been labeled $\mathbf{F_1}$ and $\mathbf{F_2}$. Because $\mathbf{W}$ is vertical and $\mathbf{F_2}$ is perpendicular to the road, the angle θ between $\mathbf{W}$ and $\mathbf{F_2}$ is equal to the angle the road makes with the horizon. Using the trigonometric functions, we have

$$F_1 = \mathbf{W} \sin \theta = 22{,}500 \sin 25° = 9{,}509 \text{ lb}$$

$$F_2 = \mathbf{W} \cos \theta = 22{,}500 \cos 25° = 20{,}392 \text{ lb}$$

Thus, the components of the truck's weight are 9,509 lb parallel to the road and 20,392 lb perpendicular to the road.

Application

EXAMPLE 10.16

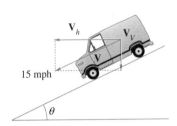

15 mph

FIGURE 10.25

If the truck in Example 10.15 rolls down the hill at 15 mph, find the magnitudes of the horizontal and vertical components of the truck's velocity.

Solution The velocity vector $\mathbf{V}$ is shown in Figure 10.25. The horizontal and vertical components of $\mathbf{V}$ are marked $\mathbf{V}_h$ and $\mathbf{V}_v$.

$$V_v = V \sin 25° = 15 \sin 25° = 6.3 \text{ mph}$$

$$V_h = V \cos 25° = 15 \cos 25° = 13.6 \text{ mph}$$

EXAMPLE 10.17

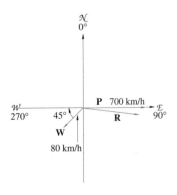

FIGURE 10.26

Application

A jet plane is traveling due east at an airspeed of 700 km/h. If a wind of 80 km/h is blowing due southwest, find the magnitude and direction of the plane's resultant velocity. (See Figure 10.26.)

Solution A table of the components for the wind's vector $\mathbf{W}$ and the plane's vector $\mathbf{P}$ allows us to quickly find the resultant vector $\mathbf{R}$.

Vector	Horizontal component		Vertical component	
$\mathbf{W}$	$80 \cos 225° =$	-56.6	$80 \sin 225° =$	-56.6
$\mathbf{P}$	$700 \cos 0° =$	700.0		0.0
$\mathbf{R}$		643.4		-56.6

The magnitude of $\mathbf{R}$ is $R = \sqrt{643.4^2 + (-56.6)^2} = 645.9$ km/h. This is the ground speed of the plane. The plane's direction is $\theta_R = \tan^{-1}\left(\frac{-56.6}{643.4}\right) = -5.03°$ or $5.03°$ south of east.

Application

EXAMPLE 10.18

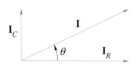

FIGURE 10.27

In a parallel RC (resistance-capacitance) circuit, the current $\mathbf{I}_C$ through the capacitance leads the current $\mathbf{I}_R$ through the resistance by $90°$, as shown in Figure 10.27. If $I_C = 0.5$ A and $I_R = 1.2$ A, find the total current $\mathbf{I}$ in the circuit and the phase angle θ of the circuit.

Solution
$$I = \sqrt{I_R^2 + I_C^2} = \sqrt{(1.2)^2 + (0.5)^2} = 1.3 \text{ A}$$

$$\theta = \tan^{-1}\left(\frac{I_C}{I_R}\right)$$

$$\theta = \tan^{-1}\left(\frac{0.5}{1.2}\right) = 22.6°$$

Application

EXAMPLE 10.19

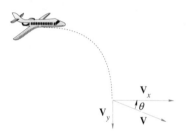

FIGURE 10.28

An airplane in level flight drops an object. The plane was traveling at 180 m/s at a height of 7 500 m. The vertical component of the dropped object is given by $\mathbf{V}_y = -9.8t$ m/s. What is the magnitude of the velocity of the object after 8 s? At what angle with the ground is the object moving at this time?

Solution   The horizontal velocity of the object will have no effect on the vertical motion. Hence, V_x will remain at 180 m/s. We are told that $\mathbf{V}_y = -9.8t$. So, when $t = 8$, $\mathbf{V}_y = -78.4$. The magnitude of the velocity of the object when $t = 8$ is

$$V = \sqrt{(-78.4)^2 + 180^2} = 196.3 \text{ m/s}$$

If θ is the angle that the object makes with the ground, then

$$\theta = \tan^{-1}\frac{V_y}{V_x} = \tan^{-1}\left(\frac{-78.4}{180}\right) \approx -23.5°$$

or $23.5°$ below the horizontal. (See Figure 10.28.)

Exercise Set 10.3

Solve Exercises 1–22.

1. *Physics*   Two forces act on an object. One force is 70 lb and the other is 50 lb. If the angle between the two forces is $35°$, find the magnitude and direction of the resultant force.

2. *Physics*   A person pulling a cart exerts a force of 35 lb on the cart at an angle of $25°$ above the horizontal. Find the horizontal and vertical components of this force.

3. *Physics*   A person pushes a lawn mower with a force of 25 lb. The handle of the lawn mower is $55°$ above the horizontal. **(a)** How much downward force is being exerted on the ground? **(b)** How much horizontal forward force is being exerted? **(c)** How do these forces change if the handle is lowered to $40°$ with the horizontal?

4. *Physics*   Two tugboats are pulling a ship. As shown in Figure 10.29, the first tug exerts a force of 1 500 N

on a cable, making an angle of 37° with the axis of the ship. The second tug pulls on a cable, making an angle of 36° with the axis of the ship. The resultant force vector is in line with the axis of the ship. What is the force being exerted by the second tug?

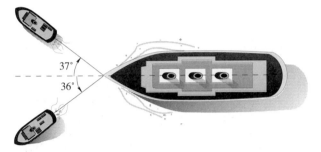

FIGURE 10.29

5. *Navigation* A tugboat is turning a barge by pushing with a force of 1 700 N at an angle of 18° with the axis of the barge. What are the forward and sideward forces in newtons exerted by the tug on the barge? (See Figure 10.30.)

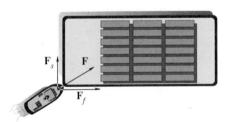

FIGURE 10.30

6. *Navigation* A ship heads due northwest at 15 km/h in a river that flows east at 7 km/h. What is the magnitude and direction of the ship's velocity relative to Earth's surface?

7. *Navigation* A plane is heading due south at 370 mph with a wind from the west at 40 mph. What are the ground speed and the true direction of the plane?

8. *Navigation* On a compass, due north is 0°, east is 90°, south 180°, and so on. A plane has a compass heading of 115° and is traveling at 420 mph. The wind is at 62 mph and blowing at 32°. What are the ground speed and true direction of the plane?

9. *Physics* A car with a mass of 1 200 kg is on a hill that makes an angle of 22° with the horizon. Which components of the car's mass are parallel and perpendicular to the road?

10. *Physics* Find the force necessary to push a 30-lb ball up a ramp that is inclined 15° with the horizon. (You want to find the component parallel to the ramp.)

11. *Construction* A guy wire runs from the top of a utility pole 35 ft high to a point on the ground 27 ft from the base of the pole. The tension in the wire is 195 lb. What are the horizontal and vertical components?

12. *Electricity* In a parallel *RC* circuit, the current I_C through the capacitance leads the current through the resistance I_R by 90°. If $I_C = 2.4$ A and $I_R = 1.6$ A, find the magnitude of the total current in the circuit and the phase angle.

13. *Electricity* If the total current I in a parallel *RC* circuit is 9.6 A and I_R is 7.5 A, what are I_C and the phase angle?

14. *Electricity* In a parallel *RL* (resistance-inductance) circuit, the current I_L through the inductance lags the current I_R through the resistance by 90°, as shown in Figure 10.31. Find the magnitude of the total current I and the negative phase angle θ, when $I_L = 6.2$ A and $I_R = 8.4$ A.

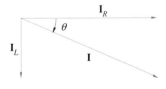

FIGURE 10.31

15. *Electricity* If the total current I in a parallel *RL* circuit is 12.4 A and I_R is 6.3 A, find I_L and the negative phase angle.

16. *Electricity* The total impedance Z of a series ac circuit is the resultant of the resistance R, the inductive reactance X_L, and the capacitive reactance X_C, as shown in Figure 10.32. Find Z and the phase angle, when $X_C = 50 \ \Omega$, $X_L = 90 \ \Omega$, and $R = 12 \ \Omega$.

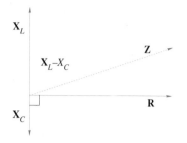

FIGURE 10.32

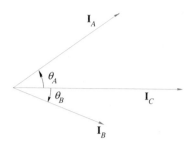

FIGURE 10.33

17. *Electricity* Find the total impedance Z and the phase angle of a series ac circuit, when $X_C = 60\ \Omega$, $X_L = 40\ \Omega$, and $R = 24\ \Omega$.

18. *Electricity* Find the total impedance and phase angle, when $X_L = 38\ \Omega$, $X_C = 265\ \Omega$, and $R = 75\ \Omega$.

19. *Electricity* In a synchronous ac motor, the current leads the applied voltage, and in an induction ac electric motor, the current lags the applied voltage. A circuit with a synchronous motor A connected in parallel with an induction motor B and a purely resistive load C has a current diagram as shown in Figure 10.33. If $I_A = 20$ A, $\theta_A = 35°$, $I_B = 15$ A, $\theta_B = -20°$, and $I_C = 25$ A, find the total current I and the phase angle between the total current and the applied voltage.

20. *Navigation* A ship is sailing at a speed of 12 km/h in the direction of 10°. A strong wind is exerting enough pressure on the ship's superstructure to move it in the direction of 270° at 2 km/h. A tidal current is flowing in the direction of 140° at the rate of 6 km/h. What is the ship's velocity and direction relative to Earth's surface?

21. *Navigation* An airplane in level flight drops an object. The plane was traveling at 120 m/s at a height of 5 000 m. The vertical component of the dropped object is $V_y = -9.8t$ m/s. What is the magnitude of the velocity of the object after 4 s and at what angle with the ground is the object moving at this time?

22. *Navigation* At any time t the object in Exercise 21 will have fallen $4.9t^2$ m. How long will it take for it to strike the ground? What is its velocity at this time? At what angle will it strike the ground?

≡ 10.4

OBLIQUE TRIANGLES: LAW OF SINES

The triangles we have examined until now have been right triangles. At first, trigonometry and the trigonometric functions dealt only with right triangles. Mathematicians quickly discovered that they needed to work with triangles that did not have a right angle. These triangles, the ones with no right angle, were named **oblique triangles**.

The trigonometric methods for solving right triangles do not work with oblique triangles. There are two methods that are usually used with oblique triangles. One of these, the Law of Sines, will be studied in this section. The other, the Law of Cosines, will be studied in Section 10.5.

The Law of Sines can be developed without too much difficulty. Consider the triangle in Figure 10.34. Select one of the vertices, in this case vertex C. Drop a perpendicular to the opposite side. From the resulting right triangles we have

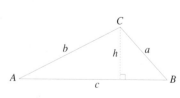

FIGURE 10.34

$$\sin A = \frac{h}{b} \quad \text{and} \quad \sin B = \frac{h}{a}$$

$$\text{or} \quad h = b \sin A \quad \text{and} \quad h = a \sin B$$

Since both of these are equal to h, then $b \sin A = a \sin B$ and $\dfrac{a}{\sin A} = \dfrac{b}{\sin B}$. If we dropped the perpendicular from vertex B we would get $\dfrac{a}{\sin A} = \dfrac{c}{\sin C}$.

Putting these together, we get the Law of Sines

The Law of Sines

The **Law of Sines** or **Sine Law** is a continued proportion and states that if $\triangle ABC$ is a triangle with sides of lengths $a, b,$ and c and opposite angles $A, B,$ and $C,$ then

$$\frac{a}{\sin A} = \frac{b}{\sin B} = \frac{c}{\sin C}$$

You may recognize that this can also be written as the continued proportion $a : b : c = \sin A : \sin B : \sin C.$

The Law of Sines can be used to solve a triangle when the parts of a triangle are known in either of the following two cases.

Case 1 (SSA)   The measure of two sides and the angle opposite one of them is known.

Case 2 (AAS)   The measure of two angles and one side is known.

As is true with any continued proportion, you work with just two of the ratios at any one time. You should also remember that the angles of a triangle add up to $180°$ or π rad.

Let's look at an example that uses the Law of Sines. This example will fit the description of a Case 1 triangle or SSA.

EXAMPLE 10.20

In a triangle, $A = 33°, a = 9.4,$ and $c = 14.3$. Solve the triangle.

Solution   We are to solve this triangle. That means that we are to find the length of the third side and the sizes of the other two angles. The following data chart shows the parts that we know and those we are to determine.

sides	angles
$a = 9.4$	$A = 33°$
$b = $ ___	$B = $ ___
$c = 14.3$	$C = $ ___

Since two of the known parts have the same letter, a, one of the ratios we should use is $\dfrac{a}{\sin A}$. The other known part has the letter c so the other ratio should be $\dfrac{c}{\sin C}$.

EXAMPLE 10.20 (Cont.)

By the Law of Sines

$$\frac{a}{\sin A} = \frac{c}{\sin C}$$

$$\frac{9.4}{\sin 33°} = \frac{14.3}{\sin C}$$

$$\sin C = \frac{(14.3)(\sin 33°)}{9.4}$$

$$\approx 0.8285466$$

$$C \approx 55.95°$$

Since $C \approx 55.95°$ and $A = 33°$, then

$$B \approx 180 - 33° - 55.95°$$

$$= 91.05°$$

We can use this information to find the length of the third side of the triangle, b.

$$\frac{a}{\sin A} = \frac{b}{\sin B}$$

$$\frac{9.4}{\sin 33°} = \frac{b}{\sin 91.05°}$$

$$b = \frac{(9.4)(\sin 91.05°)}{\sin 33°}$$

$$= 17.26$$

We can now complete the data chart and show all the parts of this triangle:

sides	angles
$a = 9.4$	$A = 33°$
$b = 17.26$	$B = 91.05°$
$c = 14.3$	$C = 55.95°$

▪

The last part of Example 10.20 was somewhat like the situation in Case 2. You knew two of the angles. Once you found the size of the third angle, you were ready to continue solving the problems.

EXAMPLE 10.21

Solve the triangle ABC, if $A = 82.17°, B = 64.43°$, and $c = 9.12$.

Solution   The beginning data chart is

sides	angles
$a = $ ___	$A = 82.17°$
$b = $ ___	$B = 64.43°$
$c = 9.12$	$C = $ ___

This is a triangle in that it satisfies Case 2. (It is an AAS triangle.) Since we know that $A + B + C = 180°$ and are given that $A + B = 82.17° + 64.43° = 146.60°$, we know that $C = 180° - 146.60° = 33.40°$.

We can now use the Sine Law to find either a or b. We will first find a.

$$\frac{a}{\sin A} = \frac{c}{\sin C}$$

$$\frac{a}{\sin 82.17°} = \frac{9.12}{\sin 33.40°}$$

$$a = \frac{(9.12)(\sin 82.17°)}{\sin 33.40°}$$

$$\approx 16.41$$

Now we use the Sine Law to find b.

$$\frac{b}{\sin B} = \frac{c}{\sin C}$$

$$\frac{b}{\sin 64.43°} = \frac{9.12}{\sin 33.40°}$$

$$b = \frac{(9.12)(\sin 64.43°)}{\sin 33.40°}$$

$$\approx 14.94$$

The completed data chart is

sides	angles
$a = 16.41$	$A = 82.17°$
$b = 14.94$	$B = 64.43°$
$c = \ \ 9.12$	$C = 33.40°$

It is possible to find 0, 1, or 2 correct solutions to the triangle if the given information includes two sides and one angle. When this ambiguous case occurs, carefully consider the practical application of the solution. We will now examine two situations that produce ambiguous results.

EXAMPLE 10.22

Solve $\triangle ABC$, if $a = 20, b = 24$, and $A = 55.4°$.

Solution   As usual, we begin with the data chart showing the given and unknown measurements.

sides	angles
$a = 20$	$A = 55.4°$
$b = 24$	$B = $ ___
$c = $ ___	$C = $ ___

We will now find B.

$$\frac{a}{\sin A} = \frac{b}{\sin B}$$

$$\frac{20}{\sin 55.4°} = \frac{24}{\sin B}$$

$$\sin B = \frac{24(\sin 55.4°)}{20}$$

$$\approx 0.9877636$$

$$B = \sin^{-1}(0.9877636)$$

So, either $B = 81.03°$ or $B = 98.97°$. Remember, the sine is positive in both the first and second quadrants. Whenever you use the Sine Law and get an equation of the form $B = \sin^{-1} n$ or $\sin B = n$, where $0 < n < 1$, then there are two possible values for B.

Will both of these answers satisfy the given parts of the triangle? Let's call the two answers for angle B, B_1, and B_2. If $B_1 = 81.03°$, and since $A = 55.4°$, then $C_1 = 180° - 81.03° - 55.4° = 43.57°$. If $B_2 = 98.97°$, then $C_2 = 180° - 98.97° - 55.4° = 25.63°$.

We will now use these angles to find the length of side c. If $C_1 = 43.57°$, then

$$\frac{a}{\sin A} = \frac{c_1}{\sin C_1}$$

$$\frac{20}{\sin 55.4°} = \frac{c_1}{\sin 43.57°}$$

$$c_1 = \frac{20(\sin 43.57°)}{\sin 55.4°}$$

$$= 16.75$$

EXAMPLE 10.22 (Cont.)

If we use $C_2 = 25.63°$, then we get

$$\frac{a}{\sin A} = \frac{c_2}{\sin C_2}$$

$$\frac{20}{\sin 55.4°} = \frac{c_2}{\sin 25.63°}$$

$$c_2 = \frac{20(\sin 25.63°)}{\sin 55.4°}$$

$$= 10.51$$

The data chart is now complete and written as two data charts.

sides	angles	sides	angles
$a = 20$	$A = 55.4°$	$a = 20$	$A = 55.4°$
$b = 24$	$B_1 = 81.03°$	$b = 24$	$B_2 = 98.97°$
$c_1 = 16.75$	$C_1 = 43.57°$	$c_2 = 10.51$	$C_2 = 25.63°$

The triangles formed with these two solutions are shown in Figure 10.35. Both are correct.

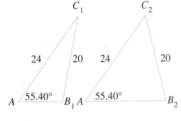

FIGURE 10.35

The other problem that can develop with the ambiguous case is that there may be no solution. Consider the situation in Example 10.23.

EXAMPLE 10.23

Solve for triangle ABC, when $a = 20, b = 27$, and $A = 70°$.

Solution   The beginning data chart is:

sides	angles
$a = 20$	$A = 70°$
$b = 27$	$B = $ __
$c = $ __	$C = $ __

By the Sine Law $\dfrac{a}{\sin A} = \dfrac{b}{\sin B}$, so

$$\frac{20}{\sin 70°} = \frac{27}{\sin B}$$

$$\sin B = \frac{27(\sin 70°)}{20}$$

$$= 1.27$$

The sine of an angle is never larger than 1. This triangle is not possible. Perhaps if you look at Figure 10.36 you can get a better idea why this is not a legitimate triangle.

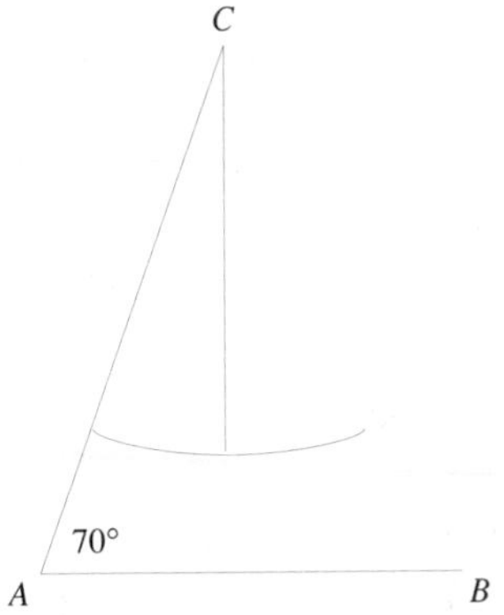

FIGURE 10.36

As with all trigonometry, there are many applications of the Law of Sines in technical areas. The problem of the technician is to recognize when the application requires trigonometry and then to select the appropriate method to solve it. The next example and the problems in the exercise set will help you make the correct selection.

Application

EXAMPLE 10.24

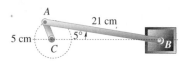

FIGURE 10.37

The crankshaft $\overline{CA}$ of an engine is 5 cm long and the connecting rod $\overline{AB}$ is 21 cm long. Find the size of $\angle ACB$ when the size of $\angle ABC$ is $5°$.

Solution A sketch of this crankshaft is in Figure 10.37. This problem falls into the SSA case. We know the lengths of two sides, AC and AB, and of the angle opposite one of them, $\angle B$. Since $AC = b$ and $AB = c$, we can use the Sine Law with

$$\frac{b}{\sin B} = \frac{c}{\sin C}$$

$$\frac{5}{\sin 5} = \frac{21}{\sin C}$$

$$\sin C = \frac{21 \sin 5°}{5}$$

$$= 0.3660541$$

$$C = 21.47° \qquad \text{or} \qquad 158.53°$$

There are two possible solutions. By looking at Figure 10.37, you can see that both are acceptable.

Exercise Set 10.4

In Exercises 1–20, solve each triangle with the given parts. Check for the ambiguous cases.

1. $A = 19.4°, B = 85.3°, c = 22.1$

2. $a = 12.4, B = 62.4°, C = 43.9°$

3. $a = 14.2, b = 15.3, B = 97°$

4. $A = 27.42°, a = 27.3, b = 35.49$

5. $A = 86.32°, a = 19.19, c = 18.42$

6. $B = 75.46°, b = 19.4, C = 44.95°$

7. $B = 39.4°, b = 19.4, c = 35.2$

8. $A = 84.3°, b = 9.7, C = 12.7°$

9. $A = 45°, a = 16.3, b = 19.4$

10. $a = 10.4, c = 5.2, C = 30°$

11. $a = 42.3, B = 14.3°, C = 16.9°$

12. $A = 105.4°, B = 68.2°, c = 4.91$

13. $b = 19.4, c = 12.5, C = 35.6°$

14. $a = 121.4, A = 19.7°, c = 63.4$

15. $a = 19.7, b = 8.5, B = 78.4°$

16. $b = 9.12, B = 1.3 \text{ rad}, C = 0.67 \text{ rad}$

17. $b = 8.5, c = 19.7, C = 1.37 \text{ rad}$

18. $b = 19.7, c = 36.4, C = 0.45 \text{ rad}$

19. $A = 0.47 \text{ rad}, b = 195.4, C = 1.32 \text{ rad}$

20. $a = 29.34, A = 1.23 \text{ rad}, C = 1.67 \text{ rad}$

Solve Exercises 21–25.

21. *Civil engineering*   Two high-tension wires are to be strung across a river. There are two towers, *A* and *B*, on one side of the river. These two towers are 360 m apart. A third tower, *C*, is on the other side of the river. If ∠*ABC* is 67.4° and ∠*BAC* is 49.3°, what are the distances between towers *A* and *C* and towers *B* and *C*?

22. *Civil engineering*   A tunnel is to be dug between points *A* and *B* on opposite sides of a hill. A point *C* is chosen 250 m from *A* and 275 m from *B*. If ∠*BAC* measures 43.62°, find the length of the tunnel.

23. *Navigation*   A plane leaves airport *A* with a heading of 313°. Several minutes later the plane is spotted from airport *B* at a heading of 27°. Airport *B* is due west of airport *A* and the two airports are 37 mi apart. How far had the airplane flown?

24. *Civil engineering*   From a point on the top of one end of a football stadium, the angle of depression to the 40-yard marker is 10.45°. The angle of depression to the 50-yard marker is 11.36°. How high is that end of the stadium above the playing field?

25. *Civil engineering*   Two technicians release a balloon containing a radio-controlled camera. In order for the camera's photographs to cover enough territory, they plan to start the camera when the balloon reaches a certain height. The technicians are 400 m apart. One study triggers the balloon when the angle of elevation at that spot is 47°. At that instant, the angle of elevation for the other technician is 67°. If the balloon is directly above the line connecting the two technicians, what is its height?

≡ 10.5

OBLIQUE TRIANGLES: LAW OF COSINES

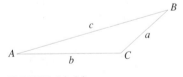

FIGURE 10.38

In Section 10.4, we learned the Law of Sines and when to use it. We were able to use the Sine Law in two cases. In Case 1, we knew the measures for two sides and the angle opposite one of them. This we named the SSA case. Case 2 existed when we knew the measures of two angles and one side of a triangle. This we called the AAS case.

There are two other cases that lead to solvable triangles. We will learn to solve the following two cases in this section.

Case 3 (SAS)   The measure of two sides and the included angle are known.

Case 4 (SSS)   The measure of three sides are known.

The Law of Cosines is used to help solve both Cases 3 and 4. Using the general oblique triangle in Figure 10.38, the Law of Cosines can be stated as follows.

> **The Law of Cosines**
>
> If △*ABC* is a triangle with sides of lengths *a*, *b*, and *c*, and opposite angles *A*, *B*, *C*, then by the **Law of Cosines** or **Cosine Law**,
> $$a^2 = b^2 + c^2 - 2bc \cos A$$
> $$b^2 = a^2 + c^2 - 2ac \cos B$$
> or $\quad c^2 = a^2 + b^2 - 2ab \cos C$

Notice that there are three versions of the Cosine Law. Each version simply restates the law so that different parts of the triangle are used.

Hint

In the Cosine Law, the side on the left-hand side of the equation has the same letter as the angle on the right-hand side of the equation.

EXAMPLE 10.25

If $b = 14.7, c = 9.3$, and $A = 46.3°$, solve the triangle.

Solution The data chart is

sides	angles
$a = \underline{\quad}$	$A = 46.3°$
$b = 14.7$	$B = \underline{\quad}$
$c = 9.3$	$C = \underline{\quad}$

This is Case 3 or the SAS type of problem. Since we know the size of angle A, we first use the Law of Cosines to find the length of side a.

$$a^2 = b^2 + c^2 - 2bc \cos A$$
$$= (14.7)^2 + (9.3)^2 - 2(14.7)(9.3) \cos 46.3°$$
$$= 216.09 + 86.49 - 273.42(0.6908824)$$
$$= 216.09 + 86.49 - 188.90107$$
$$= 113.67893$$

So, $a = 10.662032$ or $a \approx 10.7$.

At this point, you have a choice as to which method to use to solve the remainder of the problem. You know the size of one angle so you could use the Sine Law. We will use an alternate version of the Cosine Law, which takes advantage of the fact that you know the lengths of all three sides. This alternate version of the Cosine Law is used for the SSS type of problem. It will be the next example, after we state the alternate version of the Cosine Law.

The Law of Cosines (Alternate Version)

If $\triangle ABC$ is a triangle with sides of lengths a, b, and c, and opposite angles A, B, C, then by the **Law of Cosines**,

$$\cos A = \frac{b^2 + c^2 - a^2}{2bc}$$

$$\cos B = \frac{a^2 + c^2 - b^2}{2ac}$$

or $\quad \cos C = \dfrac{a^2 + b^2 - c^2}{2ab}$

You may notice that the alternate version of the Law of Cosines is just the original versions solved for $\cos A$, $\cos B$, and $\cos C$, respectively.

EXAMPLE 10.26

If $a = 10.7, b = 14.7$, and $c = 9.3$, find the sizes of the three angles.

Solution   We will use the Cosine Law to find the size of angle B.

$$b^2 = a^2 + c^2 - 2ac \cos B$$

$$\cos B = \frac{(10.7)^2 + (9.3)^2 - (14.7)^2}{2(10.7)(9.3)}$$

$$\cos B = \frac{114.49 + 86.49 - 216.09}{199.02} \approx -0.07592201793$$

$$B = 94.354201 \approx 94.4°$$

Since this is a continuation of Example 10.25, we know that $A = 46.3°$, so $C = 180° - 46.3° - 94.4° = 39.3°$. The completed data chart is:

sides	angles
$a = 10.7$	$A = 46.3°$
$b = 14.7$	$B = 94.4°$
$c = 9.3$	$C = 39.3°$

Application

EXAMPLE 10.27

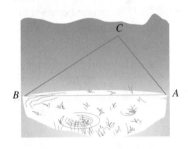

FIGURE 10.39

An electric transmission line is planned to go directly over a swamp. The power line will be supported by towers at points A and B in Figure 10.39. A surveyor measures the distance from B to C as 573 m, the distance from A to C as 347 m, and $\angle BCA$ as 106.63°. What is the distance from tower A to tower B?

Solution $BC = a = 573$ m; $AC = b = 347$ m; $\angle BCA = \angle C = 106.63°$. This is an SAS type of problem. We want to find $AB = c$. Using the Cosine Law, we have

$$c^2 = a^2 + b^2 - 2ab \cos C$$

$$= (573)^2 + (347)^2 - 2(573)(347) \cos 106.63$$

$$= 562\,544.93$$

$$c \approx 750$$

The distance between the towers will be about 750 m.

Exercise Set 10.5

In Exercises 1–20, solve each triangle with the given parts. The angles in Exercises 1–10 are in degrees and those in Exercises 11–20 are in radians.

1. $a = 9.3, b = 16.3, C = 42.3°$
2. $A = 16.25°, b = 29.43, c = 36.52$
3. $a = 47.85, B = 113.7°, c = 32.79$
4. $a = 19.52, b = 63.42, c = 56.53$
5. $a = 29.43, b = 16.37, c = 38.62$
6. $A = 121.37°, b = 112.37, c = 93.42$
7. $a = 63.92, B = 92.44°, c = 78.41$
8. $a = 19.53, b = 7.66, C = 32.56°$
9. $a = 4.527, b = 6.239, c = 8.635$
10. $A = 7.53°, b = 37.645, c = 42.635$

11. $a = 8.5, b = 15.8, C = 0.82$ rad
12. $A = 0.31, b = 15.8, c = 38.47$
13. $a = 52.65, B = 1.98, c = 35.8$
14. $a = 43.5, b = 63.4, c = 37.3$
15. $a = 36.27, b = 24.55, c = 44.26$
16. $A = 2.41, b = 153.21, c = 87.49$
17. $a = 54.8, B = 1.625, c = 38.33$
18. $a = 7.621, b = 3.429, C = 0.183$
19. $a = 2.317, b = 1.713, c = 1.525$
20. $A = 0.09, b = 40.75, c = 50.25$

Solve Exercises 21–26.

21. *Space technology*   A tracking antenna is aimed $34.7°$ above the horizon. The distance from the antenna to a spacecraft is $12\,325$ km. If the radius of the earth is $6\,335$ km, how high is the spacecraft above the surface of the earth? (See Figure 10.40.) (Hint: Find the length of BD.)

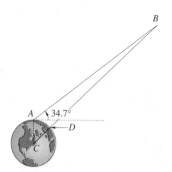

FIGURE 10.40

22. *Navigation*   A ship leaves port at noon and travels due north at 21 km/h. At 3 p.m., the ship changes direction to a heading of $37°$. How far from the port is the ship at 7 p.m.? What is the bearing of the ship from the port?

23. *Construction*   A hill makes an angle of $12.37°$ with the horizontal. A 75-ft antenna is erected on the top of the hill. A guy wire is to be strung from the top of

the antenna to a point on the hill that is 40 ft from the base of the antenna. How long is the guy wire?

24. *Physics*   Two forces are acting on an object. The magnitude of one force is 35 lb and the magnitude of the second force is 50 lb. If the angle between the two forces is $32.15°$, what is the magnitude of the resultant force?

25. *Physics*   Figure 10.41 shows two forces represented by vector $\overrightarrow{AB}$ and $\overrightarrow{BC}$. If $AB = 12$ N, $BC = 23$ N, and $\angle ABC = 121.27°$, find the magnitude of **R** and the size of θ.

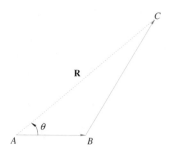

FIGURE 10.41

26. *Physics*   Two forces of 15.5 lb and 36.4 lb are acting on an object. The resultant force is 30.1 lb. What is the size of the angle between the original two forces?

☰ CHAPTER 10 REVIEW

Important Terms and Concepts

Adding vectors
  By adding components
  By the parallelogram method
Component vectors
Cosine Law
Direction vector
Initial point
Magnitude of a vector
Oblique triangle

Parallelogram method
Position vector
Resultant vector
Scalar
Sine Law
  Ambiguous case
Terminal point
Vectors

Review Exercises

In Exercises 1–4, add the given vectors by drawing the resultant vector.

1.

2.

3.

4.

In Exercises 5–8, find the horizontal and vertical components of the given vectors.

5. Magnitude 35, $\theta = 67°$

6. Magnitude 19.7, $\theta = 237°$

7. Magnitude 23.4, $\theta = 172.4°$

8. Magnitude 14.5, $\theta = 338°$

In Exercises 9–10, the horizontal and vertical components are given for a vector. Find each of the resultant vectors.

9. $A_x = 16, A_y = -8$

10. $B_x = -27, B_y = 32$

In Exercises 11–14, find the components of the indicated vectors.

11. $A = 38, \theta_A = 15°$

12. $B = 43.5, \theta_B = 127°$

13. $C = 19.4, \theta_C = 1.25$

14. $D = 62.7, \theta_D = 5.37$

In Exercises 15–18, add the given vectors by using the trigonometric functions and the Pythagorean theorem.

15. $A = 19, \theta_A = 32°, B = 32, \theta_B = 14°$

16. $C = 24, \theta_C = 57°, D = 35, \theta_D = 312°$

17. $E = 52.6, \theta_E = 2.53, F = 41.7, \theta_F = 3.92$

18. $G = 43.7, \theta_G = 4.73, H = 14.5, \theta_H = 4.42$

Solve each of the triangles in Exercises 19–26.

19. $a = 14, b = 32, c = 27$

20. $a = 43, b = 52, B = 86.4°$

21. $b = 87.4, B = 19.57°, c = 65.3$

22. $A = 121.3°, b = 42.5, c = 63.7$

23. $a = 127.35, A = 0.12, b = 132.6$

24. $b = 84.3, c = 95.4, C = 0.85$

25. $a = 67.9, b = 54.2, C = 2.21$

26. $a = 53.1, b = 63.2, c = 74.3$

Solve Exercises 27–32.

27. *Physics*   Two tow trucks are attempting to right an overturned vehicle. One truck is exerting a force of 1 650 kg. Its tow chain makes an angle of 68° with the axis of the vehicle. The other truck is exerting a force of 1 325 kg. Its chain makes an angle of 76° with the axis of the vehicle. What is the magnitude and direction of the resultant force vector? (See Figure 10.42.)

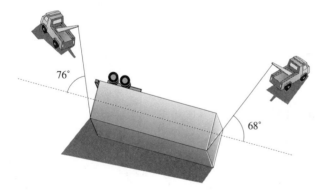

FIGURE 10.42

28. *Civil engineering*   A highway engineer has to decide whether to go over or to cut through a hill. The top of the hill makes an angle of 72.4° with the sides. One side of the hill is 2,342 ft and other side is 3,621 ft. It will cost 2.3 times as much per foot to cut through the hill and take the alternate route in Figure 10.43.

How long is the alternate route? Which route is less expensive?

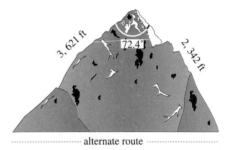

FIGURE 10.43

29. *Physics*   A block is resting on a ramp that makes an angle of 31.7° with the horizontal. The block weighs 126.5 lb. Find the components of the block's weight that are parallel and perpendicular to the road. (See Figure 10.44.)

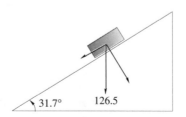

FIGURE 10.44

30. *Electricity* Find the total impedance Z and the phase angle of a series ac circuit when $X_C = 72\ \Omega$, $X_L = 52\ \Omega$, and $R = 35\ \Omega$.

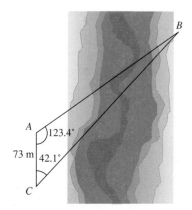

FIGURE 10.45

31. *Civil engineering* A wire is to be strung across a valley. The wire will run from Tower A to Tower B.

A surveyor is able to set up a position at a point C on the same side of the valley as Tower A, as shown in Figure 10.45. The distance from A to C is 73 m and $\angle BAC = 123.4°$ and $\angle ACB = 42.1°$. What is the distance from Tower A to Tower B?

32. *Surveying* A surveyor needs to determine the distance across a swamp. From a point C in Figure 10.46, she locates a point B on one side of the swamp. The distance from B to C is 1 235 m. Point A is directly across the swamp from B. The distance from A to C is 962 m and $\angle BCA$ is 52.57°. How far is it across the swamp from A to B?

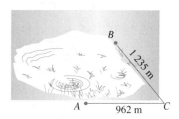

FIGURE 10.46

☰ CHAPTER 10 TEST

1. Determine the horizontal and vertical components of a vector **V** of magnitude 47 and direction $\theta = 117°$.

2. If $A_x = 12.91$ and $A_y = -14.36$, determine the magnitude and direction of vector **A**.

3. Add the given vectors by using the trigonometric functions and the Pythagorean theorem: $A = 25$, $\theta_A = 64°$, $B = 40$, $\theta_B = 112°$

4. Use the Sine Law to determine the length of side b in $\triangle ABC$, if $a = 9.42$, $\angle A = 35.6°$, and $\angle B = 67.5°$.

5. Use the Cosine Law to determine the length of side a of $\triangle ABC$, if $b = 4.95$, $c = 6.24$, and $\angle A = 113.4°$.

6. Solve $\triangle ABC$, if $A = 24°$, $b = 36.5$, and $C = 97°$.

7. Two walls meet at an angle of $97°$ to form the sides of a triangular corner cupboard. If the sides of the cupboard along each wall measure 30 in. and 36 in., what is the length of the front of the cupboard?

11

Graphs of Trigonometric Functions

The screen of an oscilloscope displays the graph of an electrical signal. This oscilloscope displays a Lissajous figure. In Section 11.6, we will see how to produce such a figure from two trigonometric functions.

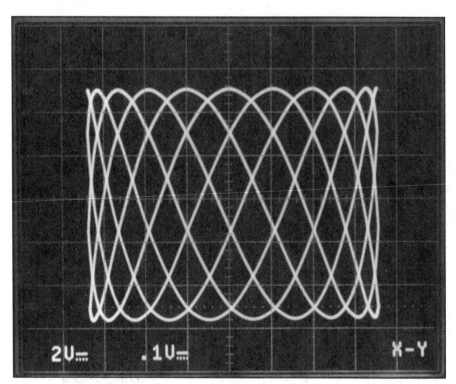

Trigonometry originally helped people solve triangles. For many years this was the main use of trigonometry. When people began graphing equations, it became possible to obtain pictures or graphs of the trigonometric functions.

Later, when people discovered light and sound waves, they found that these waves repeated at regular intervals—just like the graphs of trigonometric functions. In fact, light, sound, electricity, ocean waves, and spring action are all examples of things that can be displayed as a graph of one or more trigonometric function.

☰ 11.1
SINE AND COSINE CURVES: AMPLITUDE AND PERIOD

As you know, the trigonometric functions repeat their values on a regular basis. Thus, $\sin 15° = \sin(360° + 15°) = \sin(720° + 15°)$, and so on. This means that the graphs of the trigonometric functions repeat. This repeating nature helps us apply these graphs to many physical and technical situations. In this chapter, we will use a graphics calculator to help draw the graphs, and show how to apply these graphs to your technical area.

Sine Curves

Sine and cosine curves are used in applications of mathematics that involve circular motion. In this chapter, we will first look at ways to graph the sine and cosine functions. Later in the chapter, we will look at some applications.

When we began graphing algebraic equations, we started by making a table of values. We shall adopt the same approach in graphing trigonometric functions. The following table gives the values of x in both degrees and radians of the function $y = \sin x$. The values of x only go from $0°$ to $360°$ (0 to 2π rad), since the values of y begin to repeat after $360°$ (or 2π rad). The graph of these values is shown in Figure 11.1.

x (degrees)	0	30	45	60	90	120	135	150	180
x (radians)	0	$\dfrac{\pi}{6}$	$\dfrac{\pi}{4}$	$\dfrac{\pi}{3}$	$\dfrac{\pi}{2}$	$\dfrac{2\pi}{3}$	$\dfrac{3\pi}{4}$	$\dfrac{5\pi}{6}$	π
$y = \sin x$	0	0.5	0.707	0.866	1	0.866	0.707	0.5	0

x (degrees)	180	210	225	240	270	300	315	330	360
x (radians)	π	$\dfrac{7\pi}{6}$	$\dfrac{5\pi}{4}$	$\dfrac{4\pi}{3}$	$\dfrac{3\pi}{2}$	$\dfrac{5\pi}{3}$	$\dfrac{7\pi}{4}$	$\dfrac{11\pi}{6}$	2π
$y = \sin x$	0	-0.5	-0.707	-0.866	-1	-0.866	-0.707	-0.5	0

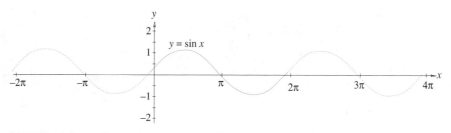

FIGURE 11.1

If we were to continue plotting the values for x larger than 2π rad or less than 0, we would get the points indicated by the dotted line in Figure 11.1. As you can see, the graph, like the values for $\sin x$, begins to repeat. The graph of the sine function has a very distinctive wave shape. A graph that has the general shape of the sine function is called a **sine wave**, a **sine curve**, or a **sinusoidal curve**.

Amplitude

One term used to help describe some functions is amplitude.

> **Amplitude**
>
> The **amplitude** A is one-half the difference between the maximum and minimum values for a periodic function.

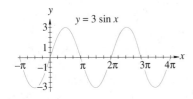

FIGURE 11.2

For $y = \sin x$, the maximum value is 1 and the minimum value is -1. The difference between these two values is $1 - (-1) = 2$ and half of that is 1. So, the amplitude of $y = \sin x$ is 1.

Let's examine another trigonometric function. Suppose we want to graph $y = 3 \sin x$. This is the same as taking each of the values for $y = \sin x$ in the previous table and multiplying it by 3. The resulting table would be as follows and the graph would be like the one in Figure 11.2.

x	0	$\frac{\pi}{6}$	$\frac{\pi}{3}$	$\frac{\pi}{2}$	$\frac{2\pi}{3}$	$\frac{5\pi}{6}$	π	$\frac{7\pi}{6}$	$\frac{4\pi}{3}$	$\frac{3\pi}{2}$	$\frac{5\pi}{3}$	$\frac{11\pi}{6}$	2π
$y = 3\sin x$	0	1.5	2.6	3	2.6	1.5	0	-1.5	-2.6	-3	-2.6	-1.5	0

Notice that the maximum value of $y = 3 \sin x$ is 3 and the minimum value is -3. The amplitude of $y = 3 \sin x$ is $\frac{3-(-3)}{2} = 3$. In fact, you should be able to see that the amplitude of $y = A \sin x$ is $|A|$.

Period and Frequency

There are some things we should notice about these first two curves. They are examples of periodic functions. Each of them repeats its shape or completes one wave every 2π units, so we say these curves each have a period of 2π rad.

> **Periodic Function**
>
> For a function f, if p is the smallest positive constant where $f(x) = f(x + p)$ for all values of x, we say that f is a **periodic function** and that it has **period** p. The portion of a graph that falls within one period is called a **cycle**.

The sine function is a periodic function with period 2π (or $360°$), because $\sin x = \sin(x + 2\pi)$.

> **Frequency**
>
> The **frequency** of a periodic function is defined as $\dfrac{1}{\text{period}}$.

Thus, the sine function with period 2π has a frequency of $\dfrac{1}{2\pi}$.

Now, let's graph a slightly different version of the sine curve, $y = \sin 4x$. Again, we will begin by using a table. This table will have values for x and $4x$, as well as for $y = \sin 4x$. The table follows, and the graph of $y = \sin 4x$ is in Figure 11.3.

x	0	$\dfrac{\pi}{24}$	$\dfrac{\pi}{12}$	$\dfrac{\pi}{8}$	$\dfrac{\pi}{6}$	$\dfrac{5\pi}{24}$	$\dfrac{\pi}{4}$	$\dfrac{7\pi}{24}$	$\dfrac{\pi}{3}$	$\dfrac{3\pi}{8}$	$\dfrac{5\pi}{12}$	$\dfrac{11\pi}{24}$	$\dfrac{\pi}{2}$
$4x$	0	$\dfrac{\pi}{6}$	$\dfrac{\pi}{3}$	$\dfrac{\pi}{2}$	$\dfrac{2\pi}{3}$	$\dfrac{5\pi}{6}$	π	$\dfrac{7\pi}{6}$	$\dfrac{4\pi}{3}$	$\dfrac{3\pi}{2}$	$\dfrac{5\pi}{3}$	$\dfrac{11\pi}{6}$	2π
$y = \sin 4x$	0	0.5	0.866	1	0.866	0.5	0	-0.5	-0.866	-1	-0.866	-0.5	0

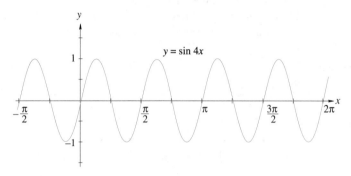

FIGURE 11.3

This graph is interesting in that it looks like a sine curve, with an amplitude of 1, but it does not have a period of 2π. In fact, its period is $\dfrac{\pi}{2}$. Notice that $\dfrac{\pi}{2} = 2\pi \div 4$. In general, we have the following statement for both the sine and the cosine functions.

Period

The **period** of $y = \sin Bx$ and $y = \cos Bx$ is $\dfrac{2\pi}{|B|}$ and the frequency is $\dfrac{|B|}{2\pi}$.

Cosine Curves

As yet, we have graphed only the sine curve. It is time to graph the cosine curve. A table of values follows and the graph of $y = \cos x$ is shown in Figure 11.4.

x	0	$\dfrac{\pi}{6}$	$\dfrac{\pi}{3}$	$\dfrac{\pi}{2}$	$\dfrac{2\pi}{3}$	$\dfrac{5\pi}{6}$	π	$\dfrac{7\pi}{6}$	$\dfrac{4\pi}{3}$	$\dfrac{3\pi}{2}$	$\dfrac{5\pi}{3}$	$\dfrac{11\pi}{6}$	2π
$y = \cos x$	1	0.87	0.5	0	-0.5	-0.87	-1	-0.87	-0.5	0	0.5	0.87	1

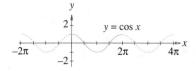

FIGURE 11.4

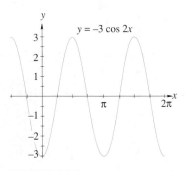

FIGURE 11.5

Here, as with the sine curve in Figure 11.1, we have shown the values graphed from the table with a solid line. The dashed line indicates what the graph would look like if we continued to plot values for $x > 2\pi$ or $x < 0$. The solid line represents one cycle. As you can see, the cosine function is periodic; $y = \cos x$ has period of 2π and amplitude of 1. A careful examination of the curves in Figures 11.1 and 11.4 shows that the shapes of the curves are exactly the same but are displaced along the horizontal axis relative to one another.

As our final example, we will graph $y = -3\cos 2x$. The period of $y = \cos x$ is 2π, so the period of $y = -3\cos 2x$ is $2\pi \div 2 = \pi$. The amplitude of $y = A\cos x$ is $|A|$, so the amplitude of $y = -3\cos 2x$ is $|-3| = 3$. To sketch this curve, we need to find only a few values of x. These x values will be at the maximum and minimum values and at the x-intercepts (when $y = 0$). This table follows and the graph of $y = -3\cos 2x$ is in Figure 11.5.

x	0	$\dfrac{\pi}{4}$	$\dfrac{\pi}{2}$	$\dfrac{3\pi}{4}$	π
$y = -3\cos 2x$	-3	0	3	0	-3

Study the graph of $y = -3\cos 2x$. This is the graph of $y = 3\cos 2x$ only "flipped" over the x-axis.

EXAMPLE 11.1

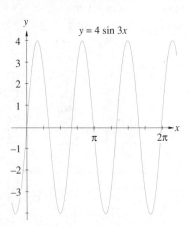

FIGURE 11.6

Sketch the graph of $y = 4\sin 3x$.

Solution We can see that the amplitude is 4 and the period is $\dfrac{2\pi}{3}$. The frequency is $\dfrac{1}{\frac{2\pi}{3}} = \dfrac{3}{2\pi}$. From this we know that $y = 0$ when $x = \dfrac{2\pi}{3}$, $x = \dfrac{\pi}{3}$, and $x = 0$. The maximum and minimum values will occur halfway between these values of x, or when $x = \dfrac{\frac{\pi}{3} + 0}{2} = \dfrac{\pi}{6}$ and $x = \dfrac{\frac{2\pi}{3} + \frac{\pi}{3}}{2} = \dfrac{\pi}{2}$. The table for these values follows.

x	0	$\dfrac{\pi}{6}$	$\dfrac{\pi}{3}$	$\dfrac{\pi}{2}$	$\dfrac{2\pi}{3}$
$y = 4\sin 3x$	0	4	0	-4	0

Based on these values and our knowledge of the shape of the sine curves, we get the graph shown in Figure 11.6. Notice how this graph is "compressed" or "shrunk" over a period of $\dfrac{2\pi}{3}$.

EXAMPLE 11.2

Sketch the graph of $y = 1.5\cos\dfrac{x}{2}$.

Solution We can see the amplitude is 1.5 and the period is $2\pi \div \dfrac{1}{2} = 4\pi$ or $720°$. Since the frequency is $\dfrac{1}{\text{period}}$, it is $\dfrac{1}{4\pi}$. The maximum values will occur

EXAMPLE 11.2 (Cont.)

when $x = 0$ and $x = 720°$. The minimum value will be halfway between these two points or when $x = 360°$. The zeros are at the midpoints between these three points or when $x = 180°$ and $x = 540°$. These findings are shown in this table.

x	$0°$	$180°$	$360°$	$540°$	$720°$
$y = 1.5 \cos \dfrac{x}{2}$	1.5	0	-1.5	0	1.5

Based on these values and our knowledge of the shape of the cosine curve, we get the graph in Figure 11.7. Notice how this curve is "stretched out" over a period of 4π or $720°$.

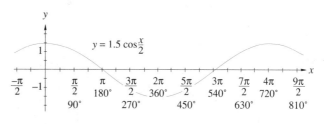

FIGURE 11.7

The following box summarizes this section's discussion on amplitude and period.

Summary

A function of the form $y = A \sin Bx$ or $y = A \cos Bx$ has

$$\text{Amplitude: } |A| \qquad \text{Period: } \frac{2\pi}{|B|} \qquad \text{Frequency: } \frac{|B|}{2\pi}$$

If $A < 0$, the graph is "flipped" over the x-axis.

Exercise Set 11.1

In Exercises 1–14, find the period, amplitude, and frequency of the given functions.

1. $y = 3 \sin 2x$

2. $y = 5 \sin 6x$

3. $y = 2 \cos x$

4. $y = 7 \cos 3x$

5. $y = 8 \cos 2\pi x$

6. $y = 7 \cos \pi x$

7. $y = \dfrac{1}{2} \sin 4x$

8. $y = 5 \cos \dfrac{1}{3} x$

9. $y = \dfrac{\cos \frac{1}{2} x}{3}$

10. $y = \dfrac{\cos 3x}{5}$

11. $y = -3 \sin \dfrac{x}{4}$

12. $y = -5 \cos \dfrac{x}{3}$

13. $y = \dfrac{-\cos x}{2}$

14. $y = \dfrac{\sin 3x}{-2}$

For Exercises 15–28, sketch one cycle of each of the functions in Exercises 1–14.

Solve the following exercise.

29. *Electronics* The current, I (in amperes), in an ac circuit is given by $I = 6.5 \sin 120\pi t$, where t is the time in seconds. **(a)** Find the period, amplitude, and frequency for this function. **(b)** Plot I as a function of t for $0 \leq t \leq 0.1$ s.

≡ 11.2
SINE AND COSINE CURVES: DISPLACEMENT OR PHASE SHIFT

The graphs of the sine and cosine curves that we have examined have all had one thing in common. That is, the sine curves all contained the point $(0, 0)$ and the cosine curves all contained $(0, A)$, where $|A|$ was the amplitude. In this section, we will shift the graphs away from these points.

EXAMPLE 11.3

Sketch the cycle of the graph of $y = \sin\left(x + \dfrac{\pi}{3}\right)$.

Solution   As usual, we make a table of values. In this table, we show three rows. The first row contains values of x, the next row contains values of $x + \dfrac{\pi}{3}$, and the third contains the values for $y = \sin\left(x + \dfrac{\pi}{3}\right)$.

x	0	$\dfrac{\pi}{6}$	$\dfrac{\pi}{3}$	$\dfrac{\pi}{2}$	$\dfrac{2\pi}{3}$	$\dfrac{5\pi}{6}$	π
$x + \dfrac{\pi}{3}$	$\dfrac{\pi}{3}$	$\dfrac{\pi}{2}$	$\dfrac{2\pi}{3}$	$\dfrac{5\pi}{6}$	π	$\dfrac{7\pi}{6}$	$\dfrac{4\pi}{3}$
$y = \sin\left(x + \dfrac{\pi}{3}\right)$	0.866	1	0.866	0.5	0	-0.5	-0.866

x	π	$\dfrac{7\pi}{6}$	$\dfrac{4\pi}{3}$	$\dfrac{3\pi}{2}$	$\dfrac{5\pi}{3}$	$\dfrac{11\pi}{6}$	2π
$x + \dfrac{\pi}{3}$	$\dfrac{4\pi}{3}$	$\dfrac{3\pi}{2}$	$\dfrac{5\pi}{3}$	$\dfrac{11\pi}{6}$	2π	$\dfrac{13\pi}{6}$	$\dfrac{8\pi}{3}$
$y = \sin\left(x + \dfrac{\pi}{3}\right)$	-0.866	-1	-0.866	-0.5	0	0.5	0.866

The graph of one cycle of this curve is given in Figure 11.8. Notice that this looks just like a sine curve except that it has to be shifted $\dfrac{\pi}{3}$ units to the left.

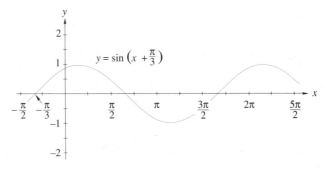

FIGURE 11.8

Let's look at a more general case.

EXAMPLE 11.4

Sketch one cycle of the graph of $y = \sin\left(2x - \dfrac{\pi}{6}\right)$.

Solution   The following table contains three rows, x, $2x - \dfrac{\pi}{6}$, and $\sin\left(2x - \dfrac{\pi}{6}\right)$.

x	0	$\dfrac{\pi}{12}$	$\dfrac{\pi}{6}$	$\dfrac{\pi}{4}$	$\dfrac{\pi}{3}$	$\dfrac{5\pi}{12}$	$\dfrac{\pi}{2}$
$2x - \dfrac{\pi}{6}$	$-\dfrac{\pi}{6}$	0	$\dfrac{\pi}{6}$	$\dfrac{\pi}{3}$	$\dfrac{\pi}{2}$	$\dfrac{2\pi}{3}$	$\dfrac{5\pi}{6}$
$y = \sin\left(2x - \dfrac{\pi}{6}\right)$	-0.5	0	0.5	0.866	1	0.866	0.5

x	$\dfrac{\pi}{2}$	$\dfrac{7\pi}{12}$	$\dfrac{2\pi}{3}$	$\dfrac{3\pi}{4}$	$\dfrac{5\pi}{6}$	$\dfrac{11\pi}{12}$	π
$2x - \dfrac{\pi}{6}$	$\dfrac{5\pi}{6}$	π	$\dfrac{7\pi}{6}$	$\dfrac{4\pi}{3}$	$\dfrac{3\pi}{2}$	$\dfrac{5\pi}{3}$	$\dfrac{11\pi}{6}$
$y = \sin\left(2x - \dfrac{\pi}{6}\right)$	0.5	0	-0.5	-0.866	-1	-0.866	-0.5

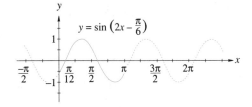

FIGURE 11.9

One cycle of this graph is shown in Figure 11.9. Notice that it has a period of π and has been shifted $\dfrac{\pi}{12}$ units to the right.

Graphs of $y = A \sin (Bx + C)$ and $y = A \cos (Bx + C)$

A curve of the form $y = A\sin(Bx + C)$ or $y = A\cos(Bx + C)$ has

Amplitude: $|A|$ Period: $\dfrac{2\pi}{|B|}$ Frequency: $\dfrac{|B|}{2\pi}$

Phase shift or displacement: $-\dfrac{C}{B}$

If the phase shift is negative, the entire curve will be shifted to the left. When the phase shift is positive, the entire curve will be shifted to the right. The phase shift gives you the x-coordinate at which to begin drawing one period of $y = A\sin(Bx + C)$ or $y = A\cos(Bx + C)$. If $A < 0$, the graph is "flipped" over the x-axis.

EXAMPLE 11.5

Sketch one cycle of the curve $y = \frac{1}{2}\cos(3x + \pi)$.

Solution Using the information contained in the equation, $A = \frac{1}{2}$, $B = 3$, and $C = \pi$, so the amplitude is $\frac{1}{2}$, the period is $\frac{2\pi}{3}$, and the phase shift is $-\frac{\pi}{3}$. Since the phase shift is $-\frac{\pi}{3}$, we start drawing the graph at $x = -\frac{\pi}{3}$. The period of $\frac{2\pi}{3}$ indicates that one complete cycle will end at $x = -\frac{\pi}{3} + \frac{2\pi}{3} = \frac{\pi}{3}$. For cosine, the maximum values occur at these two points. The minimum value will occur midway between the maximum values, or at $x = 0$. The zeros are midway between the maximum and minimum values, or at $x = -\frac{\pi}{6}$ and $x = \frac{\pi}{6}$. A sketch of $y = \frac{1}{2}\cos(3x + \pi)$ is in Figure 11.10 with the one complete cycle we just described in solid and additional cycles dashed.

EXAMPLE 11.6

Sketch one cycle of the graph of $y = -4\sin\left(\frac{1}{2}x - \frac{\pi}{8}\right)$.

Solution Here $A = -4$, $B = \frac{1}{2}$, and $C = -\frac{\pi}{8}$. The function has an amplitude of $|-4| = 4$ and a period of $2\pi \div \frac{1}{2} = 4\pi$. The phase shift is $-\left(-\frac{\pi}{8} \div \frac{1}{2}\right) = \frac{\pi}{4}$. We begin sketching the curve at $x = \frac{\pi}{4}$. The period of 4π indicates that one complete cycle will end at $x = \frac{\pi}{4} + 4\pi = \frac{17\pi}{4}$. The zeros of this sine function occur when $x = \frac{\pi}{4}$, $x = \frac{17\pi}{4}$, and midway between these two points, at $x = \frac{9\pi}{4}$. The minimum value is at $x = \frac{5\pi}{4}$ and the maximum when $x = \frac{13\pi}{4}$. The graph of $y = -4\sin\left(\frac{1}{2}x - \frac{\pi}{8}\right)$ is shown in Figure 11.11.

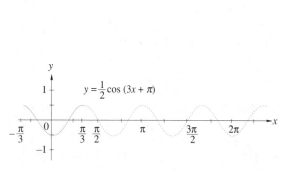

FIGURE 11.10

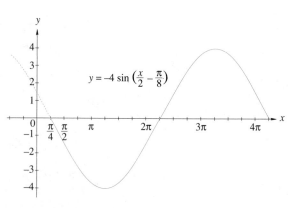

FIGURE 11.11

Exercise Set 11.2

In Exercises 1–14, give the amplitude, period, and phase shift for the given functions.

1. $y = 2\sin\left(x + \dfrac{\pi}{4}\right)$

2. $y = 1.5\cos\left(x - \dfrac{\pi}{3}\right)$

3. $y = 2.5\sin(3x - \pi)$

4. $y = 3\cos\left(2x + \dfrac{\pi}{2}\right)$

5. $y = 6\cos(1.5x + 180°)$

6. $y = 8\sin(3.5x - 90°)$

7. $y = -2\cos(3x - 90°)$

8. $y = -4\sin(5x + 600°)$

9. $y = 4.5\sin(6x - 8)$

10. $y = 9\cos(5x + 3)$

11. $y = 0.5\cos\left(\pi x + \dfrac{\pi}{8}\right)$

12. $y = -0.25\sin\left(\dfrac{x}{\pi} - \dfrac{1}{4}\right)$

13. $y = -0.75\sin\left(\pi x + \dfrac{\pi^2}{3}\right)$

14. $y = \pi\cos\left(\dfrac{1}{\pi}x + \dfrac{2}{3}\right)$

In Exercises 15–28, sketch one complete cycle of each of the functions in Exercises 1–14.

Solve Exercises 29–30.

29. *Electronics*   The current I (in amperes) in an ac circuit is given by $I = 65\cos\left(120\pi t + \dfrac{\pi}{2}\right)$, where t is the time in seconds. **(a)** Determine the amplitude, period, and phase shift for this function. **(b)** Sketch one complete cycle of this function.

30. *Oceanography*   At a certain point in the ocean, the vertical displacement of the water due to wave action is given by $y = 3\sin\left[\dfrac{\pi}{6}(t - 5)\right]$, where y is measured in meters and t is in seconds. **(a)** Determine the amplitude, period, and phase shift for the graph of this wave action. **(b)** Sketch one complete cycle of this function.

≡ 11.3
COMPOSITE SINE AND COSINE CURVES

In earlier sections, we looked at adding and subtracting functions. At that time, the functions were all algebraic. For example, when $f(x) = x^2$ and $g(x) = x$, we had little difficulty graphing $(f + g)(x)$. In a case such as this, we simply graphed $x^2 + x$ as if it were one function. Graphing composite trigonometric curves is not as easy.

In this section, we will learn how to graph trigonometric curves that combine two or more functions, using the ability to graph that we have learned earlier in this book. After we are competent with this technique, we will begin graphing trigonometric functions with the help of a calculator. Later in this chapter, we will see how these curves are used in technical areas.

The easiest way to learn how to graph a combination or composite curve is through an example.

EXAMPLE 11.7

Sketch the graph of $y = \sin x + \cos x$.

Solution   We will begin with a table of values. Both of the functions $y_1 = \sin x$ and $y_2 = \cos x$ have amplitude of 1 and period of 2π. The period of a composite curve is an integral multiple of the least common multiple of the period of each of the individual curves. In this case, since y_1 and y_2 each have periods of 2π, the least common multiple is also 2π. The following table contains values for y_1, y_2,

EXAMPLE 11.7 (Cont.)

and $y = y_1 + y_2$. $\left[\text{Notice that } \sin\dfrac{\pi}{4} = \dfrac{\sqrt{2}}{2} \approx 0.71 \text{ and } \cos\dfrac{\pi}{4} = \dfrac{\sqrt{2}}{2} \approx 0.71,\right.$

so $\sin\dfrac{\pi}{4} + \cos\dfrac{\pi}{4} = \dfrac{\sqrt{2}}{2} + \dfrac{\sqrt{2}}{2} = \sqrt{2} \approx 1.41$. However, adding the values in

the table for $\sin\dfrac{\pi}{4}$ and $\cos\dfrac{\pi}{4}$, we get $0.71 + 0.71 = 1.42$. The difference in these

two answers is the result of adding exact values $\left(\dfrac{\sqrt{2}}{2} + \dfrac{\sqrt{2}}{2}\right)$ or approximate

values $(0.71 + 0.71)$. It helps to show why you should wait until you have finished

all calculations before you round off any numbers. $\Big]$

x	0	$\dfrac{\pi}{6}$	$\dfrac{\pi}{4}$	$\dfrac{\pi}{3}$	$\dfrac{\pi}{2}$	$\dfrac{2\pi}{3}$	$\dfrac{3\pi}{4}$	$\dfrac{5\pi}{6}$	π
$y_1 = \sin x$	0.0	0.5	0.71	0.87	1.0	0.87	0.71	0.5	0.0
$y_2 = \cos x$	1.0	0.87	0.71	0.5	0.0	−0.5	−0.71	−0.87	−1.0
$y = y_1 + y_2$	1.0	1.37	1.42	1.37	1.0	0.37	0.0	−0.37	−1.0

x	π	$\dfrac{7\pi}{6}$	$\dfrac{5\pi}{4}$	$\dfrac{4\pi}{3}$	$\dfrac{3\pi}{2}$	$\dfrac{5\pi}{3}$	$\dfrac{7\pi}{4}$	$\dfrac{11\pi}{6}$	2π
$y_1 = \sin x$	0.0	−0.5	−0.71	−0.87	−1.0	−0.87	−0.71	−0.5	0.0
$y_2 = \cos x$	−1.0	−0.87	−0.71	−0.5	0.0	0.5	0.71	0.87	1.0
$y = y_1 + y_2$	−1.0	−1.37	−1.42	−1.37	−1.0	−0.37	0.0	0.37	1.0

A sketch of all three curves is shown in Figure 11.12. At any point x, the value for y
was obtained by adding the values for y_1 and y_2. The dotted line indicates how this
was done graphically.

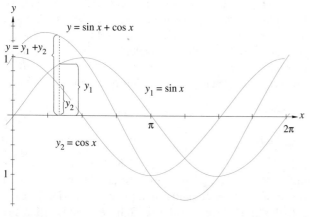

FIGURE 11.12

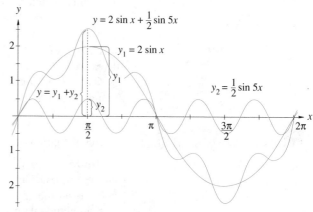

FIGURE 11.13

EXAMPLE 11.8

Sketch the graph of $y = 2 \sin x + \frac{1}{2} \sin 5x$.

Solution Once again we begin with a table of values. The function $y_1 = 2 \sin x$ has amplitude 2 and period 2π. The function $y_2 = \frac{1}{2} \sin 5x$ has amplitude $\frac{1}{2}$ and period $\frac{2\pi}{5}$. Because the least common multiple of 2π and $\frac{2\pi}{5}$ is 2π, the period of y will be 2π. The table contains values for y_1 and y_2 as well as for $y = y_1 + y_2$.

The sketch of the graphs for all three functions is in Figure 11.13. Notice at $\frac{\pi}{2}$, or 90°, the value y was obtained by adding the graphical heights of y_1 and y_2. This same method of adding the graphical distances or heights could be performed at any value of x to get the value of y at that point.

x	0	$\frac{\pi}{6}$	$\frac{\pi}{3}$	$\frac{\pi}{2}$	$\frac{2\pi}{3}$	$\frac{5\pi}{6}$	π
$y_1 = 2 \sin x$	0	1.00	1.73	2.00	1.73	1.00	0
$y_2 = \frac{1}{2} \sin 5x$	0	0.25	−0.43	0.50	−0.43	0.25	0
$y = y_1 + y_2$	0	1.25	1.30	2.50	1.30	1.25	0

x	π	$\frac{7\pi}{6}$	$\frac{4\pi}{3}$	$\frac{3\pi}{2}$	$\frac{5\pi}{3}$	$\frac{11\pi}{6}$	2π
$y_1 = 2 \sin x$	0	−1.00	−1.74	−2.00	−1.74	−1.00	0
$y_2 = \frac{1}{2} \sin 5x$	0	−0.25	0.43	−0.50	0.43	−0.25	0
$y = y_1 + y_2$	0	−1.25	−1.31	−2.50	−1.31	−1.25	0

A different graph can be obtained by changing the two functions y_1 and y_2. If we shift y_1 to the left $\frac{\pi}{6}$ rad, we would get $y_1 = 2 \sin \left(x + \frac{\pi}{6} \right)$. If y_2 remains the same, we have the graph of $y = 2 \sin \left(x + \frac{\pi}{6} \right) + \frac{1}{2} \sin 5x$. The following table shows some of the values for this curve. Not only can you add y_1 and y_2 in the table to get the value of y, but this can be done by adding values on the graph shown in Figure 11.14.

x	0	$\frac{\pi}{6}$	$\frac{\pi}{3}$	$\frac{\pi}{2}$	$\frac{2\pi}{3}$	$\frac{5\pi}{6}$	π
$y_1 = 2 \sin \left(x + \frac{\pi}{6} \right)$	1.00	1.73	2.00	1.73	1.0	0	−1.00
$y_2 = \frac{1}{2} \sin 5x$	0	0.25	−0.43	0.50	−0.43	0.25	0
$y = y_1 + y_2$	1.00	1.98	1.57	2.23	0.57	0.25	−1.00

x	π	$\dfrac{7\pi}{6}$	$\dfrac{4\pi}{3}$	$\dfrac{3\pi}{2}$	$\dfrac{5\pi}{3}$	$\dfrac{11\pi}{6}$	2π
$y_1 = 2\sin\left(x + \dfrac{\pi}{6}\right)$	-1.00	-1.74	-2.00	-1.74	-1.00	0	1.00
$y_2 = \dfrac{1}{2}\sin 5x$	0	-0.26	0.43	-0.50	0.43	-0.25	0
$y = y_1 + y_2$	-1.00	-2.00	-1.57	-2.23	-0.57	-0.25	1.00

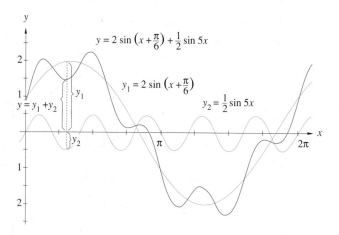

FIGURE 11.14

As the final sample graph, let's involve something more complicated than adding and subtracting two trigonometric functions. In this example, we will use a familiar curve, $y_1 = \sin 2x$, and a new curve, $y_2 = \dfrac{\sin x}{x}$.

EXAMPLE 11.9

Sketch the graph of $y = y_1 y_2$, where $y_1 = \sin 2x$, and $y_2 = \dfrac{\sin x}{x}$.

Solution The graph of y_2 is something like a sine curve. But you will notice that y_2 is not defined when $x = 0$. Also, since the values of $\sin x$ are always between $+1$ and -1, as the value of x gets larger and larger, the amplitude of y_2 will get smaller and smaller. The graph of $y_2 = \dfrac{\sin x}{x}$ is shown in Figure 11.15.

If we multiply these two functions, we get a new function, y, which is $y = (y_1)(y_2) = (\sin 2x)\left(\dfrac{\sin x}{x}\right) = \dfrac{(\sin 2x)(\sin x)}{x}$. A table of values for y_1, y_2 and $y = y_1 \cdot y_2$ follows. The curves for all three functions are shown in Figure 11.16.

x	0	$\dfrac{\pi}{3}$	$\dfrac{2\pi}{3}$	π	$\dfrac{4\pi}{3}$	$\dfrac{5\pi}{3}$	2π	$\dfrac{7\pi}{3}$	$\dfrac{8\pi}{3}$	3π
$y_1 = \sin 2x$	0	0.87	-0.87	0.02	0.87	-0.87	0.02	0.87	-0.87	0.02
$y_2 = \dfrac{\sin x}{x}$	$*$	0.83	0.41	0	-0.21	-0.17	0	0.12	0.10	0
$y = y_1 \cdot y_2$	$*$	0.71	-0.36	0	-0.18	0.14	0	0.10	-0.09	0

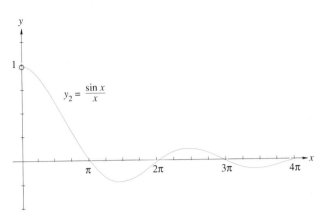

FIGURE 11.15

FIGURE 11.16

Calculator Graphics

Graphing these curves by hand takes a great deal of time. Using a calculator or a computer to draw the graphs of trigonometric curves would save time. We will briefly describe how to use a graphing calculator to graph trigonometric functions. All of the directions and illustrations are specific to a Texas Instruments *TI-81* graphics calculator. If you have another brand or model then you may have to vary the directions slightly.

While you can graph trigonometric functions with your calculator in degree mode, it is easier if you first set it in radian mode.

EXAMPLE 11.10

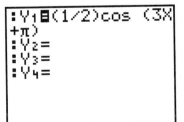

FIGURE 11.17a

Use a graphing calculator to graph $y = \frac{1}{2} \cos(3x + \pi)$.

Solution Make sure that your calculator is in radian mode. Set your calculator to the "standard" setting by pressing $\boxed{\text{ZOOM}}$ 6 $\boxed{\text{ENTER}}$. Now, press $\boxed{\text{Y=}}$ $\boxed{\text{CLEAR}}$. You are now ready to enter the function you want to graph. Press the following sequence of keys:

$$\boxed{(}\ 1\ \boxed{\div}\ 2\ \boxed{)}\ \boxed{\text{COS}}\ \boxed{(}\ 3\ \boxed{\text{X}|\text{T}}\ \boxed{+}\ \boxed{\text{2nd}}\ \boxed{\pi}\ \boxed{)}$$

The calculator screen should look like the one shown in Figure 11.17a. (Note that we could have typed 0.5 instead of $\boxed{(}\ 1\ \boxed{\div}\ 2\ \boxed{)}$.) Press $\boxed{\text{GRAPH}}$ and you should obtain the graph shown in Figure 11.17b. Compare this graph to the one in Figure 11.10.

This figure is very hard to see. We need to change the scale. We know that the amplitude of this function is $\frac{1}{2}$ and that it has a period of $\frac{2\pi}{3} \approx 2.09$. Let's change the scale as follows: Xmin $= -1$, Xmax $= 3$, Xscl $=0.7854$, Ymin $= -1$, Ymax $= 1$, Yscl $= 0.5$. Why did we select these values for Xscl and Yscl? We selected the value for Xscl because $\frac{\pi}{4} \approx 0.7854$. We selected Yscl $= 0.5$ just to help us see where the highest and lowest values of the graph are. Now press $\boxed{\text{GRAPH}}$. The result is shown in Figure 11.17c.

EXAMPLE 11.10 (Cont.)

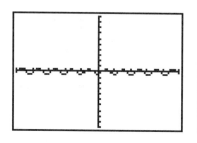

FIGURE 11.17b   FIGURE 11.17c

EXAMPLE 11.11

Use a graphing calculator to graph the product $y = (\sin 2x)\left(\dfrac{\sin x}{x}\right)$.

Solution   This is the same function we graphed in Example 11.9. We will work this so we can separately graph $y_1 = \sin 2x$ and $y_2 = \dfrac{\sin x}{x}$ and then graph their product $y = y_1 y_2$.

Based on our work in Example 11.9, we choose Xmin = 0, Xmax = 10, Xscl = 0.7854, Ymin = −1.1, Ymax = 1.1, and Yscl = 0.5. Now press $\boxed{\text{Y=}}$ $\boxed{\text{CLEAR}}$ $\boxed{\text{SIN}}$ 2 $\boxed{\text{X}\,|\,\text{T}}$ This should appear on line Y1.

To enter $y_2 = \dfrac{\sin x}{x}$, press $\boxed{\text{ENTER}}$. This moves the cursor to the next line, Y2. Now press

$$\boxed{(}\ \boxed{\text{SIN}}\ \boxed{\text{X}\,|\,\text{T}}\ \boxed{)}\ \boxed{\div}\ \boxed{\text{X}\,|\,\text{T}}$$

FIGURE 11.18a

We will put $y = y_1 y_2$ on line Y3. Rather than entering both y_1 and y_2 again, we will let the calculator do this for us. Press $\boxed{\text{ENTER}}$ to move the cursor to line Y3 and press the following key sequence:

$$\boxed{\text{2nd}}\ \boxed{\text{Y-VARS}}\ \boxed{\text{ENTER}}\ \boxed{\text{2nd}}\ \boxed{\text{Y-VARS}}\ 2$$

Notice that line Y3 now reads Y3=Y1Y2. The calculator screen should look like that in Figure 11.18a. Press $\boxed{\text{GRAPH}}$. You should see each of the functions y_1, y_2, and y graphed in sequence, with the final result like that shown in Figure 11.18b.

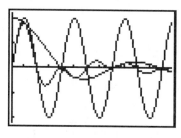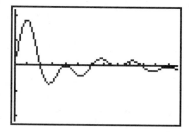

FIGURE 11.18b   FIGURE 11.18c

This is too cluttered. Let's change it so we see only the graph of $y = y_1 y_2$. Press $\boxed{\text{Y=}}$ $\boxed{\blacktriangleleft}$ to move the cursor on top of the equals sign. The equals sign is white on a black background, indicating that the calculator will graph this function. Press

EXAMPLE 11.11 (Cont.)

$\boxed{\text{ENTER}}$. Now the equals sign is black on a white background. This shows that the calculator will not graph Y1. Press $\boxed{\blacktriangledown}$ to move the cursor to the equals sign on line Y2 and press $\boxed{\text{ENTER}}$ to tell the calculator not to graph Y2. Now press $\boxed{\text{GRAPH}}$.

The result, shown in Figure 11.18c, is the graph of $y = y_1 y_2 = (\sin 2x) \left(\dfrac{\sin x}{x} \right)$.

Exercise Set 11.3

In Exercises 1–18, sketch the given curves. After you have sketched the curve, use a graphing calculator or a computer graphics program to check your sketch.

1. $y = \sin x - \cos x$

2. $y = \sin 2x + 2 \cos x$

3. $y = \dfrac{1}{3} \sin 4x + \sin 2x$

4. $y = 3 \sin \dfrac{x}{3} + \cos 2x$

5. $y = x + \sin x$

6. $y = \sin x - \sin 2x$

7. $y = \cos 2x - \cos x$

8. $y = x^2 + \cos x$

9. $y = \sin \pi x - \cos 2x$

10. $y = \cos \dfrac{\pi}{2} x + 2 \sin 3x$

11. $y = \cos 2x + \sin \left(x - \dfrac{\pi}{3} \right)$

12. $y = \dfrac{1}{4} \cos \left(x - \dfrac{\pi}{6} \right) + 4 \sin \left(x + \dfrac{\pi}{6} \right)$

13. $y = 4 \cos 2\pi x + \sin \left(\dfrac{\pi x}{2} + \dfrac{\pi}{4} \right)$

14. $y = \dfrac{\cos x}{x^2 + 1} - \sin x$

15. $y = \dfrac{\sin x}{x^2 + 1} + 2 \cos \left(\dfrac{x}{3} - \dfrac{\pi}{2} \right)$

16. $y = 2 \sin \left(3x + \dfrac{\pi}{4} \right) + 5 \cos \left(2x - \dfrac{\pi}{3} \right)$

17. $y = x \sin x$

18. $y = 3(x + \pi) \cos \left(x - \dfrac{\pi}{4} \right)$

☰ 11.4
GRAPHS OF THE OTHER TRIGONOMETRIC FUNCTIONS

We have learned how to graph two of the trigonometric functions, $y = \sin x$ and $y = \cos x$. We can change the frequency, amplitude, and displacement of each of them. In this section, we will look at the graphs of the other four trigonometric functions.

The sin and cos functions that we have graphed were each bounded. That means that there was a limit, or bound, on how large or small the value of each function could get. For the sine and cosine functions, this bound was called the amplitude.

The four remaining trigonometric functions (tangent, cotangent, secant, cosecant) are not bounded. Thus, there is no limit to how large or small each function can become.

Graph of the Tangent Function ▬▬▬▬▬

We will begin with the last of the basic trigonometric functions, the tangent function. A table of values for the tangent function, $y = \tan x$, follows. A graph based on these

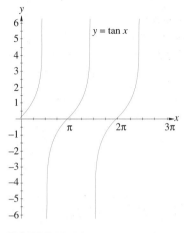

FIGURE 11.19

values is in Figure 11.19.

x	0	$\dfrac{\pi}{6}$	$\dfrac{\pi}{3}$	$\dfrac{\pi}{2}$	$\dfrac{2\pi}{3}$	$\dfrac{5\pi}{6}$	π
$y = \tan x$	0	0.58	1.73	$*$	-1.73	-0.58	0

x	π	$\dfrac{7\pi}{6}$	$\dfrac{4\pi}{3}$	$\dfrac{3\pi}{2}$	$\dfrac{5\pi}{3}$	$\dfrac{11\pi}{6}$	2π	$\dfrac{13\pi}{6}$	$\dfrac{7\pi}{3}$	$\dfrac{5\pi}{2}$
$y = \tan x$	0	0.58	1.73	$*$	-1.73	-0.58	0	0.58	1.73	$*$

*The tangent at this point is not defined.

Notice that the function is not defined when $x = \dfrac{\pi}{2}, \dfrac{3\pi}{2}, \dfrac{5\pi}{2}$, and so on. The vertical lines at each of these values of x that the graph of tangent approaches but never touches are called the **vertical asymptotes**. You might also note that the period of the tangent curve is π. (Remember, the sine and cosine have periods of 2π.)

Graphs of Cotangent, Secant, and Cosecant Functions ▬

In a similar procedure, we could graph the three remaining trigonometric functions: cotangent, secant, and cosecant. The graphs for each of these functions are given in Figures 11.20, 11.21, 11.22, respectively. Notice that the period of the cotangent is also π. The period of the secant and cosecant functions is 2π. $\Big($Remember that $\csc x = \dfrac{1}{\sin x}$ and $\sec x = \dfrac{1}{\cos x}$, and both sine and cosine have period 2π.$\Big)$

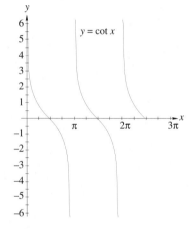

FIGURE 11.20

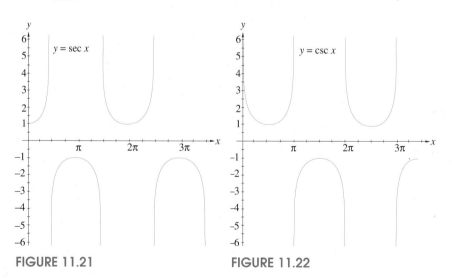

FIGURE 11.21   **FIGURE 11.22**

Variations in the Graphs of Tan, Cot, Sec, and Csc ▬▬▬

The terms period and frequency have the same meaning for these four functions as they did for the sine and cosine functions. The tangent and cotangent each have period π. The secant and cosecant have period 2π. Amplitude has no meaning for

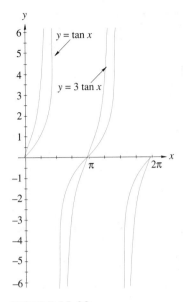

FIGURE 11.23

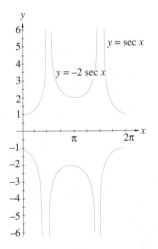

FIGURE 11.24

these functions. Each of these graphs has vertical asymptotes in which the function is not defined.

Sketching a function such as $y = 3\tan x$ or $y = -2\sec x$ is not difficult if you know the shapes of the six basic trigonometric functions. Knowing the shape means knowing where each graph crosses the x-axis (if it does), where its high and low points are, and where the function is not defined.

We will sketch both $y = 3\tan x$ and $y = -2\sec x$ to give you an idea of how to proceed. The graph for $y = 3\tan x$ is shown in Figure 11.23. The graph for $y = -2\sec x$ is shown in Figure 11.24. First, sketch the basic curve for each function. For $y = 3\tan x$, sketch $y = \tan x$. For $y = -2\sec x$, sketch $y = \sec x$. Now, multiply the y-values of each curve by the appropriate value and plot these points. For $y = \tan x$, multiply each y-value by 3 to get $y = 3\tan x$. For $y = \sec x$, multiply each y-value by -2 to get $y = -2\sec x$.

A summary of the period, amplitude, phase shift, and vertical asymptotes for all of the six trigonometric functions is given in Table 11.1.

TABLE 11.1

Function	Period	Amplitude	Phase Shift	Vertical asymptotes
$y = A\sin(Bx + C)$	$\dfrac{2\pi}{\|B\|}$	$\|A\|$	$-\dfrac{C}{B}$	none
$y = A\cos(Bx + C)$	$\dfrac{2\pi}{\|B\|}$	$\|A\|$	$-\dfrac{C}{B}$	none
$y = A\tan(Bx + C)$	$\dfrac{\pi}{\|B\|}$	*	$-\dfrac{C}{B}$	$Bx + C = n\pi + \dfrac{\pi}{2}$
$y = A\cot(Bx + C)$	$\dfrac{\pi}{\|B\|}$	*	$-\dfrac{C}{B}$	$Bx + C = n\pi$
$y = A\sec(Bx + C)$	$\dfrac{2\pi}{\|B\|}$	*	$-\dfrac{C}{B}$	$Bx + C = n\pi + \dfrac{\pi}{2}$
$y = A\csc(Bx + C)$	$\dfrac{2\pi}{\|B\|}$	*	$-\dfrac{C}{B}$	$Bx + C = n\pi$

*Not defined

Calculator Graphics

EXAMPLE 11.12

Use a graphing calculator to sketch the graph of $y = -\dfrac{1}{3}\tan\left(2x + \dfrac{\pi}{4}\right)$.

Solution   Make sure your calculator is in radian mode. From Table 11.1, we see that the period is $\dfrac{\pi}{2} \approx 1.57$. We would like to see two periods of this graph, so we set Xmin $= -2$, Xmax $= 2$, Xscl $= 0.7854$, Ymin $= -4$, Ymax $= 4$, and Yscl $= 1$. Now press [Y=] [CLEAR] and then the following sequence of keys:

$$\boxed{(-)}\ \boxed{(}\ \boxed{1}\ \boxed{\div}\ \boxed{3}\ \boxed{)}\ \boxed{\text{TAN}}\ \boxed{(}\ \boxed{2}\ \boxed{\text{X}\,|\,\text{T}}\ \boxed{+}\ \boxed{(}\ \boxed{\text{2nd}}\ \boxed{\pi}\ \boxed{\div}\ \boxed{4}\ \boxed{)}\ \boxed{)}$$

The calculator screen should now look like the one shown in Figure 11.25a. Press [GRAPH] and you should obtain the graph in Figure 11.25b.

EXAMPLE 11.12 (Cont.)

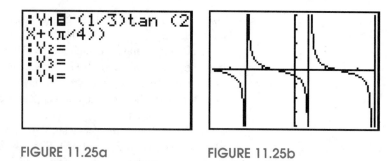

FIGURE 11.25a   FIGURE 11.25b

Notice that the calculator seems to have drawn in the vertical asymptotes. Look closely. These are not vertical lines. The portion above the x-axis is not directly above the portion below the x-axis. These apparent vertical asymptotes are put in by some graphing calculators and computer graphing programs.

EXAMPLE 11.13

Use a graphing calculator to sketch $y = 1.5 \sec (x + 2)$.

Solution   We set $Xmin = -1, Xmax = 7, Xscl = 1, Ymin = -5, Ymax = 5$, and $Yscl = .5$. Press $\boxed{Y=}$ $\boxed{CLEAR}$ to erase the previously graphed function. A calculator does not have a secant key, so before you graph this function you will need to use the reciprocal identity $\sec x = \dfrac{1}{\cos x}$ to rewrite this function as $y = \dfrac{1.5}{\cos (x + 2)}$.
Now, press the following sequence of keys:

$$1.5 \boxed{\div} \boxed{(} \boxed{COS} \boxed{(} \boxed{X\,T} \boxed{+} 2 \boxed{)} \boxed{)}$$

Press $\boxed{GRAPH}$ and you should obtain the graph in Figure 11.26. Once again, you can see that the calculator seems to have drawn in the vertical asymptotes.

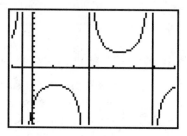

FIGURE 11.26

Exercise Set 11.4

In Exercises 1–18, sketch the given curves. Then check your sketch using a graphing calculator or a computer graphing program.

1. $y = 3 \tan x$

2. $y = -4 \tan x$

3. $y = \frac{1}{2} \sec x$

4. $y = 3 \csc x$

5. $y = -\frac{1}{2} \cot x$

6. $y = -2 \csc x$

7. $y = \tan \left(x + \frac{\pi}{4} \right)$

8. $y = \sec \left(x - \frac{\pi}{3} \right)$

9. $y = \cot \left(x + \frac{\pi}{6} \right)$

10. $y = \csc \left(x - \frac{2\pi}{5} \right)$

11. $y = \tan 2x$

12. $y = \sec 3x$

13. $y = \csc 4x$

14. $y = \cot 3x$

15. $y = 4 \tan \left(2x + \frac{\pi}{3} \right)$

16. $y = -3 \cot \left(5x - \frac{\pi}{6} \right)$

17. $y = \frac{1}{2} \sec \left(3x + \frac{\pi}{4} \right)$

18. $y = -\frac{1}{3} \csc \left(2x + \frac{\pi}{6} \right)$

≡ 11.5
APPLICATIONS OF TRIGONOMETRIC GRAPHS

There are many applications of trigonometric functions and their graphs. In this section we will examine some of them.

Periodic and Simple Harmonic Motion

One area in which period, frequency, amplitude, and displacement are applied is physics. **Periodic motion** is when an object repeats a certain motion indefinitely, always returning to its starting point after a constant time interval and then beginning a new cycle.

Simple harmonic motion is one type of periodic motion. One form of simple harmonic motion is a straight line. The acceleration is proportional to the displacement and in the opposite direction. Examples of simple harmonic motion include a weight bobbing on a spring, a simple pendulum, and a boat bobbing on water.

Simple Harmonic Motion

If a point moves so that its displacement varies with time t, measured in seconds, the point is in equilibrium when $t = 0$. Its displacement at any time t is given by

$$y = A \sin 2\pi f t$$

or $\quad y = A \sin \omega t$

where $\omega = 2\pi f$. Then the point is in **simple harmonic motion**.

As you can see from these equations, the amplitude is $|A|$. The Greek letter omega, ω, represents the angular velocity in radians per second. The period is $\frac{2\pi}{\omega}$ and the frequency f is $\frac{\omega}{2\pi}$. The frequency is measured in cycles per second and the units are hertz (Hz), kilohertz (kHz), and megahertz (MHz). One **hertz** is equivalent to 1 cycle per second.

Application

EXAMPLE 11.14

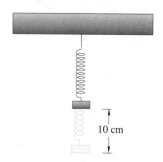

FIGURE 11.27

An object is suspended from a spring. The object is pulled down 10 cm and released. It oscillates in simple harmonic motion. If it takes 0.2 s for it to complete one oscillation, and if we let $t = 0$ when it is at rest, find the amplitude, period, frequency, and angular velocity. Write an equation to describe the motion. (See Figure 11.27.)

Solution   We are given $A = 10$ cm and period $= 0.2$ s. Since the frequency is $\dfrac{1}{\text{period}}$, we can determine that the frequency is $\dfrac{1}{0.2} = 5$ oscillations/second. Now, the angular velocity ω is $2\pi f = 2\pi(5) = 10\pi$ rad/s. Since $\omega = 10\pi$, and $A = 10$, the equation that describes this motion is $y = 10 \sin 10\pi t$.

Simple harmonic motion can also be described using the cosine function. If a point is at its maximum displacement when $t = 0$, and the rest of the previously mentioned conditions are met, then the displacement of the point at any time t can be expressed by

$$y = A \cos 2\pi f t$$

or $$y = A \cos \omega t$$

If a point is in simple harmonic motion, but is not in equilibrium or at its maximum value when $t = 0$, the sine or cosine curves can be used with the appropriate phase shift. Thus, $y = A \sin(\omega t + \phi)$ could be the equation for an object in simple harmonic motion with a phase shift of ϕ.

Alternating Current

Another common use of trigonometric curves is in the study of alternating current ac in electricity. The voltage of an ac generator varies with time t. The curve of the output voltage for a generator is a sine curve. Voltage varies with time according to these formulas.

$$V = V_{\max} \sin 2\pi f t$$

or $$V = V_{\max} \sin \omega t$$

In these formulas, $V_{\max}$ is the maximum value of the voltage and corresponds to the amplitude of the sine wave. The **angular frequency** of the alternating current is represented by $\omega = 2\pi f$ and is in rad/s. As before, f represents the frequency of the voltage in hertz.

In a similar manner, the current in an ac circuit varies with time in the same manner as the voltage. If the maximum value of the current is $I_{\max}$, then the current I is represented by

$$I = I_{\max} \sin 2\pi f t$$

or $$I = I_{\max} \sin \omega t$$

Application

EXAMPLE 11.15

Maximum voltage across a power line is 170 V at a frequency of 60 Hz. What is the amplitude, period, and angular frequency of this voltage?

Solution   The amplitude is $|V_{\max}| = 170$ V. Since the frequency is 60 Hz or 60 cycles/s and the period is $1/f$, the period is $\frac{1}{60}$ s/cycle. The angular frequency is $\omega = 2\pi f = 2\pi(60) = 120\pi$ rad/s.

Application

EXAMPLE 11.16

If household current has a frequency of 60 Hz and a maximum of 2.5 A, find the amplitude of the current.

Solution   The amplitude is $|I_{\max}| = 2.5$ A.

Application

EXAMPLE 11.17

In the household current of Example 11.16, what is the current when $t = 1.32$ s?

Solution   From Example 11.16 we know that $I_{\max} = 2.5$ A and $f = 60$ Hz. So, $I = I_{\max} \sin 2\pi f t = 2.5 \sin 120\pi t$ describes the current. When $t = 1.32$ s, $I = 2.378$ A.

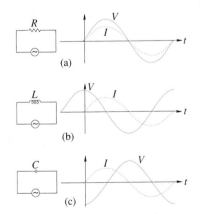

(a)

(b)

(c)

FIGURE 11.28

There are three different relationships that exist between a voltage V and a current I in an ac circuit. These three relationships are shown in Figure 11.28. In an ac circuit that contains only resistance, the current and voltage are **in phase** with each other. This means that both are 0 at the same time and both reach their maximum value at the same time. This is shown in Figure 11.28a.

In an ac circuit containing only inductance, the voltage leads the current by $\frac{1}{4}$ cycle, as shown in Figure 11.28b. In this case, the current and voltage are said to be $90° \left(\frac{\pi}{2}\right)$ rad **out of phase**, since $90°$ is $\frac{1}{4}$ cycle. This is also referred to by stating that the current **lags** the voltage by $90°$.

Figure 11.28c demonstrates the relationship between the current and the voltage in a circuit that contains only capacitance. The current **leads** the voltage by $90° \left(\frac{\pi}{2} \text{ rad}\right)$ or $\frac{1}{4}$ cycle.

In an ac circuit that contains resistance, inductance, and capacitance in series, the instantaneous voltages across the circuit elements are V_R, V_L, and V_C, respectively. At any moment, the applied voltage $V = V_R + V_L + V_C$. Because V_R, V_L, and V_C are out of phase with one another, this formula holds only for instantaneous voltages. The phase angle ϕ can be determined from the relationship $\tan \phi = \dfrac{V_L - V_C}{V_R}$ or

$$\phi = \tan^{-1}\left(\frac{V_L - V_C}{V_R}\right).$$

Application

EXAMPLE 11.18

A capacitor, inductor, and resistor are connected in series with a 120-V, 60-Hz power source. If $V_{R_{max}} = 135$ V, $V_{L_{max}} = 190$ V, and $V_{C_{max}} = 90$ V, graph the instantaneous voltage curve for the capacitance, inductance, resistance, and applied voltage and determine the phase angle.

Solution   The equation for the resistance is determined from the equation $V_R = V_{R_{max}} \sin \omega t$. Since $V_{R_{max}} = 135$ V and $\omega = 2\pi f = 120\pi$, we have $V_R = 135 \sin 120\pi t$. The inductance leads the current by 90° or $\frac{\pi}{2}$ rad, so $V_L = 190 \sin \left(120\pi + \frac{\pi}{2} \right)$.

Similarly, since the capacitance lags the current by 90°, or $\frac{\pi}{2}$, we have $V_C = 90 \sin \left(120\pi - \frac{\pi}{2} \right)$. Each of these curves and $V = V_R + V_L + V_C$ are shown in Figure 11.29. The phase angle ϕ is determined from $\tan \phi = \dfrac{V_L - V_C}{V_R} = \dfrac{190 - 90}{135} = 0.74$, so $\phi = \tan^{-1} 0.74 = 36.5°$ or 0.638 rad.

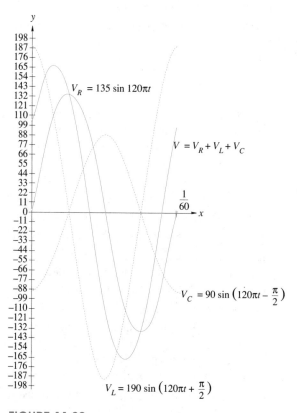

FIGURE 11.29

Exercise Set 11.5

Solve Exercises 1–20.

1. *Mechanics*   A weight vibrates on a spring in simple harmonic motion. The amplitude is 10 cm and the frequency is 4 Hz. Write an equation for the weight's position y, at any time t, if $y = 0$ when $t = 0$ s.

2. *Mechanics*   Write an equation for the position of the weight in Exercise 1 at any time t if $y = 8$ when $t = 0.1$ s.

3. *Oceanography*   A raft bobs on the water in simple harmonic motion. Write a formula for the raft's position if the amplitude is 0.8 m, $\omega = \frac{\pi}{3}$ rad/s, and $y = 0$ when $t = 0$ s.

4. *Oceanography*   What is the frequency of the raft in Exercise 3?

5. *Physics*   The acceleration a of a pendulum is given by $a = -g \sin \theta$, where g is gravitational acceleration and θ is the angular displacement from the vertical. What is the acceleration of a pendulum, when $\theta = 5°$ and $g = 32$ ft/s^2?

6. *Physics*   What is the acceleration of a pendulum when $\theta = 0.05$ rad and $g = 9.8$ m/s^2?

7. *Mechanics*   A weight hanging from a spring vibrates in simple harmonic motion according to the equation $y = 8.5 \cos 2.8t$, where y is the position of the weight in centimeters and t is time in seconds. (a) What is the amplitude of the vibrating weight? (b) What is its period? (c) What is its frequency? (d) Sketch one complete cycle of this curve.

8. *Automotive technology*   Each piston in a certain engine has stroke or total travel distance of 10.50 cm. (a) If the engine is operating at 3 000 rpm, what is the frequency of oscillation for each piston? (b) Write an equation to represent the position of the piston at any second t, if we assume the piston was at its maximum height (TDC), when $t = 0$. (c) If the engine speed is increased to 4 500 rpm, what equation would represent the position of the piston in (b) at any time t?

9. *Electricity*   In an ac circuit, the current I is given by the relation $I = 10 \sin 120\pi t$. What are the amplitude, period, frequency, and angular velocity?

10. *Electricity*   In an ac circuit, write the equation of the sine curve for the current when $I_{max} = 6.8$ A and $f = 80$ Hz.

11. *Electricity*   In the circuit in Exercise 10, what is the equation of the sine curve when the phase angle is $\frac{\pi}{3}$ rad?

12. *Electricity*   What is the equation of the sine curve for the voltage in an ac circuit, when $V_{max} = 220$ V and $f = 40$ Hz?

13. *Electricity*   What is the equation of the sine curve for the voltage in Exercise 12, when the phase angle is $-\frac{\pi}{3}$ rad?

14. *Electricity*   What is the equation of the cosine curve for the voltage in Exercise 12?

15. *Electricity*   What is the equation of the cosine curve for the voltage in Exercise 13?

16. *Medical technology*   The voltage for an x-ray can have a maximum value of 250 000 V. If the frequency of an x-ray is 10^{19} Hz, write the equation of the sine curve for the voltage.

17. *Physics*   A gamma ray can be described by the equation

$$y = 10^{-12} \sin 2\pi 10^{23} t$$

What are the frequency and amplitude of this gamma ray?

18. *Electricity*   In an ac circuit containing only a constant capacitance, the current leads the voltage by $\frac{1}{4}$ cycle. If $V = V_{max} \sin 2\pi ft$, $I_{max} = 3.6$ A and $f = 60$ Hz, write the equation for the current. Sketch one cycle of the curve.

19. *Electricity*   In a resistance-inductance circuit, $I = I_{max} \sin(2\pi ft + \phi)$. If $I_{max} = 1.2$ A, $f = 400$ Hz, and $\phi = 37°$, sketch I vs t for one cycle.

20. *Electricity*   A capacitor, inductor, and resistor are connected in series with a 120-V, 60-Hz power source. If $V_R = 120$ V, $V_L = 80$ V, and $V_C = 130$ V, graph the instantaneous voltage curves for the capacitance, inductance, resistance, and applied voltage, and determine the phase angle.

≡ 11.6
PARAMETRIC EQUATIONS

Graphing an equation in terms of two variables x and y is often done by setting up a table of values. We usually select a value for x or y and solve for the other variables. In the case of a function such as $y = x^2$ or $y = x^3 + 2x$, this has not been a problem. We have had difficulty with equations that were not functions, such as $x^2 + y^2 = 9$.

Parametric Equations

One solution is to rewrite this equation by expressing x and y as functions of a third variable, called a **parameter**. These equations are called **parametric equations**.

EXAMPLE 11.19

Describe and sketch the curve represented by the parametric equations

$$x = 2t, \quad y = t^2 - 4$$

Solution The parameter for these equations is t. We will set up a table of values for t, x, and y. The graph is shown in Figure 11.30. We connect the points in order of the increasing values of t, as indicated by the arrows in Figure 11.30.

t	-4	-3	-2	-1	0	1	2	3	4
x	-8	-6	-4	-2	0	2	4	6	8
y	12	5	0	-3	-4	-3	0	5	12

Does the curve look familiar? It looks very much like some of the equations that we graphed earlier. It is possible to eliminate the parameter and write the equation in rectangular form. In these two equations, since $x = 2t$, then $t = \frac{x}{2}$. Substituting this value for t in the equation $y = t^2 - 4$, we get $y = \frac{x^2}{4} - 4$ or $y = \frac{1}{4}x^2 - 4$.

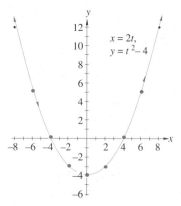

FIGURE 11.30

When eliminating the parameter, you must be very careful about the domain. Consider the following example.

EXAMPLE 11.20

Describe and sketch the curve represented by the parametric equations $x = 3 \cos t$ and $y = \cos 2t$.

Solution We will set up a table of values for t, x, and y.

t	0	$\frac{\pi}{6}$	$\frac{\pi}{4}$	$\frac{\pi}{3}$	$\frac{\pi}{2}$	$\frac{2\pi}{3}$	$\frac{3\pi}{4}$	$\frac{5\pi}{6}$	π	$\frac{7\pi}{6}$	$\frac{5\pi}{4}$	$\frac{4\pi}{3}$	$\frac{3\pi}{2}$
x	3	2.6	2.1	1.5	0	-1.5	-2.1	-2.6	-3	-2.6	-2.1	-1.5	0
y	1	0.5	0	-0.5	-1	-0.5	0	0.5	1	0.5	0	-0.5	-1

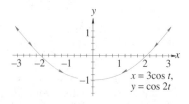

FIGURE 11.31

Notice that these values begin to repeat once we get to $t = \pi$. The sketch of this curve is shown in Figure 11.31. Again, we get a shape similar to the one in Example 11.19, except that the curve does not continue in each direction. It oscillates back and forth. Using techniques we will develop later, we can eliminate the parameter

EXAMPLE 11.20 (Cont.)

to form the rectangular equation $y = \dfrac{2x^2}{9} - 1$, where $-3 \le x \le 3$. Notice the restriction on the domain of x.

A simpler example demonstrating how the domain is often restricted when parametric equations are written in the rectangular form is shown by the next example.

EXAMPLE 11.21

Describe and sketch the curve represented by the parametric equations $x = t^2$ and $y = 2t^2$.

Solution Since $x = t^2$ and $y = 2t^2$, we see that the rectangular form is $y = 2x$. This is the equation of a straight line. Notice, however, that for all values of t, $x \ge 0$ and $y \ge 0$. The graph formed by these parametric equations is shown in Figure 11.32 as a solid line. The remainder of the curve $y = 2x$ is indicated by the dashed line.

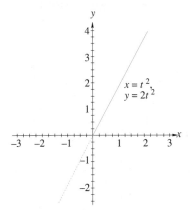

FIGURE 11.32

Lissajous Figures

When the parametric equations of a point describe simple harmonic motion, the resulting curve is called a **Lissajous figure**. When voltages of different frequencies are applied to the vertical and horizontal plates of an oscilloscope, a Lissajous figure results.

EXAMPLE 11.22

Sketch the graph of the parametric equations $x = \cos t$ and $y = \sin 3t$.

Solution The Lissajous curve for these parametric equations is shown in Figure 11.33.

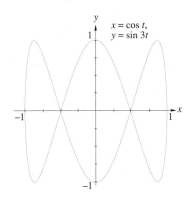

FIGURE 11.33

Examine the Lissajous figure in Figure 11.33. There are three loops along the top edge of the figure and one loop along the side. This is a ratio of 3:1. Now look at the frequencies of the parametric equations that generated this curve. The frequency of $x = \cos t$ is $\dfrac{1}{2\pi}$ and the frequency of $y = \sin 3t$ is $\dfrac{3}{2\pi}$. The ratio of the frequencies is 3:1. In Exercise Set 11.6 we will predict the number of loops on the top and side from the parametric equations and then graph the curve. While this is only an exercise in this text, it is a technique that can be used to calibrate signal generators.

Using a Calculator to Graph Parametric Equations

Before you can use a graphing calculator to draw the graphs of parametric equations, you need to put the calculator in parametric mode. On a *TI-81* this is done by pressing MODE ▼ ▼ ▼ ▶ ENTER. The screen on a *TI-81* should look like the one shown

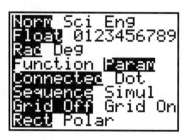

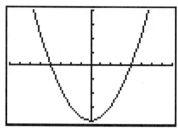

FIGURE 11.34   FIGURE 11.35

in Figure 11.34. On a Casio *fx-7700G*, press the key sequence $\boxed{\text{MODE}}$ $\boxed{\text{SHIFT}}$ $\boxed{\times}$, to set the calculator in parametric mode.

EXAMPLE 11.23

Use a graphing calculator to sketch the curve represented by the parametric equations $x = 2t$ and $y = t^2 - 4$.

Solution   These are the same parametric equations we graphed in Example 11.19. We will use the following table from that example to help graph these equations.

t	-4	-3	-2	-1	0	1	2	3	4
x	-8	-6	-4	-2	0	2	4	6	8
y	12	5	0	-3	-4	-3	0	5	12

Press $\boxed{\text{RANGE}}$. On a *TI-81*, you are first asked for Tmin, Tmax, and Tstep. Based on the table, we will let Tmin $= -4$ and Tmax $= 4$. The value of Tstep determines how much the value of t should be increased before calculating new values for x and y. We will pick Tstep $= 0.5$. You may want to try different values. Based on this table, let Xmin $= -8$, Xmax $= 8$, Xscl $= 1$, Ymin $= -4$, Ymax $= 4$, and Yscl $= 1$. A Casio *fx-7700G* calculator asks for these same values, but Tmin, Tmax, and Tptch are requested at the end, rather than at the beginning.

Now press $\boxed{\text{Y} =}$. On the first line, enter the right-hand side of the parametric equation for x. Press

$$2 \boxed{\text{X} \mid \text{T}}$$

Notice that when you press the $\boxed{\text{X} \mid \text{T}}$ key, a T is displayed on the screen. This will always happen when the calculator is in parametric mode.

Now enter the parametric equation for y. Press $\boxed{\text{ENTER}}$ to move the cursor to the line labeled Y_{1T} and press the key sequence

$$\boxed{\text{X} \mid \text{T}} \boxed{x^2} \boxed{-} 4$$

Then press $\boxed{\text{GRAPH}}$. The result is displayed in Figure 11.35.

EXAMPLE 11.24

Use a graphing calculator to sketch the curve represented by the parametric equations $x = 2\cos t$ and $y = \sin t$.

Solution Because the periods of sin and cos are 2π, we will let Tmin $= 0$, Tmax $= 6.3$, and Tstep $= 0.1$. Since the range of $x = 2\cos t$ is $[-2, 2]$, we choose Xmin $= -2$, Xmax $= 2$, and Xscl $= 1$. The range of $y = \sin t$ is $[-1, 1]$. We let Ymin $= -1.5$, Ymax $= 1.5$, and Yscl $= 1$. Now press $\boxed{\text{Y=}}$ and enter the parametric equations by pressing

$$2\ \boxed{\text{COS}}\ \boxed{\text{X\,|\,T}}\ \boxed{\text{ENTER}}\ \boxed{\text{SIN}}\ \boxed{\text{X\,|\,T}}\ \boxed{\text{GRAPH}}$$

The result is displayed in Figure 11.36.

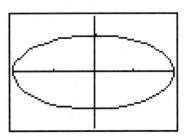

FIGURE 11.36

Exercise Set 11.6

Graph each curve given by the parametric equations in Exercises 1–14. Make a table of values for at least six values of t. Show how the figure was drawn by drawing arrows on it. Eliminate the parameters in Exercises 1–6 and write the equation as a function of y.

1. $x = t, y = 3t$

2. $x = 2t, y = 4t + 1$

3. $x = t, y = \dfrac{1}{t}$

4. $x = t + 5, y = 3t - 2$

5. $x = 3 - t, y = t^2 - 9$

6. $x = t + 2, y = t^2 - t$

7. $x = 2\sin t, y = 2\cos t$

8. $x = 5\sin t, y = 2\cos t$

9. $x = 5\sin t, y = 3\sin t$

10. $x = 2\cos t, y = 6\cos t$

11. $x = t - \sin t, y = 1 - \cos t$

12. $x = \tan t, y = 6\cot t$

13. $x = \sec t, y = 2\csc t$

14. $x = 3\sec t, y = \tan t$

Examine the pairs of parametric equations in Exercises 15–24. For each pair, (a) predict the ratio of the loops along the top to the number of loops along the side, then (b) graph each of these curves. If possible, use a graphing calculator or computer graphing program to graph the curves.

15. $x = \sin t, y = \cos t$

16. $x = \sin 2t, y = \cos t$

17. $x = \sin t, y = \cos 2t$

18. $x = \sin 3t, y = \cos t$

19. $x = \sin 4t, y = \cos t$

20. $x = \sin 4t, y = \cos 2t$

21. $x = \sin 3t, y = \cos 2t$

22. $x = \sin 4t, y = \cos 3t$

23. $x = \sin 5t, y = \cos 2t$

24. $x = \sin 5t, y = \cos 3t$

Exercises 25–30 are Lissajous curves. For these curves, we have changed the amplitude.

25. $x = 2 \sin t, y = \cos t$

26. $x = 4 \sin t, y = \cos t$

27. $x = 3 \sin 2t, y = 2 \cos t$

28. $x = 3 \sin 3t, y = \cos t$

29. $x = \sin 3t, y = 4 \cos t$

30. $x = 2 \sin 2t, y = 5 \cos 3t$

☰ 11.7
POLAR COORDINATES

Every time we have represented a point in a plane, we used the rectangular coordinate system. Each point has an x- and a y-coordinate. There is another type of coordinate system that is used to represent points. This system is called the **polar coordinate system**.

To introduce a system of polar coordinates in a plane, we begin with a fixed point O called the **pole** or **origin**. From the pole, we shall draw a half-line that has O as its endpoint. This half-line will be called the **polar axis**.

Polar Coordinates

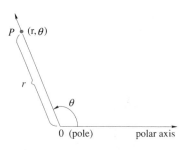

FIGURE 11.37

Consider any point in the plane different from point O. Call this point P. Let the polar axis form the initial side of an angle and $\overrightarrow{OP}$ the terminal side. This angle has measure θ. If the distance from O to P is r, we can say that the **polar coordinates** of P are (r, θ), as shown in Figure 11.37.

We have some of the same understanding about polar coordinates as we do about the angles from trigonometry. If the angle is generated by a counterclockwise rotation of the polar axis, the angle is positive. If it is generated by a clockwise rotation, θ is negative. If r is negative, the terminal side of the angle is extended in the opposite direction through the pole and is measured off r units on this extended side.

A special type of graph paper, called **polar coordinate paper**, is used to graph polar coordinates. This paper has concentric circles with their centers at the pole. The distance between any two consecutive circles is the same. Lines are drawn through the pole and correspond to some of the common angles. An example of polar coordinate paper is shown in Figure 11.38.

EXAMPLE 11.25

Plot the points with the following polar coordinates: $\left(5, \frac{\pi}{4}\right), \left(-5, \frac{\pi}{4}\right), (3, 150°),$ $(3, -70°), (6, \pi), (4, 0°),$ and $(6, 450°)$.

Solution   The points are plotted in Figure 11.39.

Notice that there is nothing unique about these points. For example, the point $(6, 450°)$ would be the same as the points $(6, 90°), (6, -270°),$ or $(-6, 270°)$. The pole has the polar coordinate $(0, \theta)$, where θ can be any angle.

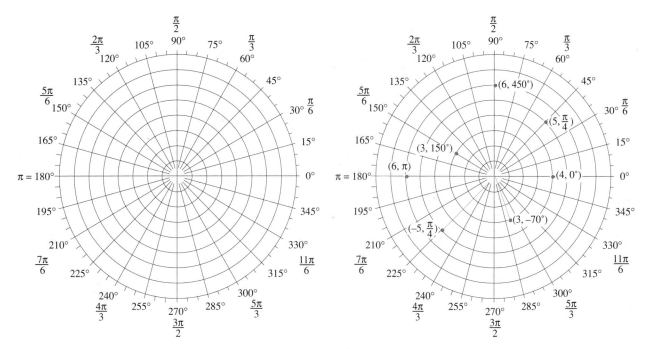

FIGURE 11.38   FIGURE 11.39

Converting Between Polar and Rectangular Coordinates ▪

Converting between the polar coordinate system and the rectangular coordinate system requires the use of trigonometry.

> **Converting Polar Coordinates to Rectangular Coordinates**
>
> If the point P has polar coordinates (r, θ) and rectangular coordinates (x, y), then
>
> $$x = r \cos \theta$$
>
> $$y = r \sin \theta$$
>
> This converts the polar coordinates to rectangular coordinates.

EXAMPLE 11.26

Find the rectangular coordinates of the point with the polar coordinates $\left(8, \dfrac{5\pi}{6} \right)$.

Solution   From the equations previously given, we have $x = r \cos \theta$ and $y = r \sin \theta$. In this example, $r = 8$ and $\theta = \dfrac{5\pi}{6}$. So, $x = 8 \left(\dfrac{-\sqrt{3}}{2} \right) = -4\sqrt{3} \approx -6.93$ and $y = 8 \left(\dfrac{1}{2} \right) = 4$. The rectangular coordinates are $\left(-4\sqrt{3}, \, 4 \right)$.

To convert from rectangular to polar coordinates, we need to use the Pythagorean theorem.

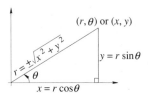

FIGURE 11.40

> **Converting Rectangular Coordinates to Polar Coordinates**
>
> The equations
>
> $$r^2 = x^2 + y^2 \qquad \text{or} \qquad r = \pm\sqrt{x^2 + y^2}$$
>
> $$\tan\theta = \frac{y}{x}, \; x \neq 0$$
>
> will convert rectangular coordinates to polar coordinates. (See Figure 11.40.)

EXAMPLE 11.27

Find polar coordinates of $\left(-4\sqrt{3},\, 4\right)$.

Solution This is the reverse of Example 11.26, so we know that the answer should be $\left(8,\, \dfrac{5\pi}{6}\right)$. Let's practice using the two conversion formulas $r^2 = x^2 + y^2$ and $\tan\theta = \dfrac{y}{x}$.

Here, $x = -4\sqrt{3}$ and $y = 4$.

$$r^2 = x^2 + y^2$$

$$= \left(-4\sqrt{3}\right)^2 + 4^2$$

$$= 48 + 16 = 64$$

and $\qquad\qquad r = \pm 8$

$$\tan\theta = \frac{4}{-4\sqrt{3}} = \frac{-1}{\sqrt{3}}$$

$$\theta = \tan^{-1}\left(\frac{-1}{\sqrt{3}}\right)$$

$$\approx 0.5236$$

and so $\qquad\qquad \approx -\dfrac{\pi}{6}$

It looks as if the answer is $\left(8,\, -\dfrac{\pi}{6}\right)$ or $\left(-8,\, -\dfrac{\pi}{6}\right)$. Now $\left(-8,\, -\dfrac{\pi}{6}\right)$ is correct, but $\left(8,\, -\dfrac{\pi}{6}\right)$ is not correct. What happened? Plot $\left(-4\sqrt{3},\, 4\right)$. It is in Quadrant II. Now plot $= \left(8,\, \dfrac{-\pi}{6}\right)$. It is in Quadrant IV. Remember that $\tan^{-1}$ will only give angles in Quadrants I and IV. You should first determine which quadrant a point is in. If it is in Quadrants II or III, you will need to add π (or $180°$) to the answer you get using the conversion formula. $\left(\text{Because } \theta \text{ may be in Quadrant II} \right.$ or III, we write $\tan\theta = \dfrac{y}{x}$, rather than $\tan^{-1}\dfrac{y}{x} = \theta.\Big)$ If we add π to $\dfrac{-\pi}{6}$, we get $\dfrac{5\pi}{6}$. Thus, two possible answers are $\left(-8,\, -\dfrac{\pi}{6}\right)$ and $\left(8,\, \dfrac{5\pi}{6}\right)$. Both of these are polar coordinates of the point with Cartesian coordinates $(-4\sqrt{3},\, 4)$.

Polar Equations

Equations can also be written using the variables r and θ. These are called **polar equations**. A polar equation states a relationship between all the points (r, θ) that satisfy the equation. In the remaining part of this section, we will graph some polar equations.

As we usually do when we graph an equation, we will use a table of values, plot the points in the table, then connect these points in order as the values of θ increase.

EXAMPLE 11.28

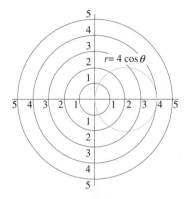

FIGURE 11.41

Graph the function $r = 4\cos\theta$.

Solution A table of values follows. The graph of the points is shown in Figure 11.41.

θ	0°	30°	60°	90°	120°	150°	180°	210°	240°	270°	300°	330°	360°
r	4	3.46	2	0	−2	−3.46	−4	−3.46	−2	0	2	3.46	4

Notice that this is a circle with radius 2 centered at $(2, 0°)$.

Circles

The graph of any equation of the type $r = a\cos\theta$ is a circle of radius $\left|\dfrac{a}{2}\right|$ and center $\left(\dfrac{a}{2}, 0°\right)$. An equation of the type $r = a\sin\theta$ is a circle with radius $\left|\dfrac{a}{2}\right|$ and center $\left(\dfrac{a}{2}, 90°\right)$.

EXAMPLE 11.29

Graph the function $r = \sin 3\theta$.

Solution The table of values follows. The graph of these points is in Figure 11.42.

θ	0°	10°	20°	30°	40°	50°	60°	70°	80°	90°
$r = \sin 3\theta$	0	0.5	0.87	1	0.87	0.5	0	−0.5	−0.87	−1

θ	90°	100°	110°	120°	130°	140°	150°	160°	170°	180°
$r = \sin 3\theta$	−1	−0.87	−0.5	0	0.5	0.87	1	0.87	0.5	0

A curve of this type is called a **rose**. This rose has three petals.

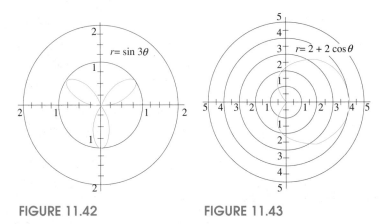

FIGURE 11.42   FIGURE 11.43

> **Roses**
>
> Any polar equation of the form $r = \sin n\theta$ or $r = \cos n\theta$, where n is a positive integer, is a rose. If n is an odd number, the rose will have n petals. If n is an even number, the rose will have $2n$ petals.

For many functions, polar equations are much simpler to work with than are rectangular equations (and vice versa). A good example of this will be seen in a later chapter.

Another type of polar graph is demonstrated by the graph of the polar equation $r = 2 + 2\cos\theta$. The graph of this equation is shown in Figure 11.43.

The curve in Figure 11.43 is called a **cardioid** because of its heart shape. In Exercise Set 11.7, we will graph other polar equations. If the curve has a special name, we will give that name.

Using a Calculator to Graph Polar Equations

You can use a graphing calculator or a computer to help you graph many of these curves. We will describe how it is done using a graphics calculator. The procedures for a *TI-81* differ quite a bit from those for a Casio *fx-7700G*. We will use a *TI-81* for the first example and a Casio *fx-7700G* for the second example.

EXAMPLE 11.30

Use a *TI-81* graphing calculator to graph $r = \sin^2 3\theta$.

Solution For a *TI-81*, you need to write this polar equation as a system of parametric equations. We use the formula for converting polar coordinates to rectangular coordinates

$$x = r\cos\theta$$

$$y = r\sin\theta$$

EXAMPLE 11.30 (Cont.)

In this example, r is the given polar equation; that is, $r = \sin^2 3\theta$. So, we will use the parametric equations

$$x = \sin^2 3\theta \cos \theta$$

$$y = \sin^2 3\theta \sin \theta$$

Set Tmin $= 0$, Tmax $= 6.3$, Tstep $= 0.1$, Xmin $= -1.5$, Xmax $= 1.5$, Xscl $= 0.5$, Ymin $= -1$, Ymax $= 1$, and Yscl $= 0.5$. Press $\boxed{\text{Y=}}$ and key in the following. [Notice that $\sin^2 3\theta$ is entered as $(\sin 3T)^2$.]

$$\boxed{(}\ \boxed{\text{SIN}}\ 3\ \boxed{\text{X}\,|\,\text{T}}\ \boxed{)}\ \boxed{x^2}\ \boxed{\text{COS}}\ \boxed{\text{X}\,|\,\text{T}}\ \boxed{\text{ENTER}}$$

$$\boxed{(}\ \boxed{\text{SIN}}\ 3\ \boxed{\text{X}\,|\,\text{T}}\ \boxed{)}\ \boxed{x^2}\ \boxed{\text{SIN}}\ \boxed{\text{X}\,|\,\text{T}}\ \boxed{\text{GRAPH}}$$

The result is a rose with 6 petals, as shown in Figure 11.44.

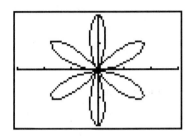

FIGURE 11.44

EXAMPLE 11.31

Use a Casio *fx-7700G* graphing calculator to graph $r = 5 \sin 2\theta - 2 \cos 3\theta$.

Solution First, put your calculator in polar mode. Press $\boxed{\text{MODE}}\ \boxed{\text{SHIFT}}\ \boxed{-}$. Next, set the range. Since the ranges of sine and cosine are both $[-1, 1]$, we know that $-7 \le r \le 7$. Because the screen is longer horizontally than it is vertically, we set Xmin -9, Xmax $= 9$, Xscl $= 1$, Ymin $= -6$, Ymax $= 6$, and Yscl $= 1$. After you press $\boxed{\text{EXE}}$, you get a new screen, where you enter the values of θ. The period of $\sin 2\theta$ is π; the period of $\cos 3\theta$ is $\frac{2}{3}\pi$. Since the least common multiple of π and $\frac{2}{3}\pi$ is 2π, the period of r is a multiple of 2π. We will set θmin $= 0$, θmax $= 2\pi$ (press 2 $\boxed{\text{SHIFT}}\ \boxed{\pi}$), and θptch $= 0.1$. Press $\boxed{\text{EXE}}\ \boxed{\text{Graph}}$ and then the following:

$$5\ \boxed{\text{SIN}}\ 2\ \boxed{\text{X}, \theta, \text{T}}\ \boxed{-}\ 3\ \boxed{\text{X}, \theta, \text{T}}\ \boxed{\text{EXE}}$$

The result is a butterfly-shaped curve as shown in Figure 11.45.

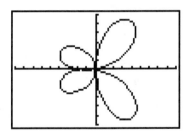

FIGURE 11.45

Exercise Set 11.7

In Exercises 1–12, plot the points with the given polar coordinates.

1. $\left(2,\ \dfrac{\pi}{4}\right)$

2. $(4,\ 60°)$

3. $(3,\ 90°)$

4. $(5,\ \pi)$

5. $\left(-4,\ \dfrac{2\pi}{3}\right)$

6. $(-6,\ 30°)$

7. $(-4,\ 0°)$

8. $\left(-2,\ \dfrac{5\pi}{6}\right)$

9. $(5,\ -30°)$

10. $\left(3,\ -\dfrac{3\pi}{4}\right)$

11. $\left(-7,\ \dfrac{11\pi}{6}\right)$

12. $(-2,\ -270°)$

In Exercises 13–24, convert each polar coordinate to its equivalent rectangular coordinates.

13. $\left(4, \dfrac{\pi}{3}\right)$

14. $(5, 75°)$

15. $(2, 135°)$

16. $\left(3, \dfrac{3\pi}{2}\right)$

17. $(-6, 20°)$

18. $\left(-3, \dfrac{5\pi}{3}\right)$

19. $(-2, 4.3)$

20. $(-5, 255°)$

21. $(3, -170°)$

22. $\left(4, -\dfrac{\pi}{8}\right)$

23. $(-6, -2.5)$

24. $(-3, -195°)$

In Exercises 25–36, convert each rectangular coordinate to an equivalent polar coordinate with $r > 0$.

25. $(4, 4)$

26. $(3, 6)$

27. $(4, 3)$

28. $(5, 12)$

29. $(-20, 21)$

30. $(-12, 5)$

31. $(-3, 4)$

32. $(9, -5)$

33. $(-7, -10)$

34. $(-8, -3)$

35. $(2, 9)$

36. $(-6, 1)$

In Exercises 37–58, graph the polar equations.

37. $r = 4$

38. $r = -6$

39. $r = 3\sin\theta$

40. $r = -5\cos\theta$

41. $r = 4 - 4\sin\theta$ (cardioid)

42. $r = 1 + 3\cos\theta$ (limacon)

43. $r = 3\cos 5\theta$ (five-petaled rose)

44. $r = 3\sin 2\theta$ (four-petaled rose)

45. $r = 5\sec\theta$

46. $r = -7\csc\theta$

47. $r = \theta$ (Let θ get larger than 4π.)

48. $r = 3^{\theta}$ (spiral)

49. $r = 4 + 4\sec\theta$

50. $r = \dfrac{1}{\theta}, \theta > 0$

51. $r = 3 + \cos\theta$

52. $r^2 = 16\sin 2\theta$

53. $r = 2 + 5\sin\theta$

54. $r = 1 + 4\sec\theta$

55. $r = \dfrac{6}{3 + 2\sin\theta}$

56. $r = \dfrac{6}{1 + 3\cos\theta}$

57. $r = \dfrac{3}{2 + 2\cos\theta}$

58. $r = \dfrac{4\sec\theta}{2\sec\theta - 1}$

▤ CHAPTER 11 REVIEW

Important Terms and Concepts

Amplitude

Cosine curve

Displacement

Frequency

Harmonic motion

Lissajous figures

Parametric equations

Period

   Periodic motion

Phase shift

Polar coordinates

Polar equations

Sine curves

Sinusoidal curve

Review Exercises

In Exercises 1–10, find the period, amplitude, frequency, and displacement of the given functions and graph one cycle of each.

1. $y = 8 \cos 4x$

2. $y = 3 \sin 2x$

3. $y = 2 \tan 3x$

4. $y = 3 \sin \left(2x + \dfrac{\pi}{2} \right)$

5. $y = \dfrac{1}{2} \cos \left(3x - \dfrac{\pi}{3} \right)$

6. $y = \dfrac{1}{3} \sec \left(2x - \dfrac{\pi}{6} \right)$

7. $y = -4 \cot \left(x + \dfrac{\pi}{4} \right)$

8. $y = -\dfrac{1}{2} \csc \left(\dfrac{x}{3} - \dfrac{\pi}{5} \right)$

9. $y = 2 \sin \left(\dfrac{2}{3} x + \dfrac{\pi}{6} \right)$

10. $y = \dfrac{-1}{4} \cos \left(-\dfrac{x}{2} + \dfrac{2\pi}{3} \right)$

In Exercises 11–20, sketch the curves of the given equations.

11. $y = \sin 2x + \cos 3x$

12. $y = 4 \sin \dfrac{x}{4} - \cos 2x$

13. $y = \cos 2x + \sin 3x$

14. $y = x + \sin \left(3x + \dfrac{\pi}{4} \right)$

15. $x = 3t, y = 5t - 2$

16. $x = t + 2, y = t^2 + t$

17. $x = 4 \sin t, y = 3 \cos t$

18. $x = \sin t, y = \cos 4t$

19. $r = \cos 7\theta$

20. $r = 3 + 2 \sin \theta$

Solve Exercises 21–24.

21. *Mechanics*   A weight vibrates on a spring in simple harmonic motion. The amplitude is 85 mm and the frequency is 5 Hz. Write an equation for the weight's position y at any time t, if $y = 0$ when $t = 0$.

22. *Oceanography*   A raft is bobbing on the water in simple harmonic motion according to the equation $y = 1.7 \sin 3.4t$, where y is the position of the raft in meters and t is the time in seconds. **(a)** What is the amplitude of the raft? **(b)** What is its period? **(c)** What is its frequency? **(d)** Sketch one complete cycle for this curve.

23. *Electronics*   In an ac circuit containing only a constant capacitance, the current leads the voltage by $\frac{1}{4}$ cycle. If $V = V_{max} \sin 2\pi ft$ V, $I_{max} = 4.8$ A, and $f = 60$ Hz, write the equation for the current. Sketch one cycle of the curve.

24. *Electronics*   In a resistance-inductance circuit, $I = I_{max} \sin(2\pi ft + \phi)$. If $I_{max} = 1.5$ A, $f = 360$ Hz, and $\phi = -31°$, sketch I vs t for one cycle.

≡ CHAPTER 11 TEST

1. For the following trigonometric functions, find the period, amplitude, frequency, and displacement of the given functions.
(a) $y = -3 \sin 5x$

(b) $y = 2.4 \cos \left(3x - \dfrac{\pi}{4} \right)$

(c) $y = 1.5 \tan \left(2x + \dfrac{\pi}{5} \right)$

2. Sketch each of the given curves.

(a) $y = -3 \sin 5x$

(b) $y = \dfrac{1}{2} \sec \left(x + \dfrac{\pi}{2} \right)$

(c) $y = \sin 2x + 2 \cos 3x$

3. Sketch $x = 2t$, $y = t^2 + 1$.

4. Sketch $r = 1 + 2 \cos \theta$.

5. Convert $(2,\ 55°)$ to rectangular coordinates.

6. Convert $(5,\ -12)$ to polar coordinates.

7. In an ac electric circuit, write the equation of the sine curve for the current, when $I_{max} = 5.7$ A and $f = 60$ Hz.

12

Exponents and Radicals

When an object is placed between two light sources so the illuminance is the same from each source, an equation involving radicals is used to determine the intensity of each light. In Section 12.4, we will learn how to determine this intensity.

Courtesy of Michael A. Gallitelli, Metroland Photo Inc.

Exponents and radicals were introduced in Chapter 1. In this chapter, we will explore the topic further and examine the close relationship between exponents and radicals.

☰ 12.1
FRACTIONAL EXPONENTS

The basic rules for exponents were given in Section 1.4. They are repeated here to refresh your memory.

Basic Rules for Exponents	
	Rule 1: $b^m b^n = b^{m+n}$
	Rule 2: $(b^m)^n = b^{mn}$
	Rule 3: $(ab)^n = a^n b^n$
	Rule 4: $\left(\dfrac{a}{b}\right)^m = \dfrac{a^m}{b^m}, b \neq 0$
	Rule 5: $\dfrac{b^m}{b^n} = b^{m-n}, b \neq 0$
	Rule 6: $b^0 = 1, b \neq 0$
	Rule 7: $b^{-n} = \dfrac{1}{b^n}, b \neq 0$

When we studied these rules in Section 1.4, m and n had to be integers. In Section 1.7, we examined the meaning of $b^{1/n}$ and found that it meant $\sqrt[n]{b}$. This makes sense, since $(b^{1/n})^n = b^{n/n} = b$. This gives us a new rule:

Rule 8 for Exponents	$b^{1/n} = \sqrt[n]{b}$

EXAMPLE 12.1

Evaluate (**a**) $27^{1/3}$, (**b**) $16^{1/2}$, and (**c**) $625^{1/4}$.

Solutions
(**a**) $27^{1/3} = \sqrt[3]{27} = 3$

(**b**) $16^{1/2} = \sqrt{16} = 4$

(**c**) $625^{1/4} = \sqrt[4]{625} = 5$

If we combine Rules 2 and 8, we obtain the following new rule.

Rule 9 for Exponents	$b^{m/n} = (b^m)^{1/n} = \sqrt[n]{b^m}$

EXAMPLE 12.2

Evaluate (**a**) $8^{2/3}$, (**b**) $64^{5/2}$, (**c**) $81^{5/4}$, and (**d**) $9^{-3/2}$.

EXAMPLE 12.2 (Cont.)

Solutions (a) $8^{2/3} = \sqrt[3]{8^2} = \sqrt[3]{64} = 4$

(b) $64^{5/2} = \sqrt{64^5} = \sqrt{1,073,741,824} = 32,768$

(c) $81^{5/4} = \sqrt[4]{81^5} = \sqrt[4]{3,486,784,401} = 243$

(d) $9^{-3/2} = \dfrac{1}{\sqrt{9^3}} = \dfrac{1}{\sqrt{729}} = \dfrac{1}{27}$

It is often easier to find the root before raising the number to a power. Thus, we could rewrite Rule 9.

Rule 9 for Exponents	$b^{m/n} = \left(b^{1/n}\right)^m = \left(\sqrt[n]{b}\right)^m$

We will rework the problems in Example 12.2 using this variation of Rule 9.

EXAMPLE 12.3

Evaluate: (a) $8^{2/3}$, (b) $64^{5/2}$, (c) $81^{5/4}$, and (d) $9^{-3/2}$

Solutions (a) $8^{2/3} = (\sqrt[3]{8})^2 = 2^2 = 4$

(b) $64^{5/2} = (\sqrt{64})^5 = 8^5 = 32,768$

(c) $81^{5/4} = (\sqrt[4]{81})^5 = 3^5 = 243$

(d) $9^{-3/2} = \dfrac{1}{(\sqrt{9})^3} = \dfrac{1}{3^3} = \dfrac{1}{27}$

All of the rules for integer exponents apply to fractional exponents. This is demonstrated in the following example.

EXAMPLE 12.4

(a) $x^{1/2}x^{2/3} = x^{1/2+2/3} = x^{7/6}$

(b) $(y^{2/3})^{4/5} = y^{2/3 \cdot 4/5} = y^{8/15}$

(c) $\left(\dfrac{x}{y}\right)^{4/5} = \dfrac{x^{4/5}}{y^{4/5}}$

(d) $\dfrac{x^{3/4}}{x^{5/3}} = x^{3/4-5/3} = x^{-11/12} = \dfrac{1}{x^{11/12}}$

(e) $\dfrac{4x^2}{x^{2/3}} = 4x^{2-2/3} = 4x^{4/3}$

(f) $(8y^3)^{5/3} = 8^{5/3}(y^3)^{5/3} = 32y^5$

(g) $\left(\dfrac{x^{15}}{y^9}\right)^{-1/3} = \left(\dfrac{y^9}{x^{15}}\right)^{1/3} = \dfrac{y^3}{x^5}$

EXAMPLE 12.4 (Cont.)

(h) $\left(\dfrac{x^{1/3}y^{2/5}}{z^{3/5}} \right)^{15} = \dfrac{(x^{1/3})^{15}(y^{2/5})^{15}}{(z^{3/5})^{15}} = \dfrac{x^5 y^6}{z^9}$

(i) $x^{2/5}x^{-1/3} = x^{2/5-1/3} = x^{1/15}$

Evaluating Exponents Using a Calculator

You might want to review Appendix A and the work with calculators, where it describes how to use the $\boxed{x^2}$ key to calculate the square, or second power, of a number and how to use the $\boxed{x^y}$, $\boxed{y^x}$, or $\boxed{\wedge}$ key for the value of any power. To use the $\boxed{x^y}$ key on an algebraic calculator, you must also use the $\boxed{=}$ key before the answer is displayed.

EXAMPLE 12.5

Evaluate $4.7^{3.42}$

Solution

	PRESS	DISPLAY
Algebraic calculator:	4.7 $\boxed{x^y}$ 3.42 $\boxed{=}$	198.8724729
RPN calculator:	4.7 $\boxed{\text{ENTER}}$ 3.42 $\boxed{y^x}$	198.8724729
TI-81 calculator:	4.7 $\boxed{\wedge}$ 3.42 $\boxed{\text{ENTER}}$	198.8724729

EXAMPLE 12.6

Evaluate $12^{5/3}$

Solution

	PRESS	DISPLAY
Algebraic calculator:	12 $\boxed{x^y}$ $\boxed{(}$ 5 $\boxed{\div}$ 3 $\boxed{)}$ $\boxed{=}$	62.89779346
RPN calculator:	12 $\boxed{\text{ENTER}}$ 5 $\boxed{\text{ENTER}}$ 3 $\boxed{\div}$ $\boxed{y^x}$	62.89779351
TI-81 calculator:	12 $\boxed{\wedge}$ $\boxed{(}$ 5 $\boxed{\div}$ 3 $\boxed{)}$ $\boxed{\text{ENTER}}$	62.89779351

Application

EXAMPLE 12.7

One study of a lake found that the light intensity was reduced 15% through a depth of 25 cm. The formula $I = (0.85)^{d/25}$ gives the approximate fraction of surface light intensity at a depth d, in centimeters. Find I at a depth of 60 cm.

Solution   Since $d = 60$ cm, we want to evaluate

$$I = (0.85)^{60/25}$$

Using an algebraic calculator, we enter

$$0.85 \boxed{x^y} \boxed{(}\ 60\ \boxed{\div}\ 25\ \boxed{)} \boxed{=}$$

and obtain 0.677026116. So, the light intensity at 60 cm is about 67.7% of the surface light intensity.

Exercise Set 12.1

In Exercises 1–20, evaluate the given expression without the use of a calculator.

1. $25^{1/2}$

2. $27^{1/3}$

3. $64^{1/3}$

4. $125^{1/3}$

5. $25^{-1/2}$

6. $81^{-1/4}$

7. $32^{-1/5}$

8. $64^{-1/3}$

9. $27^{2/3}$

10. $81^{3/4}$

11. $125^{2/3}$

12. $32^{3/5}$

13. $16^{-3/4}$

14. $(-8)^{2/3}$

15. $(-8)^{-1/3}$

16. $(-27)^{-2/3}$

17. $\left(\frac{1}{8}\right)^{1/3}$

18. $\left(\frac{1}{25}\right)^{3/2}$

19. $\left(\frac{1}{16}\right)^{-5/4}$

20. $\left(\frac{-1}{27}\right)^{4/3}$

In Exercises 21–60, express each of the given expressions in the simplest form containing only positive exponents.

21. $3^2 \cdot 3^5$

22. $5^9 5^8$

23. $7^6 7^{-2}$

24. $11^9 11^{-6}$

25. $x^4 x^6$

26. $y^7 y^9$

27. $y^6 y^{-4}$

28. $x^8 x^{-2}$

29. $(9^5)^2$

30. $(11^8)^{-5}$

31. $(x^7)^3$

32. $(p^9)^{-5}$

33. $(xy)^5$

34. $(yt)^3$

35. $(ab)^{-5}$

36. $(xyz)^{-9}$

37. $\dfrac{x^{10}}{x^2}$

38. $\dfrac{p^9}{p^3}$

39. $\dfrac{x^2}{x^8}$

40. $\dfrac{a^3}{a^{12}}$

41. $x^{1/2} x^{3/2}$

42. $a^{1/3} a^{4/3}$

43. $r^{3/4} r$

44. $a^2 a^{2/3}$

45. $a^{1/2} a^{1/3}$

46. $b^{2/3} b^{1/4}$

47. $d^{2/3} d^{-1/4}$

48. $x^{3/5} y^{-2/3}$

49. $\dfrac{a^2 b^5}{a^5 b^2}$

50. $\dfrac{x^3 y^2}{x^7 y}$

51. $\dfrac{r^5 s^2 t}{tr^3 s^5}$

52. $\dfrac{a^2 bc^3}{(abc)^3}$

53. $\dfrac{(xy^2 z)^4}{x^4 (yz^2)^2}$

54. $\dfrac{m^6 n^7}{(m^2 n)^3}$

55. $\left(\dfrac{a}{b^2}\right)^3 \left(\dfrac{a}{b^3}\right)^2$

56. $\left(\dfrac{x^2}{y}\right)^4 \left(\dfrac{x}{y^2}\right)^2$

57. $\dfrac{(xy^2 b^3)^{1/2}}{(x^{1/4} b^4 y)^2}$

58. $\dfrac{(x^{1/3} y^3)^3}{(y^{10} x^5)^{1/5}}$

59. $\left(\dfrac{2x}{p^2}\right)^{-2} \left(\dfrac{p}{4}\right)^{-1}$

60. $\left(\dfrac{5a^2}{6b}\right)^{-2} \left(\dfrac{6}{a}\right)^{-4}$

Use a calculator or computer to determine the value of each of the numbers in Exercises 61–68.

61. $8.3^{2/3}$

62. $7.3^{2/5}$

63. $92.47^{5/7}$

64. $81.94^{3/5}$

65. $(-81.52)^{2/7}$

66. $(-78.64)^{1/3}$

67. $432.61^{1/4}$

68. $(-537.15)^{2/3}$

Solve Exercises 69–72.

69. *Physics* The distance in meters traveled by a falling body starting from rest is $9.8t^2$, where t is the time, in seconds, the object has been falling. In 4.75 s, the distance fallen will be $9.8(4.75)^2$ m. Evaluate this quantity.

70. The volume of a sphere is $\frac{4}{3}\pi r^3$. If the radius r of a sphere is 19.25 in., what is the volume?

71. *Physics* When 5 m³ of helium at a temperature of 315 K and a pressure of 15 N/m² is adiabatically compressed to 0.5 m³, its new pressure p and temperature T are given by

$$p = (5)^5 \left(\frac{15}{0.5}\right)^{5/3}$$

and $$T = 315 \left(\frac{15}{0.5}\right)^{5/3}$$

Evaluate p and T.

72. *Nuclear technology* Radium has a half-life of approximately 1,600 years. It decays according to the formula $q(t) = q_0 2^{-t/1,600}$, where q_0 is the original quantity of radium and $q(t)$ is the amount left at time t. In this problem, t is given in years. If you begin with 75 mg of pure radium, how much will remain after 2,000 years?

≡ 12.2
LAWS OF RADICALS

In Section 1.3, we introduced the concept of roots. We used them in Section 12.1 for the discussion of fractional exponents. The more general name for roots is **radicals**. Fractional exponents can be used for any operation that requires radicals.

Laws of Radicals

A **radical** is any number of the form $\sqrt[n]{b}$. The number under the radical b, is called the **radicand**. The number indicating the root n is called the **order** or **index**.

In Section 1.7, we introduced four basic rules or laws of radicals. These rules are listed in the following box.

Basic Rules for Radicals	
	Rule 1: $\sqrt[n]{ab} = \sqrt[n]{a}\sqrt[n]{b}$
	Rule 2: $\sqrt[n]{\dfrac{a}{b}} = \dfrac{\sqrt[n]{a}}{\sqrt[n]{b}}$
	Rule 3: $(\sqrt[n]{b})^n = b^{n/n} = b$
	Rule 4: $\sqrt[n]{b} = b^{1/n}$

We assume in each of these rules that if n is even, neither a nor b are negative real numbers.

EXAMPLE 12.8

Simplify (a) $\sqrt[6]{27}$, (b) $\sqrt[5]{96}$, (c) $\sqrt{x^6y^8}$, (d) $\sqrt[3]{\dfrac{125y^6}{x^9}}$, and (e) $\sqrt[3]{54x^8}$

Solutions

(a) $\sqrt[6]{27} = \sqrt[6]{3^3} = 3^{3/6} = 3^{1/2} = \sqrt{3}$

(b) $\sqrt[5]{96} = \sqrt[5]{32 \cdot 3} = \sqrt[5]{32}\sqrt[5]{3} = \sqrt[5]{2^5}\sqrt[5]{3} = 2\sqrt[5]{3}$

(c) $\sqrt{x^6y^8} = (x^6y^8)^{1/2} = x^{6/2}y^{8/2} = x^3y^4$

(d) $\sqrt[3]{\dfrac{125y^6}{x^9}} = \sqrt[3]{\dfrac{5^3y^6}{x^9}} = \dfrac{5^{3/3}y^{6/3}}{x^{9/3}} = \dfrac{5y^2}{x^3}$

(e) $\sqrt[3]{54x^8} = \sqrt[3]{2 \cdot 3^3 x^6 x^2}$

$\quad = 2^{1/3}3^{3/3}x^{6/3}x^{2/3}$

$\quad = 2^{1/3} \cdot 3 \cdot x^2 x^{2/3}$

$\quad = 3x^2 2^{1/3}x^{2/3}$

$\quad = 3x^2\left(2x^2\right)^{1/3}$

$\quad = 3x^2\sqrt[3]{2x^2}$

Notice that several examples started with as many factors as possible with nth roots that could be easily found. Thus, in Example 12.8(b), we used $\sqrt[n]{ab} = \sqrt[n]{a}\sqrt[n]{b}$, when we wrote $\sqrt[5]{96} = \sqrt[5]{32} \cdot \sqrt[5]{3}$.

 Hint

In general, we try to express a radical so that the exponent of any factor in the radicand is less than the index of the radical. Thus, in Example 12.8(e), we wrote $\sqrt[3]{x^8}$ as $x^2\sqrt[3]{x^2}$.

Rationalizing Denominators

When a radicand is a fraction, a variation of Rule 2, $\sqrt[n]{\dfrac{a}{b}} = \dfrac{\sqrt[n]{a}}{\sqrt[n]{b}}$, is used to eliminate the radical in the denominator. This technique is called **rationalizing the denominator**.

> **Rationalizing the Denominator**
>
> To rationalize a denominator of the form $\sqrt[n]{x^r}$, multiply the denominator by another radical with the same radicand and the same index. In $\sqrt[n]{x^s}$, $r + s$ is a multiple of n.

The process of rationalizing the denominator makes the denominator a perfect power of x and eliminates the radical in the denominator.

 Note

Remember that whenever you multiply the denominator by something other than 1, you must also multiply the numerator by the same quantity.

EXAMPLE 12.9

Rationalize the denominators of **(a)** $\sqrt{\dfrac{3}{5}}$, **(b)** $\dfrac{1}{\sqrt[3]{2}}$, and **(c)** $\sqrt[5]{\dfrac{3}{8x^2}}$.

Solutions

(a) $\sqrt{\dfrac{3}{5}} = \dfrac{\sqrt{3}}{\sqrt{5}} = \dfrac{\sqrt{3}}{\sqrt{5}} \cdot \dfrac{\sqrt{5}}{\sqrt{5}}$ $\left(\text{Multiply by } \dfrac{\sqrt{5}}{\sqrt{5}}. \right)$

$\qquad = \dfrac{\sqrt{15}}{\sqrt{5^2}} = \dfrac{\sqrt{15}}{5}$

(b) $\dfrac{1}{\sqrt[3]{2}} = \dfrac{1}{\sqrt[3]{2}} \cdot \dfrac{\sqrt[3]{2^2}}{\sqrt[3]{2^2}}$ $\left(\text{Multiply by } \dfrac{\sqrt[3]{2^2}}{\sqrt[3]{2^2}}. \right)$

$\qquad = \dfrac{\sqrt[3]{2^2}}{\sqrt[3]{2^3}} = \dfrac{\sqrt[3]{4}}{2}$

(c) $\sqrt[5]{\dfrac{3}{8x^2}} = \dfrac{\sqrt[5]{3}}{\sqrt[5]{2^3 x^2}}$

$\qquad = \dfrac{\sqrt[5]{3}}{\sqrt[5]{2^3 x^2}} \cdot \dfrac{\sqrt[5]{2^2 x^3}}{\sqrt[5]{2^2 x^3}}$ $\left(\text{Multiply by } \dfrac{\sqrt[5]{2^2 x^3}}{\sqrt[5]{2^2 x^3}}. \right)$

$\qquad = \dfrac{\sqrt[5]{3 \cdot 2^2 x^3}}{\sqrt[5]{2^5 x^5}} = \dfrac{\sqrt[5]{12x^3}}{2x}$

Rationalizing the denominator was originally developed as a way to help computations, but the increased use of calculators and computers reduces its importance in calculations. Rationalizing the denominator, however, is often a useful way to write numbers.

Another helpful rule is used with radicals.

Rule 5 for Radicals	$\sqrt[m]{\sqrt[n]{b}} = \sqrt[mn]{b}$

EXAMPLE 12.10

(a) $\sqrt[3]{\sqrt{27}} = \sqrt[6]{27}$, **(b)** $\sqrt[4]{\sqrt[5]{914}} = \sqrt[20]{914}$, **(c)** $\sqrt[4]{\sqrt[3]{x}} = \sqrt[12]{x}$

Sometimes it is possible to reduce the index of a radical. For example,

$$\sqrt[6]{y^2} = y^{2/6} = y^{1/3} = \sqrt[3]{y}$$

Here the index was reduced from 6 to 3. Another example can be seen from 12.10(a).

$$\sqrt[3]{\sqrt{27}} = \sqrt[3]{\sqrt{3^3}} = \sqrt[6]{3^3} = 3^{3/6} = 3^{1/2} = \sqrt{3}$$

The last version, $\sqrt{3}$, is a simpler version with which to work.

EXAMPLE 12.11

Reduce the index of **(a)** $\sqrt[8]{16}$, **(b)** $\sqrt[6]{16x^2}$, and **(c)** $\sqrt[12]{27x^6y^3}$.

Solutions **(a)** $\sqrt[8]{16} = \sqrt[8]{2^4} = 2^{4/8} = 2^{1/2} = \sqrt{2}$

(b) $\sqrt[6]{16x^2} = \sqrt[6]{2^4x^2} = 2^{4/6}x^{2/6} = 2^{2/3}x^{1/3} = \sqrt[3]{4x}$

(c) $\sqrt[12]{27x^6y^3} = \sqrt[12]{3^3x^6y^3} = 3^{3/12}x^{6/12}y^{3/12} = 3^{1/4}x^{2/4}y^{1/4}$

$$= \sqrt[4]{3x^2y}$$

Notice that, in the last example, we could have factored the x^2 out of the radical using Rule 1 and written $\sqrt[4]{3x^2y} = \sqrt[4]{3y}\sqrt[4]{x^2} = \sqrt[4]{3y}\sqrt{x}$.

Simplifying Radicals

Simplifying a radical makes it easier to work with. The following three steps are used to simplify a radical.

Steps for Simplifying Radicals

A radical is simplified when all of the following steps are finished.

Step 1: All possible factors have been removed from the radicand.

Step 2: All denominators are rationalized.

Step 3: The index has been reduced as much as possible.

EXAMPLE 12.12

Simplify the following: **(a)** $\sqrt{\dfrac{x^3}{y}}$, **(b)** $\sqrt[3]{x^5y^{10}}$, **(c)** $\sqrt[5]{x^3y^{-7}}$, and **(d)** $\sqrt[3]{\dfrac{x}{y^2} + \dfrac{5y^2}{x}}$.

Solutions **(a)** $\sqrt{\dfrac{x^3}{y}} = \sqrt{\dfrac{x^3}{y}}\sqrt{\dfrac{y}{y}} = \dfrac{\sqrt{x^3y}}{\sqrt{y^2}} = \dfrac{\sqrt{x^3y}}{y} = \dfrac{x\sqrt{xy}}{y}$

(b) $\sqrt[3]{x^5y^{10}} = \sqrt[3]{x^3x^2y^9y} = \sqrt[3]{x^3y^9} \cdot \sqrt[3]{x^2y} = xy^3\sqrt[3]{x^2y}$

(c) $\sqrt[5]{x^3y^{-7}} = \sqrt[5]{\dfrac{x^3}{y^7}} = \sqrt[5]{\dfrac{x^3}{y^7}} \cdot \sqrt[5]{\dfrac{y^3}{y^3}} = \dfrac{\sqrt[5]{x^3y^3}}{\sqrt[5]{y^{10}}} = \dfrac{\sqrt[5]{x^3y^3}}{y^2}$

(d) $\sqrt[3]{\dfrac{x}{y^2} + \dfrac{5y^2}{x}} = \sqrt[3]{\dfrac{x \cdot x}{y^2x} + \dfrac{5y^2y^2}{xy^2}} = \sqrt[3]{\dfrac{x^2 + 5y^4}{xy^2}}$

$$= \sqrt[3]{\dfrac{(x^2 + 5y^4)\,x^2y}{(xy^2)\,x^2y}} = \dfrac{\sqrt[3]{x^4y + 5x^2y^5}}{\sqrt[3]{x^3y^3}}$$

$$= \dfrac{\sqrt[3]{x^4y + 5x^2y^5}}{xy}$$

Notice that in Example 12.12(d) we had to find a common denominator before we could add the fractions. After the fractions were added, we were able to rationalize the denominator.

Application

EXAMPLE 12.13

The frequency of oscillation f of a simple pendulum is given by

$$f = \frac{1}{2\pi} \sqrt{\frac{g}{L}}$$

where $g \approx 9.8$ m/s^2 is the acceleration due to gravity in the metric system and L is the length of the pendulum in meters. **(a)** Express f in simplest form, when $L = 0.35$ m. **(b)** Evaluate f to the nearest hundredth.

Solution　**(a)** Given $g \approx 9.8$ m/s^2 and $L = 0.35$ m,

$$f = \frac{1}{2\pi} \sqrt{\frac{g}{L}}$$

$$= \frac{1}{2\pi} \sqrt{\frac{9.8}{0.35}}$$

$$= \frac{1}{2\pi} \sqrt{28}$$

$$= \frac{\sqrt{7}}{\pi}$$

(b) Evaluating $\dfrac{\sqrt{7}}{\pi}$ with an algebraic calculator, we press $7\ \boxed{\sqrt{}}\ \boxed{\div}\ \boxed{\pi}\ \boxed{=}$ and obtain 0.8421687987. Thus, the pendulum oscillates about once every 0.84 s. ▪

Exercise Set 12.2

Use the rules for radicals to express each of Exercises 1–56 in simplest radical form.

1. $\sqrt[3]{16}$

2. $\sqrt[3]{81}$

3. $\sqrt{45}$

4. $\sqrt[3]{40}$

5. $\sqrt[3]{y^{12}}$

6. $\sqrt[4]{p^8}$

7. $\sqrt[5]{a^7}$

8. $\sqrt[7]{b^{10}}$

9. $\sqrt{x^2 y^7}$

10. $\sqrt[3]{x^5 y^3}$

11. $\sqrt[4]{a^5 b^3}$

12. $\sqrt[5]{p^{12} y^8}$

13. $\sqrt[3]{8x^4}$

14. $\sqrt[4]{81 y^9}$

15. $\sqrt{27 x^3 y}$

16. $\sqrt[3]{32 a^5 b^2}$

17. $\sqrt[3]{-8}$

18. $\sqrt[5]{-243}$

19. $\sqrt[3]{a^2 b^4}\, \sqrt[3]{ab^5}$

20. $\sqrt[5]{x^3 y^2 z^4}\, \sqrt[5]{x^2 y^8 z}$

21. $\sqrt[4]{p^3 q^2 r^6}\, \sqrt[4]{pq^6 r}$

22. $\sqrt[6]{m^3 n^2 e^7}\, \sqrt[6]{m^2 n^4 e^5}$

23. $\sqrt[3]{\dfrac{8x^3}{27}}$

24. $\sqrt[4]{\dfrac{81 y^8}{16}}$

25. $\sqrt[5]{\dfrac{x^5 y^{10}}{z^5}}$

26. $\sqrt[3]{\dfrac{a^3 b^9}{c^6}}$

27. $\sqrt[3]{\dfrac{16x^3y^2}{z^6}}$

28. $\sqrt{\dfrac{125a^5b^2}{c^4}}$

29. $\sqrt{\dfrac{64x^3y^4}{9z^4p^2}}$

30. $\sqrt[3]{\dfrac{8a^5b^3}{27r^6s^9}}$

31. $\sqrt{\dfrac{16}{3}}$

32. $\sqrt{\dfrac{4}{5}}$

33. $\sqrt[3]{\dfrac{27}{4}}$

34. $\sqrt[3]{\dfrac{16}{25}}$

35. $\sqrt{\dfrac{25}{2x}}$

36. $\sqrt[3]{\dfrac{8}{5y^2}}$

37. $\sqrt[4]{\dfrac{81}{32z^2}}$

38. $\sqrt[3]{\dfrac{2}{25r^2}}$

39. $\sqrt[3]{\dfrac{16x^2y}{x^5}}$

40. $\sqrt[4]{\dfrac{25a^3b^5}{a^7b}}$

41. $\sqrt[3]{\dfrac{8x^3yz}{27b^2z^4}}$

42. $\sqrt[4]{\dfrac{25a^2b^3}{16c^3b^6}}$

43. $\sqrt{4 \times 10^4}$

44. $\sqrt{9 \times 10^6}$

45. $\sqrt{25 \times 10^3}$

46. $\sqrt{16 \times 10^7}$

47. $\sqrt{4 \times 10^7}$

48. $\sqrt{9 \times 10^9}$

49. $\sqrt[3]{1.25 \times 10^{10}}$

50. $\sqrt[5]{3.2 \times 10^{14}}$

51. $\sqrt{\dfrac{x}{y} + \dfrac{y}{x}}$

52. $\sqrt{\dfrac{a}{b} - \dfrac{b}{a}}$

53. $\sqrt{a^2 + 2ab + b^2}$

54. $\sqrt{x^2 - 2xy + y^2}$

55. $\sqrt{\dfrac{1}{a^2} + \dfrac{1}{b}}$

56. $\sqrt{\dfrac{x}{y^2} + \dfrac{y}{x^2}}$

Solve Exercises 57–60.

57. *Music* Many musical instruments contain strings, which vibrate to produce music. The frequency of vibration f of a string of length L fixed at both ends and vibrating in its fundamental mode, is given by

$$f = \frac{1}{2L}\sqrt{\frac{T}{\mu}}$$

where μ is the mass per unit length and T is the tension in the string. Rationalize the right-hand side of this equation.

58. *Music* In the equation in Exercise 57, what happens to the frequency when the tension is quadrupled?

59. *Electricity* The impedance Z of a certain circuit is given by the equation

$$Z = \frac{1}{\sqrt{\dfrac{1}{x^2} + \dfrac{1}{R^2}}}$$

Rationalize the denominator in order to simplify this equation.

60. *Chemistry* The distance between ion layers of a sodium chloride crystal is given by the expression

$$\sqrt[3]{\frac{M}{2\rho N}}$$

where M is the molecular weight, N is Avogadro's number, and ρ is the density. Express this in simplest form.

☰ 12.3
BASIC OPERATIONS WITH RADICALS

In this section, we will study the basic operations of addition, subtraction, multiplication, and division of radicals. Adding and subtracting radicals is similar to adding and subtracting algebraic expressions. However, there are two cases to consider when multiplying or dividing radicals. The first case is when the radicals have the same index; the second case is when they have different indices. Each of these cases will be considered in this section.

Addition and Subtraction of Radicals

When we learned how to add and subtract algebraic expressions, we found that only like terms can be added or subtracted. Addition and subtraction of radicals is very similar.

☰ **Note** Radicals can only be added or subtracted if the radicands are identical and the indices are the same.

Trying to add or subtract two radicals such as $\sqrt{2}$ and $\sqrt{5}$ is similar to adding and subtracting x and y. Example 12.14 shows the similarity between combining radicals and combining algebraic expressions.

EXAMPLE 12.14

Combining Radicals	Combining Algebraic Expressions
(a) $\sqrt{2} + \sqrt{5} + \sqrt{2}$	$x + y + x = 2x + y$, where
$= 2\sqrt{2} + \sqrt{5}$	$x = \sqrt{2}$ and $y = \sqrt{5}$
(b) $2\sqrt{3} + 4\sqrt{7} + 6\sqrt{3}$	$2a + 4b + 6a = 8a + 4b$, where
$= 8\sqrt{3} + 4\sqrt{7}$	$a = \sqrt{3}$ and $b = \sqrt{7}$
(c) $3\sqrt[3]{6} - 7\sqrt[3]{6} = -4\sqrt[3]{6}$	$3p - 7p = -4p$, where
	$p = \sqrt[3]{6}$
(d) $\dfrac{5\sqrt{3}}{2} - \dfrac{\sqrt{3}}{2} = \dfrac{4\sqrt{3}}{2} = 2\sqrt{3}$	$\dfrac{5x}{2} - \dfrac{x}{2} = \dfrac{4x}{2} = 2x$, where
	$x = \sqrt{3}$

In each case, only the similar radicals are combined. Radicals that are not similar remain as separate terms and the addition or subtraction is only indicated. Sometimes it is possible to combine radicals that do not appear to be similar by first simplifying the radicals. For example $\sqrt{8}$ and $\sqrt{2}$ can be combined, since $\sqrt{8} = \sqrt{4}\sqrt{2} = 2\sqrt{2}$.

EXAMPLE 12.15

Simplify and combine similar radicals.
(a) $\sqrt{2} + \sqrt{8} = \sqrt{2} + 2\sqrt{2} = 3\sqrt{2}$
(b) $3\sqrt{8} + 5\sqrt{18} = 6\sqrt{2} + 15\sqrt{2} = 21\sqrt{2}$

EXAMPLE 12.15 (Cont.)

Note that $\sqrt{18} = \sqrt{9}\sqrt{2} = 3\sqrt{2}$ and $5\sqrt{18} = 5\sqrt{9}\sqrt{2} = 5 \cdot 3\sqrt{2} = 15\sqrt{2}$.

(c) $\sqrt{98x} + \sqrt{32x} = 7\sqrt{2x} + 4\sqrt{2x} = 11\sqrt{2x}$

(d) $\sqrt{20x^3} + \sqrt{8x^2} - \sqrt{45x} = 2x\sqrt{5x} + 2x\sqrt{2} - 3\sqrt{5x}$

$$= (2x - 3)\sqrt{5x} + 2x\sqrt{2}$$

Note that we can only indicate the difference of $2x\sqrt{5x} - 3\sqrt{5x}$ as $(2x - 3)\sqrt{5x}$. We factored $\sqrt{5x}$ out of each term.

(e) $\sqrt{\dfrac{5}{3}} + \dfrac{\sqrt{3}}{\sqrt{125}} = \dfrac{\sqrt{5}}{\sqrt{3}} + \dfrac{\sqrt{3}}{5\sqrt{5}} = \dfrac{\sqrt{5}\sqrt{3}}{\sqrt{3}\sqrt{3}} + \dfrac{\sqrt{3}\sqrt{5}}{5\sqrt{5}\sqrt{5}}$

$$= \dfrac{\sqrt{15}}{3} + \dfrac{\sqrt{15}}{25}$$

$$= \dfrac{25\sqrt{15}}{75} + \dfrac{3\sqrt{15}}{75}$$

$$= \dfrac{28\sqrt{15}}{75}$$

Here, we rationalized the denominators and then added the fractions by finding the common denominator. We could have rationalized the second fraction by multiplying by $\sqrt{125}$, but it was easier to first simplify the denominator to $5\sqrt{5}$ then multiply by $\sqrt{5}$.

Multiplying Radicals with the Same Index

Multiplying and dividing radicals that have the same index use two of the rules we have discussed.

$$\sqrt[n]{a}\sqrt[n]{b} = \sqrt[n]{ab}$$

and $\quad \dfrac{\sqrt[n]{a}}{\sqrt[n]{b}} = \sqrt[n]{\dfrac{a}{b}}, \qquad b \neq 0$

EXAMPLE 12.16

Multiply the following radical expressions: **(a)** $\sqrt{2}\sqrt{5}$, **(b)** $\sqrt[3]{5}\sqrt[3]{10}$, **(c)** $\sqrt{xy}\sqrt{2x}$, **(d)** $\sqrt[3]{\dfrac{3}{2}}\sqrt[3]{\dfrac{5x}{4}}$, **(e)** $\sqrt{5x}(\sqrt{5x} + \sqrt{10x^3})$, and **(f)** $(\sqrt{x} + \sqrt{y})(\sqrt{x} - \sqrt{y})$.

Solutions

(a) $\sqrt{2}\sqrt{5} = \sqrt{2 \cdot 5} = \sqrt{10}$

(b) $\sqrt[3]{5}\sqrt[3]{10} = \sqrt[3]{5 \cdot 10} = \sqrt[3]{50}$

(c) $\sqrt{xy}\sqrt{2x} = \sqrt{(xy)(2x)} = \sqrt{2x^2y} = x\sqrt{2y}$

(d) $\sqrt[3]{\dfrac{3}{2}}\sqrt[3]{\dfrac{5x}{4}} = \sqrt[3]{\left(\dfrac{3}{2}\right)\left(\dfrac{5x}{4}\right)} = \sqrt[3]{\dfrac{15x}{8}}$

$$= \dfrac{\sqrt[3]{15x}}{2}$$

EXAMPLE 12.16 (Cont.)

(e) $\sqrt{5x}\left(\sqrt{5x} + \sqrt{10x^3}\right) = \sqrt{5x}\sqrt{5x} + \sqrt{5x}\sqrt{10x^3}$

$$= \sqrt{25x^2} + \sqrt{50x^4}$$

$$= 5x + 5x^2\sqrt{2}$$

(f) $\left(\sqrt{x} + \sqrt{y}\right)\left(\sqrt{x} - \sqrt{y}\right) = \left(\sqrt{x}\right)^2 - \left(\sqrt{y}\right)^2 = x - y$ using the special product for the difference of two squares.

Multiplying Radicals with Different Indices

Radicals with different indices can be multiplied if they are rewritten so that they have the same index. The easiest way to do this is with fractional exponents.

EXAMPLE 12.17

Multiply the following: **(a)** $\sqrt[3]{2}\sqrt{7}$, **(b)** $\sqrt[4]{5x^2}\sqrt[3]{2x}$, and **(c)** $\sqrt[3]{4ab^2}\sqrt[5]{16a^4b^2}$.

Solutions   **(a)** $\sqrt[3]{2}\sqrt{7} = 2^{1/3}7^{1/2} = 2^{2/6}7^{3/6} = (2^2 7^3)^{1/6}$

$$= \sqrt[6]{2^2 7^3}$$

$$= \sqrt[6]{1,372}$$

(b) $\sqrt[4]{5x^2}\sqrt[3]{2x} = (5x^2)^{1/4}(2x)^{1/3} = (5x^2)^{3/12}(2x)^{4/12}$

$$= \sqrt[12]{(5x^2)^3}\ \sqrt[12]{(2x)^4}$$

$$= \sqrt[12]{(5x^2)^3(2x)^4}$$

$$= \sqrt[12]{5^3 x^6 2^4 x^4}$$

$$= \sqrt[12]{2,000x^{10}}$$

(c) $\sqrt[3]{4ab^2}\sqrt[5]{16a^4b^2} = (4ab^2)^{1/3}(16a^4b^2)^{1/5}$

$$= (4ab^2)^{5/15}(16a^4b^2)^{3/15}$$

$$= \sqrt[15]{(4ab^2)^5}\ \sqrt[15]{(16a^4b^2)^3}$$

$$= \sqrt[15]{(2^2ab^2)^5}\ \sqrt[15]{(2^4a^4b^2)^3}$$

$$= \sqrt[15]{2^{10}a^5b^{10}}\ \sqrt[15]{2^{12}a^{12}b^6}$$

$$= \sqrt[15]{2^{22}a^{17}b^{16}}$$

$$= 2ab\ \sqrt[15]{2^7 a^2 b}$$

$$= 2ab\ \sqrt[15]{128a^2b}$$

Division of Radicals

Division of radicals with the same index is done using Rule 2 for radicals. The result is usually simplified by rationalizing the denominator.

EXAMPLE 12.18

Divide each of the following: **(a)** $\dfrac{\sqrt{10}}{\sqrt{3}}$, **(b)** $\dfrac{\sqrt[3]{4x}}{\sqrt[3]{2x}}$, **(c)** $\dfrac{\sqrt[3]{x^2y}}{\sqrt[3]{z}}$, and **(d)** $\dfrac{\sqrt{3xy}}{\sqrt{7x^5y}}$.

Solutions

(a) $\dfrac{\sqrt{10}}{\sqrt{3}} = \dfrac{\sqrt{10}}{\sqrt{3}} \cdot \dfrac{\sqrt{3}}{\sqrt{3}} = \dfrac{\sqrt{30}}{3}$

(b) $\dfrac{\sqrt[3]{4x}}{\sqrt[3]{2x}} = \sqrt[3]{\dfrac{4x}{2x}} = \sqrt[3]{2}$

(c) $\dfrac{\sqrt[3]{x^2y}}{\sqrt[3]{z}} = \dfrac{\sqrt[3]{x^2y}}{\sqrt[3]{z}} \dfrac{\sqrt[3]{z^2}}{\sqrt[3]{z^2}} = \dfrac{\sqrt[3]{x^2yz^2}}{z}$

(d) $\dfrac{\sqrt{3xy}}{\sqrt{7x^5y}} = \dfrac{\sqrt{3xy}}{x^2\sqrt{7xy}} = \dfrac{1}{x^2} \cdot \sqrt{\dfrac{3xy}{7xy}} = \dfrac{1}{x^2}\sqrt{\dfrac{3}{7}}$

We now rationalize the denominator.

$$= \dfrac{1}{x^2}\sqrt{\dfrac{3}{7}}\sqrt{\dfrac{7}{7}} = \dfrac{\sqrt{21}}{7x^2}$$

Sometimes the denominator of a fraction is the sum of the difference of two square roots. Examples of this are $\sqrt{2}+\sqrt{5}$ and $\sqrt{x}-\sqrt{y}$. In this case, the numerator and denominator can be multiplied by a quantity that will make the denominator the difference of two squares. For example, if the denominator is $\sqrt{2}+\sqrt{5}$, then multiply the numerator and denominator by $\sqrt{2}-\sqrt{5}$. If the denominator is $\sqrt{x}-\sqrt{y}$, then multiply both numerator and denominator by $\sqrt{x}+\sqrt{y}$. Notice that in each case the denominator is then in the form $a^2 - b^2$, where a or b is a radical.

EXAMPLE 12.19

Rationalize each of these denominators: **(a)** $\dfrac{1}{\sqrt{2}+\sqrt{5}}$, **(b)** $\dfrac{a}{\sqrt{x}-\sqrt{y}}$, and **(c)** $\dfrac{\sqrt{x+y}}{1+\sqrt{x+y}}$.

Solutions

(a) $\dfrac{1}{\sqrt{2}+\sqrt{5}} = \dfrac{1}{\sqrt{2}+\sqrt{5}} \cdot \dfrac{\sqrt{2}-\sqrt{5}}{\sqrt{2}-\sqrt{5}}$

$$= \dfrac{\sqrt{2}-\sqrt{5}}{(\sqrt{2})^2 - (\sqrt{5})^2}$$

$$= \dfrac{\sqrt{2}-\sqrt{5}}{2-5}$$

$$= -\dfrac{\sqrt{2}-\sqrt{5}}{3} = \dfrac{\sqrt{5}-\sqrt{2}}{3}$$

EXAMPLE 12.19 (Cont.)

(b) $\dfrac{a}{\sqrt{x} - \sqrt{y}} = \dfrac{a}{\sqrt{x} - \sqrt{y}} \cdot \dfrac{\sqrt{x} + \sqrt{y}}{\sqrt{x} + \sqrt{y}}$

$\qquad = \dfrac{a(\sqrt{x} + \sqrt{y})}{(\sqrt{x})^2 - (\sqrt{y})^2}$

$\qquad = \dfrac{a\sqrt{x} + a\sqrt{y}}{x - y}, \ x \neq y$

(c) $\dfrac{\sqrt{x + y}}{1 + \sqrt{x + y}} = \dfrac{\sqrt{x + y}}{1 + \sqrt{x + y}} \cdot \dfrac{1 - \sqrt{x + y}}{1 - \sqrt{x + y}}$

$\qquad = \dfrac{\sqrt{x + y}(1 - \sqrt{x + y})}{(1)^2 - (\sqrt{x + y})^2}$

$\qquad = \dfrac{\sqrt{x + y} - (\sqrt{x + y})^2}{1 - (x + y)}$

$\qquad = \dfrac{\sqrt{x + y} - x - y}{1 - x - y}, \quad x + y \neq 1$

Finally, to find the quotient of two radicals with different indices, we use fractional exponents in the same manner as when we multiplied radicals with different indices.

EXAMPLE 12.20

Find the following quotients: **(a)** $\dfrac{\sqrt[3]{15}}{\sqrt{15}}$ and **(b)** $\dfrac{\sqrt{2x}}{\sqrt[4]{8x^3}}$.

Solutions

(a) $\dfrac{\sqrt[3]{15}}{\sqrt{15}} = \dfrac{(15)^{1/3}}{(15)^{1/2}} = \dfrac{(15)^{2/6}}{(15)^{3/6}} = \dfrac{\sqrt[6]{15^2}}{\sqrt[6]{15^3}} = \dfrac{\sqrt[6]{15^2}}{\sqrt[6]{15^3}} \cdot \dfrac{\sqrt[6]{15^3}}{\sqrt[6]{15^3}}$

$\qquad\qquad = \dfrac{\sqrt[6]{15^5}}{15}$

(b) $\dfrac{\sqrt{2x}}{\sqrt[4]{8x^3}} = \dfrac{(2x)^{1/2}}{(8x^3)^{1/4}} = \dfrac{(2x)^{2/4}}{(8x^3)^{1/4}} = \dfrac{\sqrt[4]{(2x)^2}}{\sqrt[4]{8x^3}} = \dfrac{\sqrt[4]{4x^2}\,\sqrt[4]{2x}}{\sqrt[4]{8x^3}\,\sqrt[4]{2x}}$

$\qquad\qquad = \dfrac{\sqrt[4]{8x^3}}{2x}$

Exercise Set 12.3

In Exercises 1–52, perform the indicated operations and express the answers in simplest form.

1. $2\sqrt{3} + 5\sqrt{3}$

2. $5\sqrt{6} - 3\sqrt{6}$

3. $\sqrt[3]{9} + 4\sqrt[3]{9}$

4. $\sqrt[4]{8} + 3\sqrt[4]{8}$

5. $2\sqrt{3} + 4\sqrt{2} + 6\sqrt{3}$

6. $5\sqrt{3} - 6\sqrt{5} - 9\sqrt{3}$

7. $\sqrt{5} + \sqrt{20}$

8. $\sqrt{8} + \sqrt{2}$

9. $\sqrt{7} - \sqrt{28}$

10. $\sqrt{8} - \sqrt{32}$

11. $\sqrt{60} - \sqrt{\dfrac{5}{3}}$

12. $\sqrt{84} + \sqrt{\dfrac{3}{7}}$

13. $\sqrt{\dfrac{1}{2}} - \sqrt{\dfrac{9}{2}}$

14. $\sqrt{\dfrac{4}{3}} - \sqrt{\dfrac{25}{3}}$

15. $\sqrt{x^3 y} + 2x\sqrt{xy}$

16. $\sqrt{a^5 b^3} - 3ab\sqrt{a^3 b}$

17. $\sqrt[3]{24p^2 q^4} + \sqrt[3]{3p^8 q}$

18. $\sqrt[4]{16a^2 b} - \sqrt[4]{81a^6 b}$

19. $\sqrt{\dfrac{x}{y^3}} - \sqrt{\dfrac{y}{x^3}}$

20. $\sqrt{\dfrac{a^3}{b^3}} + \sqrt{\dfrac{b}{a^5}}$

21. $a\sqrt{\dfrac{b}{3a}} + b\sqrt{\dfrac{a}{3b}}$

22. $x\sqrt{\dfrac{y}{5x}} - y\sqrt{\dfrac{x}{5y}}$

23. $\sqrt{5}\sqrt{8}$

24. $\sqrt[5]{-7}\sqrt[5]{11}$

25. $\sqrt{3x}\sqrt{5x}$

26. $\sqrt[3]{7x^2}\sqrt[3]{3x}$

27. $\left(\sqrt{4x}\right)^3$

28. $\left(\sqrt[3]{2x^2 y}\right)^4$

29. $\sqrt{\dfrac{5}{8}}\sqrt{\dfrac{9}{10}}$

30. $\sqrt{\dfrac{7}{6}}\sqrt{\dfrac{12}{3}}$

31. $\sqrt{2}(\sqrt{x} + \sqrt{2})$

32. $\sqrt{3}(\sqrt{12} - \sqrt{y})$

33. $(\sqrt{x} + \sqrt{y})^2$

34. $\left(\sqrt{a} + 3\sqrt{b}\right)^2$

35. $\sqrt[3]{\dfrac{5}{2}}\sqrt[3]{\dfrac{2}{7}}$

36. $\sqrt{\dfrac{ab}{5c}}\sqrt{\dfrac{abc}{5}}$

37. $(\sqrt{a} + \sqrt{b})(\sqrt{a} - \sqrt{b})$

38. $(\sqrt{x} - 2\sqrt{y})(\sqrt{x} + 5\sqrt{y})$

39. $\sqrt[3]{x}\sqrt{x}$

40. $\sqrt{5x}\sqrt[3]{2x^2}$

41. $\dfrac{\sqrt{32}}{\sqrt{2}}$

42. $\dfrac{\sqrt[3]{a^2}}{\sqrt[3]{4a}}$

43. $\dfrac{\sqrt[3]{4b^2}}{\sqrt[3]{16b}}$

44. $\dfrac{\sqrt{5a^3 b}}{\sqrt{15ab^3}}$

45. $\dfrac{1}{x + \sqrt{5}}$

46. $\dfrac{1}{x + \sqrt{3}}$

47. $\dfrac{\sqrt{5} - \sqrt{3}}{\sqrt{5} + \sqrt{3}}$

48. $\dfrac{\sqrt{5} + \sqrt{7}}{\sqrt{7} - \sqrt{5}}$

49. $\dfrac{\sqrt{x+1}}{\sqrt{x-1}} + \dfrac{\sqrt{x-1}}{\sqrt{x+1}}$

50. $\dfrac{\sqrt{x+3}}{\sqrt{x-3}} - \dfrac{\sqrt{x-3}}{\sqrt{x+3}}$

51. $\dfrac{\sqrt{x+y}}{\sqrt{x-y} - \sqrt{x}}$

52. $\dfrac{\sqrt{1+y}}{\sqrt{1-y} + \sqrt{y}}$

Solve Exercises 53–57.

53. Use the quadratic equation to find the roots of $ax^2 + bx + c = 0$. What is the sum of the two roots?

54. *Oceanography* The velocity v of a small water wave is given by

$$v = \sqrt{\frac{\pi}{4\lambda d}} + \sqrt{\frac{4\pi}{\lambda d}}$$

Simplify and combine this equation.

55. *Electricity* The equivalent resistance R of two resistors, R_1 and R_2, connected in parallel is expressed

$$R = \frac{R_1 R_2}{R_1 + R_2}$$

In a given circuit, $R_1 = x^{3/2}$ and $R_2 = \sqrt{x}$. **(a)** Express R in terms of x. Make sure you simplify the answer. **(b)** If $x = 20\ \Omega$, what is R?

56. *Sound* The theory of waves in wires uses the equation

$$\frac{\sqrt{d_1} - \sqrt{d_2}}{\sqrt{d_1} + \sqrt{d_2}}$$

Simplify this expression.

57. What is the product of the two roots of the quadratic equation $ax^2 + bx + c = 0$?

≡ **12.4**

EQUATIONS WITH RADICALS

You should now know the laws of radicals. However, the laws of radicals are rather dry without some technical and algebraic application. Because technical situations can involve solving equations that contain radicals, we now learn how to use radicals in equations and to solve those equations.

Radical Equations; Extraneous Roots

Working with radicals often means that we must work with radical equations. An equation in which the variable occurs under a radical sign or has a fractional exponent is called a **radical equation**. In order to solve radical equations, we need to eliminate the radicals or the fractional exponents. This is done by raising both sides of the equation to some power that will eliminate the radical. The equation that results may not be equivalent to the original equation. In fact, the new equation may have more roots than the original one. The roots of the new equation that are not roots of the original equation are called **extraneous roots**. Check all solutions to make sure they are actual roots. Reject any extraneous roots.

EXAMPLE 12.21

Solve the radical equation $\sqrt{x+4} - 9 = 0$.

Solution This radical equation has only one radical term. So, isolate this radical term on one side of the equation.

$$\sqrt{x+4} = 9$$

$$\left(\sqrt{x+4}\right)^2 = 9^2 \qquad \text{Square both sides.}$$

EXAMPLE 12.21 (Cont.)

$$x + 4 = 81$$

$$x = 77$$

Substituting 77 in the original equation, we get $\sqrt{77+4} - 9 = \sqrt{81} - 9 = 9 - 9 = 0$. The answer checks, so 77 is the solution.

EXAMPLE 12.22

Solve $\sqrt{x+5} - \sqrt{x-3} = 10$.

Solution This equation contains two radicals. Rewrite the equation with one radical on each side of the equals sign.

$$\sqrt{x+5} = \sqrt{x-3} + 10$$

$$\left(\sqrt{x+5}\right)^2 = \left(\sqrt{x-3} + 10\right)^2 \qquad \text{Square both sides.}$$

$$x + 5 = \left(\sqrt{x-3}\right)^2 + 20\sqrt{x-3} + 10^2 \qquad \text{"FOIL"}$$

$$x + 5 = x - 3 + 20\sqrt{x-3} + 100$$

$$x + 5 = x + 97 + 20\sqrt{x-3}$$

This is now an equation with one radical. Isolate the term with the radical so it is on one side of the equation, and use the method of squaring both sides.

$$-92 = 20\sqrt{x-3}$$

$$(-92)^2 = \left(20\sqrt{x-3}\right)^2$$

$$8{,}464 = 400\,(x-3)$$

$$8{,}464 = 400x - 1{,}200$$

$$9{,}664 = 400x$$

$$x = \frac{9{,}664}{400} = \frac{604}{25}$$

$$= 24.16$$

Replacing $x = 24.16$ in the left-hand side of the original equation gives $\sqrt{29.16} - \sqrt{21.16} = 0.8$. This does not equal the right-hand side of the original equation, 10. Since we have not made any mistakes, we can only conclude that $x = 24.16$ is an extraneous root and that there is no solution to this problem.

If we had not wanted to show how you can get extraneous roots when you square both sides, we could have solved the previous example with less work. Notice the equation $-92 = 20\sqrt{x-3}$. Since $\sqrt{x-3}$ is never negative, the product of $20\sqrt{x-3}$ can never be negative. So, there is no solution to the equation $-92 = 20\sqrt{x-3}$. The next example also involves an extraneous root.

EXAMPLE 12.23

Solve $\sqrt{2x-3} - \sqrt{x+7} = 4$.

Solution As in the last example, this equation contains two radicals, so we will first rewrite the problem with one radical on each side of the equal sign.

$$\sqrt{2x-3} = \sqrt{x+7} + 4$$

$$\left(\sqrt{2x-3}\right)^2 = \left(\sqrt{x+7}+4\right)^2 \qquad \text{Square both sides.}$$

$$\left(\sqrt{2x-3}\right)^2 = \left(\sqrt{x+7}\right)^2 + 8\sqrt{x+7} + 4^2$$

$$2x - 3 = x + 7 + 8\sqrt{x+7} + 16$$

$$x - 26 = 8\sqrt{x+7} \qquad \text{Simplify.}$$

$$(x-26)^2 = \left(8\sqrt{x+7}\right)^2 \qquad \text{Square again.}$$

$$x^2 - 52x + 676 = 64(x+7)$$

$$x^2 - 52x + 676 = 64x + 448$$

$$x^2 - 116x + 228 = 0$$

$$(x-2)(x-114) = 0$$

So, $x = 2$ or $x = 114$.

Replacing 2 for x in the original equation, we get $\sqrt{2 \cdot 2 - 3} - \sqrt{2+7} = \sqrt{1} - \sqrt{9} = 1 - 3 = -2$. Since $-2 \neq 4$, the number 2 must be an extraneous root.

Next we replace 114 for x in the original equation, and we get $\sqrt{2 \cdot 114 - 3} - \sqrt{114+7} = \sqrt{225} - \sqrt{121} = 15 - 11 = 4$. This checks. The only solution to this problem is 114.

EXAMPLE 12.24

Solve $\sqrt{x+3} = \sqrt[3]{x+21}$.

Solution This problem also involves two radicals, but the indices of the radicals are not the same. If these were written with fractional exponents, you would have $(x+3)^{1/2} = (x+21)^{1/3}$. Raise both sides of the equation to the least common denominator of these powers. In this problem, the LCD is 6, so we would have

$$\left(\sqrt{x+3}\right)^6 = \left(\sqrt[3]{x+21}\right)^6$$

$$(x+3)^3 = (x+21)^2$$

$$x^3 + 9x^2 + 27x + 27 = x^2 + 42x + 441$$

$$x^3 + 8x^2 - 15x - 414 = 0$$

As you will learn in Chapter 18, the only real number solution to this problem is $x = 6$. Check this in the original problem to see if it is a root.

EXAMPLE 12.25

Solve $\sqrt{(x+1)^3} = 64$.

Solution $\sqrt{(x+1)^3}$ can be written as $(x+1)^{3/2}$. Raising both sides to the $\frac{2}{3}$ power we get

$$\left((x+1)^{3/2}\right)^{2/3} = 64^{2/3}$$
$$x + 1 = 16$$
$$x = 15$$

Checking this in the original problem confirms that it is a root.

Application

EXAMPLE 12.26

At some point P between two light sources, the illuminance from each source is the same. If d_1 represents the distance from the light source with intensity of I_1 and d_2 is the distance from the source with intensity I_2, then

$$\frac{d_1}{d_2} = \frac{\sqrt{I_1}}{\sqrt{I_2}}$$

Find I_1 and I_2 in candelas (cd), when $d_1 = 15$ m, $d_2 = 6$ m, and $I_2 = 5I_1 - 11$.

Solution Substituting the given values in the previous equation produces

$$\frac{d_1}{d_2} = \frac{\sqrt{I_1}}{\sqrt{I_2}}$$

$$\frac{15}{6} = \frac{\sqrt{I_1}}{\sqrt{5I_1 - 11}}$$

Squaring both sides, we obtain

$$\frac{225}{36} = \frac{I_1}{5I_1 - 11}$$

or $225(5I_1 - 11) = 36I_1$

$$1\,125I_1 - 2\,475 = 36I_1$$
$$1\,125I_1 - 36I_1 = 2\,475$$
$$1\,089I_1 = 2\,475$$
$$I_1 = \frac{2\,475}{1\,089} \approx 2.27$$

So, I_1 is approximately 2.27 cd.

Substituting the exact value $I_1 = \frac{2\,475}{1\,089}$ in the equation $I_2 = 5I_1 - 11$, we determine

$$I_2 = 5\left(\frac{2\,475}{1\,089}\right) - 11 = \left(\frac{12\,375}{1\,089}\right) - 11 \approx 0.36$$

The illuminance at I_2 is approximately 0.36 cd.

Exercise Set 12.4

In Exercises 1–20, solve the radical equations. Check all roots.

1. $\sqrt{x+3} = 5$

2. $\sqrt{x-16} = 5$

3. $\sqrt{2x+4} - 7 = 0$

4. $\sqrt{3y-9} - 18 = 0$

5. $\sqrt{x^2+24} = x - 4$

6. $\sqrt{x^2-75} = x + 5$

7. $(y+12)^{1/3} = 3$

8. $(r+9)^{1/4} = 7$

9. $\left(\dfrac{x}{2}+1\right)^{1/2} = 3$

10. $\left(\dfrac{y}{3}+4\right)^{1/3} = \dfrac{5}{2}$

11. $(x-1)^{3/2} = 27$

12. $(x+1)^{3/4} = 8$

13. $\sqrt{x+1} = \sqrt{2x-1}$

14. $\sqrt{x^2-4} = \sqrt{2x-1}$

15. $\sqrt{x+1} + \sqrt{x-5} = 7$

16. $\sqrt{x-3} - \sqrt{x+2} = 19$

17. $\sqrt{x-1} + \sqrt{x+5} = 2$

18. $\sqrt{x+3} - \sqrt{2x-4} = 3$

19. $\sqrt{\dfrac{1}{x}} = \sqrt{\dfrac{4}{3x-1}}$

20. $\sqrt{\dfrac{2}{x-1}} = \sqrt{\dfrac{5}{x+1}}$

Solve Exercises 21–24.

21. *Civil engineering* The maximum speed at which it is possible to negotiate a curve without slipping is given by

$$v = \sqrt{\mu_s g R}$$

Solve this for R.

22. *Physics* The velocity v of an object falling under the influence of gravity g is given by $v = \sqrt{v_0{}^2 - 2gh}$, where v_0 is the initial velocity and h is the height fallen. Solve this equation for h.

23. *Lighting technology* The illuminance from two light sources is the same at some point P between them. Suppose d_A is the distance to P from light A and d_B is the distance from P to light B. If the luminous intensity at A is I_A and at B is I_B, then we have the equation

$$\frac{d_A}{d_B} = \frac{\sqrt{I_A}}{\sqrt{I_B}}$$

If $d_A = 11$ m, $d_B = 20$ m, and $I_A = 3.5 I_B - 15$, then find I_A and I_B.

24. *Lighting technology* The intensity of a light source can be determined with a photometer. The relationship between the luminous intensity I_x of an unknown source is found by comparing it with a standard source of known intensity I_y. The distances from each source are adjusted until a grease spot is equally illuminated by each source. If r_x is the distance to the unknown source and r_y is the distance to the standard, then the equation

$$\frac{\sqrt{I_x}}{r_x} = \frac{\sqrt{I_y}}{r_y}$$

can be used. **(a)** Solve this equation for I_x. **(b)** If $I_y = 30$ cd and $r_x = 2r_y + 3$, then solve for I_x.

☰ CHAPTER 12 REVIEW

Important Terms and Concepts

Adding radicals

Dividing radicals

Extraneous roots

Fractional exponent

Index

Multiplying radicals

Radical

Radical equations

Radicand

Rationalizing denominators

Subtracting radicals

Review Exercises

In Exercises 1–6, evaluate the given expression without the use of a calculator. Check your answers with a calculator.

1. $49^{1/2}$

2. $81^{1/4}$

3. $25^{3/2}$

4. $64^{2/3}$

5. $9^{-3/2}$

6. $125^{-4/3}$

In Exercises 7–18, express each of the given expressions in the simplest form containing only positive exponents.

7. $5^4 5^9$

8. $13^6 13^{-9}$

9. $\left(2^4\right)^8$

10. $\left(5^3\right)^6$

11. $\left(x^2 y^3\right)^4$

12. $\left(y^6 x\right)^4$

13. $\left(\dfrac{a^3}{ya^4}\right)^{-5}$

14. $\left(\dfrac{a^5}{ab^4}\right)^3$

15. $x^{1/2} x^{1/3}$

16. $y^{2/3} y^{3/4}$

17. $\dfrac{\left(xy^2 a\right)^5}{x^4 \left(ya^3\right)^2}$

18. $\dfrac{\left(ab^3 c^2\right)^4}{\left(a^2 bc\right)^3}$

Use the laws of radicals to express each of Exercises 19–42 in simplest radical form.

19. $\sqrt[3]{-27}$

20. $\sqrt[4]{81^3}$

21. $\sqrt{a^2 b^5}$

22. $\sqrt[4]{a^3 b^7}$

23. $\sqrt[3]{\dfrac{-8x^3 y^6}{z}}$

24. $\sqrt[4]{\dfrac{32a^6 b^8}{cd^2}}$

25. $\sqrt{\dfrac{a}{b^2} - \dfrac{b}{a^2}}$

26. $\sqrt{a^2 + 8a + 16}$

27. $2\sqrt{5} + 6\sqrt{5}$

28. $3\sqrt{7} - 6\sqrt{7}$

29. $3\sqrt{6} + 5\sqrt{6} - 2\sqrt{3}$

30. $\sqrt{6} + \sqrt{24}$

31. $\sqrt[4]{16a^3 b} + \sqrt[4]{81a^7 b}$

32. $\sqrt[4]{32x^5} - x^2 \sqrt[4]{2x}$

33. $a\sqrt{\dfrac{a}{7b}} - b\sqrt{\dfrac{b}{7a}}$

34. $\sqrt{\dfrac{x-2}{x+2}} + \sqrt{\dfrac{x+2}{x-2}}$

35. $\sqrt{3}\sqrt{5}$

36. $\sqrt{7}\sqrt{6}$

37. $\sqrt{2x}\sqrt{7x}$

38. $\sqrt[3]{5y}\sqrt[3]{25y}$

39. $\left(\sqrt{a} + \sqrt{2b}\right)\left(\sqrt{a} - \sqrt{2b}\right)$

40. $\left(\sqrt{x} - 3\sqrt{y}\right)\left(\sqrt{x} + 3\sqrt{y}\right)$

41. $\dfrac{\sqrt[3]{2a^2}}{\sqrt[3]{16a}}$

42. $\dfrac{\sqrt[3]{5a^2 b}}{\sqrt[3]{25ab^2}}$

In Exercises 43–50, solve the radical equations. Check your answers for extraneous roots.

43. $\sqrt{x-1} = 9$

44. $\sqrt{2x-7} - 5 = 0$

45. $\sqrt{x^2 - 7} = 2x - 4$

46. $\sqrt{\dfrac{3x}{4} - 1} = 5x - 3$

47. $\sqrt{x+1} + \sqrt{x-5} = 8$

48. $\sqrt{x+3} + \sqrt{x-2} = 20$

49. $\sqrt{\dfrac{1}{x}} = \sqrt{\dfrac{3x}{4} - 1}$

50. $\sqrt{\dfrac{3}{x-1}} = \sqrt{\dfrac{4}{x+1}}$

Solve Exercises 51–54.

51. *Nuclear technology* The half-life of tritium is 12.5yr. It decays according to the formula $q(t) = q_0 2^{-t/12.5}$, where q_0 is the original amount of tritium and $q(t)$ is the amount left at time t, where t is given in years. If you begin with 50 mg of tritium, **(a)** how much will remain after 10 years? **(b)** after 20 years?

52. *Nuclear technology* The velocity of a proton can be described in terms of its kinetic energy, KE, and mass, m, with the formula

$$v = \sqrt{\frac{2KE}{m}}$$

(a) Express this in simplest radical form. **(b)** Solve the equation for m.

53. *Navigation* Figure 12.1 shows the vector diagram for the resultant velocity v of a missile with horizontal component $\mathbf{v}_x$ and vertical component $\mathbf{v}_y$. If $\mathbf{v}_x = 3t + 2$ and $\mathbf{v}_y = 4t - 1$, express v in terms of t in simplest radical form.

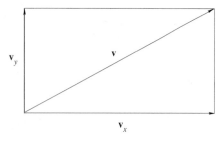

FIGURE 12.1

54. *Energy technology* Two sources, A and B, are 12 m apart as shown in Figure 12.2. Each emit waves of wavelength $\lambda = 5m$. **Constructive interference** occurs when the joint effect of the two waves results in a wave of larger amplitude. Here, constructive interference of the waves will occur at point C, located x units from A when

$$BC - AC = \lambda$$

How far is point C from A? How far from B?

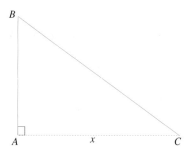

FIGURE 12.2

≡ CHAPTER 12 TEST

In Exercises 1–4, express each of the given expressions in the simplest form containing only positive exponents.

1. $8^5 8^{-7}$

2. $\left(3^5\right)^4$

3. $\left(\dfrac{x^2}{xy^3}\right)^5$

4. $x^{1/4} x^{4/3}$

In Exercises 5–13, use the laws of radicals to express each of the following in simplest form.

5. $\sqrt[3]{-64}$

6. $\sqrt[4]{81}$

7. $\sqrt{x^4 y^3}$

8. $\sqrt[3]{\dfrac{-27x^6 y}{z^2}}$

9. $\sqrt{x^2 + 6x + 9}$

10. $3\sqrt{6} + 5\sqrt{24}$

11. $\sqrt{2x}\sqrt{6x}$

12. $\left(3\sqrt{x} + \sqrt{5y}\right)\left(3\sqrt{x} - \sqrt{5y}\right)$

13. $\dfrac{\sqrt[3]{4x^2 y}}{\sqrt[3]{2x^2 y^2}}$

In Exercises 14–17, solve the radical equations.

14. $\sqrt{x+3} = 6$

15. $\sqrt{2x+1} = 5$

16. $\sqrt{2x^2 - 7} = x + 3$

17. $\sqrt{\dfrac{5}{x+3}} = \sqrt{\dfrac{4}{x-3}}$

Solve the following exercises.

18. The diameter d of a cylindrical column necessary for it to resist a crushing force F is

$$d = \left(1.12 \times 10^{-2}\right)\sqrt[3]{Fl}$$

where l is the length of the column. **(a)** Solve for F. **(b)** Determine F, when $d = 1.2$ m and $l = 5$ m.

CHAPTER

13

Exponential and Logarithmic Functions

Medical technologists and people working in microbiology often need to determine how many of a certain type of bacteria they can expect after a certain period of time. This type of growth tends to be exponential. In Section 13.2, you will learn how to determine exponential growth.

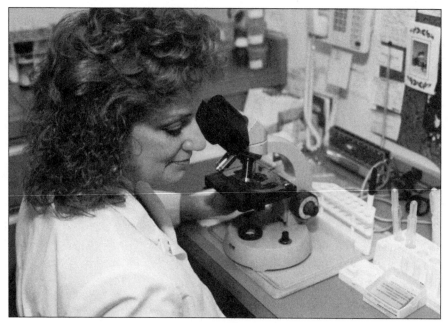

Courtesy of Ruby Gold

The first functions we studied in this text were algebraic functions. Then we introduced the trigonometric functions, examples of functions that were not algebraic. In this chapter, we will introduce two new functions. These functions are not algebraic functions, but they are very important in many technical areas, such as business, finance, nuclear technology, acoustics, electronics, and astronomy. Many of the applications in this chapter will involve growth or decay.

≣ 13.1
EXPONENTIAL FUNCTIONS

We will begin this section with the definition of an exponential function and with some examples of exponential functions.

Exponential Function

An **exponential function** is any function of the form

$$f(x) = b^x$$

where $b > 0$, $b \neq 1$, and x is any real number. The number b is called the **base**.

EXAMPLE 13.1

Some examples of exponential functions are: $f(x) = 2^x$, $g(x) = 3^x$, $h(x) = \pi^x$, $j(x) = 4.2^x$, and $k(x) = (\sqrt{3})^x$. The following exponential functions contain constants, represented by a and b: $y(x) = (5b)^{-x}$, $h(x) = (2a)^{x+1}$, and $k(x) = 3a^{x/3}$.

The functions $f(x) = (-2)^x$, $g(x) = 0^x$, and $h(x) = 1^x$ are not exponential functions. In $f(x) = (-2)^x$ the base, -2, is less than zero. Therefore, it is not an exponential function. We can see that if $x = \frac{1}{2}$, then $(-2)^{1/2} = \sqrt{-2}$ is not a real number. The function $g(x) = 0^x$ is not an exponential function, because the base is 0. Also, $h(x) = 1^x$ is not an exponential function, since the base is 1.

Graphing Exponential Functions; Asymptotes

Graphing an exponential function is done in the same manner in which we have graphed other functions. We will choose values for x and determine the corresponding values for $f(x)$. We will then plot these points and connect them in order to get the graph. Notice that, since the base b of an exponential function is positive, all powers of that base are also positive. Thus, an exponential function is positive for all values of x.

EXAMPLE 13.2

Graph the exponential function $f(x) = 2^x$.

Solution The following table gives integer values of x from -3 to 5.

x	-3	-2	-1	0	1	2	3	4	5
$f(x) = 2^x$	0.125	0.25	0.5	1	2	4	8	16	32

As x gets larger, $f(x)$ increases at a faster and faster rate. As x gets smaller, $f(x)$ keeps getting closer to 0, but never reaches 0.

The graph of $f(x) = 2^x$ is shown in Figure 13.1. Notice how the curve keeps getting closer to the negative x-axis. As a curve approaches a line when a variable approaches a certain value, the curve is said to be asymptotic to the line. The line is an **asymptote** of the curve. In this example, $y = 2^x$ is asymptotic to the negative x-axis.

Either a calculator or a computer is helpful for finding values of an exponential function. On a calculator, you will use the $\boxed{y^x}$ key. (On some calculators, this key is $\boxed{x^y}$ and on some graphics calculators it is $\boxed{\wedge}$.) On the $\boxed{y^x}$ key, the base is indicated by the letter y.

EXAMPLE 13.3

Determine $f(4.5)$, when $f(x) = 2^x$.

Solution

	PRESS	DISPLAY
Algebraic calculator:	2 $\boxed{x^y}$ 4.5 $\boxed{=}$	22.62741700
RPN calculator:	2 $\boxed{\text{ENTER}}$ 4.5 $\boxed{y^x}$	22.62741700
TI-81 calculator:	2 $\boxed{\wedge}$ 4.5 $\boxed{\text{ENTER}}$	22.627417

The most common bases are those that are larger than 1. In fact, when the base is less than 1 (and greater than 0), you get an interesting effect as shown in the next example.

EXAMPLE 13.4

Graph the exponential function $g(x) = \left(\frac{1}{2}\right)^x$.

Solution   Again, we will make a table of values.

x	-5	-4	-3	-2	-1	0	1	2	3
$g(x) = (\frac{1}{2})^x$	32	16	8	4	2	1	0.5	0.25	0.125

The graph of these points is shown in Figure 13.2. This graph would be identical to the graph of $f(x) = 2^x$, if it were reflected in the y-axis.

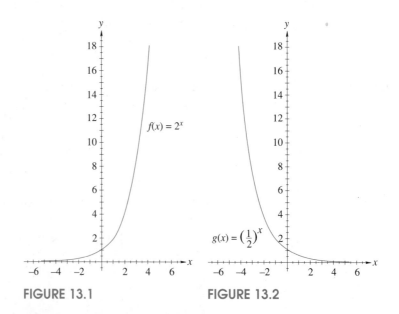

FIGURE 13.1   FIGURE 13.2

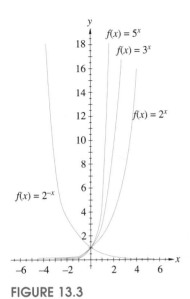

$f(x) = 5^x$

$f(x) = 3^x$

$f(x) = 2^x$

$f(x) = 2^{-x}$

FIGURE 13.3

Study the function again. Remember that $\frac{1}{2} = 2^{-1}$, so $\left(\frac{1}{2}\right)^x = 2^{-x}$. These two examples show the differences between a function of the form b^x and one of the form b^{-x}. If $b > 1$, the graph of b^x rises and the graph of b^{-x} falls, as we move from left to right.

Figure 13.3 shows the graph of several exponential functions on the same set of coordinates. These functions, and all exponential functions, have the following three features in common.

Common Features of Exponential Functions

If $f(x) = b^x$, then

1. The y-intercept is 1.
2. If $b > 1$, the negative x-axis is an asymptote; if $b < 1$, the positive x-axis is an asymptote.
3. If $b > 1$, the curves all rise as the values of x increase; if $b < 1$, the curves all fall as the values of x increase.

Business Applications

One application of exponential functions has to do with money. When an amount of money P, called the principal, is invested and interest is compounded annually (once a year) at the interest rate of r per year, then the amount of money at the end of t yr is given by the formula $S = P(1 + r)^t$. In this formula, r is expressed as a decimal. So, at 5% interest, $r = 0.05$ and at $6\frac{1}{4}\%$ interest, $r = 0.0625$.

Application

EXAMPLE 13.5

If \$800 is invested at 6% compounded annually, how much is the total value after 10 yr?

Solution   Here $P = \$800$, $r = 0.06$, and $t = 10$. As a result we have

$$S = P(1 + r)^t$$

$$= 800(1 + 0.06)^{10}$$

$$= 800(1.790847697)$$

$$\approx 1,432.68$$

After 10 yr, the total value of this investment is \$1,432.68.

If interest is compounded more than once a year, then the formula is changed. If interest is compounded semiannually, or twice a year, the formula becomes

$$S = P\left(1 + \frac{r}{2}\right)^{2t}$$

If interest is compounded quarterly, or four times a year, the formula is changed to

$$S = P\left(1 + \frac{r}{4}\right)^{4t}$$

In general, we have the following compound interest formula.

Compound Interest Formula

If an amount of money P, is invested and interest is compounded k times a year at an interest rate of r per year, then the amount of money S at the end of t yr is

$$S = P\left(1 + \frac{r}{k}\right)^{kt}$$

Application

EXAMPLE 13.6

If \$800 is invested in a savings account paying 6% interest compounded monthly, how much is this money worth after 10 yr?

Solution   In this example, $P = \$800$, $r = 0.06$, $k = 12$, and $t = 10$, so

$$S = P\left(1 + \frac{r}{k}\right)^{kt}$$

$$= 800\left(1 + \frac{0.06}{12}\right)^{12 \cdot 10}$$

$$= 800(1.005)^{120}$$

$$= 800(1.819396734)$$

$$\approx 1,455.52$$

Compare this to the result we obtained in Example 13.5. Compounding monthly increased the total by \$22.84.

In Section 13.2, we will examine what happens if you compound interest daily or continuously. We will see that there is a limit to the amount of money you will receive.

Exercise Set 13.1

In Exercises 1–6, use a calculator to approximate the given numbers. Round off each answer to four decimal places.

1. $3^{\sqrt{2}}$

2. 4^{π}

3. π^3

4. $(\sqrt{3})^{\pi}$

5. $(\sqrt{3})^{\sqrt{5}}$

6. $(\sqrt{4})^{4/3}$

In Exercises 7–16, make a table of values and draw the graph of each function.

7. $f(x) = 4^x$

8. $g(x) = 3^x$

9. $h(x) = 1.5^x$

10. $k(x) = 2.5^x$

11. $f(x) = 3^{-x}$

12. $g(x) = 5^{-x}$

13. $h(x) = 2.4^{-x}$

14. $k(x) = (\sqrt{3})^x$

15. $f(x) = 3^{(x+1/2)}$

16. $g(x) = 2^{(x-1/4)}$

Solve Exercises 17–20.

17. *Finance*   The sum of $1,000 is placed in a savings account at 6% interest. If interest is compounded **(a)** annually, **(b)** semiannually, **(c)** quarterly, and **(d)** monthly, what is the total after 5 yr?

18. *Finance*   The sum of $2,000 is placed in a savings account at 8% interest. If interest is compounded **(a)** annually, **(b)** semiannually, **(c)** quarterly, and **(d)** monthly, what is the total after 10 yr?

19. *Finance*   One bank offers 6% interest compounded semiannually. A second bank offers the same interest but compounded monthly. How much more income will result by depositing $1,000 in the second account for 5 yr than by depositing $1,000 in the first account for 5 yr?

20. *Finance*   A bank offers 8% interest compounded annually. A second bank offers 8% interest compounded quarterly. How much more income will result by depositing $2,000 in the second bank for 10 yr than in the first?

**THE EXPONENTIAL
FUNCTION e^x**

At the end of Section 13.1, we were working with an equation that determined the amount of interest gathered in an account. A variation of that formula is $\left(1 + \dfrac{1}{n}\right)^n$. Let's examine the values of $\left(1 + \dfrac{1}{n}\right)^n$ as n gets larger. Use your calculator to check these numerical values.

n	$\left(1 + \dfrac{1}{n}\right)^n$
1	2
10	2.59374246
100	2.704813829
1,000	2.716923932
10,000	2.718145927
100,000	2.718268237
1,000,000	2.718280469
10,000,000	2.718281693
100,000,000	2.718281815
1,000,000,000	2.718281827

Mathematicians have been able to prove that as the value of n gets larger, the value of $\left(1 + \dfrac{1}{n}\right)^n$ also continues to get larger, but has a limit on how large it can get. This limit is such a special number that it has been given its own symbol, e. The first nine digits of e are the same ones that we got in the previous table, 2.71828182.

As we will see later in this section, the number e is a very important number. Some calculators have two special numbers marked on them—π and e. Later in this section, we will need to determine values for the exponential function $f(x) = e^x$. Both the calculator and the computer can be used to find these values.

Finding e^x with a Calculator

Some calculators have an $\boxed{e^x}$ key. If you want to determine e^5, use this method.

	PRESS	DISPLAY
Algebraic calculator:	5 $\boxed{e^x}$	148.4131591
RPN calculator:	5 $\boxed{\text{ENTER}}$ $\boxed{e^x}$	148.4131591
TI-81 calculator:	$\boxed{\text{2nd}}$ $\boxed{e^x}$ 5 $\boxed{\text{ENTER}}$	148.4131591

Some calculators do not have a key marked e^x. On these calculators, you must use two keys. You will need to use the $\boxed{\text{INV}}$ key followed by the key $\boxed{\ln x}$. This may seem strange, but you will understand more clearly when we cover the next section. So, to determine e^5, you would use this method.

PRESS	DISPLAY
5 $\boxed{\text{INV}}$ $\boxed{\ln x}$	148.41316

Business and Finance

Let's return to a problem involving compound interest. In Section 13.1, we learned that if a certain principal amount P is invested and interest is compounded k times a year at an interest rate of r, the amount after t yr would be

$$S = P\left(1 + \frac{r}{k}\right)^{kt}$$

If we let $n = \dfrac{k}{r}$, then $k = nr$, and this formula becomes

$$S = P\left(1 + \frac{r}{nr}\right)^{nrt}$$

$$= P\left[\left(1 + \frac{1}{n}\right)^n\right]^{rt}$$

Now, if interest is compounded continuously, then the expression inside the brackets, $\left(1 + \dfrac{1}{n}\right)^n$, is equal to the number represented by e. So, the amount accumulated after t yr at the interest rate of r would be

$$S = Pe^{rt}$$

Application

EXAMPLE 13.7

If \$800 is invested in a savings account paying interest compounded continuously at 6%, how much has accumulated after 10 yr?

Solution Here $P = 800$, $r = 0.06$, and $t = 10$, so

$$S = 800e^{(0.06)10}$$

$$= 800e^{0.6}$$

$$\approx 800(1.8221188)$$

$$= 1{,}457.70$$

After 10 yr, this \$800 investment has increased to \$1,457.70.

When this is compared to the result in Examples 13.5 and 13.6, we see that, after 10 yr, continuous compounding of interest has provided an additional \$2.18.

Exponential Growth and Decay

The number represented by e is used in the two areas known as exponential growth and exponential decay.

Exponential growth can be explained by letting y represent the size of a quantity at time t. If this quantity grows or decays exponentially, it obeys the exponential growth formula given here.

Exponential Growth and Decay

The basic formula for the exponential growth or decay of a quantity is

$$y = ce^{kt}$$

where y represents the size of the quantity at time t, and c and k are positive real number constants.
If $k > 0$, this is an **exponential growth** function.
If $k < 0$, this is an **exponential decay** function.

Application

EXAMPLE 13.8

A culture of bacteria originally numbers 500. After 4 h, there are 8,000 bacteria in the culture. If we assume that these bacteria grow exponentially, how many will there be after 10 h?

Solution Since the number of bacteria grows exponentially, their growth obeys the formula $y = ce^{kt}$. When $t = 0$, we are told that $y = 500$, and so $500 = ce^{k \cdot 0} = c$. (Remember, $k \cdot 0 = 0$, $e^{k \cdot 0} = e^0 = 1$.) The formula is now $y = 500e^{kt}$. When $t = 4$ h, we have $y = 8{,}000$ and from the formula,

$$8{,}000 = 500e^{k4}$$

$$16 = e^{k4}$$

EXAMPLE 13.8 (Cont.)

However, the example asks us to determine y, when $t = 10$. It is possible to do this without determining k. (In Section 13.3, we will learn how to find the value of k.)

$$y = 500e^{k \cdot 10}$$
$$= 500(e^{k4})^{2.5}, \text{ since } 10 = (4)(2.5)$$
$$= 500(16)^{2.5}, \text{ since } e^{k4} = 16$$
$$= 500(1{,}024)$$
$$= 512{,}000$$

After 10 h, the number of bacteria increased from 500 to 512,000. ▪

Exponential decay is seen most often in radioactive substances. A common measure of the rate of decay is the **half-life** of a substance. The half-life is the amount of time needed for a substance to diminish to one-half its original size.

Application

EXAMPLE 13.9

The half-life of copper-67 is 62 h. How much of 100 g will remain after 15 days?

Solution As in Example 13.8, we have some basic information. When $t = 0$, we know that $y = 100$ g, so

$$100 = ce^{k \cdot 0}$$
$$\text{or} \quad 100 = c$$

and the exponential decay formula for copper-67 becomes

$$y = 100e^{kt}$$

Now, since the half-life is 62 h, we know that, when $t = 62$, there is only half as much copper-67, or $y = 50$ g, so

$$50 = 100e^{k(62)}$$
$$\frac{1}{2} = e^{k(62)}$$

We want to determine the amount when $t = 15$ days. Because the half-life is given in hours, we first convert 15 days to $15 \times 24 = 360$ h. Since $360 \div 62 \approx 5.8$, we can write the formula as

$$y = 100e^{k \cdot 360}$$
$$\approx 100(e^{k(62)})^{5.8}$$

EXAMPLE 13.9 (Cont.)

and since $e^{k(62)} = \frac{1}{2}$, we have

$$y \approx 100 \left(\frac{1}{2}\right)^{5.8}$$

$$= 1.79 \text{ g}$$

So, of the original 100 g of copper-67, about 1.79 g remain after 15 days.

Application

EXAMPLE 13.10

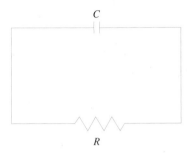

FIGURE 13.4

When a charged capacitor is discharged through a resistance, as in Figure 13.4, the decrease in charge Q is given by the formula

$$Q = Q_0 e^{-t/T}$$

where $Q_0 = CV$, the initial charge, $T = RC$, and C is the capacitance, V the battery voltage, and R the resistance. The product RC is called the time constant of the circuit. If a 15-μF capacitor is charged by being connected to a 60-V battery through a circuit with a resistance of 12 000 Ω, what is the charge on the capacitor 9 s after the battery is disconnected?

Solution   We are given $C = 15\ \mu\text{F} = 15 \times 10^{-6}$ F, $V = 60$ V, and $R = 12\,000\ \Omega$. Thus

$$Q_0 = CV$$

$$= 15 \times 10^{-6} \text{ F} \times 60 \text{ V}$$

$$= 900 \times 10^{-6}$$

$$= 9 \times 10^{-4} \text{ C}$$

and $$T = RC$$

$$= 12\,000\ \Omega \times 15 \times 10^{-6} \text{ F}$$

$$= 180\,000 \times 10^{-6} \text{ F}$$

$$= 0.18 \text{ s}$$

Substituting these values for Q_0 and T into the formula $Q = Q_0 e^{-t/T}$, we obtain

$$Q = 9 \times 10^{-4} e^{-t/0.18} \qquad (*)$$

We want to find the value of Q when $t = 9$ s, or

$$\frac{t}{T} = \frac{9 \text{ s}}{0.18 \text{ s}} = 50$$

Thus, returning to equation $(*)$, with $\frac{t}{T} = 50$, we get

$$Q = 9 \times 10^{-4} e^{-50}$$

$$= 1.74 \times 10^{-25} \text{ C}$$

So, the charge is about 1.74×10^{-25} C, 9 s after the battery is disconnected.

Exercise Set 13.2

Use a calculator or a computer to evaluate each of the numbers in Exercises 1–8.

1. e^3

2. e^7

3. $e^{4.65}$

4. $e^{5.375}$

5. e^{-4}

6. e^{-9}

7. $e^{-2.75}$

8. $e^{-0.25}$

Make a table of values and graph each of the functions in Exercises 9–12 over the given domains of x.

9. $f(x) = 4e^x$, $\{x : -2 \le x \le 4\}$

10. $g(x) = 3.5e^{5x}$, $\{x : -1 \le x \le 4\}$

11. $h(x) = 4e^{-x}$, $\{x : -4 \le x \le 2\}$

12. $k(x) = 8.5e^{-6x}$, $\{x : -4 \le x \le 2\}$

Solve Exercises 13–24.

13. *Nuclear technology*   The number of milligrams of a radioactive substance present after t yr is given by $Q = 125e^{-0.375t}$. (a) How many milligrams were present at the beginning? (b) How many milligrams were present after 1 yr? (c) How many milligrams are present after 16 yr?

14. *Nuclear technology*   Radium decays exponentially and has a half-life of 1,600 yr. How much of 100 mg will be left after 2,000 yr?

15. *Biology*   The number of bacteria in a certain culture increases from 5,000 to 15,000 in 20 h. If we assume these bacteria grow exponentially, (a) how many will be there after 10 h? (b) How many can we expect after 30 h? (c) How many can we expect after 3 days?

16. *Finance*   If $5,000 is invested in an account that pays 5% interest compounded continuously, how much can we expect to have after 10 yr?

17. *Biology*   The population of a certain city is increasing at the rate of 7% per year. The present population is 200,000. (a) What will be the population in 5 yr? (b) What can the population be expected to reach in 10 yr?

18. *Medical technology*   A pharmaceutical company is growing an organism to be used in a vaccine. The organism grows at a rate of 4.5% per hour. How many units of this organism must they begin with in order to have 1,000 units at the end of 7 days?

19. *Thermodynamics*   According to **Newton's law of cooling,** the rate at which a hot object cools is proportional to the difference between its temperature and the temperature of its surroundings. The temperature T of the object after a period of time t is

$$T = T_m + (T_0 - T_m)e^{-kt}$$

where T_0 is the initial temperature and T_m is the temperature of the surrounding medium. An object cools from $180°$F to $150°$F in 20 min when surrounded by air at $60°$F. What is the temperature at the end of 1 h of cooling?

20. *Thermodynamics*   A piece of metal is heated to $150°$C and is then placed in the outside air, which is $30°$C. After 15 min the temperature of the metal is $90°$C. What will its temperature be in another 15 min? (See Exercise 19.)

21. *Thermodynamics*   You like your drinks at $45°$F. When you arrive home from the store, the cans of drink you bought are $87°$F. You place the cans in a refrigerator. The thermostat is set at $37°$F. When you open the refrigerator 25 min later, the drinks are at $70°$F. How long will it take for the drinks to get to $45°$F? (See Exercise 19.)

22. *Electricity*   The circuit in Figure 13.5 contains a resistance R, a voltage V, and an inductance L. The current I at t s after the switch is closed is given by

$$I = I_0 \left(1 - e^{-t/T}\right)$$

where I_0 is the steady state current $\dfrac{V}{R}$ and $T = \dfrac{L}{R}$. If a circuit has $V = 120$ V, $R = 40$ Ω, and $L = 3.0$ H (henrys), determine the current in the circuit

(a) 0.01 s, (b) 0.1 s, and (c) 1.0 s after the connection is made.

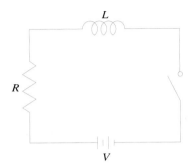

R

L

V

FIGURE 13.5

23. *Electricity* A 130-μF capacitor is charged by being connected to a 120-V circuit. The resistance is 4 500 Ω. What is the charge on the capacitor 0.5 s after the circuit is disconnected?

24. *Nuclear technology* The half-life of tritium is 12.5 yr. How much of 100 g will remain after 40 yr?

≡ 13.3
LOGARITHMIC FUNCTIONS

If you look at the graphs of exponential functions, such as the ones in Figures 13.1, 13.2, and 13.3, you can see that they would pass the horizontal line test for the inverse of a function. The inverse of the exponential function $f(x) = b^x$, $b > 0$, $b \neq 1$, is called the **logarithmic function**. The symbol for the logarithmic function is $\log_b x$. So, if $f(x) = b^x$, we have

$$f^{-1}(x) = \log_b x, \ b > 0, \ b \neq 1$$

which is read "log to the base b of x." This provides for the following definition.

Logarithmic Function

A **logarithmic function** is any function of the form

$$f(x) = \log_b x$$

where $x > 0$, $b > 0$, and $b \neq 1$. If $y = \log_b x$, then $x = b^y$. The number represented by b is called the **base**.

Remember that for any function f, with an inverse f^{-1}, we have the following relationship.

$$f\left(f^{-1}(x)\right) = x \qquad \text{and} \qquad f^{-1}(f(x)) = x$$

If $f(x) = b^x$ and $f^{-1}(x) = \log_b x$, then we see that

$$b^{\log_b x} = x \qquad \text{and} \qquad \log_b b^x = x$$

Since $y = \log_b x$ is equivalent to $b^y = x$, we can express each logarithm in exponential form.

EXAMPLE 13.11

Use the fact that $y = \log_b x$ is equivalent to $b^y = x$ to rewrite each logarithm in exponential form.

Logarithmic form	Exponential form
(a) $\log_5 125 = 3$	$5^3 = 125$
(b) $\log_7 49 = 2$	$7^2 = 49$
(c) $\log_2 128 = 7$	$2^7 = 128$
(d) $\log_5(\frac{1}{25}) = -2$	$5^{-2} = \frac{1}{25}$
(e) $\log_2 0.125 = -3$	$2^{-3} = 0.125 = \frac{1}{8}$
(f) $\log_8 16 = \frac{4}{3}$	$8^{4/3} = 16$
(g) $\log_b 1 = 0$	$b^0 = 1$

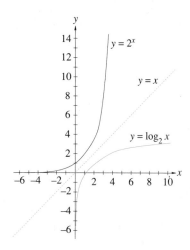

FIGURE 13.6

Graphical representation of the $\log_b x$ uses the technique we learned in Section 4.7. Figure 13.6 shows the graph of $y = 2^x$. The line $y = x$ is shown as a dashed line. The reflection of $y = 2^x$ in the line $y = x$ is shown by the colored curve and is the graph of $y = \log_2 x = f^{-1}(2^x)$.

This is a rather awkward method to graph the curve of $y = \log_b x$. Let's find another way to evaluate $\log_b x$ so that we can set up a table of values and plot points, and to connect the points to sketch the graph.

Common Logs; Natural Logs

There is another way to evaluate $\log_b x$, but the ease of doing it depends on the base of the logarithm. There are two bases that are used most often. These are 10 and e. Logarithms that have a base of 10 are called **common logs** and those with a base of e are called **natural logs**.

Because the bases 10 and e are used so often, they have special symbols. The symbol **log**, written with no indicated base, shows that common logs, or base 10 logs, are being used. If natural logs are involved, the symbol **ln** is used.

There are three major ways to find the values of a logarithm. These ways include a calculator, a computer, and a table. We will explain how to use a calculator to find the logarithm of a number. Tables are seldom used today. If, or when, you need to use a table of logarithms, you should carefully read the instructions for its use.

Finding Logs with a Calculator

Look at your calculator. It has two keys on it that we have seldom used. One key is $\boxed{\log}$ or $\boxed{\text{LOG}}$ and the other is $\boxed{\ln x}$ or $\boxed{\text{LN}}$. Now, since $\log_{10} x$ means $\log x$, we can use the $\boxed{\log}$ key to determine $\log_{10} x$. Similarly, we can use the $\boxed{\ln x}$ or $\boxed{\text{LN}}$ key to determine values of $\log_e x = \ln x$.

EXAMPLE 13.12

Use a calculator to evaluate (a) log 2, (b) log 100, (c) log 9.53, (d) ln 2, (e) ln 12.4, (f) ln 1, and (g) ln 32.4.

Solution

	PRESS	DISPLAY
(a)	2 $\boxed{\log}$	0.30103
(b)	100 $\boxed{\log}$	2
(c)	9.53 $\boxed{\log}$	0.9790929
(d)	2 $\boxed{\ln x}$	0.6931472
(e)	12.4 $\boxed{\ln x}$	2.5176965
(f)	1 $\boxed{\ln x}$	0
(g)	32.4 $\boxed{\ln x}$	3.4781584

Notice that when we wanted to determine a logarithm such as ln 32.4, we first pressed 32.4 and then the $\boxed{\ln x}$ key. On some calculators, such as a *TI-81*, you would press the appropriate logarithm key, then the number, and then $\boxed{\text{ENTER}}$. Thus, to get ln 32.4, you would press the key sequence

$$\boxed{\text{LN}} \ 32.4 \ \boxed{\text{ENTER}}$$

Logs of Different Bases

Now we see that we can use a calculator to help us get values of ln x and log x. We can also use calculators and computers to help us find the value of $\log_b x$. To do this, we use the following relationship.

Change of Base Formula	$\log_b x = \dfrac{\ln x}{\ln b}$

The relationship $\log_b x = \dfrac{\ln x}{\ln b}$ seems to be an unusual one. It uses some properties of logarithms that we will learn in Section 13.4. But, it works! In fact, it works for any base a of the logarithms on the right-hand side. So, it is true that $\log_b x = \dfrac{\log_a x}{\log_a b}$. Since $5^3 = 125$, we know $\log_5 125 = 3$. By this formula, $\log_5 125 = \dfrac{\ln 125}{\ln 5}$ should also be 3. Try it on your calculator.

PRESS	DISPLAY
125 $\boxed{\ln x}$	4.8283137
$\boxed{\div}$ 5 $\boxed{\ln x}$	1.6094379
$\boxed{=}$	3

EXAMPLE 13.13

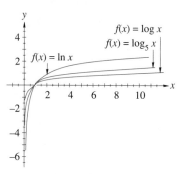

FIGURE 13.7

Plot the graphs of $\log x$, $\ln x$, and $\log_5 x$.

Solution A table of values follows and the graphs of the functions are shown in Figure 13.7.

x	0.50	1	1.50	2.00	2.50	3.00	3.50	4.00	4.50	5.00	5.50	6.00
$\log x$	−0.30	0	0.18	0.30	0.40	0.48	0.54	0.60	0.65	0.70	0.74	0.78
$\ln x$	−0.69	0	0.41	0.69	0.92	1.10	1.25	1.39	1.50	1.61	1.70	1.79
$\log_5 x$	−0.43	0	0.25	0.43	0.57	0.68	0.78	0.86	0.93	1.00	1.06	1.11

Notice that all three curves cross the x-axis at the point $(1, 0)$. If you look back at Example 13.12(f), you will see that $\log_b 1 = 0$. All logarithmic curves cross the x-axis at the point $(1, 0)$.

Exercise Set 13.3

In Exercises 1–10, rewrite each logarithm in exponential form.

1. $\log_6 216 = 3$
2. $\log_9 6{,}561 = 4$
3. $\log_4 16 = 2$
4. $\log_7 16{,}807 = 5$
5. $\log_{1/7} \dfrac{1}{49} = 2$
6. $\log_{1/2} \dfrac{1}{64} = 6$

7. $\log_2 \dfrac{1}{32} = -5$

8. $\log_3 \dfrac{1}{243} = -5$

9. $\log_9 2{,}187 = \dfrac{7}{2}$

10. $\log_8 2{,}048 = \dfrac{11}{3}$

In Exercises 11–20, rewrite each exponential in logarithmic form.

11. $5^4 = 625$
12. $3^5 = 243$
13. $2^7 = 128$
14. $4^3 = 64$
15. $7^3 = 343$
16. $11^2 = 121$

17. $5^{-3} = \dfrac{1}{125}$

18. $3^{-5} = \dfrac{1}{243}$

19. $4^{7/2} = 128$
20. $125^{5/3} = 3{,}125$

In Exercises 21–32, use a calculator or a computer to evaluate each of these logarithms.

21. $\ln 5$
22. $\ln 19$
23. $\ln 4.751$
24. $\ln 35.62$
25. $\log 4$
26. $\log 23$

27. $\log 12.67$
28. $\log 78.143$
29. $\log_5 8$
30. $\log_3 20$
31. $\log_{12} 16.4$
32. $\log_8 691.45$

In Exercises 33–39, make a table of values and sketch the graph of each function.

33. $f(x) = \ln x$

34. $g(x) = \log x$

35. $h(x) = \log_2 x$

36. $k(x) = \log_3 x$

37. $f(x) = \log_{12} x$

38. $g(x) = \log_{1/2} x$

39. $h(x) = \log_{1/4} x$

Solve Exercises 40–41.

40. *Seismology* The **Richter scale** is used to measure the magnitude of earthquakes. The formula for the Richter scale is $R = \log I$, where R is the Richter number, and I is the intensity of the earthquake. Express the Richter scale in exponential form.

41. *Acoustical engineering* The **decibel (dB) scale** is used for sound intensity. This scale is used because the response of the human ear to sound intensity is not proportional to the intensity. The intensity, $I_0 = 10^{-12}$ W/m^2 (watts/m^2), is just audible and so is given a value of 0 dB. A sound 10 times more intense is given the value 10 dB; a sound $100 = 10^2$ times more intense than 0 dB is given the value 20 dB. Continuing in this manner gives the formula

$$\beta = 10 \log \frac{I}{I_0}$$

where β is the intensity in decibels and I is the intensity in W/m^2. **(a)** Express the decibel scale in exponential notation. **(b)** What is the intensity in decibels of a sound that measures 10^{-7} W/m^2? **(c)** What is the intensity of a heavy truck passing a pedestrian at the side of a road if the sound wave intensity of the truck I is 10^{-3} W/m^2?

☰ 13.4
PROPERTIES OF LOGARITHMS

In Section 13.3, we saw that logarithms were the inverses of exponentials. We also saw that we could write each logarithm as an exponential. It is not unexpected that the properties of logarithms must be related to the rules of exponents. We will examine three properties of logarithms and mention three others. These properties used to be very important in helping to calculate logarithms. Today, we use calculators and computers for these calculations, but the properties will be important later when we need to solve equations that involve logarithms or exponents.

Property 1 ━━

Let's consider two numbers, x and y, and consider $\log_b xy$. Suppose we know that $\log_b x = m$ and $\log_b y = n$, or, in exponential form, $b^m = x$ and $b^n = y$. Then,

$$xy = b^m b^n = b^{m+n}$$

Rewriting $xy = b^{m+n}$ in logarithmic form, we get $\log_b xy = m + n = \log_b x + \log_b y$. We have established the first property of logarithms.

> **Logarithms: Property 1**
>
> If x and y are positive real numbers, $b > 0$, and $b \neq 1$, then
>
> $$\log_b xy = \log_b x + \log_b y$$

EXAMPLE 13.14

Use the first property of logarithms to rewrite each of the following: **(a)** $\log_4 35$, **(b)** $\log 21$, **(c)** $\ln 18$, **(d)** $\log_3 5x$, **(e)** $\log_8 2 + \log_8 5$, **(f)** $\log 5 + \log 2 + \log 6$, **(g)** $\ln 3 + \ln 13$, and **(h)** $\log 5 + \log a$.

Solutions

(a) $\log_4 35 = \log_4(5 \cdot 7) = \log_4 5 + \log_4 7$

(b) $\log 21 = \log(3 \cdot 7) = \log 3 + \log 7$

(c) $\ln 18 = \ln(2 \cdot 3 \cdot 3) = \ln 2 + \ln 3 + \ln 3$

(d) $\log_3 5x = \log_3 5 + \log_3 x$

(e) $\log_8 2 + \log_8 5 = \log_8(2 \cdot 5) = \log_8 10$

(f) $\log 5 + \log 2 + \log 6 = \log(5 \cdot 2 \cdot 6) = \log 60$

(g) $\ln 3 + \ln 13 = \ln(3 \cdot 13) = \ln 39$

(h) $\log 5 + \log a = \log 5a$

Property 2

For the second property, let's consider $\log_b \dfrac{x}{y}$. Again, if we let $x = b^m$ and $y = b^n$, we have $\dfrac{x}{y} = \dfrac{b^m}{b^n} = b^{m-n}$. Now, $\log_b b^{m-n} = m - n = \log_b x - \log_b y$. This gives us the following property of logarithms.

Logarithms: Property 2

If x and y are positive real numbers, $b > 0$, and $b \neq 1$, then

$$\log_b \frac{x}{y} = \log_b x - \log_b y$$

EXAMPLE 13.15

Use the second property of logarithms to rewrite each of the following: **(a)** $\log \frac{3}{5}$, **(b)** $\ln \frac{5}{4}$, **(c)** $\log_5 \frac{7}{x}$, **(d)** $\ln 8 - \ln 2$ **(e)** $\log_3 5 - \log_3 2$, and **(f)** $\log 8x^2 - \log 2x$.

Solutions

(a) $\log \dfrac{3}{5} = \log 3 - \log 5$

(b) $\ln \dfrac{5}{4} = \ln 5 - \ln 4$

(c) $\log_5 \dfrac{7}{x} = \log_5 7 - \log_5 x$

(d) $\ln 8 - \ln 2 = \ln \dfrac{8}{2} = \ln 4$

(e) $\log_3 5 - \log_3 2 = \log_3 \dfrac{5}{2}$

(f) $\log 8x^2 - \log 2x = \log \dfrac{8x^2}{2x} = \log 4x$

Property 3

Now let's consider x^p. If $m = \log_b x$, then $x = b^m$ and $x^p = (b^m)^p = b^{mp}$. So, if $x^p = b^{mp}$, then $\log_b x^p = mp$, and since $m = \log_b x$, we have the following third property of logarithms.

Logarithms: Property 3

If x and y are positive real numbers, $b > 0$, and $b \neq 1$, then

$$\log_b x^p = p \log_b x.$$

EXAMPLE 13.16

Use the third property of logarithms to rewrite each of the following: **(a)** $\log 5^3$, **(b)** $\log_2 16^5$, **(c)** $\log 100^{3.4}$, **(d)** $\log \sqrt[3]{25}$, and **(e)** $2 \log 5 + 3 \log 4 - 4 \log 2$.

Solutions

(a) $\log 5^3 = 3 \log 5$

(b) $\log_2 16^5 = 5 \log_2 16$

(c) $\log 100^{3.4} = \log(10^2)^{3.4} = \log 10^{6.8} = 6.8 \log 10 = 6.8$

(d) $\log \sqrt[3]{25} = \log 25^{1/3} = \dfrac{1}{3} \log 25$

(e) $2 \log 5 + 3 \log 4 - 4 \log 2 = \log 5^2 + \log 4^3 - \log 2^4$
$$= \log 25 + \log 64 - \log 16$$
$$= \log \frac{25 \cdot 64}{16} = \log 100$$
$$= 2$$

Properties 4, 5, and 6

In addition to these properties, there are three other properties of logarithms. These three properties are listed in the following box.

Logarithms: Properties 4, 5, and 6

If x and y are positive real numbers, $b > 0$, and $b \neq 1$, then

$$\log_b 1 = 0 \qquad \text{Property 4}$$

$$\log_b b = 1 \qquad \text{Property 5}$$

$$\log_b b^n = n \qquad \text{Property 6}$$

The properties of logarithms allow us to simplify the logarithms of products, quotients, powers, and roots.

Caution There is no way to simplify logarithms of sums or differences. You cannot change $\log_b(x + y)$ to $\log_b x + \log_b y$, except when $x = y$. This is very tempting, but do not make this error. Remember, $\log_b(x + y) \neq \log_b x + \log_b y$.

Logarithms were originally developed to help people compute. Slide rules were developed as a computational tool based on logarithms. At one time, every technician had, and knew how to use, a slide rule. Electronic calculators and the microcomputers have replaced slide rules and reduced the importance of logarithms as a help in computing.

However, logarithms are still an important tool in working many problems. In the remainder of this section, we will demonstrate the properties of logarithms. Values are sometimes obtained from a table of logarithms. We will not expect you to get values from tables. These exercises will prepare you for the next section where you will solve equations that involve logarithms.

EXAMPLE 13.17

If $\log 2 = 0.3010$, $\log 3 = 0.4771$, and $\log 5 = 0.6990$, determine each of the following: **(a)** $\log 6$, **(b)** $\log 81$, **(c)** $\log 1.5$, **(d)** $\log \sqrt{5}$, and **(e)** $\log 50$.

Solutions

(a) $\log 6 = \log(2 \cdot 3) = \log 2 + \log 3 = 0.3010 + 0.4771 = 0.7781$

(b) $\log 81 = \log 3^4 = 4 \log 3 = 4(0.4771) = 1.9084$

(c) $\log 1.5 = \log \frac{3}{2} = \log 3 - \log 2 = 0.4771 - 0.3010 = 0.1761$

(d) $\log \sqrt{5} = \log 5^{1/2} = \frac{1}{2} \log 5 = \frac{1}{2}(0.6990) = 0.3495$

(e) $\log 50 = \log(5 \cdot 10) = \log 5 + \log 10 = 0.6990 + 1 = 1.6990$

Use a calculator to check each of these answers.

Exercise Set 13.4

In Exercises 1–12, write each logarithm as the sum or difference of two or more logarithms.

1. $\log \frac{2}{3}$

2. $\log \frac{5}{4}$

3. $\log 14$

4. $\log 55$

5. $\log 12$

6. $\log 28$

7. $\log \frac{150}{7}$

8. $\log \frac{588}{5x}$

9. $\log 2x$

10. $\log \frac{1}{5x}$

11. $\log \frac{2ax}{3y}$

12. $\log \frac{4bc}{3xy}$

Express each of Exercises 13–24 as a single logarithm.

13. $\log 2 + \log 11$

14. $\log 3 + \log 13$

15. $\log 11 - \log 3$

16. $\log 17 - \log 23$

17. $\log 2 + \log 2 + \log 3$

18. $\log 3 + \log 3 + \log 5$

19. $\log 4 + \log x - \log y$

20. $\log 5 + \log 7 - \log 11$

21. $5 \log 2 + 3 \log 5$

22. $7 \log 3 - 2 \log 8$

23. $\log \dfrac{2}{3} + \log \dfrac{6}{7}$

24. $\log \dfrac{3}{24} + \log \dfrac{4}{7} - \log \dfrac{1}{3}$

If $\log 2 = 0.3010$, $\log 3 = 0.4771$, and $\log 5 = 0.6990$, determine the value of each logarithm in Exercises 25–36.

25. $\log 8$

26. $\log 9$

27. $\log 12$

28. $\log 36$

29. $\log 15$

30. $\log 30$

31. $\log \dfrac{75}{2}$

32. $\log \dfrac{45}{8}$

33. $\log 200$

34. $\log 150$

35. $\log 5,000$

36. $\log 4,500$

If $\log_b 2 = 0.3869$, $\log_b 3 = 0.6131$, and $\log_b 5 = 0.8982$, determine the value of each logarithm in Exercises 37–40. Solve Exercise 41.

37. $\log_b 16$

38. $\log_b 15$

39. $\log_b \dfrac{5}{3}$

40. $\log_b \dfrac{24}{5}$

41. Use the change of base formula to determine the base for the logarithms in Exercises 33–40. (Hint: The answer is an integer. You may not get an integer, because the values have been rounded off to four decimal places.)

☰ 13.5
EXPONENTIAL AND LOGARITHMIC EQUATIONS

We mentioned in Section 13.4 that logarithms were originally developed as an aid for computation. The wide use of electronic calculators decreased the importance of logarithms as a help for calculation. In this section, we will focus on ways we can use logarithms to help solve equations.

Exponential Equations

We will start with an equation that is relatively easy. For one thing, we already know the answer. This will help us see that the method works.

Exponential Equations

An **exponential equation** is any equation in which the variable is an exponent.

EXAMPLE 13.18

Solve $3^x = 81$.

Solution   To solve this equation, we will take the logarithm of both sides. We can use any base for the logarithm, but it is easiest to use one of the bases that is on a calculator. We will first use base 10 and then solve the same equation using base e to show that it will work either way.

$$3^x = 81$$

$$\log 3^x = \log 81$$

Since $\log 3^x = x \log 3$, we have

$$x \log 3 = \log 81$$

$$x = \frac{\log 81}{\log 3}$$

$$\approx \frac{1.908485}{0.4771213} = 4$$

≡ **Note**   Notice that $\log 81 \div \log 3 = \dfrac{\log 81}{\log 3} = 4$ and that $\log \dfrac{81}{3} = \log 81 - \log 3 \approx 1.431$. Since $4 \neq 1.431$, we see that $\log 81 \div \log 3 = \dfrac{\log 81}{\log 3} \neq \log \dfrac{81}{3}$.

Let's solve this same equation using natural logarithms.

$$3^x = 81$$

$$\ln 3^x = \ln 81 \qquad\qquad \text{Take ln of both sides.}$$

$$x \ln 3 = \ln 81 \qquad\qquad \text{Use Property 3 of logs.}$$

$$x = \frac{\ln 81}{\ln 3} \qquad\qquad \text{Divide.}$$

$$\approx \frac{4.3944492}{1.0986123} = 4.$$

EXAMPLE 13.19

Solve $5^{2x+1} = 25^{4x-1}$.

Solution   Notice that $25 = 5^2$, so we can write this equation as

$$5^{2x+1} = 25^{4x-1}$$

$$= \left(5^2\right)^{4x-1}$$

$$= 5^{8x-2}$$

Taking $\log_5$ of both sides produces

$$\log_5(5^{2x+1}) = \log_5(5^{8x-2})$$

or $\qquad (2x+1)\log_5 5 = (8x-2)\log_5 5$

EXAMPLE 13.19 (Cont.)

which simplifies to

$$2x + 1 = 8x - 2$$
$$3 = 6x$$
$$\frac{1}{2} = x$$

Logarithmic Equations

Just as we used logarithms to solve equations that involved exponentials, we can use exponentials to solve problems that involve logarithms.

Logarithmic Equations

A **logarithmic equation** is an equation that contains a logarithm of the variable.

Steps for Solving Logarithmic Equations

In solving a logarithmic equation, you should:
1. Use the properties of logarithms to combine all the logarithms into one.
2. Use the fact that $\log_b x = y$ is equivalent to $x = b^y$ to rewrite the equation.
3. Solve the resulting equation.

EXAMPLE 13.20

Solve $\ln x = 7$.

Solution The base in this equation is e, so $\ln x$ is the same as $\log_e x$ and this is equivalent to $\log_e x = 7$. Using the fact that $\log_e x = y$ is the same as $x = e^y$, we get

$$x = e^7$$
$$\approx 1{,}096.6$$

It is important that you remember to use the correct base when you solve a logarithmic equation. The next example will show you how to proceed with a more complicated situation. It also uses a different base; in this case, the base is 10.

EXAMPLE 13.21

Solve $\log(3x - 5) + 2 = \log 4x$.

Solution We will first combine this into one logarithm

$$\log(3x - 5) + 2 = \log 4x$$
$$\log(3x - 5) - \log 4x = -2$$
$$\log\left(\frac{3x - 5}{4x}\right) = -2$$

EXAMPLE 13.21 (Cont.)

We will now use the fact that $\log_b x = y$ is equivalent to $x = b^y$. Since we are using common logarithms, the base is 10.

$$\frac{3x - 5}{4x} = 10^{-2}$$

$$\frac{3x - 5}{4x} = 0.01$$

We solve this as we would any fractional equation.

$$\frac{3x - 5}{4x} = 0.01$$

$$3x - 5 = 0.04x$$

$$2.96x = 5$$

$$x = \frac{5}{2.96} \approx 1.69$$

≡ **Note**   The approximate answer 1.69 does not check, but the exact answer $\frac{5}{2.96}$ does check.

Business and Finance

An interesting equation results from the work with exponentials and we can use logarithms to solve it. If a certain amount of money P was invested at 5% compounded continuously, the investment would be worth

$$S = Pe^{rt}$$

after t yr. How long does it take for the money to double?

If it doubles, then it is worth $2P$ and we have

$$2P = Pe^{rt}$$

or $2 = e^{rt}$

Taking the natural logarithm of both sides, we obtain

$$\ln 2 = \ln e^{rt}$$

or $\ln 2 = rt$

Solving for t results in

$$t = \frac{\ln 2}{r}$$

It will take $\frac{\ln 2}{r}$ yr for the money to double. This same formula will apply to anything that is growing or decaying at an exponential rate.

Application

EXAMPLE 13.22

How long will it take for $1,000 to double at 5% compounded continuously?

Solution   We use the formula

$$t = \frac{\ln 2}{r}$$

and since

$$r = 5\% = 0.05,$$

$$t = \frac{\ln 2}{0.05} \approx 13.86$$

Application

EXAMPLE 13.23

What is the half-life of a radioactive material that decays at the rate of 1% per year?

Solution   The same formula will apply; only here $r = 1\% = 0.01$.

$$t = \frac{\ln 2}{0.01} \approx 69.31 \text{ yr}$$

Exercise Set 13.5

In Exercises 1–24, solve each equation for the indicated variable. You may want to use a calculator.

1. $5^x = 29$
2. $6^y = 32$
3. $4^{6x} = 119$
4. $3^{-7x} = 4$
5. $3^{y+5} = 16$
6. $2^{y-7} = 67$
7. $e^{2x-3} = 10$
8. $4^{3y+5} = 30$
9. $3^{4x+1} = 9^{3x-5}$
10. $2^{3-2x} = 8^{1+2x}$
11. $e^{3x-1} = 5e^{2x}$
12. $3^{2-5x} = 8(9^{x-1})$

13. $\log x = 2.3$
14. $\ln x = 5.4$
15. $\log(x - 5) = 17$
16. $\ln(y + 3) = 19$
17. $\ln 2x + \ln x = 9$
18. $2 \ln x = \ln 4$
19. $2 \ln 3x + \ln 2 = \ln x$
20. $2 \ln(x + 3) = \ln x + 4$
21. $\log(x - 1) = 2$
22. $\log(x - 4) = \log x - 4$
23. $2 \log(x - 4) = 2 \log x - 4$
24. $\ln x = 1 + 3 \ln x$

Solve Exercises 25–39.

25. *Finance*   The amount of $5,000 is placed in a savings account, where interest is compounded continuously at the rate of 6% per year. How long will it take for this amount to be doubled?

26. *Finance*   How long will it take the money in Exercise 25 to double if the rate is 8% per year?

27. *Nuclear energy* A radioactive substance decays at the rate of 0.5% per year. What is the half-life of this substance?

28. *Nuclear energy* Another radioactive substance decays at the rate of 3% per year. What is its half-life?

29. *Nuclear energy* The half-life of tritium is 12.5 yr. What is its annual rate of decay?

30. *Nuclear energy* The half-life of the sodium isotope $^{24}_{11}$Na against beta decay is 15 h. What is the rate of decay?

31. *Finance* A person has some money to invest and would like to double the investment in 8.5 yr. What annual rate of interest, compounded continuously, will be needed in order for this to be accomplished?

32. *Automotive technology* In a chrome-electroplating process, the mass m in grams of the chrome plating increases according to the formula $m = 200 - 2^{t/2}$, where t is the time in minutes. How long does it take to form 100 g of plating?

33. *Environmental science* In chemistry, the pH of a substance is defined by pH $= -\log[\text{H}^+]$, where $[\text{H}^+]$ is the concentration of hydrogen ions in the substance measured in moles per liter. The pH of distilled water is 7. A substance with a pH less than 7 is known as an **acid**. A substance with a pH greater than 7 is a **base**. Rain and snow have a natural concentration of $[\text{H}^+] = 2.5 \times 10^{-6}$ moles per liter. What is the natural pH of rain and snow?

34. *Environmental science* The pH of some acid rain is 5.3. What is the concentration of hydrogen ions in acid rain?

35. *Meteorology* The barometric equation,

$$H = (30T + 8,000) \ln \frac{P_0}{P}$$

relates the height H in meters above sea level, the air temperature T in degrees Celsius, the atmospheric pressure P_0 in centimeters of mercury at sea level, and the atmospheric pressure P in centimeters of mercury at height H. Atmospheric pressure at the summit of Mt. Whitney in California on a certain day is 43.29 cm of mercury. The average air temperature is $-5°$C and the atmospheric pressure at sea level is 76 cm of mercury. What is the height of Mt. Whitney?

36. *Environmental science* In Exercise Set 13.3, Exercise 41, we saw that the loudness in decibels β of a noise is given by the formula $\beta = 10 \log \frac{I}{I_0}$, where $I_0 = 10^{-12}$ W/m^2, and I is the intensity of the noise in W/m^2. At takeoff, a certain jet plane has a noise level of 105 dB. What is the intensity I of the sound wave produced by this airplane?

37. *Electronics* The formula for the exponential decay of electric current is given by the formula $Q = Q_0 e^{-t/T}$, where $T = RC$. (See Example 13.10.) If $Q_0 = 0.40$ A, $R = 500$ Ω, and $C = 100$ μF, what is t when $Q = 0.05$ A?

38. *Thermodynamics* The temperature T of an object after a period of time t is given by $T = T_0 + ce^{-kt}$, where T_0 is the temperature of the surrounding medium, and $c = T_I - T_0$, where T_I is the initial temperature of the object. A steel bar with a temperature of $1\,200°$C is placed in water with a temperature of $20°$C. If the rate of cooling k is 8% per hour, how long will it take for the steel to reach a temperature of $40°$C?

39. *Electronics* In an ac circuit, the current I at any time t, in seconds, is given by $I = I_0 \left(1 - e^{-Rt/L}\right)$, where I_0 is the maximum current, L is the inductance, and R the resistance. If a circuit has a 0.2 H inductor, a resistance of 4 Ω, and a maximum current of 1.5 A, at what instant does the current reach 1.4 A?

≡ 13.6

GRAPHS USING SEMILOGARITHMIC AND LOGARITHMIC PAPER

Exponential functions often produce some very large numbers. This makes graphing exponential functions difficult to show on normal graph paper. Look at the graph of $f(x) = 2^x$ in Figure 13.1. By the time x is 5, y is 32; when $x = 6$, $y = 64$. This presents a problem when trying to represent a large portion of an exponential graph. Semilogarithmic and logarithmic graph papers have been developed to allow for the plotting of a large range of values.

Semilogarithmic Graph Paper ━━━━

Semilogarithmic or **semilog graph paper** has one of the axes (usually the y-axis) marked off in distances proportional to the logarithms of numbers. Thus, the distances between lines on this axis are not equally spaced. The other axis has the distance between lines equally spaced. The result of all this is that the graph of an exponential function $y = b^x$ is a straight line when it is drawn on semilog paper.

EXAMPLE 13.24

Sketch the graph of $y = 2^x$ on semilog paper.

Solution Since this is the same curve we graphed in Figure 13.1, we will use the same table of values from that example. The table is reproduced here.

x	-3	-2	-1	0	1	2	3	4	5
$f(x) = 2^x$	0.125	0.25	0.5	1	2	4	8	16	32

Along the x-axis, we label the points -3 through 5. Along the y-axis, the points are labeled as indicated in Figure 13.8. Notice that each unit in the colored interval has been labeled $1, 2, 3, \ldots, 10$. The interval directly above it has each unit representing a multiple of 10 (10, 20, 30, etc.). In the next interval, the units would be multiples of 100 (100, 200, 300, etc.). The interval directly below the colored one has each unit represented in tenths (0.1, 0.2, 0.3, etc.), and the interval below that would be in hundredths (0.01, 0.02, 0.03, etc.).

EXAMPLE 13.25

Sketch the graph of $y = 5(4^x)$ on semilog paper.

Solution A table of values from $x = -2$ to $x = 5$ is given here.

x	-2	-1	0	1	2	3	4	5
$f(x) = 5(4^x)$	0.31	1.25	5	20	80	320	1,280	5,120

The graph of this curve on semilog paper is shown in Figure 13.9. You should study the scale of the y-axis to be sure that you understand how the units are marked.

Study Figures 13.8 and 13.9. Both of these graphs are straight lines. Yet, in both cases we started with exponential equations of the form $y = b^x$.

≡ **Note** One of the reasons for using semilog paper is that graphs of exponential functions of the form $y = b^x$ will be straight lines when they are drawn on semilogarithmic paper.

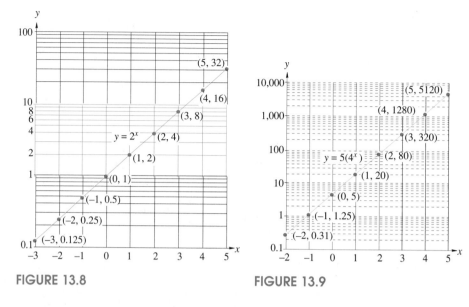

FIGURE 13.8

FIGURE 13.9

Logarithmic Graph Paper

Use semilog paper when you want to indicate a large range of values for one of the variables. When this is needed for both variables, use **logarithmic** or **log-log graph paper**. Both axes on log-log paper are marked off in logarithmic scales. A function of the type $y^b = x^a$ or $y = x^{a/b}$ graphs as a straight line on log-log paper.

EXAMPLE 13.26

Sketch the graph of $y^3 = x^2$ on log-log paper.

Solution   The equation $y^3 = x^2$ is equivalent to $y = x^{2/3}$. If we select values of x that are perfect cubes, we then get integer values for y. The following table gives some values.

x	-1	0	$\frac{1}{8} = 0.125$	1	8	27	64	125	216	1,000
$y = x^{2/3}$	1	0	$\frac{1}{4} = 0.25$	1	4	9	16	25	36	100

In Figure 13.10, we show the sketch of this curve. Notice the way in which both axes are labeled.

≣ **Note**   Negative values of x and the point (0, 0) do not appear on the graph. Since a logarithmic scale contains only positive values, the coordinates that are not positive cannot be plotted.

Semilog and log-log paper can be used to graph functions that may not graph as a straight line. This happens when a large range of values needs to be graphed, particularly when the data are exponential in nature or when the data are gathered from an experiment.

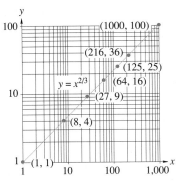

FIGURE 13.10

Both Semilogarithmic and Logarithmic Graph Paper

Sometimes it is necessary to graph the data on both semilog and log-log graph paper.

Using Graphs to Help Determine an Equation

If the graph of data is a straight line on semilog paper, then the data are related by a function of the form $y = ab^x$.

If the graph of data is a straight line with slope m on log-log paper, then the data are related by a function of the form $y = ax^m$. A function of this form, $y = ax^m$, is called a **power function**.

Application

EXAMPLE 13.27

Data from a certain experiment result in the following table.

x	1	2	3	4	5	6
y	5	40	135	320	625	1,080

Determine an equation that relates y as a function of x.

Solution   We will plot this data on both semilog graph paper and log-log graph paper. When the points are connected, we get the curves shown in Figures 13.11a and 13.11b.

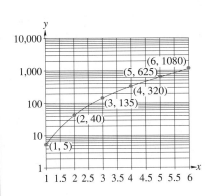

FIGURE 13.11a

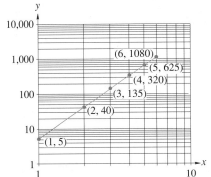

FIGURE 13.11b

The curve in Figure 13.11b is a straight line. This curve was drawn on log-log graph paper, so we conclude that these data satisfy a power function of the form $y = ax^m$.

The slope of this straight line is m, and if (x_1, y_1) and (x_2, y_2) are two points on the curve, we have

$$m = \frac{\log y_2 - \log y_1}{\log x_2 - \log x_1}$$

EXAMPLE 13.27 (Cont.)

If we use the points (1, 5) and (2, 40), we will get

$$m = \frac{\log 40 - \log 5}{\log 2 - \log 1}$$

$$\approx \frac{1.602059991 - 0.698970004}{0.301029996 - 0}$$

$$= \frac{0.903089987}{0.301029996}$$

$$= 3$$

Thus, we know that these data are of the form $y = ax^3$.

To determine a, we select one pair of values from the data (3, 135) and get

$$y = ax^3$$

$$135 = a(3)^3$$

$$= 27a$$

$$a = \frac{135}{27}$$

$$= 5$$

The data in the table satisfy the equation $y = 5x^3$.

Unlike Example 13.27, data from an actual experiment seldom fit a curve exactly. In that case, we select points on a curve that seem to "best fit" the actual data points. We will explore this in more detail in Chapter 21.

Exercise Set 13.6

In Exercises 1–12, sketch the graphs of the given functions on semilog graph paper.

1. $y = 3^x$

2. $y = 5^x$

3. $y = 3(6^x)$

4. $y = 5(7^x)$

5. $y = 2^{-x}$

6. $y = 8^{-x}$

7. $y = x^4$

8. $y = x^5$

9. $y = 8x^3$

10. $y = 5x^4$

11. $y = 3x^2 + 4x$

12. $y = 8x^3 + 2x^2 + x$

In Exercises 13–24, sketch the graphs of the given functions on log-log graph paper.

13. $y = x^{1/2}$

14. $y = x^{1/3}$

15. $y = 2x^3$

16. $y = 3x^5$

17. $y = x^2 + \sqrt{x}$

18. $y = x^3 + x$

19. $y = 5x^{-1}$

20. $y = 8x^{-3}$

21. $y^2 = 4x^3$

22. $y^4 = 6x^5$

23. $x^2y^3 = 8$

24. $x^3y = 100$

In Exercises 25–30, sketch the indicated graphs.

25. *Meteorology*   The atmospheric pressure P at a given altitude h, in feet, is given by $P = P_0e^{-kh}$, where P_0 and k are constants. On semilog paper, plot P in atmospheres and h, the altitude, in feet, for $0 \leq h \leq 10^4$, $P_0 = 1$ atmosphere, and at $0°C$, $k = 1.25 \times 10^{-4}$.

26. *Meteorology*   The atmospheric pressure in kilopascals (kPa) is given approximately by $P = 100e^{-0.3h}$, where h is the altitude in kilometers. Graph P vs h on semilog paper.

27. *Electricity*   The resistance R of a copper wire varies inversely as the square of its cross-sectional diameter D. Thus $RD^2 = k$. If $k = 0.9$ Ω·mm, graph R vs D on log-log paper, for $D = 0.1$ mm to 10 mm.

28. *Physics*   Boyle's law states that at a constant temperature, the volume V of a sample of gas is inversely proportional to the absolute pressure applied to the

gas P. Thus, $PV = C$, a constant. If $C = 5$ atm·ft^3, use log-log paper to plot the graph of P in atmospheres vs V in cubic feet. Let values of P range from 0.1 to 10.

29. An experimenter gathered the data in the following table. Plot the data on semilog and log-log papers and determine the equation that relates y as a function of x.

x	1	2	5	10
y	1,000	250	40	10

30. Repeat Exercise 29 for the data in this table.

x	1	2	3	4	5
y	16	80	400	2 000	10 000

≡ CHAPTER 13 REVIEW

Important Terms and Concepts

Exponential equation

Exponential function

Logarithmic equation

Logarithmic function

Logarithmic graph paper

Semilogarithmic graph paper

Review Exercises

In Exercises 1–8, use a calculator to evaluate each number.

1. e^5

2. e^{-7}

3. $e^{4.67}$

4. $e^{-3.91}$

5. $\log 8$

6. $\log 196.5$

7. $\ln 81.3$

8. $\ln 325.6$

In Exercises 9–16, make a table of values and graph each of these functions on ordinary graph paper over the indicated domains.

9. $f(x) = 5e^x$ $(x = -2$ to $4)$

10. $g(x) = 9e^{-0.5x}$ $(x = -4$ to $2)$

11. $h(x) = 5^x$ $(x = -3$ to $2)$

12. $k(x) = 2.1^{-x}$ $(x = -4$ to $2)$

13. $f(x) = \log x$ $(0 \leq x \leq 10)$

14. $g(x) = 3\log 2x$ $(0 \leq x \leq 10)$

15. $h(x) = \ln(2x + 1)$ $\left(-\frac{1}{2} \le x \le 10\right)$

16. $k(x) = 2 + 3\ln x$ $(0 \le x \le 10)$

In Exercises 17–24, express each as a sum, difference, or multiple of logarithms.

17. $\log \frac{3}{4}$

18. $\log 77$

19. $\log 5x$

20. $\ln \frac{7x}{3}$

21. $\ln(4x)^3$

22. $\ln 4x^3$

23. $\log \sqrt{48}$

24. $\log \frac{\sqrt[3]{2x^5}}{4x}$

In Exercises 25–30, express each as a single logarithm.

25. $\log 5 + \log 9$

26. $\log 19 - \log 11$

27. $4\log x + \log x$

28. $\log 3a + \log b - \log a$

29. $7\log x + 2\log x - \log x$

30. $\ln \frac{2}{3} + \ln \frac{15}{8}$

In Exercises 31–38, solve each equation for the indicated variable. Use a calculator if you feel it is necessary.

31. $4^x = 28$

32. $5^{3x} = 250$

33. $e^{5x} = 11$

34. $e^{5x-1} = 2e^x$

35. $\ln x = 9.1$

36. $4\log x = 49$

37. $\log(2x + 1) = 50$

38. $2\ln 4x = 100 + \ln 2x$

In Exercises 39–42, sketch the graphs of the given functions on semilog paper.

39. $y = 4(2^x)$

40. $y = 8x^4$

41. $y = 8(4^x)$

42. $y = 3(4^{x-1})$

In Exercises 43–46, sketch the graphs of the given functions on log-log paper.

43. $y = x^{1/4}$

44. $y = 5x^3$

45. $x^3y^4 = 2$

46. $y^3 = 10x^2$

Solve Exercises 47–54.

47. *Finance*   One savings institution offers 7.5% interest compounded semiannually. A second offers the same interest rate, but it is compounded monthly. A third compounds that same interest continuously. How much income will result if $1,000 is deposited in each institution for 10 yr? How long will it take for the $1,000 to double in the third bank?

48. *Nuclear engineering*   If radium has a half-life of 1,600 yr, how much of 100 mg will remain after 100 yr? after 1,000 yr?

49. *Biology*   The amount of bacteria in a culture increases from 4,000 to 25,000 in 24 h. If we assume these bacteria grow exponentially, **(a)** how many will be there after 12 h? **(b)** How many can we expect there to be after 48 h?

50. *Seismology*   The San Francisco earthquake of 1906 registered 8.3 on the Richter scale. If $R = \log I$, where I is the relative intensity of the shock, determine the number of times the 1906 earthquake was greater than **(a)** the San Francisco quake of 1979, which registered a 6.0, and **(b)** the San Francisco quake of 1989, which registered a 7.1.

51. Data from an experiment produced a straight line on semilog paper. Two of the data points were (2, 15.76) and (4, 620.84). Find an equation that approximates this scale.

52. Data from an experiment produced a straight line on log-log paper. Two of the data points were (2, 8.75) and (4, 2.1875). Find an equation that approximates this data.

53. *Astronomy* The brightness of a star perceived by the naked eye is measured in terms of a quantity called **magnitude**. The brightest stars are of magnitude 1 and the dimmest of magnitude 6. The magnitude M is given by $M = 6 - 2.5 \log \frac{I}{I_0}$, where I_0 is the

intensity of light from a just-visible star, and I is the actual intensity from a star. Even though the sun is a star, it has a magnitude of -27, and the moon has a magnitude of -12.5. **(a)** Express this equation in exponential form by solving for I. **(b)** How many times more intense is the moon than Polaris, which has a magnitude of 2.0?

54. *Biology* When the size of a colony of bacteria is limited because of a lack of room or nutrients, the growth is described by the law of logistic growth:

$$Q = \frac{mQ_0}{Q_0 + (m - Q_0)\,e^{-kmt}}$$

where Q_0 is the initial quantity present, m the maximum size, and k a positive constant. If $Q_0 = 400$, $m = 2{,}000$, and $Q = 800$ when $t = \frac{1}{2}$, find k.

≡ CHAPTER 13 TEST

In Exercises 1–3, use a calculator to evaluate each number.

1. $e^{5.3}$

2. $\log 715.3$

3. $\ln 72.35$

In Exercises 4–7, express each as a sum, difference, or multiple of logarithms.

4. $\log \dfrac{9}{15}$

5. $\log 65x$

6. $\ln(7x)^{-2}$

7. $\log \dfrac{\sqrt[5]{5x^3}}{9x}$

In Exercises 8–10, express each as a single logarithm.

8. $\log 11 + \log 3$

9. $\log 17 - 2\log x$

10. $\ln 4 - \ln 5 + \ln 15 - \ln 8$

In Exercises 11–16, solve each equation for the indicated variable. Use a calculator if you feel it is necessary.

11. $\ln x = 17.2$

12. $3 \log x = 55$

13. $3^x = 35$

14. $e^{4x+1} = 2e^x$

15. $\log(2x^2 + 4x) = 5 + \log 2x$

16. Make a table of values and graph $f(x) = 3^x$ on ordinary graph paper over the domain $x = -3$ to 3.

Solve Exercises 17 and 18.

17. Make a table of values and graph $y = 0.4(3^x)$ on semilog paper from $x = 1$ to 7.

18. Graph $y = 2x^{1/5}$ on log-log paper.

14

Complex Numbers

Complex numbers have many important applications in electronics. In Section 14.6, we will see how to use complex numbers to solve problems that involve alternating current.

Courtesy of DeVry Institutes

All the numbers we have used until now have been real numbers. Several times in the history of mathematics, problems developed that people could not solve with the number system they had. These problems led to the invention of new numbers. The first time this happened was when the number 0 had to be invented. Later, negative numbers were needed. Each time, new numbers were invented to allow people to solve new kinds of problems. At last, the set of real numbers was developed, and with it we can solve many problems.

Not all problems, however, can be solved with the real numbers. People learned that they could take the cube root -1 or of -8. We know that $\sqrt[3]{-1} = -1$ because $(-1)^3 = -1$ and that $\sqrt[3]{-8} = -2$ because $(-2)^3 = -8$. But, there is no real number for the square root of -1, or -4, or for the square root of any negative number. People had problems that could be worked only if it were possible to take the square root of a negative number. As a result, they invented the numbers we will begin using with this chapter—the complex numbers.

≡ 14.1
IMAGINARY AND COMPLEX NUMBERS

Complex numbers have many important uses in technology. They make it much easier to work with vectors and problems that involve alternating current (ac). We will see many uses for complex numbers as we work through this chapter.

Imaginary Unit

As we stated in the chapter introduction, the need for complex numbers arose because people had to solve problems that involved the square roots of negative numbers. In order to solve this dilemma, a new number was invented to correspond to the square root of -1. The name for this number is the **imaginary unit**, and it is represented by the symbol j. Thus, we have

$$j = \sqrt{-1}$$

Another popular name for the imaginary unit is the **j operator**.

≡ **Note** Many mathematics books use the symbol i instead of j. But, because i is used to represent current, people in science and technology use j for the imaginary unit.

One of the basic steps we learn in working with imaginary numbers allows us to represent the square root of a negative number as the product of a real number and the imaginary unit, j. The square root of a negative number is called a **pure imaginary number** and is defined in the following box. Remember, if b is a real number, the symbol $\sqrt{b}$ represents the **principal square root** of b and is never negative. Thus, $\sqrt{9} = 3$, $\sqrt{25} = 5$, and $\sqrt{\frac{16}{9}} = \frac{4}{3}$.

Pure Imaginary Number

If b is a real number, $b \geq 0$, then $\sqrt{-b}$ is a **pure imaginary number** and we have

$$\sqrt{-b} = \sqrt{(-1)b} = \sqrt{-1}\sqrt{b} = j\sqrt{b}$$

where $j = \sqrt{-1}$.

We call $j\sqrt{b}$ or $\sqrt{b}j$ the **standard form for a pure imaginary number**.

EXAMPLE 14.1

Simplify and express each of the following radicals in the standard form for a pure imaginary number: (a) $\sqrt{-9}$, (b) $\sqrt{-0.25}$, (c) $\sqrt{-3}$, (d) $-\sqrt{-18}$, and (e) $\sqrt{\frac{-4}{9}}$.

Solutions

(a) $\sqrt{-9} = \sqrt{9}\sqrt{-1} = 3j$

(b) $\sqrt{-0.25} = \sqrt{0.25}\sqrt{-1} = 0.5j$

(c) $\sqrt{-3} = \sqrt{3}\sqrt{-1} = \sqrt{3}j$

(d) $-\sqrt{-18} = -\sqrt{18}\sqrt{-1} = -\sqrt{9}\sqrt{2}\sqrt{-1} = -3\sqrt{2}j = -3j\sqrt{2}$

(e) $\sqrt{\frac{-4}{9}} = \sqrt{\frac{4}{9}}\sqrt{-1} = \frac{\sqrt{4}}{\sqrt{9}}\sqrt{-1} = \frac{2}{3}j$

≡ **Note**   Many people write the symbol j in front of a radical sign in order to reduce the danger of thinking that it is under the radical. Thus, you might prefer to write the answers to (c) and (d) as $j\sqrt{3}$ and $-3j\sqrt{2}$.

Since $j = \sqrt{-1}$, we have some interesting relationships.

$$j^2 = -1$$

$$j^3 = j^2 j = (-1)j = -j$$

$$j^4 = j^2 j^2 = (-1)(-1) = 1$$

Any larger power of j can be reduced to one of these basic four. Thus,

$$j^5 = j^{4+1} = j^4 j^1 = 1 \cdot j = j$$

$$j^{15} = j^{4+4+4+3} = j^4 j^4 j^4 j^3 = 1 \cdot 1 \cdot 1 \cdot (-j) = -j$$

We need to be careful when we work with imaginary numbers. Consider the problem $\sqrt{-9}\sqrt{-4}$. We know that $\sqrt{-9} = 3j$ and $\sqrt{-4} = 2j$, so $\sqrt{-9}\sqrt{-4} = (3j)(2j) = 6j^2 = -6$. But, we have gotten used to using the property $\sqrt{a}\sqrt{b} = \sqrt{ab}$. What if we tried to use it on this problem: $\sqrt{-9}\sqrt{-4} = \sqrt{(-9)(-4)} = \sqrt{36} = 6$? If we are going to have a successful set of numbers, we cannot get two different answers when we multiply imaginary numbers.

Caution   Remember that whenever you work with square roots of negative numbers, express each number in terms of j before you proceed.

Thus, the correct answer to $\sqrt{-9}\sqrt{-4}$ is -6.

> **Multiplication of Radicals**
>
> If a and b are real numbers, then
>
> $$\sqrt{a}\sqrt{b} = \sqrt{ab} \text{ if } a \geq 0 \text{ and } b \geq 0$$
>
> If either $a < 0$ or $b < 0$ (or both a and b are negative), then convert the radical to "j form" before multiplying.

EXAMPLE 14.2

Simplify the following: **(a)** $(\sqrt{-4})^2$, **(b)** $\sqrt{-3}\sqrt{-12}$, **(c)** $\sqrt{2}\sqrt{-8}$, **(d)** $\sqrt{-0.5}\sqrt{-7}$, and **(e)** $(2\sqrt{-5})(\sqrt{-7})(3\sqrt{-14})$.

Solutions

(a) $(\sqrt{-4})^2 = (2j)^2$
$= 4j^2$
$= -4$

(b) $\sqrt{-3}\sqrt{-12} = (j\sqrt{3})(2j\sqrt{3})$
$= 2j^2\sqrt{3^2}$
$= 2(-1)(3) = -6$

EXAMPLE 14.2 (Cont.)

(c) $\sqrt{2}\sqrt{-8} = (\sqrt{2})(j\sqrt{8})$
$$= j\sqrt{16}$$
$$= 4j$$

(d) $\sqrt{-0.5}\sqrt{-7} = (j\sqrt{0.5})(j\sqrt{7})$
$$= j^2\sqrt{(0.5)(7)}$$
$$= -\sqrt{3.5}$$

(e) $(2\sqrt{-5})(\sqrt{-7})(3\sqrt{-14}) = (2j\sqrt{5})(j\sqrt{7})(3j\sqrt{14})$
$$= 6j^3\sqrt{5 \cdot 7 \cdot 14}$$
$$= 42j^3\sqrt{10}$$
$$= -42j\sqrt{10}$$

Note that $\sqrt{5 \cdot 7 \cdot 14} = 7\sqrt{10}$ and that $j^3 = -j$.

Complex Numbers

When an imaginary number and a real number are added, we get a complex number. A **complex number** is of the form $a + bj$, where a and b are real numbers. When $a = 0$ and $b \neq 0$, we have a number of the form bj, which is a **pure imaginary number**. When $b = 0$, we get a number of the form a, which is a real number.

Rectangular Form of a Complex Number

The form $a + bj$ is known as the **rectangular form** of a complex number, where a is the **real part** and bj is the **imaginary part**.

Two complex numbers are equal if both the real parts are equal and the imaginary parts are equal. Symbolically, we express this as follows.

Equality of Complex Numbers

If $a + bj$ and $c + dj$ are two complex numbers, then $a + bj = c + dj$, if and only if $a = c$ and $b = d$.

EXAMPLE 14.3

Solve $4 + 3j = 7j + x + 2 + yj$ for x and y.

Solution Here we need to determine both x and y. The best way is to rearrange the terms so that the known values are on one side of the equation and the variables are on the other.

$$4 + 3j = 7j + x + 2 + yj$$

$$4 + 3j - (2 + 7j) = x + yj$$

or
$$x + yj = 4 + 3j - (2 + 7j)$$

$$x + yj = 2 - 4j$$

EXAMPLE 14.3 (Cont.)

So, $x = 2$ and $y = -4$, since the real parts must be equal and the imaginary parts must also be equal.

EXAMPLE 14.4

Simplify and express in the form $a+bj$: **(a)** $7(3+2j)$, **(b)** $j(5-3j)$, and **(c)** $\dfrac{4 - \sqrt{-12}}{2}$.

Solutions

(a) $7(3 + 2j) = 21 + 14j$

(b) $j(5 - 3j) = 5j - 3j^2 = 5j - 3(-1) = 3 + 5j$

(c) $\dfrac{4 - \sqrt{-12}}{2} = \dfrac{4 - 2j\sqrt{3}}{2} = \dfrac{4}{2} - \dfrac{2j\sqrt{3}}{2} = 2 - j\sqrt{3}$

Notice that, in this last example, we had to divide *each* term of the numerator by 2 in order to get the final answer in the form $a + bj$.

Conjugates of Complex Numbers

Every complex number has a conjugate. As you will see in Section 14.2, conjugates are particularly useful when you are dividing by a complex number.

Conjugate of a Complex Number

The conjugate of a complex number $a + bj$ is the complex number $a - bj$.

To form the conjugate of a complex number, you need to change only the sign of the imaginary part of the complex number.

EXAMPLE 14.5

(a) The conjugate of $3 + 4j$ is $3 - 4j$.

(b) The conjugate of $5 - 2j$ is $5 + 2j$.

(c) The conjugate of $-7j$ is $7j$, since $-7j = 0 - 7j$ and its conjugate is $0 + 7j = 7j$.

(d) The conjugate of 15 is 15, since $15 = 15 + 0j$ and its conjugate is $15 - 0j = 15$.

≡ **Note** The conjugate of $a + bj$ is $a - bj$ and the conjugate of $a - bj$ is $a + bj$. Thus, each number is the conjugate of the other.

EXAMPLE 14.6

Use the quadratic formula to solve $2x^2 + 3x + 5 = 0$.

Solution The quadratic formula states that the solutions of $ax^2 + bx + c = 0$ are $x = \dfrac{-b \pm \sqrt{b^2 - 4ac}}{2a}$. Here $a = 2$, $b = 3$, and $c = 5$, and so

$$x = \frac{-3 \pm \sqrt{3^2 - 4(2)(5)}}{2(2)} = \frac{-3 \pm \sqrt{9 - 40}}{4}$$

$$= \frac{-3 \pm \sqrt{-31}}{4} = \frac{-3 \pm j\sqrt{31}}{4}$$

Writing these in the form $a + bj$, we get $x = \dfrac{-3}{4} + \dfrac{\sqrt{31}}{4}j$ and $x = \dfrac{-3}{4} - \dfrac{\sqrt{31}}{4}j$. As you can see, these roots are complex conjugates of each other.

In Chapter 18, we will show that if a polynomial has only real number coefficients and has one complex root, then the conjugate of that complex number is also a root.

Exercise Set 14.1

In Exercises 1–12, simplify and express each radical in terms of j.

1. $\sqrt{-25}$

2. $\sqrt{-81}$

3. $\sqrt{-0.04}$

4. $\sqrt{-1.44}$

5. $\sqrt{-75}$

6. $-\sqrt{-72}$

7. $-3\sqrt{-20}$

8. $5\sqrt{-30}$

9. $\sqrt{-\dfrac{9}{16}}$

10. $\sqrt{-\dfrac{25}{36}}$

11. $-4\sqrt{-\dfrac{9}{16}}$

12. $-3\sqrt{-\dfrac{10}{81}}$

In Exercises 13–30, simplify each problem.

13. $(\sqrt{-11})^2$

14. $(\sqrt{-7})^2$

15. $(3\sqrt{-2})^2$

16. $(-2\sqrt{-3})^2$

17. $\sqrt{-4}\sqrt{-25}$

18. $\sqrt{-9}\sqrt{-16}$

19. $(\sqrt{-49})(2\sqrt{-9})$

20. $(3\sqrt{-16})(\sqrt{-36})$

21. $(\sqrt{-5})(-\sqrt{5})$

22. $(-\sqrt{-7})(\sqrt{7})$

23. $(\sqrt{-0.5})(\sqrt{0.5})$

24. $(\sqrt{0.8})(\sqrt{-0.2})$

25. $(-2\sqrt{2.7})(\sqrt{-3})$

26. $(-4\sqrt{1.6})(2\sqrt{-0.4})$

27. $(\sqrt{-5})(\sqrt{-6})(\sqrt{-2})$

28. $(\sqrt{-3})(\sqrt{-9})(\sqrt{-15})$

29. $(-\sqrt{-7})^2(\sqrt{-2})^2 j^3$

30. $(\sqrt{-3})^2(\sqrt{-5})^2 j^2(\sqrt{-2})$

In Exercises 31–36, solve each problem for the variables *x* and *y*.

31. $x + yj = 7 - 2j$

32. $x + yj = -9 + 2j$

33. $x + 5 + yj = 15 - 3j$

34. $6j - x + yj = 4 + 2j$

35. $x - 5j + 2 = 4 - 3j + yj$

36. $2x - 4j = 6j + 4 - yj$

In Exercises 37–52, simplify each problem and express it in the form *a + bj*.

37. $2(4 + 5j)$

38. $3(2 - 4j)$

39. $-5(2 + j)$

40. $-3(4 + 7j)$

41. $j(3 - 2j)$

42. $j(5 + 4j)$

43. $2j(4 + 3j)$

44. $-3j(2 - 5j)$

45. $\frac{1}{2}(6 - 8j)$

46. $\frac{2}{3}(6 + 9j)$

47. $\dfrac{5 - 10j}{5}$

48. $\dfrac{6 + 12j}{3}$

49. $\dfrac{6 + \sqrt{-18}}{3}$

50. $\dfrac{7 - \sqrt{-98}}{7}$

51. $\dfrac{8 - \sqrt{-24}}{4}$

52. $\dfrac{9 + \sqrt{-27}}{6}$

In Exercises 53–60, write the conjugate of the given numbers.

53. $7 + 2j$

54. $9 + \frac{1}{2}j$

55. $6 - 5j$

56. $\frac{1}{2} - 9j$

57. 19

58. $7j$

59. $-8j$

60. -11

In Exercises 61–68, use the quadratic formula to solve each of the problems. Express your answers in the form *a + bj*.

61. $x^2 + x + 2.5 = 0$

62. $x^2 + 2x + 5 = 0$

63. $x^2 + 9 = 0$

64. $x^2 + 25 = 0$

65. $2x^2 + 3x + 7 = 0$

66. $2x^2 + 7x + 9 = 0$

67. $5x^2 + 2x + 5 = 0$

68. $3x^2 + 2x + 10 = 0$

☰ 14.2
OPERATIONS WITH COMPLEX NUMBERS

As with all number systems, we want to be able to perform the four basic operations of addition, subtraction, multiplication, and division. These operations are performed after all complex numbers have been expressed in terms of *j*.

Addition and Subtraction

We will begin by giving the definitions for addition and subtraction of complex numbers. After each definition, we will provide several examples showing how to use the definition.

> **Addition of Complex Numbers**
>
> If $a + bj$ and $c + dj$ are any two complex numbers, then their sum is defined as
>
> $$(a + bj) + (c + dj) = (a + c) + (b + d)j$$

In words, this says to add the real parts of the complex numbers and add their imaginary parts.

EXAMPLE 14.7

Find each of these sums: **(a)** $(9 + 2j) + (8 + 6j)$, **(b)** $(6 + 3j) + (5 - 7j)$, **(c)** $(-2\sqrt{3} + 4j) + (5 - 6j)$, and **(d)** $(-4 + 3j) + (-1 - \sqrt{-4})$.

Solutions

(a) $(9 + 2j) + (8 + 6j) = (9 + 8) + (2 + 6)j$
$$= 17 + 8j$$

(b) $(6 + 3j) + (5 - 7j) = (6 + 5) + (3 - 7)j$
$$= 11 - 4j$$

(c) $(-2\sqrt{3} + 4j) + (5 - 6j) = (-2\sqrt{3} + 5) + (4 - 6)j$
$$= 5 - 2\sqrt{3} - 2j$$

Notice that the real part of this complex number is $5 - 2\sqrt{3}$ and the imaginary part is $-2j$.

(d) $(-4 + 3j) + (-1 - \sqrt{-4}) = (-4 + 3j) + (-1 - 2j)$
$$= (-4 - 1) + (3 - 2)j$$
$$= -5 + j$$

Application

EXAMPLE 14.8

In an ac circuit, if two sections are connected in series and have the same current in each section, the voltage V is given by $V = V_1 + V_2$. Find the total voltage in a given circuit if the voltages in the individual sections are $8.9 - 2.4j$ and $11.2 + 6.3j$.

Solution To find the total voltage in this circuit, we need to add the voltages in the individual sections.

$$V = V_1 + V_2$$
$$= (8.9 - 2.4j) + (11.2 + 6.3j)$$
$$= (8.9 + 11.2) + (-2.4 + 6.3)j$$
$$= 20.1 + 3.9j$$

The voltage in this circuit is $20.1 + 3.9j$ V.

Subtraction of Complex Numbers

If $a + bj$ and $c + dj$ are complex numbers, then their difference is defined as

$$(a + bj) - (c + dj) = (a - c) + (b - d)j$$

EXAMPLE 14.9

Find each of the following differences: **(a)** $(3 + 4j) - (2 + j)$, **(b)** $(5 + 7j) - (3 - 10j)$, **(c)** $(-8 + 4j) - (3 + 10j)$, and **(d)** $(9 + \sqrt{-18}) - (6 + \sqrt{-2})$.

Solutions

(a) $(3 + 4j) - (2 + j) = (3 - 2) + (4 - 1)j$
$$= 1 + 3j$$

(b) $(5 + 7j) - (3 - 10j) = (5 - 3) + [7 - (-10)]j$
$$= 2 + 17j$$

(c) $(-8 + 4j) - (3 + 10j) = (-8 - 3) + (4 - 10)j$
$$= -11 - 6j$$

(d) $(9 + \sqrt{-18}) - (6 + \sqrt{-2}) = (9 + 3j\sqrt{2}) - (6 + j\sqrt{2})$
$$= (9 - 6) + (3\sqrt{2} - \sqrt{2})j$$
$$= 3 + 2j\sqrt{2}$$

Notice that in Examples 14.7(d) and 14.9(d), we had to first write some of the numbers in the $a + bj$ form. In Example 14.7(d), we changed $\sqrt{-4}$ to $2j$ and in Example 14.9(d) we changed $\sqrt{-18}$ to $3j\sqrt{2}$ and $\sqrt{-2}$ to $j\sqrt{2}$. Do not overlook this step.

Multiplication

Multiplication of complex numbers uses the FOIL method, which was introduced in Chapter 2. As you can see, you will have to replace j^2 with -1 to obtain a simplified answer.

$$(a + bj)(c + dj) = ac + adj + bcj + bdj^2$$
$$= ac + adj + bcj - bd$$
$$= (ac - bd) + (ad + bc)j$$

Multiplication of Complex Numbers

If $a + bj$ and $c + dj$ are any two complex numbers, then their product $(a + bj)(c + dj)$ is defined as

$$(a + bj)(c + dj) = (ac - bd) + (ad + bc)j$$

EXAMPLE 14.10

Multiply and write each answer in the form $a + bj$: **(a)** $(2 + 5j)(3 - 4j)$, **(b)** $(5 + 3j)^2$, and **(c)** $(7 + 3j)(7 - 3j)$.

Solutions   We have used the FOIL method rather than the definition to determine these products. By using the FOIL method, you do not have to remember the rule. However, remember that $j^2 = -1$.

(a) $\begin{aligned}(2 + 5j)(3 - 4j) &= 2 \cdot 3 + 2(-4)j + 5j(3) + 5(-4)j^2\\ &= 6 - 8j + 15j - 20(-1)\\ &= 6 - 8j + 15j + 20\\ &= 26 + 7j\end{aligned}$

(b) $\begin{aligned}(5 + 3j)^2 &= (5 + 3j)(5 + 3j)\\ &= 5^2 + 2(5)(3)j + 9j^2\\ &= 25 + 30j + 9(-1)\\ &= 25 + 30j - 9\\ &= 16 + 30j\end{aligned}$

(c) Here we can use the difference of squares.

$\begin{aligned}(7 + 3j)(7 - 3j) &= 7^2 - 3^2 j^2\\ &= 49 - 9(-1)\\ &= 49 + 9\\ &= 58\end{aligned}$

Application

EXAMPLE 14.11

In an ac circuit, the formula $V = ZI$ relates the voltage V, to impedance Z and current I. Find the voltage in a given circuit if the impedance is $8 - 2j \ \Omega$ and the current is $11 + 6j$ A.

Solution   To find the voltage, we need to multiply the given values of Z and I. As before, we will use the FOIL method.

$$V = ZI$$
$$= (8 - 2j)(11 + 6j)$$
$$= 8 \cdot 11 + 8(6j) + (-2j)(11) + (-2j)6j$$
$$= 88 + 48j - 22j + 12$$
$$= 100 + 26j$$

The voltage in this circuit is $100 + 26j$ V.

Division

In Example 14.10(c) we multiplied a complex number and its conjugate. The following note will be helpful when we divide complex numbers.

≡ **Note**   The product of a complex number and its conjugate is a real number.

Division of Complex Numbers

If $a + bj$ and $c + dj$ are complex numbers, then the quotient $(a + bj) \div (c + dj) = \dfrac{a + bj}{c + dj}$, is defined as

$$\frac{a + bj}{c + dj} = \frac{a + bj}{c + dj} \cdot \frac{c - dj}{c - dj} = \frac{(ac + bd) + (bc - ad)j}{c^2 + d^2}$$

$$= \frac{ac + bd}{c^2 + d^2} + \frac{bc - ad}{c^2 + d^2} j$$

This looks very complicated, but the following hint gives an important idea to remember whenever you divide by a complex number.

Hint

In division of complex numbers, you multiply both the numerator and the denominator by the conjugate of the denominator.

The four problems in Example 14.12 show how to use the conjugate to divide.

EXAMPLE 14.12

Divide and express each answer in the form $a + bj$: (a) $\dfrac{10 - 4j}{1 + j}$, (b) $\dfrac{8 + 6j}{2 - j}$, (c) $\dfrac{3 - 2j}{4 + 2j}$, and (d) $\dfrac{0.5 + j\sqrt{3}}{2.4j}$.

Solutions

(a) $\dfrac{10 - 4j}{1 + j} = \dfrac{10 - 4j}{1 + j} \cdot \dfrac{1 - j}{1 - j}$

$= \dfrac{10 - 10j - 4j + 4j^2}{1 + 1}$

$= \dfrac{10 - 10j - 4j - 4}{2}$

$= \dfrac{6 - 14j}{2} = 3 - 7j$

(b) $\dfrac{8 + 6j}{2 - j} = \dfrac{8 + 6j}{2 - j} \cdot \dfrac{2 + j}{2 + j}$

$= \dfrac{16 + 8j + 12j + 6j^2}{4 + 1}$

$= \dfrac{16 + 8j + 12j - 6}{5}$

$= \dfrac{10 + 20j}{5} = 2 + 4j$

(c) $\dfrac{3 - 2j}{4 + 2j} = \dfrac{3 - 2j}{4 + 2j} \cdot \dfrac{4 - 2j}{4 - 2j}$

$= \dfrac{12 - 6j - 8j - 4}{16 + 4}$

$= \dfrac{8 - 14j}{20} = \dfrac{8}{20} - \dfrac{14}{20}j$

$= \dfrac{2}{5} - \dfrac{7}{10}j = 0.4 - 0.7j$

EXAMPLE 14.12 (Cont.)

(d) $\dfrac{0.5 + j\sqrt{3}}{2.4j} = \dfrac{0.5 + j\sqrt{3}}{2.4j} \cdot \dfrac{-2.4j}{-2.4j}$

$\qquad = \dfrac{(0.5)(-2.4j) + (j\sqrt{3})(-2.4j)}{(2.4j)(-2.4j)}$

$\qquad = \dfrac{-1.2j + 2.4\sqrt{3}}{5.76}$

$\qquad = \dfrac{2.4\sqrt{3} - 1.2j}{5.76}$

$\qquad = \dfrac{2.4\sqrt{3}}{5.76} - \dfrac{1.2}{5.76}j = \dfrac{\sqrt{3}}{2.4} - \dfrac{1}{4.8}j$

Application

EXAMPLE 14.13

In an ac circuit, use the formula $V = ZI$ to find the impedance in a given circuit if the voltage is $95 + 9j$ V and the current is $7 - 3j$ A.

Solution We want to find the impedance Z, given V and I. Using the formula $V = ZI$, we see that $Z = \dfrac{V}{I}$.

$$Z = \dfrac{V}{I}$$

$$= \dfrac{95 + 9j}{7 - 3j}$$

$$= \dfrac{95 + 9j}{7 - 3j} \cdot \dfrac{7 + 3j}{7 + 3j}$$

$$= \dfrac{95(7) + 95(3j) + (9j)7 + (9j)(3j)}{7^2 - (3j)^2}$$

$$= \dfrac{665 + 285j + 63j + 27j^2}{49 - 9j^2}$$

$$= \dfrac{638 + 348j}{58}$$

$$= 11 + 6j$$

The impedance in this circuit is $11 + 6j$ Ω.

Using Calculators with Complex Numbers

Some calculators can be used for working with complex numbers. One way to tell if your calculator can work with complex numbers is to look in the index of the owner's manual. Another way is to see if your calculator has an $\boxed{\text{Img}}$, an $\boxed{\text{I}}$, or a $\boxed{\theta}$ key. If it does, then it can probably be used to calculate with complex numbers.

If you are going to use complex numbers on your algebraic calculator, you must first put the calculator in the complex mode. Turn on your calculator and press the $\boxed{\text{Img}}$ key. The display should show 0 on the right and CMPLX in the lower left corner.

CMPLX

FIGURE 14.1

(See Figure 14.1.) The calculator is now in the complex mode. To get the calculator out of the complex mode press $\boxed{\text{INV}}$ $\boxed{\text{Img}}$.

To enter a complex number of the form $a + bj$ into an algebraic calculator, first enter the imaginary part b and press $\boxed{\text{Img}}$. Next, enter the real part, a. With an RPN calculator, first enter the real part a, press $\boxed{\text{ENTER}}$, and then enter the imaginary part b and press $\boxed{\text{2nd}}$ $\boxed{\text{I}}$.

EXAMPLE 14.14

Enter **(a)** $9 + 2j$ and **(b)** $7 - 4j$ into a calculator.

Solutions

		PRESS	DISPLAY
(a)	Algebraic calculator:	2 $\boxed{\text{Img}}$	0
		9	9
	RPN calculator:	9 $\boxed{\text{ENTER}}$	9
		2 $\boxed{\text{2nd}}$ $\boxed{\text{I}}$	9
(b)	Algebraic calculator:	4 $\boxed{\text{Img}}$ $\boxed{^+/_-}$	0
		7	7
	RPN calculator:	7 $\boxed{\text{ENTER}}$	9
		4 $\boxed{\text{CHS}}$ $\boxed{\text{2nd}}$ $\boxed{\text{I}}$	7

If you want to determine the value stored in the imaginary part of the algebraic calculator, press $\boxed{\text{EXC}}$ $\boxed{\text{Img}}$ and, in the case of Example 14.14(b), the display will show −4 and the symbol CMPLX will blink. The blinking indicates that this is the imaginary part of the number. Enter $\boxed{\text{EXC}}$ $\boxed{\text{Img}}$ again and you will see the real part, 7, of the number. The CMPLX symbol stops blinking and is a steady display to show that this is the real part of the display. You must return to the real part of the display before you can perform any calculations.

To view the value stored in the imaginary part of an RPN calculator, press $\boxed{\text{2nd}}$ $\boxed{\text{(i)}}$. In the case of Example 14.14(b), the display will show −4 for about two seconds and then the real part of the complex number, 7, will again be displayed. To see the imaginary value for longer than two seconds, press the $\boxed{\text{2nd}}$ $\boxed{\text{(i)}}$ keys and hold down the $\boxed{\text{(i)}}$ key until you are finished viewing the imaginary part of the display.

Calculations with a calculator are done in the same manner in which they are done with real numbers. The following example shows how each of these is done.

EXAMPLE 14.15

Use a calculator to perform each of the following: **(a)** $(3 + 5j) + (7 - 8j)$, **(b)** $(2 + j)(-6 - 3j)$, and **(c)** $\dfrac{10 - 4j}{1 + j}$.

Solutions

PRESS	DISPLAY
(a) Algebraic calculator:	
5 $\boxed{\text{Img}}$ 3 $\boxed{+}$	3
8 $\boxed{^+/_-}$ $\boxed{\text{Img}}$ 7 $\boxed{=}$	10 steady CMPLX
$\boxed{\text{EXC}}$ $\boxed{\text{Img}}$	−3 blinking CMPLX

EXAMPLE 14.15 (Cont.)

RPN calculator:

3 [ENTER] 5 [2nd] [I]	3
7 [ENTER] 8 [CHS] [2nd] [I]	7
[+]	10
[2nd] [(i)]	−3

So, $(3 + 5j) + (7 − 8j) = 10 − 3j$

(b) Algebraic calculator:

1 [Img] 2 [×]	2
3 [+/−] [Img] 6 [+/−] [=]	−9 steady CMPLX
[EXC] [Img]	−12 blinking CMPLX

RPN calculator:

2 [ENTER] 1 [2nd] [I]	2
6 [CHS] [ENTER] 3 [2nd] [I]	−6
[×]	−9
[2nd] [(i)]	−12

From this we see that $(2 + j)(−6 − 3j) = −9 − 12j$.

(c) Algebraic calculator:

4 [+/−] [Img] 10 [÷]	10
1 [Img] 1 [=]	3 steady CMPLX
[EXC] [Img]	−7 blinking CMPLX

RPN calculator:

10 [ENTER] 4 [CHS] [2nd] [I]	10
1 [ENTER] 1 [2nd] [I]	1
[÷]	3
[2nd] [(i)]	−7

And, as we saw in Example 14.12(a), $(10 − 4j) ÷ (1 + j) = 3 − 7j$.

Exercise Set 14.2

In Exercises 1–48, perform the indicated operations. Express all answers in the form $a + bj$. If possible, use a calculator to check your answers.

1. $(5 + 2j) + (−6 + 5j)$

2. $(9 − 7j) + (6 − 8j)$

3. $(11 − 4j) + (−6 + 2j)$

4. $(21 + 3j) + (−7 − 6j)$

5. $(4 + \sqrt{−9}) + (3 − \sqrt{−16})$

6. $(−11 + \sqrt{−4}) + (9 + \sqrt{−36})$

7. $(2 + \sqrt{−9}) + (8j − \sqrt{5})$

8. $(3 + \sqrt{−8}) + (3 − \sqrt{8})$

9. $(14 + 3j) − (6 + j)$

10. $(−8 + 3j) − (4 − 3j)$

11. $(9 − \sqrt{−4}) − (\sqrt{−16} + 6)$

12. $(\sqrt{−25} − 3) − (3 − \sqrt{−25})$

13. $(4 + 2j) + j + (3 − 5j)$

14. $(2 − 3j) − j − (6 + \sqrt{−81})$

15. $(2 + j)3j$

16. $(5 − 3j)2j$

17. $(9 + 2j)(-5j)$

18. $(11 - 4j)(-3j)$

19. $(2 + j)(5 + 3j)$

20. $(3 - 2j)(4 + 5j)$

21. $(6 - 2j)(5 + 3j)$

22. $(4 - 2j)(7 - 3j)$

23. $(2\sqrt{-9} + 3)(5\sqrt{-16} - 2)$

24. $(6\sqrt{-25} - 4)(-3 - 2\sqrt{-49})$

25. $(\sqrt{-3})^4$

26. $(\sqrt{-9})^3$

27. $(1 + 2j)^2$

28. $(3 + 4j)^2$

29. $(7 - j)^2$

30. $(4 - 3j)^2$

31. $(5 + 2j)(5 - 2j)$

32. $(7 + 3j)(7 - 3j)$

33. $\dfrac{6 - 4j}{1 + j}$

34. $\dfrac{4 - 8j}{2 - 2j}$

35. $\dfrac{6 - 3j}{1 + 2j}$

36. $\dfrac{5 - 10j}{1 - 2j}$

37. $\dfrac{4 + 2j}{1 - 2j}$

38. $\dfrac{9 + 5j}{3 + j}$

39. $\dfrac{2j}{5 + j}$

40. $\dfrac{5j}{6 - j}$

41. $\dfrac{\sqrt{3} - \sqrt{-6}}{\sqrt{-3}}$

42. $\dfrac{\sqrt{5} + \sqrt{-10}}{\sqrt{-5}}$

43. $\dfrac{(5 + 2j)(3 - j)}{4 + j}$

44. $\dfrac{(6 - j)(2 + 3j)}{-1 + 3j}$

45. $(1 + j)^4$

46. $(1 - j)^4$

47. $\left(\dfrac{1}{2} + \dfrac{\sqrt{3}}{2}j\right)^2$

48. $\dfrac{4 + j}{(3 - 2j) + (4 - 3j)}$

Solve Exercises 49–65.

49. Show that the sum of a complex number and its conjugate is a real number.

50. Show that the difference of a complex number $a + bj$, with $b \neq 0$ and its conjugate is a pure imaginary number.

51. Show that the product of a complex number and its conjugate is a real number.

52. *Electricity* In an ac circuit, if two sections are connected in series and have the same current in each section, the voltage V is given by $V = V_1 + V_2$. Find the total voltage in a given circuit if the voltages in the individual sections are $9.32 - 6.12j$ and $7.24 + 4.31j$.

53. *Electricity* Find the total voltage in an ac series circuit that has the same current in each section, if the voltages in the individual sections are $6.21 - 1.37j$ and $4.32 - 2.84j$.

54. *Electricity* If two sections of an ac series circuit have the same current in each section, what is the

voltage in one section if the total voltage is $19.2 - 3.5j$ and the voltage in the other section is $12.4 + 1.3j$?

55. *Electricity* If two sections of an ac series circuit have the same current in each section, what is the voltage in one section if the total voltage is $7.42 + 1.15j$ and the voltage in the other section is $2.34 - 1.73j$?

56. *Electricity* The total impedance Z of an ac circuit containing two impedances Z_1 and Z_2 in series is $Z = Z_1 + Z_2$. If $Z_1 = 0.25 + 0.20j \, \Omega$ and $Z_2 = 0.15 - 0.25j \, \Omega$, what is Z?

57. *Electricity* If the total impedance of an ac series circuit containing two impedances is $Z = 9.13 - 4.27j \, \Omega$ and one of the impedances is $3.29 - 5.43j \, \Omega$, what is the impedance of the other circuit?

58. *Electricity* If an ac circuit contains two impedances Z_1 and Z_2 in parallel, then the total impedance Z is

given by

$$Z = \frac{Z_1 Z_2}{Z_1 + Z_2}$$

What is Z when $Z_1 = 6\ \Omega$ and $Z_2 = 3j\ \Omega$?

59. *Electricity*   What is the total impedance in an ac circuit that contains two impedances Z_1 and Z_2 in parallel, if $Z_1 = 20 + 10j\ \Omega$ and $Z_2 = 10 - 20j\ \Omega$?

60. *Electricity*   In an ac circuit the voltage V, current I, and impedance Z are related by $V = IZ$. If $I = 12.3 + 4.6j$ A and $Z = 16.4 - 9.0j\ \Omega$, what is the voltage?

61. *Electricity*   What is the current when $V = 5.2 + 3j$ V and $Z = 4 - 2j\ \Omega$? (See Exercise 60.)

62. *Electricity*   What is the impedance when $V = 10.6 - 6.0j$ V and $I = 4 + j$ A?

 63. Write a computer program that will add or subtract two complex numbers.

64. Write a computer program that will multiply two complex numbers.

65. Write a computer program that will divide two complex numbers.

☰ 14.3
GRAPHING COMPLEX NUMBERS; POLAR FORM OF A COMPLEX NUMBER

We have been able to graph real numbers since Chapter 4. It would be helpful if we could also represent complex numbers as points in a plane. The fact that each complex number has a real part and an imaginary part makes it possible to graph complex numbers.

Complex Plane

Each complex number can be written in the form $a + bj$. We can represent this as the point (a, b) in the plane, as shown in Figure 14.2. Notice that the origin corresponds to the point $(0, 0)$, or $0 + 0j$. Since the points in the plane are representing complex numbers, this is called the **complex plane**. It is also referred to as the **Argand plane**, after the French mathematician Argand (1768–1822). In the complex plane, the horizontal axis acts as the real axis and the vertical axis as the **imaginary axis**.

EXAMPLE 14.16

Graph the following complex numbers: **(a)** $4+2j$, **(b)** $-3+5j$, **(c)** $-5-10j$, **(d)** $9-4j$, **(e)** 12, and **(f)** $-3j$.

Solutions   The solutions are shown in Figure 14.3.

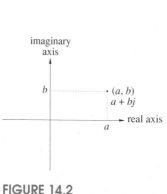

FIGURE 14.2

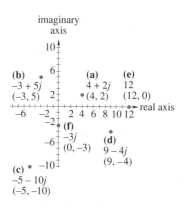

FIGURE 14.3

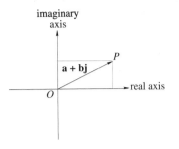

FIGURE 14.4

The complex number $a + bj$ can also be represented in the plane by the position vector **OP** from the origin to the point $P(a, b)$. We now have three correct and interchangeable ways to refer to the complex number: $a + bj$, the point (a, b) on the complex (Argand) plane, and the vector $\mathbf{a} + \mathbf{bj}$. (See Figure 14.4.)

If we have two complex numbers on a graph, their sum or difference can be represented in the same way in which we add or subtract vectors. To add two complex numbers graphically, locate the point corresponding to one of them and draw the position vector for that point. Repeat the process for the second point. Finally, use the parallelogram method to add these two vectors. The sum will be the diagonal of the parallelogram that has the origin as an endpoint.

EXAMPLE 14.17

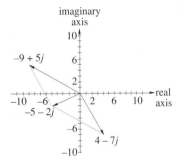

FIGURE 14.5

Graphically add the complex numbers $-9 + 5j$ and $4 - 7j$.

Solution   The solution is shown in Figure 14.5.

As you can see, the graphical solution agrees with the method given in Section 14.2.

$$(-9 + 5j) + (4 - 7j) = (-9 + 4) + (5 - 7)j$$
$$= -5 - 2j$$

Polar Form of a Complex Number

The rectangular form is not the only way to represent complex numbers. In Figure 14.6, the angle θ that the vector **OP** makes with the positive real axis is called the **argument** or **amplitude** of the complex number $a + bj$. The length r of **OP** is called the **absolute value** or **modulus** of $a + bj$. The absolute value is a real number and is never negative. The absolute value is always positive or 0. Using the Pythagorean theorem, we see that the length of r is $\sqrt{a^2 + b^2}$.

Since $a + bj$ can be considered a vector in the complex plane, $|a + bj|$ is the magnitude of the vector.

Absolute Value of a Complex Number

The **absolute value of a complex number** $a + bj$ is denoted $|a + bj|$ and has the value

$$|a + bj| = \sqrt{a^2 + b^2}$$

A careful examination of Figure 14.6 reveals four useful relationships. First, from our definitions of the trigonometric functions, we see that $\cos \theta = \dfrac{a}{r}$ and $\sin \theta = \dfrac{b}{r}$. From these, we see that

$$a = r \cos \theta \qquad (1)$$

and $\qquad b = r \sin \theta \qquad (2)$

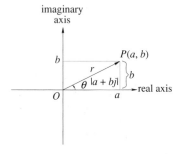

FIGURE 14.6

The other two relationships require not only our knowledge of trigonometry but of the Pythagorean theorem. These two relationships state that

$$\tan \theta = \frac{b}{a} \qquad (3)$$

and

$$r = \sqrt{a^2 + b^2} \qquad (4)$$

These four equations will be very valuable to us. From the first two, we see that

$$a + bj = r \cos \theta + jr \sin \theta = r(\cos \theta + j \sin \theta)$$

The expression $r(\cos \theta + j \sin \theta)$ is often abbreviated as r cis θ or $r\underline{/\theta}$. In the abbreviation r cis θ, the c represents cosine, the s represents sine, and the i represents the mathematician's symbol for j. The symbol $r\underline{/\theta}$ is read "r at angle θ." The right-hand side of the previous equation, $r(\cos \theta + j \sin \theta)$, is called the **polar** or **trigonometric form** of a complex number.

Changing Complex Numbers from Polar to Rectangular Form

A complex number written in polar form as

$$r\underline{/\theta} \qquad \text{or} \qquad r \text{ cis } \theta \qquad \text{or} \qquad r(\cos \theta + j \sin \theta)$$

has the rectangular coordinates $a + bj$, where

$$a = r \cos \theta \qquad \text{and} \qquad b = r \sin \theta$$

EXAMPLE 14.18

Locate the point $5(\cos 120° + j \sin 120°)$ in the complex plane and convert the number to rectangular form.

Solution   The graphical representation is given in Figure 14.7.

In this example, $r = 5$ and $\theta = 120°$, and so

$$a = r \cos \theta = 5 \cos 120° = 5(-0.5) = -2.5$$
$$b = r \sin \theta = 5 \sin 120° \approx 5(0.8660) = 4.3301$$

Thus, 5 cis $120° \approx -2.5 + 4.3301j$.

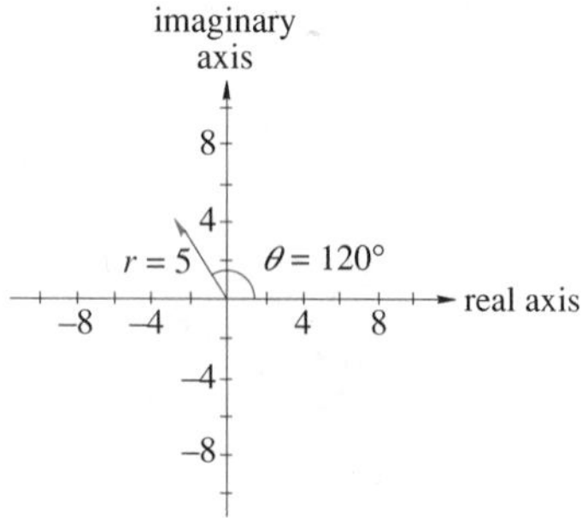

FIGURE 14.7

The previous example was worked using a calculator. The exact value for b would have been represented by $\sin \theta = \frac{\sqrt{3}}{2}$, and so $b = \frac{5\sqrt{3}}{2}$. Even though a calculator gives values to more than four decimal places, we will give degrees to the nearest tenth and trigonometric functions to four decimal places.

  Do not round off numbers until all calculations have been finished. Rounding off numbers before you complete the problem can make your final results different from the correct results.

EXAMPLE 14.19

Locate the point $8\big/\,215°$ in the complex plane and convert the number to rectangular form.

Solution   The graphical representation is given in Figure 14.8.

In this example, $r = 8$ and $\theta = 215°$, so

$$a = r\cos\theta = 8\cos 215° \approx 8(-0.8192) = -6.5532$$

$$b = r\sin\theta = 8\sin 215° \approx 8(-0.5736) = -4.5886$$

Thus, $8\big/\,215° \approx -6.5532 - 4.5886j$.

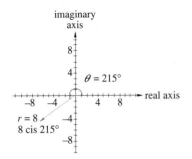

FIGURE 14.8

Changing Complex Numbers from Rectangular to Polar Form

A complex number written in rectangular form as

$$a + bj$$

has the polar coordinate forms

$$r\big/\,\theta, \qquad r\text{ cis }\theta, \qquad \text{or} \qquad r(\cos\theta + j\sin\theta),$$

where

$$r = \sqrt{a^2 + b^2} \qquad \text{and} \qquad \tan\theta = \frac{b}{a}$$

≡ Note   While there may be many values for θ that satisfy the given conditions, we will normally select the smallest positive value.

EXAMPLE 14.20

Represent $5 - 12j$ graphically and convert it to polar form.

Solution   Graphically, this point is shown as a vector in Figure 14.9.

Here $a = 5$ and $b = -12$, so

$$r = \sqrt{a^2 + b^2}$$

$$= \sqrt{5^2 + (-12)^2} = \sqrt{25 + 144}$$

$$= \sqrt{169} = 13$$

From the graph we can see that this point is in Quadrant IV.

$$\tan\theta = \frac{b}{a}$$

$$= \frac{-12}{5}$$

so

$$\theta = \tan^{-1}\left(\frac{-12}{5}\right)$$

$$\approx -67.4°$$

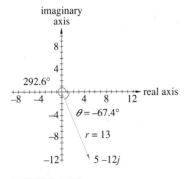

FIGURE 14.9

EXAMPLE 14.20 (Cont.)

Notice that we used the arctan function. We can let θ be either $-67.4°$ or $-67.4° + 360° = 292.6°$. So, $5 - 12j \approx 13\underline{/\,292.6°} = 13\underline{/\,-67.4°}$. But, as we just stated, we will usually use the smallest positive value for θ that satisfies the given conditions. In this case, $\theta = 292.6°$.

EXAMPLE 14.21

Locate the point $-5 + 7.5j$ in the complex plane and convert it to polar form.

Solution   The point $-5 + 7.5j$ is shown in the graph in Figure 14.10 and we see that it is in Quadrant II.

In this example $a = -5$ and $b = 7.5$, so

$$r = \sqrt{(-5)^2 + 7.5^2}$$

$$= \sqrt{25 + 56.25}$$

$$= \sqrt{81.25} \approx 9.01$$

$$\tan \theta = \frac{7.5}{-5} = -1.5$$

so $\theta \approx 123.7°$

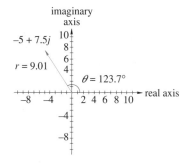

FIGURE 14.10

Thus, $-5 + 7.5j \approx 9.01\underline{/\,123.7°}$.

Caution   Remember that using a calculator to work with inverse trigonometric functions may not give the desired angle. You must determine which quadrant the point is in.

In Example 14.21, we had $\tan \theta = -1.5$. If you use a calculator (in degree mode) to determine $\theta = \tan^{-1}(-1.5)$, you get

	PRESS	DISPLAY
Algebraic calculator:	1.5 $\boxed{+/-}$ $\boxed{\text{INV}}$ $\boxed{\text{tan}}$	-56.309932
Graphics calculator:	$\boxed{\text{2nd}}$ $\boxed{\text{tan}}$ $\boxed{(-)}$ 1.5	-56.30993247

This value for θ, $-56.3°$, is an angle in the fourth quadrant. We can see, from Figure 14.10, that the point is in Quadrant II. So, $\theta = -56.3° + 180° = 123.7°$

Finding the polar form of a real number or a pure imaginary number is relatively easy, as shown in Example 14.22.

EXAMPLE 14.22

Express each of the following complex numbers in polar form: **(a)** 8, **(b)** -9, **(c)** $3j$, and **(d)** $-4j$.

Solutions

(a) $8 = 8 + 0j$ lies on the positive real axis, so $\theta = 0°$ and $r = 8$; thus, $8 = 8(\cos 0° + j \sin 0°) = 8 \operatorname{cis} 0°$.

(b) $-9 = -9 + 0j$ lies on the negative real axis so, $\theta = 180°$ and $r = 9$; thus, $-9 = 9 \operatorname{cis} 180°$.

EXAMPLE 14.22 (Cont.)

(c) $3j = 0 + 3j$ lies on the positive imaginary axis, so $\theta = 90°$ and $r = 3$; $3j = 3\underline{/90°}$.

(d) $-4j = 0 - 4j$ lies on the negative imaginary axis, so $\theta = 270°$ and $r = 4$; $-4j = 4(\cos 270° + j \sin 270°) = 4 \text{ cis } 270°$.

Exercise Set 14.3

For each number in Exercises 1–16, locate the point in the complex plane and express the number in rectangular form.

1. $4(\cos 30° + j \sin 30°)$

2. $10(\cos 45° + j \sin 45°)$

3. $5(\cos 135° + j \sin 135°)$

4. $8(\cos 305° + j \sin 305°)$

5. $7 \text{ cis } 260°$

6. $3 \text{ cis } 340°$

7. $2 \text{ cis } 115°$

8. $5 \text{ cis } 285°$

9. $3\underline{/25°}$

10. $4\underline{/240°}$

11. $5\underline{/340°}$

12. $6\underline{/90°}$

13. $4.5\underline{/245°}$

14. $6.8\underline{/10°}$

15. $2.5\underline{/180°}$

16. $5.9\underline{/270°}$

For each number in Exercises 17–32, locate the point in the complex plane and express the number in polar form.

17. $5 + 5j$

18. $6 + 3j$

19. $4 - 8j$

20. $8 - 2j$

21. $-4 + 7j$

22. $-9 + 3j$

23. $-6 - 2j$

24. $-10 - 2j$

25. 6

26. $1.2 + 7.3j$

27. $4.2 - 6.3j$

28. $9j$

29. $-5.8 + 0.2j$

30. $-7j$

31. -2.7

32. $-4.7 - 1.1j$

In Exercises 33–40, perform the indicated operations graphically.

33. $(5 + j) + (3 + 2j)$

34. $(3 - 4j) + (4 - 2j)$

35. $(-4 + 2j) + (2 - 8j)$

36. $(8 - j) + (9 + 6j)$

37. $(-8 + 7j) + (-5 - 3j)$

38. $(4 + 2j) + (-3 - 9j)$

39. $(-5 + 3j) + (4 - 8j)$

40. $(-3 + 2j) + (-8 - 5j)$

Solve Exercise 41.

41. Write a computer program that will change a complex number from the rectangular form to polar form, or vice versa.

≡ 14.4
EXPONENTIAL FORM OF A COMPLEX NUMBER

There is yet another way in which complex numbers are often represented. It is called the **exponential form of a complex number**, because it involves exponents of the number e. (Remember from Section 13.2, that $e \approx 2.718281828....$) If $z = r\underline{/\theta}$ is a complex number, then we know that $z = r(\cos\theta + j\sin\theta)$.

Exponential Form of a Complex Number

The **exponential form** of a complex number uses **Euler's formula**, $e^{j\theta} = \cos\theta + j\sin\theta$, and states that

$$z = re^{j\theta}$$

where θ is in radians.

While θ can have any value, we will express answers with $0 \le \theta < 2\pi$.

Caution When using the exponential form, you must remember that θ is in radians.

EXAMPLE 14.23

Write the complex number $6(\cos 180° + j\sin 180°)$ in exponential form.

Solution   In this example, $r = 6$ and $\theta = 180°$. The exponential form requires that θ be expressed in radians: $180° = \pi$ rad and so $\theta = \pi$. Thus $6(\cos 180° + j\sin 180°) = 6e^{j\pi}$.

EXAMPLE 14.24

Write the complex number $8\underline{/225°}$ in exponential form.

Solution   In this example, $r = 8$ and $\theta = 225° = \dfrac{5\pi}{4} \approx 3.927$ rad.

$$8\underline{/225°} = 8e^{\frac{5\pi}{4}j} = 8e^{5j\pi/4} \approx 8e^{3.927j}$$

All of the last three versions are correct. Because you will probably be using a calculator to convert from degrees to radians, the last version is the one that you will most likely use.

EXAMPLE 14.25

Express $-4 + 3j$ in exponential form.

Solution   This example is in the form $a + bj$. We must first determine r and θ. Now, $r = \sqrt{a^2 + b^2} = \sqrt{(-4)^2 + 3^2} = 5$ and $\tan\theta = \frac{3}{-4} = -0.75$. With an algebraic calculator in radian mode, we can determine θ.

PRESS	DISPLAY
.75 $\boxed{+/_-}$ $\boxed{\text{INV}}$ $\boxed{\text{tan}}$	-0.6435011
$\boxed{+}$ $\boxed{\pi}$	3.1415927
$\boxed{=}$	2.4980915

EXAMPLE 14.25 (Cont.)

The last two steps were needed because the original answer for θ, -0.6435011, was for an angle in Quadrant IV. The point $-4 + 3j$ is in Quadrant II, so we added π to the original answer. Thus, we can see that $\theta \approx 2.4981$ and $-4 + 3j \approx 5e^{2.4981j}$.

EXAMPLE 14.26

Express $-5 - 8j$ in exponential form.

Solution $r = \sqrt{a^2 + b^2} = \sqrt{(-5)^2 + (-8)^2} = \sqrt{25 + 64} = \sqrt{89} \approx 9.4340$.
$\tan \theta = \frac{-8}{-5} = 1.6$ and so $\theta \approx 4.1538$. (Note that $-5 - 8j$ is in Quadrant III.) Thus, we see that $-5 - 8j = 9.4340e^{4.1538j}$.

EXAMPLE 14.27

Express $4.6e^{5.7j}$ in polar and rectangular form.

Solution Here, $r = 4.6$ and $\theta = 5.7$, so we write the given expression in polar form as $4.6e^{5.7j} = 4.6(\cos 5.7 + j \sin 5.7)$. With a calculator in radian mode, we get

$$a = 4.6 \cos 5.7 \approx 3.8397$$

$$b = 4.6 \sin 5.7 \approx -2.5332$$

We have determined that $4.6e^{5.7j} = a + bj \approx 3.8397 - 2.5332j$.

Application

EXAMPLE 14.28

The voltage in an ac circuit is represented by the complex number $V = -85.6 + 72.3j$ V. Express this complex number in exponential form.

Solution First, we find r.

$$r = \sqrt{(-85.6)^2 + 72.3^2} = \sqrt{7327.36 + 5227.29} \approx 112.0$$

Next, we find the angle θ.

$$\tan \theta = \frac{72.3}{-85.6}$$

$$\theta \approx -0.70137$$

Since V is in the second quadrant, $\theta = -0.70137 + \pi \approx 2.4402$.

So, in exponential form, we find that voltage $V = -85.6 + 72.3j \approx 112e^{2.4402j}$ V.

Application

EXAMPLE 14.29

Express the current $I = 2.5 \underline{/-50°}$ A in exponential and rectangular forms.

Solution Here, $r = 2.5$ and $\theta = -50° \approx -0.8727$ rad, so in exponential form we have $I = re^{j\theta} = 2.5e^{-0.8727j}$.

EXAMPLE 14.29 (Cont.)

To convert the given number to rectangular form, we have $x = 2.5\cos(-50°) \approx 1.6070$ and $y = 2.5\sin(-50°) \approx -1.9151$. Thus, in rectangular form, the current is $I = 1.6070 - 1.9151j$ A.

Multiplying and Dividing in Exponential Form

One advantage of using the exponential form for complex numbers is that complex numbers written in exponential form obey the laws of exponents. There are three properties of exponents that are of interest. These three properties concern multiplication, division, and powers, and all of them use four basic rules introduced in Section 1.4.

$$b^m b^n = b^{m+n}$$

$$\frac{b^m}{b^n} = b^{m-n}$$

$$(b^m)^n = b^{mn}$$

$$(ab)^m = a^m b^m$$

If we have two complex numbers $z_1 = r_1 e^{j\theta_1}$ and $z_2 = r_2 e^{j\theta_2}$, we can then multiply, divide, or take powers of them using the preceding rules. For example,

$$z_1 z_2 = (r_1 e^{j\theta_1})(r_2 e^{j\theta_2}) = r_1 r_2 e^{j(\theta_1 + \theta_2)}$$

and $\qquad \dfrac{z_1}{z_2} = \dfrac{r_1 e^{j\theta_1}}{r_2 e^{j\theta_2}} = \dfrac{r_1}{r_2} e^{j(\theta_1 - \theta_2)}$

EXAMPLE 14.30

Multiply $7e^{4.2j}$ and $2e^{1.5j}$.

Solution $(7e^{4.2j})(2e^{1.5j}) = 7 \cdot 2e^{(4.2+1.5)j} = 14e^{5.7j}$

If we want to divide, then

$$\frac{z_1}{z_2} = \frac{r_1 e^{j\theta_1}}{r_2 e^{j\theta_2}} = \frac{r_1}{r_2} e^{j(\theta_1 - \theta_2)}$$

EXAMPLE 14.31

Divide $9e^{3.2j}$ by $2e^{4.3j}$.

Solution $\dfrac{9e^{3.2j}}{2e^{4.3j}} = \dfrac{9}{2} e^{(3.2-4.3)j} = 4.5e^{-1.1j}$

If you want to express the angle in the answer between 0 and 2π ($0 \leq \theta < 2\pi$), then let $\theta = 2\pi - 1.1$, or, using a calculator,

PRESS	DISPLAY
2 $\boxed{\times}$ $\boxed{\pi}$ $\boxed{-}$	6.2831853
1.1 $\boxed{=}$	5.1831853

Thus, $4.5e^{-1.1j}$ could be expressed as $4.5e^{5.2j}$.

The last property involves raising a complex number to a power and the properties $(b^m)^n = b^{mn}$ and $(ab)^m = a^m b^m$.

$$z^n = (re^{j\theta})^n = r^n e^{jn\theta}$$

EXAMPLE 14.32

Calculate $(4e^{2.3j})^5$.

Solution
$$(4e^{2.3j})^5 = 4^5 e^{5(2.3)j}$$
$$= 1{,}024 e^{11.5j}$$
$$\approx 1{,}024 e^{5.2j}$$

Since 11.5 is greater than 2π, we subtracted 2π to get an angle of 5.2, which is between 0 and 2π.

Application

EXAMPLE 14.33

Given that $I = 5.8e^{-0.4363j}$ A and $Z = 8.5e^{1.047j}$ Ω, calculate $V = IZ$.

Solution Since $V = IZ$, we want to determine the product of the given numbers.
$$V = (5.8e^{-0.4363j})(8.5e^{1.047j}) = (5.8)(8.5)e^{-0.4363j+1.047j}$$
$$= 49.3e^{0.6107j}$$

The voltage is $49.3e^{0.6107j}$ V.

Application

EXAMPLE 14.34

If $V_C = 78.3e^{-.3725j}$ V and $X_C = 87.0e^{-1.6500j}$ Ω, and $V_C = IX_C$, determine I.

Solution Since $V_C = IX_C$, then $I = \dfrac{V_C}{X_C}$, and so we need to divide.
$$I = \frac{V_C}{X_C} = \frac{78.3e^{-.3725j}}{87.0e^{-1.6500j}}$$
$$= \frac{78.3}{87.0}e^{-.3725j-(-1.6500j)}$$
$$= 0.9e^{1.2775j}$$

So, we have determined that $I = 0.9e^{1.2775j}$ A.

Exercise Set 14.4

In Exercises 1–12, express each complex number in exponential form.

1. $3\left(\cos\frac{3\pi}{2} + j\sin\frac{3\pi}{2}\right)$

2. $7(\cos 1.4 + j\sin 1.4)$

3. $2(\cos 60° + j\sin 60°)$

4. $11(\cos 320° + j\sin 320°)$

5. $1.3(\cos 5.7 + j\sin 5.7)$

6. $9.5(\cos 2.1 + j\sin 2.1)$

7. $3.1(\cos 25° + j\sin 25°)$

8. $10.5(\cos 195° + j\sin 195°)$

9. $8 + 6j$

10. $12 - 5j$

11. $-9 + 12j$

12. $-8 - 12j$

In Exercises 13–16, express each number in rectangular and polar form.

13. $5e^{0.5j}$

14. $8e^{1.9j}$

15. $2.3e^{4.2j}$

16. $4.5e^{\frac{7\pi}{6}j}$

In Exercises 17–34, perform each of the indicated operations.

17. $2e^{3j} \cdot 6e^{2j}$

18. $3e^{j} \cdot 4e^{2j}$

19. $e^{1.3j} \cdot 2.4e^{4.6j}$

20. $1.5e^{4.1j} \cdot 0.2e^{1.7j}$

21. $7e^{4.3j} \cdot 4e^{5.7j}$

22. $3.6e^{5.4j} \cdot 2.5e^{6.1j}$

23. $8e^{3j} \div 2e^{j}$

24. $28e^{5j} \div 7e^{2j}$

25. $17e^{4.3j} \div 4e^{2.8j}$

26. $8.5e^{3.4j} \div 2e^{5.3j}$

27. $(3e^{2j})^4$

28. $(4e^{3j})^5$

29. $(2.5e^{1.5j})^4$

30. $(7.2e^{2.3j})^5$

31. $(4e^{6j})^{1/2}$

32. $(16e^{6j})^{1/4}$

33. $(6.25e^{4.2j})^{1/2}$

34. $(1.728e^{2.1j})^{1/3}$

Solve Exercises 35–42.

35. *Electricity* The voltage in an ac circuit is represented by the complex number $V = 56.5 + 24.1j$ V. Express this complex number in exponential form.

36. *Electricity* Express the current $I = 4.90 - 4.11j$ A in exponential form.

37. *Electricity* Express the impedance $Z = 135 \times \underline{/-52.5°}\ \Omega$ in exponential and rectangular forms.

38. *Electricity* Express the capacitive reactance $X_C = 40.5\underline{/-\pi/2}\ \Omega$ in exponential and rectangular forms.

39. *Electricity* Given that $I = 12.5e^{-0.7256j}$ A and $Z = 6.4e^{1.4285j}\ \Omega$, find $V = IZ$.

40. *Electricity* Given that $I = 4.24e^{0.5627j}$ A and $X_L = 28.5e^{-1.5708j}\ \Omega$, find $V_L = IX_L$.

41. *Electricity* If $V = 115e^{-0.2145j}$ V and $Z = 2.5e^{0.5792j}\ \Omega$, find I, given that $V = IZ$.

42. *Electricity* If $V_R = 35.1e^{1.3826j}$ V and $I = 0.78e^{1.3826j}$ A, find R, given that $V_R = IR$.

☰ 14.5

OPERATIONS IN POLAR FORM; DeMOIVRE'S FORMULA

Multiplication, division, and powers of complex numbers are easily performed when the numbers are written in exponential form. They are just as easily performed when the numbers are in polar form. Remember the relationship between the exponential and polar forms.

$$re^{j\theta} = r(\cos\theta + j\sin\theta)$$

Multiplication

Again, we will let $z_1 = r_1 e^{j\theta_1}$ and $z_2 = r_2 e^{j\theta_2}$ be two complex numbers. We know that

$$z_1 z_2 = r_1 r_2 e^{j(\theta_1 + \theta_2)}$$

so in polar form, this would be

$$r_1(\cos\theta_1 + j\sin\theta_1) \cdot r_2(\cos\theta_2 + j\sin\theta_2) = r_1 r_2 [\cos(\theta_1 + \theta_2) + j\sin(\theta_1 + \theta_2)]$$

(In Chapter 19, you will develop skills with trigonometry that will allow you to verify this trigonometric identity.)

Using the alternative way of writing the polar form of a complex number, we have

$$r_1 \underline{/\theta_1} \cdot r_2 \underline{/\theta_2} = r_1 r_2 \underline{/(\theta_1 + \theta_2)}$$

Notice that the angles do not have to be written in radians. Angles can be in either degrees or radians.

Product of Two Complex Numbers

The product of two complex numbers $z_1 = r_1 \text{ cis } \theta_1 = r_1 \underline{/\theta_1} = r_1 e^{j\theta_1}$ and $z_2 = r_2 \text{ cis } \theta_2 = r_2 \underline{/\theta_2} = r_2 e^{j\theta_2}$ is

$$z_1 z_2 = (r_1 \text{ cis } \theta_1)(r_2 \text{ cis } \theta_2) = r_1 r_2 \text{ cis } (\theta_1 + \theta_2)$$

$$= r_1 \underline{/\theta_1} r_2 \underline{/\theta_2} = r_1 r_2 \underline{/\theta_1 + \theta_2}$$

$$= r_1 r_2 e^{j(\theta_1 + \theta_2)}$$

EXAMPLE 14.35

Find each of the following products.

(a) $2(\cos 15° + j\sin 15°) \cdot 5(\cos 80° + j\sin 80°)$, **(b)** $6\underline{/30°} \cdot 3\underline{/60°}$, **(c)** $4\underline{/2.3} \cdot 1.5\underline{/0.5}$, and **(d)** $(3e^{1.2j})(5e^{0.3j})$.

Solutions

(a) $2(\cos 15° + j\sin 15°) \cdot 5(\cos 80° + j\sin 80°)$
$$= 2 \cdot 5[(\cos(15° + 80°) + j\sin(15° + 80°)]$$
$$= 10(\cos 95° + j\sin 95°)$$

(b) $6\underline{/30°} \cdot 3\underline{/60°} = 6 \cdot 3\underline{/(30° + 60°)} = 18\underline{/90°} = 18j$

EXAMPLE 14.35 (Cont.)

(c) $4\underline{/\,2.3} \cdot 1.5\underline{/\,0.5} = 4(1.5)\underline{/\,(2.3+0.5)} = 6\underline{/\,2.8}$

(d) $(3e^{1.2j})(5e^{0.3j}) = 3(5)e^{(1.2+0.3)j} = 15e^{1.5j}$

Division

Division uses a similar process. Again, we will use the exponential form from Section 14.4 to show that

$$\frac{z_1}{z_2} = \frac{r_1 e^{j\theta_1}}{r_2 e^{j\theta_2}} = \frac{r_1}{r_2}e^{j(\theta_1-\theta_2)}$$

The results for both the exponential and polar forms are summarized as follows.

Quotient of Two Complex Numbers

The quotient of two complex numbers $z_1 = r_1 \operatorname{cis} \theta_1 = r_1\underline{/\,\theta_1} = r_1 e^{j\theta_1}$ and $z_2 = r_2 \operatorname{cis} \theta_2 = r_2\underline{/\,\theta_2} = r_2 e^{j\theta_2}$ is

$$\frac{z_1}{z_2} = \frac{r_1(\cos\theta_1 + j\sin\theta_1)}{r_2(\cos\theta_2 + j\sin\theta_2)}$$

$$= \frac{r_1}{r_2}[\cos(\theta_1 - \theta_2) + j\sin(\theta_1 - \theta_2)]$$

or
$$= \frac{r_1\underline{/\,\theta_1}}{r_2\underline{/\,\theta_2}}$$

$$= \frac{r_1}{r_2}\underline{/\,\theta_1 - \theta_2}$$

or
$$= \frac{r_1 e^{j\theta_1}}{r_2 e^{j\theta_2}} = \frac{r_1}{r_2}e^{j(\theta_1-\theta_2)}$$

EXAMPLE 14.36

Find each of the following quotients.

(a) $\dfrac{12(\cos 45° + j\sin 45°)}{2(\cos 15° + j\sin 15°)}$, (b) $\dfrac{15\underline{/\,135°}}{3\underline{/\,75°}}$, and (c) $4e^{2.1j} \div 8e^{1.7j}$.

Solutions

(a) $\dfrac{12(\cos 45° + j\sin 45°)}{2(\cos 15° + j\sin 15°)} = \dfrac{12}{2}[\cos(45° - 15°) + j\sin(45° - 15°)]$

$\qquad\qquad\qquad\qquad\qquad = 6(\cos 30° + j\sin 30°)$

(b) $\dfrac{15\underline{/\,135°}}{3\underline{/\,75°}} = \dfrac{15}{3}\underline{/\,(135° - 75°)} = 5\underline{/\,60°}$

(c) $4e^{2.1j} \div 8e^{1.7j} = \frac{4}{8}e^{(2.1-1.7)j} = \frac{1}{2}e^{0.4j}$

Powers, Roots, and DeMoivre's Formula ▬▬▬▬

Finding powers of complex numbers in polar form also uses the same process we developed for powers in the exponential form.

DeMoivre's Formula

For any complex number $z = r \operatorname{cis} \theta = r\underline{/\theta} = re^{j\theta}$

$$z^n = (re^{j\theta})^n = r^n e^{jr\theta}$$

or $\quad [r(\cos\theta + j\sin\theta)]^n = r^n(\cos n\theta + j\sin n\theta)$

This formula is known as **DeMoivre's formula**.

EXAMPLE 14.37

Use DeMoivre's formula to find each of the following powers. Convert your answers to rectangular form.

(a) $[3(\cos 30° + j\sin 30°)]^6$, **(b)** $(2\underline{/135°})^5$, **(c)** $(16\underline{/225°})^{1/4}$, and **(d)** $(2e^{0.4j})^7$.

Solutions

(a) $[3(\cos 30° + j\sin 30°)]^6 = 3^6(\cos 6 \cdot 30° + j\sin 6 \cdot 30°)$
$$= 729(\cos 180° + j\sin 180°)$$
$$= -729$$

(b) $\qquad (2\underline{/135°})^5 = 2^5\underline{/5 \cdot 135°} = 32\underline{/675°}$
$$a = 32\cos 675° \approx 22.6274$$
$$b = 32\sin 675° \approx -22.6274$$

so, $\qquad (2\underline{/135°})^5 = 22.6274 - 22.6274j$

(c) $\qquad (16\underline{/225°})^{1/4} = 16^{1/4}\underline{/\frac{225}{4}°} = 2\underline{/56.25°}$
$$a = 2\cos 56.25° \approx 1.1111$$
$$b = 2\sin 56.25° = 1.6629$$

so, $\qquad (16\underline{/225°})^{1/4} \approx 1.1111 + 1.6629j$

There are three other 4th roots of $16\underline{/255°}$. We will soon see how to use DeMoivre's formula to find these other three roots.

(d) $(2e^{0.4j})^7 = (2)^7 e^{(0.4)7j} = 128e^{2.8j}$
$$\approx -120.6045 + 42.8785j$$

DeMoivre's formula can be used to help find all of the roots of a complex number. For example, the equation $z^3 = -1$ has three roots. One of the roots is -1. What are the other two? DeMoivre's formula can be used to find those roots.

First, we will write -1 in polar form.

$$-1 = 1(\cos 180° + j\sin 180°)$$

Using DeMoivre's formula with $n = \frac{1}{3}$, we get

$$(-1)^{1/3} = 1^{1/3}\left(\cos\frac{180°}{3} + j\sin\frac{180°}{3}\right)$$

$$= 1(\cos 60° + j\sin 60°)$$

$$= 0.5000 + \frac{\sqrt{3}}{2}j$$

$$\approx 0.5000 + 0.8660j$$

We can see that this is not -1, the answer that we had before. A check would verify that $(0.5 + 0.8660j)^3 = -1$. So, this is a correct answer.

Why did we get a different answer? If you divide any number between $0°$ and $1{,}080°$ (or 0 and 6π) by 3, you find an angle between $0°$ and $360°$ (or between 0 and 2π). Now, $180°$, $540°$, and $900°$ all have the same terminal side. So, we could have written -1 as $1(\cos 540° + j\sin 540°)$ or as $1(\cos 900° + j\sin 900°)$. Let's find the cube root of each of these numbers.

$$[1(\cos 540° + j\sin 540°)]^{1/3} = 1^{1/3}\left(\cos\frac{540°}{3} + j\sin\frac{540°}{3}\right)$$

$$= 1(\cos 180° + j\sin 180°)$$

$$= -1$$

$$[1(\cos 900° + j\sin 900°)]^{1/3} = 1(\cos 300° + j\sin 300°)$$

$$= 0.5000 - 0.8660j$$

We have found three different cube roots of -1. The first and last are conjugates of each other. We have also given an example of a process we can use to find all of the nth roots of a number, where n is a positive integer.

Roots of a Complex Number

If $z = r(\cos\theta + j\sin\theta)$, then the nth roots of z are given by the formula

$$w_k = \sqrt[n]{r}\left[\cos\left(\frac{\theta}{n} + \frac{360° \cdot k}{n}\right) + j\sin\left(\frac{\theta}{n} + \frac{360° \cdot k}{n}\right)\right] \qquad (*)$$

where $k = 0, 1, 2, \ldots, n-1$.

If θ is in radians, then substitute 2π for $360°$.

EXAMPLE 14.38

Find the five fifth roots of $32j$.

Solution The five fifth roots will be called w_0, w_1, w_2, w_3, and w_4. Using $32j = 32(\cos 90° + j\sin 90°)$ and applying formula $(*)$, we obtain the following.

$$w_0 = \sqrt[5]{32}\left[\cos\left(\frac{90°}{5} + \frac{360° \cdot 0}{5}\right) + j\sin\left(\frac{90°}{5} + \frac{360° \cdot 0}{5}\right)\right]$$

EXAMPLE 14.38 (Cont.)

$$= 2(\cos 18° + j \sin 18°)$$

$$\approx 1.9021 + 0.6180j$$

$$w_1 = \sqrt[5]{32} \left[\cos \left(\frac{90°}{5} + \frac{360° \cdot 1}{5} \right) + j \sin \left(\frac{90°}{5} + \frac{360° \cdot 1}{5} \right) \right]$$

$$= 2(\cos 90° + j \sin 90°)$$

$$= 2j$$

$$w_2 = \sqrt[5]{32} \left[\cos \left(\frac{90°}{5} + \frac{360° \cdot 2}{5} \right) + j \sin \left(\frac{90°}{5} + \frac{360° \cdot 2}{5} \right) \right]$$

$$= 2(\cos 162° + j \sin 162°)$$

$$\approx -1.9021 + 0.6180j$$

$$w_3 = \sqrt[5]{32} \left[\cos \left(\frac{90°}{5} + \frac{360° \cdot 3}{5} \right) + j \sin \left(\frac{90°}{5} + \frac{360° \cdot 3}{5} \right) \right]$$

$$= 2(\cos 234° + j \sin 234°)$$

$$\approx -1.1756 - 1.6180j$$

$$w_4 = \sqrt[5]{32} \left[\cos \left(\frac{90°}{5} + \frac{360° \cdot 4}{5} \right) + j \sin \left(\frac{90°}{5} + \frac{360° \cdot 4}{5} \right) \right]$$

$$= 2(\cos 306° + j \sin 306°)$$

$$\approx 1.1756 - 1.6180j$$

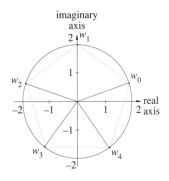

FIGURE 14.11

The five roots are shown in Figure 14.11. Notice the symmetry of these five roots around the circle. Any time you graph the n roots of a number, they should be equally spaced around a circle with the center at the origin in the complex plane. ▪

EXAMPLE 14.39

Find the three cube roots of $2\sqrt{11} + 10j$.

Solution   We first write this number in polar form. $r = \sqrt{(2\sqrt{11})^2 + 10^2} = \sqrt{144} = 12$ and $\tan \theta = \dfrac{10}{2\sqrt{11}} \approx 1.5075567$, so $\theta \approx 56.4°$. We will find the three cube roots w_0, w_1, and w_2.

$$w_0 = \sqrt[3]{12} \left[\cos \left(\frac{56.4°}{3} + \frac{360° \cdot 0}{3} \right) + j \sin \left(\frac{56.4°}{3} + \frac{360° \cdot 0}{3} \right) \right]$$

$$= \sqrt[3]{12}(\cos 18.8° + j \sin 18.8°)$$

$$\approx 2.1673 + 0.7378j$$

$$w_1 = \sqrt[3]{12} \left[\cos \left(\frac{56.4°}{3} + \frac{360° \cdot 1}{3} \right) + j \sin \left(\frac{56.4°}{3} + \frac{360° \cdot 1}{3} \right) \right]$$

$$= \sqrt[3]{12}(\cos 138.8° + j \sin 138.8°)$$

$$\approx -1.7230 + 1.5076j$$

EXAMPLE 14.39 (Cont.)

$$w_2 = \sqrt[3]{12}\left[\cos\left(\frac{56.4°}{3} + \frac{360° \cdot 2}{3}\right) + j\sin\left(\frac{56.4°}{3} + \frac{360° \cdot 2}{3}\right)\right]$$

$$= \sqrt[3]{12}(\cos 258.8° + j\sin 258.8°)$$

$$\approx -0.4447 - 2.2459j$$

Geometrically, these three points are shown on the graph in Figure 14.12.

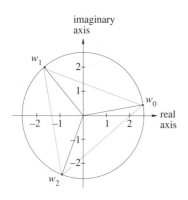

FIGURE 14.12

Exercise Set 14.5

In Exercises 1–20, perform the indicated operations and give the answers in polar form.

1. $3(\cos 46° + j\sin 46°) \cdot 5(\cos 23° + j\sin 23°)$

2. $4(\cos 135° + j\sin 135°) \cdot 5(\cos 63° + j\sin 63°)$

3. $2.5(\cos 1.43 + j\sin 1.43) \cdot 4(\cos 2.67 + j\sin 2.67)$

4. $6.4(\cos 0.25 + j\sin 0.25) \cdot 3.5(\cos 1.1 + j\sin 1.1)$

5. $\dfrac{8(\cos 85° + j\sin 85°)}{2(\cos 25° + j\sin 25°)}$

6. $\dfrac{6(\cos 273° + j\sin 273°)}{3(\cos 114° + j\sin 114°)}$

7. $\dfrac{9\,/\,137°}{2\,/\,26°}$

8. $\dfrac{18\,/\,3.52}{5\,/\,2.14}$

9. $(3\,/\,2.7)(4\,/\,5.3)$

10. $[3(\cos 20° + j\sin 20°)]^4$

11. $[5(\cos 84° + j\sin 84°)]^6$

12. $[2.5(\cos 118° + j\sin 118°)]^3$

13. $[10.4(\cos 3.42 + j\sin 3.42)]^3$

14. $(2\,/\,1.38)^5$

15. $(3.4\,/\,5.3)^4$

16. $(4.41\,/\,124°)^{1/2}$

17. $(4e^{2.1j})(3e^{1.7j})$

18. $6e^{1.5j} \div 4e^{0.9j}$

19. $(0.5e^{0.3j})^3$

20. $(0.0625e^{4.2j})^{1/2}$

In Exercises 21–28, use DeMoivre's formula to find the indicated roots. Give the answers in rectangular form.

21. cube roots of 1

22. cube roots of j

23. cube roots of $-8j$

24. fourth roots of -16

25. fourth roots of $-16j$

26. fourth roots of $1 - j$

27. fifth roots of $1 + j$

28. sixth roots of $-1 - j$

In Exercises 29–32, solve the given equations. Express your answers in rectangular form.

29. $x^3 = -j$

30. $x^3 = 125j$

31. $x^6 - 64j = 0$

32. $x^6 - 1 = j$

Solve Exercises 33–37.

33. *Electricity* Ohm's law for alternating current states that for a current with voltage V, current I, and impedance Z, $V = IZ$. If the current is $12\underline{/-23°}$ and the impedance is $9\underline{/42°}$, what is the voltage?

34. *Electricity* If an ac circuit has a voltage of $20\underline{/30°}$ and a current of $5\underline{/40°}$, what is the impedance?

 35. Modify your program for multiplying complex numbers to also work when the numbers are entered in polar form. (See Exercise Set 14.2, Exercise 64.)

 36. Modify your program for dividing two complex numbers to also work when the numbers are entered in polar form. (See Exercise Set 14.2, Exercise 65.)

 37. Write a computer program to take the power, or find all the roots, of a complex number by using DeMoivre's formula.

≡ 14.6
COMPLEX NUMBERS IN AC CIRCUITS

In direct current (dc) circuits, the basic relation between voltage and current is given by Ohm's law. If V is the voltage across a resistance R, and I is the current flowing through the resistor, then Ohm's law states that

$$V = IR$$

In alternating current (ac) circuits, there is a very similar equation. If V is the voltage across an impedance Z, and I is the current flowing through the impedance, then we have the relationship

$$V = IZ$$

The main difference in these equations is that the dc circuits are expressed as real numbers. Using complex numbers with the ac equations allows them to take the same simple form as the dc equations, except that all quantities are complex numbers. We will examine some of those relationships in this section.

Impedance is the opposition to alternating current produced by a resistance R, an inductance L, a capacitance C, or any combination of these. When a sinusoidal voltage V of a given frequency f is applied to a circuit of constant resistance R, constant capacitance C, and constant inductance L, the circuit, like the one in Figure 14.13, is an *RLC* circuit.

In Figure 14.13, the opposition to the current produced by the inductance is called the **inductive reactance** X_L and the opposition to the current produced by the capacitance is the **capacitive reactance** X_C. The **reactance** X is a measure of how much the capacitance and inductance retard the flow of current in an ac circuit and is the difference between the capacitive reactance and the inductive reactance. Thus,

$$X = X_L - X_C$$

The impedance, resistance, and reactance can be represented by a **vector impedance triangle**. The angle ϕ between Z and R is the **phase angle**. In the

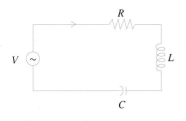

FIGURE 14.13

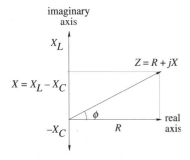

FIGURE 14.14

complex plane, we can represent the resistance along the real axis and the reactance along the imaginary axis as shown in Figure 14.14. Thus,

$$Z = R + jX = R + j(X_L - X_C)$$

$$= Z\underline{/\phi}$$

From our study of complex numbers, you can see that the **magnitude** of the impedance is $|Z| = \sqrt{R^2 + X^2}$ and $\tan\phi = \dfrac{X}{R}$.

Application

EXAMPLE 14.40

A circuit has a resistance of 8 Ω in series with a reactance of 5 Ω. What are the magnitude of the impedance and its phase angle?

Solution Using a vector impedance triangle, we can see that the impedance can be represented by

$$Z = 8 + 5j$$

The magnitude of the impedance is

$$|Z| = \sqrt{R^2 + X^2} = \sqrt{8^2 + 5^2} = \sqrt{89} \approx 9.43\ \Omega$$

The phase angle ϕ is given by

$$\phi = \arctan\frac{5}{8} = 32° \text{ or } 0.56 \text{ rad}$$

Application

EXAMPLE 14.41

In the *RLC* circuit in Figure 14.14, $R = 60\ \Omega$, $X_L = 75\ \Omega$, $X_C = 30\ \Omega$, and $I = 2.25$ A. Find **(a)** the magnitude and phase angle of Z, and **(b)** the voltage across the circuit.

Solutions

(a) The reactance $X = X_L - X_C = 75 - 30 = 45\ \Omega$. The impedance $Z = 60 + 45j$, so the magnitude of the impedance is

$$|Z| = \sqrt{R^2 + X^2} = \sqrt{60^2 + 45^2} = 75\ \Omega$$

The phase angle ϕ is given by

$$\phi = \arctan\frac{45}{60} \approx 36.87° \text{ or } 0.64 \text{ rad}$$

(b) Since the current is 2.25 A and the impedance is 75 Ω, the voltage $V = IZ = (2.25)(75) = 168.75$ V, or approximately 169 V.

An alternating current is produced by a coil of wire rotating through a magnetic field. If the angular velocity of the wire is ω, the capacitive reactance X_C and the inductive reactance X_L are given by the formulas

$$X_C = \frac{1}{\omega C} \qquad \text{and} \qquad X_L = \omega L$$

Since $\omega = 2\pi f$, where f is the frequency of the current, these are also expressed as

$$X_C = \frac{1}{2\pi fC} \quad \text{and} \quad X_L = 2\pi fL$$

From these formulas you can see that if C, L, and either ω or f are known, the reactance of the circuit may be determined.

Application

EXAMPLE 14.42

If $R = 40\ \Omega$, $L = 0.1$ H, $C = 50\ \mu$F, and $f = 60$ Hz, determine the impedance and phase difference between the current and voltage.

Solution   Converting $C = 50\ \mu$F to farads produces $C = 50 \times 10^{-6}$ F.

$$X_C = \frac{1}{2\pi fC} = \frac{1}{2\pi(60)(50 \times 10^{-6})} \approx 53\ \Omega$$

$$X_L = 2\pi fL = 2\pi(60)(0.1) \approx 38\ \Omega$$

$$Z = R + jX = R + j(X_L - X_C) = 40 + (38 - 53)j = 40 - 15j$$

$$|Z| = \sqrt{R^2 + X^2} = \sqrt{40^2 + (-15)^2} \approx 42.7\ \Omega$$

$$\phi = \arctan \frac{-15}{40} \approx -20.6°$$

So, the impedance is $42.7\ \Omega$ and the phase difference is $-20.6°$.

In the study of dc circuits, you learn that if several resistors are connected in series, their total resistance is the sum of the individual resistances. Thus, if two resistors R_1 and R_2 are connected in series, their total resistance, R equals $R_1 + R_2$. If R_1, R_2, and R_3 are connected in series, then $R = R_1 + R_2 + R_3$.

If the resistors are connected in parallel, the relationship is more complicated. The total resistance is the reciprocal of the sum of the reciprocals of the resistances. What this means is that if two resistors R_1 and R_2 are connected in parallel, then the total resistance $R = \dfrac{1}{\dfrac{1}{R_1} + \dfrac{1}{R_2}}$. This can be simplified to $R = \dfrac{R_1 R_2}{R_1 + R_2}$. If three resistors, R_1, R_2, and R_3 are connected in parallel, then $R = \dfrac{1}{\dfrac{1}{R_1} + \dfrac{1}{R_2} + \dfrac{1}{R_3}}$. This can be rewritten as

$$R = \frac{R_1 R_2 R_3}{R_1 R_2 + R_1 R_3 + R_2 R_3}$$

Corresponding formulas hold for complex impedances. Thus, if two impedances Z_1 and Z_2 are connected in series, the total impedance is $Z = Z_1 + Z_2$. If there are three impedances in series, Z_1, Z_2, and Z_3, then $Z = Z_1 + Z_2 + Z_3$.

If complex impedances are connected in parallel, we then have the more complicated formulas. If two impedances are connected in parallel, then the total impedance

$$Z = \cfrac{1}{\cfrac{1}{Z_1} + \cfrac{1}{Z_2}} = \frac{Z_1 Z_2}{Z_1 + Z_2}.$$ If three impedances are connected in parallel, then the

total impedance is $Z = \cfrac{1}{\cfrac{1}{Z_1} + \cfrac{1}{Z_2} + \cfrac{1}{Z_3}} = \dfrac{Z_1 Z_2 Z_3}{Z_1 Z_2 + Z_1 Z_3 + Z_2 Z_3}.$

Application

EXAMPLE 14.43

If $Z_1 = 2 + 3j \ \Omega$ and $Z_2 = 1 - 6j \ \Omega$, what is the total impedance if these are connected **(a)** in series and **(b)** in parallel?

Solutions

(a) If they are connected in series,

$$Z = Z_1 + Z_2$$
$$= (2 + 3j) + (1 - 6j)$$
$$= 3 - 3j \ \Omega$$

(b) If they are connected in parallel,

$$Z = \cfrac{1}{\cfrac{1}{Z_1} + \cfrac{1}{Z_2}} = \frac{Z_1 Z_2}{Z_1 + Z_2}$$

$$= \frac{(2 + 3j)(1 - 6j)}{(2 + 3j) + (1 - 6j)}$$

$$= \frac{20 - 9j}{3 - 3j} \qquad \text{multiply by } \frac{3 + 3j}{3 + 3j}$$

$$= \frac{87 + 33j}{18}$$

$$= \frac{87}{18} + \frac{33}{18} j \ \Omega$$

$$= \frac{29}{6} + \frac{11}{6} j \ \Omega$$

Application

EXAMPLE 14.44

If $Z_1 = 3.16 \underline{/\ 18.4°} \ \Omega$ and $Z_2 = 4.47 \underline{/\ 63.4°} \ \Omega$, what is the total impedance if these are connected **(a)** in series and **(b)** in parallel?

Solutions

(a) If they are connected in series,

$$Z = Z_1 + Z_2$$

EXAMPLE 14.44 (Cont.)

Since we cannot add complex numbers in polar form, we need to change these to rectangular form.

$$Z_1 = 3.16(\cos 18.4° + j\sin 18.4°) \approx 2.998 + 0.997j$$

$$Z_2 = 4.47(\cos 63.4° + j\sin 63.4°) \approx 2.001 + 3.997j$$

$$Z = (2.998 + 0.997j) + (2.001 + 3.997j)$$

$$= 4.999 + 4.994j \ \Omega$$

$$\approx 7.07\underline{/45.0°} \ \Omega$$

(b) If they are in parallel,

$$Z = \frac{Z_1 Z_2}{Z_1 + Z_2}$$

$$= \frac{(3.16\underline{/18.4°})(4.47\underline{/63.4°})}{7.07\underline{/45.0°}}$$

Using our knowledge of multiplying and dividing complex numbers in polar form, we get

$$Z = \frac{(3.16)(4.47)}{7.07}\underline{/18.4° + 63.4° - 45.0°}$$

$$= 2.0\underline{/36.8°} \ \Omega$$

As you can see, some problems are easier to work if you use the rectangular form and some are easier if you use the polar form. They would all be easier to work if you had a computer program to solve these problems, such as the one in Exercise Set 14.6, Exercise 35.

Exercise Set 14.6

In Exercises 1–12, find the total impedance if the given impedances are connected (a) in series and (b) in parallel.

1. $Z_1 = 2 + 3j, Z_2 = 1 - 5j$

2. $Z_1 = 4 - 7j, Z_2 = -3 + 4j$

3. $Z_1 = 1 - j, Z_2 = 3j$

4. $Z_1 = 2 + j, Z_2 = 4 - 3j$

5. $Z_1 = 2.19\underline{/18.4°}, Z_2 = 5.16\underline{/67.3°}$

6. $Z_1 = 2\sqrt{3}\underline{/30°}, Z_2 = 2\underline{/120°}$

7. $Z_1 = 3\sqrt{5}\underline{/\frac{\pi}{7}}, Z_2 = 1.5\underline{/0.45}$

8. $Z_1 = 2.57\underline{/0.25}, Z_2 = 1.63\underline{/1.38}$

9. $Z_1 = 4 + 3j, Z_2 = 3 - 2j, Z_3 = 5 + 4j$

10. $Z_1 = 3 - 4j, Z_2 = 1 + 5j, Z_3 = -2j$

11. $Z_1 = 1.64\underline{/38.2°}, Z_2 = 2.35\underline{/43.7°}, Z_3 = 4.67\underline{/-39.6°}$

12. $Z_1 = 0.15\underline{/0.95}, Z_2 = 2.17\underline{/1.39}, Z_3 = 1.10\underline{/0.40}$

In Exercises 13–18, use the formula $V = IZ$ to determine the missing unit.

13. $I = 4 - 3j$ A, $Z = 8 - 15j$ Ω

14. $V = 5 + 5j$ V, $I = 4 + 3j$ A

15. $Z = 1 - j$ Ω, $I = 1 + j$ A

16. $V = 3 + 4j$ V, $Z = 5 - 12j$ Ω

17. $V = 7\underline{/\ 36.3°}$ V, $I = 2.5\underline{/\ 12.6°}$ A

18. $V = 3\underline{/\ 1.37}$ V, $Z = 4\underline{/\ 0.16}$ Ω

In Exercises 19–24, determine the inductive reactance, capacitive reactance, impedance, and phase difference between the current and voltage.

19. $R = 38$ Ω, $L = 0.2$ H, $C = 40$ μF, and $f = 60$ Hz

20. $R = 35$ Ω, $L = 0.15$ H, $C = 80$ μF, and $f = 60$ Hz

21. $R = 20$ Ω, $L = 0.4$ H, $C = 60$ μF, and $f = 60$ Hz

22. $R = 12$ Ω, $L = 0.3$ H, $C = 250$ μF, and $\omega = 80$ rad/s

23. $R = 28$ Ω, $L = 0.25$ H, $C = 200$ μF, and $\omega = 50$ rad/s

24. $R = 2\,000$ Ω, $L = 3.0$ H, $C = 0.5$ μF, and $\omega = 1\,000$ rad/s

In the *RLC* circuits in Exercises 25–28, find (a) the magnitude and phase angle of *Z* and (b) the voltage across the circuit.

25. $R = 75$ Ω, $X_L = 60$ Ω, $X_C = 40$ Ω, and $I = 3.50$ A

26. $R = 40$ Ω, $X_L = 30$ Ω, $X_C = 60$ Ω, and $I = 7.50$ A

27. $R = 3.0$ Ω, $X_L = 6.0$ Ω, $X_C = 5.0$ Ω, and $I = 2.85$ A

28. $R = 12.0$ Ω, $X_L = 11.4$ Ω, $X_C = 2.4$ Ω, and $I = 0.60$ A

Solve Exercises 29–36.

29. *Electronics* Figure 14.15 indicates part of an electrical circuit. Kirchhoff's law implies that $I_2 = I_1 + I_3$. If $I_1 = 7 + 2j$ A and $I_2 = 9 - 7j$ A, find I_3.

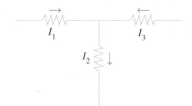

FIGURE 14.15

30. *Electronics* **Resonance** in an *RLC* circuit occurs when $X_L = X_C$. Under these conditions, the current and voltage are in phase. Resonance is required for the tuning of radio and television receivers. If $\omega = 100$ rad/s and $L = 0.500$ H, what is the value of C if the system is in resonance?

31. *Electronics* What is the frequency in Hertz for a circuit in resonance if $L = 2.5$ H and $C = 20.0$ μF?

32. *Electronics* The **admittance** Y is the reciprocal of the impedance. If $Z = 3 - 2j$ Ω, what is the admittance?

33. *Electronics* If the admittance is $4 + 3j$ Ω, what is the impedance?

34. *Electronics* The **susceptance** B of an ac circuit of reactance X and impedance Z is defined as $B = \dfrac{X}{Z^2} j$. If the reactance is 4 Ω and $Z = 8 + 7j$, what is the susceptance?

35. Write a computer program that will compute the total impedance of a group of impedances in an ac circuit that are connected either in series or in parallel. The program should allow the impedance to be entered in either rectangular or polar form.

36. Using the computer program from Exercise 35, rework Exercises 1–12.

☰ CHAPTER 14 REVIEW

Important Terms and Concepts

Changing complex numbers from
  Polar form to rectangular form
  Rectangular form to polar form

Complex numbers
  Absolute value
  Addition
  Conjugate
  Division
  Exponential form
  Imaginary part

Multiplication
Polar form
Power
Real part
Rectangular part
Roots
Subtraction
Complex plane
DeMoivre's formula
Imaginary unit

Review Exercises

In Exercises 1–6, simplify each number in terms of j.

1. $\sqrt{-49}$

2. $\sqrt{-36}$

3. $\sqrt{-54}$

4. $(2j^3)^3$

5. $\sqrt{-2}\sqrt{-18}$

6. $\sqrt{-9}\sqrt{-27}$

In Exercises 7–14, perform the indicated operation. Express each answer in the form $a + bj$.

7. $(2 - j) + (7 - 2j)$

8. $(9 + j)(4 + 7j)$

9. $(5 + j) - (6 - 3j)$

10. $\dfrac{1}{11 - j}$

11. $(6 + 2j)(-5 + 3j)$

12. $\dfrac{2 - 5j}{6 + 3j}$

13. $(4 - 3j)(4 + 3j)$

14. $\dfrac{-4}{\sqrt{3} + 2j}$

In Exercises 15–20, graph each complex number and change each number from rectangular form to polar form, or vice versa.

15. $9 - 6j$

16. $-8 + 2j$

17. $4 - 4j$

18. $4 \operatorname{cis} 60°$

19. $6.5 / 2.3$

20. $10 / 20°$

In Exercises 21–28, perform the indicated operation and express each answer in rectangular form.

21. $(2 \operatorname{cis} 30°)(5 \operatorname{cis} 150°)$

22. $\dfrac{3 \operatorname{cis} 20°}{6 \operatorname{cis} 80°}$

23. $\left(3 \operatorname{cis} \dfrac{5\pi}{4}\right)^{14}$

24. $(324 \operatorname{cis} 225°)^{1/5}$

25. $\left(3 / \dfrac{\pi}{4}\right)\left(9 / \dfrac{2\pi}{3}\right)$

26. $44 / 125° \div 4 / 97°$

27. $\left(2 / \dfrac{\pi}{6}\right)^{12}$

28. $(2048 / 330°)^{1/11}$

For Exercises 29–32, find all roots and express in rectangular form.

29. $\sqrt[3]{-j}$

30. $\sqrt[4]{16}$

31. $\sqrt{16 \text{ cis } 120°}$

32. $\sqrt[3]{27j}$

In Exercises 33–36, change each number to the exponential form and perform the indicated operation.

33. $(3 + 2j)(5 - j)$

34. $\dfrac{4 - 7j}{3 + j}$

35. $(5 + 3j)^5$

36. $(-7 - 2j)^{1/3}$

Solve Exercises 37–40.

37. *Physics*   A force vector in the complex plane is given by $7.3 - 1.4j$. What is the magnitude and direction (argument) of this vector?

38. *Electronics*   Given an *RLC* circuit with $R = 3.0 \ \Omega$, $X_L = 7.0 \ \Omega$, $X_C = 4.5 \ \Omega$, and $I = 1.5$ A, **(a)** what is the magnitude and phase angle of Z, and **(b)** what is the voltage across the circuit?

39. *Electronics*   What are the inductive reactance, capacitive reactance, impedance, and phase difference between the current and voltage, if $R = 55 \ \Omega$, $L = 0.3$ H, $C = 50 \ \mu$F, and $f = 60$ Hz?

40. *Electronics*   If $Z_1 = 3 + 5j$ and $Z_2 = 6 - 3j$, then what is Z, if Z_1 and Z_2 are connected **(a)** in series and **(b)** in parallel?

▤ CHAPTER 14 TEST

1. Write $\sqrt{-80}$ in terms of j.

2. Change $7 - 2j$ from rectangular form to polar form.

3. Change 8 cis 150° from polar form to rectangular form.

In Exercises 4–11, perform the indicated operation. Express each answer in the form a + bj.

4. $(5 + 2j) + (8 - 6j)$

5. $(-5 + 2j) - (8 - 6j)$

6. $(2 + 3j)(4 - 5j)$

7. $\dfrac{6 + 5j}{-3 - 4j}$

8. $(7 \text{ cis } 75°)(2 \text{ cis } 105°)$

9. $\dfrac{4 \text{ cis } 115°}{3 \text{ cis } 25°}$

10. $\left(9 \text{ cis } \dfrac{2\pi}{3}\right)^{5/2}$

11. $(27 \underline{/\ 129°})^{1/3}$

12. Find all four roots of $\sqrt[4]{j}$.

13. If $Z_1 = 4 + 2j$ and $Z_2 = 5 - 3j$, what is Z, if Z_1 and Z_2 are connected in **(a)** series and **(b)** parallel?

15

An Introduction to Plane Analytic Geometry

A lithotripter is a medical instrument. Its design is based on an ellipse: In Section 15.4, we will see how the ellipse makes this instrument effective.

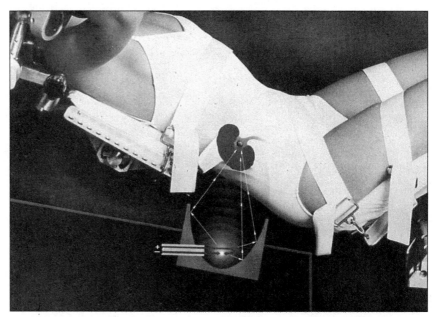

Courtesy of Dornier Medical Systems Inc.

Many scientific and technical applications make use of four special curves: the circle, parabola, ellipse, and hyperbola. These are four of the seven conic sections that can be formed by a plane intersecting a cone. (The others are a point, a line, and two intersecting lines.)

One medical instrument that uses an ellipse is a lithotripter, an instrument that uses sound waves to break up kidney stones. An ellipse has two foci. (Foci is the plural of focus.) A sound wave transmitter is placed at one focus and the patient's kidney is placed at the other. The elliptical shape of the lithotripter causes sound waves to be reflected off the ellipse and focused on the kidney stones.

In Chapter 4, the rectangular coordinate system was introduced. We began by graphing equations on the Cartesian coordinate system. The process of using algebra to describe geometry is called analytic geometry. In this chapter, we first review the line and then look at circles, parabolas, ellipses, and hyperbolas.

≡ 15.1
BASIC DEFINITIONS AND STRAIGHT LINES

We introduced graphing in Chapter 4 and have worked with graphs in almost every chapter since then. In this section, we will introduce some new ideas and review some of those that were introduced earlier.

If we have two points in a plane, we often want to know some things about them. Besides the location of the points, some of the things we might want to know include the distance between them, the point halfway between them, and the equation of the line through those two points, including its slope and intercepts. Let's look at these and a few other ideas, one at a time.

The Distance Formula

For most of this discussion we will be using two points P_1 and P_2. The point P_1 has coordinates (x_1, y_1) and P_2 has coordinates (x_2, y_2). Using the Pythagorean theorem, we can find the distance d between these points.

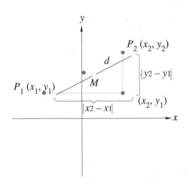

FIGURE 15.1

> **The Distance Formula**
>
> The distance d between any two points $P_1(x_1, y_1)$ and $P_2(x_2, y_2)$ in a plane is given by the **distance formula**
>
> $$d = \sqrt{(x_1 - x_2)^2 + (y_1 - y_2)^2}$$

If there are several points, the symbol $d(P_1, P_2)$ would then represent the two points that are being used. (See Figure 15.1.)

EXAMPLE 15.1

Find the lengths of the sides of the triangle with vertices $A(-1, -3)$, $B(6, 1)$, and $C(2, -7)$.

Solutions

$$d(C, A) = \sqrt{(-1-2)^2 + (-3+7)^2} = \sqrt{(-3)^2 + 4^2}$$

$$= \sqrt{9 + 16} = \sqrt{25} = 5$$

$$d(B, A) = \sqrt{(-1-6)^2 + (-3-1)^2} = \sqrt{(-7)^2 + (-4)^2}$$

$$= \sqrt{49 + 16} = \sqrt{65}$$

$$d(C, B) = \sqrt{(6-2)^2 + (1+7)^2} = \sqrt{4^2 + 8^2}$$

$$= \sqrt{16 + 64} = \sqrt{80} = 4\sqrt{5}$$

The Midpoint Formula

If you want to find the coordinates of the point M halfway between two points P_1 and P_2, then you use the **midpoint formula**.

The Midpoint Formula

The coordinates of the midpoint M between any two points $P_1(x_1, y_1)$ and $P_2(x_2, y_2)$ are given by the **midpoint formula**

$$M = \left(\frac{x_1 + x_2}{2}, \frac{y_1 + y_2}{2} \right)$$

The midpoint formula gives you the coordinates of the point on the segment $\overline{P_1 P_2}$ halfway between P_1 and P_2.

EXAMPLE 15.2

Find the midpoints of the sides of the triangle with vertices $A(-1, -3)$, $B(6, 1)$, and $C(2, -7)$. Note that this is the same triangle we used in Example 15.1.

Solutions $M(A, B) = \left(\dfrac{-1+6}{2}, \dfrac{-3+1}{2} \right) = \left(\dfrac{5}{2}, \dfrac{-2}{2} \right) = \left(\dfrac{5}{2}, -1 \right)$

$M(A, C) = \left(\dfrac{-1+2}{2}, \dfrac{-3-7}{2} \right) = \left(\dfrac{1}{2}, \dfrac{-10}{2} \right) = \left(\dfrac{1}{2}, -5 \right)$

$M(B, C) = \left(\dfrac{6+2}{2}, \dfrac{1-7}{2} \right) = \left(\dfrac{8}{2}, \dfrac{-6}{2} \right) = (4, -3)$

Slope

As we have discussed before, the slope of a line measures the steepness of the line. The slope also indicates whether, as values of x increase, a line is rising (positive slope), falling (negative slope), or constant (slope of 0). A formula for determining the slope of the line through two points follows.

Slope of a Line

If a line through two points $P_1(x_1, y_1)$ and $P_2(x_2, y_2)$ is not vertical, then the **slope**, m, of this line is

$$m = \frac{y_2 - y_1}{x_2 - x_1}$$

All lines except vertical lines have a slope. The slope of a vertical line is undefined. The slope of a horizontal line is 0.

EXAMPLE 15.3

What is the slope of the line through the points $A(2, -5)$ and $B(-4, 7)$?

Solution

$$m = \frac{7 - (-5)}{-4 - 2} = \frac{7 + 5}{-4 - 2} = \frac{12}{-6} = -2$$

Any line that is not parallel to the x-axis must eventually cross the x-axis. The angle measure in a positive direction from the x-axis to a line is called the **inclination** of the line. The inclination of a line parallel to the x-axis is defined as 0. The inclination provides us with an alternative definition for the slope. If α is the inclination that a line makes with the x-axis, as shown in Figures 15.2a and 15.2b, then the slope is

$$m = \tan \alpha, \qquad 0° \leq \alpha < 180° \qquad \text{or} \qquad 0 \leq \alpha < \pi$$

Of course, since $\tan 90°$ and $\tan \frac{\pi}{2}$ are undefined, the inclination of a vertical line is undefined.

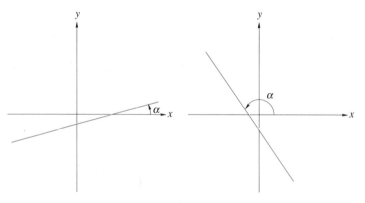

FIGURE 15.2a   FIGURE 15.2b

EXAMPLE 15.4

What is the slope of a line with an inclination of **(a)** 30°, **(b)** 0.85 rad, and **(c)** 115°?

Solutions

(a) $m = \tan 30° = 0.5773503$

(b) $m = \tan 0.85 \text{ rad} = 1.1383327$

(c) $m = \tan 115° = -2.1445069$

EXAMPLE 15.5

The line in Example 15.3 has a slope of -2. What is the inclination of this line?

Solution

$$m = \tan \alpha = -2 \text{ so } \alpha = \tan^{-1}(-2) \approx 116.56505°$$

[If you used a calculator, $\tan^{-1}(-2) \approx -63.43498°$. To obtain the inclination, you need to add 180°, with the result $180° + (-63.43495°) = 116.56505°$.]

If two lines are parallel, they have the same inclination, and so parallel lines have the same slope. If two lines are perpendicular, they intersect at a 90° angle. This means that their inclinations must differ by 90°. It also means that the slope of one of these lines is the negative reciprocal of the other.

Slopes of Perpendicular Lines

If m_1 is the slope of a line and m_2 is the slope of a line perpendicular to the first line, then

$$m_1 = -\frac{1}{m_2}$$

This can also be written as $m_1 m_2 = -1$.

 Caution

The above rule does not apply if one of the lines is horizontal because the other line would then be vertical and not have a slope.

EXAMPLE 15.6

What is the slope of a line that is perpendicular to a line with a slope of -2?

Solution If $m_1 = -2$, then the slope of the perpendicular line is $m_2 = -\frac{1}{-2} = \frac{1}{2}$

In Chapter 5, we introduced several formulas for the equation of a line. If b represents the y-intercept of a line, then the line crosses the y-axis at the point $(0, b)$. If the slope of this line is m, then the **slope-intercept form for the equation of a line** is

$$y = mx + b$$

EXAMPLE 15.7

Write an equation for a line that has a slope of $\frac{1}{4}$ and a y-intercept of 5.

Solution We have $m = \frac{1}{4}$ and $b = 5$, so $y = \frac{1}{4}x + 5$. Writing this without fractions, we get $4y = x + 20$.

EXAMPLE 15.8

What are the slope and y-intercept of the line $4x + 6y - 3 = 0$?

Solution We need to rewrite this equation in the form $y = mx + b$.

$$4x + 6y - 3 = 0$$

$$6y = -4x + 3$$

$$y = -\frac{4}{6}x + \frac{3}{6}$$

$$= -\frac{2}{3}x + \frac{1}{2}$$

The slope is $-\frac{2}{3}$ and the y-intercept is $\frac{1}{2}$.

If you know the slope of a line is m and $P(x_1, y_1)$ is a point on the line, then the equation for the line can be written in the **point-slope form for the equation of a line**.

$$y - y_1 = m(x - x_1)$$

EXAMPLE 15.9

What is the equation of the line through the point $(2, -3)$ and with a slope of 5?

Solution   In this example, $x_1 = 2$, $y_1 = -3$, and $m = 5$. Substitute these values into the equation $y - y_1 = m(x - x_1)$.

$$y - (-3) = 5(x - 2)$$
$$y + 3 = 5x - 10$$
$$y = 5x - 13$$

Every straight line can be written in the **general form for the equation of a line**. This form is represented by the equation

$$Ax + By + C = 0$$

where A, B, and C represent constants and A and B are not both 0.

EXAMPLE 15.10

What are the slope and intercepts of the line $4x + 3y - 24 = 0$?

Solution   We know that the intercepts are the points where the line crosses the axes. The line crosses the x-axis at $(a, 0)$ and the y-axis at $(0, b)$.

If we let $y = 0$, we then get $x = 6$, and if $x = 0$, then $y = 8$; so the x-intercept is $(6, 0)$ and the y-intercept is $(0, 8)$. We will use these two points to determine the following slope.

$$m = \frac{8 - 0}{0 - 6} = \frac{8}{-6} = -\frac{4}{3}$$

The three forms for the equation of a line are summarized in the following box.

Different Forms for the Equation of a Line

If a line has a slope of m, a y-intercept of $(0, b)$, and $P(x_1, y_1)$ is a point on the line, then the equation for the line can be written in any of these three forms:

Type of Equation	Equation for Line
Point-slope form	$y - y_1 = m(x - x_1)$
Slope-intercept form	$y = mx + b$
General form	$Ax + By + C = 0$

Application

EXAMPLE 15.11

A mechanic charges $96 for a job that takes 2 h to finish and $135 if the job is completed in 5 h. **(a)** Find a linear equation that describes how much the mechanic should charge for a job of x hours. **(b)** Use your equation to determine how much should be charged for a job that takes 7 h 15 min.

Solutions **(a)** We begin by thinking of the given information as points on a line. We give each point as an ordered pair of the form (x, C), where x represents the number of hours a job takes to complete and C stands for the amount the mechanic charges for this job.

A 2-h job with a charge of $96 can be thought of as the point (2, 96), and the 5-h job as the point (5, 135). From this information we determine that the slope of this line is

$$m = \frac{135 - 96}{5 - 2} = \frac{39}{3} = 13$$

Using this value for the slope and selecting the point (2, 96) we can use the point-slope form for the equation of a line to determine the needed equation.

$$C - 96 = 13(x - 2)$$

$$C - 96 = 13x - 26$$

or $\qquad C = 13x + 70$

Thus the desired equation is $C = 13x + 70$.

(b) A 7 h 15 min job can be thought of as a 7.25-h job. Substituting 7.25 for x in the equation $C = 13x + 70$, we obtain $C = 13(7.25) + 70 = 94.25 + 70 = 164.25$. The mechanic should charge $164.25 for a job that will take 7 h and 15 min.

Application

EXAMPLE 15.12

The resistance of a circuit element varies directly with its temperature. The resistance at 0°C is found to be 5.00 Ω. Further tests show that the resistance increases by 1.5 Ω for every 1°C increase in temperature. **(a)** Determine the equation for the resistance R in terms of the temperature T. **(b)** Use the equation from part (a) to find the resistance at 27°C.

Solutions **(a)** In order to find R in terms of T, we should think of R as the dependent variable. Since resistance varies directly with temperature, the two quantities have a linear relationship. From the given information, we notice that since $R = 5.00$ when $T = 0$, we have the R-intercept. As a result, we shall use the slope-intercept form for the equation of a line.

$$R = mT + b$$

$$R = mT + 5.00$$

EXAMPLE 15.12 (Cont.)

To find the slope m, we will use

$$m = \frac{\text{Change in } R}{\text{Change in } T} = \frac{1.5 \ \Omega}{1°C} = 1.5 \ \Omega/°C$$

Substituting 1.5 for m in the slope-intercept equation, we are able to complete the equation.

$$R = 1.5T + 5.00$$

(b) To find the resistance at $27°C$, we will substitute $27°C$ for T in the equation from part **(a)**.

$$R = 1.5(27) + 5.00 = 45.5 \ \Omega$$

Exercise Set 15.1

In Exercises 1–8, find the distance between the given pairs of points.

1. $(2, 4)$ and $(7, -9)$

2. $(-3, 5)$ and $(0, 1)$

3. $(5, -6)$ and $(-3, -5)$

4. $(-8, -4)$ and $(6, 10)$

5. $(12, 1)$ and $(3, -13)$

6. $(11, -2)$ and $(4, -8)$

7. $(-2, 5)$ and $(5, 5)$

8. $(-4, -4)$ and $(-4, 7)$

In Exercises 9–16, find the midpoints of the given pairs of points in Exercises 1–8.

In Exercises 17–24, find the slopes of the lines through the points in Exercises 1–8.

In Exercises 25–32, use the point-slope form of the equation for the line to write the equation of the lines through the points in Exercises 1–8.

In Exercises 33–36, find the slopes of the lines with the given inclinations.

33. $45°$

34. $175°$

35. 0.15 rad

36. 1.65 rad

In Exercises 37–40, find the inclinations of the lines with the given slopes.

37. 2.5

38. 0.30

39. -0.50

40. -1.475

In Exercises 41–44, determine the slope of a line that is perpendicular to a line of the given slope.

41. 3

42. $\dfrac{2}{5}$

43. $-\dfrac{1}{2}$

44. -5

In Exercises 45–56, find the equation of each line with the given properties.

45. Passes through $(2, -5)$ with a slope of 6

46. Passes through $(6, -2)$ with a slope of $-\dfrac{1}{2}$

47. Passes through $(2, 3)$ and $(4, -2)$

48. Passes through $(-5, 2)$ and $(-3, -4)$

49. Passes through $(-2, -4)$ with an inclination of $60°$

50. Passes through $(5, 1)$ with an inclination of 0.75 rad

51. Has an inclination of $20°$ and a y-intercept of $(0, 3)$

52. Has an inclination of 2.5 rad and a y-intercept of $(0, -2)$

53. Passes through $(2, 5)$ and is parallel to $2x - 3y + 4 = 0$

54. Passes through $(-3, 2)$ and is parallel to $3x + 4y = 12$

55. Passes through $(4, -1)$ and is perpendicular to $2x + 5y = 20$

56. Passes through $(-1, -6)$ and is perpendicular to $8x - 3y = 24$

In Exercises 57–60, determine the slope and intercepts of each line.

57. $3x + 2y = 12$

58. $5x - 3y = 15$

59. $x - 3y = 9$

60. $6x + y = 9$

Solve Exercises 61–64.

61. *Physics* The instantaneous velocity v of an object under constant acceleration a during an elapsed time t is given by $v = v_0 + at$, where v_0 is its initial velocity. If an object has an initial velocity of 2.6 m/s and a velocity of 5.8 m/s after 8 s of constant acceleration, write the equation relating velocity to time.

62. *Physics* The **coefficient of linear expansion** α is the change in length of a solid due to the change in temperature.

(a) Determine the coefficient of linear expansion of a copper rod that is 1.000 000 cm at $10°$C and expands to 1.000 084 cm at $15°$C.

(b) Determine the coefficient of linear expansion of a copper rod that is 4.000 000 cm at $10°$C and is 4.000 336 cm at $15°$C.

(c) The coefficient of linear expansion for any specific solid is a constant. This means that the answers for parts (a) and (b) should have been the same. Assume that the answer in part (a) is correct. What changes need to be made in the way you determine the coefficient of linear expansion?

(d) Determine the coefficient of linear expansion of a glass rod that expands from 72.000 024 cm at $10°$C to 72.005 208 cm at $18°$C.

63. *Electronics* In a dc circuit, when the internal resistance of the voltage source is taken into account, the voltage E is a linear function of the current I and is given by $E = IR + Ir$, where R is the circuit resistance and r is the internal resistance. If the resistance of the circuit $R = 4.0\ \Omega$ and $I = 2.5\ A$ when $E = 12.0$ V, find r.

64. *Electronics* The resistance of a circuit element is found to increase by $0.006\ \Omega$ for every $1°$C increase in temperature over a wide range of temperatures. If the resistance is $7.000\ \Omega$ at $0°$C, (a) write the equation relating the resistance R to the temperature T, and (b) find R, when $T = 17°$C.

≡ 15.2
THE CIRCLE

In Section 15.1, we developed a general equation for a straight line. In this section, we will work with the circle.

A **circle** is the set of all points in a plane that are at some fixed distance from a fixed point in the plane. The fixed point is called the **center** of the circle and the fixed distance is called the **radius**. In general, suppose we have a circle with radius r and center (h, k). An seen in Figure 15.3, if (x, y) is any point on the circle, then

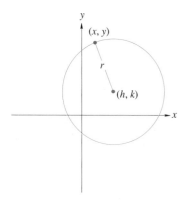

FIGURE 15.3

the distance from (x, y) to (h, k) must be r. Using the distance formula, we have the equation

$$\sqrt{(x - h)^2 + (y - k)^2} = r$$

Squaring both sides, we have

$$(x - h)^2 + (y - k)^2 = r^2$$

Thus, we have established the following standard equation of a circle.

Standard Equation of a Circle

The standard equation of a circle with center at (h, k) and radius r is

$$(x - h)^2 + (y - k)^2 = r^2$$

EXAMPLE 15.13

The equation $(x - 5)^2 + (y + 3)^2 = 36$ is a circle with center at $(5, -3)$ and a radius of 6. Remember, since the equation states that $y + 3 = y - k$, you obtain $k = -3$. Note very carefully the signs of the numbers in the equation and the signs of the coordinates of the center. This circle is shown in Figure 15.4.

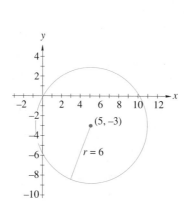

FIGURE 15.4

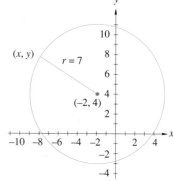

FIGURE 15.5

EXAMPLE 15.14

Write the equation for the circle with center at $(-2, 4)$ and radius 7.

Solution   The circle is shown in Figure 15.5. Since the center is $(-2, 4)$, we have $h = -2$ and $k = 4$, and since the radius is 7, then $r = 7$. So,

$$(x - h)^2 + (y - k)^2 = r^2$$

$$[x - (-2)]^2 + (y - 4)^2 = 7^2$$

$$(x + 2)^2 + (y - 4)^2 = 49$$

Notice that if the center is at the origin, then $h = 0$ and $k = 0$. The equation becomes

$$x^2 + y^2 = r^2$$

If we expand the general equation for a circle, we get

$$(x - h)^2 + (y - k)^2 = r^2$$
$$x^2 - 2hx + h^2 + y^2 - 2ky + k^2 = r^2$$

Since h^2, k^2, and r^2 are all constants, we will let $h^2 + k^2 - r^2 = F$. If we let the constants $-2h = D$ and $-2k = E$, then the equation for a circle can be written as the following general equation.

General Equation of a Circle

If A, D, E, and F are any real number constants, $A \neq 0$, then the general equation for a circle is

$$Ax^2 + Ay^2 + Dx + Ey + F = 0$$

EXAMPLE 15.15

Write $x^2 + y^2 + 4x - 8y + 4 = 0$ in the standard form for the equation of a circle and determine the center and radius.

Solution   To convert the equation to the standard form, we need to complete the square. We first write the constant on the right-hand side and group the terms containing x together and then group those containing y. The $\bigcirc$ and $\square$ show the missing constants that we must determine in order to complete the square. (See Section 8.3.)

$$x^2 + y^2 + 4x - 8y + 4 = 0$$
$$\left(x^2 + 4x + \bigcirc\right) + \left(y^2 - 8y + \square\right) = -4 + \bigcirc + \square$$

The coefficient of the x-term is 4. If we take half of that and square it, we can add the result, 4, to both sides of the equation. Similarly, half of -8 is -4, and we also add $(-4)^2 = 16$ to both sides. The equation then becomes

$$\left(x^2 + 4x + \boxed{4}\right) + \left(y^2 - 8y + \boxed{16}\right) = -4 + 4 + 16$$
$$(x + 2)^2 + (y - 4)^2 = 16$$

Since $16 = 4^2$, the radius is 4, and the center is $(-2, 4)$.

EXAMPLE 15.16

Write the equation $x^2 + y^2 + x - 5y + 2 = 0$ in the standard form for the equation of a circle and determine the center and radius.

Solution Again we will complete the square and use a ◯ and a ▢ to show where the constants need to be placed.

$$x^2 + y^2 + x - 5y + 2 = 0$$

$$\left(x^2 + x + \bigcirc\right) + \left(y^2 - 5y + \square\right) = -2 + \bigcirc + \square$$

$$\left(x^2 + x + \frac{1}{4}\right) + \left(y^2 - 5y + \frac{25}{4}\right) = -2 + \frac{1}{4} + \frac{25}{4}$$

$$\left(x + \frac{1}{2}\right)^2 + \left(y - \frac{5}{2}\right)^2 = \frac{18}{4}$$

The center is $\left(-\frac{1}{2}, \frac{5}{2}\right)$ and the radius is $\sqrt{\frac{18}{4}} = \frac{3\sqrt{2}}{2}$.

EXAMPLE 15.17

Write the equation $4x^2 + 4y^2 - 8x - 24y - 9 = 0$ in the standard form for the equation of a circle and determine the center and radius.

Solution Again, we will complete the square. First, move the constant to the right-hand side and group the x-terms and the y-terms.

$$4x^2 - 8x + 4y^2 - 24y = 9$$

Next, factor the 4 out of the x-terms and the y-terms.

$$4(x^2 - 2x) + 4(y^2 - 6y) = 9$$

We now use a ◯ and a ▢ to indicate the missing constants. Notice that a $4\bigcirc$ and a $4\square$ are added on the right-hand side of the equation to show where the constants need to be placed.

$$4(x^2 - 2x + \bigcirc) + 4(y^2 - 6y + \square) = 9 + 4\bigcirc + 4\square$$

$$4(x^2 - 2x + 1) + 4(y^2 - 6y + 9) = 9 + 4(1) + 4(9)$$

$$4(x - 1)^2 + 4(y - 3)^2 = 49$$

$$(x - 1)^2 + (y - 3)^2 = \frac{49}{4}$$

The center is $(1, 3)$ and the radius $r = \sqrt{\frac{49}{4}} = \frac{7}{2}$.

While we could have started by dividing this entire equation by 4, we waited until the end in order to delay working with fractions. This example uses a process that we will need in the sections on ellipses and hyperbolas.

A circle with an equation of the form $(x - h)^2 + y^2 = r^2$ has its center on the x-axis [at $(h, 0)$] and thus is symmetrical with the x-axis. In the same way, a circle with an equation of the form $x^2 + (y - k)^2 = r^2$ has its center on the y-axis [at $(0, k)$] and is symmetrical with the y-axis.

If $(x - h)^2 + (y - k)^2 = 0$, the circle has a radius of 0. This is sometimes called a **point circle**. If the radius is 1, the circle is often referred to as a **unit circle**.

Finally, if $r^2 < 0$, then the equation does not define a circle and this equation would have no graph on the Cartesian plane.

Application

EXAMPLE 15.18

A square concrete post is reinforced axially with eight rods arranged symmetrically around a circle, much like those shown in the photo in Figure 15.6a. The cross section in Figure 15.6b provides a schematic drawing of the post and rods. **(a)** Determine the equation of the circle using the upper left-hand corner of the post as the origin. **(b)** Find the coordinates of each rod's location.

FIGURE 15.6

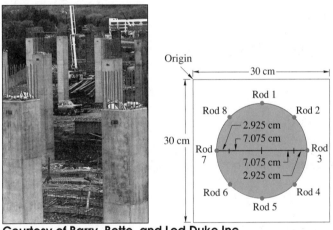

Courtesy of Barry, Bette, and Led Duke Inc.
(a)   (b)

Solution **(a)** While it is not specifically stated, we will assume that the eight rods are placed around a circle that shares the center of the square post. This would place the center at the point $(15, -15)$. The radius of the circle is $2.925 + 7.075$ cm $= 10$ cm. Thus, the equation for this circle is $(x - 15)^2 + (y + 15)^2 = 100$.

(b) The coordinates of rods 1, 3, 5, and 7 should be fairly easy to determine from the given information. Determining the coordinates of the four remaining rods is a little more difficult. We will show how to determine the coordinates for Rod 2 and then use the symmetry of the rod's placement to determine the coordinates of the other three rods.

The radius of the circle on which the rods are placed is 10 cm. Thus, each rod is located 10 cm from the center of the post. Rod 2 is located directly above a point that is 7.075 cm from the center of the post. Using the Pythagorean theorem, we find that this rod is placed $\sqrt{10^2 - 7.075^2} = \sqrt{100 - 50.055\,625} = \sqrt{49.944\,375} \approx 7.067$ cm above the line through Rods 7 and 3. Then the y-coordinate of Rod 2 is $-15 + 7.067 = -7.933$. This places Rod 2 at the coordinates $(22.075, -7.933)$.

EXAMPLE 15.18 (Cont.)	The eight rods have the coordinates $(15, -5)$, $(22.075, -7.933)$, $(25, -15)$, $(22.075, -22.067)$, $(15, -25)$, $(7.925, -22.067)$, $(5, -15)$, and $(7.925, -7.933)$.

Exercise Set 15.2

In Exercises 1–8, find the standard and general form of an equation for the circle with the given center C and radius r.

1. $C = (2, 5), r = 3$

2. $C = (4, 1), r = 8$

3. $C = (-2, 0), r = 4$

4. $C = (0, -7), r = 2$

5. $C = (-5, -1), r = \dfrac{5}{2}$

6. $C = (-4, -5), r = \dfrac{5}{4}$

7. $C = (2, -4), r = 1$

8. $C = (5, -2), r = \sqrt{3}$

In Exercises 9–14, give the center and radius of the circle described by each equation. Sketch the circle.

9. $(x - 3)^2 + (y - 4)^2 = 9$

10. $(x - 7)^2 + (y + 5)^2 = 25$

11. $\left(x + \dfrac{1}{2}\right)^2 + \left(y + \dfrac{13}{4}\right)^2 = 7$

12. $\left(x + \dfrac{7}{2}\right)^2 + \left(y - \dfrac{7}{3}\right)^2 = \dfrac{11}{6}$

13. $x^2 + \left(y - \dfrac{7}{3}\right)^2 = 6$

14. $(x + 3)^2 + y^2 = 1.21$

In Exercises 15–26, describe the graph of each equation. If it is a circle, give the center and radius.

15. $x^2 + y^2 + 4x - 6y + 4 = 0$

16. $x^2 + y^2 - 10x + 2y + 22 = 0$

17. $x^2 + y^2 + 10x - 6y - 47 = 0$

18. $x^2 + y^2 + 2x - 12y - 27 = 0$

19. $x^2 + y^2 - 2x + 2y + 3 = 0$

20. $x^2 + y^2 + 2x + 2y - 3 = 0$

21. $x^2 + y^2 + 6x - 16 = 0$

22. $x^2 + y^2 - 8y - 9 = 0$

23. $x^2 + y^2 + 5x - 9y = 9.5$

24. $x^2 + y^2 - 7x + 3y + 2.5 = 0$

25. $9x^2 + 9y^2 + 18x - 15y + 27 = 0$

26. $25x^2 + 25y^2 - 10x + 30y + 1 = 0$

Solve Exercises 27–33.

27. *Physics* When a particle with mass m and charge q enters a magnetic field of induction β with a velocity v at right angle to β, the path of the particle is a circle. The radius of the path is $\sigma = \dfrac{mv}{\beta q}$. If a proton of mass 1.673×10^{-27} kg and charge 1.5×10^{-19} C enters a magnetic field with induction 4×10^{-4} T (tesla $= V \cdot s \cdot m^2$) with a velocity of 1.186×10^7 m/s, find the equation of the path of the electron.

28. *Astronomy* The earth's orbit around the sun is approximately a circle of radius 1.495×10^8 km. The moon's orbit around earth is approximately a circle of radius 3.844×10^5 km. If the sun is placed at the center of a coordinate system, what is the equation of the earth's orbit?

29. *Astronomy* If the sun is placed at the center of a coordinate system and the earth on the positive x-axis, what is the equation of the moon's orbit around the earth?

30. *Industrial design* A drafter is drawing two gears that intermesh. They are represented by two intersecting circles, as shown in Figure 15.7. The first circle has a radius of 6 in. and the second has a radius of 5 in. The section of intersection has a maximum depth of 1 in. What is the equation of each circle, if the center of the first circle is at the origin and the positive x-axis passes through the center of the second circle?

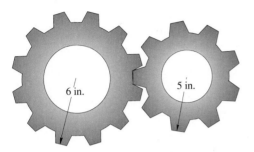

FIGURE 15.7

31. *Industrial design* In designing a tool, an engineer calls for a hole to be drilled with its center 0.9 cm directly above a specified origin. If the hole must have a diameter of 1.1 cm, find the equation of the hole.

32. *Industrial engineering* In designing dies for blanking a sheet-metal piece, the configuration shown in Figure 15.8 is used. Write the equation of the small circle with its center at B, using the point marked A as the origin. Use the following values: $r = 3$ in., $a = 3.25$ in., $b = 1.0$ in., and $c = 0.25$ in.

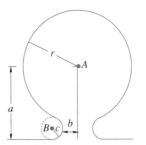

FIGURE 15.8

33. *Industrial engineering* A flywheel that is 26 cm in diameter is to be mounted so that its shaft is 5 cm above the floor, as shown in Figure 15.9. **(a)** Write an equation for the path followed by a point on the rim. Use the surface of the floor as the horizontal axis and the perpendicular line through the center as the vertical axis. **(b)** Find the width of the opening in the floor, allowing a 2-cm clearance on both sides of the wheel.

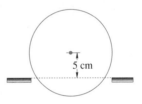

FIGURE 15.9

≣ 15.3
THE PARABOLA

The second curve we will study in this chapter is the parabola. A **parabola** is the set of points in a plane for which the distance from a fixed line is equal to its distance from a fixed point not on the line. The given line is called the **directrix** and the given point is the **focus**.

In Figure 15.10, we have indicated the line ℓ is the directrix, and F is the focus. The line through F perpendicular to the directrix is called the **axis** of the parabola. Examination of the parabola in Figure 15.10 indicates that the axis of the parabola is the only axis of symmetry for the parabola. The point V on the axis that is halfway between the focus and the directrix is the **vertex**.

Let P be any point on the parabola. Draw a line through P that is perpendicular to the directrix at a point P'. According to the definition, the distance from P to P' is the same as the distance from P to F.

Let's set up a similar drawing on a coordinate system so that we can develop a formula for a parabola. We will let the vertex be at the origin and the focus F at $(0, p)$. Since the vertex is midway between the focus and the directrix, the directrix

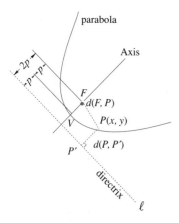

FIGURE 15.10

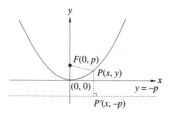

FIGURE 15.11

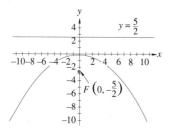

FIGURE 15.12

is the line $y = -p$, as shown in Figure 15.11. If $P(x, y)$ is any point on the parabola, then the distance from F to P is, by the distance formula,

$$\sqrt{(x - 0)^2 + (y - p)^2}$$

Since P' is on the directrix, its coordinates are $(x, -p)$. So the distance from P to P' is

$$\sqrt{(x - x)^2 + (y + p)^2}$$

Since these two distances are equal, we have

$$\sqrt{(x - 0)^2 + (y - p)^2} = \sqrt{(x - x)^2 + (y + p)^2}$$

Squaring both sides and simplifying, we obtain

$$x^2 + (y - p)^2 = (y + p)^2$$
$$x^2 + y^2 - 2py + p^2 = y^2 + 2py + p^2$$
$$x^2 = 4py$$

We have established the following standard equation of a vertical parabola.

Standard Equation of a Vertical Parabola

The standard equation of a vertical parabola with focus F at $(0, p)$ and directrix $y = -p$ is

$$x^2 = 4py$$

If $p > 0$, the parabola opens upward, as shown in Figure 15.11.
If $p < 0$, the parabola opens downward.

The term vertical parabola is used because the axis is a vertical line. Notice that $|p|$ is the distance from the focus to the vertex.

EXAMPLE 15.19

Find the focus and directrix of the parabola with the equation $x^2 = -10y$. Sketch its graph.

Solution   Since $x^2 = 4py$, then $4p = -10$ and $p = -\frac{5}{2}$. The focus is at $\left(0, -\frac{5}{2}\right)$. The equation of the directrix is $y = -p$ or $y = \frac{5}{2}$. The graph is sketched in Figure 15.12. Notice that $p < 0$ and that this parabola opens downward.

EXAMPLE 15.20

Find the equation of the parabola that has its vertex at the origin, opens upward, and passes through $(-5, 9)$.

Solution   The general form of the equation is $x^2 = 4py$. Since it opens upward, $p > 0$, and since $(-5, 9)$ is on the parabola, then $(-5)^2 = 4p(9)$. Solving for p, we

EXAMPLE 15.20 (Cont.)

get $p = \frac{25}{36}$. Substituting this value for p in the equation, we get

$$x^2 = 4\left(\frac{25}{36}\right)y$$

$$x^2 = \frac{25}{9}y$$

or $\quad 9x^2 = 25y$

If we had used the x-axis as the axis of the parabola, and the vertex was still at the origin, the focus would then have been at $(p, 0)$, and the directrix would have had the equation $x = -p$. Using the same method that we used earlier, we would get the following standard equation for a horizontal parabola. It is called a horizontal parabola because the axis is horizontal.

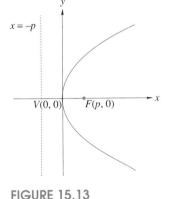

FIGURE 15.13

Standard Equation of a Horizontal Parabola

The standard equation of a horizontal parabola with focus F at $(p, 0)$ and directrix $x = -p$ is

$$y^2 = 4px$$

If $p > 0$, the parabola opens to the right, as shown in Figure 15.13.
If $p < 0$, the parabola opens to the left.

EXAMPLE 15.21

Find the equation of the parabola with its vertex at the origin and focus at $(5, 0)$.

Solution Since the focus is at $(5, 0)$, $p = 5$. A parabola with focus on the x-axis and vertex at the origin is of the form $y^2 = 4px$. Since $p = 5$, the equation is $y^2 = 20x$.

EXAMPLE 15.22

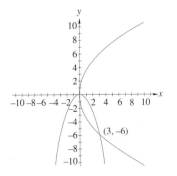

FIGURE 15.14

Sketch the graphs of one horizontal and one vertical parabola, each of which has its vertex at the origin and passes through the point $(3, -6)$. Determine the equation for each parabola.

Solution The graphs are in Figure 15.14. One graph is of the form $x^2 = 4py$ and the other is of the form $y^2 = 4px$. Substituting 3 for x and -6 for y in each equation, we get

$$x^2 = 4py \qquad\qquad y^2 = 4px$$

$$3^2 = 4p(-6) \qquad\qquad (-6)^2 = 4p(3)$$

$$9 = -24p \qquad\qquad 36 = 12p$$

$$p = -\frac{9}{24} = -\frac{3}{8} \qquad\qquad p = 3$$

EXAMPLE 15.22 (Cont.)

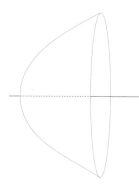

FIGURE 15.15

$$x^2 = 4\left(-\frac{3}{8}\right)y \qquad\qquad y^2 = 4(3)x$$

$$x^2 = -\frac{3}{2}y \qquad\qquad y^2 = 12x$$

$$\text{or}\quad 2x^2 = -3y$$

The vertical parabola has the equation $x^2 = -\frac{3}{2}y$ and the horizontal parabola has the equation $y^2 = 12x$.

Reflective Properties of Parabolas

A three-dimensional object called a **paraboloid of revolution** is formed when a parabola is revolved around its axis of symmetry. (See Figure 15.15.) Paraboloids of revolution have many uses based on the following two facts.

1. Rays entering a parabola along lines parallel to its axis are all reflected through its focus. Many examples exist for the different types of energy rays. Radio telescopes, radar antennae, and satellite television dishes used as downlinks are all examples of paraboloids of revolution. Parabolic reflectors are sometimes used at sports events to pick out specific sounds from background noises. In each of these cases, a ray or wave that is directed toward the dish is reflected off the sides of the paraboloid through its focus.

2. Rays drawn through a parabola's focus are all reflected along lines parallel to the parabola's axis. This uses rays in the reverse of the first property. Applications include flashlights, automobile headlights, search lights, and satellite dishes used as uplinks.

Paraboloids are sometimes used to both send and receive information. For example, a radar transmitter alternately sends and receives waves.

Application

EXAMPLE 15.23

The television satellite dish shown in Figure 15.16a is in the shape of a paraboloid of revolution measuring 0.4 m deep and 2.5 m across at its opening. Where should the receiver be placed in order to pick up the incoming signals?

Solution A cross-section through the axis of this satellite dish has been drawn over the photograph of the dish in Figure 15.16a. Because the cross-section is a parabola, the receiver should be placed at the parabola's focus. We will next determine the location of that focus.

If we place the drawing of this cross-section on a coordinate system so that the vertex is at the origin and the focus is placed on the x-axis, we get Figure 15.16b. We know that the parabola in Figure 15.16b has an equation of the form $y^2 = 4px$. From the given data, the point $(-0.4,\ 1.2)$ is on the parabola. This means that $p = \dfrac{y^2}{4x} = \dfrac{1.2^2}{-1.6} = -0.9$. Since the focus is at the point $(0,\ -p)$, this means that we can place the focus 0.9 m from the vertex.

FIGURE 15.16

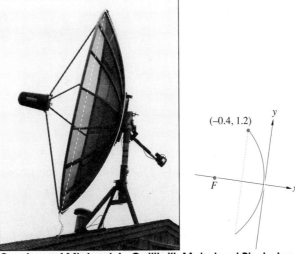

Courtesy of Michael A. Gallitelli, Metroland Photo Inc.
(a)   (b)

Using a Graphing Calculator

Graphing calculators can be used to sketch the graph of any of the curves in this chapter. You must remember, however, that these calculators will only sketch the graphs of functions. This means that you will first need to solve the equation algebraically for y and then graph the one or two equations that result. The next two examples will demonstrate how this is done.

EXAMPLE 15.24

Use a graphing calculator to sketch the graphs of the parabolas (**a**) $y = \frac{3}{4}x^2$ and (**b**) $y = \frac{3}{4}x^2 - 3$.

Solution As usual, after you turn on the calculator, clear the screen of any previous graphs.

Now you are ready to graph $y = \frac{3}{4}x^2$. Since this equation has already been solved for y, we need only to press the following key sequence on a Casio *fx-7700G*.

$$\boxed{\text{Graph}}\ \boxed{(}\ \boxed{3}\ \boxed{\div}\ \boxed{4}\ \boxed{)}\ \boxed{\text{X}, \theta, \text{T}}\ \boxed{\text{SHIFT}}\ \boxed{x^2}\ \boxed{\text{EXE}}$$

The result is the graph in Figure 15.17a.

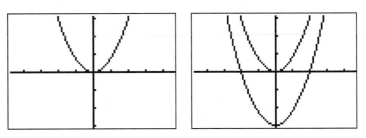

FIGURE 15.17a   FIGURE 15.17b

EXAMPLE 15.24 (Cont.)

If you want to graph $y = \frac{3}{4}x^2 - 3$, enter the key sequence

$$\boxed{\text{Graph}}\ \boxed{(}\ 3\ \boxed{\div}\ 4\ \boxed{)}\ \boxed{\text{X},\theta,\text{T}}\ \boxed{\text{SHIFT}}\ \boxed{x^2}\ \boxed{-}\ 3\ \boxed{\text{EXE}}$$

The result is the graph in Figure 15.17b. (If you did not clear the screen after you graphed $y = \frac{3}{4}x^2$, you will see both graphs on your screen.) Notice how the graph of $y = \frac{3}{4}x^2 - 3$ is simply the graph of $y = \frac{3}{4}x^2$ shifted down 3 units.

EXAMPLE 15.25

Use a graphing calculator to sketch the graph of the circle $x^2 + y^2 + 4x - 8y + 4 = 0$.

Solution   This is the same equation we used in Example 15.15. This equation can be simplified to

$$(x+2)^2 + (y-4)^2 = 16$$

and represents the equation of a circle with center $(-2,\ 4)$ and radius $r = 4$.

In order to use the graphing calculator, we need to solve the equation for y.

$$(x+2)^2 + (y-4)^2 = 16$$

$$(y-4)^2 = 16 - (x+2)^2$$

$$y - 4 = \pm\sqrt{16 - (x+2)^2}$$

$$y = 4 \pm \sqrt{16 - (x+2)^2}$$

Since this result has two solutions, we will have to graph the two equations $y = 4 + \sqrt{16 - (x+2)^2}$ and $y = 4 - \sqrt{16 - (x+2)^2}$.

Now, we know that since this circle has its center at $(-2,\ 4)$ and a radius of 4, the x-values will be from $x = -2 - 4 = -6$ to $x = -2 + 4 = 2$ and the y-values will be from $y = 4 - 4 = 0$ to $y = 4 + 4 = 8$. We will want to set the calculator screen's "range" values so that both of these appear on the screen. We will let x be over the interval $[-8, 4]$ and y be over the interval $[-2, 10]$.

The graph of $y = 4 + \sqrt{16 - (x+2)^2}$ is obtained by first pressing $\boxed{\text{Graph}}$ and then pressing the key sequence

$$4\ \boxed{+}\ \boxed{\sqrt{}}\ \boxed{(}\ 16\ \boxed{-}\ \boxed{(}\ \boxed{\text{X},\theta,\text{T}}\ \boxed{+}\ 2\ \boxed{)}\ \boxed{\text{SHIFT}}\ \boxed{x^2}\ \boxed{)}\ \boxed{\text{EXE}}$$

The result is the graph in Figure 15.18a. Now graph $y = 4 - \sqrt{16 - (x+2)^2}$ by pressing $\boxed{\text{Graph}}$ followed by the sequence

$$4\ \boxed{-}\ \boxed{\sqrt{}}\ \boxed{(}\ 16\ \boxed{-}\ \boxed{(}\ \boxed{\text{X},\theta,\text{T}}\ \boxed{+}\ 2\ \boxed{)}\ \boxed{\text{SHIFT}}\ \boxed{x^2}\ \boxed{)}\ \boxed{\text{EXE}}$$

This graph is displayed over the earlier graph, as shown in Figure 15.18b.

This does not look like a circle. First, the top and bottom parts do not touch and it does not have a circular shape. The fact that parts do not touch is a common error of graphing calculators and computer graphing programs. The noncircular shape is caused by the fact that the calculator screen's ranges for both the x- and y-values are 12 units long. But, the calculator does not have a square screen. In fact, the ratio of

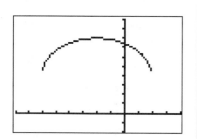

FIGURE 15.18a

EXAMPLE 15.25 (Cont.)

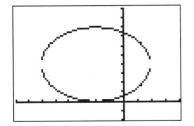

FIGURE 15.18b

the screen's horizontal to vertical size is 3 : 2. So, if we want this graph of a circle to look like a circle, then the ratio of the calculator screen's range of x-values to the range of y-values must be 3 : 2.

Let's leave the y-values at $[-2, 10]$. This is 12 units, so the x-values should cover 18 units. We will change them to $[-10, 8]$. Now press EXE and the graphs should be redrawn as shown in Figure 15.18c.

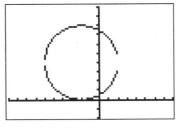

FIGURE 15.18c

FIGURE 15.18d

If you use a Texas Instruments *TI-81* graphics calculator, the following procedure makes the units on the axes the same length. After you have graphed the two functions and obtain a graph like the one in Figure 15.18b, press the ZOOM button on the top row of the calculator. You should see a screen similar to the one shown in Figure 15.18d. This zoom menu presents you with seven options. Pressing a 5 selects option 5: Square. This option adjusts the current range values so that the width of the dots on the x- and y-axes are equalized.

Exercise Set 15.3

In Exercises 1–12, determine the coordinates of the focus and the equation of the directrix of each parabola. State the direction in which each parabola opens and sketch each curve.

1. $x^2 = 4y$
2. $x^2 = 12y$
3. $y^2 = -4x$
4. $y^2 = -20x$
5. $x^2 = -8y$
6. $x^2 = -16y$

7. $y^2 = 10x$
8. $y^2 = -14x$
9. $x^2 = 2y$
10. $x^2 = -17y$
11. $y^2 = -21x$
12. $y^2 = 5x$

In Exercises 13–20, determine the equation of the parabola satisfying the given conditions.

13. Focus $(0, 4)$, directrix $y = -4$
14. Focus $(0, -5)$, directrix $y = 5$
15. Focus $(-6, 0)$, directrix $x = 6$
16. Focus $(3, 0)$, directrix $x = -3$

17. Focus $(2, 0)$, vertex $(0, 0)$
18. Focus $(0, -7)$, vertex $(0, 0)$
19. Vertex $(0, 0)$, directrix $x = -\frac{3}{2}$
20. Vertex $(0, 0)$, directrix $y = \frac{3}{4}$

Solve Exercises 21–24.

21. *Civil engineering*   In a suspension bridge, the main cables are in a parabolic shape. This is because a parabola is the only shape that will bear the total weight load evenly. The twin towers of a certain bridge extend 90 m above the road surface and are 360 m apart, as shown in Figure 15.19. The cables are suspended from the tops of the towers and are tangent to the road surface at a point midway between the towers. What is the height of the cable above the road surface at a point 100 m from the center of the bridge?

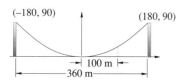

FIGURE 15.19

22. *Optics*   A parabolic reflector is a mirror formed by rotating a parabola around its axis. If a light is placed at the focus of the mirror, all the light rays starting from the focus will be reflected off the mirror in lines parallel to the axis. A radar antenna is constructed in the shape of a parabolic reflector. The receiver is placed at the focus. If the reflector has a diameter of 2 m and a depth of 0.4 m, what is the location of the receiver?

23. *Electronics*   When rays from a distant source strike a parabolic reflector, they will be reflected to a single point—the focus. A parabolic antenna is used to catch the television signals from a satellite. The antenna is 5 m across and 1.5 m deep. If the receiver is located at the focus, what is its location?

24. *Energy*   The rate at which heat is dissipated in an electric current is referred to as the power loss. This loss P is given by the relationship $P = I^2 R$, where I is the current and R is the resistance. If the resistance is 10 Ω, sketch a graph of the power loss as a function of the current.

☰ 15.4
THE ELLIPSE

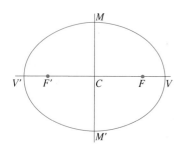

FIGURE 15.20

The third curve we will study in this chapter is the ellipse. An **ellipse** is the set of points in a plane that have the sum of their distances from two fixed points a constant. Each of the fixed points is called a **focus**. The **major axis** of an ellipse is the line segment through the two foci with endpoints on the ellipse. The endpoints of the major axis are called the **vertices**. In Figure 15.20, the foci are labeled F and F' and the vertices are V and V'. The **center** C of the ellipse is the midpoint of the segment joining the foci. The segment through the center and perpendicular to the major axis is called the **minor axis**. The endpoints of the minor axis are on the ellipse. In Figure 15.20, the endpoints of the minor axis are M and M'. The major axis is always longer than the minor axis.

If we were to draw an ellipse with its center at the origin and its major axis along the x-axis, we would get a figure like the one in Figure 15.21. We will let the vertices have the coordinates $V(a, 0)$ and $V'(-a, 0)$. The foci will have the coordinates $F(c, 0)$ and $F'(-c, 0)$. If $P(x, y)$ is any point on the ellipse, then the distance from P to F, d_1, plus the distance from P to F', d_2, is a constant, which we will call $2a$. The sum of these distances, $d_1 + d_2 = (a - c) + (a + c) = 2a$.

The points F, F', and M form an isosceles triangle with the lengths of two equal sides, $\overline{FM}$ and $\overline{F'M}$, being a. Thus, F, M, and C form a right triangle with legs of lengths c and b and hypotenuse of length a. As a result, we see that $a^2 = b^2 + c^2$.

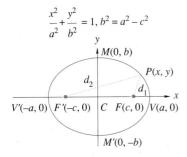

$$\frac{x^2}{a^2} + \frac{y^2}{b^2} = 1, b^2 = a^2 - c^2$$

FIGURE 15.21

Applying the distance formula, we see that $d_1 = \sqrt{(x-c)^2 + y^2}$ and $d_2 = \sqrt{(x+c)^2 + y^2}$. Since $d_1 + d_2 = 2a$, we have $\sqrt{(x-c)^2 + y^2} + \sqrt{(x+c)^2 + y^2} = 2a$. Using the techniques from Chapter 12 for solving radical equations, we get

$$\sqrt{(x-c)^2 + y^2} = 2a - \sqrt{(x+c)^2 + y^2}$$

$$\left[\sqrt{(x-c)^2 + y^2}\right]^2 = \left[2a - \sqrt{(x+c)^2 + y^2}\right]^2 \qquad \text{Square both sides.}$$

$$(x-c)^2 + y^2 = (2a)^2 - 4a\sqrt{(x+c)^2 + y^2} + (x+c)^2 + y^2$$

$$4a\sqrt{(x+c)^2 + y^2} = 4a^2 + 4cx$$

$$a\sqrt{(x+c)^2 + y^2} = a^2 + cx$$

$$\left[a\sqrt{(x+c)^2 + y^2}\right]^2 = \left[a^2 + cx\right]^2 \qquad \text{Square again.}$$

$$a^2\left[(x+c)^2 + y^2\right] = a^4 + 2a^2cx + c^2x^2$$

$$a^2x^2 + 2a^2cx + a^2c^2 + a^2y^2 = a^4 + 2a^2cx + c^2x^2$$

$$(a^2 - c^2)x^2 + a^2y^2 = a^2(a^2 - c^2)$$

But, $a^2 = b^2 + c^2$, so $a^2 - c^2 = b^2$. The equation becomes

$$b^2x^2 + a^2y^2 = a^2b^2$$

Dividing both sides by a^2b^2 and simplifying results in one of the standard equations for an ellipse:

$$\frac{x^2}{a^2} + \frac{y^2}{b^2} = 1$$

Standard Equation: Ellipse with a Horizontal Major Axis

The standard equation of an ellipse with center at (0, 0) and the major axis on the x-axis is

$$\frac{x^2}{a^2} + \frac{y^2}{b^2} = 1$$

where $a > b$.
The vertices are $(-a, 0)$ and $(a, 0)$.
The endpoints of the minor axis are $(0, -b)$ and $(0, b)$.
The foci are at $(-c, 0)$ and $(c, 0)$, where $c^2 = a^2 - b^2$.

EXAMPLE 15.26

Sketch the ellipse $4x^2 + 25y^2 = 100$. Give the coordinates of the vertices, foci, and endpoints of the minor axis.

Solution   This equation is not in the standard form, since the right-hand side is not 1. To get the equation into the standard form, we must divide both sides of the equation by 100. This produces the equation

$$\frac{x^2}{25} + \frac{y^2}{4} = 1$$

From this, we see that $a^2 = 25$, so $a = 5$ and $b^2 = 4$ or $b = 2$. Since $c^2 = a^2 - b^2 = 25 - 4 = 21$, we see that $c = \sqrt{21}$. So, the vertices are $V(5, 0)$ and $V'(-5, 0)$, and the foci are $F(\sqrt{21}, 0)$ and $F'(-\sqrt{21}, 0)$. The endpoints of the minor axis are $M(0, 2)$ and $M'(0, -2)$. A sketch of this ellipse is shown in Figure 15.22.

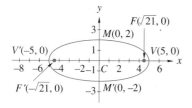

FIGURE 15.22

Not all ellipses have their major axis along the x-axis. For some situations, it is more convenient if the center is at the origin and the major axis is along the y-axis. When that is the case, the vertices are $V(0, a)$ and $V'(0, -a)$, the foci are $F(0, c)$ and $F'(0, -c)$, and the endpoints of the minor axis are $M(b, 0)$ and $M'(-b, 0)$. In this case, we have the following standard form of the equation for an ellipse. A typical graph of an ellipse with center at the origin and a vertical major axis is shown in Figure 15.23.

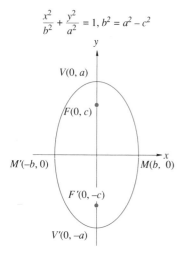

$$\frac{x^2}{b^2} + \frac{y^2}{a^2} = 1, \; b^2 = a^2 - c^2$$

FIGURE 15.23

Standard Equation: Ellipse with a Vertical Major Axis

The standard equation of an ellipse with center at $(0, 0)$ and the major axis on the y-axis is

$$\frac{x^2}{a^2} + \frac{y^2}{b^2} = 1$$

where $a < b$.
The vertices are $(0, -b)$ and $(0, b)$.
The endpoints of the minor axis are $(-a, 0)$ and $(a, 0)$.
The foci are at $(0, -c)$ and $(0, c)$, where $c^2 = b^2 - a^2$.

If we put these two types of ellipses together, we see that the equation of an ellipse with center at the origin and foci on a coordinate axis can always be written in the form

$$\frac{x^2}{p^2} + \frac{y^2}{q^2} = 1 \qquad \text{or} \qquad q^2x^2 + p^2y^2 = p^2q^2$$

where p and q are both positive. If $p^2 > q^2$, then the major axis is along the x-axis. If $p^2 < q^2$, then the major axis is along the y-axis.

≡ **Note**   The *x*-axis is the major, or longer, axis if the denominator of x^2 is the larger denominator, and the *y*-axis is the major axis if the denominator of y^2 is the larger denominator.

 Hint   Remember, c^2 is found by subtracting the smaller denominator from the larger. You may want to remember this as $c^2 = |a^2 - b^2|$.

EXAMPLE 15.27

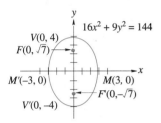

FIGURE 15.24

Discuss and graph the equation $16x^2 + 9y^2 = 144$.

Solution   Dividing both sides of the equation by 144, we get the standard equation

$$\frac{16x^2}{144} + \frac{9y^2}{144} = \frac{144}{144}$$

$$\frac{x^2}{9} + \frac{y^2}{16} = 1$$

So, $a^2 = 16$, $b^2 = 9$, and $c^2 = a^2 - b^2 = 7$. Since $a = 3$ and $b = 4$, we see that $a < b$, so the major axis is along the *y*-axis. The vertices are $V(0, 4)$ and $V'(0, -4)$, the foci are $F(0, \sqrt{7})$ and $F'(0, -\sqrt{7})$, and the endpoints of the minor axis are $M(3, 0)$ and $M'(-3, 0)$. The sketch of this ellipse is shown in Figure 15.24.

EXAMPLE 15.28

Find the equation of the ellipse with center $(0, 0)$, one focus at $(5, 0)$, and a vertex at $(-8, 0)$.

Solution   The second focus is at $(-5, 0)$ and the other vertex is at $(8, 0)$. Since the major axis is horizontal (along the *x*-axis), the equation is of the form

$$\frac{x^2}{a^2} + \frac{y^2}{b^2} = 1$$

where $c^2 = a^2 - b^2$, $a > b$.

Since one focus is at $(5, 0)$, $c = 5$ and the fact that one vertex is at $(8, 0)$ tells us that $a = 8$. So, $5^2 = 8^2 - b^2$ or $b^2 = 64 - 25 = 39$, and $b = \sqrt{39}$. The equation of this ellipse is $\dfrac{x^2}{64} + \dfrac{y^2}{39} = 1$.

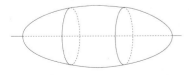

FIGURE 15.25

Reflective Properties of Ellipses

A three-demensional object called an **ellipsoid of revolution** is formed when an ellipse is revolved around one of its axes of symmetry. In Figure 15.25, the ellipse has been revolved around its longer axis.

These ellipsoids of revolution have uses based on the fact that rays leaving or passing through one focus of an ellipse are all reflected by the ellipse so that they pass through the other focus of the ellipse. One architectural application of these ellipsoids of revolution is in the construction of whispering galleries, such as the one in Statuary Hall of the Capitol in Washington, D.C. or St. Paul's Cathedral in London, England. If you position yourself at one focus in a whispering gallery, you will be able to hear anything said at the other focus, regardless of the direction in which the speakers address their remarks.

Application

EXAMPLE 15.29

A lithotripter is a medical device that is used to break up kidney stones with shock waves through water. An ellipsoid is cut in half and careful measurements are used to place the patient's kidney stones at one focus of the ellipse. A source that produces ultra–high-frequency shock waves is placed at the other focus. (See Figure 15.26.) The shock waves are reflected by the ellipsoid to the other focus, where they break up the kidney stones.

If the endpoints of the major axis are located 6 in. from the center of the ellipse that is rotated to make a lithotripter and one end of the minor axis is 2.5 in. from the center, what are the locations of the lithotripter's foci?

FIGURE 15.26

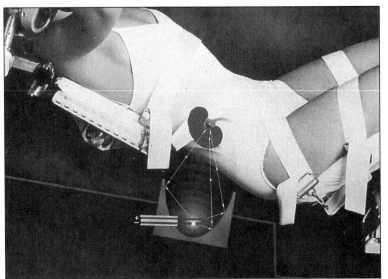

(a)

(b)

Courtesy of Dornier Medical Systems Inc.

Solution If we place the drawing of the cross-section of the ellipse in Figure 15.26a on a coordinate system so that the center is at the origin and the major axis is on the x-axis, we get Figure 15.26b. From the given information, we know that the ellipse in Figure 15.26b has endpoints on the major axis at $(-6, 0)$ and $(6, 0)$, and that the endpoints of the minor axis are $(0, -2.5)$ and $(0, 2.5)$. From the given data, we know that the desired ellipse has the equation $\dfrac{x^2}{a^2} + \dfrac{y^2}{b^2} = \dfrac{x^2}{6^2} + \dfrac{y^2}{2.5^2} = 1$.

Thus, $a = 6$ and $b = 2.5$, and since $c^2 = a^2 - b^2$, we have $c^2 = 6^2 - 2.5^2 = 29.75$, so $c = \sqrt{29.75} \approx 5.45$. Thus, the foci are at $(-5.45, 0)$ and $(5.45, 0)$.

Exercise Set 15.4

In Exercises 1–8, find the equation of the ellipse with the stated properties. Each ellipse has its center at (0, 0).

1. Focus at (4, 0), vertex at (6, 0)

2. Focus at (9, 0), vertex at (12, 0)

3. Focus at (0, 2), vertex at (0, −4)

4. Focus at (0, −3), vertex at (0, 5)

5. Focus at (3, 0), vertex at (5, 0)

6. Focus at (0, −2), vertex at (0, −3)

7. Length of the minor axis 6, vertex at (4, 0)

8. Length of the minor axis 10, vertex at (0, −6)

In Exercises 9–16, give the coordinates of the vertices, foci, and endpoints of the minor axis, and sketch each curve.

9. $\dfrac{x^2}{4} + \dfrac{y^2}{9} = 1$

10. $\dfrac{x^2}{9} + \dfrac{y^2}{4} = 1$

11. $\dfrac{x^2}{4} + \dfrac{y^2}{1} = 1$

12. $\dfrac{x^2}{16} + \dfrac{y^2}{36} = 1$

13. $4x^2 + y^2 = 4$

14. $9x^2 + 4y^2 = 36$

15. $25x^2 + 36y^2 = 900$

16. $9x^2 + y^2 = 18$

Solve Exercises 17–21.

17. *Astronomy*   The orbit of the earth is an ellipse with the sun at one focus. The major axis has a length of 2.992×10^8 km and the ratio of $\dfrac{c}{a} = \dfrac{1}{60}$. What is the length of the minor axis? $\left(\text{The ratio } \dfrac{c}{a} \text{ is called the \textbf{eccentricity.}}\right)$

18. *Civil engineering*   In order to support a bridge, an arch in the shape of the upper half of an ellipse is built. As shown in Figure 15.27, the bridge is to span a river 80 ft wide and the center of the arch is to be 24 ft above the water. If the water level is used as the *x*-axis and the *y*-axis passes through the center of the bridge, write an equation for the ellipse that forms the arch of the bridge.

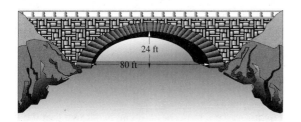

FIGURE 15.27

19. *Mechanical engineering*   A circular pipe has an inside diameter of 10 in. One end of the pipe is cut at an angle of 45°, as shown in Figure 15.28. What are the lengths of the major and minor axes of the elliptical opening?

FIGURE 15.28

20. *Mechanical engineering*   Elliptical gears are used in some machinery to provide a quick-return mechanism or a slow power stroke (for heavy cutting) in each revolution. Figure 15.29, on page 538, shows two congruent gears that remain in contact as they rotate around the indicated foci. If the distance between the foci used at the centers of rotation is 6 cm and the shortest distance from a focus to the edge of a gear is 1 cm, what is the length of the minor axis?

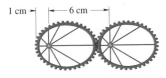

FIGURE 15.29

21. *Astronomy*   The comet Kahoutek has major and minor axes of lengths 3 600 and 44 astronomical units. One astronomical unit is about 1.495×10^8 km. What is the eccentricity of the comet's orbit?

☰ 15.5
THE HYPERBOLA

The fourth and last conic section that we will study is the hyperbola. A **hyperbola** is the set of points in a plane for which the difference of the distances of the points from two fixed points is a constant. As with the ellipse, each of the fixed points is called a **focus**. The **transverse axis** is the line segment through the two foci with its endpoints on the hyperbola. The endpoints of the transverse axis are called the **vertices**.

A hyberbola is shown by the blue curve in Figure 15.30. In Figure 15.30, the foci are labeled F and F' and the vertices V and V'. The **center** C of the hyperbola is the midpoint of the foci. The line segment through the center that is perpendicular to the transverse axis is called the **conjugate axis**. The endpoints of the conjugate axis, labeled W and W' in Figure 15.30, are not on the hyperbola. We will learn how to determine the length of the conjugate axis later in this section.

If we draw a hyperbola with its center at the origin and its transverse axis along the x-axis, we get a figure like the one in Figure 15.31. If $P(x, y)$ is any point on the hyperbola, the distance from P to F, d_1, minus the distance from P to F', d_2, is a constant. The difference of the distances is $|d_1 - d_2| = |(c - a) - (c + a)| = 2a$, $a > 0$.

Here we will use the distance formulas $d_1 = \sqrt{(x - c)^2 + y^2}$ and $d_2 = \sqrt{(x + c)^2 + y^2}$. As a result, we compute the difference of the distances as

$$\left| \sqrt{(x - c)^2 + y^2} - \sqrt{(x + c)^2 + y^2} \right| = 2a$$

or $\sqrt{(x - c)^2 + y^2} - \sqrt{(x + c)^2 + y^2} = \pm 2a$

Using the same technique we used on the ellipse, we get

$$c^2x^2 - a^2x^2 - a^2y^2 = a^2c^2 - a^4$$

or $x^2(c^2 - a^2) - a^2y^2 = a^2(c^2 - a^2)$

If we let $b^2 = c^2 - a^2$, this can be written as $b^2x^2 - a^2y^2 = a^2b^2$. Dividing both sides by a^2b^2, the equation becomes

$$\frac{x^2}{a^2} - \frac{y^2}{b^2} = 1$$

which is one of the standard equations for a hyperbola. It can be shown that the lines $y = \dfrac{b}{a}x$ and $y = -\dfrac{b}{a}x$ are the hyperbola's asymptotes.

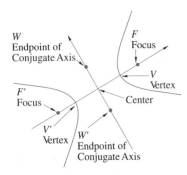

W
Endpoint of
Conjugate Axis
F
Focus
V
Vertex
F'
Focus
Center
V'
Vertex
W'
Endpoint of
Conjugate Axis

FIGURE 15.30

$$\frac{x^2}{a^2} - \frac{y^2}{b^2} = 1, \, b^2 = c^2 - a^2$$

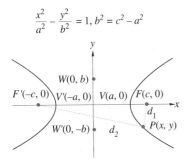

$W(0, b)$
$F'(-c, 0)$ $V'(-a, 0)$ $V(a, 0)$ $F(c, 0)$
d_1
$W'(0, -b)$ d_2 $P(x, y)$

FIGURE 15.31

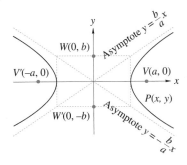

FIGURE 15.32

> **Standard Equation: Hyperbola with a Horizontal Major Axis**
>
> The standard equation of a hyperbola with center at (0, 0) and the major axis on the x-axis is
>
> $$\frac{x^2}{a^2} - \frac{y^2}{b^2} = 1$$
>
> The vertices are $(-a, 0)$ and $(a, 0)$.
> The endpoints of the conjugate axis are $(0, -b)$ and $(0, b)$.
> The foci are at $(-c, 0)$ and $(c, 0)$, where $c^2 = a^2 + b^2$.
> The lines $y = \dfrac{b}{a}x$ and $y = -\dfrac{b}{a}x$ are **asymptotes** of this hyperbola.

The asymptotes provide convenient guidelines for drawing hyperbolas. Asymptotes are easily drawn as the diagonals of a rectangle. The midpoints of the sides of the rectangles are the vertices and the endpoints of the conjugate axis, as shown in Figure 15.32.

If the transverse axis is along the y-axis and the conjugate axis is along the x-axis, the hyperbola has a second standard form.

> **Standard Equation: Hyperbola with a Vertical Major Axis**
>
> The standard equation of a hyperbola with center at (0, 0) and the major axis on the y-axis is
>
> $$\frac{y^2}{a^2} - \frac{x^2}{b^2} = 1$$
>
> The vertices are $(0, -a)$ and $(0, a)$.
> The endpoints of the conjugate axis are $(-b, 0)$ and $(b, 0)$.
> The foci are at $(0, -c)$ and $(0, c)$, where $c^2 = a^2 + b^2$.
> The lines $y = \dfrac{a}{b}x$ and $y = -\dfrac{a}{b}x$ are **asymptotes** of this hyperbola.

Caution

The formulas relating a, b, and c for the hyperbola are different from the corresponding formulas for the ellipse.

For the ellipse

$$c^2 = |a^2 - b^2| \quad \text{(See Figure 15.33a.)} \qquad a > c \quad \text{(See Figure 15.33b.)}$$

For the hyperbola

$$c^2 = a^2 + b^2 \quad \text{(See Figure 15.34a.)} \qquad c > a \quad \text{(Figure 15.34b.)}$$

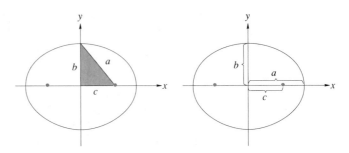

FIGURE 15.33a   FIGURE 15.33b

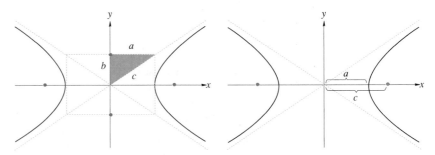

FIGURE 15.34a   FIGURE 15.34b

EXAMPLE 15.30

Discuss and graph the equation $16x^2 - 9y^2 = 144$.

Solution   To get the equation in standard form, we divide both sides by 144 and obtain the equation

$$\frac{x^2}{9} - \frac{y^2}{16} = 1$$

This is the equation of a hyperbola with center $(0, 0)$. The transverse axis is the x-axis. Also, $a^2 = 9$ and $b^2 = 16$, so $c^2 = a^2 + b^2 = 9 + 16 = 25$. The foci are at $F(5, 0)$ and $F'(-5, 0)$, the vertices at $V(3, 0)$ and $V'(-3, 0)$, and the endpoints of the conjugate axis at $W(0, 4)$ and $W'(0, -4)$. If we form the rectangle with midpoints of its sides V, V', W, and W' and draw its diagonals, we get the asymptotes $y = \frac{4}{3}x$ and $y = -\frac{4}{3}x$ shown by the dotted lines in Figure 15.35. The sketch of the graph is shown by the solid curve in Figure 15.35.

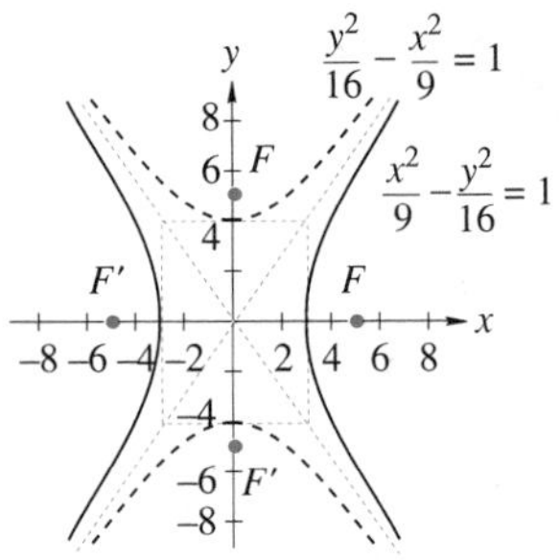

FIGURE 15.35

EXAMPLE 15.31

Discuss and graph the equation $9y^2 - 16x^2 = 144$.

Solution   Again, we divide both sides by 144 to get

$$\frac{y^2}{16} - \frac{x^2}{9} = 1$$

This is the equation of a hyperbola with center $(0, 0)$. The transverse axis is the y-axis. Also, $a^2 = 16$ and $b^2 = 9$, so $c^2 = 25$. The foci are at $F(0, 5)$ and $F'(0, -5)$,

EXAMPLE 15.31 (Cont.)

the vertices at $V(0, 4)$ and $V'(0, -4)$, and the endpoints of the conjugate axis at $W(3, 0)$ and $W'(-3, 0)$. This hyperbola has the same asymptotes as the hyperbola in Example 15.30. The sketch of this hyperbola is shown by the dotted curves in Figure 15.35. The hyperbolas in Examples 15.30 and 15.31 are called **conjugate hyperbolas**, because they have the same set of asymptotes.

EXAMPLE 15.32

Find the equation of a hyperbola with center $(0, 0)$, a focus at $(6, 0)$, and a vertex at $(-3, 0)$.

Solution   The second focus is at $(-6, 0)$, and the second vertex at $(3, 0)$. Since the foci and vertices are on the x-axis, it is the transverse axis. The equation is

$$\frac{x^2}{a^2} - \frac{y^2}{b^2} = 1$$

Since $a = 3$ and $c = 6$, then $b^2 = c^2 - a^2 = 36 - 9 = 27$. The equation of this hyperbola is

$$\frac{x^2}{9} - \frac{y^2}{27} = 1$$

or $27x^2 - 9y^2 = 243$.

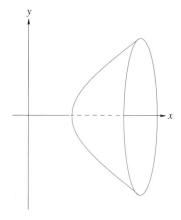

FIGURE 15.36

Reflective Properties of Hyperbolas

Like the other conic sections, hyperbolas are used in many of today's technological applications. A three-dimensional object called a **hyperboloid of revolution** is formed when a hyperbola is revolved around its transverse axis of symmetry. (See Figure 15.36.) Notice that a hyperbola of revolution normally uses only one branch of the hyperbola. Uses of these hyperboloids of revolution are based on the fact that rays aimed at one focus of a hyperbolic reflector are reflected so that they pass through the other focus.

Hyperboloids are frequently used to locate the source of a sound or radio signal. In Exercise Set 15.5, Exercise 18 describes how a system called LORAN (for LOng RAnge Navigation) uses hyperbolas as aids for navigation.

A telescope can use both a hyperbolic lens and a parabolic lens. The main lens is parabolic with its focus at F_1 and vertex at F_2. A second lens is hyperbolic with its foci at F_1 and F_2, as shown in Figure 15.37. The eye is positioned at F_2. The key to this arrangement is that the same point, F_1, is both the focus of the parabola and a focus of the hyperbola.

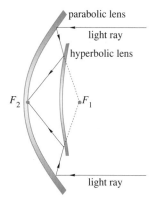

FIGURE 15.37

Application

EXAMPLE 15.33

As we will see in Section 15.6, many telescopes use both a parabolic reflector and a hyperbolic reflector. In this example, we shall examine a simple hyperbolic reflector. In Section 15.6, we shall consider an application that involves both parabolic and hyperbolic reflectors.

EXAMPLE 15.33 (Cont.)

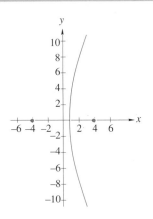

FIGURE 15.38

A hyperbolic reflector is designed to fit in a confined area. It is possible to get the foci so they are 8 ft apart and one vertex is 3 ft from a focus. What is the equation for this hyperbolic reflector?

Solution We shall arrange this reflector on a coordinate system, so that the transverse axis is on the x-axis and the center is at the origin. A cross-section through the axis of this hyperbolic reflector has been drawn on the desired coordinate system in Figure 15.38.

Since the foci are 8 ft apart, this means that the foci are at $(-4, 0)$ and $(4, 0)$. The given vertex is 3 ft from one of the foci (and hence, 5 ft from the other), so it is at $(-1, 0)$ or $(1, 0)$. We have $a = 1$ and $c = 4$, so $b^2 = c^2 - a^2 = 4^2 - 1^2 = 15$. Thus, the desired equation is $\dfrac{x^2}{1} - \dfrac{y^2}{15} = 1$.

Exercise Set 15.5

In Exercises 1–8, find the equation of the hyperbola with the stated properties. Each hyperbola has its center at (0, 0).

1. Focus at $(6, 0)$, vertex at $(4, 0)$

2. Focus at $(12, 0)$, vertex at $(9, 0)$

3. Focus at $(0, 5)$, vertex at $(0, -3)$

4. Focus at $(0, -4)$, vertex at $(0, 2)$

5. Focus at $(5, 0)$, vertex at $(3, 0)$

6. Focus at $(0, -3)$, vertex at $(0, -2)$

7. Focus at $(4, 0)$, length of conjugate axis, 6

8. Focus at $(0, -6)$, length of conjugate axis, 10

In Exercises 9–16, give the coordinates of the vertices, foci, and endpoints of the conjugate axis and sketch each curve.

9. $\dfrac{x^2}{4} - \dfrac{y^2}{9} = 1$

10. $\dfrac{x^2}{9} - \dfrac{y^2}{4} = 1$

11. $\dfrac{y^2}{4} - \dfrac{x^2}{1} = 1$

12. $\dfrac{y^2}{36} - \dfrac{x^2}{16} = 1$

13. $4x^2 - y^2 = 4$

14. $9x^2 - 4y^2 = 36$

15. $36y^2 - 25x^2 = 900$

16. $y^2 - 9x^2 = 18$

Solve Exercises 17–20.

17. The **eccentricity** e of the hyperbola is defined as $\dfrac{c}{a}$. Since $c > a$, then $e > 1$. (Remember, for the ellipse, $0 < e < 1$.) What is the equation of a hyperbola with center at the origin, vertex at $(7, 0)$, and $e = 1.5$?

18. *Navigation* Hyperbolas are used in long-range navigation as part of the LORAN system of navigation. A

transmitter is located at each focus and radio signals are sent to the navigator simultaneously from each station. The difference in time at which the signals are received allows the navigator to determine the position. Suppose the transmitting towers are 1 000 km apart on an east-west line. A ship on the line between two towers determines its position to be 250 km from the west tower.

(a) Sketch the signal hyperbola on which the ship is located. Place the center at (0, 0) and let the *x*-axis be the transverse axis.

(b) Find the equation of the hyperbola.

19. A hyperbola for which $a = b$ is called an **equilateral hyperbola**. Find the eccentricity of an equilateral hyperbola.

20. *Civil engineering* A proposed design for a cooling tower at a nuclear power plant is a branch of a hyperbola rotated about a conjugate axis. Find the equation of the hyperbola passing through the point (238, −616) with center (0, 0) and one vertex at (153, 0).

≡ 15.6
TRANSLATION OF AXES

So far, the equations that we have used for the ellipse and the hyperbola have all been centered at the origin. You may remember from our study of circles that some circles had their center at the origin and some did not. In this section, we will look at cases where one of the axes of an ellipse is parallel to one of the coordinate axes. To do this, we will use translation axes.

The method we will use to translate axes will be very similar to the one we used when we were solving the equation for a circle. We took each equation that was not centered at the origin and completed the square. This told us the location of the center and the radius.

The equations that we will be working with are all in the form

$$Ax^2 + Cy^2 + Dx + Ey + F = 0$$

where *A* and *C* cannot both be zero at the same time. The process can best be explained by following the next example.

EXAMPLE 15.34

Discuss and sketch the graph of $9x^2 + 4y^2 - 36x + 40y + 100 = 0$.

Solution As we just mentioned, we will complete the square. First we will group the terms, and use ◯ and ☐ to indicate the missing constants that we must determine in order to complete the square.

$$(9x^2 - 36x) + (4y^2 + 40y) + 100 = 0$$

$$9(x^2 - 4x + ◯) + 4(y^2 + 10y + ☐) = -100 + 9◯ + 4☐$$

Next, we complete the square of each expression within parentheses by adding 4 to the *x* terms and 25 to the *y* terms. Remember that on the right-hand side of the equation, each of these numbers (4 and 25) must be multiplied by the number outside the parentheses (9 and 4):

$$9(x^2 - 4x + 4) + 4(y^2 + 10y + 25) = -100 + 9(4) + 4(25)$$

or $$9(x - 2)^2 + 4(y + 5)^2 = 36$$

Dividing both sides by 36, we get

$$\frac{(x - 2)^2}{4} + \frac{(y + 5)^2}{9} = 1$$

EXAMPLE 15.34 (Cont.)

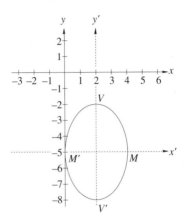

FIGURE 15.39

This looks like the equation for an ellipse. If we let $x' = x - 2$ and $y' = y + 5$, then the equation can be written as

$$\frac{(x')^2}{4} + \frac{(y')^2}{9} = 1$$

This is an ellipse with center at $(x', y') = (0', 0')$. If we draw a new coordinate system, the $x'y'$-system, with its origin at $(2, -5)$, we can draw our ellipse on this new coordinate system. Thus, this ellipse is centered at $(0', 0')$, the vertices are at $V(0', 3')$ and $V'(0', -3')$, the endpoints of the minor axis at $M(2', 0')$ and $M'(-2', 0')$, and the foci are at $F(0', \sqrt{5}')$ and $F'(0', -\sqrt{5}')$. Notice that we have used $3', 0', 2'$, and so on, to show that these are points on the $x'y'$-axes. A sketch of this graph is shown in Figure 15.39. The $x'y'$-axes have been traced over the xy-system.

We can use a table to show the coordinates in both systems. Since we know the coordinates in the $x'y'$-system and we know that $x' = x - 2$ and $y' = y + 5$, to find the coordinates in the xy-system, we solve these equations for x and y: $x = x' + 2$ and $y = y' - 5$. The table follows.

	$x'y'$-system	xy-system
center	$(0', 0')$	$(2, -5)$
vertices	$(0', \pm 3')$	$(2, -2), (2, -8)$
foci	$(0', \pm\sqrt{5}')$	$(2, \sqrt{5} - 5), (2, -\sqrt{5} - 5)$
endpoints of minor axis	$(\pm 2', 0')$	$(4, -5), (0, -5)$

In general, if we want the point (h, k) in the xy-coordinate system to become identified as the origin of the $x'y'$-coordinate system, we use the substitution

$$x' = x - h \qquad \text{and} \qquad y' = y - k$$

This procedure is called a **translation of axes**. As a result of a translation of axes, new axes are formed that are parallel to the old ones.

EXAMPLE 15.35

Discuss and sketch the graph of $x^2 - 18y^2 + 6x - 36y - 45 = 0$.

Solution Again, we will complete the square.

$$(x^2 + 6x) + (-18y^2 - 36y) = 45$$

$$(x^2 + 6x + \bigcirc) - 18(y^2 + 2y + \square) = 45 + \bigcirc - 18\square$$

We add 9 to complete the square of $x^2 + 6x$ and 1 to complete $y^2 + 2y$.

$$(x^2 + 6x + 9) - 18(y^2 + 2y + 1) = 45 + 9 - 18(1)$$

$$(x + 3)^2 - 18(y + 1)^2 = 36$$

EXAMPLE 15.35 (Cont.)

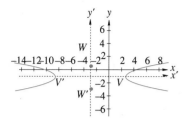

FIGURE 15.40

Dividing both sides by 36, we get

$$\frac{(x+3)^2}{36} - \frac{(y+1)^2}{2} = 1$$

If $x' = x + 3$ and $y' = y + 1$, then the equation becomes

$$\frac{x'^2}{36} - \frac{y'^2}{2} = 1$$

which is a hyperbola with center at $(0', 0')$ and vertices at $(\pm 6', 0')$ of the $x'y'$-system. The origin of the $x'y'$-system is found at $(-3, -1)$ of the xy-system. The sketch of this hyperbola on both coordinate systems is shown in Figure 15.40. Corresponding coordinates in the two systems are shown in the following table.

	$x'y'$-system	xy-system
center	$(0', 0')$	$(-3, -1)$
vertices	$(\pm 6', 0')$	$(-9, -1), (3, -1)$
foci	$(\pm\sqrt{38}, 0')$	$(-3 + \sqrt{38}, -1), (-3 - \sqrt{38}, -1)$
endpoints of conjugate axis	$(0', \pm\sqrt{2})$	$(-3, -1 + \sqrt{2}), (-3, -1 - \sqrt{2})$

EXAMPLE 15.36

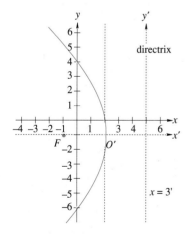

FIGURE 15.41

Discuss and sketch the graph of $y^2 + 12x + 2y - 23 = 0$.

Solution There is no x^2-term, so this is the equation for a parabola. We will complete the square on just the y-terms. We will also put the y-terms on one side of the equation and all the other terms on the other side.

$$y^2 + 2y = -12x + 23$$

$$y^2 + 2y + 1 = -12x + 23 + 1$$

$$(y + 1)^2 = -12(x - 2)$$

If $y' = y + 1$ and $x' = x - 2$, the equation becomes

$$y'^2 = -12x'$$

which is a parabola with vertex at $(0', 0')$, a horizontal axis, and that opens to the left. Since $4p = -12$, $p = -3$ and the directrix is the line $x = 3'$. The focus is at $(-3', 0')$. The sketch of this curve is shown in Figure 15.41.

EXAMPLE 15.37

Find the equation of the hyperbola with vertices $(3, 1)$ and $(-5, 1)$, and focus $(5, 1)$.

Solution The center of a hyperbola is at the midpoint between the vertices. Using the midpoint formula, we can determine that the center is at $\left(\frac{3 + (-5)}{2}, \frac{1 + 1}{2}\right) = (-1, 1)$. The distance from the vertex to the center is 4, so $a = 4$. The distance

EXAMPLE 15.37 (Cont.)

from the focus to the center is 6, so $c = 6$. This is a hyperbola, so $b^2 = c^2 - a^2 = 6^2 - 4^2 = 20$. Finally, this is a hyperbola with foci on the x'-axis and the equation is

$$\frac{x'^2}{16} - \frac{y'^2}{20} = 1$$

Since $x' = x - h$ and $y' = y - k$, we have $x' = x + 1$ and $y' = y - 1$, so the equation becomes

$$\frac{(x + 1)^2}{16} - \frac{(y - 1)^2}{20} = 1$$

Application

EXAMPLE 15.38

A telescope is shown in Figure 15.42a. The focus of one branch of the hyperbolic lens is 20.4 cm from the vertex of the other branch of this same lens. The parabolic lens is 63 cm deep and measures 168 cm across. Determine **(a)** the location of the focus for the parabolic lens **(b)** an equation for the parabola, and **(c)** an equation for the hyperbola.

Solution A cross-section through the axis of this telescope has been drawn on a coordinate system in Figure 15.42b. Because the focus of one branch of the hyperbolic lens is also the vertex of the parabolic lens, we have placed the given focus at the origin. The given vertex of the other branch has been located at $(20.4, 0)$. The parabolic lens is 63 cm deep and 168 cm across and so the point $(63, 84)$ is on this lens.

FIGURE 15.42

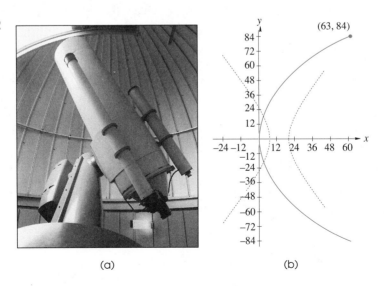

(a)   (b)

This is a parabola. Its equation is of the form $y^2 = 4px$. We know that $(63, 84)$ is a point on the parabola, hence we have $p = \dfrac{y^2}{4x} = \dfrac{84^2}{4(63)} = 28$. Thus, the

EXAMPLE 15.38 (Cont.)

parabola's focus is at (28, 0) or 28 cm from the vertex and the equation of the parabola is $y^2 = 4(28)x = 112x$.

The foci of the hyperbola are at (0, 0) and (28, 0). Thus, the center of the hyperbola is at (14, 0), and so $c = 28 - 14 = 14$. Since one of the hyperbola's vertices is at (20.4, 0), $a = 20.4 - 14 = 6.4$. Using these values of a and c, we can determine that $b^2 = c^2 - a^2 = 14^2 - 6.4^2 = 155.04$. Thus, the standard equation of the hyperbola for this lens is

$$\frac{(x - 14)^2}{6.4^2} - \frac{y^2}{155.04} = 1$$

Exercise Set 15.6

Sketch the graphs in Exercises 1–12, after making suitable translations of axes.

1. $\dfrac{(x - 4)^2}{9} + \dfrac{(y + 3)^2}{4} = 1$

2. $\dfrac{(x + 5)^2}{16} - \dfrac{(y + 3)^2}{25} = 1$

3. $\dfrac{(y - 3)^2}{36} - (x + 4)^2 = 1$

4. $(y + 5)^2 = 12(x + 1)$

5. $100(x + 5)^2 - 4y^2 = 400$

6. $(y + 3)^2 = 8(x - 2)$

7. $16x^2 + 4y^2 + 64x - 12y + 57 = 0$

8. $x^2 + y^2 - 2x + 2y - 2 = 0$

9. $25x^2 + 4y^2 - 250x - 16y + 541 = 0$

10. $100x^2 - 180x - 100y + 81 = 0$

11. $2x^2 - y^2 - 16x + 4y + 24 = 0$

12. $9x^2 - 36x + 16y^2 - 32y - 524 = 0$

In Exercises 13–22, determine the equation of each of the curves described by the given information.

13. Parabola, vertex at (2, −3), $p = 8$, axis parallel to y-axis

14. Hyperbola, vertex at (2, 1), foci at (−6, 1) and (8, 1)

15. Ellipse, center at (4, −3), focus at (8, −3), vertex at (10, −3)

16. Ellipse, center at (−2, 0), focus at (6, 0), vertex at (9, 0)

17. Hyperbola, center at (−3, 2), focus at (−3, 7), transverse axis 6 units

18. Ellipse, center at (3, 5), focus at (3, 8), minor axis 2 units

19. Parabola, vertex at (−5, 1), $p = -4$, axis parallel to x-axis

20. Parabola, vertex at (−3, −6), $p = -12$, axis parallel to y-axis

21. Ellipse, center at (−4, 1), focus at (−4, 9), minor axis 12 units

22. Hyperbola, center at (2, 8), vertex at (6, 8), conjugate axis 10 units

Solve Exercises 23–27.

23. *Physics* The height s of a ball thrown vertically upward is given by the equation $s = 29.4t - 4.9t^2$, where s is in meters and t is the elapsed time in seconds. Graph this curve. Discuss the curve. Determine the maximum height of the ball.

24. *Astronomy* Satellites often orbit earth in an elliptical path with the center of the earth at one of the foci. For a certain satellite, the maximum altitude is 140 mi above the surface of the earth and the minimum is

90 mi. If the radius of earth is approximately 4,000 mi, find the equation of the orbit.

25. *Navigation* Two navigational transmitting towers A and B are 1 000 km apart along an east-west line. Radio signals are sent (traveling at 300 m/μs) simultaneously from both towers. An airplane is located somewhere north of the line joining the two towers. The signal from A arrives at the plane 600 μs after the signal from B. The signal sent from B and reflected by the plane takes a total of 8 000 μs to reach A. What is the location of the plane?

26. *Astronomy* The orbit of Halley's comet is an ellipse with the sun at one focus. The major and minor semi-axes of this orbit are 18.09 A.U. and 4.56 A.U., respectively. (A **semi-axis** is half an axis. One A.U. is one astronomical unit or about 1.496×10^8 km.) What is the equation of this orbit, if the sun is at the origin and the major axis is along the x-axis? What are the maximum and minimum distances from the sun to Halley's comet?

27. *Mechanical engineering* A cantilever beam is a beam fixed at one end. Under a uniform load, the beam assumes a parabolic curve with the fixed end the vertex. For a cantilever beam 2.00 m long, the equation of the load parabola is approximately $x^2 + 200y - 4.00 = 0$, where x is the distance in meters from the fixed end and y is the displacement. See Figure 15.43.

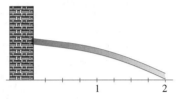

FIGURE 15.43

(a) Change this equation to the standard equation for a parabola.
(b) How far is the free end of the beam displaced from its no-load position?

≡ 15.7
ROTATION OF AXES; THE GENERAL SECOND-DEGREE EQUATION

The conic sections that we have considered so far have all had their axes on a coordinate axis or parallel to a coordinate axis. All of these could be written as second-degree equations of the form

$$Ax^2 + Cy^2 + Dx + Ey + F = 0$$

Notice that this equation does not contain an xy-term.

In this section, we will work with curves in which an axis is not parallel to a coordinate axis. However, the axes are still perpendicular to each other. These equations will all have an xy-term. In order to recognize its graph, we will change to a new coordinate system. In Section 15.6, we changed to a new coordinate system through a translation of axes. In this section, we will use a rotation of axes and we will work with the general second-degree equation.

General Form of a Second-Degree Equation

The **general form of a second-degree equation** is

$$Ax^2 + Bxy + Cy^2 + Dx + Ey + F = 0$$

where A, B, and C are not all 0.

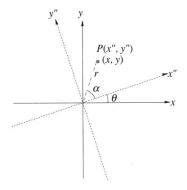

FIGURE 15.44

In a rotation, the origin remains fixed, while the x-axis and y-axis are rotated through a positive acute angle θ. These rotated axes will be labeled the x''-axis and the y''-axis. (See Figure 15.44.) If $P(x, y)$ is any general point in the xy-coordinate system, then that same point would have the coordinates $P(x'', y'')$ in the $x''y''$-coordinate system.

We can convert from one coordinate system to the other with the help of a pair of equations that express x and y in terms of x'' and y''.

$$x = x'' \cos \theta - y'' \sin \theta$$

$$y = y'' \cos \theta + x'' \sin \theta$$

These appear to be terrible equations to work with, but the following examples should show that they are not as difficult to use as they seem.

EXAMPLE 15.39

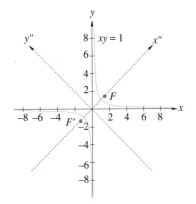

FIGURE 15.45

Transform the equation $xy = 1$ by rotating the axes through an angle of $45°$.

Solution Since $\theta = 45°$, the conversion equations become

$$x = x'' \cos 45° - y'' \sin 45° = 0.7071x'' - 0.7071y''$$

$$y = y'' \cos 45° + x'' \sin 45° = 0.7071y'' + 0.7071x''$$

Substituting these into the original equation, $xy = 1$, we get

$$[0.7071(x'' - y'')] \cdot [0.7071(y'' + x'')] = 1$$

$$0.5(x''^2 - y''^2) = 1$$

$$\frac{x''^2}{2} - \frac{y''^2}{2} = 1$$

This is an equation of a hyperbola with center at the origin of the $x''y''$-coordinate system and transverse axis along the x''-axis. Here $a = b = \sqrt{2}$ and $c = 2$. The asymptotes of this hyperbola happen to be the original xy-axes. A graph of this curve is shown in Figure 15.45. ▪

Any equation of the type in Example 15.39 is called an **equilateral hyperbola** or a **rectangular hyperbola** because the asymptotes are perpendicular. Equilateral hyperbolas are of the form $xy = k$, where k is a constant.

Now, suppose we have an equation that is a general second-degree equation and $B \neq 0$. What we want to do is rotate the axes so that we will get an equation of the form

$$A''(x^2) + B''xy + C''(y^2) + D''x + E''y + F'' = 0$$

where $B'' \neq 0$. Then we will be able to put the equation in the $x''y''$-coordinate system in a recognizable form of one of the conic sections.

We can cause B'' to be 0 if we let θ be the unique acute angle where

$$\cot 2\theta = \frac{A - C}{B} \qquad 0 < \theta < 90° \qquad 0 < \theta < \frac{\pi}{2}$$

If you use a calculator, you have to be careful. First there is no $\boxed{\text{COT}}$ key on your calculator, so you will have to take the reciprocal of the value. Since $\dfrac{1}{\cot\theta} = \tan\theta$, you can then use the $\boxed{\text{INV}}$ $\boxed{\text{tan}}$ keys. But, the arctan function will give answers between $-\dfrac{\pi}{2}$ and $\dfrac{\pi}{2}$. If you get a negative angle, you will have to add π rad or $180°$. Let's see how it works in the next example.

EXAMPLE 15.40

Determine the graph of the equation $29x^2 + 24xy + 36y^2 - 54x - 72y - 135 = 0$.

Solution We will first determine the angle of rotation using the formula $\cot 2\theta = \dfrac{A - C}{B}$. Here $A = 29$, $B = 24$, and $C = 36$. The following description shows how to use a calculator to determine θ.

PRESS	DISPLAY	
29 $\boxed{-}$ 36 $\boxed{=}$	-7	
$\boxed{\div}$ 24 $\boxed{=}$	-0.2916667	
$\boxed{1/x}$	-3.4285714	Changes from $\cot 2\theta$ to $\tan 2\theta$.
$\boxed{\text{INV}}$ $\boxed{\text{tan}}$	-73.739795	2θ
$\boxed{+}$ 180	106.2602	Add $180°$, because 2θ is negative.
$\boxed{\div}$ 2	53.130102	This is θ.

So, using our conversion formulas and the fact that $\sin\theta = 0.8$ and $\cos\theta = 0.6$, we get

$$x = x'' \cos\theta - y'' \sin\theta = 0.6x'' - 0.8y''$$

$$y = y'' \cos\theta + x'' \sin\theta = 0.6y'' + 0.8x''$$

Substituting these values for x and y in the given equation, $29x^2 + 24xy + 36y^2 - 54x - 72y - 135 = 0$, we obtain

$$29(0.6x'' - 0.8y'')^2 + 24(0.6x'' - 0.8y'')(0.6y'' + 0.8x'')$$
$$+ 36(0.6y'' + 0.8x'')^2 - 54(0.6x'' - 0.8y'')$$
$$- 72(0.6y'' + 0.8x'') - 135 = 0$$
$$29(0.36x''^2 - 0.96x''y'' + 0.64y''^2)$$
$$+ 24(0.48x''^2 + 0.36x''y'' - 0.48y''^2 - 0.64x''y'')$$
$$+ 36(0.36y''^2 + 0.96x''y'' + 0.64x''^2)$$
$$- 54(0.6x'' - 0.8y'') - 72(0.6y'' + 0.8x'') - 135 = 0$$
$$10.44x''^2 - 27.84x''y'' + 18.56y''^2 + 11.52x''^2 + 8.64x''y'' - 11.52y''^2$$
$$- 15.36x''y'' + 12.96y''^2 + 34.56x''y'' + 23.04x''^2 - 32.4x''$$
$$+ 43.2y'' - 43.2y'' - 57.6x'' - 135 = 0$$
$$45x''^2 + 0x''y'' + 20y''^2 - 90x'' + 0y'' - 135 = 0$$
$$45x''^2 + 20y''^2 - 90x'' - 135 = 0$$

EXAMPLE 15.40 (Cont.)

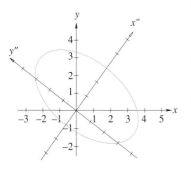

FIGURE 15.46

Completing the square on the x''-terms, we get

$$45(x''^2 - 2x'' + 1) + 20y''^2 = 135 + 45$$

$$45(x'' - 1)^2 + 20y''^2 = 180$$

Dividing both sides by 180 we obtain

$$\frac{(x'' - 1)^2}{4} + \frac{y''^2}{9} = 1$$

This is the equation of an ellipse with a major axis of 6 ($a = 3$) and a minor axis of 4 ($b = 2$) with the major axis along the vertical axis. The center of this ellipse is at ($1''$, $0''$). The graph of this ellipse is shown in Figure 15.46.

The Discriminant

There is an easy method for determining the nature of the curve described by a general second-degree equation. It turns out that the **discriminant**, $B^2 - 4AC$, will tell the type of curve described by the equation.

Classifying the Graph of a General Quadratic Equation

The graph of a general quadratic equation

$$Ax^2 + Bxy + Cy^2 + Dx + Ey + F = 0$$

is either a conic or a degenerate conic. If it is a conic, then the discriminant, $B^2 - 4AC$, can be used to classify the graph of the equation by using the following:

If $B^2 - 4AC > 0$, the curve is a hyperbola.

If $B^2 - 4AC = 0$, the curve is a parabola.

If $B^2 - 4AC < 0$, the curve is an ellipse.

The majority of the problems we will work are not as long as the last example. We are more likely to get problems similar to the one in the next example.

Application

EXAMPLE 15.41

When the power P in an electric circuit is constant, the voltage V is inversely proportional to the current I, as indicated by the equation $P = IV$. If the power is 120 W, sketch the relationship of I vs V.

Solution   The equation is $IV = 120$. A table of values is

I	1	2	4	6	8	10	12	15	20	30	60	120
V	120	60	30	20	15	12	10	8	6	4	2	1

EXAMPLE 15.41 (Cont.)

Negative values do not apply. The graph is one branch of an equilateral hyperbola, as shown in Figure 15.47.

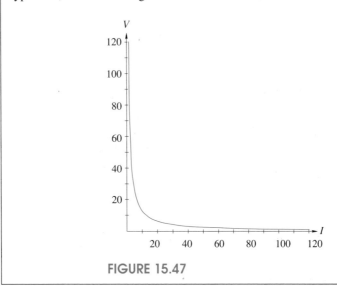

FIGURE 15.47

Exercise Set 15.7

In Exercises 1–12, (a) use the discriminant to identify the graph of the given equation, (b) determine the angle θ needed to rotate the coordinate axes to remove the xy-term, (c) rotate the axes through the angle θ, (d) determine the equation of the conic in the $x''y''$-coordinate system, and (e) sketch the graph.

1. $xy = -9$
2. $xy = 9$
3. $x^2 - 6xy + y^2 - 8 = 0$
4. $x^2 + 4xy - 2y^2 - 6 = 0$
5. $52x^2 - 72xy + 73y^2 - 100 = 0$
6. $11x^2 + 10\sqrt{3}xy + y^2 - 4 = 0$

7. $x^2 - 2xy + y^2 + x + y = 0$
8. $3x^2 + 2\sqrt{3}xy + y^2 - 2x + 2\sqrt{3}y = 0$
9. $2x^2 + 12xy - 3y^2 - 42 = 0$
10. $7x^2 - 20xy - 8y^2 + 52 = 0$
11. $6x^2 - 5xy + 6y^2 - 26 = 0$
12. $9x^2 - 6xy + y^2 - 12\sqrt{10}x - 36\sqrt{10}y = 0$

Solve Exercises 13–16.

13. *Physics*   Boyle's law states that the volume of a gas is inversely proportional to its pressure, provided that the mass and temperature are constant. Thus, if P is the pressure and V the volume, $PV = k$, where k is a constant. Draw the graph of P vs V, if 15 mm³ of gas is under a pressure of 400 kPa at a constant temperature.

14. *Electricity*   For a given alternating current circuit, the capacitance C and the capacitive reactance X_C are related by the equation $X_C = \dfrac{1}{\omega C}$, where ω is the angular frequency. What kind of curve is this? Sketch a graph of the equation if $\omega = 280$ rad/s.

15. *Thermodynamics*   When a cross-section of a pipe is suddenly enlarged, as shown in Figure 15.48, the loss

of heat of the fluid through the pipe is related by the formula $19.62h_L = (\overline{v_1} - \overline{v_2})^2$, where h_L is the heat loss and $\overline{v_1}$ and $\overline{v_2}$ are the average velocities in the two pipes. If h_L is to be held to less that $5°C$, describe this curve. Sketch a graph of the curve with $(\overline{v_1} - \overline{v_2})$ on one axis and h_L on the other.

16. The sides of a rectangle are $3x$ and y and the diagonal is $x + 10$. What kind of curve is represented by the equation relating x and y? Sketch the curve.

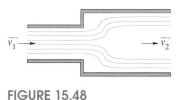

FIGURE 15.48

≡ 15.8
CONIC SECTIONS IN POLAR COORDINATES

In Section 11.7, we studied graphs in the polar coordinate system. Some of the special curves we were able to draw were the rose, the cardioid, and the limaçon. In this section, we will return to the polar coordinate system and see how we can graph the conic sections using polar coordinates.

As we progressed through this chapter, you saw that the equations of the conic sections have very simple forms if the center or vertex is at the origin. But, we had to learn a different form of each conic section. Polar coordinates will allow us to use one equation to represent each conic except the circle. Each of these conics will have a focus at the origin and one axis will be a coordinate axis.

When we defined the parabola, we said that it was the set of points that was an equal distance from the focus and the directrix. The ratio of these two distances, since the distances are the same, is 1. This ratio is called the **eccentricity**. In the problems for the ellipse and the hyperbola, we also used the eccentricity. Each conic can be defined in terms of a point on the curve and the ratio of its distance from a focus and a line called the directrix. This ratio is the eccentricity, e.

Each type of conic is determined by the eccentricity and leads to the general definition of a conic section that follows.

General Definition of a Conic Section

Let ℓ be a fixed line (the directrix) and F a fixed point (**focus**) not on ℓ. A **conic section** is the set of all points P in the plane such that

$$\frac{d(P, F)}{d(P, \ell)} = e$$

where $d(P, F)$ is the distance from P to F, and $d(P, \ell)$ is the perpendicular distance from P to ℓ. The constant e is called the **eccentricity** and if

$$0 < e < 1, \text{ the conic is an ellipse}$$

$$e = 1, \text{ the conic is a parabola}$$

$$e > 1, \text{ the conic is a hyperbola}$$

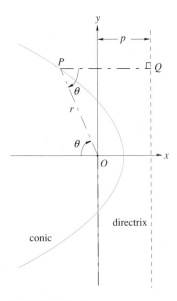

FIGURE 15.49

Now, if $P(r, \theta)$ is a point on a conic with focus O, directrix $x = p$ $(p > 0)$, and eccentricity e (see Figure 15.49), then

$$\frac{d(P, O)}{d(P, Q)} = e$$

where Q is a point on the directrix.

But, $d(P, O) = r$ and $d(P, Q) = p + r \cos \theta$, and so

$$\frac{r}{p + r \cos \theta} = e$$

Solving this for r, we obtain

$$r = \frac{pe}{1 - e \cos \theta}$$

If we had chosen the directrix as $x = -p$ $(p > 0)$, we would have obtained the equation

$$r = \frac{pe}{1 + e \cos \theta}$$

If the directrix is $y = \pm p(p > 0)$, the equations would be

$$r = \frac{pe}{1 \pm e \sin \theta}$$

Thus, we have two sets of equations, and by determining e, we can determine the type of curve. The previous results are summarized in the following box.

Polar Equations of Conic Sections

A polar equation that has one of the four forms

$$r = \frac{pe}{1 \mp e \cos \theta} \qquad r = \frac{pe}{1 \pm e \sin \theta}$$

is a conic section. The conic is a parabola if $e = 1$, an ellipse if $0 < e < 1$, or a hyperbola if $e > 1$.

EXAMPLE 15.42

Describe and sketch the graph of the equation

$$r = \frac{10}{5 + 3 \cos \theta}$$

Solution   Since the constant term in the denominator must be 1, we divide both numerator and denominator by 5. The equation then becomes

$$r = \frac{2}{1 + \frac{3}{5} \cos \theta}$$

From this, we see that $e = \frac{3}{5}$. Since $\frac{3}{5} < 1$, the conic is an ellipse. The denominator contains the cosine function and so the major axis of this conic section is horizontal. The vertices can be determined by setting θ equal to 0 and π. When $\theta = 0$, $r = \frac{2}{8/5} = \frac{10}{8} = 1.25$ and when $\theta = \pi$, $r = \frac{2}{2/5} = 5$. So $2a = 5 + 1.25 = 6.25$, or

EXAMPLE 15.42 (Cont.)

$a = 3.125$. The eccentricity $e = \dfrac{c}{a}$, so $\dfrac{3}{5} = \dfrac{c}{3.125}$ and we get $c = 1.875$. Finally, $b^2 = a^2 - c^2 = 6.25$, thus $b = 2.5$. The sketch of this ellipse is given in Figure 15.50.

EXAMPLE 15.43

Describe and sketch the graph of the equation

$$r = \frac{8}{2 + 2 \sin \theta}$$

Solution   We divide the numerator and denominator by 2, obtaining

$$r = \frac{4}{1 + 1 \sin \theta}$$

From this we see that $e = 1$, and the curve is a parabola. Also, since $pe = 4, p = 4$. This curve has a vertical axis, so the directrix is $y = 4$. If we plot the points that correspond to the x- and y-intercepts, we get the following table and the curve in Figure 15.51.

θ	0	$\dfrac{\pi}{2}$	π	$\dfrac{3\pi}{2}$
r	4	2	4	Not defined

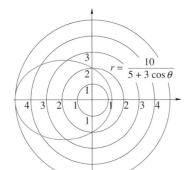

FIGURE 15.50

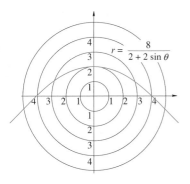

FIGURE 15.51

In Section 15.7, we learned a very complicated process for rotating conic sections on a rectangular coordinate system. The polar coordinate equation of a rotated conic is quite simple. Consider the ellipse in Figure 15.52. It has been rotated through a positive angle α about its focus at the origin O. In the rotated system with polar coordinates (r', θ'), the ellipse has the equation

$$r' = \frac{pe}{1 - e \cos \theta'}$$

But, $r' = r$ and $\theta' = \theta - \alpha$, so the equation of this ellipse in the original (unrotated) polar coordinate system is

$$r = \frac{pe}{1 - e \cos(\theta - \alpha)}$$

Since, in this example, the conic is an ellipse, $0 < e < 1$. But if $e = 1$, the conic would be a parabola, and if $e > 1$, the conic would be a hyperbola.

EXAMPLE 15.44

Discuss and sketch the graph of the equation

$$r = \frac{1.30}{1 + 0.65 \cos \left(\theta + \dfrac{\pi}{6} \right)}$$

Solution   This equation has a denominator that begins with the number 1, so we can immediately see that $e = 0.65$. Since $e < 1$, the conic is an ellipse. The angle is $\theta + \dfrac{\pi}{6}$, so the major axis is rotated $\alpha = -\dfrac{\pi}{6}$. The denominator contains the cosine

EXAMPLE 15.44 (Cont.)

function, thus the major axis is the horizontal axis, which has been rotated $-\frac{\pi}{6}$ rad. Again, the vertices can be determined by setting $\theta = 0$ and $\theta = \pi$. This gives $V = 0.83$ and $V' = 2.97$, so $2a = 3.80$ and $a = 1.90$. The eccentricity $e = \frac{c}{a}$, so $c = ea = (0.65)(1.90) = 1.235$. Thus, $b^2 = a^2 - c^2 = 2.085$ or $b \approx 1.444$. The sketch of this ellipse is shown in Figure 15.53.

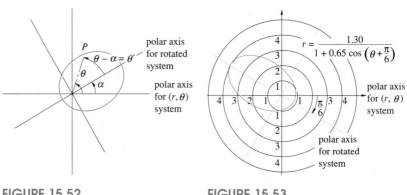

FIGURE 15.52   FIGURE 15.53

We will see some applications of this polar equation for a rotated conic section in the following exercise set and again in Chapter 20.

Exercise Set 15.8

In Exercises 1–10, identify and sketch the conic section with the given equation.

1. $r = \dfrac{6}{1 + 3\cos\theta}$

2. $r = \dfrac{8}{1 - 2\sin\theta}$

3. $r = \dfrac{12}{3 - \cos\theta}$

4. $r = \dfrac{15}{3 + 9\sin\theta}$

5. $r = \dfrac{12}{4 - 4\cos\theta}$

6. $r = \dfrac{8}{6 + 6\sin\theta}$

7. $r = \dfrac{12}{3 + 2\cos\theta}$

8. $r = \dfrac{12}{2 - 3\cos\theta}$

9. $r = \dfrac{12}{2 - 4\cos\left(\theta + \dfrac{\pi}{3}\right)}$

10. $r = \dfrac{6}{2 + \sin\left(\theta + \dfrac{\pi}{6}\right)}$

In Exercises 11–16, find a polar equation of the conic with focus at the origin and the given eccentricity and directrix.

11. Directrix $x = 4$; $e = \dfrac{3}{2}$

12. Directrix $x = -2$; $e = \dfrac{3}{4}$

13. Directrix $y = -5$; $e = 1$

14. Directrix $y = 3$; $e = 2$

15. Directrix $x = 1$; $e = \dfrac{2}{3}$

16. Directrix $x = 5$; $e = 1$

Solve Exercises 17–19.

17. *Astronomy* The planet Mercury travels around the sun in an elliptical orbit given approximately by

$$r = \frac{3.442 \times 10^7}{1 - 0.206 \cos \theta}$$

where r is measured in miles and the sun is at the pole. Determine Mercury's greatest and shortest distance from the sun.

18. *Mechanical engineering* An engine gear is tested for dynamic balance by rotating it and tracing a polar graph of a point on its rim. Draw the graph given by the equation $r = \dfrac{4}{1 - 0.05 \sin \theta}$ and determine the state of balance of the gear.

19. *Mechanical engineering* A cam is shaped such that the equation of the upper half is given by $r = 2 + \cos \theta$ and the equation of the lower half is given by $r = \dfrac{3}{2 - \cos \theta}$. Sketch the shape of this cam.

☰ CHAPTER 15 REVIEW

Important Terms and Concepts

Angle of inclination
Circle
  Center
  Radius
Degenerate conics
Distance
Eccentricity
Ellipse
  Center
  Foci
  Major axis
  Minor axis
  Vertices
Hyperbola
  Asymptotes
  Center

Conjugate axis
Foci
Transverse axis
Vertices
Midpoint
Parabola
  Axis
  Directrix
  Foci
  Vertices
Polar equations for conic sections
Rotation of axes
Slope
Translation of axes
x-intercept
y-intercept

Review Exercises

For each pair of points in Exercises 1–4, find (a) the distance between them, (b) their midpoint, (c) the slope of the line through the points, and (d) the equation of the line through the points.

1. $(2, 5)$ and $(-1, 9)$

2. $(-2, -5)$ and $(10, -10)$

3. $(1, -4)$ and $(3, 6)$

4. $(2, -5)$ and $(-6, 3)$

5. For each line in Exercises 1–4, find the slope of one of its perpendiculars.

6. For each pair of points in Exercises 1–4, write the equation for the line passing through the midpoint of each pair and perpendicular to the line through them.

7. What is the equation of the line that passes through the point $(-3, 5)$ and is parallel to $2y + 4x = 9$?

8. What is the equation of the line through $(2, -7)$ with a slope of 4?

Sketch each of the conic sections in Exercises 9–28.

9. $x^2 + y^2 = 16$

10. $x^2 - y^2 = 16$

11. $x^2 + 4y^2 = 16$

12. $y^2 = 16x$

13. $(x - 2)^2 + (y + 4)^2 = 16$

14. $(x - 2)^2 - (y + 4)^2 = 16$

15. $(x - 2)^2 + 4(y + 4)^2 = 16$

16. $(y + 4)^2 = 16(x - 2)$

17. $x^2 + y^2 + 6x - 10y + 18 = 0$

18. $x^2 - 4y^2 + 6x + 40y - 107 = 0$

19. $x^2 + 4y^2 + 6x - 40y + 93 = 0$

20. $x^2 + 6x - 4y + 29 = 0$

21. $2x^2 + 12xy - 3y^2 - 42 = 0$

22. $5x^2 - 4xy + 8y^2 - 36 = 0$

23. $3x^2 + 2\sqrt{3}xy + y^2 + 8x - 8\sqrt{3}y = 32$

24. $2x^2 - 4xy - y^2 = 6$

25. $r = \dfrac{16}{5 - 3\cos\theta}$

26. $r = \dfrac{9}{3 - 5\cos\theta}$

27. $r = \dfrac{9}{3 + 3\sin\theta}$

28. $r = \dfrac{2}{1 + \cos\theta}$

Solve Exercises 29–32.

29. *Civil engineering* A concrete bridge for a highway overpass is constructed in the shape of half an elliptic arch, as shown in Figure 15.54. The arch is 100 m long. In order to have enough clearance for tall vehicles, the arch must be 6.1 m high at a point 5 m from the end of the arch. What is the equation of the ellipse for this arch, if the origin is at the midpoint of the major axis?

30. *Civil engineering* The main cables on a suspension bridge approximate a parabolic shape. The twin towers of a suspension bridge are to be 120 m above the road surface and are 400 m apart. If the cable's lowest point is 10 m above the road surface, what is the equation of the parabola for the main cable?

31. *Navigation* An airplane sends out an impulse that travels at the speed of sound (320 m/μs). The plane is 50 km south of the line connecting two receiving stations. The stations are on an east-west line with station A 400 km west of station B. Station A receives the signal from the plane 500 μs after station B. What is the location of the plane?

32. *Astronomy* A map of the solar system is drawn so that the surface of earth is represented by the equation $x^2 + y^2 - 2x + 4y - 6{,}361 = 0$. A satellite orbits earth in a circular orbit 0.8 units above earth. What is the equation of the satellite's orbit on this map?

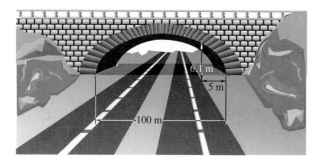

FIGURE 15.54

CHAPTER 15 TEST

1. Find the focus and the directrix of the parabola $y^2 = -18x$, and sketch the graph.

2. Graph the hyperbola $9x^2 - 4y^2 = 36$. Specify the foci, vertices, and endpoints of the conjugate axis.

3. Determine the distance between the points $(-2, 4)$ and $(5, -6)$.

4. What is the equation of the line that is perpendicular to the line through the points $(4, -5)$ and $(2, -8)$ and passes through their midpoint?

5. The x-intercept of a line is 3, and its angle of inclination is $30°$. Write the equation of the line in slope-intercept form.

6. **(a)** Write the equation $3x^2 + 4y^2 - 6x + 16y + 7 = 0$ in the appropriate standard form. **(b)** Determine the type of conic section described by this equation. **(c)** Determine the coordinates of all "significant" points, such as vertices and foci. **(d)** Graph the equation.

7. **(a)** Determine the angle needed to rotate the coordinate axes so that the transformed equation of $x^2 + xy + y^2 = 4$ has no xy-term. **(b)** Identify the graph.

8. Identify and sketch the graph of the polar equation $r = \dfrac{2}{1 + 4\cos\theta}$.

9. A cable hangs in a parabolic curve between two vertical supports that are 120 ft apart. At a distance 48 ft in from each support, the cable is 3.0 ft above its lowest point. How high up is the cable attached on each support?

Systems of Equations and Inequalities

Every week, an electronics company manufactures CD players and television sets. In Section 16.5, we will learn how to analyze variables, such as cost of materials and the profit on each manufactured item, to determine at which point the company will make the most profit.

In Chapter 5, we solved systems of linear equations using four methods: graphing, elimination by substitution, elimination by addition and subtraction, and Cramer's rule. In this chapter, we will use some of these same methods to solve systems, where one or both equations are of second degree.

We will also begin to explore inequalities. Many problems in technology, industry, business, science, engineering, and mathematics require the use of inequalities. Some of the same techniques we used on equations will be used to solve inequalities. Finally, we will study linear programming and use this powerful technique in applied situations.

16.1
SOLUTIONS OF NONLINEAR SYSTEMS OF EQUATIONS

We will look at two types of nonlinear systems of equations in this section. A **nonlinear system** is a system of equations in which one or more of the equations is not linear. The first nonlinear system we will look at contains one linear and one quadratic equation. Next, we will look at a nonlinear system of two quadratic equations. With each type, we will begin with a graphical method of solution and then consider algebraic techniques.

For our first system of nonlinear equations, we will use the following:

$$\begin{cases} x^2 + y^2 = 25 \\ x - y = 1 \end{cases}$$

Graphically, we can see in Figure 16.1 that these two curves seem to intersect at the points $(4,\ 3)$ and $(-3,\ -4)$. In the next example, we will see that most graphical solutions are not solved as accurately as this one.

We will now examine how these equations can be solved algebraically using elimination by substitution. When you have a nonlinear system of equations that contains one linear equation, first solve the linear equation for one of the variables and substitute this solution into the other equation.

EXAMPLE 16.1

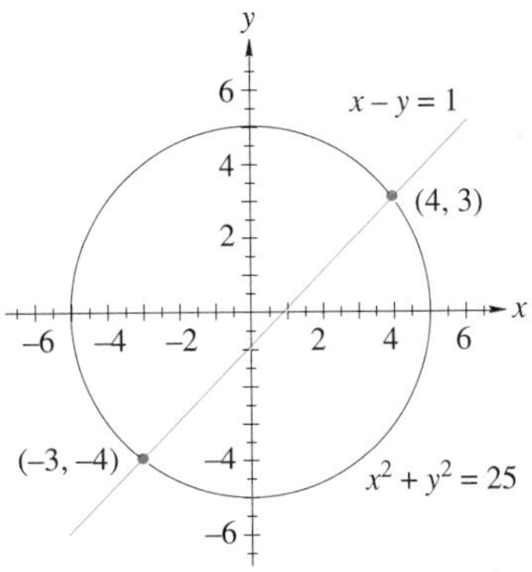

Use the substitution method to solve the system

$$\begin{cases} x^2 + y^2 = 25 & (1) \\ x - y = 1 & (2) \end{cases}$$

Solution Solve equation (2) for one of the variables. If you solve equation (2) for x, you get

$$x = y + 1 \qquad (3)$$

Substitute the value of x from equation (3) into equation (1) and then solve for y.

$$(y + 1)^2 + y^2 = 25$$

$$y^2 + 2y + 1 + y^2 = 25$$

$$2y^2 + 2y = 24$$

$$y^2 + y - 12 = 0$$

$$(y + 4)(y - 3) = 0$$

So, $\qquad y = -4 \qquad$ or $\qquad y = 3$

Substituting these values for y in equation (3), we get the corresponding values for x.

$$\text{If } y = -4, \text{ then } x = -4 + 1$$

$$x = -3$$

$$\text{and if } y = 3, \text{ then } x = 3 + 1$$

$$x = 4$$

FIGURE 16.1

The graph shows the line $x - y = 1$, the circle $x^2 + y^2 = 25$, and the intersection points $(4, 3)$ and $(-3, -4)$.

EXAMPLE 16.1 (Cont.)

The solutions are the points $(-3, -4)$ and $(4, 3)$. These are the same solutions we got when we graphed this system of equations.

Now, let's try this technique on another system of nonlinear equations that contains one linear equation.

EXAMPLE 16.2

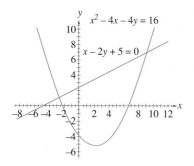

FIGURE 16.2

Solve.

$$\begin{cases} x^2 - 4x - 4y = 16 & (1) \\ x - 2y + 5 = 0 & (2) \end{cases}$$

Solution The graphs of these two curves are shown in Figure 16.2. We need to determine where they intersect.

We solve equation (2) for x, and obtain

$$x = 2y - 5$$

Substituting this value of x into equation (1), we get

$$(2y - 5)^2 - 4(2y - 5) - 4y = 16$$
$$4y^2 - 20y + 25 - 8y + 20 - 4y = 16$$
$$4y^2 - 32y + 29 = 0$$

Using the quadratic formula, $y = \dfrac{-b \pm \sqrt{b^2 - 4ac}}{2a}$, we find that

$$y = \frac{32 \pm \sqrt{32^2 - 4(4)(29)}}{2(4)}$$

$$= \frac{32 \pm \sqrt{1024 - 464}}{8}$$

$$= \frac{32 \pm \sqrt{560}}{8}$$

$$\approx \frac{32 \pm 23.66}{8}$$

So, $y \approx \dfrac{32 + 23.66}{8} \approx 6.96$

and $y \approx \dfrac{32 - 23.66}{8} \approx 1.04$

Substituting these values of y into equation (2) we get

$$x - 2(6.96) + 5 = 0 \quad \text{or} \quad x = 8.92$$

and $x - 2(1.04) + 5 = 0 \quad \text{or} \quad x = -2.92$

The approximate solutions are $(-2.92, 1.04)$ and $(8.92, 6.96)$.

When neither of the equations is a linear equation, you may then use either of the elimination methods: substitution, or addition and subtraction.

EXAMPLE 16.3

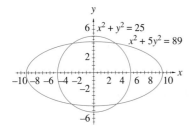

FIGURE 16.3

Solve.

$$\begin{cases} x^2 + 5y^2 = 89 & (1) \\ x^2 + y^2 = 25 & (2) \end{cases}$$

Solution The graphical solution of these two equations is shown in Figure 16.3.

Algebraically, we will use elimination by addition and subtraction. Since both equations have an x^2 term, we will subtract equation (2) from equation (1) to get

$$4y^2 = 64 \qquad (3)$$

$$y^2 = 16$$

$$y = \pm 4$$

Since $y = \pm 4$, we will substitute these values in equation (2) to determine x. For both substitutions we get

$$x^2 + 16 = 25$$

$$x^2 = 9$$

$$x = \pm 3$$

This gives a total of four solutions: $(3, 4)$, $(3, -4)$, $(-3, 4)$, and $(-3, -4)$, and these appear to be the four points in Figure 16.3 where the two curves intersect. ▪

In Example 16.3, a circle and an ellipse intersected in four points. It is also possible that a circle and an ellipse would not intersect at any points, at one point, at two points, or at three points.

EXAMPLE 16.4

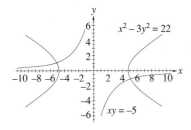

FIGURE 16.4

Solve.

$$\begin{cases} x^2 - 3y^2 = 22 & (1) \\ xy = -5 & (2) \end{cases}$$

Solution The graphical solution is shown in Figure 16.4.

Algebraically, we will use elimination by substitution. We will solve equation (2) for y and substitute this value into equation (1).

$$y = \frac{-5}{x} \qquad (3)$$

$$x^2 - 3\left(\frac{-5}{x}\right)^2 = 22$$

$$x^2 - \frac{75}{x^2} = 22$$

EXAMPLE 16.4 (Cont.)

Multiplying both sides of the equation by x^2, we get

$$x^4 - 75 = 22x^2$$

or $$x^4 - 22x^2 - 75 = 0$$

which factors into

$$(x^2 - 25)(x^2 + 3) = 0$$

Now, $x^2 - 25 = 0$ means that $x = \pm 5$ and $x^2 + 3 = 0$ means that $x = \pm j\sqrt{3}$. Substituting these values of x into equation (3), we get $y = \pm 1$ (when $x = \pm 5$) and $y = \pm \dfrac{5}{j\sqrt{3}} = \pm \dfrac{5}{3}j\sqrt{3}$ (when $x = \pm j\sqrt{3}$).

The two real solutions are $(5, -1)$ and $(-5, 1)$. These are shown on the graph in Figure 16.4. There are two imaginary roots $\left(j\sqrt{3}, \frac{5}{3}j\sqrt{3}\right)$ and $\left(-j\sqrt{3}, -\frac{5}{3}j\sqrt{3}\right)$. These two solutions are not on the graph, because this graph is in the real number plane and not the complex plane.

In our examples, all the real roots were at integer values. However, it would not be unexpected if the solutions were any real (or complex) number, as shown in the next example.

Application

EXAMPLE 16.5

Two satellites are in orbit in the same plane. One orbit is described by $5(x + 10)^2 + 3y^2 = 3\,000$. The other is described by $5x^2 + 4y^2 = 4\,000$. What are the points where the two orbits intersect?

Solution   We want to solve the system that contains the equations of two ellipses.

$$\begin{cases} 5(x + 10)^2 + 3y^2 = 3\,000 & (1) \\ 5x^2 + 4y^2 = 4\,000 & (2) \end{cases}$$

We will multiply equation (1) by 4 and equation (2) by 3. This will give the y^2-terms the coefficient of 12.

$$\begin{cases} 20(x + 10)^2 + 12y^2 = 12\,000 \\ 15x^2 + 12y^2 = 12\,000 \end{cases}$$

Subtracting, we get

$$20(x + 10)^2 - 15x^2 = 0$$

or $$5x^2 + 400x + 2\,000 = 0$$

$$x^2 + 80x + 400 = 0$$

EXAMPLE 16.5 (Cont.)

Using the quadratic formula we get

$$x = \frac{-80 \pm \sqrt{80^2 - 4(400)}}{2} = \frac{-80 \pm \sqrt{4800}}{2}$$

$$= -40 \pm 20\sqrt{3}$$

Substituting $x = -40 + 20\sqrt{3} \approx -5.36$ into equation (2), we get

$$143.65 + 4y^2 = 4\,000$$

$$y^2 = 964.09$$

$$y \approx \pm 31.05$$

Substituting $x = -40 - 20\sqrt{3} \approx -74.64$ into equation (2), we get

$$27\,855.65 + 4y^2 = 4\,000$$

$$y^2 \approx -5\,963.91$$

Since $y^2 < 0$, these roots are imaginary.

The satellites will intersect at two points: $(-5.36, 31.05)$ and $(-5.36, -31.05)$.

Exercise Set 16.1

In Exercises 1–10, graph the systems of equations and then solve each system algebraically for x and y. Be sure to include real and complex roots.

1. $\begin{cases} x - 2y = 5 \\ x^2 - 4y^2 = 45 \end{cases}$

2. $\begin{cases} x^2 + y^2 = 13 \\ 2x - y = 4 \end{cases}$

3. $\begin{cases} x^2 + 4y^2 = 32 \\ x + 2y = 0 \end{cases}$

4. $\begin{cases} x - y = 4 \\ x^2 - y^2 = 32 \end{cases}$

5. $\begin{cases} x^2 + y^2 = 4 \\ x^2 - 2y = 1 \end{cases}$

6. $\begin{cases} x^2 - y^2 = 4 \\ 2x - 3y = 10 \end{cases}$

7. $\begin{cases} x^2 + y^2 = 7 \\ y^2 = 6x \end{cases}$

8. $\begin{cases} x^2 - y^2 + 6 = 0 \\ y^2 = 5x \end{cases}$

9. $\begin{cases} xy = 3 \\ 2x^2 - 3y^2 = 15 \end{cases}$

10. $\begin{cases} y = x^2 - 4 \\ x^2 + 3y^2 + 4y - 6 = 0 \end{cases}$

Solve Exercises 11–18.

11. *Transportation* A truck driver travels the first 100 mi of a 120-mi trip in light traffic. For the last 20 mi, traffic is heavy enough that the average speed is reduced by 10 mph. The total time for the trip is 3 h. What was the average speed for each part of the trip?

12. *Agricultural technology*   The perimeter of a rectangular field is 160 m and the area is 1 200 m². What are the dimensions of the field?

13. *Electricity*   In a direct current (dc) circuit, the equivalent resistance R of two resistors R_1 and R_2 connected in series is $R = R_1 + R_2$. If the resistors are connected in parallel, then $R = \dfrac{R_1 R_2}{R_1 + R_2}$. When two resistors are connected in series, the equivalent resistance is 100 Ω. When they are connected in parallel, $R = 24$ Ω. Find R_1 and R_2.

14. *Electricity*   When two resistors are connected in series, the equivalent resistance is 10 Ω. When they are connected in parallel, the equivalent resistance is 2.1 Ω. What are each of the resistances? (See Exercise 13.)

15. *Physics*   When a 40-kg object traveling at a velocity of v_1 m/s collides with a 60-kg object on a frictionless surface traveling in the same direction at v_2 m/s, the final total momentum is 450 kg · m/s and is given by the formula

$$40v_1 + 60v_2 = 450$$

The relationship between the initial and final kinetic energy ($KE = \frac{1}{2}mv^2$) is given by the equation

$$20v_1{}^2 + 30v_2{}^2 = 1\,087.5 \text{ J}$$

where J stands for joule. Determine the velocity of each object.

16. *Space technology*   Two satellites are in orbits in the same plane. One orbit is described by $20x^2 + 40y^2 = 8\,000$. The other is described by $30x^2 + 20y^2 = 6\,000$. What are the points where these two orbits intersect?

17. *Physics*   A 200-kg object 0.8 m from the fulcrum of a lever can be lifted by a certain minimum force at the other end of the lever. If that same force is applied 1 m closer to the fulcrum, an additional 50 kg are needed to lift the object. If the minimum force is F and its original distance from the fulcrum is ℓ, then we know that

$$F\ell = (200)(0.8)$$

and $(F + 50)(\ell - 1) = (200)(0.8)$

Find F and ℓ.

18. *Physics*   The vertical distance in meters that an object falls is given by $y = 15t - 4.9t^2$ m. When the horizontal distance equals twice the vertical distance in meters then $y = 10t$. What values of y and t satisfy these conditions?

≡ 16.2

PROPERTIES OF INEQUALITIES; LINEAR INEQUALITIES

In Section 1.1, we introduced the inequality symbols. Since that time, we have not used them except to indicate when conditions were being placed on a group of numbers. The fact that we have not used the inequality symbols should not indicate that they are seldom used. In fact, in some parts of mathematics, inequalities are used as often as equations. In the remainder of this chapter, we will focus on inequalities.

An **inequality** is formed whenever two expressions are separated by one of the inequality symbols: $>$, $<$, $\geq$, and $\leq$. The two expressions are called the **sides**, or **members** of the inequality. A **compound inequality** has more than two expressions separated by inequality symbols.

EXAMPLE 16.6

$$2x < 5,$$

$$3x + 7 \geq 8,$$

$$16x - 8 \leq 5x - 2,$$

and $5x^2 - 3x - 2 > x - 7$

EXAMPLE 16.6 (Cont.)

are all inequalities.

$$3x < 4x - 5 < 9,$$

$$x < y \leq 2z,$$

$$-8 \leq 3x + 1 \leq 4,$$

and $\quad 1 > 2x - 3 \geq 17$

are all compound inequalities.

A compound inequality can always be written as a combination of simple inequalities.

EXAMPLE 16.7

The compound inequality $3x < 4x - 5 < 9$ is the same as the two simple inequalities $3x < 4x - 5$ and $4x - 5 < 9$.

$-8 \leq 3x + 1 \leq 4$ is the same as the two simple inequalities $-8 \leq 3x + 1$ and $3x + 1 \leq 4$.

There are three types of inequalities. Most of the inequalities we will use are conditional inequalities. A **conditional inequality** is true for some real numbers and false for others. For example, $x \geq 5$ is a conditional inequality. It is true if x is $5, 6\frac{1}{2}$, 19, or any other number larger than 5. On the other hand, it is false for values of x such as $4, -3, 2\frac{1}{2}$, or any number smaller than 5.

An **absolute inequality** is true for all real numbers. An example of an absolute inequality is $x + 1 > x$. No matter which real number you select, this is a true statement. The opposite of an absolute inequality is a contradictory inequality. A **contradictory inequality** is false for all real numbers. For example $x + 1 < x$ is a contradictory inequality, because it is never true.

The **solution** of an inequality consists of those values of the variable that make the inequality true. To **solve an inequality** is to determine the solutions of the inequality.

Properties of Inequalities

There are six basic properties or rules of inequalities that are used to solve an inequality. In all of the explanations and examples of these properties, we will use the $<$ symbol. We could just as easily have used any of the other three inequality symbols.

Property 1 of Inequalities

If a, b, and c are real numbers, with $a < b$, then $a + c < b + c$.

In words, this property says that the same algebraic expression can be added or subtracted to both sides of an inequality without changing the direction of the inequality symbol.

EXAMPLE 16.8

(a) $x + 3 < 7$ is equivalent to $x < 4$.   Subtract 3 from both sides.

(b) $3x + 4 < 6 - 2x$ is equivalent to $5x < 2$.   Add $2x$ to both sides, and subtract 4 from both sides.

Property 2 of Inequalities

If a, b, and c are real numbers, with $a < b$ and $c > 0$, then $ac < bc$ or $\dfrac{a}{c} < \dfrac{b}{c}$

In words, this states that you can multiply or divide both sides of an inequality by the same *positive* expression without changing the direction of the inequality symbol.

EXAMPLE 16.9

(a) $\dfrac{x}{3} - \dfrac{4}{3} < 2x + 1$ is equivalent to $x - 4 < 6x + 3$.   Multiply both sides by 3.

(b) $3x < 21$ is equivalent to $x < 7$.   Divide both sides by 3.

Property 3 of Inequalities

If a, b, and c are real numbers, with $a < b$ and $c < 0$, then $ac > bc$ or $\dfrac{a}{c} > \dfrac{b}{c}$

≡ Note   If you multiply or divide both sides of an inequality by the same *negative* expression, then the direction of the inequality symbol is reversed.

EXAMPLE 16.10

(a) $-3 < 7$ is equivalent to $6 > -14$.   Multiply both sides by -2.

(b) $\dfrac{-x}{3} < 5$ is equivalent to $x > -15$.   Multiply both sides by -3.

(c) $-4x < 20$ is equivalent to $x > -5$.   Divide both sides by -4.

Property 4 of Inequalities

If a, b, and n are positive real numbers and $a < b$, then $a^n < b^n$ and $\sqrt[n]{a} < \sqrt[n]{b}$

If both sides of an inequality are positive and n is a positive number, then the nth power or root of both sides retains the direction of the inequality.

EXAMPLE 16.11

(a) If $3 < 5$ and $n = 6$, then $3^6 < 5^6$.

(b) If $2 < 10$ and $n = 8$, then $\sqrt[8]{2} < \sqrt[8]{10}$.

Property 5 of Inequalities

If x and a are real numbers, $a > 0$, and $|x| < a$, then $-a < x < a$

EXAMPLE 16.12

(a) $|x + 3| < 5$ is equivalent to $-5 < x + 3 < 5$.

(b) $|2x - 7| < 10$ is equivalent to $-10 < 2x - 7 < 10$.

Property 6 of Inequalities

If x and a are real numbers, $a > 0$, and $|x| > a$, then $x > a$ or $x < -a$

EXAMPLE 16.13

(a) $|x - 5| > 2$ is equivalent to $x - 5 > 2$ or $x - 5 < -2$.

(b) $|7x + 4| > 9$ is equivalent to $7x + 4 > 9$ or $7x + 4 < -9$.

Solving Linear Inequalities

We can use these six properties to help us solve problems that use inequalities. As is true when you solve an equation, you often use several of the properties in each problem. Each of the problems is an example of a linear inequality.

Linear Inequality

A **linear inequality** in one variable, x, is an inequality of the form $ax + b < 0$, where a and b are constants and $a \neq 0$.

First, we will solve each of the linear inequalities in the following examples algebraically and then graph the solution. Since these are inequalities in one variable, each solution will be graphed on a number line.

EXAMPLE 16.14

Solve $6x + 3 \geq 21$.

Solution

$$6x + 3 \geq 21$$

$$6x \geq 18 \qquad \text{Subtract 3 from both sides.}$$

$$x \geq 3 \qquad \text{Divide both sides by 6.}$$

EXAMPLE 16.14 (Cont.)

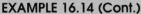
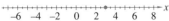

FIGURE 16.5

The solution to $6x + 3 \geq 21$ is all numbers greater than or equal to 3. This is shown by the shaded section of the number line in Figure 16.5. The solid circle at 3 indicates that 3 is included in the solution.

EXAMPLE 16.15

Solve $\frac{3}{4} - \frac{5}{6}x > \frac{-1}{2}x - \frac{11}{12}$.

Solution

$$\frac{3}{4} - \frac{5}{6}x > \frac{-1}{2}x - \frac{11}{12}$$

$9 - 10x > -6x - 11$ Multiply by 12, the LCD of the denominators.

$9 - 4x > -11$ Add $6x$ to both sides.

$-4x > -20$ Subtract 9 from both sides.

$x < 5$ Divide both sides by -4 and reverse the inequality sign.

FIGURE 16.6

The solution is $x < 5$. This is shown graphically in Figure 16.6. The open circle at 5 indicates that the solution does not include 5.

EXAMPLE 16.16

Solve $|3x - 5| < 13$.

Solution

$$|3x - 5| < 13$$

$-13 < 3x - 5 < 13$ Apply Property 5.

$-8 < 3x < 18$ Add 5 to all three expressions.

$-\frac{8}{3} < x < 6$ Divide by 3.

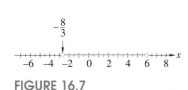

FIGURE 16.7

This solution is shown graphically in Figure 16.7. Notice that $-\frac{8}{3} < x < 6$ is equivalent to $-\frac{8}{3} < x$ and $x < 6$.

EXAMPLE 16.17

Solve $|3x + 2| \geq 8$.

Solution

$$|3x + 2| \geq 8$$

$3x + 2 \geq 8$ or $3x + 2 \leq -8$ Apply Property 6.

$3x \geq 6$ or $3x \leq -10$ Subtract 2.

$x \geq 2$ or $x \leq -\frac{10}{3}$ Divide by 3.

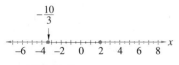

FIGURE 16.8

The solution is the numbers less than or equal to $-\frac{10}{3}$ or those greater than or equal to 2. This is shown graphically in Figure 16.8.

Application

EXAMPLE 16.18

A technician determines that an electronic circuit fails to operate because the resistance between A and B, 600 Ω, exceeds the specifications. See Figure 16.9. The specifications state that the resistance must be between 200 Ω and 500 Ω. If a shunt resistor of R Ω, $R > 0$, is added, the circuit will satisfy the specifications. The equivalent resistance for the parallel connection is $\dfrac{1}{\dfrac{1}{600} + \dfrac{1}{R}}$ or $\dfrac{600R}{600 + R}$ Ω. What are the possible values for R?

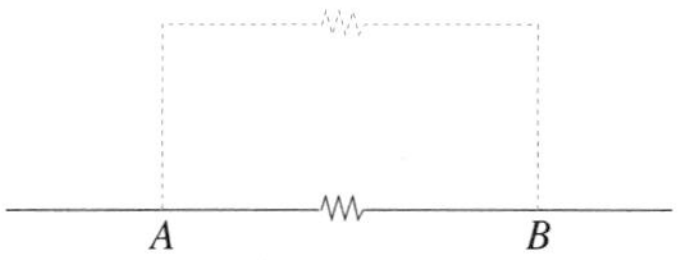

Solution   To satisfy the specifications, the following inequality must be true.

$$200 < \frac{600R}{600 + R} < 500$$

Since $R > 0$, $600 + R > 0$, and we can multiply the inequality by $600 + R$ without changing the direction of the inequality sign.

$$200(600 + R) < 600R < 500(600 + R)$$

$$120\,000 + 200R < 600R < 300\,000 + 500R$$

We will now write this compound inequality as two simple inequalities:

$120\,000 + 200R < 600R$	and	$600R < 300\,000 + 500R$
$120\,000 < 400R$	and	$100R < 300\,000$
$300 < R$	and	$R < 3\,000$

Thus, $300 < R < 3\,000$ and the shunt resistance must be between 300 Ω and 3 000 Ω.

FIGURE 16.9

Exercise Set 16.2

In Exercises 1–8, using the inequality $x < 10$, state the inequality that results when the operations given are performed on each side. Assume $x > 0$.

1. Add 5.

2. Subtract 7.

3. Multiply by 3.

4. Multiply by -4.

5. Divide by -2.

6. Divide by 5.

7. Square both.

8. Take the square root of both.

In Exercises 9–32, solve each of the given inequalities algebraically and graphically.

9. $3x > 6$

10. $4x < -8$

11. $2x + 5 > -7$

12. $3x - 4 \leq 8$

13. $2x - 5 < x$

14. $3x + 4 \geq x$

15. $4x - 7 \geq 2x + 5$

16. $7 - 3x \leq 3 + 5x$

17. $\frac{2}{3}x - 4 < \frac{1}{3}x + 2$

18. $\frac{1}{5}x - \frac{2}{5} > \frac{3}{5}x + \frac{4}{5}$

19. $\frac{x-2}{4} \leq \frac{3}{8}$

20. $\frac{x+2}{3} \geq \frac{5}{6}$

21. $\frac{x-3}{4} \leq \frac{2x}{3}$

22. $\frac{2x+5}{3} > \frac{3x-1}{2}$

23. $|x + 1| < 5$

24. $|2x - 1| < 7$

25. $|x + 4| > 6$

26. $|3x + 9| \geq 6$

27. $|x + 5| < -3$

28. $|3x + 2| < 12$

29. $-7 \leq 3x + 5 < 26$

30. $-6 < 5 - 3x \leq 17$

31. $3x + 1 < 5 < 2x - 3$

32. $4x - 6 \leq 11 < 9x + 1$

For Exercises 33–40, express the answer in terms of an inequality.

33. *Astronomy* The radius of the Earth at the equator is 6 378 km and the polar radius is 6 357 km. What is the range of the radius r?

34. *Astronomy* The moon has a maximum distance of 407 000 km from the Earth and a minimum distance of 357 000 km. What is the range of the moon's distance, d, from the Earth?

35. *Automotive technology* The company specifications state that the camber angle c of the wheels should be $+0.60° \pm 0.50°$. Express the range for c as an inequality.

36. *Automotive technology* The amount of R-12 in an automotive air-conditioning system may vary from 0.9 kg to 1.8 kg (2 to 4 lb). Knowing that there are 16 oz/lb, express the range of R-12 in ounces.

37. *Electricity* If the resistance between A and B in Example 16.18 had been 800 Ω, what are the possible values for the shunt resistance?

38. *Machine technology* A certain welding operation must be performed at temperatures between 1 800° and 2 200°C. What is the temperature range in degrees Fahrenheit? [Use $C = \frac{5}{9}(F - 32)$.]

39. *Business* The weekly cost of manufacturing x microcomputers of a certain type is given by $C = 1{,}500 + 25x$. The revenue from selling these is given by $R = 60x$. At least how many microcomputers must be made and sold each week to produce a profit?

40. *Energy technology* A rectangular solar collector is to have a height of 1.5 m. The collector will supply 600 W per m^2 and is to supply a total between 2 400 and 4 000 W. What range of values can the length of the collector have in order to provide this voltage?

☰ 16.3
NONLINEAR INEQUALITIES

In this section, we will learn how to solve nonlinear inequalities in one variable. There are two generally accepted methods for solving these kinds of inequalities. We will briefly show you one of those methods and then concentrate on the other method. You can use the one that seems easier.

Let's consider the inequality $x^2 + x - 6 > 0$. This is not linear, since it has an x^2-term. It is an inequality in one variable, x.

The first method for solving this inequality is to graph the equation $y = f(x) = x^2 + x - 6$, as shown in Figure 16.10. Look at the graph. When is $y > 0$? That is, when is the graph of $y = x^2 + x - 6$ above the x-axis? You can see that $y > 0$ when $x < -3$ or $x > 2$. Thus, the solution of $x^2 + x - 6 > 0$ is $x < -3$ or $x > 2$. This is a fairly easy approach. The fact that you have to graph the function requires time. While a graphing calculator can speed up the graphing, it is not always easy to determine

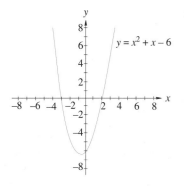

FIGURE 16.10

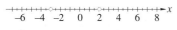

FIGURE 16.11a

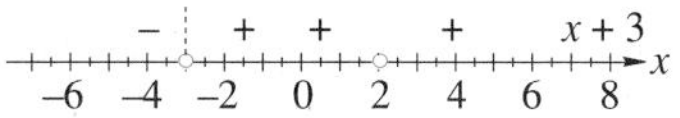

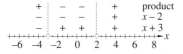

FIGURE 16.11b

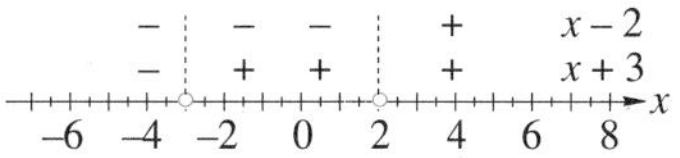

FIGURE 16.11c

FIGURE 16.11d

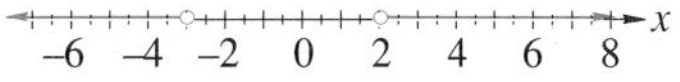

FIGURE 16.11e

exactly where the graph crosses the x-axis. We need to find a faster and more accurate method.

We will use the same inequality to demonstrate the second method. We begin by finding the roots of the corresponding equation $x^2 + x - 6 = 0$. Since $x^2 + x - 6 = (x + 3)(x - 2) = 0$, the roots are -3 and 2. Draw a number line and mark the roots on the number line, as in Figure 16.11a.

At one of these roots, put an open circle on top of the mark for that root. For example, in Figure 16.11a we have put an open circle on top of the mark for -3. Draw a dashed line above the circle at -3. The line is dashed because -3 is not a solution of this inequality. On the right side of this circle $x + 3 > 0$, and on the other side $x + 3 < 0$. Put a row of $+$ signs on the side where $x + 3$ is positive and a row of $-$ signs on the other side. Write $x + 3$ at one end to remind you which factor is indicated by that row. Your number line should now look like the one in Figure 16.11b.

At the second root, put an open circle at the mark for the root at 2 and draw a dashed line upward from this circle. Indicate the side where $x - 2$ is positive and the side where $x - 2$ is negative, as in Figure 16.11c.

We need one more row of signs to complete the process. The last row is the product row. The original inequality is $x^2 + x - 6 > 0$ or $(x+3)(x-2) > 0$. The product row will indicate the sign of $(x + 3)(x - 2)$. Remember, whenever you multiply an even number of negative numbers together, you get a positive answer. If you multiply an odd number of negative numbers, you get a negative number.

In Figure 16.11d, to the right of 2 both the $x - 2$ and the $x + 3$ rows are positive, so their product is positive. We have put $+$ signs in the product row to the right of the dashed line above the 2. The next interval is the section between -3 and 2. Here $x - 2$ is negative, and $x + 3$ is positive, so the product is negative. Thus, we have put $-$ signs in the product row between the dashed lines above -3 and 2. The last interval is to the left of -3 or the numbers less than -3. In this interval both factors are negative, so the product is positive. This is indicated by the $+$ signs in the product row to the left of the dashed line above the -3, as shown in Figure 16.11d.

By inspecting the product row in Figure 16.11d, we can see that the solution is $x < -3$ or $x > 2$. This is shown graphically in Figure 16.11e. This is the same answer we got using the other method. This seems like a longer and more complicated method, but it is actually easier. If you are solving an inequality that has the sign $\leq$ or $\geq$, you would fill in the circles that mark the roots and draw solid lines instead of dashed lines above the circles.

EXAMPLE 16.19

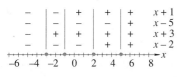

FIGURE 16.12a

Solve $(x - 2)(x + 3)(x - 5)(x + 1) \leq 0$.

Solution This is already factored. The roots are $2, -3, 5,$ and -1. The sign rows for each of these factors are given in Figure 16.12a. The product row is given in Figure 16.12b. Because this inequality has the sign $\leq$, we have filled in the circles to indicate that the roots are included. From inspection, we see that the product row is negative when $-3 \leq x \leq -1$ or $2 \leq x \leq 5$. This is the solution, and it is shown graphically in Figure 16.12c.

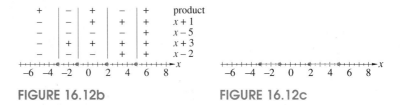

FIGURE 16.12b   FIGURE 16.12c

Do not be misled into thinking that the signs in each row are always positive to the right of the root. You need to solve each linear inequality to determine the signs, as the next example demonstrates.

EXAMPLE 16.20

Solve: $x(x - 2)(5 - x) > 0$.

Solution The roots are 0, 2, and 5. The sign row for each root is given in Figure 16.13a. The circles are not filled in because the inequality sign does not include an equal sign. Also, notice the sign row for $5 - x$. The product in Figure 16.13b indicates that the solution is $x < 0$ or $2 < x < 5$. Figure 16.13c shows the answer graphically.

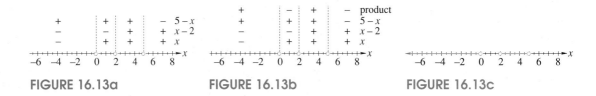

FIGURE 16.13a   FIGURE 16.13b   FIGURE 16.13c

The procedure for finding the solutions to a division problem can be determined the same way. You must check to ensure that any numbers that make the quotient zero are not in the solutions.

EXAMPLE 16.21

Solve: $\dfrac{x^2 + 6x + 5}{x + 2} \leq 0$.

Solution The numerator factors into $(x + 5)(x + 1)$ and so the numerator is zero at -5 and -1; the denominator is zero at $x = -2$. These three points are indicated in Figure 16.14a with the sign rows for $x + 5$, $x + 1$, and $x + 2$. Notice that the circles are filled at -5 and -1 because the inequality includes an equal sign. The circle at -2 is not filled in because the denominator is zero at -2. Thus, -2 is not a solution. If we remember that $\dfrac{x^2 + 6x + 5}{x + 2} = (x + 5)(x + 1)\left(\dfrac{1}{x + 2}\right)$ and realize that $x + 2$ and $\dfrac{1}{x + 2}$ have the same signs, then the signs of the answer are given by a product row. The product row is shown in Figure 16.14b and the graphical solution is given in Figure 16.14c. From this, we see that the solution is $x \leq -5$ or $-2 < x \leq -1$.

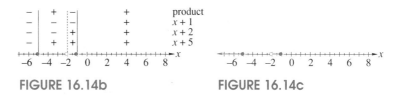

FIGURE 16.14b   FIGURE 16.14c

Application

EXAMPLE 16.22

From the top of a 120-ft building an object is thrown upward with an initial velocity of 68 ft/s. Its distance d above the ground at any time t is given by the equation $d = 120 + 68t - 16t^2$. For what period of time is the object higher than the building?

Solution   Since the building is 120 ft high, this is really asking when

$$120 + 68t - 16t^2 > 120$$

or $$-16t^2 + 68t > 0$$

This factors into $-16t(t - 4.25) > 0$. This inequality is zero at $t = 0$ and $t = 4.25$. As shown in Figure 16.15, the sign row for $-16t$ is negative for $t > 0$ and positive for $t < 0$. According to the product row, the object is above the building for the first 4.25 s after it is thrown.

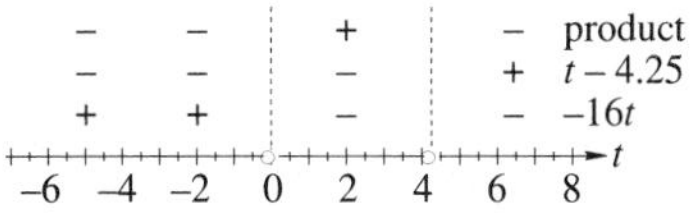

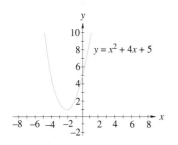

FIGURE 16.15

In each of these examples, we were concerned with solutions in the real numbers. If the nonlinear equation does not have any real roots, the problem will be an absolute inequality and every real number will satisfy the inequality or it will be a contradictory inequality and no real number will satisfy the inequality. The next example illustrates one of these situations.

EXAMPLE 16.23

Solve: $x^2 + 4x + 5 > 0$.

Solution   Using the quadratic formula, we see that the roots are $\dfrac{-4 \pm \sqrt{16 - 20}}{2} = -2 \pm j$. Both of these are complex numbers. Since there are only two roots to $x^2 + 4x + 5 = 0$, and neither of them is a real number, this inequality is either absolute or contradictory. If it is absolute, any real number will satisfy it. Select any real number, say 0 and we get $0^2 + 4(0) + 5 = 5 > 0$. So, 0 is a solution. This is an absolute inequality and all real numbers satisfy it.

The graph of $y = x^2 + 4x + 5$ is shown in Figure 16.16. From this graph, you can see that y is always larger than 0. This confirms our solution.

FIGURE 16.16

EXAMPLE 16.24

Solve: $x^2 + 4x + 5 < 0$.

Solution   Check the graph of $y = x^2 + 4x + 5$ in Figure 16.16. As we saw in the previous example, y is always positive. This is a contradictory inequality and no real numbers will make it true.

We could have also solved this in the same manner as described in Example 16.23. Select a real number, say 2. Since $2^2 + 4(2) + 5 = 17$, this inequality has no real number solutions.

The techniques described in Examples 16.23 and 16.24 depend on first determining that there are no real roots. At present, we can only do this for quadratics. In Chapter 18, we will learn how to determine the roots of higher degree polynomials.

Exercise Set 16.3

Solve the inequalities in Exercises 1–40.

1. $(x + 1)(x - 3) > 0$
2. $(x - 4)(x + 2) \geq 0$
3. $(x - 1)(x + 4) \leq 0$
4. $(x - 2)(x + 5) < 0$
5. $x^2 - 1 < 0$
6. $x^2 \leq 16$
7. $x^2 - 2x - 15 \leq 0$
8. $x^2 + x - 2 > 0$
9. $x^2 - x - 2 \geq 0$
10. $x^2 - 3x - 10 < 0$
11. $x^2 - 5x > -6$
12. $x^2 + 7x \leq -12$
13. $2x^2 + 7x + 3 < 0$
14. $2x^2 - 5x + 2 \geq 0$
15. $2x^2 - x < 1$
16. $2x^2 - x \leq 3$
17. $4x^2 + 2x > x^2 - 1$
18. $2x^2 + 5x < 2 - x^2$
19. $(x + 1)(x - 2)(x + 3) < 0$
20. $(x - 1)(x + 2)(x - 3) \leq 0$
21. $(x + 3)(2x - 5)(x + 4) > 0$
22. $(x + 4)(3x - 1)(x - 5) \geq 0$
23. $(x - 2)^2(x + 4) < 0$

24. $(x + 3)^2(x - 5) \geq 0$
25. $x^3 - 4x > 0$
26. $x^4 - 9x^2 \leq 0$
27. $x^2 + 2x + 3 \leq 0$
28. $x^2 + x + 1 > 0$
29. $\dfrac{(x - 2)(x - 5)}{x + 1} < 0$
30. $\dfrac{(x + 2)(x - 4)}{x - 2} \geq 0$
31. $\dfrac{x}{(x + 1)(x - 2)} > 0$
32. $\dfrac{x(x - 3)}{(x + 1)(x - 4)} \leq 0$
33. $\dfrac{(3x + 1)(x - 3)}{2x - 1} \leq 0$
34. $\dfrac{(x - 1)(x + 6)}{(x + 1)(x - 3)} > 0$
35. $\dfrac{4}{x - 1} < \dfrac{5}{x + 1}$
36. $\dfrac{-3}{x + 2} \geq \dfrac{6}{x - 3}$
37. $\dfrac{x^2 + 2x + 3}{x - 1} \geq 0$
38. $\dfrac{x^2 + x + 1}{x + 2} < 0$
39. $|x^2 + 3x + 2| \leq 4$
40. $|x^2 - 3x + 2| > 5$

Solve Exercises 41–44.

41. *Lighting technology*   The intensity I in candelas (cd) of a certain light is $I = 75d^2$, where d is the distance in meters from the source. For what range of distances will the intensity be between 75 and 450 cd?

42. *Physics*   An object is thrown straight upward from the ground with an initial velocity of 34.3 m/s. Its distance d above the ground at any time t is given by the equation $d = 34.3t - 4.9t^2$. For what time period is the object more than 49 m above the ground?

43. *Architecture* The deflection of a beam d is given by $x^2 - 1.1x + 0.2$, where x is the distance from one end. For what values of x is $d > 0.08$?

44. *Architecture* The load L that can be safely supported by a wooden beam of length ℓ with rectangular cross-section of width w and depth d is given by the formula

$$L = \frac{kwd^2}{\ell}$$

where k is a constant, depending on the type of wood. If $k = 90$ and the beam is to be 18 ft long and 6 in. wide, what are the acceptable values for d, if the beam must support at least 3,000 lb and $d < 12$ in.? (Do not change the length measurements to the same units.)

≣ 16.4
INEQUALITIES IN TWO VARIABLES

One variable is not enough to describe some real problems. As a result, we often work with equations of two or more variables. In this section, we will consider inequalities with two variables.

Solutions to equations and inequalities in one variable are represented on a number line. Solutions to equations are represented by points and solutions to inequalities by intervals.

Equations in two variables are represented by a graph in a plane. Inequalities in two variables are also represented graphically in a plane. The graph, however, is not a line or a curve.

We will begin by looking at linear inequalities in two variables.

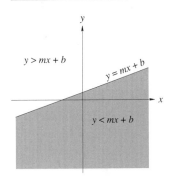

FIGURE 16.17

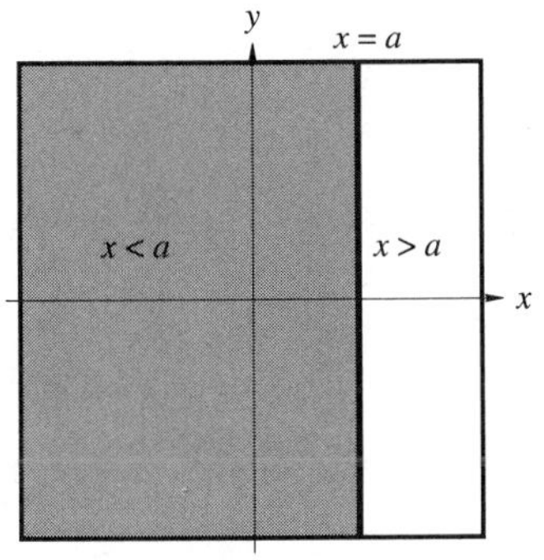

> **Linear Inequalities in Two Variables**
>
> A **linear inequality in two variables**, x and y, is an inequality of the form $ax + by + c < 0$, where $a, b,$ and c are constants and a and b are not both zero.

≣ **Note**

The use of the $<$ symbol in this definition is not meant to restrict this to only inequalities that use $<$. This same definition applies for $>, \leq,$ and $\geq$.

Every nonvertical line can be written in the slope-intercept form, $y = mx + b$. This line divides the plane into three separate regions, as shown in Figure 16.17.

1. The points that satisfy $y = mx + b$ (the line)
2. The region that satisfies $y < mx + b$ (the points below the line)
3. The region that satisfies $y > mx + b$ (the points above the line)

A vertical line is of the form $x = a$, where a is a constant. Vertical lines also divide the plane into three separate regions, as shown in Figure 16.18: the line ($x = a$), the points to the right of the line ($x > a$), and the points to the left of the line ($x < a$).

To illustrate how this works, we will solve $2x + y < 4$. We begin by solving the inequality for y, getting $y < -2x + 4$. Next, we graph the line $y = -2x + 4$. This inequality does not include the points where $y = -2x + 4$, so we will make this a dotted line rather than a solid line. The plane is now divided into three regions. We want the region below the line. This area has been shaded in Figure 16.19.

It is always a good idea to check your results. Select any point in the shaded region. For example, select the point $(-1, 4)$. Substitute these values in the original inequality and, on the left-hand side, you get $2(-1) + 4 = -2 + 4 = 2$. This is certainly

FIGURE 16.18

less than 4, so this point checks. Obviously checking one value does not mean that you have not made a mistake. This answer has an infinite number of solutions and we checked only one. But, checking one value at least *helps* us see if we may be correct.

≡ Note   Notice that if the inequality does not include equality, the curve dividing the region will be a broken line. A solid line will indicate that the points on the line are included in the solution.

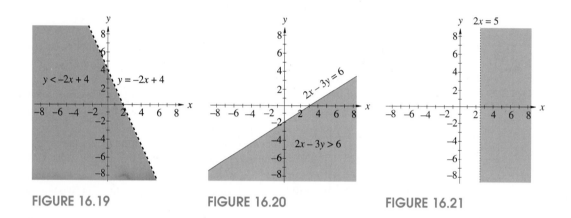

FIGURE 16.19   FIGURE 16.20   FIGURE 16.21

EXAMPLE 16.25

Find the region described by $2x - 3y \geq 6$.

Solution We first solve the inequality for y.

$$2x - 3y \geq 6$$

$$-3y \geq -2x + 6$$

$$y \leq \frac{2}{3}x - 2$$

Notice that, when we divided by -3, the direction of the inequality symbol reversed.

Next, graph the line $y = \frac{2}{3}x - 2$, as shown in Figure 16.20. Because this problem involves equality ($\leq$), we make it a solid line. When we solved the inequality for y, we got $y \leq \frac{2}{3}x - 2$ and this indicates that we want the region below the line, as is shaded in Figure 16.20. This shaded region and the line are the solutions for this problem.

EXAMPLE 16.26

Find the solution to $2x > 5$.

Solution If you solve this for x, you get $x > \frac{5}{2}$. Graph the line $x = \frac{5}{2}$ with a dotted line. The desired region is to the right of this line. The solution is the shaded region in Figure 16.21.

Not all equations in two variables are equations of a line. In the same way, not all inequalities in two variables are of regions separated by a line.

EXAMPLE 16.27

Find the solution to $y \leq x^2 - 4$.

Solution The graph of $y = x^2 - 4$ is the parabola in Figure 16.22. All points on the parabola or below it satisfy the given inequality, as indicated in Figure 16.22.

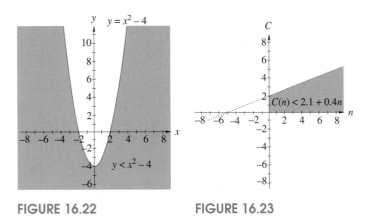

FIGURE 16.22   **FIGURE 16.23**

Application

EXAMPLE 16.28

A company has determined that it can make a profit on its microcomputers if the production costs C satisfies the inequality $C(n) < 2.1 + 0.4n$, where n is the number of computers produced. Graph this inequality.

Solution We will let n be on the horizontal axis and C the vertical axis. We graph the line $C(n) = 2.1 + 0.4n$, and make the line dashed to show that the solution does not include the points on the line. Since we want $C(n) < 2.1 + 0.4n$, we shade the points below the line. We only shade points in the first quadrant, because $n > 0$. (After all, the company cannot make less than 0 microcomputers.) We have also $C(n) > 0$, because it will cost something to produce the computers. The solution is shown by the graph in Figure 16.23.

Exercise Set 16.4

Graph each inequality in Exercises 1–30.

1. $y > 2$

2. $y \leq -3$

3. $x \geq 5$

4. $y < 2$

5. $x + y > 3$

6. $y - x \leq 4$

7. $2x + y < 4$

8. $3x + y > -4$

9. $x + 2y \leq 4$

10. $2y - x > -3$

11. $2x + 3y < 6$

12. $3x + 2y \geq 6$

13. $4x - 3y \leq -6$

14. $3y + 4x \geq 9$

15. $0.5x + 1.5y > 2.5$

16. $0.3x - 0.9y \leq 1.2$

17. $1.4x - 0.7y < 1.0$

18. $0.2x + 0.6y \leq 2.0$

19. $y \leq x^2 + 2$

20. $y \geq x^2 - 3$

21. $y \geq 3x^2 - 1$

22. $y \leq 4x^2 - 5$

23. $y < x^2 + 2x + 1$

24. $y \geq x^2 - 4x + 4$

25. $x^2 + y^2 \leq 16$

26. $x^2 + y^2 > 9$

27. $4x^2 + 9y^2 \leq 36$

28. $9x^2 + y^2 > 9$

29. $|x + 1| < 4$

30. $|x - 5| \leq 3$

Solve Exercises 31–32.

31. *Chemistry* The temperature (°C) and pressure (kPa) of a controlled chemical reaction must satisfy $1.8P + T < 270$. Graph this inequality for $T > 0$ and $P \geq 0$.

32. *Business*   In order to make a profit, the cost C of producing a computer chip in a certain factory must satisfy the inequality $C(n) < n^2 + 10n + 35$, where n is the number of chips produced. Graph this inequality. Remember that $n \geq 0$.

≡ 16.5
SYSTEMS OF INEQUALITIES; LINEAR PROGRAMMING

Several times in this book we have worked with systems of equations. In fact, the first section of this chapter was devoted to solving systems of quadratic equations. You can just as easily have systems of inequalities. In this section, you will learn how to solve linear inequalities and then learn how to apply this to a technique called linear programming.

The **graph of a system of inequalities** consists of all the points in the plane that satisfy every inequality in the system. To get this graph you will graph on the same coordinate system the region determined by each of the inequalities. The region where all these regions overlap is the graph of the system of inequalities.

EXAMPLE 16.29

Graph the solution of the following system.

$$\begin{cases} x + y \leq 5 \\ x - 3y < 3 \end{cases}$$

Solution   This system is equivalent to

$$\begin{cases} y \leq -x + 5 \\ y > \dfrac{x}{3} - 1 \end{cases}$$

As we did in Section 16.4, we first sketch the lines $y = -x + 5$ and $y = \dfrac{x}{3} - 1$. Use broken lines when you sketch them. Next, lightly shade the regions that satisfy each inequality, as shown in Figures 16.24a and 16.24b.

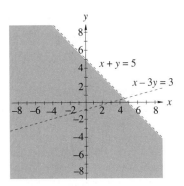

FIGURE 16.24a

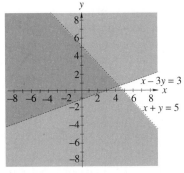

FIGURE 16.24b

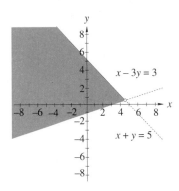

FIGURE 16.24c

EXAMPLE 16.29 (Cont.)

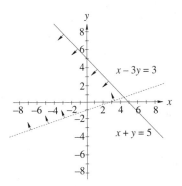

FIGURE 16.24d

The part where both of the shaded regions overlap is part of the solution. Shade this overlapped region more heavily. The first inequality used the $\leq$ symbol, so the part of the line $y = -x + 5$ that borders the answer region should be made solid. The area you have shaded is the solution and should look like the one in Figure 16.24c. Remember to work with broken lines until you can determine which section of the line is included in the solution.

Notice the "hole" where the two boundaries intersect. This point of intersection $(4.5,\ 0.5)$ does not satisfy the second inequality and so it is not part of the solution for this linear system. In essence, the intersection of a solid line and a broken line results in a "hole."

Another approach is just to place arrows on the boundaries with the arrows pointing toward the region that satisfies the inequality. This is shown in Figure 16.24d.

EXAMPLE 16.30

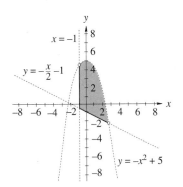

FIGURE 16.25

Graph the solution of this system.

$$\begin{cases} y + x^2 < 5 \\ x \geq -1 \\ x + 2y \geq -2 \end{cases}$$

Solution   This system is equivalent to

$$\begin{cases} y < -x^2 + 5 \\ x \geq -1 \\ y \geq -\dfrac{1}{2}x - 1 \end{cases}$$

Each curve is sketched using broken lines. The region defined by each inequality is lightly shaded and the region common to all three is then shaded more darkly. The borders of this common region that were formed by the last two inequalities are changed from broken lines to solid lines. The final region is shown in Figure 16.25. If you want to verify that this is the correct region, then select a point in the region and verify that it satisfies all of these inequalities. A good point to test is $(0,\ 0)$ and you can readily see that it satisfies all three inequalities.

One very important area of discipline that uses graphs of systems of inequalities is called **linear programming**. With the use of linear programming it is possible to solve problems in which a quantity is to be maximized or minimized, subject to certain limitations called **constraints**.

Linear programming is used with functions of two or more variables. We will work only with functions or equations of two variables. In linear programming, you establish a system of linear inequalities and graph its solution. This region is the set of **feasible solutions** to the problem and is called the **feasible region**. Any point in this region is called a **feasible point**. Finally, the **objective function** is formed, which describes the quantity to be maximized or minimized. The maximum and minimum values of the objective function will occur on the boundary of the feasible region. This will always be a vertex or a segment that connects two vertices of the feasible region.

Application

EXAMPLE 16.31

A company manufactures two types of microcomputers: Model A with 256K RAM and Model B with 128K RAM. The company can make a maximum of 75 model A micros per day and 50 Model B micros per day. It takes 8 h to manufacture a Model A micro and 3 h to manufacture a Model B machine. The number of employees can provide a total of 630 h of work each day. The profit on each Model A computer is $20 and the profit on each Model B is $27. How many of each type should be made each day to give the maximum profit?

Solution   If x is the number of Model A microcomputers made each day and y the number of Model B, we then have the following constraints.

$0 \leq x \leq 75$ They can make no more than 75 of Model A.

$0 \leq y \leq 50$ They can make a maximum of 50 Model B.

$8x + 3y \leq 630$ They can only work 630 h a day.

The profit P is given by $P = 20x + 27y$. This is the objective function.

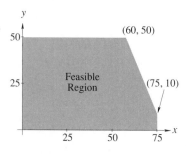

FIGURE 16.26

The set of feasible points is the shaded region in Figure 16.26. While it is too lengthy to prove here, it can be shown that the solution to the problem lies on the boundary of the feasible region. Thus, since this region is bordered by a polygon, the solution will be a vertex or one side of the polygon.

There are two ways to approach this solution. We can test the object value at each vertex or we can select a value for P and draw its graph, a test line, for that equation. The answer will be found by drawing a line parallel to the test line so that every point in the feasible region is either on the line or on one side of the line. We will begin with the first method.

EXAMPLE 16.31 (Cont.)

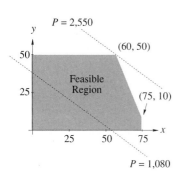

P = 2,550

(60, 50)

Feasible
Region

(75, 10)

P = 1,080

FIGURE 16.27

The feasible region has five vertices. If we evaluate the objective function $P = 20x + 27y$, at each of these vertices we get the following table of values.

Vertex	P
(0, 0)	0
(0, 50)	1,350
(60, 50)	2,550
(75, 10)	1,770
(75, 0)	1,500

From this table, we can clearly see that the most profit will occur when they make 60 Model A and 50 Model B microcomputers each day.

With the second method, you can select a value for P, say 1,080, and graph $1,080 = 20x + 27y$, as shown in Figure 16.27. Now find a line parallel to this, where the entire feasible region is on the same side of the line except for the point, or points, where the line intersects the region. (We know that this occurs when $P = 2,550$.) This happens when $y = 50$ and $8x + 3y = 630$ intersect. Solving these two simultaneous equations gives us the same solution, $x = 60$ and $y = 50$, and the maximum daily profit of $2,550.

Application

EXAMPLE 16.32

Each week, an electronics company makes home entertainment sets. Among those that they make are compact disc players (CDs) and television sets (TVs). The number of hours it takes to manufacture each set, the cost of the materials, and the profit are shown in the following table. The last line of the table shows the maximum number of each item that are available each week. How many of each type of set should the company make in order to make the largest profit?

	Manufacturing time (h)	Material Cost ($)	Profit ($)
CD	3.5	112	67
TV	4.0	76	58
Available	488.0	13,500	

Solution Since the company makes a profit of $67 on each CD and only $58 on each TV, you might suggest that they make only CD players. Since there are 488 h of work time available each week, and since each CD takes 3.5 h to make, they could make a total of

$$488 \div 3.5 = 139.4 \text{ CD players}$$

It requires $112 of materials to make each CD, so a total of 139 CDs would require $112 \times 139 = $15,568. Unfortunately, there is only $13,500 available for materials each week.

Using a similar argument, we can see that it would not be wise to make as many CDs as possible with the available money for materials. Since each CD requires

EXAMPLE 16.32 (Cont.)

$112 worth of materials and we have $13,500 available, then why not make

$$\$13,500 \div \$112 = 120.5 \text{ CD players}$$

Thus, we could make 120 CD players each week (and make a weekly profit of $120 \times \$67 = \$8,040$. It would require $120 \times 3.5 = 420$ h to make these sets. That would leave 68 h when no manufacturing was taking place, but no more money to make any TV sets. The company is committed to making both CD players and TV sets, so this solution is not acceptable.

Now, let's use linear programming to see how many CDs and TVs should be made in order to make the largest profit.

If we let c represent the number of CDs and t the number of TVs we can make in one week, then our weekly profit P will be

$$P = 67c + 58t$$

This equation, $P = 67c + 58t$, is the objective function.

The manufacturing time cannot be more than 488 h, so

$$3.5c + 4t \leq 488$$

The cost of materials cannot be more than $13,500, so

$$112c + 76t \leq 13,500$$

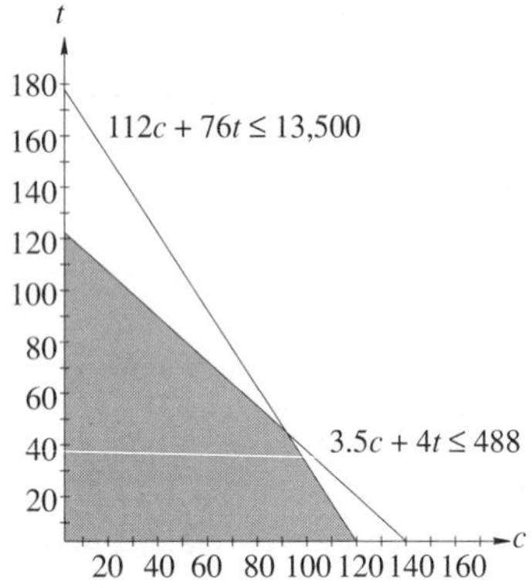

FIGURE 16.28

Finally, we assume that $c \geq 0$ and $t \geq 0$. Both of these last two inequalities contain the equal signs, because we want to allow for the possible (but unexpected) answer that we should make all CD players and no TVs or all TVs and no CDs. We now have our objective function and four constraints. The constraints are graphed in Figure 16.28.

Three vertices of the feasible region are $(0, 0)$, $(0, 120)$ and $(120, 0)$. We find the fourth vertex of the feasible region by simultaneously solving the systems of equations

$$\begin{cases} 3.5c + 4t = 488 \\ 112c + 76t = 13,500 \end{cases}$$

Using the techniques from Chapter 5, such as Cramer's rule, we see that $c \approx 92.92$ and $t \approx 40.69$. These are not integers, so the solution is still not clear. If we round both of these numbers off to the nearest whole number, we would conclude that we should make 93 CDs and 41 TVs. But if we do, this would require 489.5 h of manufacturing time and $13,532 of materials a week. Both of these are more than we are allowed.

The alternative is to round one of these numbers up and the other one down. This gives the following two possibilities.

$$c = 92 \quad \text{and} \quad t = 41$$

$$\text{or} \quad c = 93 \quad \text{and} \quad t = 40$$

In order to determine the answer, we substitute each pair of values in the objective function $P = 67c + 58t$ and see which pair of numbers produces the higher profit.

EXAMPLE 16.32 (Cont.)	When

$$c = 92 \text{ and } t = 41, P = 67(92) \text{ and } 58(41)$$

$$= 8,542$$

and when $c = 93$ and $t = 40, P = 67(93)$ and $58(40)$

$$= 8,551$$

Thus, the company will make the most profit, $8,551, when they manufacture 93 CDs and 40 TVs.

For many companies, other factors might have entered into this example. One factor is the size of the company's warehouse. Suppose that the company in Example 16.32 only ships once a week. That means that they must store all the work from one week in the warehouse. Further suppose that there was only enough room for 125 boxes in the warehouse (assuming that the TV and CD boxes are the same size). Then our answer of 93 CDs and 40 TVs, or 133 total products might not be acceptable. Other factors might be considered, such as the amount of time available to test each new product or the number of people that are trained to assemble each type of equipment.

Exercise Set 16.5

In Exercises 1–14, give the graphical solution to each system of inequalities.

1. $\begin{cases} x - y > 0 \\ y - 2x < 4 \end{cases}$

2. $\begin{cases} x + y \geq 6 \\ y - 3x < 6 \end{cases}$

3. $\begin{cases} x + y \geq 5 \\ y > 1 \end{cases}$

4. $\begin{cases} x + 2y \leq 4 \\ y < 1 \end{cases}$

5. $\begin{cases} 2x + 3y \leq 7 \\ 3x - y \leq 5 \end{cases}$

6. $\begin{cases} -2x + y > 6 \\ x + 3y \leq 9 \end{cases}$

7. $\begin{cases} x + 2y - 3 \leq 0 \\ 2x + y - 4 > 0 \end{cases}$

8. $\begin{cases} -2x - 3y < 6 \\ -4x + 3y \geq 12 \end{cases}$

9. $\begin{cases} 3x + y \geq 4 \\ y - 2x \geq -1 \end{cases}$

10. $\begin{cases} -4x + 7y \leq 28 \\ 2x + 3y < 6 \\ y > -3 \end{cases}$

11. $\begin{cases} x + 2y < 4 \\ x - 2y < 4 \\ y \leq 3 \end{cases}$

12. $\begin{cases} x > 1 \\ y < 2 \\ 8x + 3y \geq 24 \end{cases}$

13. $\begin{cases} -x + 2y \leq 6 \\ 3x + y \leq 9 \\ x > -1 \\ y \geq 1 \end{cases}$

14. $\begin{cases} x - 2y > -10 \\ 4x + 3y \geq 26 \\ 3x - 4y < 7 \\ x + y \leq 14 \end{cases}$

In Exercises 15–20, find the maximum or minimum value (as specified) of the objective function that is subject to the given constraints.

15. Maximum $P = 2x + y$ and $\begin{cases} 3x + 4y \leq 24 \\ x \geq 2 \\ y \geq 3 \end{cases}$

16. Minimum $C = 9x + 5y$ and $\begin{cases} 3x + 4y \geq 25 \\ x + 3y \geq 15 \\ x \geq 0 \\ y \geq 0 \end{cases}$

17. Maximum $F = 15x + 10y$ and $\begin{cases} 3x + 2y \leq 80 \\ 2x + 3y \leq 70 \\ x \geq 0 \\ y \geq 0 \end{cases}$

18. Minimum $F = 6x + 9y$ and $\begin{cases} 2x + 5y \leq 50 \\ x + y \leq 12 \\ 2x + y \leq 20 \\ y - x \geq 2 \\ x \geq 0 \end{cases}$

19. Maximum $P = 8x + 10y$ and $\begin{cases} 5x + 10y \leq 180 \\ 10x + 5y \leq 180 \\ x \geq 0 \\ y \geq 3 \end{cases}$

20. Minimum $M = 10x - 8y + 15$ and $\begin{cases} y \geq -3 \\ x - y \geq -5 \\ x + y \geq -5 \\ x \leq 2 \end{cases}$

Solve Exercises 21–24.

21. *Business*   A company manufactures two products. Each product must pass two inspection points, A and B. Each unit of Product X requires 30 min at A and 45 min at B. Product Y requires 15 min at A and 10 min at B. There are enough trained people to provide 100 h at A and 80 h at B. The company makes a profit of $10 and $8 on each of X and Y, respectively. What numbers of each should be manufactured to make the most profit? How much is the most profit? If more people could be added at A or B (but not both), where should they be added to increase profits?

22. *Business*   An electronics company manufactures two models of computer chips. Model A requires 1 unit of labor and 4 units of parts. Model B requires 1 unit of labor and 3 units of parts. If 120 units of labor and 390 units of parts are available, and if the company makes a profit of $7.00 on each model A chip and $5.50 on each model B, how many should it manufacture to maximize its profits?

23. *Business*   The company in Exercise 22 raises its prices so that it makes a profit of $8.00 on each model A chip and $9.50 on each model B. Now, how many should it manufacture to maximize profits?

24. *Business*   A computer company manufactures a personal computer (PC) and a business computer (BC). Each computer uses two types of chips, an AB chip and an EP chip. The number of chips needed for each computer and the profit for each are given in the following table.

Computer	AB	EP	Profit
PC	2	3	145
BC	3	8	230

Because of a shortage in chips, the company has 200 AB chips and 450 EP chips in stock. **(a)** How many of each computer should be made in order to maximize profits? **(b)** What is that maximum profit?

≡ CHAPTER 16 REVIEW

Important Terms and Concepts

Feasible region
Feasible solution
Inequalities

Linear
Linear system
Nonlinear

Properties

Linear programming

Nonlinear system of equations

Objective function

Substitution method

Review Exercises

In Exercises 1–8, solve each inequality both algebraically and graphically.

1. $3x < -12$

2. $4x + 5 < 6$

3. $2x - 7 \geq 15$

4. $7 - 5x < 2 + 3x$

5. $\dfrac{2x + 5}{4} < \dfrac{4x - 1}{3}$

6. $|2x - 1| \leq 5$

7. $|3x + 2| > 7$

8. $2x + 3 < 13 \leq 3x - 9$

In Exercises 9–13, solve each inequality graphically.

9. $-2x + y < 5$

10. $2x + 3y \geq 6$

11. $4x - 3y < 3$

12. $y > 2x^2 - 5$

13. $y \leq x^2 - 6x + 9$

In Exercises 14–21, graph each system of equations or inequalities.

14. $\begin{cases} x + y = 8 \\ x^2 + y^2 = 49 \end{cases}$

15. $\begin{cases} x^2 + y^2 = 9 \\ x^2 = 5y \end{cases}$

16. $\begin{cases} 2x^2 - y^2 = 8 \\ x^2 + 2y^2 = 4 \end{cases}$

17. $\begin{cases} 2x + 4y^2 = 16 \\ xy = 10 \end{cases}$

18. $\begin{cases} x + 3y < 5 \\ x - 4y \leq 8 \end{cases}$

19. $\begin{cases} 4x - y \leq 4 \\ y + x > -2 \\ y - x < 1 \end{cases}$

20. $\begin{cases} 2x + y - 3 < 0 \\ x - 2y - 4 \geq 0 \end{cases}$

21. $\begin{cases} 3x - y \leq -2 \\ 4 - x - y \geq 0 \\ x > -3 \\ 2x + y > -4 \end{cases}$

In Exercises 22–23, find the maximum or minimum value (as specified) of the objective function that is subject to the given constraints.

22. Maximum $P = 3x + 5y$ and $\begin{cases} x + y \leq 5 \\ y - x \geq -2 \\ x \geq -1 \\ x \leq 3 \end{cases}$

23. Minimum $F = 4x - 3y$ and $\begin{cases} 2x + y \leq 6 \\ y \geq \dfrac{x}{2} - 3 \\ x \geq -1 \\ y \leq 4x - 1 \end{cases}$

Solve Exercises 24–27.

24. The length of a computer chip is given as $7.0\,\text{mm} < \ell < 7.5\,\text{mm}$ and the width as $1.4\,\text{mm} < w < 1.6\,\text{mm}$. What is the range in the area?

25. *Electricity*   Two resistances R_1 and R_2 connected in parallel must have a total resistance R of at least $4\,\Omega$. If

$R_1R_2 = 20 \, \Omega$, what are acceptable values for $R_1 + R_2$?
Remember $R = \dfrac{R_1R_2}{R_1 + R_2}$.

26. *Transportation*   A trucker was delayed 30 min because of trouble loading the shipment. To make up for the delay, the trucker increased the speed and took advantage of interstate highways. The speed was increased by an average of 4 mph, and as a result, the trucker was on time, 7 h and 30 min after leaving. Find the usual average speed of the trucker and the distance the truck traveled.

27. *Business*   A robotics manufacturing company makes two types of robots: a cylindrical coordinate robot (Model C) and a spherical (polar) coordinate robot (Model S). Each robot has two assembly stations. The number of hours needed at each station and the profit of each robot are given in the following table.

Robot	Station 1 (hours)	Station 2 (hours)	Profit ($)
C	12	15	1,250
S	18	10	1,620

During the next month, the workers will take their vacations. As a result, the company will only have enough workers for 220 h at Station 1 and 180 h at Station 2. How many of each robot should be manufactured in order to make the most profit?

CHAPTER 16 TEST

1. Solve the inequality $5x > 30$, both algebraically and graphically.

2. Solve $|4x + 5| < 17$, both algebraically and graphically.

3. Solve $\dfrac{4x - 5}{2} \leq \dfrac{7x - 2}{3}$, both algebraically and graphically.

4. Solve $(x - 2)(x - 5) > 0$, both algebraically and graphically.

5. Solve $5x^2 + 3x \leq 2x^2 - x - 1$, both algebraically and graphically.

6. Graph this system of inequalities .

$$\begin{cases} x + 2y > 4 \\ 3x - y \leq 3 \end{cases}$$

7. A brand X multivitamin tablet contains 12 mg of iron and 10 mg of zinc. Each brand Y tablet contains 5 mg of iron and 8 mg of zinc. A nutritionist suggests that a patient take at least 80 mg of iron and 90 mg of zinc each day. The brand X tablets cost 4¢ each and brand Y pills are 6¢ each.

(a) How many of each pill should the patient take each day to satisfy the suggestion at the minimum cost?

(b) What is the minimum cost?

17

Matrices

Kirchhoff's laws can be used to write a system of linear equations. In Section 17.4, we will learn how to use matrices to solve this linear system and determine the currents in each electric circuit's path.

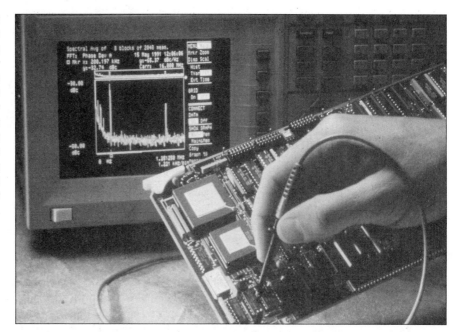

Courtesy of Hewlett-Packard Company

In Chapter 5, we studied systems of equations and we learned that determinants can be used to help solve systems of linear equations. In this chapter, another technique for solving systems of equations will be introduced. This technique uses matrices. The introduction of computers has led to increased applications of matrices in engineering, physics, biology, information science, transportation, and other technical areas. We will use calculators and computer programs to help us work with matrices.

≡ 17.1
MATRICES

Dimension

A **matrix** is a rectangular array of numbers arranged in rows and columns. A matrix with m rows and n columns is an $m \times n$ matrix. ($m \times n$ is read m by n.) The **dimensions** of a matrix are given in the form $m \times n$, where m represents the number of rows and n the number of columns.

EXAMPLE 17.1

$$\begin{bmatrix} 4 & 3 & 7 \\ 8 & 19 & -11 \end{bmatrix}$$ is a matrix with dimension 2×3.

$$\begin{bmatrix} 5 & 8 & -9 & 12 \\ 4 & 0 & -3 & 5.7 \\ 9 & 3 & 21 & -6 \end{bmatrix}$$ is a 3×4 matrix.

$$\begin{bmatrix} 2 & 4 & 6 & 8 & 9 \end{bmatrix}$$ is a 1×5 matrix.

$$\begin{bmatrix} 10 \\ 8 \\ 7 \\ 6 \\ 3 \end{bmatrix}$$ is a 5×1 matrix.

A matrix is often used to present numerical data in a condensed form. For example, suppose a company makes two types of robots, C and S. Model C requires 78 h to assemble and contains 14 Type I computer chips, 7 Type II chips, and 16 m of wiring. Model S requires 65 h to assemble and contains 12 Type I chips, 9 Type II chips, and 11 m of wiring. The company makes a profit of $1,250 on each Model C it sells and $1,500 on each Model S. All this information can be presented in a rectangular array or matrix like the following.

	Hours	Type I Chip	Type II Chip	Wiring (meters)	Profit ($)
Model C	78	14	7	16	1,250
Model S	65	12	9	11	1,500

The numbers in the first row indicate the data for the Model C robot and those in the second row indicate the Model S data. This is a simple example of a matrix that has two rows and five columns.

Row and Column Vectors

A matrix of dimension $1 \times n$ is a **row vector** and a matrix of dimension $m \times 1$ is a **column vector**.

EXAMPLE 17.2

Some row vectors are:

$$[2 \quad 4 \quad 6] \qquad [1 \quad -3 \quad 2 \quad 8]$$

$$[4 \quad 6 \quad 0 \quad -7 \quad 0] \qquad [9]$$

Some column vectors are:

$$\begin{bmatrix} 2 \\ 4 \\ 6 \end{bmatrix} \quad \begin{bmatrix} 1 \\ -3 \\ 2 \\ 8 \end{bmatrix} \quad \begin{bmatrix} 4 \\ 6 \\ 0 \\ -7 \\ 0 \end{bmatrix} \quad [9]$$

Notice that [9] is both a row vector and a column vector.

Zero Matrix

A matrix is a **zero matrix** if all the elements are zero. The matrix

$$\begin{bmatrix} 0 & 0 \\ 0 & 0 \\ 0 & 0 \end{bmatrix}$$

is a zero matrix of dimension 3×2.

Square Matrix

A matrix with the same number, n, of rows and columns is a **square matrix of order** n.

EXAMPLE 17.3

The matrix

$$\begin{bmatrix} 2 & 1 \\ 5 & 7 \end{bmatrix}$$

is a square matrix of order 2 and the matrix

$$\begin{bmatrix} 5 & -7 & 8 \\ 2 & -1 & 0 \\ 0 & 3 & 5 \end{bmatrix}$$

is a square matrix of order 3.

A double-subscript notation has been developed to allow us to refer to specific elements. The first numeral in the subscripts refers to the row in which the element lies and the second numeral refers to the column. An example of this notation is matrix A.

$$A = \begin{bmatrix} a_{11} & a_{12} & a_{13} & a_{14} \\ a_{21} & a_{22} & a_{23} & a_{24} \\ a_{31} & a_{32} & a_{33} & a_{34} \\ a_{41} & a_{42} & a_{43} & a_{44} \end{bmatrix}$$

The elements $a_{11}, a_{22}, a_{33}, a_{44}$ are the **main diagonal** of this square matrix of order 4.

Coefficient Matrix

When working with a system of linear equations, a **coefficient matrix** can be formed from the coefficients of the equations. For example, the system

$$\begin{cases} 2x + 3y - 5z = 8 \\ 4x - 7y + 22z = 15 \end{cases}$$

has the coefficient matrix

$$\begin{bmatrix} 2 & 3 & -5 \\ 4 & -7 & 22 \end{bmatrix}$$

Augmented Matrix

If the column vector formed by the constants to the right of the equal symbols, $\begin{bmatrix} 8 \\ 15 \end{bmatrix}$, is adjoined to the right of the coefficient matrix, we get a new matrix called the **augmented matrix** of the system of linear equations. For example, for the system used previously,

$$\begin{cases} 2x + 3y - 5z = 8 \\ 4x - 7y + 22z = 15 \end{cases}$$

the augmented matrix would be

$$\left[\begin{array}{ccc:c} 2 & 3 & -5 & 8 \\ 4 & -7 & 22 & 15 \end{array}\right]$$

The dashed line between columns 3 and 4 is not needed but is often used to indicate an augmented matrix.

Two matrices are **equal**, if they have the same dimension and their corresponding elements are equal.

EXAMPLE 17.4

If $\begin{bmatrix} 6 & 4 & 3 & x \\ 2 & 9 & y & -3 \end{bmatrix} = \begin{bmatrix} z & 4 & 3 & -11 \\ 2 & w & 7 & -3 \end{bmatrix}$, then $x = -11$, $y = 7$, $z = 6$, and $w = 9$.

EXAMPLE 17.5

For the matrices $A = \begin{bmatrix} 1 & 2 & 6 \\ 7 & 8 & 9 \end{bmatrix}$, $B = \begin{bmatrix} 1 & 2 & 6 \\ 7 & 8 & 5 \end{bmatrix}$, and $C = \begin{bmatrix} 1 & 2 \\ 7 & 8 \end{bmatrix}$, $A \neq B$ because the corresponding elements $a_{23} = 9$ and $b_{23} = 5$ are not equal. $A \neq C$ because they have different dimensions.

Addition and Subtraction of Matrices

If A and B are two matrices, each of dimension $m \times n$, then their sum (or difference) is defined to be another matrix, C, also of dimension $m \times n$, where every element of C is the sum (or difference) of the corresponding elements A and B.

EXAMPLE 17.6

(a) $\begin{bmatrix} 2 & 3 & 5 \\ 6 & 7 & 8 \end{bmatrix} + \begin{bmatrix} 10 & 11 & 12 \\ 14 & 16 & 22 \end{bmatrix} = \begin{bmatrix} 2+10 & 3+11 & 5+12 \\ 6+14 & 7+16 & 8+22 \end{bmatrix}$

$= \begin{bmatrix} 12 & 14 & 17 \\ 20 & 23 & 30 \end{bmatrix}$

(b) $\begin{bmatrix} 4 & 9 \\ -2 & 8 \\ 6 & 1 \end{bmatrix} + \begin{bmatrix} 16 & 21 \\ 2 & -8 \\ 9 & 14 \end{bmatrix} = \begin{bmatrix} 4+16 & 9+21 \\ -2+2 & 8+(-8) \\ 6+9 & 1+14 \end{bmatrix}$

$= \begin{bmatrix} 20 & 30 \\ 0 & 0 \\ 15 & 15 \end{bmatrix}$

(c) $\begin{bmatrix} 1 & -5 \\ 4 & 3 \\ 2 & -1 \end{bmatrix} - \begin{bmatrix} 4 & -6 \\ 2 & 1 \\ -3 & 9 \end{bmatrix} = \begin{bmatrix} 1-4 & -5-(-6) \\ 4-2 & 3-1 \\ 2-(-3) & -1-9 \end{bmatrix}$

$= \begin{bmatrix} -3 & 1 \\ 2 & 2 \\ 5 & -10 \end{bmatrix}$

Caution You can only add or subtract two matrices if they have the same dimension.

EXAMPLE 17.7

If $A = \begin{bmatrix} 4 & 2 & 7 & 9 \\ 5 & 4 & 6 & 1 \end{bmatrix}$ and $B = \begin{bmatrix} 5 & 3 & 8 & 1 \\ 4 & 6 & 2 & 4 \end{bmatrix}$, then determine $A+B$, $A-B$, and $B-A$.

Solutions

$A+B = \begin{bmatrix} 4+5 & 2+3 & 7+8 & 9+1 \\ 5+4 & 4+6 & 6+2 & 1+4 \end{bmatrix}$

$= \begin{bmatrix} 9 & 5 & 15 & 10 \\ 9 & 10 & 8 & 5 \end{bmatrix}$

EXAMPLE 17.7 (Cont.)

$$A - B = \begin{bmatrix} 4-5 & 2-3 & 7-8 & 9-1 \\ 5-4 & 4-6 & 6-2 & 1-4 \end{bmatrix}$$

$$= \begin{bmatrix} -1 & -1 & -1 & 8 \\ 1 & -2 & 4 & -3 \end{bmatrix}$$

$$B - A = \begin{bmatrix} 5-4 & 3-2 & 8-7 & 1-9 \\ 4-5 & 6-4 & 2-6 & 4-1 \end{bmatrix}$$

$$= \begin{bmatrix} 1 & 1 & 1 & -8 \\ -1 & 2 & -4 & 3 \end{bmatrix}$$

Application

EXAMPLE 17.8

At the beginning of a laboratory experiment, two groups of four mice were timed to see how long it took them to go through a maze. The mice in the group that went through the maze at night took 8, 6, 5, and 7 s. The mice in the group that went through the maze during the day took 22, 14, 18, and 12 s. **(a)** Write a 2×4 matrix using this information.

A week later, the mice were sent through the maze again. They went in the same order as they did the week before. This time the night group completed the maze in 4, 3, 5, and 3 s and the day group took 9, 8, 10, and 8 s. **(b)** Write this information as a 2×4 matrix and use matrix subtraction to write a matrix that shows the amount of change each mouse made in its time to complete the maze.

Solutions

(a) We will put the night group in the top row and the day group of mice in the bottom row. The result is the following matrix.

$$\begin{matrix} \text{Night group} \\ \text{Day group} \end{matrix} \begin{bmatrix} 8 & 6 & 5 & 7 \\ 22 & 14 & 18 & 12 \end{bmatrix}$$

(b) The matrix for the data at the end of the week is

$$\begin{matrix} \text{Night group} \\ \text{Day group} \end{matrix} \begin{bmatrix} 4 & 3 & 5 & 3 \\ 9 & 8 & 10 & 8 \end{bmatrix}$$

To find the change in the amount of time each mouse took going through the maze, we subtract the last matrix from the matrix in **(a)**.

$$\begin{matrix} \text{Night group} \\ \text{Day group} \end{matrix} \begin{bmatrix} 8 & 6 & 5 & 7 \\ 22 & 14 & 18 & 12 \end{bmatrix} - \begin{bmatrix} 4 & 3 & 5 & 3 \\ 9 & 8 & 10 & 8 \end{bmatrix}$$

$$= \begin{bmatrix} 4 & 3 & 0 & 4 \\ 13 & 6 & 8 & 4 \end{bmatrix}$$

Scalar Multiplication

If A is a $m \times n$ matrix and k a real number, then kA is an $m \times n$ matrix B, where $b_{ij} = ka_{ij}$ for each element of A. This is referred to as **scalar multiplication** and the real number k is called a **scalar**.

EXAMPLE 17.9

If $A = \begin{bmatrix} 3 & 5 \\ 9 & 2 \end{bmatrix}$ and $k = 4$, then

$$kA = 4 \begin{bmatrix} 3 & 5 \\ 9 & 2 \end{bmatrix} = \begin{bmatrix} 4 \cdot 3 & 4 \cdot 5 \\ 4 \cdot 9 & 4 \cdot 2 \end{bmatrix}$$

$$= \begin{bmatrix} 12 & 20 \\ 36 & 8 \end{bmatrix}$$

This means that $A + A = 2A, B + B + B = 3B$, and so on.

Application

EXAMPLE 17.10

A computer retailer sells three types of computers: the personal computer (PC), the business computer (BC), and the industrial computer (IC). The retailer has two stores. The matrix below shows the number of each computer model in stock at each store.

$$\begin{array}{c} \\ \text{Store A} \\ \text{Store B} \end{array} \begin{array}{ccc} \text{PC} & \text{BC} & \text{IC} \\ \begin{bmatrix} 6 & 18 & 12 \\ 10 & 22 & 28 \end{bmatrix} \end{array}$$

The store plans to have a sale in the near future. In order to have enough computers in stock at each store when the sale begins, it plans to order 1.5 times the current number. How many of each computer should be ordered for each store?

Solution Here the scalar is 1.5, so we want to multiply each element of this matrix by 1.5. The result is

$$1.5 \left(\begin{array}{c} \\ \text{Store A} \\ \text{Store B} \end{array} \begin{array}{ccc} \text{PC} & \text{BC} & \text{IC} \\ \begin{bmatrix} 6 & 18 & 12 \\ 10 & 22 & 28 \end{bmatrix} \end{array} \right) = \begin{array}{c} \\ \text{Store A} \\ \text{Store B} \end{array} \begin{array}{ccc} \text{PC} & \text{BC} & \text{IC} \\ \begin{bmatrix} 9 & 27 & 18 \\ 15 & 33 & 42 \end{bmatrix} \end{array}$$

Thus, we see that the retailer needs to order 9 PCs for Store A and 15 for Store B, 27 BC for Store A and 33 for Store B, and 18 ICs for Store A and 42 for Store B.

We can summarize the work in this section by listing the following properties of matrices.

> **Matrix Properties**
>
> If A, B, and C are three matrices, all of the same dimension, 0 is the zero matrix, and k is a real number, then
>
> $A + B = B + A$ (commutative law)
> $A + 0 = 0 + A$ (identity for addition)
> $A + (B + C) = (A + B) + C$ (associative law)
> $k(A + B) = kA + kB$

Exercise Set 17.1

Give the dimension of the matrices in Exercises 1–6. Solve Exercises 7–10.

1. $\begin{bmatrix} 2 & 4 & 5 \\ 3 & 2 & 1 \end{bmatrix}$

2. $\begin{bmatrix} 3 & 4 & 6 \\ 2 & 5 & 9 \\ 8 & 7 & 2 \end{bmatrix}$

3. $\begin{bmatrix} 2 & 1 & 0 & 7 & 9 & 6 \\ 3 & 2 & 4 & 8 & 7 & 2 \\ 9 & 6 & 4 & 5 & 0 & 2 \end{bmatrix}$

4. $\begin{bmatrix} 3 \\ 2 \\ 1 \\ 4 \end{bmatrix}$

5. $\begin{bmatrix} 3 & 2 \\ 4 & 1 \\ 5 & 3 \\ 7 & 9 \end{bmatrix}$

6. $\begin{bmatrix} 1 & 2 & 4 & 6 & 9 & 11 & 12 \end{bmatrix}$

7. Given matrix A, determine the values of elements a_{11}, a_{24}, a_{21}, and a_{32}.

$$A = \begin{bmatrix} 1 & 2 & 5 & 7 \\ 8 & 9 & 11 & 13 \\ 4 & 6 & 12 & 15 \end{bmatrix}$$

8. Given matrix B, determine the values of elements b_{12}, b_{21}, b_{23}, and b_{32}.

$$B = \begin{bmatrix} -5 & 0 & 2 \\ 4 & 7 & 9 \\ 5 & -8 & 11 \end{bmatrix}$$

9. Given that $\begin{bmatrix} 4 & 3 & 2 & x \\ 5 & 9 & 7 & 11 \\ y & 8 & 3 & 4 \end{bmatrix} = \begin{bmatrix} 4 & 3 & z & 12 \\ w & 9 & 7 & 11 \\ -6 & 8 & 3 & 4 \end{bmatrix}$, determine the values of x, y, z, and w.

10. Given that

$$\begin{bmatrix} 3 & 2 & x-2 \\ 4 & y+5 & 9 \\ w+6 & 11 & 10 \end{bmatrix} = \begin{bmatrix} 3z & 2 & 7 \\ 4 & 3 & 9 \\ 7 & 11 & 2p \end{bmatrix},$$

determine the values of x, y, z, w, and p.

In Exercises 11–14, find the indicated sum or difference.

11. $\begin{bmatrix} 3 & 2 & 1 & -4 \\ 9 & 7 & 6 & -1 \\ 4 & 3 & 5 & 2 \end{bmatrix} + \begin{bmatrix} -6 & 4 & 5 & -4 \\ 3 & 2 & 11 & 1 \\ 5 & -3 & 0 & 8 \end{bmatrix}$

12. $\begin{bmatrix} 2 & 4 \\ -5 & 1 \\ 3 & 4 \end{bmatrix} + \begin{bmatrix} 5 & -6 \\ 2 & 9 \\ 8 & 2 \end{bmatrix}$

13. $\begin{bmatrix} 3 & 2 & -1 \\ 9 & 12 & 4 \end{bmatrix} - \begin{bmatrix} 1 & 3 & -5 \\ 4 & 3 & 6 \end{bmatrix}$

14. $\begin{bmatrix} 3 & 2 & -1 & 5 \\ 4 & 3 & 11 & 9 \\ 16 & 4 & 3 & -8 \\ 12 & 5 & 4 & 6 \\ 2 & 9 & 18 & 7 \end{bmatrix} - \begin{bmatrix} 1 & 2 & -4 & 2 \\ 1 & 3 & 2 & 8 \\ 2 & -4 & 3 & 6 \\ 2 & -5 & 2 & -4 \\ 2 & 3 & -5 & 3 \end{bmatrix}$

In Exercises 15–22, use matrices A, B, and C to determine the indicated matrix, where

$$A = \begin{bmatrix} 4 & 3 & 2 \\ 5 & 0 & 7 \end{bmatrix}, \quad B = \begin{bmatrix} -5 & 4 & 2 \\ 3 & 1 & 8 \end{bmatrix}, \quad \text{and} \quad C = \begin{bmatrix} 5 & 8 & 4 \\ 6 & -2 & 9 \end{bmatrix}.$$

15. $A + B$

16. $B - C$

17. $3A$

18. $4B$

19. $2B + C$

20. $4A - 3C$

21. $2B - 3A$

22. $4C + 2B$

In Exercises 23–30, use matrices D, E, and F to determine the indicated matrix, where

$$D = \begin{bmatrix} 4 \\ 2 \\ -1 \\ 5 \end{bmatrix}, \quad E = \begin{bmatrix} 7 \\ 9 \\ -4 \\ 10 \end{bmatrix}, \quad \text{and} \quad F = \begin{bmatrix} 14 \\ 18 \\ -8 \\ 5 \end{bmatrix}.$$

23. $D - E$

24. $E + F$

25. $3E$

26. $F - 2E$

27. $3D + 2E$

28. $2F - 7D$

29. $7D - 2F$

30. $5E - 2F$

Solve Exercises 31–39.

31. *Business* In keeping inventory records, a company uses a matrix. At one storage point, they have 18 of computer chip A, 7 of computer chip B, 9 EPROMS, 7 keyboards, 11 motherboards, and 4 disk drives. This is represented by the matrix $\begin{bmatrix} 18 & 7 & 9 \\ 7 & 11 & 4 \end{bmatrix}$. At a second storage point, the inventory is $\begin{bmatrix} 9 & 5 & 6 \\ 11 & 4 & 12 \end{bmatrix}$. How many of each item do they have on hand?

32. *Business* Wrecker's Auto Supply Company has three stores, each of which carries five sizes of a certain tire model. The following matrix represents the present inventories.

Store	195/70SR-14	205/70SR-14	185/70SR-15	205/70SR-15	215/70SR-15
A	12	15	28	7	16
B	25	19	11	40	11
C	15	29	21	17	17

At the beginning of the next month, they would like to have the following inventory.

Store	195/70SR-14	205/70SR-14	185/70SR-15	205/70SR-15	215/70SR-15
A	20	48	60	24	44
B	36	64	24	40	32
C	44	72	48	60	40

Write a matrix that represents the number of each size of tire that must be ordered for each store, so as to achieve their goal. (Assume that no more tires are sold.)

33. *Medical technology* A drug company tested 400 patients to see if a new medicine is effective. Half of the patients received the new drug and half received a placebo. The results for the first 200 patients are shown in the following matrix.

	New drug	Placebo
Effective	70	40
Not effective	30	60

Using the same matrix format, the results for a second 200 patients were $\begin{bmatrix} 65 & 42 \\ 35 & 58 \end{bmatrix}$. What were the results for the entire test group?

34. *Business* A computer supply company has its inventory of four types of computer chips in three warehouses. At the beginning of the month they had the following inventory.

Chip type:	286	386	486	586
Warehouse A	1,200	3,200	4,800	900
Warehouse B	1,650	4,580	7,200	700
Warehouse C	1,120	5,100	6,200	1,200

The sales for the month were

Chip type:	286	386	486	586
Warehouse A	270	2,130	3,210	265
Warehouse B	1,120	4,230	3,124	75
Warehouse C	320	3,126	2,743	1,012

Write a matrix that shows the inventory at the end of the month.

35. *Business* The computer supply company in Exercise 34 expects sales to increase by 10% during the next month. What are next month's projected sales? (Round off any fractional answers.) Remember, last month the company sold

Chip type:	286	386	486	586
Warehouse A	270	2,130	3,210	265
Warehouse B	1,120	4,230	3,124	75
Warehouse C	320	3,126	2,743	1,012

36. *Business* The following matrix represents the normal monthly order of a retail store for three models of exercise suits in four different sizes.

	S	M	L	XL
Jogging	5	4	3	4
Sweating	7	12	15	7
Walking	4	8	12	14

During its spring sale, the store expects to do four times the usual volume of business. Determine the matrix that represents their order for the month of the sale.

37. Write a computer program to add or subtract two $m \times n$ matrices.

38. Write a computer program to multiply a matrix by a scalar.

39. Write a computer program that will allow you to multiply two matrices by a scalar and then add or subtract the results.

≡ 17.2
MULTIPLICATION OF MATRICES

Learning to add and subtract matrices was straightforward. We had to be careful that each matrix had the same number of rows and the same number of columns. Multiplying matrices is more involved.

Multiplying Matrices

If A and B are two matrices, then in order to define the product AB, the number of columns in matrix A must be the same as the number of rows in matrix B. Thus, matrix A must be of dimension $m \times n$ and matrix B dimension $n \times p$. The product will have dimension $m \times p$.

We will begin by multiplying a row vector and a column vector.

Product of a Row Vector and a Column Vector

If $A = [a_{11} \quad a_{12} \quad a_{13} \quad \cdots \quad a_{1n}]$ is a row vector, and $B = \begin{bmatrix} b_{11} \\ b_{21} \\ b_{31} \\ \vdots \\ b_{n1} \end{bmatrix}$ is a column vector, then

$$AB = [a_{11}b_{11} + a_{12}b_{21} + a_{13}b_{31} + \cdots + a_{1n}b_{n1}]$$

EXAMPLE 17.11

If $A = [3 \quad 4]$ and $B = \begin{bmatrix} 5 \\ 1 \end{bmatrix}$, then

$$AB = [3 \cdot 5 + 4 \cdot 1] = [15 + 4] = [19]$$

EXAMPLE 17.12

If $C = [1 \quad 2 \quad 3 \quad x]$, $D = \begin{bmatrix} 3 \\ -1 \\ 1 \\ 2 \end{bmatrix}$, and $CD = [22]$, find x.

Solution

$$CD = [1(3) + 2(-1) + 3(1) + 2x]$$
$$= [3 - 2 + 3 + 2x]$$
$$= [4 + 2x] = [22]$$

These two matrices are equal only if $4 + 2x = 22$ or $2x = 18$, so $x = 9$.

We will use the following method of multiplying a row vector and a column vector to multiply larger matrices.

Matrix Multiplication

If A is an $m \times n$ matrix and B is an $n \times p$ matrix, then the product matrix $C = AB$ is an $m \times p$ matrix, where element c_{ij} in the ith row and jth column is formed by multiplying the elements in the ith row of A by the corresponding elements in the jth row of B and adding the results. Thus,

$$c_{ij} = a_{i1}b_{1j} + a_{i2}b_{2j} + \cdots + a_{in}b_{nj}$$

We will try to illustrate this definition by showing how you get an element in the following product. Suppose. $A = \begin{bmatrix} 1 & 2 & -1 & 0 \\ 3 & 5 & 2 & -3 \end{bmatrix}$ and

$B = \begin{bmatrix} 2 & 1 & 3 \\ 0 & 4 & -2 \\ 1 & 2 & 0 \\ 1 & 3 & -2 \end{bmatrix}$. You can see that A is a 2×4 matrix and B is a 4×3

matrix, so AB will have dimension 2×3. If we let $C = AB$, we will show how to find c_{12} or the element in row 1, column 2 of matrix C. To get this element, we multiply row 1 of matrix A and column 2 of matrix B. This is the same as multiplying a row vector and a column vector. Thus,

$$c_{12} = [\,1 \quad 2 \quad -1 \quad 0\,]\begin{bmatrix} 1 \\ 4 \\ 2 \\ 3 \end{bmatrix}$$

$$= 1(1) + 2(4) - 1(2) + 0(3)$$

$$= 7$$

EXAMPLE 17.13

Multiply $\begin{bmatrix} 1 & 2 & -1 & 0 \\ 3 & 5 & 2 & -3 \end{bmatrix}\begin{bmatrix} 2 & 1 & 3 \\ 0 & 4 & -2 \\ 1 & 2 & 0 \\ 1 & 3 & -2 \end{bmatrix}$.

Solution

$$\begin{bmatrix} 1 & 2 & -1 & 0 \\ 3 & 5 & 2 & -3 \end{bmatrix}\begin{bmatrix} 2 & 1 & 3 \\ 0 & 4 & -2 \\ 1 & 2 & 0 \\ 1 & 3 & -2 \end{bmatrix}$$

$$= \begin{bmatrix} 1(2) + 2(0) - 1(1) + 0(1) & 1(1) + 2(4) - 1(2) + 0(3) \\ 3(2) + 5(0) + 2(1) - 3(1) & 3(1) + 5(4) + 2(2) - 3(3) \end{bmatrix}$$

$$\begin{matrix} 1(3) + 2(-2) - 1(0) + 0(-2) \\ 3(3) + 5(-2) + 2(0) - 3(-2) \end{matrix}$$

$$= \begin{bmatrix} 2 + 0 - 1 + 0 & 1 + 8 - 2 + 0 & 3 - 4 - 0 + 0 \\ 6 + 0 + 2 - 3 & 3 + 20 + 4 - 9 & 9 - 10 + 0 + 6 \end{bmatrix}$$

EXAMPLE 17.13 (Cont.)

$$= \begin{bmatrix} 1 & 7 & -1 \\ 5 & 18 & 5 \end{bmatrix}$$

≡ **Note**   In general, $AB \neq BA$. In fact, if AB is defined, then BA may not be defined. The only time that both AB and BA are defined is when both A and B are square matrices with the same dimensions. Thus, the commutative law does not hold for multiplication of matrices.

The following properties will hold, assuming that the dimensions of the matrices allow the products or sums to be defined.

$$A(BC) = (AB)C \qquad \text{associative law}$$

$$A(B + C) = AB + AC \qquad \text{left distributive law}$$

$$(B + C)A = BA + CA \qquad \text{right distributive law}$$

Because the commutative law does not hold, the left and right distributive laws usually give different results.

Application

EXAMPLE 17.14

A computer supply company has its inventory of four types of computer chips in three warehouses. The sales of each chip from each warehouse are shown in matrix S.

$$S = \begin{array}{r} \text{Chip type:} \\ \text{Warehouse A} \\ \text{Warehouse B} \\ \text{Warehouse C} \end{array} \begin{array}{cccc} 286 & 386 & 486 & 586 \\ \begin{bmatrix} 270 & 2{,}130 & 3{,}210 & 265 \\ 1{,}120 & 4{,}230 & 3{,}124 & 75 \\ 320 & 3{,}126 & 2{,}743 & 1{,}012 \end{bmatrix} \end{array}$$

Matrix P below shows the selling price and the profit for each type of chip.

$$P = \begin{array}{r} \text{Chip} \\ 286 \\ 386 \\ 486 \\ 586 \end{array} \begin{array}{cc} \text{Selling Price} & \text{Profit} \\ \begin{bmatrix} 25 & 10 \\ 35 & 13 \\ 52 & 17 \\ 97 & 33 \end{bmatrix} \end{array}$$

(a) How much money in sales did each warehouse generate during this month?
(b) How much profit did each warehouse make?

EXAMPLE 17.14 (Cont.)

Solution Since matrix S is a 3×4 matrix and P is a 4×2, we can determine the solution from the 3×2 matrix that results when we multiply SP.

$$SP = \begin{bmatrix} 270 & 2{,}130 & 3{,}210 & 265 \\ 1{,}120 & 4{,}230 & 3{,}124 & 75 \\ 320 & 3{,}126 & 2{,}743 & 1{,}012 \end{bmatrix} \cdot \begin{bmatrix} 25 & 10 \\ 35 & 13 \\ 52 & 17 \\ 97 & 33 \end{bmatrix}$$

$$SP = \begin{matrix} & \text{Sales} & \text{Profit} \\ \text{Warehouse A} \\ \text{Warehouse B} \\ \text{Warehouse C} \end{matrix} \begin{bmatrix} 273{,}925 & 93{,}705 \\ 345{,}773 & 121{,}773 \\ 358{,}210 & 123{,}865 \end{bmatrix}$$

From this matrix we can see, for example, that Warehouse A sold \$273,925 in chips for a profit of \$93,705.

Exercise Set 17.2

In Exercises 1–10, find the indicated products.

1. $\begin{bmatrix} 2 & 3 \\ 1 & -4 \end{bmatrix} \begin{bmatrix} 4 & -3 \\ 6 & 5 \end{bmatrix}$

2. $\begin{bmatrix} 1 & -2 & 4 \end{bmatrix} \begin{bmatrix} 3 \\ 9 \\ -7 \end{bmatrix}$

3. $\begin{bmatrix} 5 & 1 & 6 & 2 \end{bmatrix} \begin{bmatrix} 2 \\ 9 \\ -8 \\ 3 \end{bmatrix}$

4. $\begin{bmatrix} 2 & 3 & 1 \\ -4 & 2 & 0 \end{bmatrix} \begin{bmatrix} 4 & 1 \\ -5 & 2 \\ 3 & 0 \end{bmatrix}$

5. $\begin{bmatrix} 1 & 2 & 4 \\ 2 & 3 & 6 \end{bmatrix} \begin{bmatrix} 5 & -7 & 4 \\ 4 & 6 & 0 \\ -4 & 2 & 1 \end{bmatrix}$

6. $\begin{bmatrix} 4 & 7 \\ 2 & 3 \\ 6 & 2 \\ 9 & 1 \end{bmatrix} \begin{bmatrix} 3 & 7 & 6 & 1 & -5 \\ 4 & -3 & 2 & 0 & 10 \end{bmatrix}$

7. $\begin{bmatrix} 1 & 2 & 3 \\ 6 & 5 & 4 \\ -1 & 0 & 2 \end{bmatrix} \begin{bmatrix} 9 & 8 & 7 \\ -2 & 3 & -1 \\ 0 & 1 & 0 \end{bmatrix}$

8. $\begin{bmatrix} 9 & 8 & 7 \\ -2 & 3 & -1 \\ 0 & 1 & 0 \end{bmatrix} \begin{bmatrix} 1 & 2 & 3 \\ 6 & 5 & 4 \\ -1 & 0 & 2 \end{bmatrix}$

9. $\begin{bmatrix} 3 & 4 & -2 \\ 7 & 8 & 10 \end{bmatrix} \begin{bmatrix} 0 & 4 & 1 \\ 1 & 0 & 2 \\ 8 & -1 & 0 \end{bmatrix}$

10. $\begin{bmatrix} 1 & 2 & -1 \\ 2 & -1 & 3 \end{bmatrix} \begin{bmatrix} 0 & 1 & 0 & 2 \\ 1 & 0 & 2 & -1 \\ 3 & 4 & -1 & 0 \end{bmatrix}$

In Exercises 11–20, use matrices A, B, and C to determine the indicated answer, where $A = \begin{bmatrix} 1 & 2 \\ 3 & 4 \end{bmatrix}$, $B = \begin{bmatrix} 3 & 6 \\ 2 & 5 \end{bmatrix}$, **and** $C = \begin{bmatrix} 4 & -2 \\ -6 & 2 \end{bmatrix}$.

11. AB

12. BA

13. AC

14. BC

15. $A(BC)$

16. $(AB)C$

17. $A(B + C)$

18. $(B + C)A$

19. $(2A)(3B)$

20. $\left(2A + \dfrac{1}{2}C\right)B$

Solve Exercises 21–29.

21. Suppose $A = \begin{bmatrix} 2 & 10 \\ 3 & 15 \end{bmatrix}$ and $B = \begin{bmatrix} -10 & 25 \\ 2 & -5 \end{bmatrix}$.
What is AB? Can you conclude that if $AB = 0$, then either $A = 0$ or $B = 0$?

22. If $\begin{bmatrix} 2 & x \\ y & 5 \end{bmatrix} \begin{bmatrix} 4 & 6 \\ 9 & -1 \end{bmatrix} = \begin{bmatrix} 35 & 9 \\ 49 & 1 \end{bmatrix}$ determine x and y.

23. *Business* A computer supply company has its inventory of four types of computer chips in three warehouses. One month they had the following sales.

Chip type:	286	386	486	586
Warehouse A	270	2,130	3,210	265
Warehouse B	1,120	4,230	3,124	75
Warehouse C	320	3,126	2,743	1,012

The 286 chips sell for $25, the 386 chips sell for $35, the 486 chips sell for $55, and the 586 chips sell for $95. How much money in sales did each warehouse generate during this month?

24. *Construction* A contractor builds three sizes of houses (ranch, bi-level, and two story) in two different models, A and B. The contractor plans to build 100 new homes in a subdivision. Matrix P shows the number of each type of house planned for the subdivision.

$$P = \begin{array}{c} \text{Model A} \\ \text{Model B} \end{array} \begin{bmatrix} \overset{\text{Ranch}}{30} & \overset{\text{Bi-level}}{20} & \overset{\text{Two story}}{15} \\ 10 & 10 & 15 \end{bmatrix}$$

The amounts of each type of exterior material needed for each house are shown in matrix A. Here, concrete is in cubic yards, lumber in 1,000 board feet, bricks in 1,000s, and shingles in units of 100 ft^2.

$$A = \begin{array}{c} \text{Ranch} \\ \text{Bi-level} \\ \text{Two story} \end{array} \begin{bmatrix} \overset{\text{Concrete}}{10} & \overset{\text{Lumber}}{2} & \overset{\text{Bricks}}{2} & \overset{\text{Shingles}}{3} \\ 15 & 3 & 4 & 4 \\ 25 & 5 & 6 & 3 \end{bmatrix}$$

The cost of each of the units for each kind of material is given by matrix C.

$$C = \begin{array}{c} \text{Concrete} \\ \text{Lumber} \\ \text{Brick} \\ \text{Shingles} \end{array} \begin{bmatrix} \overset{\text{Cost per unit}}{25} \\ 210 \\ 75 \\ 40 \end{bmatrix}$$

(a) How much of each type of material will the contractor need for each house model?

(b) What will it cost to build each size of house?

(c) What will it cost to build each house model?

(d) What will it cost to build the entire subdivision?

25. *Physics*   The **Pauli spin matrices** in quantum mechanics are $A = \begin{bmatrix} 0 & 1 \\ 1 & 0 \end{bmatrix}$, $B = \begin{bmatrix} 0 & -i \\ i & 0 \end{bmatrix}$, and $C = \begin{bmatrix} 1 & 0 \\ 0 & -1 \end{bmatrix}$, where $i = \sqrt{-1}$. Show that $A^2 = B^2 = C^2 = I$, where $I = \begin{bmatrix} 1 & 0 \\ 0 & 1 \end{bmatrix}$.

26. *Physics*   Show that the Pauli spin matrices **anticommute**. That is, show that $AB = -BA$, $AC = -CA$, and $BC = -CB$.

27. *Physics*   The **commutator** of two matrices P and Q is $PQ - QP$. For the Pauli spin matrices, show that

(a) the commutator of A and B is $2iC$, (b) the commutator of A and C is $-2iB$, and (c) the commutator of B and C is $2iA$, where $i = \sqrt{-1}$.

28. Show, by multiplying the matrices, that the following equation represents an ellipse.

$$[x \quad y] \begin{bmatrix} 5 & -7 \\ 7 & 3 \end{bmatrix} \begin{bmatrix} x \\ y \end{bmatrix} = [30]$$

29. Write a computer program to multiply two matrices.

☰ 17.3
INVERSES OF MATRICES

One use of matrices is to help solve a system of linear equations. For example, if you had the system of linear equations

$$\begin{cases} a_1 x_1 + a_2 x_2 + a_3 x_3 = k_1 \\ b_1 x_1 + b_2 x_2 + b_3 x_3 = k_2 \\ c_1 x_1 + c_2 x_2 + c_3 x_3 = k_3 \end{cases}$$

you could write these as the matrices:

$$\begin{bmatrix} a_1 & a_2 & a_3 \\ b_1 & b_2 & b_3 \\ c_1 & c_2 & c_3 \end{bmatrix} \begin{bmatrix} x_1 \\ x_2 \\ x_3 \end{bmatrix} = \begin{bmatrix} k_1 \\ k_2 \\ k_3 \end{bmatrix}$$

Another way to look at this would be to let the coefficients of the linear equations be represented by the matrix

$$A = \begin{bmatrix} a_1 & a_2 & a_3 \\ b_1 & b_2 & b_3 \\ c_1 & c_2 & c_3 \end{bmatrix}$$

If X represents the column vector of variables, and K represents the column vector of constants, you have

$$X = \begin{bmatrix} x_1 \\ x_2 \\ x_3 \end{bmatrix} \text{ and } K = \begin{bmatrix} k_1 \\ k_2 \\ k_3 \end{bmatrix}$$

and the system of equations could be represented by $AX = K$.

With normal equations you would divide both sides by A to get the values of X, but these are matrices and there is no commutative law for multiplication. We will return to this dilemma after we introduce a new matrix.

In Section 17.2, we introduced matrix multiplication. A square $n \times n$ matrix with a 1 in each position of the main diagonal and zeros elsewhere is called the $n \times n$

identity matrix and is denoted as I_n. Thus,

$$I_2 = \begin{bmatrix} 1 & 0 \\ 0 & 1 \end{bmatrix} \quad \text{and} \quad I_3 = \begin{bmatrix} 1 & 0 & 0 \\ 0 & 1 & 0 \\ 0 & 0 & 1 \end{bmatrix}$$

are both identity matrices.

If the size of the identity matrix is understood, we simply write I.

Now, back to our problem of $AX = K$. What we need is a matrix A^{-1}, where $A^{-1}A = I$. We would then find that

$$X = A^{-1}K$$

When A is an $n \times n$ matrix, then A is an **invertible** or **nonsingular matrix**, if there exists another $n \times n$ matrix, A^{-1}, where

$$A^{-1}A = AA^{-1} = I$$

If A^{-1} exists, it is called the **inverse** of matrix A.

There are two procedures for finding the inverse of a matrix. The first method may be easier on 2×2 matrices.

Steps for Finding the Inverse of a 2 × 2 Matrix

If $A = \begin{bmatrix} a & b \\ c & d \end{bmatrix}$ is a 2 × 2 matrix, then use the following four steps to determine A^{-1}:

1. Interchange the elements on the main diagonal.
2. Change the signs of the elements that are not on the main diagonal.
3. Find the determinant of the original matrix.
4. Divide each element at the end of Step 2 by the determinant of the original matrix.

≡ **Note**   Symbolically, we can summarize the four steps for finding the inverse of a 2 × 2 matrix as

$$A^{-1} = \frac{1}{ad - bc} \begin{bmatrix} d & -b \\ -c & a \end{bmatrix}$$

EXAMPLE 17.15

Find the inverse of the matrix $A = \begin{bmatrix} 1 & 2 \\ 4 & 10 \end{bmatrix}$.

Solution   **Step 1:** $\begin{bmatrix} 10 & 2 \\ 4 & 1 \end{bmatrix}$ Interchange elements on main diagonal.

Step 2: $\begin{bmatrix} 10 & -2 \\ -4 & 1 \end{bmatrix}$ Change signs of elements not on main diagonal.

Step 3: $\begin{vmatrix} 1 & 2 \\ 4 & 10 \end{vmatrix} = 10 - 8 = 2$ Find the determinant of the original matrix.

EXAMPLE 17.15 (Cont.)

Step 4: $\begin{bmatrix} \dfrac{10}{2} & \dfrac{-2}{2} \\ \dfrac{-4}{2} & \dfrac{1}{2} \end{bmatrix}$ Divide each element from Step 2 by the determinant (Step 3).

The inverse is $A^{-1} = \begin{bmatrix} 5 & -1 \\ -2 & \dfrac{1}{2} \end{bmatrix}$. Check your answer. Is $A^{-1}A = I$?

$$\begin{bmatrix} 5 & -1 \\ -2 & \dfrac{1}{2} \end{bmatrix} \begin{bmatrix} 1 & 2 \\ 4 & 10 \end{bmatrix} = \begin{bmatrix} 5-4 & 10-10 \\ -2+2 & -4+5 \end{bmatrix} = \begin{bmatrix} 1 & 0 \\ 0 & 1 \end{bmatrix} = I$$

The second method for determining the inverse of a matrix is slightly more complicated for a 2×2 matrix, but it will work for a matrix of any size.

With the second method, we are going to transform the given matrix into the identity matrix. At the same time, we will change an identity matrix into the inverse of the given matrix.

Steps for Finding the Inverse for Any $n \times n$ Matrix

The idea in this method for finding A^{-1} is to begin with the matrix $[\,A \ \vdots \ I\,]$ and use the following two steps to change it to $[\,I \ \vdots \ A^{-1}\,]$.

1. Multiply or divide all elements in a row by a nonzero constant.
2. Add a constant multiple of the elements of one row to the corresponding elements of another row.

We demonstrate this method in Example 17.16 with the same matrix used in Example 17.15.

EXAMPLE 17.16

Find the inverse of $A = \begin{bmatrix} 1 & 2 \\ 4 & 10 \end{bmatrix}$.

Solution First, we form the augmented matrix $[\,A \ \vdots \ I\,]$.

$$\begin{bmatrix} 1 & 2 & \vdots & 1 & 0 \\ 4 & 10 & \vdots & 0 & 1 \end{bmatrix}$$

In the following demonstration, we let R_1 mean Row 1 and R_2 mean Row 2 for this 2×4 matrix. The arrows point to the row that we changed and that row is always listed last. Everything we do is designed to change the left half of the matrix to the identity matrix.

$$\begin{bmatrix} 1 & 2 & \vdots & 1 & 0 \\ 4 & 10 & \vdots & 0 & 1 \end{bmatrix}$$

First, multiply R_1 by -4 and add that to R_2 to get a new Row 2.

$$\begin{bmatrix} 1 & 2 & \vdots & 1 & 0 \\ 0 & 2 & \vdots & -4 & 1 \end{bmatrix} \quad \longleftarrow -4R_1 + R_2$$

EXAMPLE 17.16 (Cont.)

Next, multiply R_2 by -1 and add that to R_1. This is the new row 1.

$$\begin{bmatrix} 1 & 0 & \vdots & 5 & -1 \\ 0 & 2 & \vdots & -4 & 1 \end{bmatrix} \quad \leftarrow -R_2 + R_1$$

Now, all that is left is to divide Row 2 by 2 (or multiply it by $\frac{1}{2}$).

$$\begin{bmatrix} 1 & 0 & \vdots & 5 & -1 \\ 0 & 1 & \vdots & -2 & \frac{1}{2} \end{bmatrix} \quad \leftarrow \frac{1}{2} R_2$$

The left half of the matrix is the identity matrix and the right half is the inverse that we wanted.

$$A^{-1} = \begin{bmatrix} 5 & -1 \\ -2 & \frac{1}{2} \end{bmatrix}$$

This is the same answer we got for Example 17.15.

We will work another 2×2 example and then work a 3×3 example. Before beginning, you should always check to see if the matrix is invertible. It will not be invertible if every element in any row or column is zero. Thus

$$\begin{bmatrix} 1 & 2 & 3 \\ 0 & 0 & 0 \\ 4 & 5 & 6 \end{bmatrix} \quad \text{and} \quad \begin{bmatrix} 1 & 2 & 0 \\ 4 & 5 & 0 \\ 7 & 8 & 0 \end{bmatrix}$$

cannot be inverted. A matrix cannot be inverted if every element in one row (or column) is a constant multiple of every corresponding element in another row (or column). Thus $\begin{bmatrix} 4 & 2 \\ 12 & 6 \end{bmatrix}$ cannot be inverted, because each element in row 2 is 3 times its corresponding element in row 1. Notice that a matrix that cannot be inverted has a determinant of 0.

EXAMPLE 17.17

Find the inverse of $B = \begin{bmatrix} 8 & 4 \\ 3 & 2 \end{bmatrix}$.

Solution

$$\begin{bmatrix} 8 & 4 & \vdots & 1 & 0 \\ 3 & 2 & \vdots & 0 & 1 \end{bmatrix}$$

$$\begin{bmatrix} 8 & 4 & \vdots & 1 & 0 \\ 0 & 4 & \vdots & -3 & 8 \end{bmatrix} \quad \leftarrow -3R_1 + 8R_2$$

$$\begin{bmatrix} 8 & 0 & \vdots & 4 & -8 \\ 0 & 4 & \vdots & -3 & 8 \end{bmatrix} \quad \leftarrow -R_2 + R_1$$

$$\begin{bmatrix} 1 & 0 & \vdots & \frac{1}{2} & -1 \\ 0 & 1 & \vdots & -\frac{3}{4} & 2 \end{bmatrix} \quad \begin{matrix} \leftarrow R_1 \div 8 \\ \leftarrow R_2 \div 4 \end{matrix}$$

The inverted matrix is $B^{-1} = \begin{bmatrix} \frac{1}{2} & -1 \\ -\frac{3}{4} & 2 \end{bmatrix}$.

EXAMPLE 17.18

Find the inverse of $C = \begin{bmatrix} 7 & -8 & 5 \\ -4 & 5 & -3 \\ 1 & -1 & 1 \end{bmatrix}$.

Solution

$\begin{bmatrix} 7 & -8 & 5 & | & 1 & 0 & 0 \\ -4 & 5 & -3 & | & 0 & 1 & 0 \\ 1 & -1 & 1 & | & 0 & 0 & 1 \end{bmatrix}$

$\begin{bmatrix} 7 & -8 & 5 & | & 1 & 0 & 0 \\ -4 & 5 & -3 & | & 0 & 1 & 0 \\ 0 & -1 & -2 & | & 1 & 0 & -7 \end{bmatrix}$ ← $R_1 - 7R_3$

$\begin{bmatrix} 7 & -8 & 5 & | & 1 & 0 & 0 \\ 0 & 3 & -1 & | & 4 & 7 & 0 \\ 0 & -1 & -2 & | & 1 & 0 & -7 \end{bmatrix}$ ← $4R_1 + 7R_2$

$\begin{bmatrix} 7 & -8 & 5 & | & 1 & 0 & 0 \\ 0 & 3 & -1 & | & 4 & 7 & 0 \\ 0 & 0 & -7 & | & 7 & 7 & -21 \end{bmatrix}$ ← $R_2 + 3R_3$

$\begin{bmatrix} 7 & -8 & 5 & | & 1 & 0 & 0 \\ 0 & -21 & 0 & | & -21 & -42 & -21 \\ 0 & 0 & -7 & | & 7 & 7 & -21 \end{bmatrix}$ ← $R_3 - 7R_2$

$\begin{bmatrix} 7 & -8 & 5 & | & 1 & 0 & 0 \\ 0 & 1 & 0 & | & 1 & 2 & 1 \\ 0 & 0 & 1 & | & -1 & -1 & 3 \end{bmatrix}$ ← $R_2 \div -21$
   ← $R_3 \div -7$

$\begin{bmatrix} 7 & -8 & 0 & | & 6 & 5 & -15 \\ 0 & 1 & 0 & | & 1 & 2 & 1 \\ 0 & 0 & 1 & | & -1 & -1 & 3 \end{bmatrix}$ ← $-5R_3 + R_1$

$\begin{bmatrix} 7 & 0 & 0 & | & 14 & 21 & -7 \\ 0 & 1 & 0 & | & 1 & 2 & 1 \\ 0 & 0 & 1 & | & -1 & -1 & 3 \end{bmatrix}$ ← $8R_2 + R_1$

$\begin{bmatrix} 1 & 0 & 0 & | & 2 & 3 & -1 \\ 0 & 1 & 0 & | & 1 & 2 & 1 \\ 0 & 0 & 1 & | & -1 & -1 & 3 \end{bmatrix}$ ← $R_1 \div 7$

The inverse of C is $C^{-1} = \begin{bmatrix} 2 & 3 & -1 \\ 1 & 2 & 1 \\ -1 & -1 & 3 \end{bmatrix}$.

Exercise Set 17.3

In Exercises 1–20, if the given matrix is invertible find its inverse.

1. $\begin{bmatrix} 6 & 1 \\ 5 & 1 \end{bmatrix}$

2. $\begin{bmatrix} 5 & 1 \\ 9 & 2 \end{bmatrix}$

3. $\begin{bmatrix} 10 & 4 \\ 8 & 3 \end{bmatrix}$

4. $\begin{bmatrix} 9 & 6 \\ 4 & 3 \end{bmatrix}$

5. $\begin{bmatrix} 8 & -6 \\ -6 & 4 \end{bmatrix}$

6. $\begin{bmatrix} 12 & 9 \\ -9 & -7 \end{bmatrix}$

7. $\begin{bmatrix} 15 & 10 \\ 4 & 3 \end{bmatrix}$

8. $\begin{bmatrix} 15 & 10 \\ 3 & 4 \end{bmatrix}$

9. $\begin{bmatrix} 0 & -3 & 0 \\ 1 & 0 & 0 \\ 0 & 0 & 4 \end{bmatrix}$

10. $\begin{bmatrix} 1 & 0 & 0 \\ 0 & 4 & 0 \\ 0 & 0 & 2 \end{bmatrix}$

11. $\begin{bmatrix} 1 & 2 & 6 \\ 0 & 0 & 2 \\ -3 & -6 & -9 \end{bmatrix}$

12. $\begin{bmatrix} 4 & 5 & 1 \\ 1 & 0 & 1 \\ 4 & 5 & 1 \end{bmatrix}$

13. $\begin{bmatrix} 8 & 7 & -1 \\ -5 & -5 & 1 \\ -4 & -4 & 1 \end{bmatrix}$

14. $\begin{bmatrix} 1 & 2 & 3 \\ 2 & 5 & 7 \\ 1 & 1 & 1 \end{bmatrix}$

15. $\begin{bmatrix} 3 & -1 & 0 \\ -6 & 2 & 0 \\ 1 & 0 & 5 \end{bmatrix}$

16. $\begin{bmatrix} 1 & -1 & 1 \\ 7 & -8 & 5 \\ -4 & 5 & -3 \end{bmatrix}$

17. $\begin{bmatrix} 2 & 0 & 0 \\ 2 & 2 & 0 \\ 2 & 2 & 2 \end{bmatrix}$

18. $\begin{bmatrix} 3 & 1 & -1 \\ 1 & -2 & 0 \\ 0 & 3 & 1 \end{bmatrix}$

19. $\begin{bmatrix} 1 & -1 & 1 \\ 0 & 2 & -1 \\ 2 & 3 & 0 \end{bmatrix}$

20. $\begin{bmatrix} 1 & 2 & -1 \\ 2 & -2 & 1 \\ 6 & 4 & 3 \end{bmatrix}$

Solve Exercises 21 and 22.

21. Check your answers to 1–20 by multiplying the given matrix and its inverse. Remember, if the matrix in the problem is called A and your answer A^{-1}, then $AA^{-1} = I$.

22. Write a computer program to determine the inverse of an $n \times n$ matrix.

☰ 17.4
MATRICES AND LINEAR EQUATIONS

Earlier in this chapter, we said that a system of equations could be represented by matrices. For example, the system

$$\begin{cases} 2x + 3y = 5 \\ 3x + 5y = 9 \end{cases}$$

would be represented by three matrices $A, X,$ and K, where $AX = K$. To do this, let A be the matrix formed by the coefficients of the variables, let X be a column vector of the variables, and let K be a column vector of the constants. Thus, for this system, we have

$$A = \begin{bmatrix} 2 & 3 \\ 3 & 5 \end{bmatrix} \qquad X = \begin{bmatrix} x \\ y \end{bmatrix} \qquad K = \begin{bmatrix} 5 \\ 9 \end{bmatrix}$$

Now that we can invert a matrix, we can find A^{-1}. Since $A^{-1}A = I$ then $A^{-1}(AX) = (A^{-1}A)X = IX = X$. With this knowledge, we can find the solution to our system of equations.

$$AX = K$$

$$A^{-1}(AX) = A^{-1}K$$

$$X = A^{-1}K$$

EXAMPLE 17.19

Use the inverse of the coefficient matrix to solve the linear system

$$\begin{cases} 2x + 3y = 5 \\ 3x + 5y = 9 \end{cases}$$

Solution   Since $A = \begin{bmatrix} 2 & 3 \\ 3 & 5 \end{bmatrix}$, and $|A| = \begin{vmatrix} 2 & 3 \\ 3 & 5 \end{vmatrix} = 1$, we have, by the first method of inverting a matrix, $A^{-1} = \begin{bmatrix} 5 & -3 \\ -3 & 2 \end{bmatrix}$. So

$$X = A^{-1}K = \begin{bmatrix} 5 & -3 \\ -3 & 2 \end{bmatrix} \cdot \begin{bmatrix} 5 \\ 9 \end{bmatrix}$$

$$= \begin{bmatrix} 5(5) - 3(9) \\ -3(5) + 2(9) \end{bmatrix}$$

$$= \begin{bmatrix} 25 - 27 \\ -15 + 18 \end{bmatrix}$$

$$= \begin{bmatrix} -2 \\ 3 \end{bmatrix}$$

and $x = -2$ and $y = 3$.

Let's try this method for solving a system of linear equations on a system of three equations with three variables.

EXAMPLE 17.20

Use the inverse of the coefficient matrix to solve the linear system

$$\begin{cases} 7x - 8y + 5z = 18 \\ -4x + 5y - 3z = -11 \\ x - y + z = 1 \end{cases}$$

Solution   This linear system can be represented by the three matrices A, X, and K, where $AX = K$. Again, A is the matrix formed by the coefficients, X is the column vector of the variables, and K is the column vector of the constants. From this system we have

$$A = \begin{bmatrix} 7 & -8 & 5 \\ -4 & 5 & -3 \\ 1 & -1 & 1 \end{bmatrix} \qquad X = \begin{bmatrix} x \\ y \\ z \end{bmatrix} \qquad K = \begin{bmatrix} 18 \\ -11 \\ 1 \end{bmatrix}$$

EXAMPLE 17.20 (Cont.)

We will first need A^{-1}. In Example 17.18, we found that

$$A^{-1} = \begin{bmatrix} 2 & 3 & -1 \\ 1 & 2 & 1 \\ -1 & -1 & 3 \end{bmatrix}$$

Thus,

$$X = A^{-1}K = \begin{bmatrix} 2 & 3 & -1 \\ 1 & 2 & 1 \\ -1 & -1 & 3 \end{bmatrix} \begin{bmatrix} 18 \\ -11 \\ 1 \end{bmatrix}$$

$$= \begin{bmatrix} 2(18) + 3(-11) - 1(1) \\ 1(18) + 2(-11) + 1(1) \\ -1(18) - 1(-11) + 3(1) \end{bmatrix}$$

$$= \begin{bmatrix} 36 - 33 - 1 \\ 18 - 22 + 1 \\ -18 + 11 + 3 \end{bmatrix}$$

$$= \begin{bmatrix} 2 \\ -3 \\ -4 \end{bmatrix}$$

From this, we see that $x = 2, y = -3$, and $z = -4$.

This is a lot of work. It would be nice if we had a computer program to help solve a system of equations. If fact, the design of such a program is included as an exercise in the following exercise set.

Application

EXAMPLE 17.21

Apply Kirchhoff's laws to the circuit in Figure 17.1 and determine the currents in I_1, I_2, and I_3.

Solution   If we apply Kirchhoff's laws to the circuit in Figure 17.1, we obtain the following system of linear equations.

$$\begin{cases} I_1 - I_2 + I_3 = 0 \\ 5I_1 + 8I_2 = 4 \\ 8I_2 + 2I_3 = 38 \end{cases}$$

FIGURE 17.1

EXAMPLE 17.21 (Cont.)

The coefficient matrix for this system is $A = \begin{bmatrix} 1 & -1 & 1 \\ 5 & 8 & 0 \\ 0 & 8 & 2 \end{bmatrix}$. (Notice that the coefficients of the missing variables are 0.) The inverse of A is

$$A^{-1} = \begin{bmatrix} \dfrac{8}{33} & \dfrac{5}{33} & -\dfrac{4}{33} \\ -\dfrac{5}{33} & \dfrac{1}{33} & \dfrac{5}{66} \\ \dfrac{20}{33} & -\dfrac{4}{33} & \dfrac{13}{66} \end{bmatrix} = \dfrac{1}{66} \begin{bmatrix} 16 & 10 & -8 \\ -10 & 2 & 5 \\ 40 & -8 & 13 \end{bmatrix}$$

If the constant matrix is $B = \begin{bmatrix} 0 \\ 4 \\ 38 \end{bmatrix}$ and $X = \begin{bmatrix} I_1 \\ I_2 \\ I_3 \end{bmatrix}$, then

$$X = A^{-1}B$$

$$= \dfrac{1}{66} \begin{bmatrix} 16 & 10 & -8 \\ -10 & 2 & 5 \\ 40 & -8 & 13 \end{bmatrix} \begin{bmatrix} 0 \\ 4 \\ 38 \end{bmatrix}$$

$$= \dfrac{1}{66} \begin{bmatrix} -264 \\ 198 \\ 462 \end{bmatrix}$$

$$= \begin{bmatrix} -4 \\ 3 \\ 7 \end{bmatrix}$$

Thus, the three currents are $I_1 = -4$ A, $I_2 = 3$ A, and $I_3 = 7$ A.

Exercise Set 17.4

In Exercises 1–10, solve the given systems of equations by using the inverse of the coefficient matrix. The numbers in parentheses refer to exercises in Exercise Set 17.3, where you determined the inverse of the coefficient matrix.

1. $\begin{cases} 6x + y = -4 \\ 5x + y = -3 \end{cases}$ (Exercise Set 17.3, Exercise 1)

2. $\begin{cases} 10x + 4y = 8 \\ 8x + 3y = 7 \end{cases}$ (Exercise Set 17.3, Exercise 3)

3. $\begin{cases} 8x - 6y = -27 \\ -6x + 4y = 19 \end{cases}$ (Exercise Set 17.3, Exercise 5)

4. $\begin{cases} 15x + 10y = -5 \\ 4x + 3y = 0 \end{cases}$ (Exercise Set 17.3, Exercise 7)

5. $\begin{cases} x + 2y + 6z = 12 \\ 2z = 2 \\ -3x - 6y - 9z = -27 \end{cases}$ (Exercise Set 17.3, Exercise 11)

6. $\begin{cases} 8x + 7y - z = 9 \\ -5x - 5y + z = -1 \\ -4x - 4y + z = 0 \end{cases}$ (Exercise Set 17.3, Exercise 13)

7. $\begin{cases} x + 2y + 3x = 4 \\ 2x + 5y + 7z = 7.5 \\ x + y + z = 1 \end{cases}$ (Exercise Set 17.3, Exercise 14)

9. $\begin{cases} 2x = 11 \\ 2x + 2y = 4 \\ 2x + 2y + 2z = -9 \end{cases}$ (Exercise Set 17.3, Exercise 17)

8. $\begin{cases} x - y + z = -6.6 \\ 7x - 8y + 5z = -43 \\ -4x + 5y - 3z = 26.9 \end{cases}$ (Exercise Set 17.3, Exercise 16)

10. $\begin{cases} x - y + z = 22 \\ 2y - z = -23 \\ 2x + 3y = -11.2 \end{cases}$ (Exercise Set 17.3, Exercise 19)

In Exercises 11–20, solve each system of equations by using the inverse of the coefficient matrix.

11. $\begin{cases} 2x + y = 1 \\ -3x + 2y = 16 \end{cases}$

12. $\begin{cases} 4x + 5y = 2 \\ 3x - 2y = 13 \end{cases}$

13. $\begin{cases} 2x + 2y = 4 \\ 4x + 3y = 1 \end{cases}$

14. $\begin{cases} 3x + y = -5 \\ 4x + 3y = 2 \end{cases}$

15. $\begin{cases} 1.5x + 2.5y = 0.3 \\ 3.2x + 2.6y = 7.2 \end{cases}$

16. $\begin{cases} 7x + 2y + z = 2 \\ 3x - 2y + 4z = 13 \\ 4x + 5y - z = 1 \end{cases}$

17. $\begin{cases} 5x + 2y + 3z = 1 \\ -3x + 2y - 8z = 6 \\ 4x - 2y + 9z = -7 \end{cases}$

18. $\begin{cases} 2x + 4y + z = 10 \\ 4x + 2y + z = 8 \\ 6x + 4y + 7z = -2 \end{cases}$

19. $\begin{cases} x + 2y + 4z = 7 \\ 3x + y + 4z = -2 \\ 2x + 9y - 2z = 10 \end{cases}$

20. $\begin{cases} 2x - y + z = 7 \\ 4x - 2y + z = 5 \\ 6x - 3y + 5z = -3 \end{cases}$

Solve Exercises 21–25.

21. *Electricity*   If Kirchhoff's laws are applied to the circuit in Figure 17.2, the following equations are obtained. Determine the indicated currents.

$$I_A - I_B - I_C = 0$$

$$16I_A + 4I_C = 100$$

$$12I_B - 4I_C = 60$$

22. *Electricity*   Applying Kirchhoff's laws to the circuit in Figure 17.3 results in the following equations. Determine the indicated currents.

$$I_3 - I_1 - I_2 = 0$$

$$20I_1 + 0.5I_1 - 15I_2 - 0.4I_2 = 120 - 80$$

$$15I_2 + 0.4I_2 + 10I_3 + 0.6I_3 = 140$$

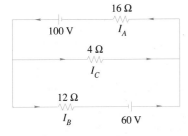

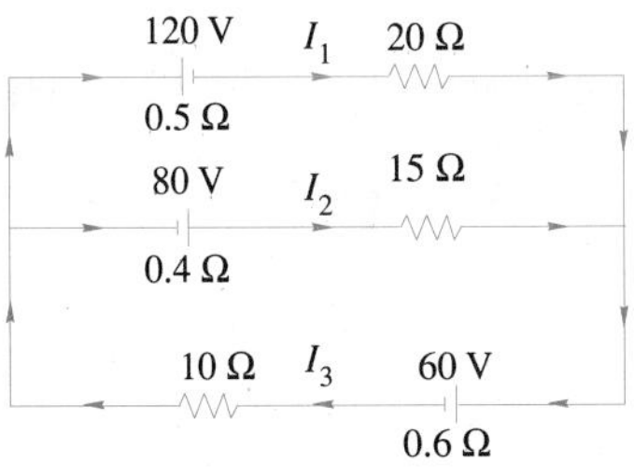

FIGURE 17.2

FIGURE 17.3

23. *Electricity*   The currents through the resistors in Figure 17.4 produce the following equations. Determine the currents.

$$7.18I_1 - I_2 + 2.2I_3 = 10$$

$$-I_1 + 5.8I_2 + 1.5I_3 = 15$$

$$2.2I_1 + 1.5I_2 + 8.4I_3 = 20$$

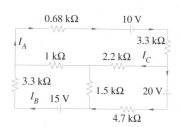

FIGURE 17.4

24. *Business*   A company manufactures robotic controls. Their current models are the *RC*-1 and *RC*-2. Each *RC*-1 unit requires 8 transistors and 4 integrated circuits. Each *RC*-2 unit uses 9 transistors and 5 integrated circuits. Each day the company receives 1,596 transistors and 860 integrated circuits. How many units of each model can be made if all parts are used?

 25.   Write a computer program that uses matrices to solve a system of *n* linear equations with *n* variables.

☰ CHAPTER 17 REVIEW

Important Terms and Concepts

Augmented matrix

Coefficient matrix

Cofactor

Column vector

Dimension

Identity matrix

Inverse of a matrix

Invertible matrix

Matrix
   Addition

Multiplication

Scalar multiplication

Subtraction

Nonsingular matrix

Row vector

Scalar

Singular matrix

Square matrix

Zero matrix

Review Exercises

In Exercises 1–6, use the following matrices to determine the indicated matrix.

$$A = \begin{bmatrix} 4 & 3 & 2 & 5 \\ 6 & 7 & -1 & 4 \\ 9 & 10 & -8 & 3 \end{bmatrix} \quad B = \begin{bmatrix} 3 & -2 & 1 & 0 \\ 5 & -1 & 2 & 0 \\ 4 & 3 & -2 & 0 \end{bmatrix} \quad C = \begin{bmatrix} 0 & 1 & 0 & 2 \\ 3 & 0 & 4 & 0 \\ 0 & -5 & 0 & -6 \end{bmatrix}$$

1. $A + C$

2. $B + C$

3. $A - B$

4. $C - B$

5. $2A - 3C$

6. $4A + B - 2C$

In Exercises 7–10, find the indicated products.

7. $\begin{bmatrix} 3 & 2 \\ 4 & 5 \\ 1 & 0 \end{bmatrix} \begin{bmatrix} 1 & 2 \\ 0 & 1 \end{bmatrix}$

9. $\begin{bmatrix} 3 \\ 2 \\ -1 \end{bmatrix} [4 \quad 5 \quad 1]$

8. $\begin{bmatrix} 1 & -4 \\ 5 & 1 \end{bmatrix} \begin{bmatrix} 3 & 4 & 1 \\ 2 & 5 & 0 \end{bmatrix}$

10. $[4 \quad 5 \quad 1] \begin{bmatrix} -2 \\ 7 \\ -3 \end{bmatrix}$

In Exercises 11–14, if the given matrix is invertible, find its inverse.

11. $\begin{bmatrix} 2 & 3 \\ -4 & -5 \end{bmatrix}$

13. $\begin{bmatrix} -2 & 1 & 0 \\ 0 & 4 & 0 \\ 1 & 0 & 1 \end{bmatrix}$

12. $\begin{bmatrix} 2 & -1 \\ 0 & 4 \end{bmatrix}$

14. $\begin{bmatrix} 2 & 3 & -1 \\ 1 & 2 & 1 \\ -1 & -1 & 3 \end{bmatrix}$

In Exercises 15–18, solve the system of equations by using the inverse of the coefficient matrix.

15. $\begin{cases} 12x + 5y = -2 \\ 3x + y = 1.1 \end{cases}$

17. $\begin{cases} 2x + 3y + 5z = 20 \\ -2x + 3y + 5z = 12 \\ 5x - 3y - 2z = 9 \end{cases}$

16. $\begin{cases} 4x + y = -4 \\ -3x + 2y = 14 \end{cases}$

18. $\begin{cases} x + y + 6z = -3 \\ -2x + 2y + 4z = -2 \\ 3x + 2y + 4z = 14 \end{cases}$

Solve Exercises 19–24.

19. *Physics* Masses of 9 kg and 11 kg are attached to a cord that passes over a frictionless pulley. When the masses are released, the acceleration a of each mass, in meters per second squared (m/s²), and the tension T in the cord, in newtons (N), are related by the system

$$\begin{cases} T - 75.4 = 9a \\ 100.0 - T = 11a \end{cases}$$

Find a and T.

20. *Business* A computer company makes three types of computers—a personal computer (PC), a business computer (BC), and a technical computer (TC). There are three parts in each computer that they have difficulty getting: RAM chips, EPROMS, and transistors. The number of each part needed by each computer is shown in this table.

	RAM	EPROM	Transistor
PC	4	2	7
BC	8	3	6
TC	12	5	11

If the company is guaranteed 1,872 RAM chips, 771 EPROMS, and 1,770 transistors each week, how many of each computer can be made?

21. Given the equations

$$\begin{cases} x' = \dfrac{1}{2}\left(x + y\sqrt{3}\right) \\ y' = \dfrac{1}{2}\left(-x\sqrt{3} + y\right) \end{cases}$$

and

$$\begin{cases} x'' = \frac{1}{2}\left(-x' + y'\sqrt{3}\right) \\ y'' = -\frac{1}{2}\left(x'\sqrt{3} + y'\right) \end{cases}$$

write each set as a matrix equation and solve for x'' and y'' in terms of x and y by multiplying matrices.

22. The equations in Exercise 21 represent rotations of axes in two directions. In Section 15.7, we found that, if the angle of rotation is θ, then

$$\begin{cases} x'' = x\cos\theta + y\sin\theta \\ y'' = -x\sin\theta + y\cos\theta \end{cases}$$

What was the rotation angle for the equations in Exercise 21?

23. *Optics* The following matrix product is used in discussing two thin lenses in air:

$$M = \begin{bmatrix} 1 & -\dfrac{1}{f_2} \\ 0 & 1 \end{bmatrix} \cdot \begin{bmatrix} 1 & 0 \\ d & 1 \end{bmatrix} \cdot \begin{bmatrix} 1 & -\dfrac{1}{f_1} \\ 0 & 1 \end{bmatrix}$$

where f_1 and f_2 are the focal lengths of the lenses and d is the distance between them. Evaluate M.

24. *Optics* In Exercise 23, element M_{12} of M is $-\dfrac{1}{f}$, where f is the focal length of the combination. Determine $\dfrac{1}{f}$.

CHAPTER 17 TEST

1. Given $A = \begin{bmatrix} 8 & 0 & -4 \\ 16 & -6 & 2 \end{bmatrix}$ and $B = \begin{bmatrix} -1 & 5 & -3 \\ 3 & 0 & 4 \end{bmatrix}$, find

(a) $A + B$

(b) $3A - 2B$

2. Calculate the product $[\,1 \quad -2 \quad 3\,]\begin{bmatrix} -4 \\ -6 \\ 8 \end{bmatrix}$.

3. If $C = \begin{bmatrix} 4 & 6 \\ -10 & 4 \end{bmatrix}$ and $D = \begin{bmatrix} \dfrac{1}{2} & 0 \\ -\dfrac{3}{2} & 4 \end{bmatrix}$, calculate

(a) CD and (b) DC.

4. If $E = \begin{bmatrix} 2 & -3 \\ 7 & 9 \end{bmatrix}$, (a) find E^{-1}. (b) Use E^{-1} to solve $EX = F$, where $F = \begin{bmatrix} -31 \\ 28 \end{bmatrix}$.

5. Solve the following system of equations by using the inverse of the coefficient matrix.

$$\begin{cases} x + 3y + z = -2 \\ 2x + 5y + z = -5 \\ x + 2y + 3z = 6 \end{cases}$$

18

Higher Degree Equations

Package designers often need to design a box that will hold a specific volume. In Section 18.3, we will see how to design such a box made from a given sheet of material.

The majority of equations that we have solved have been linear or quadratic equations, or systems of linear equations. We were able to solve higher degree equations when they could be factored. It seems, however, that most equations do not factor. When you get to equations of degree higher than two, there is no easy formula, such as the quadratic formula, to help solve them. In this chapter, we will explore some methods for solving polynomial equations of degree higher than two.

≡ 18.1
THE REMAINDER AND FACTOR THEOREMS

We will begin this section with the definition of a polynomial.

> **Polynomial**
>
> A **polynomial** is a function of the form
>
> $$P(x) = a_n x^n + a_{n-1} x^{n-1} + \cdots + a_2 x^2 + a_1 x + a_0$$
>
> where $a_0, a_1, a_2, \ldots, a_n$ are numbers, n is a non-negative integer, and $a_n \neq 0$. This polynomial has **degree** n.
>
> a_n is called the **leading coefficient** of the polynomial.

EXAMPLE 18.1

The following are some examples of polynomials.

$$P(x) = 3x^2 + 2x - 1 \text{ has degree 2.}$$

$$P(x) = 5x^6 + \frac{7}{3}x^3 - 4x + 1 \text{ has degree 6.}$$

$$P(x) = -\sqrt{7} \text{ has degree 0.}$$

$$P(x) = \frac{2}{3}x^4 - \frac{1}{2}x^3 + x^2 \text{ has degree 4.}$$

EXAMPLE 18.2

The following are not polynomials.

   $P(x) = 4x^{5/2} + 2x^2 - 3$, because one of the exponents, $\frac{5}{2}$, is not an integer.

   $P(x) = 5x^{-3}$, because the exponent is not positive.

The third polynomial in Example 18.1, $P(x) = -\sqrt{7}$, is an example of a constant polynomial.

> **Constant and Zero Polynomials**
>
> A **constant polynomial function** is of the form $P(x) = c$, where c is a real number and $c \neq 0$. The degree of a constant polynomial function is zero.
>
> $P(x) = 0$ is called the **zero polynomial**. It has no degree.

The roots, solutions, or zeros of a polynomial function $P(x)$ are the values of x for which $P(x) = 0$. In earlier chapters, we learned that, if $P(x) = ax + b$, then $P(x) = 0$, when $x = \dfrac{-b}{a}$. We also learned the quadratic formula, which states that if $P(x)$ is of degree 2 and $P(x) = ax^2 + bx + c$, then $P(x) = 0$, when $x = \dfrac{-b \pm \sqrt{b^2 - 4ac}}{2a}$.

EXAMPLE 18.3

Find the roots of **(a)** $P(x) = 7x - \sqrt{2}$ and **(b)** $P(x) = 4x^2 - 3x - 5$.

Solutions

(a) The roots occur when $P(x) = 7x - \sqrt{2} = 0$. That is, when $7x = \sqrt{2}$ or $x = \frac{1}{7}\sqrt{2}$.

(b) Here, we want the solution to be $4x^2 - 3x - 5 = 0$. Using the quadratic formula, we get

$$x = \frac{-(-3) \pm \sqrt{(-3)^2 - 4(4)(-5)}}{2(4)}$$

$$= \frac{3 \pm \sqrt{9 + 80}}{8} = \frac{3 \pm \sqrt{89}}{8}$$

The roots are $x = \frac{3}{8} + \frac{1}{8}\sqrt{89}$ and $x = \frac{3}{8} - \frac{1}{8}\sqrt{89}$.

We will now look at some ways we can solve polynomial functions of degree larger than two.

Remainder Theorem

Whenever we divide a polynomial $P(x)$ by $x - a$ we get a quotient $Q(x)$ that is a polynomial and a remainder R that is a constant. In fact, when $P(x)$ is divided by $x - a$, the remainder $R = P(a)$. This is the **remainder theorem**.

Remainder Theorem

If a polynomial $P(x)$ is divided by $x - a$ until a remainder that does not contain x is obtained, then the remainder $R = P(a)$.

EXAMPLE 18.4

Determine the remainder, when $P(x) = x^3 - 4x^2 + 2x + 5$ is divided by $x - 3$ and $x + 2$.

Solution   When $P(x) = x^3 - 4x^2 + 2x + 5$ is divided by $x - 3$, then $a = 3$. So, $P(3) = 3^3 - 4(3)^2 + 2(3) + 5 = 27 - 36 + 6 + 5 = 2$. When $P(x)$ is divided by $x - 3$, the remainder is 2.

When we divide $P(x)$ by $x + 2$, we have $a = -2$, so $P(-2) = (-2)^3 - 4(-2)^2 + 2(-2) + 5 = -8 - 16 - 4 + 5 = -23$. So, when $P(x)$ is divided by $x + 2$, the remainder is -23.

We can check to see if these are true by dividing $P(x)$ by $x - 3$ and by $x + 2$.

$$
\begin{array}{r}
x^2 - x - 1 \\
x - 3 \overline{\smash{\big)}\ x^3 - 4x^2 + 2x + 5} \\
\underline{x^3 - 3x^2} \\
-x^2 + 2x + 5 \\
\underline{-x^2 + 3x} \\
-x + 5 \\
\underline{-x + 3} \\
2
\end{array}
\qquad
\begin{array}{r}
x^2 - 6x + 14 \\
x + 2 \overline{\smash{\big)}\ x^3 - 4x^2 + 2x + 5} \\
\underline{x^3 + 2x^2} \\
-6x^2 + 2x + 5 \\
\underline{-6x^2 - 12x} \\
14x + 5 \\
\underline{14x + 28} \\
-23
\end{array}
$$

From this we see that, when $P(x) = x^3 - 4x^2 + 2x + 5$ is divided by $x - 3$, we get a quotient $Q(x) = x^2 - x - 1$ and a remainder of $R = 2$. When $P(x)$ is divided by $x + 2$, the quotient is $Q(x) = x^2 - 6x + 14$ with a remainder of $R = -23$. These remainders are the same ones we found in Example 18.4.

Factor Theorem

In some cases, $R = 0$, and we see that $P(x) = Q(x)(x - a) + R$ can be rewritten as

$$P(x) = Q(x)(x - a)$$

That is, $x - a$ is a factor of $P(x)$. This leads to our second theorem for this section.

Factor Theorem

The **factor theorem** states that a polynomial $P(x)$ contains $x - a$ as a factor if and only if $P(a) = 0$.

This means that a is a root of $P(x)$ if $x - a$ is a factor of $P(x)$.

Synthetic Division

What we need is a shorter method to determine the quotient and remainder of a polynomial. Both of the divisions we worked earlier took a lot of time and a lot of space. A process has been developed that allows us to shorten the division. This process is called **synthetic division**.

Consider the problem we worked earlier when $x^3 - 4x^2 + 2x + 5$ was divided by $x + 2$. We had the following work.

$$
\begin{array}{r}
x^2 - 6x + 14 \\
x+2\overline{\smash{\big)}\,x^3 - 4x^2 + 2x + 5} \\
\underline{x^3 + 2x^2} \\
-6x^2 + 2x + 5 \\
\underline{-6x^2 - 12x} \\
14x + 5 \\
\underline{14x + 28} \\
-23
\end{array}
$$

In synthetic division, we do not write the x's and we do not repeat terms. So, if we erase all the x's and the terms that repeat, we get a group of coefficients that look like this.

$$
\begin{array}{r}
1 - 6 \quad 14 \\
2\overline{\smash{\big)}\,1 - 4 \quad\; 2 \quad\;\; 5} \\
2 \\
\underline{} \\
-6 \\
\underline{-12} \\
14 \\
\underline{28} \\
-23
\end{array}
$$

We can write all the numbers below the dividend in two lines and get

$$
\begin{array}{r}
1 - 6 \quad 14 \\
2\overline{\smash{\big)}\,1 - 4 \quad\; 2 \quad\;\; 5} \\
2 - 12 \quad 28 \\
\underline{} \\
-6 \quad 14 - 23
\end{array}
$$

If we repeat the leading coefficient on the bottom line, we see that the numbers on the third line represent the quotient and the remainder. Finally, we will change the sign of the divisor. Changing the sign of the divisor forces us to change the signs in the second row. This allows us to add the first two rows.

$$
\begin{array}{r}
-2\,\underline{\smash{\big|}\,1 - 4 \quad\;\; 2 \quad\;\; 5} \\
-2 \quad 12 - 28 \\
\underline{} \\
1 - 6 \quad 14 - 23
\end{array}
$$

$$\underbrace{}_{\text{quotient}} \quad \underbrace{}_{\text{remainder}}$$

quotient   remainder

The next example shows you how to use synthetic division.

EXAMPLE 18.5

Use synthetic division to determine the quotient and remainder when $x^4 - 3x^2 + 10x - 5$ is divided by $x + 3$.

Solution

$$\underline{-3}\,|\,1 \quad 0 \quad -3 \quad 10 \quad -5$$

Step 1: If the divisor is $x - a$, write a in the box. Arrange the coefficients of the dividend by descending powers of x. Use a zero coefficient when a power is missing. In this example, there is no x^3 term and $a = -3$.

$$\underline{-3}\,|\,1 \quad 0 \quad -3 \quad 10 \quad -5$$
$$\overline{\,1}$$

Step 2: Copy the leading coefficient in the third row.

$$\underline{-3}\,|\,1 \quad 0 \quad -3 \quad 10 \quad -5$$
$$-3$$
$$\overline{\,1 \quad -3}$$

Step 3: Multiply the last entry in the third row by the number in the box and write the result in the second row under the second coefficient. Add the numbers in that column.

$$\underline{-3}\,|\,1 \quad 0 \quad -3 \quad 10 \quad -5$$
$$-3 \quad 9$$
$$\overline{\,1 \quad -3 \quad 6}$$

Step 4a: Repeat the process from Step 3, but write the results under the third coefficient.

$$\underline{-3}\,|\,1 \quad 0 \quad -3 \quad 10 \quad -5$$
$$-3 \quad 9 \quad -18 \quad 24$$
$$\overline{\,1 \quad -3 \quad 6 \quad -8 \quad 19}$$
$$\underbrace{}_{\text{quotient}} \quad \underbrace{}_{\text{remainder}}$$

Step 4b: Repeat the process from Step 3 until there are as many entries in row 3 as there are in row 1. The last number in row 3 is the remainder. The other numbers are the coefficients of the quotient.

From this we see that the quotient is $Q(x) = x^3 - 3x^2 + 6x - 8$ and the remainder is $R = 19$.

EXAMPLE 18.6

Use synthetic division to determine the quotient and remainder when $2x^6 - 11x^4 + 17x^2 - 20$ is divided by $x - 2$.

Solution Since there are no x^5, x^3, or x terms, we will replace them with 0 coefficients. The synthetic division looks like this.

$$\underline{2}\,|\,2 \quad 0 \quad -11 \quad 0 \quad 17 \quad 0 \quad -20$$
$$4 \quad 8 \quad -6 \quad -12 \quad 10 \quad 20$$
$$\overline{\,2 \quad 4 \quad -3 \quad -6 \quad 5 \quad 10 \quad 0}$$
$$\underbrace{}_{\text{quotient}} \quad \underbrace{}_{\text{remainder}}$$

The quotient is $2x^5 + 4x^4 - 3x^3 - 6x^2 + 5x + 10$ and the remainder is 0. So, $x - 2$ is a factor of the polynomial $2x^6 - 11x^4 + 17x^2 - 20$.

EXAMPLE 18.7

Use synthetic division to determine if $3x - 2$ and $x - 1$ are factors of $18x^4 - 12x^3 - 45x^2 + 57x - 18$.

Solution   The remainder and factor theorems both refer to division by $x - a$. We are to check $3x - 2$. But $3x - 2 = 3\left(x - \frac{2}{3}\right)$, and if $x - \frac{2}{3}$ divides the polynomial, then $3x - 2$ will also divide it. Our synthetic division follows.

$$\frac{2}{3} \big| \begin{array}{ccccc} 18 & -12 & -45 & 57 & -18 \\ & 12 & 0 & -30 & 18 \\ \hline 18 & 0 & -45 & 27 & 0 \end{array}$$

Since the remainder is zero, $3x - 2$ is a factor of $18x^4 - 12x^3 - 45x^2 - 57x - 18$. We will now see if $x - 1$ is a factor. Instead of checking the original equation, we will use the result from the synthetic division.

When we divided $18x^4 - 12x^3 - 45x^2 + 57x - 18$ by $x - \frac{2}{3}$, we got 18 0 −45 27 in the third row. This told us that $18x^4 - 12x^3 - 45x^2 + 57x - 18 = \left(x - \frac{2}{3}\right)(18x^3 - 45x + 27)$. The factor $18x^3 - 45x + 27$ is called a **depressed equation** of the original equation. If $x - 1$ is a factor of the depressed equation $18x^3 - 45x + 27$, it is a factor of the original equation. Thus we can use the third row from our earlier synthetic division as the first row when we check $x - 1$.

$$\frac{2}{3} \big| \begin{array}{ccccc} 18 & -12 & -45 & 57 & -18 \\ & 12 & 0 & -30 & 18 \\ \end{array}$$
$$1 \big| \begin{array}{ccccc} 18 & 0 & -45 & 27 & 0 \\ & 18 & 18 & -27 & \\ \hline 18 & 18 & -27 & 0 \end{array}$$

Thus, $x - 1$ is a factor of the depressed equation $18x^3 - 45x + 27$, so $x - 1$ is also a factor of $18x^4 - 12x^3 - 45x^2 + 57x - 18$.

So,

$$18x^4 - 12x^3 - 45x^2 + 57x - 18$$

$$= \left(x - \frac{2}{3}\right)(x - 1)(18x^2 + 18x - 27)$$

$$= 9\left(x - \frac{2}{3}\right)(x - 1)(2x^2 + 2x - 3)$$

We can use the quadratic formula on the last depressed equation to find that the remaining roots are $x = \dfrac{-1 + \sqrt{7}}{2}$ and $x = \dfrac{-1 - \sqrt{7}}{2}$.

Exercise Set 18.1

In Exercises 1–8, find the value of $P(x)$ for the given value of x.

1. $P(x) = 3x^2 - 2x + 1; x = 2$

2. $P(x) = 2x^3 - 4x + 5; x = -1$

3. $P(x) = x^4 + x^3 + x^2 - x + 1; x = -1$

4. $P(x) = x^4 - 2x^2 + x; x = -2$

5. $P(x) = 5x^3 - 4x + 7; x = 3$

6. $P(x) = 7x^4 - 5x^2 + x - 7; x = -2$

7. $P(x) = x^5 - x^4 + x^3 + x^2 - x + 1; x = -1$

8. $P(x) = 3x^5 + 4x^2 - 3; x = -3$

In Exercises 9–16, use the remainder theorem to find the remainder R, when $P(x)$ is divided by $x - a$.

9. $P(x) = x^3 + 2x^2 - x - 2; x - 1$

10. $P(x) = x^3 + 2x^2 - 12x - 9; x - 3$

11. $P(x) = x^3 - 3x^2 + 2x + 5; x - 3$

12. $P(x) = x^3 - 9x^2 + 23x - 15; x - 1$

13. $P(x) = 4x^4 + 13x^3 - 13x^2 - 40x + 12; x + 2$

14. $P(x) = 2x^4 - 2x^3 - 6x^2 - 14x - 7; x - 7$

15. $P(x) = 3x^4 - 12x^3 - 60x + 4; x - 5$

16. $P(x) = 4x^3 - 4x^2 - 10x + 8; x - \frac{1}{2}$

In Exercises 17–24, use the factor theorem to determine whether or not the second factor is a factor of the first.

17. $x^3 + 2x^2 - 12x - 9; x - 3$

18. $x^4 - 9x^3 + 18x^2 - 3; x + 1$

19. $2x^5 - 6x^3 + x^2 + 4x - 1; x + 1$

20. $2x^5 - 6x^3 + x^2 + 4x - 1; x - 1$

21. $3x^5 + 3x^4 - 14x^3 + 4x^2 - 24x; x + 3$

22. $3x^5 + 3x^4 - 14x^3 + 4x^2 - 24x; x - 2$

23. $6x^4 - 15x^3 - 8x^2 + 20x; 2x - 5$

24. $20x^4 + 12x^3 + 10x + 9; 5x + 3$

In Exercises 25–32, use synthetic division to determine the quotient and remainder when each polynomial is divided by the given $x - a$.

25. $x^5 - 17x^3 + 75x + 9; x - 3$

26. $2x^5 - x^2 + 8x + 44; x + 2$

27. $5x^3 + 7x^2 + 9; x + 3$

28. $x^3 + 3x^2 - 2x - 4; x - 2$

29. $8x^5 - 4x^3 + 7x^2 - 2x; x - \frac{1}{2}$

30. $9x^5 + 3x^4 - 6x^3 - 2x^2 + 6x + 1; x + \frac{1}{3}$

31. $4x^4 - 12x^3 + 9x^2 - 8x + 12; 2x - 3$

32. $4x^3 + 7x^2 - 3x - 15; 4x - 5$

≡ 18.2
ROOTS OF AN EQUATION

In Section 18.1, we learned the factor theorem to help us determine if a number is a root of a polynomial. We also learned how to use synthetic division to quickly find the quotient and remainder when a polynomial is divided by a first degree polynomial, $x - a$. In this section, we shall learn some theorems that determine the number of roots of the equation $P(x) = 0$.

In working with first degree polynomials, we were always able to find one root. With second degree polynomials, we could find two roots. At times, as in $P(x) = x^2 + 6x + 9$, both of these roots were the same. (Both roots of $x^2 + 6x + 9 = 0$ were -3.) As you might expect, every polynomial of degree n has exactly n roots. The **fundamental theorem of algebra** states that every polynomial equation of degree $n > 0$ has at least one (real or complex) root.

Combining the fundamental theorem of algebra with the factor theorem leads to the **linear factorization theorem**, which states the following:

Linear Factorization Theorem

If $P(x)$ is a polynomial function of degree $n > 1$, then there is a non-zero number a and there are numbers, $r_1, r_2, \ldots, r_n$, such that

$$P(x) = a(x - r_1)(x - r_2) \cdots (x - r_n)$$

The proof of this theorem is fairly easy. If $P(x)$ is a polynomial and $P(x) = 0$, then by the fundamental theorem there is a number r_1 such that $P(r_1) = 0$. From the factor theorem, we know $P(x) = (x - r_1)P_1(x)$ where $P_1(x)$ is a polynomial.

Again, the fundamental theorem states that there is a number r_2 such that $P_1(r_2) = 0$, so $P_1(x) = (x - r_2)P_2(x)$ and $P(x) = (x - r_1)(x - r_2)P_2(x)$.

We continue until one of the quotients is a constant a. At that time, we have

$$P(x) = a(x - r_1)(x - r_2) \cdots (x - r_n)$$

A linear factor appears each time a root is found. Since the degree of $P(x)$ is n, there are n linear factors and n roots. These roots are $r_1, r_2, r_3, \ldots, r_n$. Now, as we have seen, these roots may not all be different. Yet, even if they are not distinct, each root is counted.

EXAMPLE 18.8

Use the linear factorization theorem to determine the roots of the polynomial $P(x) = 3x^4 - 8x^3 - 11x^2 + 28x - 12$.

Solution

$$
\begin{aligned}
3x^4 - 8x^3 - 11x^2 + 28x - 12 &= (x - 1)(3x^3 - 5x^2 - 16x + 12) \\
&= (x - 1)(x - 3)(3x^2 + 4x - 4) \\
&= (x - 1)(x - 3)(x + 2)(3x - 2) \\
&= 3(x - 1)(x - 3)(x + 2)\left(x - \frac{2}{3}\right)
\end{aligned}
$$

The roots are 1, 3, -2, and $\frac{2}{3}$.

Since $P(x)$ was of degree 4, it should have four roots. It does. Notice that the constant factor is the leading coefficient.

Just as it is not necessary that all roots be distinct, it is not necessary that all roots are real numbers. Remember that complex numbers can be roots to quadratic equations.

EXAMPLE 18.9

Determine the roots of $P(x) = x^2 + 6x + 25$.

Solution   Using the quadratic formula, we determine that $P(x) = x^2 + 6x + 25$ has the following roots.

$$x = \frac{-6 \pm \sqrt{6^2 - 4(25)}}{2}$$

$$= \frac{-6 \pm \sqrt{36 - 100}}{2}$$

$$= \frac{-6 \pm \sqrt{-64}}{2}$$

$$= \frac{-6 \pm 8j}{2}$$

$$= -3 \pm 4j$$

The roots are $r_1 = -3 + 4j$ and $r_2 = -3 - 4j$.

EXAMPLE 18.10

Determine the roots of $P(x) = (x - 2)^3(x^2 + 6x + 25)$.

Solution   From the previous example we know that the roots of $x^2 + 6x + 25$ are $r_1 = -3 + 4j$ and $r_2 = -3 - 4j$. Thus, the roots of $P(x) = (x - 2)^3(x^2 + 6x + 25)$ would be 2, 2, 2, $-3 + 4j$, and $-3 - 4j$. Notice that there are five roots, but three of them are the same.

One basic property of complex roots follows.

> **Complex Roots Theorem**
>
> If $P(x)$ is a polynomial with real coefficients and $a + bj$ is a root of $P(x)$, then its conjugate, $a - bj$, is also a root.

This would not be true if $P(x)$ was a polynomial with complex coefficients, but it is true when all the coefficients of P are real numbers.

 Hint

Remember, when solving polynomials, if you find enough roots so the remaining factor is quadratic, you can always find the last two roots by using the quadratic formula.

EXAMPLE 18.11

Solve the equation $4x^3 - 9x^2 - 25x - 12$, given the fact that $-\frac{3}{4}$ is a root.

Solution   Using synthetic division, we get

$$-\tfrac{3}{4} \underline{\big|\, 4 \quad -9 \quad -25 \quad -12}$$
$$\phantom{-\tfrac{3}{4}\big|\,4\quad} \underline{-3 \quad\;\; 9 \quad\;\; 12}$$
$$\phantom{-\tfrac{3}{4}\big|\,} 4 \quad -12 \quad -16 \quad\;\; 0$$

EXAMPLE 18.11 (Cont.)	So,

$$4x^3 - 9x^2 - 25x - 12 = \left(x + \frac{3}{4}\right)(4x^2 - 12x - 16)$$

$$= 4\left(x + \frac{3}{4}\right)(x^2 - 3x - 4)$$

We can factor $x^2 - 3x - 4$ as $(x - 4)(x + 1)$, and so

$$4x^3 - 9x^2 - 25x - 12 = 4\left(x + \frac{3}{4}\right)(x - 4)(x + 1)$$

The roots of $P(x) = 4x^3 - 9x^2 - 25x - 12$ are $-\frac{3}{4}$, 4, and -1.

EXAMPLE 18.12

Solve $P(x) = x^4 + 5x^3 + 10x^2 + 20x + 24$ if you know that $2j$ is a root.

Solution   Since $2j$ is a root and the coefficients of P are all real, its conjugate $-2j$ is also a root. So, two of the linear factors are $x - 2j$ and $x + 2j$. We can use synthetic division twice or divide the original polynomial by $(x - 2j)(x + 2j) = x^2 + 4$. We will use synthetic division twice.

$$
\begin{array}{r|rrrrr}
2j & 1 & 5 & 10 & 20 & 24 \\
   &   & +2j & -4 + 10j & -20 + 12j & -24 \\
\hline
-2j & 1 & (5 + 2j) & (6 + 10j) & (0 + 12j) & 0 \\
    &   & -2j & -10j & -12j & \\
\hline
    & 1 & 5 & 6 & 0 &
\end{array}
$$

Thus,

$$x^4 + 5x^3 + 10x^2 + 20x + 24 = (x - 2j)(x + 2j)(x^2 + 5x + 6)$$

$$= (x - 2j)(x + 2j)(x + 3)(x + 2)$$

The four roots are $2j$, $-2j$, -3, and -2.

Notice that the second time we performed synthetic division it was done on the depressed equation that resulted from the first synthetic division.

EXAMPLE 18.13

Solve $2x^4 - 17x^3 + 49x^2 - 51x + 9$ if 3 is a double root.

Solution   Since 3 is a double root, we know that two of the linear factors are $x - 3$ and $x - 3$. Using synthetic division twice, we get

$$
\begin{array}{r|rrrrr}
3 & 2 & -17 & 49 & -51 & 9 \\
  &   & 6 & -33 & 48 & -9 \\
\hline
3 & 2 & -11 & 16 & -3 & 0 \\
  &   & 6 & -15 & 3 & \\
\hline
  & 2 & -5 & 1 & 0 &
\end{array}
$$

EXAMPLE 18.13 (Cont.)

The last factor is $2x^2 - 5x + 1$. Using the quadratic formula, we see that its roots are

$$x = \frac{5 \pm \sqrt{(-5)^2 - 4(2)(1)}}{2(2)}$$

$$= \frac{5 \pm \sqrt{17}}{4}$$

Thus, the roots are 3, 3, $\dfrac{5 + \sqrt{17}}{4}$, and $\dfrac{5 - \sqrt{17}}{4}$.

Exercise Set 18.2

In Exercises 1–22, solve the equations using synthetic division and the given roots.

1. $5x^3 - 8x + 3, r_1 = 1$

2. $2x^3 + 5x^2 - 11x + 4, r_1 = -4$

3. $9x^3 - 3x^2 - 81x + 27, r_1 = \dfrac{1}{3}$

4. $10x^3 - 4x^2 - 40x + 16, r_1 = \dfrac{2}{5}$

5. $x^4 - 3x^2 - 4, r_1 = j$

6. $3x^4 + 6x^2 - 189, r_1 = -3j$

7. $x^4 + 2x^3 - 4x^2 - 18x - 45, r_1 = -1 + 2j$

8. $2x^4 + 4x^3 + 2x^2 - 16x - 40, r_1 = -1 - 2j$

9. $x^4 - 3x^3 - 3x^2 + 7x + 6, r_1 = 2, r_2 = 3$

10. $x^4 - 3x^3 - 12x^2 + 52x - 48, r_1 = 2, r_2 = -4$

11. $3x^4 + 12x^3 + 6x^2 - 12x - 9, r_1 = r_2 = -1$ (a double root)

12. $2x^4 + 6x^3 - 12x^2 - 24x + 16, r_1 = 2, r_2 = -2$

13. $6x^4 + 25x^3 + 33x^2 + x - 5, r_1 = -\dfrac{1}{2}, r_2 = \dfrac{1}{3}$

14. $12x^4 - 47x^3 + 55x^2 + 9x - 5, r_1 = \dfrac{1}{4}, r_2 = -\dfrac{1}{3}$

15. $3x^4 - 2x^3 - 3x + 2, r_1 = \dfrac{2}{3}, r_2 = 1$

16. $4x^4 + 3x^3 - 32x - 24, r_1 = 2, r_2 = -\dfrac{3}{4}$

17. $x^5 - x^4 + x^3 - 7x^2 + 10x - 4, r_1 = r_2 = r_3 = 1$ (a triple root)

18. $x^5 - 5x^4 + 7x^3 - 2x^2 + 4x - 8, r_1 = r_2 = r_3 = 2$

19. $3x^5 - 2x^4 - 24x^3 + x^2 + 28x - 12, r_1 = 3, r_2 = -2, r_3 = \dfrac{2}{3}$

20. $3x^5 - 4x^4 + 5x^3 - 18x^2 - 28x - 8, r_1 = -\dfrac{2}{3}, r_2 = 2j$

21. $x^6 - 6x^5 + 3x^4 - 60x^3 - 61x^2 - 54x - 63, r_1 = j, r_2 = -3j$

22. $6x^6 - 19x^5 + 63x^4 - 152x^3 + 216x^2 - 304x + 240, r_1 = r_2 = 2j$

☰ 18.3
RATIONAL ROOTS

In Section 18.2, we determined the number of roots of a polynomial. We also learned that complex roots come in pairs and that once we are able to reduce the nonlinear factor to degree 2, we can use the quadratic formula.

What we need is some help in locating roots when the degree is larger than two. Those roots can be complex numbers, irrational numbers, or rational numbers. In this section, we will focus on ways to determine the rational roots of a polynomial.

Rational Root Test

Let's consider the general polynomial:

$$P(x) = a_n x^n + a_{n-1} x^{n-1} + \cdots + a_2 x^2 + a_1 x + a_0$$

We begin with a test to help determine if a rational number is a root of P.

Rational Root Test

If the rational number $\dfrac{c}{d}$ is in lowest terms and is a root of $P(x) = a_n x^n + a_{n-1} x^{n-1} + \cdots + a_2 x^2 + a_1 x + a_0$, then c is a factor of a_0 and d is a factor of a_n.

One consequence of this test is that if $a_n = 1$, then all rational roots are integers. The next three examples show how to use the rational root test.

EXAMPLE 18.14

Find all rational roots of $P(x) = 4x^3 - 5x^2 - 2x + 3$.

Solution   In this polynomial, $4x^3 - 5x^2 - 2x + 3$, $n = 3$, so $a_n = a_3 = 4$ and $a_0 = 3$. The factors of $a_n = 4$ are $d : \pm 1, \pm 2, \pm 4$, and the factors of $a_0 = 3$ are $c : \pm 1, \pm 3$. Any rational roots must be among these combinations of $\dfrac{c}{d} : \pm 1, \pm 3,$ $\pm \frac{1}{2}, \pm \frac{3}{2}, \pm \frac{1}{4}, \pm \frac{3}{4}$. A check will verify that the roots are $-\frac{3}{4}$, 1, and 1.

EXAMPLE 18.15

Find all rational roots of $P(x) = 5x^6 - 4x^5 - 41x^4 + 32x^3 + 43x^2 - 28x - 7$.

Solution   In this polynomial, $n = 6$, so $a_n = a_6 = 5$ and $a_0 = -7$. The factors of a_6 are $d : \pm 1, \pm 5$. The factors of a_0 are $c : \pm 1, \pm 7$. Any rational roots must be among these combinations of $\dfrac{c}{d} : \pm 1, \pm 7, \pm \dfrac{1}{5}$, and $\pm \dfrac{7}{5}$. A check will verify that the rational roots are $-1, -\frac{1}{5}$, 1, and 1. The other two roots, $\sqrt{7}$ and $-\sqrt{7}$, are not rational numbers.

EXAMPLE 18.16

Find all rational roots of $P(x) = x^5 - 4x^4 - 5x^3 + 8x^2 - 32x - 40$.

Solution   In this polynomial, $n = 5$, so $a_n = a_5 = 1$ and $a_0 = -40$. Since $a_5 = 1$, any rational roots will be integers. The factors of a_0 are $c : \pm 1, \pm 2, \pm 4, \pm 5, \pm 8,$ $\pm 10, \pm 20, \pm 40$. And these are the only 16 possible rational roots. A check will verify that the rational roots are $-2, -1$, and 5 and that the complex roots are $1 + j\sqrt{3}$ and $1 - j\sqrt{3}$.

In Example 18.14, we were able to narrow the number of possible rational roots to 12, in Example 18.15 we had only eight possible rational roots, and in Example 18.16 we had 16 possible rational roots. It would be beneficial if there were some additional methods we could use to help us in our search for roots.

Descartes' Rule of Signs

Descartes' rule of signs gives us a little more help. This rule gives an idea of the number of positive and negative roots there are in a polynomial. The rule applies to all real number roots.

> **Descartes' Rule of Signs**
>
> **Descartes' rule of signs** states that the number of positive roots of a polynomial $P(x) = 0$ cannot be more than the number of changes of sign in the terms of $P(x)$ and, if there are less, it will be reduced by a multiple of two. The equation $P(x) = 0$ has at most as many negative roots as the number of changes in sign of $P(-x)$ and, if there are less, it will be reduced by a multiple of two.

EXAMPLE 18.17

What information does Descartes' rule of signs tell us about the roots of $P(x) = 3x^5 - 2x^4 + x^3 - x - 7$?

Solution   The signs of the terms of this polynomial equation are marked below each term.

$$P(x) = 3x^5 - 2x^4 + x^3 - x - 7$$

There are three places where the signs change. By Descartes' rule of signs, there can be no more than three positive real roots. If there are less than three, then there is one. So, this equation has either one or three positive real roots.

To find the number of negative real roots, we perform the same procedure on $P(-x)$.

$$P(-x) = 3(-x)^5 - 2(-x)^4 + (-x)^3 - (-x) - 7$$
$$= -3x^5 - 2x^4 - x^3 + x - 7$$

There are two sign changes and so the equation has either two or zero negative real roots.

Thus, this polynomial equation $3x^5 - 2x^4 + x^3 - x - 7 = 0$ has no more than three positive and two negative roots.

Caution   Remember, a polynomial may have complex number roots. This means that you cannot find the number of negative roots by subtracting the number of positive roots from the total number of roots.

Upper and Lower Bounds

We spent a lot of time in Section 18.2 learning synthetic division because it is a quick and easy way to determine if a given number is a root. What we will develop next is a way to tell when to stop looking for roots larger or smaller than the root we just checked.

Before we give this next technique, we need to learn two new terms. A real number U is an **upper bound** of the real roots of a polynomial if no root is larger than the number U. A real number L is a **lower bound** of the real roots of a polynomial if no root is smaller than L.

Let $P(x) = a_n x^n + a_{n-1} x^{n-1} + \cdots + a_2 x^2 + a_1 x + a_0$ be a polynomial with real coefficients and $a_n > 0$. If $k_1 > 0$ and if the terms in the third row of synthetic division of $P(x)$ by $x - k_1$ are all positive or zero, then k_1 is an upper bound of the real roots of $P(x)$. If $k_2 \leq 0$ and if the terms in the third row of synthetic division of $P(x)$ by $x - k_2$ alternate in sign, then k_2 is a lower bound of the real roots of $P(x)$.

EXAMPLE 18.18

Find an upper and lower bound of $P(x) = 4x^3 + 7x^2 + x - 9$.

Solution If $\dfrac{c}{d}$ is a rational root, then c must be a factor of -9 and d must be a factor of 4. The factors of -9 are ± 1, ± 3, ± 9; and the factors of 4 are ± 1, ± 2, and ± 4. The possible rational roots are ± 1, ± 3, ± 9, $\pm \frac{1}{2}$, $\pm \frac{3}{2}$, $\pm \frac{9}{2}$, $\pm \frac{1}{4}$, $\pm \frac{3}{4}$, and $\pm \frac{9}{4}$.

If we use synthetic division to determine if 1 is a root, we get

$$
\begin{array}{r|rrrr}
1 & 4 & 7 & 1 & -9 \\
  &   & 4 & 11 & 12 \\
\hline
  & 4 & 11 & 12 & 3
\end{array}
\qquad \text{All are positive.}
$$

From this, we can see that 1 is not a root, but it is an upper bound. Thus, none of the numbers larger than 1 $(3, 9, \frac{3}{2}, \frac{9}{2},$ and $\frac{9}{4})$ can be roots of this polynomial.

Now, we will check to see if -3 is a root.

$$
\begin{array}{r|rrrr}
-3 & 4 & 7 & 1 & -9 \\
   &   & -12 & 15 & -48 \\
\hline
   & 4 & -5 & 16 & -57
\end{array}
\qquad \text{Signs alternate.}
$$

Again, -3 is not a root, but the numbers in the third row alternate sign. So, -3 is a lower bound and none of the numbers less than -3 $(-9$ and $-\frac{9}{2})$ can be roots.

Summary

Now let's summarize what we have learned in the last two sections and then work an example that uses everything. We have developed a list of six hints at the top of the next page to help you determine the roots of a polynomial. See the box that follows.

EXAMPLE 18.19

Find the roots of $P(x) = 6x^4 + 23x^3 + 3x^2 - 32x + 12$.

Solution $P(x)$ has degree 4, so there are four roots. There are two sign changes in $P(x)$, so there are two or zero positive real roots. Now, $P(-x) = 6x^4 - 23x^3 + 3x^2 + 32x + 12$ also has two sign changes, so $P(x)$ has two or zero negative real roots.

The leading coefficient $a_4 = 6$ and the constant $a_0 = 12$. The factors of a_4 are ± 1, ± 2, ± 3, and ± 6. The factors of a_0 are ± 1, ± 2, ± 3, ± 4, ± 6, and ± 12. The

Hints for Finding Roots of Polynomials

If $P(x) = 0$ is a polynomial equation of degree n:

1. There are n roots.
2. Complex roots appear in conjugate pairs.
3. Any rational root of the form $\frac{c}{d}$, when in lowest terms, has c as a factor of the constant term, a_0, and d as a factor of the leading coefficient, a_n.
4. The maximum number of positive roots is the number of sign changes in $f(x)$; the maximum number of negative roots is the number of sign changes in $f(-x)$.
5. If we use synthetic division to determine whether or not a non-negative number c is a root and the third row has all non-negative numbers, then c is an upper bound; if c is a non-positive number and the third row alternates signs, then c is a lower bound.
6. Once we determine $n - 2$ roots, the remaining roots can be found by using the quadratic formula or by factoring.

EXAMPLE 18.19 (Cont.)

possible rational roots of $P(x)$ are ± 1, ± 2, ± 3, ± 4, ± 6, ± 12, $\pm \frac{1}{2}$, $\pm \frac{1}{3}$, $\pm \frac{1}{6}$, $\pm \frac{2}{3}$, $\pm \frac{3}{2}$, and $\pm \frac{4}{3}$.

We will now check these to see if any are roots of $P(x)$. Using synthetic division we first check 1.

$$\begin{array}{r|rrrrr} 1 & 6 & 23 & 3 & -32 & 12 \\ & & 6 & 29 & 32 & 0 \\ \hline & 6 & 29 & 32 & 0 & 12 \end{array}$$

We found two pieces of information from this synthetic division. First, 1 is not a root. But, since the numbers in the third row are all nonnegative, 1 is an upper bound. Thus, none of the numbers larger than 1 (2, 3, 4, 6, 12, $\frac{3}{2}$, and $\frac{4}{3}$) are roots. The only possible positive rational roots must be selected from $\frac{1}{6}$, $\frac{1}{3}$, $\frac{1}{2}$, and $\frac{2}{3}$.

Instead of continuing to look for positive roots, we will check for negative roots. We will first check -1.

$$\begin{array}{r|rrrrr} -1 & 6 & 23 & 3 & -32 & 12 \\ & & -6 & -17 & 14 & 18 \\ \hline & 6 & 17 & -14 & -18 & 30 \end{array}$$

Since -1 is neither a root nor a lower bound, we will check -2.

$$\begin{array}{r|rrrrr} -2 & 6 & 23 & 3 & -32 & 12 \\ & & -12 & -22 & 38 & -12 \\ \hline & 6 & 11 & -19 & 6 & 0 \end{array}$$

We have found our first root, -2. If you check -2 again, you will see it is not a double root. It is also not a lower bound. We know that there is another negative

EXAMPLE 18.19 (Cont.)

root. We next check -3, but this time on the depressed equation in the third row.

$$
\begin{array}{r|rrrr}
-3 & 6 & 11 & -19 & 6 \\
   &   & -18 & 21 & -6 \\
\hline
   & 6 & -7 & 2 & 0
\end{array}
$$

We see that -3 is also a root. The depressed equation from the third row is $6x^2 - 7x + 2$, which factors into $(3x - 2)(2x - 1)$. The roots of $P(x) = 6x^4 + 23x^3 + 3x^2 - 32x + 12$ are -3, -2, $\frac{1}{2}$, and $\frac{2}{3}$.

Application

EXAMPLE 18.20

A company needs a box like the one shown in Figure 18.1a. The box is to be made from a sheet of metal that measures 45 cm by 30 cm by cutting a square from each corner and bending up the sides. If the box is to hold 3 500 cm³, what are the lengths of the sides of the squares that are cut out of each corner?

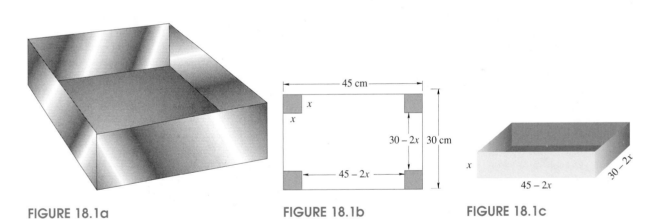

FIGURE 18.1a   FIGURE 18.1b   FIGURE 18.1c

Solution   A sketch of this situation is shown in Figure 18.1b. The lengths of the sides of the squares have been labeled x. Once these squares have been removed, the part of the remaining sheet that will form the length of the box is $45 - 2x$ cm. Similarly, the box's width is $30 - 2x$ cm. The width of the box must be at least 0 and less than 30 cm. So, $0 < 30 - 2x < 30$ or $0 < x < 15$.

When the metal is bent to form the sides of the box, a box like the one in Figure 18.1c is obtained. This box is a rectangular prism, so its volume is $V = (45 - 2x)(30 - 2x)x$. We are told that this box is to have a volume of 3 500 cm³, so

$$(45 - 2x)(30 - 2x)x = 3\,500$$

EXAMPLE 18.20 (Cont.)

Multiplying the factors on the left-hand side produces

$$1\,350x - 150x^2 + 4x^3 = 3\,500$$

or $\quad 4x^3 - 150x^2 + 1\,350x - 3\,500 = 0$

There are three sign changes, so there is either 1 or 3 positive real roots. $P(-x) = -4x^3 - 150x^2 - 1\,350x - 3\,500$ has no sign changes, so there are zero negative real roots.

The leading coefficient is $a_3 = 4$ and the constant is $a_0 = 3\,500$. Since we know there are no negative real roots (and since x must be positive), we can eliminate the negative factors. The factors of a_3 are $d : 1, 2, 4$. The positive factors of a_0 are $c : 1, 2, 4, 5, 7, 10, 14, 20, 25, 28 \ldots$. (Since $x < 15$, we can disregard any potential roots of a_0 that are 15 or larger.) Thus, possible roots less than 15 include $\frac{c}{d} : 1, 2,$ $4, 5, 7, 10, 14, \frac{1}{2}, \frac{1}{4}, \frac{5}{2}, \frac{5}{4}, \frac{7}{2}, \frac{7}{4}, \frac{25}{2},$ and $\frac{25}{4}$.

Since $P(1) = -2\,296$, 1 is not a root. Similarly, $P(2) = -1\,368$ and $P(4) = -244$ indicate that neither 2 nor 4 are roots. But, $P(5) = 0$, so 5 is a root.

Using synthetic division, we see that

$$
\begin{array}{r|rrrr}
5 & 4 & -150 & 1\,350 & -3\,500 \\
  &   & 20 & -650 & 3\,500 \\
\hline
  & 4 & -130 & 700 & 0
\end{array}
$$

and so

$$4x^3 - 150x^2 + 1\,350x - 3\,500 = (x - 5)(4x^2 - 130x + 700)$$

Using the quadratic formula on the depressed equation $4x^2 - 130x + 700$, we obtain the other two roots.

$$x = \frac{130 \pm \sqrt{130^2 - 4(4)(700)}}{8}$$

$$= \frac{130 \pm \sqrt{5\,700}}{8}$$

So, $\quad x = \dfrac{130 + \sqrt{5\,700}}{8} \approx 25.7$

or $\quad x = \dfrac{130 - \sqrt{5\,700}}{8} \approx 6.8$

One of these possible solutions, 25.7 is too large since $x < 15$.

Thus, there are two possible ways to cut this metal to obtain a box with the desired volume. If $x = 5$ cm, the box has a length of $45 - 2(5) = 35$ cm and a width of $30 - 2(5) = 20$ cm. Checking, we see that $(35)(20)5 = 3\,500$.

If $x \approx 6.812\,7$ cm, the length is approximately $31.374\,6$ cm and the width is about $16.374\,6$ cm. Multiplying, we get a volume of $(31.374\,6)(16.374\,6)(6.812\,7) = 3\,500.000\,952 \approx 3\,500$ cm^3.

This problem has two correct solutions. The solution the company uses will depend on other factors, such as which is easier (or less expensive) to make, which

| EXAMPLE 18.20 (Cont.) | one does a better job of holding the product, and which shape is more appealing to the customer. |

Exercise Set 18.3

In Exercises 1–10, use Descartes' rule of signs to determine the number of possible positive and negative real roots of the polynomial equation. List the possible rational roots of each equation.

1. $x^3 - 3x - 2 = 0$
2. $x^3 - 2x^2 - x + 2 = 0$
3. $x^3 - 8x^2 - 17x + 6 = 0$
4. $x^3 + 2x^2 - x - 2 = 0$
5. $x^3 - 2x^2 - 5x + 6 = 0$
6. $3x^3 - x^2 - 2x + 2 = 0$
7. $2x^4 - x^3 - 5x^2 + x + 3 = 0$
8. $2x^4 - 7x^3 - 10x^2 + 33x + 18 = 0$
9. $6x^4 - 7x^3 - 13x^2 + 4x + 4 = 0$
10. $5x^6 - x^5 - 5x^4 + 6x^3 - x^2 - 5x + 1 = 0$

In Exercises 11–26, find all rational roots of the polynomial equation. If possible, find all roots.

11. $x^3 - 9x^2 + 23x - 15 = 0$
12. $x^3 - 3x^2 - 4x + 12 = 0$
13. $x^3 - 3x - 2 = 0$
14. $x^3 - x^2 - 8x + 12 = 0$
15. $4x^3 - 5x^2 + 10x + 12 = 0$
16. $3x^3 - 4x^2 - 17x + 6 = 0$
17. $x^4 - 4x^3 + 7x^2 - 12x + 12 = 0$
18. $4x^4 + 4x^3 + 9x^2 + 8x + 2 = 0$
19. $4x^4 - 20x^3 + x^2 + 18x + 6 = 0$
20. $x^5 - x^3 + 27x^2 - 27 = 0$
21. $x^4 - 8x^3 + 39x^2 + 2x - 10 = 0$
22. $8x^3 - 36x^2 + 54x - 27 = 0$
23. $2x^5 - 13x^4 + 26x^3 - 22x^2 + 24x - 9 = 0$
24. $8x^5 - 4x^4 + 6x^3 - 3x^2 - 2x + 1 = 0$
25. $x^5 - 5x^4 + 6x^3 + 11x^2 - 43x + 30 = 0$
26. $2x^5 - x^4 - 6x^3 - 18x^2 + 4x + 40 = 0$

Solve Exercises 27–34.

27. *Sheet metal technology*   A box was made from a rectangular sheet of metal by cutting identical squares from the four corners and bending up the sides. If the piece of sheet metal originally measured 8.0 in. by 10.0 in. and the volume of the box is 48 in.^{3}, what was the length of each side of the squares that were removed?

28. *Drafting*   A rectangular box is constructed so that its width is 2.5 cm longer than its height and the length is 4 cm longer than the width. If the box has a volume of 210 cm^3, what are its dimensions?

29. *Agriculture*   A grain silo has the shape of a right circular cylinder with a hemisphere on top, as shown in Figure 18.2. The total height of the silo is 34 ft. Determine the radius of the cylinder if the total volume is 2,511π ft^3.

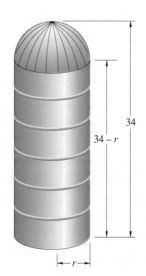

34

34 − r

r

FIGURE 18.2

30. *Petroleum engineering* A propane gas storage tank is in the shape of a right circular cylinder of height 6 m with a hemisphere attached at each end. Determine the radius r so that the volume of the tank is 18π m^3.

31. *Electricity* Three electric resistors are connected in parallel. The second resistor is 4 Ω more than the first and the third resistor is 1 Ω larger than the first. The total resistance is 1 Ω. In order to find the first resistance R, we must solve the equation

$$\frac{1}{R} + \frac{1}{R+4} + \frac{1}{R+1} = 1$$

What are the values of the resistances?

32. *Electricity* Suppose that the resistors in Exercise 31 had been related such that the second was 4 Ω more than the first and the third 9 Ω more than the first. If the total resistance was 3 Ω, then

$$\frac{1}{R} + \frac{1}{R+4} + \frac{1}{R+9} = \frac{1}{3}$$

What are the values of these resistances?

33. *Industrial technology* A rectangular box is made from a piece of metal 12 cm by 19 cm by cutting a square from each corner and bending up the sides and welding the seams. If the volume of the box is 210 cm^3, what is the size of the square that is cut from each corner?

34. *Architecture* The bending moment of a beam is given by $M(d) = 0.1d^4 - 2.2d^3 + 15.2d^2 - 32d$, where d is the distance in meters from one end. Find the values of d, where the bending moment is zero. (Hint: First multiply by 10 to eliminate the decimals.)

≣ 18.4
IRRATIONAL ROOTS

In Section 18.3, we concentrated on rational roots of polynomials. In this section, we expand our search for roots to include the remaining real numbers—the irrational numbers.

We have no difficulty in locating irrational roots of quadratic equations, because we can use the quadratic formula. Difficulties arise with polynomials of degrees larger than two.

In Section 18.3, we learned some rules that apply to all real numbers, not just rational numbers. Descartes' rule of signs applies to all real numbers. The upper- and lower-bound tests also apply to all real numbers. Suppose that we have a polynomial of degree 4 and Descartes' rule of signs tells us that there are some real roots. We have checked all possible rational roots and none of them satisfy the equation. The real roots must be irrational numbers. How can we find out what they are?

Locator Theorem

Because we are working with polynomials, the function is continuous. There is a theorem, called the **locator theorem**, that we can use.

Locator Theorem

If $P(x)$ is a real polynomial and a and b are two real numbers such that $P(a)$ and $P(b)$ have different signs, then there is at least one real root between a and b. Thus, there is a real number c between a and b, where $P(c) = 0$.

EXAMPLE 18.21

Locate the roots of $P(x) = x^3 + x^2 - 7x + 3$.

Solution Descartes' rule of signs tells us there are 0 or 2 positive real roots and one negative real root. The only possible rational roots are ± 1 and ± 3. By using synthetic division, we see that none of these are roots, but 3 is an upper bound.

EXAMPLE 18.21 (Cont.)

A table of values for $P(x)$ will let us use the locator theorem. Since 3 is an upper bound we do not have to check values of x larger than 3.

x	-5	-4	-3	-2	-1	0	1	2	3
$P(x)$	-62	-17	6	13	10	3	-2	1	18

According to the locator theorem, there is a root between -4 and -3, because $P(-4) = -17 < 0$ and $P(-3) = 6 > 0$. There is also a root between 0 and 1 and one between 1 and 2. Notice that while -3 was the smallest possible rational root, there is an irrational root that is smaller.

Linear Interpolation

Once the locator theorem has given us an idea where to look for roots, we need a method to help get approximations of their values. We will use a process called **linear interpolation**. There are other methods that are more efficient, but they all use calculus. This method is easy to remember, but, as you will see, using a calculator or a computer is a definite advantage.

We have three roots. We will demonstrate linear interpolation for the root between -4 and -3. We will approximate the graph of $P(x)$ by drawing the line joining $(-4, -17)$ and $(-3, 6)$, as shown in Figure 18.3. The x-intercept of this line approximates the root of $P(x)$. The slope of this line is 23, $\left(\frac{6+17}{-3+4} \right)$, and its equation is

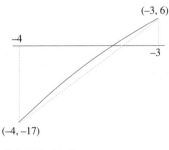

(−3, 6)

−4

−3

(−4, −17)

FIGURE 18.3

$$y + 17 = 23(x + 4)$$

$$\text{or} \qquad x = -4 + \frac{y + 17}{23}$$

The x-intercept occurs where $y = 0$ or $x = -4 + \frac{17}{23} \approx -3.261$. $P(-3.261) \approx 1.7806$. Thus, by the locator theorem, the actual root is between -4 and -3.261. We can repeat this process on these two values to find a closer approximation.

The process can be summarized as follows. Suppose we found that a root of $P(x)$ lies between a and b. The line joining $(a, P(a))$ and $(b, P(b))$, has the equation

$$y - P(a) = m(x - a)$$

$$\text{where} \qquad m = \frac{P(b) - P(a)}{b - a}.$$

The x-intercept of this line provides an approximation of the root. This intercept occurs where $y = 0$ in the previous equation or where

$$x = a - \frac{P(a)}{m}$$

We will call this value c. Repeating the process will give closer approximations to the roots. You can continue the process until one of two things happens: (1) two consecutive values of c have the same value to the nearest tenth, hundredth, or whatever you select; or (2) the absolute value of $f(c)$ is below a certain value. Remember, since each value of c is closer to the actual root, the values of $f(c)$ are getting closer to zero.

The following table shows the approximate values that we found for the root between $x = -4$ and $x = -3$ for $P(x) = x^3 + x^2 - 7x + 3$, when we continued until $|P(x)| < 0.01$. Notice that in each step b assumes the value of c from the previous step.

a	$P(a)$	b	$P(b)$	$c = a - \dfrac{P(a)}{m}$	$P(c)$
-4	-17	-3.000	6.000	-3.2609	1.7806
-4	-17	-3.261	1.786	-3.3311	0.4508
-4	-17	-3.331	0.451	-3.3484	0.1090
-4	-17	-3.348	0.109	-3.3526	0.0261
-4	-17	-3.353	0.026	-3.3535	0.0062

Application

EXAMPLE 18.22

The stability of a molecule can be determined by solving its characteristic polynomial, $P(x)$. The roots of the equation $P(x) = 0$ determine the molecule's pi electrons. The characteristic polynomial of naphthalene $C_{10}H_8$ is

$$P(x) = x^{10} - 11x^8 + 41x^6 - 65x^4 + 43x^2 - 9$$

Determine the energies of naphthalene's 10 pi electrons.

Solution   Descartes' rule of signs tell us that there are 1, 3, or 5 positive real roots and 1, 3, or 5 negative real roots. In fact, P is an even function and so it is symmetric about the y-axis. Thus, for each positive root, its additive inverse is also a root.

The only possible rational roots are ± 1, ± 3, and ± 9. Using synthetic division, we would determine that $x = 1$, and so $x = -1$, are roots. The resulting depressed polynomial is

$$P_1(x) = x^8 - 10x^6 + 31x^4 - 34x^2 + 9$$

A table of values of $P_1(x)$ will let us use the locator theorem. Because P_1 is even, we only need to check positive values of x.

x	0	0.5	1	1.5	2.0	2.5
$P_1(x)$	9	2.3	-3	1.2	-1.5	91.9

According to the locator theorem, there is a root of P_1 (and so also a root of P) between 0.5 and 1, a root between 1 and 1.5, a root between 1.5 and 2, and a root between 2 and 2.5.

Using linear interpolation to approximate the root between $x = 0.5$ and $x = 1$, we obtain the following table.

a	$P_1(a)$	b	$P_1(b)$	$c = a - \dfrac{P_1(a)}{m}$	$P_1(c)$
0.5	2.2852	1.000000	-3.0000	0.716186	-1.5638
0.5	2.2852	0.716186	-1.5638	0.628351	-0.1827
0.5	2.2852	0.628351	-0.1827	0.618847	-0.0145
0.5	2.2852	0.618847	-0.0145	0.618096	-0.0011
0.5	2.2852	0.618096	-0.0011	0.618039	-0.0001

EXAMPLE 18.22 (Cont.)

One of the roots is $x \approx 0.618039$, so $x \approx -0.618039$ is another root. Similarly, we can determine that the other six roots are approximately ± 1.303776, ± 1.618034, and ± 2.302776.

Notice that we could have obtained similar results to Example 18.22 if we had used a graphing calculator or a computer graphing program with its zoom and trace features.

Exercise Set 18.4

In Exercises 1–6, approximate the specified irrational root to the nearest 0.01.

1. The positive root of $x^3 + 5x - 3 = 0$
2. The largest root of $x^3 - 3x + 1 = 0$
3. The root of $x^4 + 2x^3 - 5x^2 + 1 = 0$, which is between 0 and 1
4. The root of $x^4 + 2x^3 - 5x^2 + 1 = 0$, which is between 1 and 2

5. The root of $x^3 + x^2 - 7x + 3 = 0$, which is between 0 and 1
6. The root of $x^3 + x^2 - 7x + 3 = 0$, which is between 1 and 2

In Exercises 7–16, find to two decimal places all rational and irrational roots of the given polynomial equation.

7. $x^4 - x - 2 = 0$
8. $x^5 - 2x^2 + 4 = 0$
9. $x^4 + x^3 - 2x^2 - 7x - 5 = 0$
10. $x^4 - 2x^3 - 3x + 4 = 0$
11. $2x^4 + 3x^3 - x^2 - 2x - 2 = 0$

12. $3x^4 + 3x^3 + x^2 + 4x + 3 = 0$
13. $2x^5 - 5x^3 + 2x^2 + 4x - 1 = 0$
14. $x^5 + 3x^4 - 5x^3 - 2x^2 + x + 2 = 0$
15. $8x^4 + 6x^3 - 15x^2 - 12x - 2 = 0$
16. $9x^4 + 15x^3 - 20x^2 - 20x + 16 = 0$

Solve Exercises 17–22.

17. *Petroleum engineering* The pressure drop P in pounds per square inch (psi) in a particular oil reservoir is a function of the number of years t that the reservoir has been in operation. The pressure drop is approximated by the equation

$$P = 150t - 20t^2 + t^3$$

How many years will it take for the pressure to drop 400 psi?

18. *Industrial design* A cylindrical storage tank 12 ft high contains 674 ft³. Determine the thickness of the tank, if the outside radius is 4.5 ft and the sides, top, and bottom have the same thickness.

19. *Nuclear technology* A cylindrical container for storing radioactive waste is to be constructed from lead.

The sides, top, and bottom of the cylinder of the container must be at least 15.5 cm thick.
(a) If the volume of the outside cylinder is $1\,000\,000\pi$ cm³, and the height of the inside cylinder is twice the radius of the inside cylinder (as shown in Figure 18.4), determine the radius of the inside cylinder.
(b) What is the volume of the inside cylinder?

FIGURE 18.4

20. *Mechanical engineering*   The characteristic polynomial for a certain material is

$$S^3 - 6S^2 - 78S + 108 = 0$$

Find the approximate value(s) of the stress that lies
between 1 and 2 psi.

21. *Chemistry*   The reference polynomial for naphthalene $C_{10}H_8$ is given by

$$R(x) = x^{10} - 11x^8 + 41x^6 - 61x^4 + 31x^2 - 3$$

Solve this reference polynomial.

22. Write a program that uses linear interpolation
to approximate a root of a polynomial.

≡ 18.5
RATIONAL FUNCTIONS

When you multiply two polynomials, you get another polynomial. In this section, we
will study what happens when we divide two polynomials.

Rational Function

If $P(x)$ and $Q(x)$ are two polynomials and if $Q(x) \neq 0$, then a function of the form

$$R(x) = \frac{P(x)}{Q(x)}$$

is called a **rational function**.

EXAMPLE 18.23

Each of the following are rational functions. Notice the restrictions placed on x so
that the denominator is not zero.

(a) $R(x) = \dfrac{2x^2 + 3x - 5}{x + 2}$, $x \neq -2$

(b) $f(x) = \dfrac{4x^3 - 7x + 3}{3x - 4}$, $x \neq \dfrac{4}{3}$

(c) $g(x) = \dfrac{4}{x + 3}$, $x \neq -3$

(d) $h(x) = \dfrac{9x^2 + 4x - 5}{x^2 + 6x + 5}$, $x \neq -1$, $x \neq -5$

Let's first look at solving these functions and then at graphing them. In order to
solve a rational function, we need to remember that any fraction $\dfrac{a}{b} = 0$, if and only
if $a = 0$ and $b \neq 0$. Using this rule, we will determine when each of the functions in
Example 18.23 is zero.

EXAMPLE 18.24

Solve: $\dfrac{2x^2 + 3x - 5}{x + 2} = 0$.

Solution   This equation will be true only when $2x^2 + 3x - 5 = 0$ and $x + 2 \neq 0$.
The left-hand equation factors to $2x^2 + 3x - 5 = (2x + 5)(x - 1) = 0$, and so $x = 1$,

EXAMPLE 18.24 (Cont.)

or $x = -\frac{5}{2}$. Neither of these make $x + 2 = 0$, so 1 and $-\frac{5}{2}$ are both roots of the original equation.

EXAMPLE 18.25

Solve: $\dfrac{4x^3 - 7x + 3}{3x - 4} = 0$.

Solution   This equation will be true only when $4x^3 - 7x + 3 = 0$ and $3x - 4 \neq 0$. The left-hand equation has a root at 1, so $4x^3 - 7x + 3 = (x - 1)(4x^2 + 4x - 3) = (x - 1)(2x - 1)(2x + 3)$; thus $x = 1, \frac{1}{2}, -\frac{3}{2}$. None of these make $3x - 4 = 0$, so all three are roots of the original equation.

EXAMPLE 18.26

Solve: $\dfrac{4}{x + 3} = 0$.

Solution   Since the numerator, 4, is never zero, this equation has no solution. That is, there are no numbers that make it true.

EXAMPLE 18.27

Solve: $\dfrac{9x^2 + 4x - 5}{x^2 + 6x + 5} = 0$.

Solution   This equation will be true when $9x^2 + 4x - 5 = 0$, if $x^2 + 6x + 5 \neq 0$ for the same values of x. The numerator factors to $9x^2 + 4x - 5 = (9x - 5)(x + 1) = 0$, so the numerator is zero when $x = -1$ or $x = \frac{5}{9}$. The denominator is zero when $x^2 + 6x + 5 = (x + 5)(x + 1) = 0$, or $x = -5$ or $x = -1$. Notice that -1 makes both the numerator and the denominator zero. Since the denominator cannot be zero, -1 is not a solution. The only solution of the original equation is $\frac{5}{9}$.

Graphing Rational Functions

Some interesting things occur when we graph rational functions. Let's graph some of the examples we just worked.

EXAMPLE 18.28

Sketch the graph of $g(x) = \dfrac{4}{x + 3}$.

Solution   We already know that this is defined everywhere except when $x = -3$. We also know that it has no real roots, so it does not cross the x-axis. Some of the values for $g(x)$ are shown in this table.

x	-11	-10	-9	-8	-7	-6	-5	-4
$g(x)$	-0.5	-0.57	$-\frac{2}{3}$	-0.8	-1	$-\frac{4}{3}$	-2	-4

x	-2	-1	0	1	2	3	4	5
$g(x)$	4	2	$\frac{4}{3}$	1	0.8	$\frac{2}{3}$	0.57	0.5

EXAMPLE 18.28 (Cont.)

Notice that $x = -3$ is not in the table. Let's see what happens when we get values of x closer to -3, as shown in this table.

x	-4	-3.8	-3.6	-3.4	-3.2	-3.1
$g(x)$	-4	-5	$-6\frac{2}{3}$	-10	-20	-40

x	-2.9	-2.8	-2.6	-2.4	-2.2	-2.0
$g(x)$	40	20	10	$6\frac{2}{3}$	5	4

From this second table, you can see that as x gets closer to -3 from the left, $x + 3$ gets closer to zero and $g(x)$ gets smaller. As x gets closer to -3 from the right, $x + 3$ again gets closer to zero, but this time $g(x)$ gets larger. Because $g(x)$ is not defined at $x = -3$, the graph never crosses the line $x = -3$. Thus $x = -3$ is a **vertical asymptote** for this graph.

In the same way, the x-axis is a horizontal asymptote for the graph. The graph of $g(x)$ is shown in Figure 18.5.

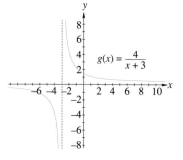

$g(x) = \dfrac{4}{x + 3}$

FIGURE 18.5

Vertical Asymptotes

The graph of any rational function $R(x) = \dfrac{P(x)}{Q(x)}$, $Q(x) \neq 0$, where P and Q have no common factors, will have vertical asymptotes at the values of x where $Q(x) = 0$.

In Example 18.28, the denominator was 0 when $x = -3$ and the line $x = -3$ was a vertical asymptote.

EXAMPLE 18.29

Determine the vertical asymptotes, if any, for the graph of $R(x) = \dfrac{2x^2 + 3x - 5}{x + 2}$.

Solution   The denominator is zero when $x = -2$ and this is the vertical asymptote, as shown in Figure 18.6.

EXAMPLE 18.30

Sketch the graph of $h(x) = \dfrac{9x^2 + 4x - 5}{x^2 + 6x + 5}$.

Solution   The denominator is zero when $x = -5$ or $x = -1$. These are not both vertical asymptotes. Both the numerator and denominator have a factor of $x + 1$, so, $\dfrac{9x^2 + 4x - 5}{x^2 + 6x + 5} = \dfrac{9x - 5}{x + 5}$, except when $x = -1$. The graph of these functions will be the same, except one is defined at $x = -1$ and the other is not. Thus, the vertical asymptote is $x = -5$. The following is a table of values and the graph is given in

EXAMPLE 18.30 (Cont.)

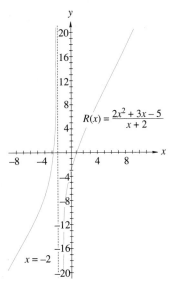

$R(x) = \dfrac{2x^2 + 3x - 5}{x + 2}$

$x = -2$

FIGURE 18.6

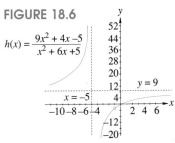

$h(x) = \dfrac{9x^2 + 4x - 5}{x^2 + 6x + 5}$

$x = -5$ $y = 9$

FIGURE 18.7

Figure 18.7.

x	-10	-9	-8	-7	-6	-4	-3	-2	-1
$h(x)$	19	21.5	$25\frac{2}{3}$	34	59	-41	-16	$-7\frac{2}{3}$	$\boxed{-3.5}$

x	0	1	2	3	4	5	10
$h(x)$	-1	$\frac{2}{3}$	1.86	2.75	3.44	4	$5\frac{2}{3}$

The value of $h(x) = -3.5$ is boxed because the function is not defined at $x = -1$. The value of -3.5 is the value of the simplified version of the function, $h(x) = \dfrac{9x - 5}{x + 5}$, when $x \neq -1$. In Figure 18.7, an open circle at the point $(-1, -3.5)$ emphasizes the fact that $h(x)$ is not defined at $x = -1$.

It appears as if the function h shown in Figure 18.7, has another asymptote, a horizontal one; and, in fact, it does. This asymptote is $y = 9$.

Horizontal Asymptote

If two polynomials, $P(x)$ and $Q(x)$, $Q(x) \neq 0$, are of the same degree, then the rational function $R(x) = \dfrac{P(x)}{Q(x)}$ has a **horizontal asymptote** at $y = \dfrac{a}{b}$, where a is the leading coefficient of $P(x)$ and b is the leading coefficient of $Q(x)$. If the degree of P is less than the degree of Q, then $y = 0$ is a horizontal asymptote.

In Example 18.30, $P(x) = 9x^2 + 4x - 5$ with a leading coefficient of 9 and $Q(x) = x^2 + 6x + 5$ with a leading coefficient of 1. So, $y = \frac{9}{1} = 9$ is the horizontal asymptote.

EXAMPLE 18.31

Determine the asymptotes of $f(x) = \dfrac{8x^2 - 22x - 6}{3x^2 - 2x - 1}$.

Solution   The numerator and denominator have the same degree, 2, so there is a horizontal asymptote. The leading coefficient of the numerator, $8x^2 - 22x - 6$, is 8; the leading coefficient of the denominator, $3x^2 - 2x - 1$, is 3. The horizontal asymptote is $y = \frac{8}{3}$.

The denominator factors into $(3x + 1)(x - 1)$, so it is zero when $x = -\frac{1}{3}$ or $x = 1$. Since neither of these makes the numerator zero, they are both vertical asymptotes. The graph of f is shown in Figure 18.8 with the asymptotes indicated by dashed lines.

EXAMPLE 18.32

Determine the asymptotes and sketch the graph of $g(x) = \dfrac{5x^2 + 10}{x^2 + 7x + 10}$.

Solution   The numerator and denominator are both degree 2, so there is a horizontal asymptote at $y = 5$.

EXAMPLE 18.32 (Cont.)

The denominator factors as $x^2 + 7x + 10 = (x + 5)(x + 2)$, so it is zero when $x = -5$ and $x = -2$. Neither of these makes the numerator zero, so they are vertical asymptotes. The graph of g is shown in Figure 18.9 with the asymptotes indicated by dashed lines. Notice that the graph actually crosses the asymptote $y = 5$ at $x \approx -1.143$.

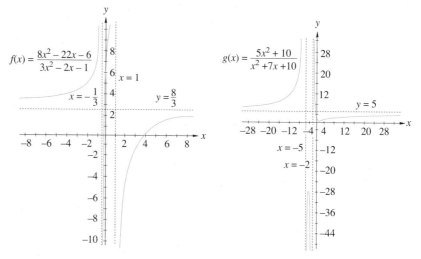

FIGURE 18.8   **FIGURE 18.9**

Solving Rational Functions

Now let's return to solving rational functions. We want to solve an equation of the form $R(x) = T(x)$, where R and T are both rational functions. We solve this in the same way that we work with fractions. First we write $R(x) - T(x)$ as a fraction and then solve $R(x) - T(x) = 0$.

EXAMPLE 18.33

Solve $\dfrac{6}{x + 1} = \dfrac{4}{x + 2}$.

Solution   First, we get both rational functions on the same side of the equal sign.

$$\frac{6}{x + 1} = \frac{4}{x + 2}$$

or $\qquad \dfrac{6}{x + 1} - \dfrac{4}{x + 2} = 0$

The lowest common denominator (LCD) of these two rational functions is $(x + 1)(x + 2)$. Writing each of these rational functions with this common denominator produces

$$\frac{6}{x + 1} = \frac{6(x + 2)}{(x + 1)(x + 2)}$$

EXAMPLE 18.33 (Cont.)

and $\dfrac{4}{x+2} = \dfrac{4(x+1)}{(x+1)(x+2)}$

So,

$$\dfrac{6}{x+1} - \dfrac{4}{x+2} = \dfrac{6(x+2)}{(x+1)(x+2)} - \dfrac{4(x+1)}{(x+1)(x+2)}$$

$$= \dfrac{(6x+12) - (4x+4)}{(x+1)(x+2)}$$

$$= \dfrac{2x+8}{(x+1)(x+2)} = 0$$

This is zero when $2x + 8 = 0$ or $x = -4$. Since $x = -4$ does not make the denominator zero, the solution is $x = -4$.

Exercise Set 18.5

In Exercises 1–10, determine the horizontal and vertical asymptotes, if any, of the function. Sketch the graph.

1. $f(x) = \dfrac{3}{x+2}$

2. $g(x) = \dfrac{4}{x^2 - 9}$

3. $g(x) = \dfrac{2x}{x+4}$

4. $k(x) = \dfrac{5x}{x^2 + x - 6}$

5. $j(x) = \dfrac{3x^2}{x^2 + 6x + 8}$

6. $f(x) = \dfrac{x^2 - 4}{x^2 - 16}$

7. $g(x) = \dfrac{x(x+1)}{(x+1)(x+2)}$

8. $h(x) = \dfrac{x^2 + 1}{2x^2}$

9. $j(x) = \dfrac{4x^2 - 12x - 27}{x^2 - 6x + 8}$

10. $k(x) = \dfrac{4x^2 - 9}{2x^2 + 7x + 3}$

In Exercises 11–20, solve the given equation.

11. $\dfrac{(x+2)(x-3)}{(x+5)(x+4)} = 0$

12. $\dfrac{x^2 - 2x - 1}{x(x+1)} = 0$

13. $\dfrac{x^2 - 1}{x^2 + 2x + 1} = 0$

14. $\dfrac{3x - 4}{6x^2 - 5x - 4} = 0$

15. $\dfrac{4}{x-3} = \dfrac{2}{x+1}$

16. $\dfrac{5}{x+7} = \dfrac{2}{x-8}$

17. $\dfrac{5}{2x} - \dfrac{3}{4x} = \dfrac{2}{3}$

18. $\dfrac{2x - 3}{x+5} = 2$

19. $\dfrac{3}{x-2} + \dfrac{5}{x+2} = \dfrac{20}{x^2 - 4}$

20. $\dfrac{4}{x+3} - \dfrac{6}{x-3} = \dfrac{10}{x^2 - 9}$

Solve Exercise 21.

21. *Medical technology*   When medicine is injected into the bloodstream, its concentration t h after injection is given by

$$C(t) = \dfrac{3t^2 + t}{t^3 + 50}$$

The concentration is greatest when $3t^4 + 2t^3 - 300t - 50 = 0$. Approximate this time to the nearest hundredth of an hour.

≡ CHAPTER 18 REVIEW

Important Terms and Concepts

Complex roots theorem
Constant polynomial
Descartes' rule of signs
Factor of a polynomial
Factor theorem
Fundamental theorem of algebra
Horizontal asymptote
Imaginary root
Irrational root
Linear factorization theorem
Linear interpolation
Lower bound

Multiple roots
Polynomial function
Rational function
Rational root test
Remainder theorem
Root of an equation
Synthetic division
Upper bound
Vertical asymptote
Zero of a function
Zero polynomial

Review Exercises

In Exercises 1–4, find the value of $P(x)$ for the given value of x and give the degree of $P(x)$.

1. $P(x) = 9x^2 - 4x + 2, x = 2$

2. $P(x) = 2x^3 - 4x^2 + 7, x = 3$

3. $P(x) = 4x^3 - 5x^2 + 7x + 3, x = -1$

4. $P(x) = 5x^2 + 6x - 3, x = -3$

In Exercises 5–8, find the quotient and remainder when $P(x)$ is divided by $x - a$.

5. $P(x) = 4x^2 - 6x - 3; x + 3$

6. $P(x) = 6x^3 - 2x + 1; x - 1$

7. $P(x) = 9x^4 - 5x^3 + 2x^2 - 7x + 1; x - 1$

8. $P(x) = 7x^3 - 3x^2 - 12; x + 2$

In Exercises 9–16, find all the roots of the given equation. Approximate irrational roots to 3 decimal places.

9. $x^3 - 3x^2 + 3x - 1 = 0$

10. $x^4 - 5x^3 + 9x^2 - 7x + 2 = 0$

11. $x^4 - 6x^2 - 8x - 3 = 0$

12. $x^4 + 10x^3 + 25x^2 - 16 = 0$

13. $x^4 + x^3 - 6x^2 + x + 1 = 0$

14. $2x^5 - 3x^4 - 18x^3 + 75x^2 - 104x + 48 = 0$

15. $x^6 + 2x^5 + 3x^4 + 4x^3 + 3x^2 + 2x + 1 = 0$

16. $x^6 - 2x^5 - 4x^4 + 6x^3 - x^2 + 8x + 4 = 0$

In Exercises 17–20, determine the horizontal and vertical asymptotes, if any, and solve the equation.

17. $\dfrac{5x}{x^2 + 3x - 4} = 0$

18. $\dfrac{7x^3}{x^3 + 1} = 0$

19. $\dfrac{4}{x - 2} = \dfrac{5}{x + 2}$

20. $\dfrac{3}{x - 1} + \dfrac{2}{x + 1} = \dfrac{4}{x^2 - 1}$

Solve Exercises 21–22.

21. *Medical technology* Psychologists have developed mathematical models, called **learning curves**, that are used to predict the performance as a function of the number of trials n for a certain task. One learning curve is

$$P(n) = \frac{0.5 + 0.8(n - 1)}{1.5 + 0.8(n - 1)}, \quad n > 0$$

where P is the percentage, expressed as a decimal, of the correct responses after n trials. According to this model, what is the limiting percentage of correct responses as n increases; that is, what is the horizontal asymptote?

22. *Medical technology* An experiment on learning retention is conducted by the training people at a certain company. Each trainee is given 1 day to memorize a list of 30 rules. At the end of the day, the list is turned in. Each day after that, each trainee is asked to turn in a written list of as many rules as he or she can remember. It is found that

$$N(t) = \frac{5t^2 + 6t + 15}{t^2}, \quad t > 0$$

provides a close approximation of the average number of rules, $N(t)$, remembered after t days. Determine the limiting number of rules that the trainees remembered; that is, find the horizontal asymptote of N.

☰ CHAPTER 18 TEST

1. What is the degree of $P(x) = 7x^5 - 2x^4 - 5$?

2. If $P(x) = 4x^3 - 5x^2 + 2x - 8$, what is $P(-2)$?

3. Find the quotient and remainder when $P(x) = x^4 - 8x^3 + 3x^2 + 44x + 75$ is divided by $x - 6$.

4. List the possible rational roots of $P(x) = 3x^4 - 7x^2 + 2x - 2$.

5. Find all the roots of $3x^3 + 5x^2 - 4x - 4 = 0$.

6. Find all the roots of $x^6 + 2x^5 - 4x^4 - 10x^3 - 41x^2 - 72x - 36 = 0$.

7. Determine the horizontal and vertical asymptotes, if any, and solve the equation

$$\frac{3x^2 - 2x - 1}{x^2 + 5x + 6} = 0$$

19

Sequences, Series, and the Binomial Formula

How much insulation is needed to reduce fuel consumption by 20%? The solution to this question requires knowledge of geometric sequences, a topic we will study in Section 19.2.

Courtesy of Ruby Gold

Sequences have played an important role in advanced mathematics. Many natural and physical patterns can be described by a sequence of numbers. In this chapter, we will study two specific kinds of sequences—arithmetic and geometric. We will also study the sum of a sequence, which is called a series. Both sequences and series are studied in more detail in calculus.

≡ **19.1**
SEQUENCES

A **sequence** is a set of numbers arranged in some order. Each number in the sequence is labeled with a variable, such as a. The variable is indexed with a natural number that tells its position in the sequence. The numbers a_1, a_2, a_3, and so on are the **terms** of the sequence. So, the first term in the sequence is a_1, the second term a_2, the third term a_3, and so on.

EXAMPLE 19.1

The sequence $3, 7, 11, 15, 19, 23$ has $a_1 = 3$, $a_2 = 7$, $a_3 = 11$, $a_4 = 15$, $a_5 = 19$, and $a_6 = 23$. This sequence has six terms. ∎

Many sequences follow some sort of pattern. The pattern is usually described by the nth term of the sequence. This term, a_n, is called the **general term** of the sequence. A **finite sequence** has a specific number of terms and so it has a last term. An **infinite sequence** does not have a last term. The notation $\{a_n\}$ is often used to present the nth term of a sequence. The $\{\ \}$ indicate that it is a sequence.

EXAMPLE 19.2

Find the first six terms of the sequence $a_n = 4n + 1$.

Solution

$$a_1 = 4(1) + 1 = 5$$
$$a_2 = 4(2) + 1 = 9$$
$$a_3 = 4(3) + 1 = 13$$
$$a_4 = 4(4) + 1 = 17$$
$$a_5 = 4(5) + 1 = 21$$
$$a_6 = 4(6) + 1 = 25$$

The first six terms of this sequence are 5, 9, 13, 17, 21, and 25. ∎

EXAMPLE 19.3

Find the first five terms of the sequence $\{n^2 - 3\}$.

Solution

$$a_1 = 1^2 - 3 = 1 - 3 = -2$$
$$a_2 = 2^2 - 3 = 4 - 3 = 1$$
$$a_3 = 3^2 - 3 = 9 - 3 = 6$$
$$a_4 = 4^2 - 3 = 16 - 3 = 13$$
$$a_5 = 5^2 - 3 = 25 - 3 = 22$$

The first five terms of this sequence are -2, 1, 6, 13, and 22. ∎

EXAMPLE 19.4

Find the first seven terms of the sequence $\left\{ \dfrac{n}{n+2} \right\}$.

Solution

$$a_1 = \frac{1}{1+2} = \frac{1}{3}$$

$$a_2 = \frac{2}{2+2} = \frac{2}{4} = \frac{1}{2}$$

$$a_3 = \frac{3}{3+2} = \frac{3}{5}$$

$$a_4 = \frac{4}{4+2} = \frac{4}{6} = \frac{2}{3}$$

$$a_5 = \frac{5}{5+2} = \frac{5}{7}$$

$$a_6 = \frac{6}{6+2} = \frac{6}{8} = \frac{3}{4}$$

$$a_7 = \frac{7}{7+2} = \frac{7}{9}$$

The first seven terms of this sequence are $\frac{1}{3}, \frac{1}{2}, \frac{3}{5}, \frac{2}{3}, \frac{5}{7}, \frac{3}{4}$, and $\frac{7}{9}$.

Recursion Formula

A **recursion formula** defines a sequence in terms of one or more previous terms. A sequence that is specified by giving the first term, or the first few terms, and a recursion formula is said to be **defined recursively**.

EXAMPLE 19.5

Find the first five terms of the sequence defined recursively by $a_1 = 2$ and $a_n = na_{n-1}$.

Solution We are told that $a_1 = 2$. By the recursion formula

$$a_2 = 2(a_1) = 2(2) = 4$$

$$a_3 = 3(a_2) = 3(4) = 12$$

$$a_4 = 4(a_3) = 4(12) = 48$$

$$a_5 = 5(a_4) = 5(48) = 240$$

The first five terms are 2, 4, 12, 48, and 240.

EXAMPLE 19.6

Give the first six terms of the sequence defined by $a_1 = 1$, $a_n = -2a_{n-1}$.

Solution

$$a_1 = 1$$

$$a_2 = -2(a_1) = -2(1) = -2$$

$$a_3 = -2(a_2) = -2(-2) = 4$$

$$a_4 = -2(a_3) = -2(4) = -8$$

EXAMPLE 19.6 (Cont.)

$$a_5 = -2(a_4) = -2(-8) = 16$$

$$a_6 = -2(a_5) = -2(16) = -32$$

The first six terms are $1, -2, 4, -8, 16,$ and -32.

Application

EXAMPLE 19.7

A contractor is preparing a bid for constructing an office building. The first floor will cost $275,000. Each floor above the first will cost $15,000 more than the floor below it. **(a)** How much will the fifth floor cost? **(b)** What is the total cost for the first five floors?

Solutions

(a) Notice that this is a recursive sequence with $a_1 = 275,000$, $a_n = a_{n-1} + 15,000$. The first five terms of the sequence are

$$a_1 = 275,000$$

$$a_2 = a_1 + 15,000 = 275,000 + 15,000 = 290,000$$

$$a_3 = a_2 + 15,000 = 290,000 + 15,000 = 305,000$$

$$a_4 = a_3 + 15,000 = 305,000 + 15,000 = 320,000$$

$$a_5 = a_4 + 15,000 = 320,000 + 15,000 = 335,000$$

The cost of the fifth floor is $335,000.

(b) The total cost for the first 5 floors is

$$a_1 + a_2 + a_3 + a_4 + a_5 = 275,000 + 290,000 + 305,000$$

$$+ 320,000 + 335,000$$

$$= 1,525,000$$

The cost for the first five floors in $1,525,000.

Exercise Set 19.1

In Exercises 1–12, find the first six terms with the specified general term.

1. $a_n = \dfrac{1}{n+1}$

2. $b_n = \dfrac{(-1)^n}{n}$

3. $a_n = (n+1)^2$

4. $a_n = \dfrac{1}{n(n+1)}$

5. $a_n = (-1)^n n$

6. $a_n = \dfrac{1 + (-1)^n}{1 + 4n}$

7. $\left\{ \dfrac{2n-1}{2n+1} \right\}$

8. $\left\{ \left(\dfrac{-1}{2} \right)^n \right\}$

9. $\left\{\left(\frac{2}{3}\right)^{n-1}\right\}$

10. $\{n \cos n\pi\}$

11. $\left\{\left(\frac{n-1}{n+1}\right)^2\right\}$

12. $\left\{\frac{n^2-1}{n^2+1}\right\}$

In Exercises 13–24, find the first six terms of the recursively defined sequence.

13. $a_1 = 1, a_n = na_{n-1}$

14. $a_1 = 3, a_n = a_{n-1} + n$

15. $a_1 = 5, a_n = a_{n-1} + 3$

16. $a_1 = 2, a_n = (a_{n-1})^n$

17. $a_1 = 2, a_n = (a_{n-1})^{n-1}$

18. $a_1 = 1, a_n = \left(\frac{-1}{n}\right)a_{n-1}$

19. $a_1 = 1, a_n = n^{a_{n-1}}$

20. $a_1 = \frac{1}{2}, a_n = (a_{n-1})^{-n}$

21. $a_1 = 1, a_2 = \frac{1}{2}, a_n = (a_{n-1})(a_{n-2})$

22. $a_1 = 1, a_2 = 1, a_n = a_{n-2} + a_{n-1}$

23. $a_1 = 0, a_2 = 2, a_n = a_{n-1} - a_{n-2}$

24. $a_1 = 1, a_2 = 2, a_n = (a_{n-1})(a_{n-2})$

Solve Exercises 25–28.

25. *Environmental Science* An ocean beach is eroding at the rate of 3 in. per year. If the beach is currently 25 ft wide, how wide will the beach be in 25 y? (Hint: $a_1 = 25$ ft -3 in.; find a_{25}.)

26. *Construction* A contractor is preparing a bid for constructing an office building. The foundation and basement will cost $750,000. The first floor will cost $320,000. The second floor will cost $240,000. Each floor above the second will cost $12,000 more than the floor below it.

(a) How much will the 10th floor cost?
(b) How much will the 20th floor cost?

27. *Business* You are offered a job with a starting salary of $26,000 per year with a guaranteed raise of $1,100 per year. What can you expect as an annual salary in your (a) 5th year? (b) 10th year?

28. *Architecture* The first row of an auditorium has 60 seats. Each row after the first has 4 more seats than the row in front of it. How many seats are in the 10th row?

☰ 19.2
ARITHMETIC AND GEOMETRIC SEQUENCES

In Section 19.1, we began our study of sequences. In this section, we will learn about two special sequences, the arithmetic and geometric sequences.

Arithmetic Sequence

The first special sequence we will study is the arithmetic sequence. Its definition depends on a recursion formula.

> **Arithmetic Sequence**
>
> An **arithmetic sequence,** or **arithmetic progression**, is a sequence where each term is obtained from the preceding term by adding a fixed number called the **common difference**. If the common difference is d, then an arithmetic sequence follows the recursion formula
>
> $$a_n = a_{n-1} + d$$

The terms of an arithmetic sequence follow the pattern $a_1, a_2 = a_1 + d, a_3 = a_2 + d = a_1 + 2d, a_4 = a_3 + d = a_1 + 3d, \ldots, a_n = a_{n-1} + d = a_1 + (n-1)d$. We have developed a formula for finding the nth term of an arithmetic sequence.

Terms of an Arithmetic Sequence

If a_1 is the first term of an arithmetic sequence, a_n the nth term and d the common difference, then

$$a_n = a_1 + (n-1)d$$

If a_{n-1} and a_n are consecutive terms of an arithmetic sequence, then

$$d = a_n - a_{n-1}$$

EXAMPLE 19.8

Find the 15th term of $3, 7, 11, 15 \ldots$.

Solution   Since $a_1 = 3$ and $a_2 = 7$, then $d = 7 - 3 = 4$. We want the 15th term, so $n = 15$, and we have

$$a_{15} = a_1 + (15 - 1)d$$
$$= 3 + 14 \cdot 4$$
$$= 59$$

The 15th term of this sequence is 59.

EXAMPLE 19.9

If the 20th term is 122 and the first term is 8, what is the common difference?

Solution   Since we know the 20th term and the first term, we let $n = 20$. Then

$$a_n = a_1 + (n-1)d$$
$$a_{20} = a_1 + (20 - 1)d$$
$$122 = 8 + (19)d$$
$$114 = 19d$$
$$6 = d$$

The common difference is 6.

EXAMPLE 19.10

If the first term is 5, the last (nth) term is -139, and $d = -6$, how many terms are there?

Solution
$$a_n = a_1 + (n-1)d$$
$$-139 = 5 + (n-1)(-6)$$
$$-144 = -6n + 6$$

EXAMPLE 19.10 (Cont.)

$$-150 = -6n$$

$$25 = n$$

There are 25 terms.

Geometric Sequence

The second special sequence that we will study is the geometric sequence. Like the arithmetic sequence, it follows a recursion formula.

Geometric Sequence

A **geometric sequence**, or **geometric progression**, is a sequence where each term is obtained by multiplying the preceding term by a fixed number called the **common ratio**. If the common ratio is r, then a geometric sequence follows the recursion formula:

$$a_n = ra_{n-1}$$

The terms of a geometric sequence follow the pattern $a_1, a_2 = ra_1, a_3 = ra_2 = r^2a_1, a_4 = ra_3 = r^3a_1, \ldots, a_n = ra_{n-1} = r^{n-1}a_1$.

Terms of a Geometric Sequence

If a_1 is the first term of a geometric sequence, a_n the nth term, and r the common ratio, then

$$a_n = r^{n-1}a_1$$

If a_{n-1} and a_n are consecutive terms of a geometric sequence, then

$$r = \frac{a_n}{a_{n-1}}$$

EXAMPLE 19.11

Find the eighth term of the geometric sequence 4, 20, 100, ...

Solution   Since we know that this is a geometric sequence, the common ratio is

$$r = \frac{a_2}{a_1} = \frac{20}{4} = 5$$

and since $a_1 = 4$, then

$$a_8 = r^{8-1}a_1$$

$$= 5^7 \cdot 4$$

$$= 312{,}500$$

The eighth term is 312,500.

EXAMPLE 19.12

If the 12th term is $\frac{-3}{2,048}$ and the first term is -3, what is the common ratio?

Solution Here we let $n = 12$, and so

$$a_{12} = r^{11}a_1$$

$$\frac{-3}{2,048} = r^{11}(-3)$$

$$\frac{1}{2,048} = r^{11}$$

$$r = \sqrt[11]{\frac{1}{2,048}}$$

$$= \frac{1}{2}$$

The common ratio is $\frac{1}{2}$.

EXAMPLE 19.13

In a geometric sequence, if the first term is -2, the last (nth) term is $-1,062,882$, and $r = 3$, how many terms are there?

Solution

$$a_n = r^{n-1}a_1$$

$$-1,062,882 = 3^{n-1}(-2)$$

$$531,441 = 3^{n-1}$$

$$\ln 531,441 = (n - 1)\ln 3$$

$$n - 1 = \frac{\ln 531,441}{\ln 3}$$

$$= 12$$

$$n = 13$$

There are 13 terms in this sequence.

Application

EXAMPLE 19.14

Each 2-in. layer of insulation like that shown in the photograph in Figure 19.1, reduces energy consumption by 4%. How many layers would be needed to reduce consumption by 20%, from 850 kW to 680 kW?

Solution This is a geometric sequence with $a_1 = 850$ and $a_n = 680$. Each 2-in. layer reduces energy consumption by 4%, so $r = 1 - 4\% = 1 - 0.04 = 0.96$. We need to determine n, the number of 2-in. layers of insulation.

Substituting the given values into the formula

$$a_n = a_1r^{n-1}$$

we get $680 = 850(0.96)^{n-1}$

FIGURE 19.1

Courtesy of Owens-Corning Fiberglass

EXAMPLE 19.14 (Cont.)

or $0.80 = (0.96)^{n-1}$

Using logarithms, we obtain

$$\ln 0.80 = \ln \left[(0.96)^{n-1} \right]$$

$$= (n-1) \ln 0.96$$

$$\frac{\ln 0.80}{\ln 0.96} = n-1$$

$$5.47 \approx n-1$$

so, $n \approx 6.47$

This means that 14 in. of insulation (7 layers of 2-in. insulation) are needed to reduce energy consumption by 20%.

You should notice in Example 19.14, that even though the answer (6.47) would round off to 6, we would have to use 7 layers of insulation. The actual energy used after adding 7 layers of insulation is

$$a_7 = a_1 r^6$$

$$= 850(0.96)^6$$

$$\approx 665.34 \text{ kW}$$

This represents a savings of $850 - 665.34 = 184.66$ kW, which is $\frac{184.66}{850} \approx 0.217$ or about 21.7%.

You may have noticed in the last two solutions that we used a technique that we had not used for several chapters—the technique of taking the logarithm of both sides of an equation.

Exercise Set 19.2

In Exercises 1–20, determine whether the given sequence is an arithmetic sequence, a geometric sequence, or neither. For the arithmetic and geometric sequences, find the common difference or ratio and the indicated term.

1. $1, 9, 17, \ldots$ (9th term)
2. $3, -5, -13, \ldots$ (7th term)
3. $8, -4, 2, \ldots$ (10th term)
4. $4, 1, \frac{1}{4}, \ldots$ (8th term)
5. $-1, 4, -16, \ldots$ (6th term)
6. $-1, 4, 9, \ldots$ (12th term)
7. $3, -2, -7, \ldots$ (8th term)
8. $1, -3, 9, \ldots$ (7th term)
9. $3, -2, 1, \ldots$ (10th term)
10. $1, \sqrt{2}, 2, \ldots$ (12th term)
11. $0.4, 0.8, 1.2, \ldots$ (9th term)

12. $4, 2, \sqrt{2}, \ldots$ (11th term)
13. $\frac{2}{3}, \frac{1}{2}, \frac{1}{3}, \ldots$ (8th term)
14. $\frac{2}{3}, \frac{1}{3}, \frac{1}{6}, \ldots$ (8th term)
15. $5, 0.5, 0.05, \ldots$ (6th term)
16. $6, -2, \frac{2}{3}, \ldots$ (7th term)
17. $1 + \sqrt{3}, 3 + \sqrt{3}, 5 + \sqrt{3}, \ldots$ (9th term)
18. $1.02, 10.2, 102, \ldots$ (10th term)
19. $4.3, -3.2, 2.1, \ldots$ (6th term)
20. $3.4, -6.8, 13.6, \ldots$ (9th term)

Solve Exercises 21–35.

21. For an arithmetic sequence $a_5 = 9$ and $a_6 = 24$, what is a_1?

22. For an arithmetic sequence $a_7 = 11$ and $a_9 = 15$, what is a_1?

23. For a geometric sequence $a_4 = 9$ and $a_5 = 3$, what is a_1?

24. For a geometric sequence $a_8 = \frac{1}{3}$ and $a_9 = 5$, what is a_1?

25. In an arithmetic sequence $a_1 = 5$, $a_n = 81$, and $d = 4$, what is n?

26. In an arithmetic sequence $a_1 = 20$, $a_n = -31$, and $d = -3$, what is n?

27. In a geometric sequence $a_1 = 3$, $a_n = \frac{1}{559{,}872}$ and $r = \frac{1}{6}$, find n.

28. In a geometric sequence $a_1 = -4$, $a_n = \frac{-1}{4{,}096}$, and $r = \frac{1}{2}$, what is n?

29. If the first term of an arithmetic sequence is -5 and the 7th term is 4, what is the second term?

30. If the first term of a geometric sequence is $\frac{1}{3}$ and the 8th term is 729, what is the third term?

31. *Medical technology* A group of people make out an exercise program. The first day they will jog $\frac{1}{2}$ mi. After that they will jog a certain amount more each day until on the 61st day, 8 mi are jogged. How much was the distance increased each day?

32. *Physics* A ball is dropped to the ground from a height of 80 m. Each time it bounces it goes $\frac{7}{8}$ as high as the previous bounce. How high does it go on the sixth bounce?

33. *Physics* A pendulum swings 15 cm on its first swing. Each subsequent swing is reduced by 0.3 cm. How far is the 10th swing? How many times will the pendulum swing before it comes to rest?

34. *Aeronautical engineering* The atmospheric pressure at sea level is approximately 100 kPa and decreases 12.5% for each km increase in altitude. What is the atmospheric pressure at the top of Mt. Everest, which is about 8.8 km high?

35. *Environmental science* A chemical spill pollutes a river. A monitor located 1 mi downstream from the spill measures 940 parts of the chemical for every million parts of water. (This is written as 940 ppm.) The readings decrease 16.2% for each mile farther downstream.

(a) How far downstream from the spill will it be before the concentration is reduced to 100 ppm?
(b) The water is considered safe for human consumption at 1.5 ppm. How far downstream from the spill will you have to go before the water is considered safe for humans to drink?

≡ 19.3
SERIES

There are many times when we need to add the terms of a sequence. The sum of the terms of a sequence is called a **series**. The series

$$a_1 + a_2 + a_3 + a_4 + a_5$$

is a **finite series** with five terms. A finite series can have 1, 2, 5, 19, 437, or any number of terms as long as the number is a natural number. A series of the form

$$a_1 + a_2 + a_3 + a_4 + \cdots$$

is an **infinite series**. An infinite series has an infinite or endless number of terms.

Summation Notation

A more compact notation is often used to indicate a series. This notation is referred to as **summation** or **sigma notation** because it uses the capital Greek letter sigma ($\sum$). In general, the sigma notation means

$$\sum_{k=1}^{n} a_k = a_1 + a_2 + a_3 + \cdots + a_n$$

Here $\sum$ indicates a sum. The letter k is called an **index of summation**. The summation begins with $k = 1$ as is indicated below the $\sum$ and ends with $k = n$ as is indicated above the $\sum$.

Sometimes it is useful to indicate that a series begins with the zeroth, second, or other term. If it starts with the 0th term, then

$$\sum_{k=0}^{n} a_k = a_0 + a_1 + a_2 + \cdots + a_n$$

and if it starts with the second term,

$$\sum_{k=2}^{n} a_k = a_2 + a_3 + a_4 + \cdots + a_n$$

The numbers below and above the $\sum$ are the **limits of summation**. Here they are 2 and n.

EXAMPLE 19.15

Evaluate $\sum_{k=1}^{5} (-3)k$.

Solution

$$\sum_{k=1}^{5} (-3)k = (-3)1 + (-3)2 + (-3)3 + (-3)4 + (-3)5$$

EXAMPLE 19.15 (Cont.)

$$= -3 + (-6) + (-9) + (-12) + (-15)$$

$$= -45$$

EXAMPLE 19.16

Evaluate $\displaystyle\sum_{k=1}^{4} (2k - 3)$.

Solution

$$\sum_{k=1}^{4} (2k - 3) = (2 \cdot 1 - 3) + (2 \cdot 2 - 3) + (2 \cdot 3 - 3) + (2 \cdot 4 - 3)$$

$$= (2 - 3) + (4 - 3) + (6 - 3) + (8 - 3)$$

$$= -1 + 1 + 3 + 5$$

$$= 8$$

EXAMPLE 19.17

Evaluate $\displaystyle\sum_{k=0}^{3} \dfrac{2^k}{3k - 5}$.

Solution

$$\sum_{k=0}^{3} \frac{2^k}{3k - 5} = \frac{2^0}{3 \cdot 0 - 5} + \frac{2^1}{3 \cdot 1 - 5} + \frac{2^2}{3 \cdot 2 - 5} + \frac{2^3}{3 \cdot 3 - 5}$$

$$= \frac{1}{0 - 5} + \frac{2}{3 - 5} + \frac{4}{6 - 5} + \frac{8}{9 - 5}$$

$$= \frac{1}{-5} + \frac{2}{-2} + \frac{4}{1} + \frac{8}{4}$$

$$= 4\frac{4}{5}$$

Partial Sums

Informally speaking, the expression $\displaystyle\sum_{k=0}^{n} a_k$ directs us to find the "sum" of the terms $a_1, a_2, a_3, \ldots, a_n$. To carry out this process, we proceed as follows. We let S_n denote the sum of the first n terms of the series. Thus,

$$S_1 = a_1$$

$$S_2 = a_1 + a_2$$

$$S_3 = a_1 + a_2 + a_3$$

$$\vdots$$

$$S_n = a_1 + a_2 + a_3 + \cdots + a_n = \sum_{k=1}^{n} a_k$$

The number S_n is called the **nth partial sum** of the series. The sequence $S_1, S_2, S_3, \ldots, S_n$ is called the **sequence of partial sums**.

EXAMPLE 19.18

Find the first five partial sums of the series $\sum\limits_{k=1}^{n} k^2$.

Solution

$$S_1 = 1^2 = 1$$

$$S_2 = 1^2 + 2^2 = 5$$

$$S_3 = 1^2 + 2^2 + 3^2 = 14$$

$$S_4 = 1^2 + 2^2 + 3^2 + 4^2 = 30$$

$$S_5 = 1^2 + 2^2 + 3^2 + 4^2 + 5^2 = 55$$

Arithmetic Series

Now let's look at the series formed from the two special sequences that we studied in Section 19.2. An arithmetic sequence has a first term of a_1 and a common difference d. The sum of the first n terms of an arithmetic sequence is the nth partial sum of the sequence:

$$S_n = a_1 + (a_1 + d) + (a_1 + 2d) + (a_1 + 3d) + \cdots + [a_1 + (n-1)d] \qquad (1)$$

If we write the nth term a_n first, then S_n could be written as

$$S_n = a_n + (a_n - d) + (a_n - 2d) + (a_n - 3d) + \cdots + [a_n - (n-1)d] \qquad (2)$$

If we add the corresponding terms of equations (1) and (2), we get

$$2S_n = (a_1 + a_n) + (a_1 + a_n) + (a_1 + a_n) + (a_1 + a_n) + \cdots + (a_1 + a_n)$$

The right-hand side has n terms that are all the same, $a_1 + a_n$. Thus, we have the following sum S_n of the first n terms of an arithmetic sequence.

Sum of First n Terms of an Arithmetic Sequence

The sum S_n of the first n terms of an arithmetic sequence is

$$S_n = \frac{n(a_1 + a_n)}{2}$$

where a_1 is the first term and a_n is the nth term.

EXAMPLE 19.19

Find the sum of the first eight terms of the arithmetic sequence 3, 8, 13, 18, 23, 28, 33, 38, ...

Solution We will use the formula $S_n = \dfrac{n(a_1 + a_n)}{2}$ with $n = 8, a_1 = 3$, and $a_8 = 38$.

$$S_8 = \frac{8(3 + 38)}{2} = 164$$

EXAMPLE 19.20

Find the sum of the first 14 terms of the arithmetic sequence $9, 3, -3, \ldots$.

Solution In this sequence, $d = -6$, $a_1 = 9$, and $n = 14$. From Section 19.2, we know that

$$a_{14} = a_1 + (14 - 1)d$$

$$= 9 + 13(-6)$$

$$= -69$$

We can now find the sum of the first 14 terms.

$$S_{14} = \frac{14(a_1 + a_n)}{2}$$

$$= \frac{14(9 - 69)}{2}$$

$$= -420$$

The sum of the first 14 terms of this sequence is -420.

Geometric Series

Now let's see if we can develop a similar formula for the first n terms of a geometric series. The sum of the first n terms of a geometric sequence is

$$S_n = a_1 + a_1 r + a_1 r^2 + a_1 r^3 + \cdots + a_1 r^{n-1} \tag{3}$$

Multiplying both sides of equation (3) by r, we get

$$rS_n = a_1 r + a_1 r^2 + a_1 r^3 + a_1 r^4 + \cdots + a_1 r^{n-1} + a_1 r^n \tag{4}$$

If we subtract equation (4) from equation (3), we get

$$S_n - rS_n = a_1 - a_1 r^n$$

Factoring each side $S_n(1 - r) = a_1(1 - r^n)$ and solving for S_n, we have the following formula for the first n terms of a geometric sequence.

Sum of First n Terms of a Geometric Sequence

The sum S_n of the first n terms of a geometric sequence is

$$S_n = a_1 \frac{1 - r^n}{1 - r}$$

where a_1 is the first term and r is the common ratio.

EXAMPLE 19.21

Find the sum of the first 10 terms of the geometric sequence $4, -8, 16, -32, \ldots$.

Solution In this sequence, $a_1 = 4, r = -2$, and $n = 10$.

$$S_{10} = a_1 \frac{1 - r^{10}}{1 - r}$$

$$= 4 \left(\frac{1 - (-2)^{10}}{1 - (-2)} \right)$$

$$= 4 \left(\frac{1 - 1,024}{3} \right)$$

$$= 4 \left(\frac{-1,023}{3} \right)$$

$$= -1,364$$

The sum of the first 10 terms of $4 - 8 + 16 - 32 + \cdots$ is $-1,364$.

EXAMPLE 19.22

Find the sum of the geometric series $\frac{1}{2} + \frac{1}{3} + \frac{2}{9} + \cdots + \frac{1}{2}(\frac{2}{3})^8$.

Solution In this series, $a_1 = \frac{1}{2}, r = \frac{2}{3}$, and $n = 9$.

$$S_9 = \frac{1}{2} \left(\frac{1 - (\frac{2}{3})^9}{1 - \frac{2}{3}} \right) = \frac{1}{2} \left(\frac{1 - \frac{512}{19,683}}{1 - \frac{2}{3}} \right) = \frac{1}{2} \left(\frac{\frac{19,171}{19,683}}{\frac{1}{3}} \right) = \frac{57,513}{39,366}$$

$$\approx 1.46$$

An Application of the Geometric Series

One application of the geometric series deals with compound interest. Suppose you deposit $100 in a savings account that pays 8% compounded annually. The amount of money after 1 year would be

$$100 + 100(0.08) = 100 + 8$$

$$= \$108$$

After 2 years the amount would be

$$108 + 108(0.08) = 108 + 8.64$$

$$= \$116.64$$

As you can see, the amount after each year forms a geometric sequence

$$100 + 108 + 116.64 + \cdots$$

where $a_1 = 100$ and $r = 1.08$. Thus $r = 1 + i$, where i is the interest rate. From this, we get the formula for compound interest.

> **Compound Interest**
>
> $$A = P(1 + i)^n$$
>
> where A is the amount after n interest periods, P is the principal or initial amount invested, and i is the interest rate per interest period expressed as a decimal.

Thus A is similar to S_n. Since the initial investment is P, this formula begins with $n = 0$.

The interest rate is given for the interest period. This means that in order to determine i, you must divide the annual interest rate by the number of interest periods. An 8% interest rate compounded quarterly (four times a year) would have $i = \frac{0.08}{4} = 0.02$. (Remember 8% = 0.08.) If the 8% interest rate was compounded monthly, or 12 times a year, then $i = \frac{0.08}{12} = 0.0067$.

Application

EXAMPLE 19.23

If $100 is deposited in a savings account paying 6% annually, what is the amount after 1 year if interest is compounded **(a)** quarterly, or **(b)** monthly?

Solutions We will use $A = P(1 + i)^n$ in both parts with $P = 100$.

(a) Since interest is compounded quarterly

$$i = \frac{0.06}{4} = 0.015 \quad \text{and} \quad n = 4$$

$$A = 100(1 + 0.015)^4$$

$$= 106.14$$

The total is $106.14 at the end of 1 year.

(b) In this example, the interest is compounded monthly, so

$$i = \frac{0.06}{12} = 0.005 \quad \text{and} \quad n = 12$$

$$A = 100(1.005)^{12}$$

$$= 106.17$$

The total is $106.17 at the end of 1 year.

Application

EXAMPLE 19.24

Suppose the money in Example 19.23 were left in the savings account for 4 more years (a total of 5). What would it then be worth?

Solutions

(a) At quarterly compounding, i is still 0.015, but n is now 4×5 (4 times a year for 5 years).

$$A = 100(1.015)^{20}$$

$$= 134.69$$

EXAMPLE 19.24 (Cont.)

(b) At monthly compounding, $i = 0.005$ and $n = 12 \times 5 = 60$.

$$A = 100(1.005)^{60}$$

$$= 134.89$$

The totals are $134.69 if the money is compounded quarterly for 5 years and $134.89 if it is compounded monthly.

Some older calculators have a $\boxed{\sum}$ or $\boxed{\sum +}$ key. This key is used to sum a group of numbers. The total is placed in memory and the number of terms is displayed. On some calculators, it is possible to recall this total from memory and on others it is not. The $\boxed{\sum +}$ key on a calculator is primarily used for statistical problems.

Caution Check the owner's manual for your calculator to see if you can use your calculator to sum a sequence.

Exercise Set 19.3

In Exercises 1–4, write the first four terms of the indicated series.

1. $\displaystyle\sum_{k=1}^{20} 3\left(\frac{1}{2}\right)^{k}$

2. $\displaystyle\sum_{n=1}^{50} [4 + (n-1)3]$

3. $\displaystyle\sum_{n=0}^{20} (-1)^{n}\frac{3^{n}}{n+1}$

4. $\displaystyle\sum_{k=1}^{60} (-1)^{k}$

In Exercises 5–12, evaluate the given sum.

5. $\displaystyle\sum_{k=1}^{5} (k-3)$

6. $\displaystyle\sum_{k=0}^{4} (2k+1)$

7. $\displaystyle\sum_{k=1}^{6} k^{2}$

8. $\displaystyle\sum_{k=2}^{8} \frac{k+2}{(k-1)k}$

9. $\displaystyle\sum_{n=1}^{5} n^{3}$

10. $\displaystyle\sum_{n=1}^{8} \frac{(-1)^{n}(n^{2}+1)}{n}$

11. $\displaystyle\sum_{i=3}^{8} \frac{2^{i}}{3i+1}$

12. $\displaystyle\sum_{i=0}^{7} \frac{(-1)^{i+1}(i+1)^{2}}{2i+1}$

In Exercises 13–28, determine whether the terms of the given series form an arithmetic or geometric series and find the indicated sum.

13. $3 + 6 + 9 + 12 + \cdots$; S_{10}

14. $5 + 1 - 3 + \cdots$; S_{12}

15. $1 + 2 + 4 + \cdots$; S_{8}

16. $\frac{1}{2} + 2 + 8 + \cdots$; S_{9}

17. $-6 - 2 + 2 + \cdots$; S_{12}

18. $1 - \frac{1}{2} + \frac{1}{4} + \cdots$; S_{10}

19. $0.5 + 0.75 + 1.0 + \cdots$; S_{20}

20. $0.5 + 0.2 + 0.08 + \cdots$; S_{15}

21. $\dfrac{3}{4} - \dfrac{1}{4} + \dfrac{1}{12} + \cdots ; S_{16}$

22. $0.4 + 0.04 + 0.004 + \cdots ; S_6$

23. $0.4 + 1.6 + 2.8 + \cdots ; S_{14}$

24. $16 + 8 + 4 + \cdots ; S_{10}$

25. $2\sqrt{3} + 6 + 6\sqrt{3} + \cdots ; S_{15}$

26. $\sqrt{5} + 10 + 20\sqrt{5} + \cdots ; S_{12}$

27. $3 + 9 + 27 + \cdots ; S_8$

28. $5 - 25 + 125 + \cdots ; S_{10}$

Solve Exercises 29–35.

29. *Physics* A ball is dropped to the ground from a height of 80 m, and each time it bounces 80% as high as it did on the previous bounce. How far has it traveled when it hits the ground the fourth time? The tenth time?

30. *Physics* A pendulum swings a distance of 50 m initially from one side to the other. After the first swing, each swing is only 0.8 of the distance of the previous swing. What is the total distance covered by the pendulum in 10 swings?

31. *Construction* The end of a gable roof is in the shape of a triangle. If the span is 32 ft (as shown in Figure 19.2), the height of the gable is 8 ft, and the studs are placed every 16 in., what is the total length of the studs?

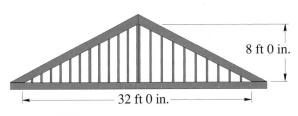

FIGURE 19.2

32. *Finance* How much will $2,000 earn in 3 years if it is invested in a savings account that pays 8% compounded quarterly?

33. *Finance* You have the opportunity to invest the $2,000 in Exercise 32 in a savings account that pays 8% compounded monthly. How much would this total after 3 years? How much more was this than the amount in Exercise 32?

34. *Finance* Your final chance is to invest the money in Exercise 32 at 8% compounded continuously. What would this total after 3 years? (See Section 13.2.)

35. *Construction* In Exercise 26, Exercise Set 19.1, a contractor was preparing a bid for the construction of an office building. The foundation and basement will cost $750,000 and the first floor will cost $320,000. The second floor will cost $240,000 and each floor above the second will cost $12,000 more than the floor below it.
(a) If the 10th floor will cost $336,000, how much will a 10-story building cost?
(b) If the 20th floor will cost $456,000, how much will it cost to build a 20-story building?

≡ 19.4
INFINITE GEOMETRIC SERIES

In this section, we will continue discussing geometric series. All of the series that we discussed in Section 19.3 were finite series. A finite series has a first term and a last term. In this section, we will discuss series that are not finite. In particular, those that do not have a last term. A series that does not have a last term is called an **infinite series**.

If we have a sequence of the form

$$a_1, a_2, a_3, \ldots, a_n, \ldots$$

then this is an **infinite sequence** because it does not have a last term. The corresponding infinite series is designed by

$$\sum_{k=1}^{\infty} a_k = a_1 + a_2 + a_3 + \cdots + a_n + \cdots$$

The symbol ∞ is read, "infinity."

We have seen examples of infinite numbers before and have always found ways of working with these numbers. But now we want to find a way to determine the sum of an infinite series.

One way that we can determine the sum of an infinite series is with our earlier definition of partial sum. Remember, each partial sum S_n is the sum of the first n term in the series. That means that each partial sum is the sum of a finite number of terms. So,

$$S_1 = a_1$$

$$S_2 = a_1 + a_2$$

$$S_3 = a_1 + a_2 + a_3$$

$$\vdots$$

$$S_n = a_1 + a_2 + a_3 + \cdots + a_n$$

S_n is called the ***n*th partial sum** of the series $\sum\limits_{k=1}^{\infty} a_k$. The **sequence of partial sums** is $S_1, S_2, S_3, \ldots, S_n$. If it happens that, as n gets larger and larger, S_n seems to be approaching some number that we will call S, we then say that S is the sum of the infinite series.

The idea of the number S is much the same as when we studied asymptotes. If you remember a curve kept getting closer and closer to an asymptote, but never reached it. In the same way, as n gets larger, S_n will get closer and closer to S, but never reach it. We say that S is the limit of the sequence S_n and write it symbolically as

$$S = \lim_{n\to\infty} S_n = \sum_{n=1}^{\infty} a_n$$

The symbol $\lim\limits_{n\to\infty} S_n$ is read "the limit of S_n as n goes to infinity." Limits are studied in detail in calculus.

Now, let's consider a couple of infinite geometric sequences. The first one we will consider is

$$1 + \frac{1}{2} + \frac{1}{4} + \frac{1}{8} + \cdots + \frac{1}{2^{n-1}} + \cdots$$

The partial sums are

$$S_1 = 1$$

$$S_2 = 1 + \tfrac{1}{2} = \tfrac{3}{2}$$

$$S_3 = 1 + \tfrac{1}{2} + \tfrac{1}{4} = \tfrac{7}{4}$$

$$S_4 = 1 + \tfrac{1}{2} + \tfrac{1}{4} + \tfrac{1}{8} = \tfrac{15}{8}$$

$$S_5 = 1 + \tfrac{1}{2} + \tfrac{1}{4} + \tfrac{1}{8} + \tfrac{1}{16} = \tfrac{31}{16}$$

From the Section 19.3 we know that

$$S_n = a_1 \left(\frac{1 - r^n}{1 - r} \right) = 1 \left[\frac{1 - (\frac{1}{2})^n}{1 - \frac{1}{2}} \right] = 2 \left[1 - \left(\frac{1}{2} \right)^n \right]$$

The sequence of partial sums

$$1, \frac{3}{2}, \frac{7}{4}, \frac{15}{8}, \frac{31}{16}, \ldots, 2 \left[1 - \left(\tfrac{1}{2} \right)^n \right], \ldots$$

seems to be approaching 2 as a limit. Since $\left(\frac{1}{2} \right)^n$ keeps getting closer and closer to 0 as n gets larger, it seems that

$$\lim_{n \to \infty} S_n = \lim_{n \to \infty} 2 \left[1 - \left(\tfrac{1}{2} \right)^n \right] = 2(1 - 0) = 2$$

Not all partial sums of infinite series approach a finite limit. If the partial sums of an infinite series approach a finite limit, we say that the series **converges** or is a **convergent series**. A series that does not converge is said to **diverge** or to be a **divergent series**. All arithmetic series diverge.

In general, an infinite geometric series of the form

$$\sum_{n=1}^{\infty} a_1 r^{n-1} = a_1 + a_1 r + a_1 r^2 + a_1 r^3 + \cdots$$

converges under certain circumstances. We know from Section 19.3 that the nth partial sum of this series is

$$S_n = \frac{a_1 (1 - r^n)}{1 - r}$$

When does this have a limit? The key is in the term r^n. Pick a number, such as 2, and take larger powers of 2. What happens? $2^2 = 4$, $2^4 = 16$, $2^8 = 256$, ... Notice that as n gets larger, 2^n keeps getting larger. Now select a smaller number such as 0.5. $(0.5)^2 = 0.25$, $(0.5)^4 = 0.0625$, $(0.5)^8 = 0.00390625$. As n gets larger, $(0.5)^n$ gets smaller and smaller. In fact, for any value r that is between -1 and 1, as n gets larger r^n gets closer to zero. In symbols, using our new limit notation, this is

$$\text{If } |r| < 1, \text{then } \lim_{n \to \infty} r^n = 0$$

This means that for an infinite geometric series, if $|r| < 1$, then

$$\lim_{n \to \infty} S_n = \lim_{n \to \infty} a_1 \left(\frac{1 + r^n}{1 - r} \right) = \frac{a_1}{1 - r}$$

This provides us with a formula for the sum of an infinite geometric series.

Sum of an Infinite Geometric Series

If $|r| < 1$, then the infinite geometric series

$$a_1 + a_1 r + a_1 r^2 + a_1 r^3 + \cdots + a_1 r^n + \cdots$$

has the sum $S = \dfrac{a_1}{1 - r}$.

If $|r| \geq 1$, then the series diverges.

EXAMPLE 19.25

Show that the series $\sum\limits_{n=1}^{\infty} \left(\frac{-2}{5}\right)^{n}$ converges and find its sum.

Solution   This is a geometric series with $r = \frac{-2}{5}$. Since $|r| = \left|\frac{-2}{5}\right| = \frac{2}{5}$, and $|r|$ is less than one, the sum of this geometric series is

$$\frac{a_1}{1-r} = \frac{-\frac{2}{5}}{1 - \frac{-2}{5}} = \frac{\frac{-2}{5}}{\frac{7}{5}} = \frac{-2}{7}$$

Always remember that we can never reach $-\frac{2}{7}$ exactly, no matter how many terms in the series we add; but, by definition, $-\frac{2}{7}$ is the sum of this series.

EXAMPLE 19.26

Show that each of the following series is divergent:

(a) $1 + \frac{4}{3} + \frac{16}{9} + \cdots$; **(b)** $\sum\limits_{k=1}^{\infty} (-1)^{k}$

Solutions

(a) Since $r = \frac{4}{3}$, we can see that $r > 1$ and so this series is divergent.

(b) The expanded series is $-1 + 1 - 1 + 1 - 1 + 1 - 1 + \cdots$ which has partial sums

$$S_1 = -1, S_2 = 0, S_3 = -1, S_4 = 0, S_5 = -1, \ldots$$

These partial sums do not get arbitrarily large, but they also do not approach some specific limit. Since $r = -1$, we see that $|r| = |-1| = 1$, and so this series diverges.

EXAMPLE 19.27

Find the fraction that has the repeating decimal form $0.232323\ldots$.

Solution   This decimal can be thought of as the series $0.23 + 0.0023 + 0.000023 + \cdots$, which is the geometric series $\sum\limits_{n=0}^{\infty} 0.23(0.01)^{n}$. In this series $a_1 = 0.23$ and $r = 0.01$, so

$$S = \frac{a_1}{1-r} = \frac{0.23}{1 - 0.01} = \frac{0.23}{0.99} = \frac{23}{99}$$

Thus, the decimal $0.232323\ldots$ is equivalent to the fraction $\frac{23}{99}$.

EXAMPLE 19.28

Express $4.2315315315\ldots$ as a rational number.

Solution   The number $4.2315315315\ldots$ repeats the digits 315. We can write the number as $4.2 + (0.0315 + 0.0000315 + 0.0000000315 + \cdots)$. The portion inside the parentheses is an infinite geometric series with $a_1 = 0.0315$ and $r = 0.001$. So, $4.2315315\ldots = 4.2 + S$, where

$$S = \frac{0.0315}{1 - 0.001} = \frac{0.0315}{0.999} = \frac{315}{9,990} = \frac{35}{1,110}$$

EXAMPLE 19.28 (Cont.) Since $4.2 = \frac{42}{10}$, we have

$$4.2315315\ldots = \frac{42}{10} + \frac{35}{1{,}110} = \frac{4{,}662 + 35}{1{,}110} = \frac{4{,}697}{1{,}110}.$$

Exercise Set 19.4

Determine which of the infinite geometric series in Exercises 1–22 converge and which diverge. For those that converge, find the sum.

1. $\dfrac{1}{2} - \dfrac{1}{4} + \dfrac{1}{8} - \dfrac{1}{16} + \cdots$

2. $1 + \dfrac{3}{4} + \dfrac{9}{16} + \cdots$

3. $1 - \dfrac{2}{3} + \dfrac{4}{9} - \dfrac{8}{27} + \cdots$

4. $1.5 + 2.25 + 3.375 + \cdots$

5. $3 + \dfrac{3}{4} + \dfrac{3}{16} + \cdots$

6. $4 - 1 + \dfrac{1}{4} + \cdots$

7. $\dfrac{5}{4} + \dfrac{1}{8} + \dfrac{1}{80} + \cdots$

8. $\dfrac{3}{4} + \dfrac{3}{4^2} + \dfrac{3}{4^3} + \cdots$

9. $0.03 + 0.003 + 0.0003 + \cdots$

10. $1.4 + 0.014 + 0.00014 + \cdots$

11. $\displaystyle\sum_{n=1}^{\infty} (-0.8)^{n-1}$

12. $\displaystyle\sum_{n=1}^{\infty} \left(\dfrac{3}{4}\right)^{n}$

13. $\displaystyle\sum_{n=1}^{\infty} \left(\dfrac{5}{4}\right)^{n+1}$

14. $\displaystyle\sum_{n=1}^{\infty} (-1.1)^{n}$

15. $\displaystyle\sum_{n=1}^{\infty} 0.3(10)^{n}$

16. $\displaystyle\sum_{n=1}^{\infty} 2^{-n}$

    $\left[\text{Hint: Let } 2^{-n} = (2^{-1})^{n}.\right]$

17. $\displaystyle\sum_{n=1}^{\infty} \left(\dfrac{3}{2}\right)^{-n}$

18. $\displaystyle\sum_{n=1}^{\infty} 6\left(-\dfrac{2}{3}\right)^{n}$

19. $\displaystyle\sum_{n=1}^{\infty} \left(\dfrac{1}{\sqrt{2}}\right)^{n}$

20. $\displaystyle\sum_{n=1}^{\infty} \left(\dfrac{\sqrt{3}}{3}\right)^{n}$

21. $\displaystyle\sum_{n=1}^{\infty} \left(\sqrt{5}\right)^{n}$

22. $\displaystyle\sum_{n=1}^{\infty} \left(\sqrt{7}\right)^{1-n}$

In Exercises 23–30, use infinite series to find the rational number corresponding to the given decimal number.

23. $0.444\ldots$

24. $0.777\ldots$

25. $0.575757\ldots$

26. $0.848484\ldots$

27. $1.352135213521\ldots$

28. $4.12341234123\ldots$

29. $6.3021021021\ldots$

30. $2.1906906906\ldots$

Solve Exercises 31–34.

31. *Physics*   A ball is dropped from a height of 5 m. After the first bounce, the ball reaches a height of 4 m, after the second, 3.2 m, and so on. What is the total distance traveled by the ball before it comes to rest?

32. *Physics*   An object suspended on a spring is oscillated up and down. The first oscillation was 100 mm and each oscillation after that was $\frac{9}{10}$ that of the preceding one. What is the total distance that the object traveled?

33. *Machine technology*   When a motor is turned off, a flywheel attached to the motor coasts to a stop. In the first second, the flywheel makes 250 revolutions. Each of the following seconds, it revolves $\frac{8}{10}$ of the number in the preceding second. What are the total number of revolutions made by the flywheel when it stops?

34. *Physics*   A pendulum swings a distance of 50 cm initially from one side to the other. After the first swing, each swing is 0.85 the distance of the previous swing. What is the total distance covered by the pendulum when it comes to a rest?

≡ 19.5
THE BINOMIAL THEOREM

In Section 7.1, we learned several special products. Among them were

$$(x+y)^2 = x^2 + 2xy + y^2$$

$$(x+y)^3 = x^3 + 3x^2y + 3xy^2 + y^3$$

$$(x-y)^2 = x^2 - 2xy + y^2$$

and $$(x-y)^3 = x^3 - 3x^2y + 3xy^2 - y^3$$

To find the larger powers of $x + y$ or $x - y$ would require repeated multiplication by $x + y$ or $x - y$. In this section, we will develop the binomial theorem. This theorem allows us to expand $x + y$ or $x - y$ to any power without direct multiplication.

Expansions of $(x + y)^n$ and Pascal's Triangle ────────

By direct multiplication, we can obtain the following expansions of $x + y$:

$$(x+y)^0 = 1$$

$$(x+y)^1 = x + y$$

$$(x+y)^2 = x^2 + 2xy + y^2$$

$$(x+y)^3 = x^3 + 3x^2y + 3xy^2 + y^3$$

$$(x+y)^4 = x^4 + 4x^3y + 6x^2y^2 + 4xy^3 + y^4$$

$$(x+y)^5 = x^5 + 5x^4y + 10x^3y^2 + 10x^2y^3 + 5xy^4 + y^5$$

$$(x+y)^6 = x^6 + 6x^5y + 15x^4y^2 + 20x^3y^3 + 15x^2y^4 + 6xy^5 + y^6$$

If you look at these expansions you may notice some patterns.
1. There are $n + 1$ terms in a binomial raised to the nth power.
2. The powers of x decrease by 1, each term beginning with x^n and ending with x^0.
3. The powers of y increase by 1, each term beginning with y^0 and ending with y^n.
4. The first term is x^n and the last term is y^n.

5. In each term, the sum of the exponents of x and y is n. For example, the third term of $(x + y)^6$ contains $x^4 y^2$ and $4 + 2 = 6$, and the fifth term contains $xy^5 = x^1 y^5$ and $1 + 5 = 6$.
6. The coefficients of terms equidistant from the ends are equal.
7. If the coefficient of any term is multiplied by the exponent of x in that term, and this product is divided by the exponent of y in the next term, we get the coefficient of the next term.
8. The coefficients form a pattern known as **Pascal's triangle** in which each coefficient is the sum of the two nearest coefficients in the row above.

$$
\begin{array}{rccccccccccccccc}
n = 0 & & & & & & & & 1 & & & & & & & \\
n = 1 & & & & & & & 1 & & 1 & & & & & & \\
n = 2 & & & & & & 1 & & 2 & & 1 & & & & & \\
n = 3 & & & & & 1 & & 3 & & 3 & & 1 & & & & \\
n = 4 & & & & 1 & & 4 & & 6 & & 4 & & 1 & & & \\
n = 5 & & & 1 & & \underbrace{5 + 10} & & 10 & & 5 & & 1 & & & \\
n = 6 & & 1 & & 6 & & 15 & & 20 & & \underbrace{15 + 6} & & 1 & & \\
n = 7 & 1 & & 7 & & 21 & & 35 & & 35 & & 21 & & 7 & & 1 \\
\end{array}
$$

You can see that the first and last number in each row is 1 and the second and next to last numbers are n. The pattern in Pascal's triangle can be continued forever. It would take a long time to develop row 26 if you needed the coefficients for $(x + y)^{26}$. That is why pattern number 7 is easier to use.

EXAMPLE 19.29

Use Pascal's triangle to expand $(3a - 2b)^6$.

Solution   Here $x = 3a$ and $y = -2b$. The coefficients from row 6 of Pascal's triangle are 1, 6, 15, 20, 15, 6, and 1, and so we get

$$1(3a)^6 + 6(3a)^5(-2b) + 15(3a)^4(-2b)^2 + 20(3a)^3(-2b)^3$$
$$+ 15(3a)^2(-2b)^4 + 6(3a)(-2b)^5 + 1(-2b)^6$$
$$= 729a^6 - 2{,}916a^5 b + 4{,}860a^4 b^2 - 4{,}320a^3 b^3 + 2{,}160a^2 b^4$$
$$- 576ab^5 + 64b^6$$

Notice that the signs alternate when the second term in the binomial is negative.

Binomial Formula

If we were to apply pattern number 7 to a general binomial expansion $(x + y)^n$, we would get the following expansion known as the **binomial formula**.

$$(x + y)^n = x^n + nx^{n-1}y + \frac{n(n-1)}{2}x^{n-2}y^2 + \frac{n(n-1)(n-2)}{2 \cdot 3}x^{n-3}y^3 + \cdots + y^n$$

Mathematicians have developed a way to abbreviate these coefficients. They use $\binom{n}{r}$ to represent $\dfrac{n!}{r!(n-r)!}$, where $n!$, read "n factorial," is the product

$$n! = n(n-1)(n-2)(n-3)\cdots(3)(2)(1)$$

and $\quad 0! = 1$

EXAMPLE 19.30

Determine 5! and 8!

Solutions

$$5! = 5 \cdot 4 \cdot 3 \cdot 2 \cdot 1 = 120$$

$$8! = 8 \cdot 7 \cdot 6 \cdot (5!) = 40,320$$

Some calculators have a key marked $\boxed{n!}$ or $\boxed{x!}$. On some of these calculators you have to use the $\boxed{\text{2nd}}$ key in order to use the $\boxed{x!}$. As you can see from comparing 5! and 8! in the last example, factorials get very large, very quickly. To determine 10! with a calculator you would

PRESS	DISPLAY
10 $\boxed{n!}$	3628800

Now,

$$\binom{n}{2} = \frac{n!}{2!(n-2)!}$$

$$= \frac{n(n-1)(n-2)\cdots(3)(2)(1)}{(2)(1)(n-2)(n-3)\cdots(2)(1)}$$

$$= \frac{n(n-1)}{2 \cdot 1}$$

This is the coefficient we got for the $x^{n-2}y^2$ term. A similar pattern holds for each term. This allows us to rewrite the binomial formula as follows.

Binomial Formula

$$(x+y)^n = x^n + \binom{n}{1}x^{n-1}y + \binom{n}{2}x^{n-2}y^2 + \binom{n}{3}x^{n-3}y^3 + \cdots$$

$$+ \binom{n}{n-1}xy^{n-1} + y^n$$

$$= \sum_{r=0}^{n} \binom{n}{r}x^{n-r}y^r$$

where the general term is $\binom{n}{r}x^{n-r}y^r$ and

$$\binom{n}{r} = \frac{n!}{(n-r)!r!} = \frac{n(n-1)(n-2)\cdots(n-r+1)}{r!}$$

EXAMPLE 19.31

Find the first four terms $(x + y)^{15}$.

Solution   The first two coefficients are 1 and 15. The next two terms are

$$\binom{15}{2} = \frac{15!}{2!13!}$$

$$= \frac{15 \cdot 14 \cdot 13 \cdots 3 \cdot 2 \cdot 1}{(2 \cdot 1)(13 \cdot 12 \cdots 3 \cdot 2 \cdot 1)}$$

$$= \frac{15 \cdot 14}{2} = 105$$

and

$$\binom{15}{3} = \frac{15!}{3!12!}$$

$$= \frac{15 \cdot 14 \cdot 13 \cdots 3 \cdot 2 \cdot 1}{(3 \cdot 2 \cdot 1)(12 \cdot 11 \cdots \cdot 2 \cdot 1)}$$

$$= \frac{15 \cdot 14 \cdot 13}{3 \cdot 2 \cdot 1} = 455$$

So, the first four terms of $(x + y)^{15}$ are

$$x^{15} + 15x^{14}y + 105x^{13}y^2 + 455x^{12}y^3$$

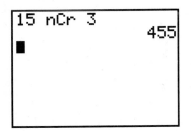

FIGURE 19.3a

FIGURE 19.3b

You could have used the $\boxed{n!}$ key on a calculator to determine these coefficients. For example, to determine $\binom{15}{3}$

PRESS	DISPLAY
15 $\boxed{n!}$	1.3076744 12
$\boxed{\div}$ 12 $\boxed{n!}$	4.790016 08
$\boxed{\div}$	2730
3 $\boxed{n!}$	6
$\boxed{=}$	455

Some calculators have a $\boxed{Cy, x}$ or $\boxed{nC_r}$ key that can be used to determine this value. For example, on an RPN calculator, $\binom{15}{3}$ could be evaluated by using the keystroke sequence $\boxed{\text{ENTER}}$ 3 $\boxed{\text{2nd}}$ $\boxed{Cy, x}$ and the answer 455 will be displayed.

On a *TI-81*, you first press the top number in this expression, 15, then press $\boxed{\text{MATH}}$ to access that Math menu. Next press $\boxed{\blacktriangleright}$ three times so that the PRB at the top of the Math menu is highlighted. (PRB stands for probability.) There are three operations listed on the PRB screen. The third operation is listed as "3: nCr," as shown in Figure 19.3a. Press "3" (or $\boxed{\blacktriangledown}$ three times) and the second number, 3. Press $\boxed{\text{ENTER}}$. The screen now displays the answer, 455, as shown in Figure 19.3b.

EXAMPLE 19.32

Find the eighth term of $(x - 2)^{17}$.

Solution The eighth term will be

$$\binom{17}{7}x^{10}(-2)^7 = \frac{17!}{7!10!}x^{10}(-2)^7$$

$$= 19,448x^{10}(-128)$$

$$= -2,489,344x^{10}$$

Binomial Series

If we rewrite the binomial in the form $1 + x$, then we obtain the **binomial series**

$$(1 + x)^n = 1 + nx + \binom{n}{2}x^2 + \binom{n}{3}x^3 + \cdots + \binom{n}{n-1}x^{n-1} + x^n$$

It can be shown that the binomial series is valid for any real number n if $|x| < 1$. If n is negative or a rational number, we then get an infinite series.

$$(1 + x)^n = 1 + nx + \frac{n(n-1)}{2!}x^2 + \frac{n(n-1)(n-2)}{3!}x^3 \cdots$$

Normally, with an infinite binomial series, we calculate as many terms as are needed. One application of the binomial series is to find numerical approximations, particularly if you want more accuracy than you can get from a calculator or computer.

EXAMPLE 19.33

Find the value of $\sqrt[7]{0.875}$ to three significant digits.

Solution We can rewrite $\sqrt[7]{0.875} = (1 - 0.125)^{1/7}$. We will use the binomial series with $n = \frac{1}{7}$ and $x = -0.125$.

$$\sqrt[7]{0.875} = 1 + \frac{1}{7}(-0.125) + \frac{\frac{1}{7}(\frac{-6}{7})}{2}(-0.125)^2$$

$$+ \frac{\frac{1}{7}(\frac{-6}{7})(\frac{-13}{7})}{6}(-0.125)^3 + \cdots$$

We can omit the terms after the fourth one, because they are not significant.

$$\sqrt[7]{0.875} \approx 1 - 0.0179 - 0.0010 - 0.0001 = 0.981.$$

Checking this on a calculator, you get $0.9811\ldots$, which is the same to three significant digits.

EXAMPLE 19.34

Show that $\lim\limits_{n\to\infty} \left(1 + \dfrac{1}{n}\right)^n = e$.

Solution   This is a binomial series and

$$\left(1 + \frac{1}{n}\right)^n = 1 + n\left(\frac{1}{n}\right) + \frac{n(n-1)}{2!}\left(\frac{1}{n}\right)^2$$

$$+ \frac{n(n-1)(n-2)}{3!}\left(\frac{1}{n}\right)^3 + \cdots$$

$$= 1 + 1 + \frac{1}{2!}\left(\frac{n-1}{n}\right) + \frac{1}{3!}\left(\frac{n-1}{n}\right)\left(\frac{n-2}{n}\right) + \cdots$$

$$= 1 + 1 + \frac{1}{2!}\left(1 - \frac{1}{n}\right) + \frac{1}{3!}\left(1 - \frac{1}{n}\right)\left(1 - \frac{2}{n}\right) + \cdots$$

Since $\lim\limits_{n\to\infty} \dfrac{1}{n} = 0$, this is equivalent to

$$1 + 1 + \frac{1}{2!} + \frac{1}{3!} + \frac{1}{4!} + \cdots$$

If you calculate these values, you will get 2.7083333. Including more values will show that

$$\lim_{n\to\infty} 1 + 1 + \frac{1}{2!} + \frac{1}{3!} + \cdots + \frac{1}{n!} + \cdots = 2.71828\cdots = e.$$

Exercise Set 19.5

In Exercises 1–8, expand and simplify the given expression by use of Pascal's triangle.

1. $(a+1)^4$

2. $(2x+b)^5$

3. $(3x-1)^4$

4. $(x-2y)^5$

5. $\left(\dfrac{x}{2} + d\right)^6$

6. $\left(xy - \dfrac{a}{3}\right)^6$

7. $\left(\dfrac{a}{2} - \dfrac{4}{b}\right)^5$

8. $\left(\dfrac{x}{3} + \dfrac{2}{y}\right)^4$

In Exercises 9–16, expand and simplify the given expression by use of the binomial formula.

9. $(a+b)^7$

10. $(a+b)^9$

11. $(t-a)^8$

12. $(x-2a)^7$

13. $(2a-1)^6$

14. $(a^2-3)^9$

15. $\left(x^2y + \dfrac{a}{2}\right)^7$

16. $\left(\dfrac{a}{3} - \dfrac{xy}{2}\right)^6$

In Exercises 17–22, find the first four terms of each binomial expansion.

17. $(x + y)^{12}$

18. $(x - 5)^{10}$

19. $(1 - a)^{-2}$

20. $(1 + x)^{1/2}$

21. $(1 + b)^{-1/3}$

22. $(1 - y)^{-1/4}$

In Exercises 23–30, approximate the values of the given expression to three decimal places using the binomial series. Check your result with a calculator.

23. $(1.1)^4 = (1 + 0.1)^4$

24. $(0.85)^5 = (1 - 0.15)^5$

25. $\sqrt{1.1}$

26. $\sqrt[3]{1.01}$

27. $\sqrt[5]{1.04}$

28. $\sqrt[6]{0.98}$

29. $\sqrt[3]{0.95}$

30. $\sqrt[4]{0.925}$

In Exercises 31–36, find the indicated term of the given binomial expression.

31. The sixth term of $(x^2 + y)^{15}$

32. The eighth term of $(a + b)^{20}$

33. The fifth term of $(2x - y)^{12}$

34. The fourth term of $(3x - 2y)^{10}$

35. The term involving b^4 in $(a + b)^{14}$

36. The term involving x^5 in $(x + y)^{15}$

Solve Exercises 37–40.

37. *Nuclear physics*   The energy of an electron traveling at speed v in special relativity theory, is $mc^2 \left(1 - \dfrac{v^2}{c^2}\right)^{-1/2}$, where m is the electron mass and c is the speed of light. The factor mc^2 is called the **rest mass energy** (or the energy when $v = 0$).
   (a) Find the first three terms of the expansion of $\left(1 - \dfrac{v^2}{c^2}\right)^{-1/2}$.
   (b) Multiply the result of **(a)** by mc^2 to get the energy at speed v.

38. *Nuclear physics*   The velocity v of electrons from a high-energy accelerator is very near the velocity of light, c. Given the voltage V of the accelerator, we often need to calculate the ratio v/c. This ratio can be calculated from the formula
$$\frac{v}{c} = \sqrt{1 - \frac{1}{4V^2}}, V = \text{number of million volts}$$
   (a) Find the first two terms of the expansion of
$$\sqrt{1 - \frac{1}{4V^2}}$$

   (b) Use your answer from (a) to determine $1 - \dfrac{v}{c}$, given the following values of V. (Remember, V is the number of million volts.)
   (i) 100 million volts
   (ii) 500 MV (1 MV = 1 megavolt = 1 million volts)
   (iii) 125 billion volts
   (iv) 250 GV (1 GV = 1 gigavolt = 1 billion volts)

39. *Physics*   At a point in a magnetic field, the field strength is given by
$$H = \frac{2m\ell}{(r^2 + \ell^2)^{3/2}}$$
   Use the binomial series to find the first three terms of the expression $(r^2 + \ell^2)^{3/2}$ by expressing it as
$$(r^2)^{3/2} \left(1 + \frac{\ell^2}{r^2}\right)^{3/2}$$

40. In the formula for a derivative in calculus, the following expression may occur:
$$\frac{(x + h)^4 - x^4}{h}$$
   Expand and simplify this expression.

CHAPTER 19 REVIEW

Important Terms and Concepts

Arithmetic sequence
   Common difference
   Sum of first n terms

Binomial theorem

Geometric sequence
   Common ratio
   Sum of first n terms

Partial sum

Pascal's triangle

Recursion formula

Sequence
   Arithmetic
   Finite
   Geometric
   Infinite

Series
   Arithmetic
   Binomial
   Finite
   Geometric
   Infinite

Summation notation

Review Exercises

In Exercises 1–4, find the first six terms with the specified general term.

1. $a_n = \dfrac{1}{n+2}$

2. $b_n = \dfrac{3}{2n-1}$

3. $\left\{ \dfrac{(-1)^n}{n(2n+1)} \right\}$

4. $\left\{ \dfrac{n^2+1}{3n-1} \right\}$

In Exercises 5–8, find the first six terms of the recursively defined sequence.

5. $a_1 = 1, a_n = \dfrac{a_{n-1}}{n}$

6. $a_1 = 5, a_n = a_{n-1} - n$

7. $a_1 = 1, a_2 = 3, a_n = a_{n-1} + a_{n-2}$

8. $a_1 = 1, a_2 = 2, a_n = (a_{n-1})(a_{n-2})$

In Exercises 9–14, determine whether the given sequence is an arithmetic sequence, a geometric sequence, or neither. For the arithmetic and geometric sequences, find the common difference or ratio and the indicated term.

9. $10, 7, 4, \ldots$ (10th term)

10. $8, 2, \dfrac{1}{2}, \ldots$ (8th term)

11. $-1, 3, 7, \ldots$ (7th term)

12. $7, 4, 0, -5, \ldots$ (9th term)

13. $3, -\dfrac{1}{2}, \dfrac{1}{12}, \ldots$ (10th term)

14. $2.05, 20.5, 205, \ldots$ (7th term)

Solve Exercises 15–17.

15. For an arithmetic sequence, $a_7 = 12$ and $a_8 = 18$. What is a_1?

16. For a geometric sequence, $a_6 = 9$ and $a_7 = \frac{9}{7}$. What is a_8?

17. Write the first four terms of

(a) $\displaystyle\sum_{k=1}^{30} 4\left(\dfrac{1}{3}\right)^k$ and   (b) $\displaystyle\sum_{k=0}^{25} \dfrac{k+2}{k+1}$.

In Exercises 18–20, evaluate the given sum.

18. $\sum_{k=1}^{4} (k+1)$

20. $\sum_{n=1}^{4} \frac{(-1)^n n}{2^n + 1}$

19. $\sum_{k=0}^{5} (3k-1)$

In Exercises 21–24, determine whether the terms of the given series form an arithmetic or geometric series and find the indicated sum.

21. $4 + 9 + 14 + \cdots ; S_{10}$

23. $1 + \frac{1}{3} + \frac{1}{9} + \cdots ; S_{14}$

22. $2 - 5 - 12, \cdots ; S_{12}$

24. $\sqrt{5} + 1 + 2\sqrt{5} + 2 + 3\sqrt{5} + 3 \cdots ; S_8$

In Exercises 25–30, determine which of the infinite geometric series converge and which diverge. For those that converge, find the sum.

25. $\frac{1}{3} + \frac{1}{6} + \frac{1}{12} + \cdots$

29. $\sum_{n=1}^{\infty} \left(\frac{1}{\sqrt{3}} \right)^n$

26. $8 + 6 + 4.5 + \cdots$

27. $0.03 + 0.3 + 3 + \cdots$

30. $\sum_{n=0}^{\infty} \left(\sqrt{2} \right)^{-n}$

28. $1.5 - 0.15 + 0.015 + \cdots$

In Exercises 31 and 32, use infinite series to find the rational numbers corresponding to the given decimal number.

31. $0.185185\ldots$

32. $0.611111\ldots$

In Exercises 33–36, expand and simplify the given binomial expression.

33. $(a+2)^5$

36. $\left(\frac{2a^3}{5} + \frac{5b}{2} \right)^5$

34. $(3x - y)^6$

35. $\left(\frac{x}{2} - 3y^2 \right)^6$

In Exercises 37–40, find the first four terms of the given binomial expansion.

37. $(2x + y)^{15}$

39. $(1-x)^{-5}$

38. $(1 + ax^2)^{10}$

40. $(1+b)^{-1/4}$

Solve Exercises 41–50.

41. What is the seventh term of $(x + 2y)^{20}$?

42. What is the fifth term of $(a - 3b)^{12}$?

43. Approximate $\sqrt[5]{1.02}$ to three decimal places using the binomial expansion. Check your work with a calculator.

44. Approximate $\sqrt[7]{0.98}$.

45. *Finance* If you invest \$400 at 6% compounded monthly for 10 years, how much will you have at the end of that time?

46. Some copying machines make reduced copies. Suppose that you copy a page 20 cm wide and it comes out $\frac{3}{4}$ as wide. Then you copy this, and so on. How wide is the fifth copy?

47. If you lay the original and five copies from Exercise 46 side by side on the floor, how far will they extend?

48. Suppose that you continued the copying process that you began in Exercise 46 indefinitely. If you lay the original and all the copies side by side on the floor, how far would they extend?

49. *Physics* A ball starting from rest rolls down a uniform incline so that it covers 10 in. during the first second, 25 in. during the second second, 40 in. during the third second, and so on. How long will it take until it covers 250 ft in 1 s?

50. *Business* A company had sales of $250,000 during its first year of operation. Sales increased by $40,000 per year during each successive year. What were the sales of the company in the 10th year? What were the total sales of the company for the first 10 years?

≡ CHAPTER 19 TEST

1. Determine the first six terms of $a_n = \dfrac{5}{2n-3}$.

2. Determine the first six terms of $a_1 = -2$, $a_n = 3 + na_{n-1}$.

3. Determine whether the given sequence is an arithmetic sequence, geometric sequence, or neither. If appropriate, find the common difference or ratio and the indicated term.

$$15, 12.5, 10, \ldots \text{ (10th term)}$$

4. For an arithmetic sequence with $a_5 = 10$ and $a_6 = 14$, what is a_1?

5. Evaluate $\displaystyle\sum_{k=1}^{5} (2k-1)$.

6. Determine whether this infinite geometric series converges or diverges. If it converges, find its sum.

$$\frac{1}{5} - \frac{1}{10} + \frac{1}{20} + \cdots$$

7. Determine the rational number that corresponds to $0.435435\ldots$.

8. Find the first four terms of $(3x - 2y)^{12}$.

9. A computer software company predicts that it will sell 2,000 copies of a new type of software during the first month and that each month sales will be 1,500 copies more than the sales of the previous month. How many copies can they expect to sell during the first year?

CHAPTER

20

Trigonometric Formulas, Identities, and Equations

A highway engineer is designing the curve at an intersection where two highways intersect at angle θ. In Section 20.3, we will learn how to use trigonometry to help design this curve.

Courtesy of Ruby Gold

In Chapters 9 through 11, we discussed the fundamental reciprocal and quotient identities of trigonometry. We also learned how to use trigonometry to solve both right and oblique triangles and how to graph trigonometric functions. In this chapter, we will return to the study of trigonometry, establishing the remaining standard trigonometric identities. Among these will be identities for the sums, differences, and multiples of angles. The identities you will learn in this chapter are used in advanced mathematics, particularly calculus, and in engineering, physics, and technical areas to simplify complicated expressions and to help solve equations that involve trigonometry.

680

☰ 20.1
BASIC IDENTITIES

An **identity** is an equation that is true for all values of the variable. In Section 9.2, we introduced two groups of trigonometric identities. One group was called **reciprocal identities** and consisted of

$$\csc \theta = \frac{1}{\sin \theta} \qquad \sec \theta = \frac{1}{\cos \theta} \qquad \cot \theta = \frac{1}{\tan \theta}$$

The second group was the **quotient identities**.

$$\tan \theta = \frac{\sin \theta}{\cos \theta} \qquad \cot \theta = \frac{\cos \theta}{\sin \theta}$$

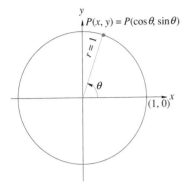

FIGURE 20.1

Pythagorean Identity

Now, suppose that we have a unit circle as shown in Figure 20.1. (Remember, a unit circle is a circle with radius 1.) If this circle is centered at the origin and $P(x, y)$ is any point on the circle on the terminal side of an angle θ in standard position, then $x = \cos \theta$ and $y = \sin \theta$. By the Pythagorean theorem, $x^2 + y^2 = r^2$, and since $r = 1$, we get the first of the **Pythagorean identities.**

$$\sin^2 \theta + \cos^2 \theta = 1$$

☰ **Note** The term $\sin^2 \theta$ is an abbreviation for $(\sin \theta)^2$. Similarly, $\cos^2 \theta$ is an abbreviation for $(\cos \theta)^2$ and $\tan^2 \theta$ is an abbreviation for $(\tan \theta)^2$. While we write $\sin^2 \theta$, $\cos^2 \theta$, or $\tan^2 \theta$, you must enter them in a calculator as $(\sin \theta)^2$, $(\cos \theta)^2$, or $(\tan \theta)^2$.

If we divide both sides of this identity by $\cos^2 \theta$, we get

$$\frac{\sin^2 \theta}{\cos^2 \theta} + \frac{\cos^2 \theta}{\cos^2 \theta} = \frac{1}{\cos^2 \theta}$$

or $\qquad \tan^2 \theta + 1 = \sec^2 \theta$

This is the second Pythagorean identity. The third, and last, Pythagorean identity is produced by dividing $\sin^2 \theta + \cos^2 \theta = 1$ by $\sin^2 \theta$, with the result

$$1 + \cot^2 \theta = \csc^2 \theta$$

Thus, there are three Pythagorean identities.

Pythagorean Identities	$\sin^2 \theta + \cos^2 \theta = 1$
	$\tan^2 \theta + 1 = \sec^2 \theta$
	$1 + \cot^2 \theta = \csc^2 \theta$

Proving Identities

The eight basic identities, the reciprocal, quotient, and Pythagorean identities, can be used to develop and prove other identities. Your ability to prove an identity depends greatly on your familiarity with the eight basic identities.

To prove that an identity is true, you change either side, or both sides, until the sides are the same. Each side must be worked separately. Since you do not know that the two sides are equal (which is what you are trying to prove), you cannot transpose terms from one side to the other. Some people draw a vertical line between the two sides until they can show that the sides are equal. The vertical line acts as a reminder that you should work on each side separately.

EXAMPLE 20.1

Prove the identity $\csc \theta = \dfrac{\cot \theta}{\cos \theta}$.

Solution We will change the right-hand side of the identity until it looks like the left-hand side.

$\csc \theta$	$\dfrac{\cot \theta}{\cos \theta}$	
	$\dfrac{\frac{\cos \theta}{\sin \theta}}{\cos \theta}$	Change $\cot \theta$ to $\dfrac{\cos \theta}{\sin \theta}$.
	$\dfrac{\cos \theta}{\sin \theta} \cdot \dfrac{1}{\cos \theta}$	Change the division problem to a multiplication problem.
	$\dfrac{1}{\sin \theta}$	Multiply.
	$\csc \theta$	Reciprocal identity.

So, $\csc \theta = \dfrac{\cot \theta}{\cos \theta}$.

EXAMPLE 20.2

Prove the identity: $|\sin x| = \dfrac{|\tan x|}{\sqrt{1 + \tan^2 x}}$.

Solution The right-hand side is more complicated than the left-hand side, so we will simplify the right-hand side until it matches the left-hand side.

$|\sin x|$	$\dfrac{|\tan x|}{\sqrt{1 + \tan^2 x}}$	
	$\dfrac{|\tan x|}{\sqrt{\sec^2 x}}$	Use the Pythagorean identity to replace $1 + \tan^2 x$ with $\sec^2 x$.
	$\left| \dfrac{\tan x}{\sec x} \right|$	Take the square root. Notice that $\sqrt{\sec^2 x} = |\sec x|$.
	$\left| \dfrac{\frac{\sin x}{\cos x}}{\frac{1}{\cos x}} \right|$	Express $\tan x$ and $\sec x$ in terms of $\sin x$ and $\cos x$.
	$\left| \dfrac{\sin x}{\cos x} \cdot \dfrac{\cos x}{1} \right|$	Change the division problem to a multiplication problem.

EXAMPLE 20.2 (Cont.)

$$| \sin x| \qquad \text{Multiply.}$$

Thus, we have shown that $|\sin x| = \dfrac{|\tan x|}{\sqrt{1 + \tan^2 x}}$.

Notice that we had to use the properties of absolute value to prove this identity.

EXAMPLE 20.3

Prove the identity $\sec \theta - \sec \theta \sin^2 \theta = \cos \theta$.

Solution   In this example, we start with the more complicated left-hand side and simplify it until it matches the right-hand side.

$\sec \theta - \sec \theta \sin^2 \theta$	$\cos \theta$
$\sec \theta(1 - \sin^2 \theta)$	Factor.
$\sec \theta(\cos^2 \theta)$	Pythagorean identity.
$\dfrac{1}{\cos \theta}(\cos^2 \theta)$	Reciprocal identity.
$\cos \theta$	Multiply.

And so, $\sec \theta - \sec \theta \sin^2 \theta = \cos \theta$.

In this example, we used a different version of a Pythagorean identity. We gave you the identity $\sin^2 \theta + \cos^2 \theta = 1$. You should also recognize the two variations of this identity: $\sin^2 \theta = 1 - \cos^2 \theta$ and $\cos^2 \theta = 1 - \sin^2 \theta$. There are two variations of each of the other Pythagorean identities.

EXAMPLE 20.4

Prove the identity: $\sec^2 \theta \csc^2 \theta = \sec^2 \theta + \csc^2 \theta$.

Solution   Here we simplify the right-hand side until it matches the left-hand side.

$\sec^2 \theta \csc^2 \theta$	$\sec^2 \theta + \csc^2 \theta$
	$\dfrac{1}{\cos^2 \theta} + \dfrac{1}{\sin^2 \theta}$ Reciprocal identities.
	$\dfrac{\sin^2 \theta}{\cos^2 \theta \sin^2 \theta} + \dfrac{\cos^2 \theta}{\cos^2 \theta \sin^2 \theta}$ Rewrite with a common denominator.
	$\dfrac{\sin^2 \theta + \cos^2 \theta}{\cos^2 \theta \sin^2 \theta}$ Add.
	$\dfrac{1}{\cos^2 \theta \sin^2 \theta}$ Pythagorean identity.
	$\dfrac{1}{\cos^2 \theta} \cdot \dfrac{1}{\sin^2 \theta}$ Factor.

EXAMPLE 20.4 (Cont.)

$$\left| \ \sec^2 \theta \csc^2 \theta \right. \hspace{4cm} \text{Reciprocal identities.}$$

So, $\sec^2 \theta \csc^2 \theta = \sec^2 \theta + \csc^2 \theta$.

Application

EXAMPLE 20.5

Malus's law concerns light incident on a polarizing plate and describes the amount of light transmitted I in terms of the angle of incidence θ and the maximum intensity of light transmitted, M. Malus's law can be written as

$$I = M - M \tan^2 \theta \cos^2 \theta$$

Express the right-hand side of this equation in terms of the $\cos \theta$.

Solution   We have

$$I = M - M \tan^2 \theta \cos^2 \theta$$

$$= M - M \left(\frac{\sin^2 \theta}{\cos^2 \theta} \right) \cos^2 \theta$$

$$= M - M \sin^2 \theta$$

$$= M(1 - \sin^2 \theta)$$

$$= M \cos^2 \theta$$

So, Malus's law can be more simply expressed as

$$I = M \cos^2 \theta$$

Exercise Set 20.1

1. Prove the Pythagorean identity $1 + \cot^2 \theta = \csc^2 \theta$ from the identity $\sin^2 \theta + \cos^2 \theta = 1$.

Prove each of the identities in Exercises 2–30.

2. $\tan x \cot x = 1$

3. $\sin \theta \sec \theta = \tan \theta$

4. $\cos \theta(\tan \theta + \sec \theta) = \sin \theta + 1$

5. $\dfrac{\sin \theta}{\cot \theta} = \sec \theta - \cos \theta$

6. $\tan x = \dfrac{\sec x}{\csc x}$

7. $(1 - \sin^2 \theta)(1 + \tan^2 \theta) = 1$

8. $\dfrac{\sin A}{\csc A} + \dfrac{\cos A}{\sec A} = 1$

9. $1 - \dfrac{\sin A}{\csc A} = \cos^2 A$

10. $(1 + \tan \theta)(1 - \tan \theta) = 2 - \sec^2 \theta$

11. $(1 + \cos x)(1 - \cos x) = \sin^2 x$

12. $\sec^4 x - \sec^2 x = \tan^4 x + \tan^2 x$

13. $2 \csc \theta = \dfrac{\sin \theta}{1 + \cos \theta} + \dfrac{1 + \cos \theta}{\sin \theta}$

14. $\cos x = \sin x \cot x$

15. $(\sin \theta + \cos \theta)^2 = 1 + 2 \sin \theta \cos \theta$

16. $\csc^2 x(1 - \cos^2 x) = 1$

17. $\dfrac{\tan \theta + \cot \theta}{\tan \theta - \cot \theta} = \dfrac{\tan^2 \theta + 1}{\tan^2 \theta - 1}$

18. $\dfrac{1 - \sin x}{\cos x} = \dfrac{\cos x}{1 + \sin x}$

19. $\dfrac{\sec \theta - \csc \theta}{\sec \theta + \csc \theta} = \dfrac{\tan \theta - 1}{\tan \theta + 1}$

20. $(\sin A + \cos A)^2 + (\sin A - \cos A)^2 = 2$

21. $\tan^2 x \cos^2 x + \cot^2 x \sin^2 x = 1$

22. $\tan \theta + \dfrac{\cos \theta}{1 + \sin \theta} = \sec \theta$

23. $\sec^4 x - \sec^2 x = \tan^2 x \sec^2 x$

24. $\cos^2 A - \sin^2 A = 2 \cos^2 A - 1$

25. $\dfrac{\tan \theta - \sin \theta}{\sin^3 \theta} = \dfrac{\sec \theta}{1 + \cos \theta}$

26. $\dfrac{\sin x - \cos x + 1}{\sin x + \cos x - 1} = \dfrac{\sin x + 1}{\cos x}$

27. $\tan^2 \theta \csc^2 \theta \cot^2 \theta \sin^2 \theta = 1$

28. $\tan x \sin x + \cos x = \sec x$

29. $\dfrac{\sec A + \csc A}{\tan A + \cot A} = \sin A + \cos A$

30. $\dfrac{\sin^3 x + \cos^3 x}{\sin x + \cos x} = 1 - \sin x \cos x$

To show that something is not an identity, all you need is one counterexample. A counterexample is an example that shows that something is not true. In Exercises 31–35, use the indicated angles as a counterexample to show that the relation is not an identity.

31. $2 \sin \theta \neq \sin 2\theta$; $\theta = 90°$

32. $\dfrac{\tan A}{2} \neq \tan \left(\dfrac{A}{2} \right)$; $A = 60°$

33. $\cos(\theta^2) \neq (\cos \theta)^2$; $\theta = \pi$

34. $\sin(x - y) \neq \sin x - \sin y$; $x = 60°$, $y = 30°$

35. $\sin x \neq \dfrac{\tan x}{\sqrt{1 + \tan^2 x}}$; $x = 120°$

36. In finding the rate of change of $\cot x$, you get the expression $\dfrac{(\sin x)(- \sin x) - (\cos x)(\cos x)}{\sin^2 x}$. Show that this is equal to $- \csc^2 x$.

37. In finding the rate of change of $\cot^2 x$, you get the expression $-2 \cot x \csc^2 x$. Show that this is equal to $-2 \cos x \csc^3 x$.

38. In calculus, in order to determine the integral of $\sin^5 x$, we need to show that it is identical to $(1 - 2 \cos^2 x + \cos^4 x) \sin x$. Prove this identity.

39. *Electricity*   In electric circuit theory, we use the expression

$$\dfrac{(1.2 \sin \omega t - 1.6 \cos \omega t)^2 + (1.6 \sin \omega t + 1.2 \cos \omega t)^2}{2L}$$

Show that this is identical to $\dfrac{2.0}{L}$.

☰ 20.2
THE SUM AND DIFFERENCE IDENTITIES

As we saw in Exercises 31 and 34 in Exercise Set 20.1, $2 \sin \theta \neq \sin 2\theta$ and $\sin(x - y) \neq \sin x - \sin y$. In this section, we will develop some identities for the sum and differences of the trigonometric functions. These identities are important for further studies in mathematics, such as in calculus. They are also important in wave mechanics, electric circuit theory, and in theory for other technical areas.

Cos($\theta + \phi$) ▬▬▬▬▬▬▬▬▬▬▬▬▬▬▬▬▬▬▬▬▬▬▬▬▬▬▬

We will begin with a rather lengthy development of the identity for the $\cos(\theta + \phi)$. Even this lengthy proof does not include all cases, but it will serve our purposes.

In Figures 20.2a and 20.2b, we have drawn two unit circles. In Figure 20.2a, $\angle AOB$ is θ and $\angle BOC$ is ϕ, so $\angle AOC$ is $\theta + \phi$. The coordinates of A, B, and C are also given. Since A is on the x-axis, its coordinates are $(1, 0)$. The coordinates of B are given in terms of θ and those of C are given in terms of $\theta + \phi$.

In Figure 20.2b, we have rotated $\triangle AOC$ through the angle $-\theta$ to get $\triangle DOF$. The coordinates of D and F are given in terms of θ and ϕ. Now, since $\triangle AOC$ is congruent to $\triangle DOF$, the distance from A to C must be the same as the distance from

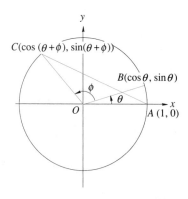

FIGURE 20.2a

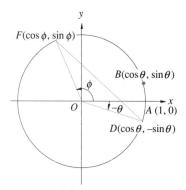

FIGURE 20.2b

D to F. According to the distance formula from Section 15.1, the distance from A to C is

$$d(A, C) = \sqrt{(\cos(\theta + \phi) - 1)^2 + (\sin(\theta + \phi) - 0)^2}$$

Squaring both sides we get

$$[d(A, C)]^2 = (\cos(\theta + \phi) - 1)^2 + (\sin(\theta + \phi) - 0)^2$$
$$= \cos^2(\theta + \phi) - 2\cos(\theta + \phi) + 1 + \sin^2(\theta + \phi)$$
$$= 2 - 2\cos(\theta + \phi)$$

In a similar manner, the distance from D to F is

$$d(D, F) = \sqrt{(\cos\phi - \cos\theta)^2 + (\sin\phi + \sin\theta)^2}$$

Again, squaring both sides, we get

$$[d(D, F)^2] = (\cos\phi - \cos\theta)^2 + (\sin\phi + \sin\theta)^2$$
$$= \cos^2\phi - 2\cos\theta\cos\phi$$
$$+ \cos^2\theta + \sin^2\phi + 2\sin\theta\sin\phi + \sin^2\theta$$
$$= (\cos^2\phi + \sin^2\phi) + (\cos^2\theta + \sin^2\theta)$$
$$- 2\cos\theta\cos\phi + 2\sin\theta\sin\phi$$
$$= 2 - 2\cos\theta\cos\phi + 2\sin\theta\sin\phi.$$

Since $d(A, C) = d(D, F)$, we have

$$2 - 2\cos(\theta + \phi) = 2 - 2\cos\theta\cos\phi + 2\sin\theta\sin\phi$$

or $$\cos(\theta + \phi) = \cos\theta\cos\phi - \sin\theta\sin\phi$$

Cos($\theta - \phi$)

If we substitute $-\phi$ for ϕ in the previous formula and remember the two identities $\cos(-\phi) = \cos\phi$ and $\sin(-\phi) = -\sin\phi$, we could show that

$$\cos(\theta - \phi) = \cos\theta\cos\phi + \sin\theta\sin\phi$$

Identities for the sum and difference of the sine and tangent functions can also be developed. When we are done, there will be a total of six sum and difference identities.

Sum and Difference Identities	$\sin(\theta + \phi) = \sin\theta\cos\phi + \cos\theta\sin\phi$
	$\sin(\theta - \phi) = \sin\theta\cos\phi - \cos\theta\sin\phi$
	$\cos(\theta + \phi) = \cos\theta\cos\phi - \sin\theta\sin\phi$
	$\cos(\theta - \phi) = \cos\theta\cos\phi + \sin\theta\sin\phi$
	$\tan(\theta + \phi) = \dfrac{\tan\theta + \tan\phi}{1 - \tan\theta\tan\phi}$
	$\tan(\theta - \phi) = \dfrac{\tan\theta - \tan\phi}{1 + \tan\theta\tan\phi}$

EXAMPLE 20.6

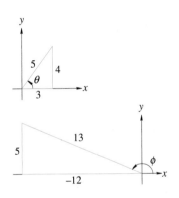

FIGURE 20.3

If $\sin\theta = \frac{4}{5}$, $\cos\phi = \frac{-12}{13}$, θ is in Quadrant I, and ϕ is in Quadrant II, find **(a)** $\sin(\theta + \phi)$, **(b)** $\cos(\theta - \phi)$, and **(c)** $\tan(\theta + \phi)$.

Solutions   If we draw reference triangles for θ and ϕ, we can determine the values of $\cos\theta$, $\tan\theta$, $\sin\phi$, and $\tan\phi$. These triangles are shown in Figure 20.3. From them, we can determine that $\cos\theta = \frac{3}{5}$, $\tan\theta = \frac{4}{3}$, $\sin\phi = \frac{5}{13}$, and $\tan\phi = \frac{-5}{12}$. We are now ready to apply the formulas.

(a) $\sin(\theta + \phi) = \sin\theta\cos\phi + \cos\theta\sin\phi$

$$= \left(\frac{4}{5}\right)\left(\frac{-12}{13}\right) + \left(\frac{3}{5}\right)\left(\frac{5}{13}\right)$$

$$= \frac{-48}{65} + \frac{15}{65} = -\frac{33}{65}$$

(b) $\cos(\theta - \phi) = \cos\theta\cos\phi + \sin\theta\sin\phi$

$$= \left(\frac{3}{5}\right)\left(\frac{-12}{13}\right) + \left(\frac{4}{5}\right)\left(\frac{5}{13}\right)$$

$$= \frac{-36}{65} + \frac{20}{65} = -\frac{16}{65}$$

(c) $\tan(\theta + \phi) = \dfrac{\tan\theta + \tan\phi}{1 - \tan\theta\tan\phi}$

$$= \frac{\frac{4}{3} + \frac{-5}{12}}{1 - \left(\frac{4}{3}\right)\left(\frac{-5}{12}\right)}$$

$$= \frac{\frac{16}{12} - \frac{5}{12}}{\frac{56}{36}} = \frac{\frac{11}{12}}{\frac{56}{36}}$$

$$= \frac{11}{12} \cdot \frac{36}{56} = \frac{33}{56}$$

So, $\sin(\theta + \phi) = -\frac{33}{65}$, $\cos(\theta - \phi) = -\frac{16}{65}$, and $\tan(\theta + \phi) = \frac{33}{56}$. Since $\sin(\theta + \phi)$ is negative and $\tan(\theta + \phi)$ is positive, $\theta + \phi$ is in the third quadrant.

EXAMPLE 20.7

If $\sin\alpha = 0.25$ and $\cos\beta = 0.65$, α in Quadrant II and β in Quadrant I, find **(a)** $\sin(\alpha - \beta)$ and **(b)** $\tan(\alpha + \beta)$.

Solutions Using the Pythagorean theorem and the fact that α is in Quadrant II, we determine that $\cos\alpha = -\sqrt{1 - 0.25^2} \approx -0.97$. Similarly, we find that $\sin\beta \approx 0.76$. We can now apply the formulas.

(a) $\sin(\alpha - \beta) = \sin\alpha\cos\beta - \cos\alpha\sin\beta$

$$\approx (0.25)(0.65) - (-0.97)(0.76)$$

$$= 0.1625 + 0.7372$$

$$= 0.8997$$

$$\approx 0.90$$

(b) $\tan(\alpha + \beta) = \dfrac{\tan\alpha + \tan\beta}{1 - \tan\alpha\tan\beta}$

$$\approx \frac{\dfrac{0.25}{-0.97} + \dfrac{0.76}{0.65}}{1 - \left(\dfrac{0.25}{-0.97}\right)\left(\dfrac{0.76}{0.65}\right)}$$

$$\approx \frac{-0.2577 + 1.1692}{1 - (-0.2577)(1.1692)}$$

$$= 0.7005$$

EXAMPLE 20.8

Find $\sin 75°$ by using the trigonometric values for $30°$ and $45°$.

Solution Since $75° = 30° + 45°$, we will use $\sin 75° = \sin(30° + 45°)$. Now $\sin 30° = \frac{1}{2}$, $\cos 30° = \frac{\sqrt{3}}{2}$, and $\sin 45° = \cos 45° = \frac{\sqrt{2}}{2}$. This gives

$$\sin 75° = \sin 30°\cos 45° + \cos 30°\sin 45°$$

$$= \frac{1}{2} \cdot \frac{\sqrt{2}}{2} + \frac{\sqrt{3}}{2} \cdot \frac{\sqrt{2}}{2}$$

$$= \frac{\sqrt{2}}{4} + \frac{\sqrt{6}}{4}$$

$$= \frac{\sqrt{2} + \sqrt{6}}{4}$$

 Note We realize that the use of calculators makes it very unlikely that you will use such procedures to evaluate a given trigonometric value. We did these examples and have included exercises to give you practice using the sum and difference identities with numbers that you can verify on your calculator. This practice will also help you remember the identities later when you need them.

EXAMPLE 20.9

Simplify $\sin\left(x + \dfrac{\pi}{2}\right)$.

Solution
$$\sin\left(x + \frac{\pi}{2}\right) = \sin x \cos \frac{\pi}{2} + \cos x \sin \frac{\pi}{2}$$
$$= \sin x \cdot 0 + \cos x \cdot 1$$
$$= \cos x$$

EXAMPLE 20.10

Verify that $\sin(\alpha + \beta) + \sin(\alpha - \beta) = 2\sin\alpha\cos\beta$.

Solution

$\sin(\alpha + \beta) + \sin(\alpha - \beta)$	$2\sin\alpha\cos\beta$
$\sin\alpha\cos\beta + \cos\alpha\sin\beta$ $+ \sin\alpha\cos\beta - \cos\alpha\sin\beta$	Expand $\sin(\alpha + \beta)$ and $\sin(\alpha - \beta)$.
$2\sin\alpha\cos\beta$	Collect terms.

So, $\sin(\alpha + \beta) + \sin(\alpha - \beta) = 2\sin\alpha\cos\beta$.

 Caution

Remember, $\sin(\alpha + \beta) \neq \sin\alpha + \sin\beta$. Make sure that you rewrite $\sin(\alpha + \beta)$ as $\sin\alpha\cos\beta + \cos\alpha\sin\beta$. In a similar way, you can show that $\cos(\alpha + \beta) \neq \cos\alpha + \cos\beta$ and $\tan(\alpha + \beta) \neq \tan\alpha + \tan\beta$.

Application

EXAMPLE 20.11

The displacement d of an object oscillating in simple harmonic motion can be determined by the expression
$$d = a\sin 2\pi ft \cos\beta + a\cos 2\pi ft \sin\beta$$

Express the right-hand side as a single term.

Solution If we factor an a out of both terms, then
$$d = a(\sin 2\pi ft \cos\beta + \cos 2\pi ft \sin\beta)$$

If we let $\alpha = 2\pi ft$, then the expression in parentheses is in the form $\sin\alpha\cos\beta + \cos\alpha\sin\beta = \sin(\alpha + \beta)$. So, the desired expression is
$$d = a\sin(2\pi ft + \beta)$$

Exercise Set 20.2

In Exercises 1–8, use the fact that $\sin 30° = \cos 60° = \frac{1}{2}$, $\sin 60° = \cos 30° = \frac{\sqrt{3}}{2}$, and $\sin 45° = \cos 45° = \frac{\sqrt{2}}{2}$, along with the other facts you know from trigonometry to determine the following.

1. $\sin 15°$

2. $\cos 75°$

3. $\sin 120°$

4. $\cos(-15°)$

5. $\tan 15°$

6. $\tan 135°$

7. $\sin 150°$

8. $\cos 105°$

In Exercises 9–16, simplify the given expression.

9. $\sin(x + 90°)$

10. $\cos(x + \pi)$

11. $\cos(x + \frac{\pi}{2})$

12. $\sin(\frac{\pi}{2} - x)$

13. $\cos(\pi - x)$

14. $\tan(x - \frac{\pi}{4})$

15. $\sin(180° - x)$

16. $\tan(180° + x)$

If α and β are first quadrant angles, $\sin \alpha = \frac{3}{4}$, and $\cos \beta = \frac{7}{8}$, evaluate the given expressions in Exercises 17–24.

17. $\sin(\alpha + \beta)$

18. $\cos(\alpha + \beta)$

19. $\tan(\alpha + \beta)$

20. $\sin(\alpha - \beta)$

21. $\cos(\alpha - \beta)$

22. $\tan(\alpha - \beta)$

23. In what quadrant is $\alpha + \beta$?

24. In what quadrant is $\alpha - \beta$?

If α is a second quadrant angle, and β is a third quadrant angle with $\sin \alpha = \frac{3}{4}$ and $\cos \beta = \frac{-7}{8}$, determine each of the given expressions in Exercises 25–32.

25. $\sin(\alpha + \beta)$

26. $\cos(\alpha + \beta)$

27. $\tan(\alpha + \beta)$

28. $\sin(\alpha - \beta)$

29. $\cos(\alpha - \beta)$

30. $\tan(\alpha - \beta)$

31. In what quadrant is $\alpha + \beta$?

32. In what quadrant is $\alpha - \beta$?

Simplify the given expression in Exercises 33–40.

33. $\sin 47° \cos 13° + \cos 47° \sin 13°$

34. $\sin 47° \sin 13° + \cos 47° \cos 13°$

35. $\cos 32° \cos 12° - \sin 32° \sin 12°$

36. $\dfrac{\tan 40° + \tan 15°}{1 - \tan 40° \tan 15°}$

37. $\cos(\alpha + \beta) \cos \beta + \sin(\alpha + \beta) \sin \beta$

38. $\sin(x - y) \cos y + \cos(x - y) \sin y$

39. $\cos(x + y) \cos(x - y) - \sin(x + y) \sin(x - y)$

40. $\sin A \cos(-B) + \cos A \sin(-B)$

Prove each of the identities in Exercises 41–46.

41. $\sin(x + y) \sin(x - y) = \sin^2 x - \sin^2 y$

42. $(\sin A \cos B - \cos A \sin B)^2 + (\cos A \cos B + \sin A \sin B)^2 = 1$

43. $\cos \theta = \sin(\theta + 30°) + \cos(\theta + 60°)$

44. $\dfrac{\sin(x + y)}{\cos(x - y)} = \dfrac{\tan x + \tan y}{1 + \tan x \tan y}$

45. $\tan x - \tan y = \dfrac{\sin(x - y)}{\cos x \cos y}$

46. $\cos(A + B)\cos(A - B) = 1 - \sin^2 A - \sin^2 B$

In Exercises 47–50, use a calculator to show that the statements are true.

47. $\sin(20° + 37°) = \sin 20° \cos 37° + \cos 20° \sin 37°$

48. $\cos(15° + 63°) = \cos 15° \cos 63° - \sin 15° \sin 63°$

49. $\tan(0.2 + 1.3) = \dfrac{\tan 0.2 + \tan 1.3}{1 - (\tan 0.2)(\tan 1.3)}$

50. $\sin(2.3 - 1.1) = (\sin 2.3)(\cos 1.1) - (\cos 2.3)(\sin 1.1)$

In Exercises 51–54, with the help of a calculator use the given angles as a counterexample to show that each relation is not an identity.

51. $\sin(x + y) \neq \sin x + \sin y; x = 55°, y = 37°$

52. $\cos(x - y) \neq \cos x - \cos y; x = 68°, y = 24°$

53. $\cos(x + y) \neq \cos x + \cos y; x = 40°, y = 35°$

54. $\tan(x - y) \neq \tan x - \tan y; x = 76°, y = 37°$

Solve Exercises 55–58.

55. In Chapter 14, we learned that when two complex numbers are written in polar form their product is

$$[r_1(\cos \theta_1 + j \sin \theta_1)][r_2(\cos \theta_2 + j \sin \theta_2)]$$

$$= r_1 r_2[\cos(\theta_1 + \theta_2) + j \sin(\theta_1 + \theta_2)]$$

Prove this formula.

56. *Physics* A spring vibrating in harmonic motion described by the equation $y_1 = A_1 \cos(\omega t + \pi)$ is subjected to another harmonic motion described by $y_2 = A_2 \cos(\omega t - \pi)$. Show that $y_1 + y_2 = -(A_1 + A_2) \cos \omega t$.

57. *Optics* When a light beam passes from one medium through another medium and exists in a third medium

of the same density as the first, the displacement d of the light beam is $d = \dfrac{h}{\cos \theta_r} \sin(\theta_i - \theta_r)$, where θ_r is the angle of refraction, θ_i is the angle of incidence, and h is the thickness of the medium. Show that $d = h(\sin \theta_i - \cos \theta_i \tan \theta_r)$.

58. *Electricity* The angle between voltage and current in an RC circuit is 45°. Develop an expression in terms of ω for $i(t)$ in milliamperes (mA) using $i(t) = I_p \sin(\theta - \omega t)$, if $I_p = 14.8$ mA.

☰ 20.3
THE DOUBLE- AND HALF-ANGLE IDENTITIES

In the Section 20.2, we studied the sum and difference identities. We can use these identities to develop double-angle identities. The double-angle identities can then be used to develop some half-angle identities.

Double-Angle Identities

The identity for $\sin(\theta + \phi)$ can be used to develop an identity for $\sin 2\theta$. To do this, calculate $\sin(\theta + \theta)$.

$$\sin 2\theta = \sin(\theta + \theta)$$

$$= \sin \theta \cos \theta + \cos \theta \sin \theta$$

$$= 2 \sin \theta \cos \theta$$

In the same manner, we can develop $\cos 2\theta$.

$$\cos 2\theta = \cos(\theta + \theta)$$
$$= \cos \theta \cos \theta - \sin \theta \sin \theta$$
$$= \cos^2 \theta - \sin^2 \theta$$

This last identity has two other forms. If we use the Pythagorean identity, $\sin^2 \theta + \cos^2 \theta = 1$, we get

$$\cos 2\theta = \cos^2 \theta - \sin^2 \theta$$
$$= (1 - \sin^2 \theta) - \sin^2 \theta$$
$$= 1 - 2\sin^2 \theta$$

We can replace $\sin^2 \theta$ with $1 - \cos^2 \theta$ and get a third version of this formula.

$$\cos 2\theta = \cos^2 \theta - \sin^2 \theta$$
$$= \cos^2 \theta - (1 - \cos^2 \theta)$$
$$= 2\cos^2 \theta - 1$$

Once again, if we evaluate $\tan(\theta + \theta)$, we get the third double-angle identity:

$$\tan 2\theta = \tan(\theta + \theta)$$
$$= \frac{\tan \theta + \tan \theta}{1 - \tan \theta \tan \theta}$$
$$= \frac{2\tan \theta}{1 - \tan^2 \theta}$$

This completes the list of double-angle identities.

Double-Angle Identities	$\sin 2\theta = 2\sin \theta \cos \theta$
	$\cos 2\theta = \cos^2 \theta - \sin^2 \theta$
	$\qquad = 2\cos^2 \theta - 1$
	$\qquad = 1 - 2\sin^2 \theta$
	$\tan 2\theta = \dfrac{2\tan \theta}{1 - \tan^2 \theta}$

EXAMPLE 20.12

If $\sin x = 0.60$ and x is in the second quadrant, then determine **(a)** $\sin 2x$, **(b)** $\cos 2x$, and **(c)** $\tan 2x$.

Solutions   We will first determine the value of $\cos x$. Since $\sin x = 0.60$, $\cos^2 x = 1 - \sin^2 x = 1 - (0.60)^2 = 0.64$ and $\cos x = \pm\sqrt{0.64} = \pm 0.80$. Since x is in

EXAMPLE 20.12 (Cont.)

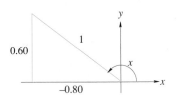

FIGURE 20.4

Quadrant II (see Figure 20.4), $\cos x = -0.80$.

(a) $\sin 2x = 2 \sin x \cos x$

$$= 2(0.60)(-0.80) = -0.96$$

(b) $\cos 2x = \cos^2 x - \sin^2 x$

$$= (-0.8)^2 - (0.6)^2$$

$$= 0.64 - 0.36 = 0.28$$

(c) Since $\sin x = 0.60$, $\cos x = -0.8$, and $\tan x = \dfrac{\sin x}{\cos x}$, we have $\tan x = \dfrac{0.60}{-0.80} = -\dfrac{3}{4}$. Thus,

$$\tan 2x = \frac{2 \tan x}{1 - \tan^2 x}$$

$$= \frac{2(-\frac{3}{4})}{1 - (-\frac{3}{4})^2}$$

$$= \frac{-\frac{3}{2}}{1 - \frac{9}{16}}$$

$$= \frac{-\frac{3}{2}}{\frac{7}{16}}$$

$$= \frac{-24}{7}$$

$$\approx -3.43$$

Caution Don't forget that $\sin 2\alpha \neq 2 \sin \alpha$ and $\cos 2\alpha \neq 2 \cos \alpha$. We now know that

$$\sin 2\alpha = 2 \sin \alpha \cos \alpha$$

and that

$$\cos 2\alpha = \cos^2 \alpha - \sin^2 \alpha$$

EXAMPLE 20.13

Rewrite $\cos 4x$ in terms of $\cos x$.

Solution   Using the double-angle identity for $\cos 2\theta$, if we let $\theta = 2x$, then we get $\cos 4x = 2 \cos^2 2x - 1$. Now using the double-angle identity $\cos 2x = 2 \cos^2 x - 1$, we get

$$\cos 4x = 2(2 \cos^2 x - 1)^2 - 1$$

$$= 2(4 \cos^4 x - 4 \cos^2 x + 1) - 1$$

$$= 8 \cos^4 x - 8 \cos^2 x + 1$$

Half-Angle Identities

If we solve $\cos 2x = 2\cos^2 x - 1$ for the $\cos x$, we get another identity—a half-angle identity.

$$\cos 2x = 2\cos^2 x - 1$$

or

$$\cos^2 x = \frac{1 + \cos 2x}{2}$$

and, taking the square root of both sides,

$$\cos x = \pm\sqrt{\frac{1 + \cos 2x}{2}}$$

If we let $x = \dfrac{\theta}{2}$, then

$$\cos \frac{\theta}{2} = \pm\sqrt{\frac{1 + \cos \theta}{2}}$$

If we solve $\cos^2 x = 1 - 2\sin^2 x$ for $\sin x$, we obtain

$$\sin^2 x = \frac{1 - \cos 2x}{2}$$

or

$$\sin x = \pm\sqrt{\frac{1 - \cos 2x}{2}}$$

Again, if $x = \dfrac{\theta}{2}$, then

$$\sin \frac{\theta}{2} = \pm\sqrt{\frac{1 - \cos \theta}{2}}$$

Since $\tan x = \dfrac{\sin x}{\cos x}$, we can show that

$$\tan \frac{\theta}{2} = \pm\sqrt{\frac{1 - \cos \theta}{1 + \cos \theta}}$$

$$= \frac{\sin \theta}{1 + \cos \theta} = \frac{1 - \cos \theta}{\sin \theta}$$

We have developed the following three half-angle identities.

Half-Angle Identities	$\sin \dfrac{\theta}{2} = \pm\sqrt{\dfrac{1 - \cos \theta}{2}}$
	$\cos \dfrac{\theta}{2} = \pm\sqrt{\dfrac{1 + \cos \theta}{2}}$
	$\tan \dfrac{\theta}{2} = \pm\sqrt{\dfrac{1 - \cos \theta}{1 + \cos \theta}} = \dfrac{\sin \theta}{1 + \cos \theta}$
	$= \dfrac{1 - \cos \theta}{\sin \theta}$

EXAMPLE 20.14

If $\cos \theta = \frac{-5}{13}$ and θ is in the third quadrant, find the values of **(a)** $\sin \frac{\theta}{2}$, **(b)** $\cos \frac{\theta}{2}$, and **(c)** $\tan \frac{\theta}{2}$.

Solution We need to determine which quadrant $\frac{\theta}{2}$ is in. We know θ is in Quadrant III, or $\pi < \theta < \frac{3\pi}{2}$, and so $\frac{\pi}{2} < \frac{\theta}{2} < \frac{1}{2}\left(\frac{3\pi}{2}\right)$ or $\frac{\pi}{2} < \frac{\theta}{2} < \frac{3\pi}{4}$. This means that $\frac{\theta}{2}$ is in Quadrant II. In Quadrant II, the sine is positive, cosine is negative, and tangent is negative.

(a) $\sin \frac{\theta}{2} = +\sqrt{\dfrac{1 - \cos \theta}{2}} = +\sqrt{\dfrac{1 - \left(-\frac{5}{13}\right)}{2}} = \sqrt{\dfrac{9}{13}} \approx 0.832$

(b) $\cos \frac{\theta}{2} = -\sqrt{\dfrac{1 + \cos \theta}{2}} = -\sqrt{\dfrac{1 + \left(-\frac{5}{13}\right)}{2}} = -\sqrt{\dfrac{4}{13}} \approx -0.555$

(c) $\tan \frac{\theta}{2} = -\sqrt{\dfrac{1 - \cos \theta}{1 + \cos \theta}} = -\sqrt{\dfrac{1 - \left(-\frac{5}{13}\right)}{1 + \left(-\frac{5}{13}\right)}} = -\sqrt{\dfrac{\frac{18}{13}}{\frac{8}{13}}} = -\sqrt{\dfrac{9}{4}}$

$= -\dfrac{3}{2} = -1.500$

We could have determined $\tan \frac{\theta}{2}$ by using $\dfrac{\sin \frac{\theta}{2}}{\cos \frac{\theta}{2}}$. With the values above, we would have gotten -1.499. The error of 0.001 was caused by the use of approximations in parts **(a)** and **(b)**.

EXAMPLE 20.15

Prove the identity $2 \sin \frac{x}{2} \cos \frac{x}{2} = \sin x$.

Solution

$2 \sin \dfrac{x}{2} \cos \dfrac{x}{2}$	$\sin x$
$2\left(\pm\sqrt{\dfrac{1 - \cos x}{2}}\right)\left(\pm\sqrt{\dfrac{1 + \cos x}{2}}\right)$	Replace with half-angle identities.
$2\sqrt{\dfrac{1 - \cos^2 x}{4}}$	Multiply.
$2\sqrt{\dfrac{\sin^2 x}{4}}$	Pythagorean identity.

EXAMPLE 20.15 (Cont.)

$$2\left(\frac{\sin x}{2}\right)$$ Take the square root.

$$\sin x$$ Simplify.

Caution Be careful when you use the double- and half-angle formulas. Begin by calculating the necessary values of θ, 2θ, and $\frac{\theta}{2}$ before they are substituted into the formula.

Application

EXAMPLE 20.16

A highway engineer is designing the curve at an intersection like the one shown by the photograph in Figure 20.5a. These two highways intersect at an angle θ. The edge of the highway is to join the two points A and B with an arc or a circle that is tangent to the highways at these two points. Determine the relationship between the radius of the arc r, the distance d of A and B from the intersection, and angle θ.

FIGURE 20.5

Courtesy of Michael A. Gallitelli, Metroland Photo Inc.
(a)

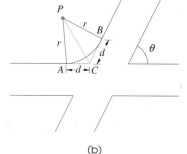

(b)

Solution   We begin by noticing in Figure 20.5b, that $\angle BCA$ and θ are supplementary angles. So, $m\angle BCA = 180° - \theta$. If the center of the circle is at P, then $\overline{PA} \perp \overline{AC}$, because a tangent to a circle, in this case $\overline{AC}$, is perpendicular to a radius at the point of tangency. Now, $\overline{PC}$ bisects $\angle BCA$, so $m\angle PCA = \frac{1}{2}m\angle BCA = 90° - \frac{\theta}{2}$. Since $\triangle PAC$ is a right triangle with right angle at A, $\tan \angle PCA = \frac{r}{d}$; so $d = \frac{r}{\tan \angle PCA} = r \cot \angle PCA = r \cot\left(90° - \frac{\theta}{2}\right) = r \tan \frac{\theta}{2}$. Thus, we have shown that $d = r \tan \frac{\theta}{2}$.

Application

EXAMPLE 20.17

Two highways meet at an angle of 34°. The curb is to join points A and B located 45 ft from the beginning of the intersection.

(a) Approximate the radius of the arc joining A and B.

(b) Determine the length of the arc.

Solutions (a) From Example 20.16, we have the formula

$$d = r \tan \frac{\theta}{2}$$

In this example, $d = 45$ ft and $\theta = 34°$. We are to determine r.

$$r = \frac{d}{\tan \dfrac{\theta}{2}}$$

$$= \frac{45}{\tan\left(\frac{34}{2}\right)^\circ}$$

$$= \frac{45}{\tan 17°}$$

$$\approx \frac{45}{0.3057}$$

$$\approx 147.19 \text{ ft}$$

(b) We want the length of the arc that forms the curb. As we saw in Section 9.6, the arc length s of a circle with radius r, formed by an angle θ, is

$$s = r\theta$$

provided that θ is in radians.

In this example, $\theta = 34° = \dfrac{34\pi}{180} = \dfrac{17\pi}{90}$ rad. So,

$$s = r\theta$$

$$= (147.19)\left(\frac{17\pi}{90}\right)$$

$$\approx 87.34 \text{ ft}$$

Exercise Set 20.3

In Exercises 1–14, if $\sin 30° = \cos 60° = \frac{1}{2}$, $\sin 60° = \cos 30° = \frac{\sqrt{3}}{2}$, and $\sin 45° = \cos 45° = \frac{\sqrt{2}}{2}$, determine the exact values of the given trigonometric function.

1. $\cos 15°$

2. $\sin 75°$

3. $\sin 15°$

4. $\cos 105°$

5. $\sin 105°$

6. $\cos 210°$

7. $\cos 7\frac{1}{2}^{\circ}$

8. $\tan 15^{\circ}$

9. $\tan 22\frac{1}{2}^{\circ}$

10. $\cos 67\frac{1}{2}^{\circ}$

11. $\cos 75^{\circ}$

12. $\sin 37\frac{1}{2}^{\circ}$

13. $\sin 127\frac{1}{2}^{\circ}$

14. $\tan(-15^{\circ})$

Use the information given in each of Exercises 15–20 to determine $\sin 2x$, $\cos 2x$, $\tan 2x$, $\sin \dfrac{x}{2}$, $\cos \dfrac{x}{2}$, **and** $\tan \dfrac{x}{2}$.

15. $\sin x = \dfrac{7}{25}$, x in Quadrant II

16. $\cos x = \dfrac{8}{17}$, x in Quadrant IV

17. $\sec x = \dfrac{29}{20}$, x in Quadrant I

18. $\csc x = \dfrac{-41}{9}$, x in Quadrant III

19. $\tan x = \dfrac{35}{12}$, x in Quadrant III

20. $\cot x = \dfrac{-45}{28}$, x in Quadrant II

Prove the identities in Exercises 21–32 for all angles in the domains of the functions.

21. $\cos^2 x = \sin^2 x + \cos 2x$

22. $\cos^2 4x - \sin^2 4x = \cos 8x$

23. $\cos 4x = 1 - 8 \sin^2 x \cos^2 x$

24. $\cos 3x = 4 \cos^3 x - 3 \cos x$

25. $\dfrac{1 + \tan^2 \alpha}{1 - \tan^2 \alpha} = \sec 2\alpha$

26. $\sin 2x \cos 2x = \frac{1}{2} \sin 4x$

27. $1 - 2 \sin^2 3x = \cos 6x$

28. $\dfrac{2 \tan 3x}{1 - \tan^2 3x} = \tan 6x$

29. $\sin^2 x \cos^2 x = \frac{1}{4} \sin^2 2x$

30. $1 + \tan \beta \tan \dfrac{\beta}{2} = \dfrac{1}{\cos \beta}$

31. $\tan \left(\dfrac{\alpha + \beta}{2} \right) \cot \left(\dfrac{\alpha - \beta}{2} \right) = \dfrac{(\sin \alpha + \sin \beta)^2}{\sin^2 \alpha - \sin^2 \beta}$

32. $\cot \theta = \dfrac{1}{2} \left(\cot \dfrac{\theta}{2} - \tan \dfrac{\theta}{2} \right)$

In Exercises 33–36, use a calculator and the given angles as counterexamples to show that the following are not identities.

33. $\cos 2\theta \neq 2 \cos \theta$; $\theta = 45^{\circ}$

34. $\tan \dfrac{\theta}{2} \neq \dfrac{\tan \theta}{2}$; $\theta = 80^{\circ}$

35. $\cot 2\alpha \neq 2 \cos \alpha$; $\alpha = 150^{\circ}$

36. $\sin \dfrac{x}{2} \neq \dfrac{\sin x}{2}$; $x = 210^{\circ}$

Solve Exercises 37–40.

37. *Optics* The index of refraction n of a prism whose apex angle is α and whose angle of minimum deviation is ϕ is given by

$$n = \dfrac{\sin(\dfrac{\alpha + \phi}{2})}{\sin \dfrac{\alpha}{2}} \quad \text{with } n > 0$$

Show that

$$n = \sqrt{\dfrac{1 - \cos \alpha \cos \phi + \sin \alpha \sin \phi}{1 - \cos \alpha}}$$

38. *Optics* Show that an equivalent expression for the index of refraction described in Exercise 37 is

$$n = \sqrt{\dfrac{1 + \cos \phi}{2}} + \left(\cot \dfrac{\alpha}{2} \right) \sqrt{\dfrac{1 - \cos \phi}{2}}$$

39. *Electricity* In an ac circuit containing reactance, the instantaneous power is given by

$$P = V_{max} I_{max} \cos \omega t \sin \omega t$$

Show that $P = \dfrac{V_{max} I_{max}}{2} \sin 2\omega t$.

40. *Physics* A cable vibrates with a decreased amplitude that is given by $A = \sqrt{e^{-2x}(1 + \sin 2x)}$. Show that $A = e^{-x}(\sin x + \cos x)$.

▦ **20.4**
TRIGONOMETRIC EQUATIONS

For Sections 20.1 through 20.3, we studied different types of trigonometric identities. Many people find proving and developing trigonometric identities to be very interesting. Our main interest in them, however, was to give you some skills for solving equations that involve trigonometric functions.

A **trigonometric equation** is an equation involving trigonometric functions of unknown angles. If these equations have been true for all angles, then we have called them identities. A trigonometric equation that is not an identity is a **conditional equation**. A conditional equation is true for some values for the angle and not true for others. To **solve** a conditional trigonometric equation means to find all values of the angle for which the equation is true. To solve a trigonometric equation, you must use both algebraic and trigonometric identities.

Solving a trigonometric equation of the type $2 \tan x = 1$ would produce an infinite number of answers. As you remember from our earlier study, the trigonometric functions are periodic. Thus, the solution to this equation would not only be true when $x = 26.565°$ but also for $x = 26.565° + 180°n$, where n is any integer. Usually, it is sufficient to give only the **primary solutions** or **principal values**, which are the solutions for x, where $0° \leq x < 360°$ or $0 \leq x < 2\pi$.

EXAMPLE 20.18

Solve $2 \tan x = 1$.

Solution
$$2 \tan x = 1$$
$$\tan x = \frac{1}{2}$$

We know that $x = \arctan \frac{1}{2} = \tan^{-1}(\frac{1}{2})$. Using a calculator, we see that $x \approx 26.565°$. But, we know that the tangent function is also positive in Quadrant III, so $x = 26.565° + 180° = 206.565°$. The primary solutions are $26.565°$ and $206.565°$.

EXAMPLE 20.19

Solve $\cos 4x = \frac{\sqrt{2}}{2}$, where $0 \leq x < 2\pi$.

Solution   A natural way to proceed would be to use the double-angle identities to rewrite this as an equation in x.

A little foresight will save a lot of work. We will let $\theta = 4x$ and solve for θ. Once we have the value for θ we can then solve for x. But, since $x = \frac{\theta}{4}$ and $0 \leq x < 2\pi$, we must solve for $0 \leq \theta < 8\pi$.

$$\cos \theta = \frac{\sqrt{2}}{2}$$

$$\theta = \cos^{-1}\left(\frac{\sqrt{2}}{2}\right) = \frac{\pi}{4}$$

Since the cosine is also positive in Quadrant IV, we see that $\theta = \frac{7\pi}{4}$. If we keep adding 2π to each of these answers until we exceed 8π, we will get the other solutions for θ:

$$\frac{\pi}{4} + 2\pi = \frac{9\pi}{4},$$

EXAMPLE 20.19 (Cont.)

$$\frac{\pi}{4} + 4\pi = \frac{17\pi}{4},$$

$$\frac{\pi}{4} + 6\pi = \frac{25\pi}{4};$$

and we also get

$$\frac{7\pi}{4} + 2\pi = \frac{15\pi}{4},$$

$$\frac{7\pi}{4} + 4\pi = \frac{23\pi}{4},$$

and $$\quad \frac{7\pi}{4} + 6\pi = \frac{31\pi}{4}$$

The solutions then for $0 \le \theta < 8\pi$ are $4x = \theta = \frac{\pi}{4}, \frac{7\pi}{4}, \frac{9\pi}{4}, \frac{15\pi}{4}, \frac{17\pi}{4},$ $\frac{23\pi}{4}, \frac{25\pi}{4},$ and $\frac{31\pi}{4},$ and so the values of x are $x = \frac{\pi}{16}, \frac{7\pi}{16}, \frac{9\pi}{16}, \frac{15\pi}{16}, \frac{17\pi}{16}, \frac{23\pi}{16}, \frac{25\pi}{16},$ and $\frac{31\pi}{16}.$

EXAMPLE 20.20

Solve $\sin\theta \tan\theta = \sin\theta$ for $0 \le \theta < 360°$.

Solution We begin by collecting terms and factoring:

$$\sin\theta \tan\theta = \sin\theta$$

$$\sin\theta \tan\theta - \sin\theta = 0$$

$$\sin\theta(\tan\theta - 1) = 0$$

We now determine when each of these factors can be 0.

So $\qquad\qquad \sin\theta = 0 \qquad\qquad$ or $\qquad \tan\theta - 1 = 0$

$$\sin\theta = 0 \qquad\qquad\qquad \tan\theta - 1 = 0$$

$$\theta = 0°, 180° \qquad\qquad\qquad \tan\theta = 1$$

$$\theta = 45°, 225°$$

The solutions are $0°, 45°, 180°,$ and $225°.$

EXAMPLE 20.21

Solve $2\sin^2 x - \cos x - 1 = 0$ for $0 \le x < 2\pi$.

Solution We will use one of the Pythagorean identities to replace $\sin^2 x$ in the given expression.

$$2(1 - \cos^2 x) - \cos x - 1 = 0$$

$$2 - 2\cos^2 x - \cos x - 1 = 0$$

$$-2\cos^2 x - \cos x + 1 = 0$$

or $\qquad\qquad 2\cos^2 x + \cos x - 1 = 0$

EXAMPLE 20.21 (Cont.)

Factoring, we get

$$(2\cos x - 1)(\cos x + 1) = 0$$

Solving each factor, we have

$$2\cos x - 1 = 0$$

$$\cos x = \frac{1}{2}$$

$$x = \frac{\pi}{3}, \frac{5\pi}{3}$$

and,

$$\cos x + 1 = 0$$

$$\cos x = -1$$

$$x = \pi$$

The solutions are $\frac{\pi}{3}$, π, and $\frac{5\pi}{3}$.

EXAMPLE 20.22

Solve $\tan\theta + \sec\theta = 1$ for $0 \le \theta < 2\pi$.

Solution   We will follow a procedure similar to the one we followed when solving radical equations with two radicals.

This equation has two different trigonometric functions and there does not seem to be any identity that relates them. We will rewrite the equation with the functions on different sides of the equation and then square both sides. When we solved identities, we could not square both sides. However, in solving a conditional equation, it is possible to square both sides, if you check for extraneous roots later.

$$\tan\theta + \sec\theta = 1$$

$\sec\theta = 1 - \tan\theta$	Rewrite.
$\sec^2\theta = (1 - \tan\theta)^2$	Square both sides.
$\sec^2\theta = 1 - 2\tan\theta + \tan^2\theta$	Simplify.
$1 + \tan^2\theta = 1 - 2\tan\theta + \tan^2\theta$	Pythagorean identity.
$0 = -2\tan\theta$	Subtract $1 + \tan^2\theta$.
$0 = \tan\theta$	Divide by -2.

The solutions to $\tan\theta = 0$ are 0 and π. Now we need to check each of these in the *original* equation to see if they are actual roots or extraneous roots.

We first check $\theta = 0$:

$$\tan 0 + \sec 0 = 0 + 1 = 1$$

This checks so $\theta = 0$ is a root.

EXAMPLE 20.22 (Cont.)

Next we check $\theta = \pi$:

$$\tan \pi + \sec \pi = 0 + -1 = -1$$

This is not equal to 1, so π is not a solution. (π is an extraneous root.)

Thus, $\theta = 0$ is the only solution.

Application

EXAMPLE 20.23

The range r of a projectile thrown at an angle of elevation θ at a velocity v is given by

$$r = \frac{2v^2 \cos \theta \sin \theta}{g}$$

If v is in ft/s, then g is 32 ft/s^2 and if v is in m/s, then g is 9.8 m/s^2. A projectile is fired with a velocity of 750 m/s with the purpose of hitting an object 20 000 m away. Determine the angle θ at which the projectile should be fired.

Solution   Here $v = 750$ m/s, $g = 9.8$ m/s^2, and $r = 20\,000$ m. Substituting these values in the given equation, we obtain

$$20\,000 = \frac{2(750)^2 \cos \theta \sin \theta}{9.8}$$

Now, $2 \cos \theta \sin \theta = 2 \sin \theta \cos \theta = \sin 2\theta$, and the equation becomes

$$20\,000 = \frac{(750)^2 \sin 2\theta}{9.8}$$

$$\text{or} \quad \sin 2\theta = \frac{20\,000(9.8)}{(750)^2}$$

$$\approx 0.3484$$

so

$$2\theta \approx 20.39° \quad \text{or} \quad 159.61°$$

$$\text{and} \quad \theta \approx 10.195° \quad \text{or} \quad 79.805°$$

So, the projectile should be fired at an angle of 10.195° or 79.805°.

Exercise Set 20.4

Solve each equation in Exercises 1–30 for nonnegative angles less than 360° or 2π. You may want to use a calculator.

1. $2 \cos \theta = 0$

2. $2 \sin \theta = -1$

3. $\sqrt{3} \tan x = 1$

4. $\sqrt{3} \sec x = -2$

5. $4 \sin \theta = -3$

6. $2 \cos x = 3$

7. $4 \tan \alpha = 5$

8. $3 \csc x = 1$

9. $\cos 2x = -1$

10. $\sin 2x = \frac{1}{2}$

11. $\tan \dfrac{\theta}{4} = 1$

12. $\cos \dfrac{\theta}{3} = -1$

13. $\sin^2 \alpha = \sin \alpha$

14. $\cos^2 \beta = \frac{1}{2} \cos \beta$

15. $\sin x \cos x = 0$

16. $\dfrac{\sec \theta}{\csc \theta} = -1$

17. $3 \tan^2 x = 1$

18. $\sec^2 \theta = 2$

19. $4 \sin \alpha \cos \alpha = 1$

20. $\sin^2 \beta = \frac{1}{2} \sin \beta$

21. $\sin \theta - \cos \theta = 0$

22. $\tan \theta = \csc \theta$

23. $\sin 6\theta + \sin 3\theta = 0$

24. $4 \tan^2 x = 3 \sec^2 x$

25. $2 \cos^2 x - 3 \cos 2x = 1$

26. $\sin^2 4\alpha = \sin 4\alpha + 2$

27. $\sec^2 \theta + \tan \theta = 1$

28. $\tan 2x + \sec 2x = 1$

29. $\sin 2x = \cos x$

30. $\csc^2 \theta - \cot \theta = 1$

Solve Exercises 31–36.

31. *Optics*   The second law of refraction (Snell's law) states that as a light ray passes from one medium to a second, the ratio of the sine of the angle of incidence θ_i to the sine of the angle of refraction θ_r is a constant, μ, called the **index of refraction**, with respect to the two mediums. (See Figure 20.6.) Thus, we have

$$\frac{\sin \theta_i}{\sin \theta_r} = \mu$$

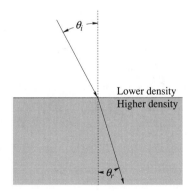

FIGURE 20.6

The index of refraction of general epoxy relative to air is $\mu = 1.61$. Determine the angle of refraction θ_r of a ray of light that strikes some general epoxy with an angle of incidence $\theta_i = 35°$.

32. *Optics*   The index of refraction of glass silicone relative to air is 1.43. If the angle of incidence is $27°$, what is the angle of refraction?

33. *Electronics*   An oscillating signal voltage is given by $E = 125 \cos(\omega t - \phi)$ millivolts, where the angular frequency is $\omega = 120\pi$, phase angle is $\phi = \dfrac{\pi}{2}$, and t is time in seconds. The triggering mechanism of an oscilloscope starts the sweep when $E = 60$ mV. What is the smallest positive value of t for which the triggering occurs?

34. *Agriculture*   As shown in Figure 20.7, an irrigation ditch has a cross-section in the shape of an isosceles trapezoid with the smaller base on the bottom. The area A of the trapezoid is given by

$$A = a \sin \theta (b + a \cos \theta)$$

If $a = 3$ m, $b = 3.4$ m, and $A = 10$ m^2, find θ to the nearest tenth of a degree if $\sin \theta \cos \theta \approx 0.4675$.

FIGURE 20.7

35. *Automotive technology*   The displacement d of a piston is given by

$$d = \sin \omega t + \frac{1}{2} \sin 2\omega t$$

For what primary solutions of ωt less than 2π is $d = 0$?

36. *Optics*   Refraction causes a submerged object in a liquid to appear closer to the surface than it actually is. The relation between the true depth a and the apparent depth b is

$$\frac{a}{b} = \sqrt{\frac{\mu^2 - \sin^2 \theta_i}{\cos^2 \theta_i}}$$

where μ is the index of refraction for the two media and θ_i is the angle of incidence. An object that is 14 ft under water appears to be only 10 ft under water. If the index of refraction of water is 1.333, what is the angle of incidence?

▬ CHAPTER 20 REVIEW

Important Terms and Concepts

Double-angle identities

Half-angle identities

Pythagorean identities

Quotient identities

Reciprocal identities

Sum and difference identities

Trigonometric equations

Review Exercises

Prove the identities in Exercises 1–10.

1. $\dfrac{\sin(x + y)}{\cos x \cos y} = \tan x + \tan y$

2. $(\sin x + \cos x)^2 = 1 + \sin 2x$

3. $\dfrac{\sin 3x}{\sin x} - \dfrac{\cos 3x}{\cos x} = 2$

4. $\cos(\theta + \phi) \cos(\theta - \phi) = \cos^2 \phi - \sin^2 \theta$

5. $\sin(\alpha - \beta) \cos \beta - \cos(\alpha + \beta) \sin \beta = \sin \alpha$

6. $\tan 2x = \dfrac{2 \cos x}{\csc x - 2 \sin x}$

7. $\cos^4 x - \sin^4 x = \cos 2x$

8. $\dfrac{\sin 2x - \sin x}{\cos 2x + \cos x} = \tan \dfrac{x}{2}$

9. $\sin 3\theta = 2 \sin \theta \cos 2\theta + \sin \theta$

10. $\dfrac{\cos 3x - \cos 5x}{\sin 3x + \sin 5x} = \tan x$ (Hint: Let $3x = 4x - x$ and $5x = 4x + x$.

Solve the equations in Exercises 11–20 for nonnegative values less than 360° or 2π.

11. $2 \tan x = -\sqrt{3}$

12. $3 \sin x = -2$

13. $\cos 2x + \cos x = -1$

14. $\cos x - \sin 2x - \cos 3x = 0$

15. $\sin 4x - 2 \sin 2x = 0$

16. $\sin(30° + x) - \cos(60° + x) = -\dfrac{\sqrt{3}}{2}$

17. $2 \sin \theta = \sin 2\theta$

18. $\sin^2 \alpha + 5 \cos^2 \alpha = 3$

19. $\sin^2 x = 1 + \sin x$

20. $\sin \theta - 2 \csc \theta = -1$

In Exercises 21–24, use the facts that $\sin x = \frac{5}{13}$, $\cos x = -\frac{12}{13}$, and x is in Quadrant II to determine the exact value of the given function.

21. $\sin 2x$

22. $\cos \dfrac{x}{2}$

23. $\sin \dfrac{x}{2}$

24. $\tan 2x$

In Exercises 25–30, use the facts that x is in Quadrant III, $\sin x = \frac{-5}{13}$, and $\cos x = \frac{-12}{13}$, and y is in Quadrant II, $\sin y = \frac{8}{17}$, and $\cos y = \frac{-15}{17}$ to determine the exact value of the given function.

25. $\sin(x + y)$

26. $\cos(x - y)$

27. $\cos(x + y)$

28. $\tan(x + y)$

29. $\cos(y - x)$

30. $\sin(x - y)$

Solve Exercises 31–33.

31. *Automotive technology*   The acceleration of a piston is given by

$$a = 5.0(\sin \omega t + \cos 2\omega t)$$

For what primary solutions of ωt does $a = 0$?

32. *Acoustical engineering*   When two sinusoidal sound waves that are close together in frequency are superimposed, the resultant disturbance exhibits beats. The two waves are represented by $y_1 = A_1 \cos \omega_1 t$ and $y_2 = A_2 \cos(\omega_2 t + \phi)$, where ω_2 is slightly larger than ω_1 and ϕ is a phase constant. If $\phi = 0$, $\alpha = \dfrac{\omega_1 + \omega_2}{2}$, and $\beta = \dfrac{\omega_2 - \omega_1}{2}$, show that

$$y = y_1 + y_2 = A_1 \cos(\alpha - \beta)t + A_2 \cos(\alpha + \beta)t$$

33. *Mechanics*   Consider a machine that is mounted on four springs with a known stiffness and on four dampers with a known damping constant . If this system is initially at rest and a certain force is applied, then under certain conditions the system has a time-displacement equation of $x = 0.01e^{-6t}(\cos 8t + \sin 8t)$. Another version gives the time-displacement equation at $x = \frac{\sqrt{2}}{100} e^{-6t} \cos(8t - \frac{\pi}{4})$. Show that these two equations are identical.

≡ CHAPTER 20 TEST

In Exercises 1–8, use the fact that $\sin \alpha = \frac{4}{5}$ and α is in Quadrant II and that $\cos \beta = -\frac{12}{13}$, with β in Quadrant III to determine the exact value of the given function.

1. $\sin(\alpha + \beta)$

2. $\cos(\alpha + \beta)$

3. $\sin(\alpha - \beta)$

4. $\cos(\alpha - \beta)$

5. $\sin 2\alpha$

6. $\cos 2\beta$

7. $\sin \dfrac{\alpha}{2}$

8. $\cos \dfrac{\beta}{2}$

Solve Exercises 9–12.

9. Write $8\cos 6x \sin 6x$ using a single trigonometric function.

10. Prove the identity $\tan x = \dfrac{\sec x}{\csc x}$

11. Prove the identity $\dfrac{\sin x}{1 - \cos x} + \dfrac{\sin x}{1 + \cos x} = 2\csc x$.

12. Solve $6\cos^2 x + \cos x = 2$ for x with either $0° \leq x < 360°$ or $0 \leq x < 2\pi$.

21

Statistics and Empirical Methods

A police officer is measuring car speed in a residential area. In this chapter, we will see some ways in which statistics can be used to help interpret the information gathered by this officer.

In this chapter, we will begin exploring a new area of mathematics—the world of probability, statistics, and empirical methods. Probability theory has many applications in the physical sciences and technology. It is of basic importance in statistical mechanics. Probability is needed in any problem dealing with large numbers of variables where it is impossible or impractical to have complete information.

In many technological settings, information is needed about an operation. When it is not possible to gather information about the entire operation, information is gathered on a part, or sample, of the operation. This information is then analyzed and decisions are made as a result of this analysis. Statistics are the basis of this analysis, and in this chapter we will learn how to conduct some statistical analyses and how to use that information to help us make decisions.

≡ 21.1
PROBABILITY

The word "probably" is used often in everyday life. For example, you might tell someone that "the test will probably be hard" or that "it will probably rain today." The word "probably" indicates that we do not know what will happen. The theory of probability tries to express more precisely just how much we do know.

In the theory of probability, a numerical value between 0 and 1 is given to the likelihood that some particular event will happen. When we do this, we assume that all events are equally likely to occur, unless we have some special information that indicates otherwise.

Outcome and Sample Space

In working with probability, we will be concerned with the possible outcomes of experiments. An experiment can be something as simple as tossing a coin or something more complicated such as determining the number of bad computer chips at a production station. A result of an experiment is called an **outcome**. The group of all possible outcomes for an experiment is the **sample space**. Each performance of an experiment is called a **trial**.

EXAMPLE 21.1

(a) If a coin is tossed, the sample space contains the two possible outcomes of heads (H) or tails (T) and can be written as $\{H, T\}$.

(b) If a die is rolled, the sample space of the number of dots on the upper face is $\{1, 2, 3, 4, 5, 6\}$. Each trial will produce one of these numbers.

(c) If two coins are tossed, the sample space has the following four possible outcomes $\{HH, HT, TH, TT\}$.

Event

An event is something that may or may not occur from the possible sample space. In particular, an **event** is some part (or all) of the sample space. For example, in rolling a die we might consider the event that an odd number is obtained. Another possible event would be that a 6 is obtained. A third possible event would be that a number larger than 2 is obtained. The outcomes for which an event occurs are said to be **favorable** to the event. For example, the outcomes 1, 3, and 5 are favorable to the event of "rolling an odd number."

Three Types of Probability

If E is an event from a sample space with n equally likely elements and k is the number of ways in which this event can happen, then the **probability of E, $P(E)$**, is

$$P(E) = \frac{k}{n}$$

This definition is known as the **classical** or **a priori** definition. In this approach, no experimenting is conducted and the probability is based on our knowledge of the nature of the event.

A second approach is the **empirical** or **frequency** approach, and it is based on the results of an actual experiment or of previous experience. In this definition, a series of N trials are made and the event is observed to occur K times. The empirical definition says that the probability of this event occurring would be

$$P(E) = \frac{K}{N}$$

There is a third approach to probability called the subjective approach, and it is based on a person's belief that an event will occur. The subjective approach is normally employed when there is no past history to use for an empirical approach and no basis for using the classical approach.

EXAMPLE 21.2

Find the probability of drawing a 9 or 10 from a well-shuffled deck of cards.

Solution   A deck has 52 cards. Since this is well shuffled, the likelihood of getting any one card is the same as the likelihood of getting any other card. There are four 9s and four 10s in the deck. There are 8 cards in this event and so $P(9 \text{ or } 10) = \frac{8}{52} = \frac{2}{13}$. This is an example of the classical approach to probability.

EXAMPLE 21.3

Samples have shown that, out of every 1,000 discs it copies, a certain machine will produce 75 discs with errors. The probability of a disc being bad if copied on this machine is $\frac{75}{1,000} = \frac{3}{40}$. This is an example of the empirical approach to probability.

As you can see, since $0 \leq K \leq N$, we can make the following five points about the probability of any event:
- The probability that a given event will happen ranges from 0 to 1 or $0 \leq P(E) \leq 1$.
- The higher the probability (the closer to 1), the more likely it is that the event will happen.
- If the probability is $\frac{1}{2}$, then it is equally likely that an event will happen (or will not happen).
- A probability of 1 indicates that the event is certain to occur.
- A probability of 0 means that the event will not happen.
    If E represents a certain event, then E' represents all the elements in the sample space that are not in E. So, if the probability that E will happen is $P(E)$, then the probability that E will not happen (or that E' will happen) is $P(E') = 1 - P(E)$.

EXAMPLE 21.4

What is the probability that the disc-copying machine in Example 21.3 will make a good copy?

Solution   The probability that a copy is bad is $\frac{3}{40}$, so the probability that a copy is good is $1 - \frac{3}{40} = \frac{37}{40}$.

Sometimes we are interested in the probability of an event when we know that another event has already occurred. If the probability that the second event will occur is not affected by what happens to the first event, we then say that the events are **independent**.

> **Probability of Two Independent Events**
>
> If events A and B are independent, then the probability that both A and B will occur is $P(A \text{ and } B) = P(A) \cdot P(B)$.

EXAMPLE 21.5

Two cards are to be drawn from a well-shuffled deck of cards. After the first card is drawn, it is replaced and the deck is shuffled again. What is the probability that we will draw two diamonds?

Solution   There are 13 diamonds in the deck, so the probability of getting a diamond is $\frac{13}{52} = \frac{1}{4}$. Since the first card is replaced and the deck is shuffled, the probability of getting a diamond on the second draw is independent of what was drawn on the first card. The probability of getting a diamond on the second card is $\frac{1}{4}$ and the probability of getting two diamonds is $\frac{1}{4} \cdot \frac{1}{4} = \frac{1}{16}$.

How would the results of Example 21.5 have been different if we had not replaced the first card? In this instance, we are working with **conditional probability**. If A is an event that has already occurred, and we want to know the probability that a second event B will occur, we call this the **probability of B given A**. This is often written $P(B|A)$. To determine $P(B|A)$, we will use the following reasoning: We know that event A has already occurred. This reduces the sample space to the elements in A. We now count the number of times that event B is in this reduced sample space. Then $P(B|A)$ are the number of elements in B that are also in A divided by the number of elements in A. Thus,

$$P(A \text{ and } B) = P(A) \cdot P(B \mid A) \qquad \text{or} \qquad P(B \mid A) = \frac{P(A \text{ and } B)}{P(A)}$$

EXAMPLE 21.6

Two cards are going to be drawn from a well-shuffled deck of cards. The first card will not be replaced after it is drawn. What is the probability that both cards will be diamonds?

Solution   Let A be the event that the first card is a diamond. Let B be the event that the second card is a diamond. We want the probability of A and B. We know from Example 21.5 that $P(A) = \frac{1}{4}$. Since the first card is not replaced, the deck now has 51 cards. Of these 51 cards, 12 of them are diamonds. So, $P(B|A) = \frac{12}{51} = \frac{4}{17}$ and

$$P(A \text{ and } B) = P(A) \cdot P(B|A)$$

$$= \frac{1}{4} \cdot \frac{4}{17} = \frac{1}{17}$$

In this section, we have given you a very brief introduction to probability. As you might imagine, there are many more complicated situations that could happen, and determining their probabilities is much more involved. We have, however, given you a foundation for the work with statistics that is in the remainder of this chapter.

Exercise Set 21.1

In Exercises 1–10, let the sample space be an ordinary deck of well-shuffled playing cards. What is the probability of drawing each of the following?

1. a heart
2. a six
3. a queen
4. a two or a seven
5. a face card (jack, queen, or king)
6. a black card
7. two black cards on successive draws, if the first card is replaced before the second card is drawn
8. a black card and then a face card, if the first card is replaced before the second card is drawn
9. a red card and then a black card, if the first card is not replaced before the second draw
10. a face card and then an ace, if the first card is not replaced before the second draw

Assume that you are rolling a single die. What is the probability of rolling the following?

11. 6
12. 1 or 6
13. even number
14. 4 or more
15. two successive 6s
16. not a 5
17. at least one 5 on two successive rolls (First, determine the probability of not getting a 5.)
18. three successive 6s

In Exercises 19–20, assume that you are rolling two dice. There are 36 possible ways for the dice to fall.

19. What is the sample space?
20. What is the probability of getting two 6s?

When rolling a pair of dice, what is the probability of obtaining each of the following totals?

21. 12
22. 2
23. 1
24. 3
25. 7
26. 7 or 11
27. less than 5
28. 7 on two successive rolls
29. 12 on two successive rolls
30. at least 7

In Exercises 31–34, use the fact that a loaded (or unfair) die has probabilities of $\frac{1}{21}$, $\frac{2}{21}$, $\frac{3}{21}$, $\frac{4}{21}$, $\frac{5}{21}$, and $\frac{6}{21}$ on showing a 1, 2, 3, 4, 5, and 6, respectively.

31. What is the probability of rolling two 6s in succession?

32. What is the probability of rolling a 3 and then a 4?

33. What is the probability that the number on the die is even?

34. What is the probability of a total of 12 when two dice are rolled?

Solve Exercises 35–40.

35. *Medical technology*   A certain medication is known to cure a specific illness for 75% of the people who have the illness. If two people with the illness are selected at random and take the medicine, what is the probability that
   (a) both will be cured?
   (b) neither will be cured?
   (c) only one will be cured?

36. *Aeronautics*   On a two-engine jet plane, the probability of either engine failing is 0.001. If one engine fails, then the probability that the second will fail is 0.005. What is the probability that both engines will fail?

37. *Insurance*   Insurance company tables show that for a married couple of a certain age group the probability that the husband will be alive in 25 years is 0.7. The probability that the wife will be alive in 25 years is 0.8.

   (a) What is the probability that both are alive in 25 years?
   (b) What is the probability that both are dead in 25 years?
   (c) What is the probability that in 25 years only one is alive?

38. *Industrial technology*   A machine produces 25 defective parts out of every 1,000. What is the probability of a defective part being produced?

39. *Industrial technology*   What is the probability of testing two parts from the machine in Exercise 38 and finding that one of them is defective?

40. *Industrial technology*   What is the probability of testing four parts from the machine in Exercise 38 and finding that all four are not defective?

≡ 21.2
MEASURES OF CENTRAL TENDENCY

In Section 21.1, we stated that there were three types of probability. One of these was the empirical or frequency approach. This approach is based on past experience or on the results of an actual experiment. The experiment might consist of a series of trials. For each of these trials, the person directing the experiment will collect data, analyze the data, and then interpret it. In the remainder of this chapter, we will look at some of the ways in which the experimenter might gather and analyze this data.

In statistics, we deal with numbers. Sometimes it is possible to deal with entire sets of numbers. For example, if your teacher wants to determine how much your class has learned about probability, a test could be given to the class. Businesses also are sometimes able to test an entire group of items. For example, a company might be able to test each television set to see that it works.

There are times when it is not possible to check each item and so a sample of items must be selected. For example, it is not possible to check every electric fuse to see that it provides the proper protection. It also might not be practical to check every part made on a machine that produces 10,000 parts a day. In these last two cases, a sample is selected and tested. They might check every 100th fuse or every 1,000th part made by a machine. No matter how the data are gathered, it needs to be organized in a way that makes it easier to understand.

Frequency Distribution ━━━━━━━━━━━━━

One way to organize the information is in a **frequency distribution**. In a frequency distribution, one line contains a list of possible values and a second line contains the number of times each value was observed in a particular time.

Application

EXAMPLE 21.7

People who live near an interstate highway have complained about traffic noise. A measuring instrument is placed along the highway and every 15 min it measures the noise level in decibels. A frequency distribution is made of the readings for the first day.

Decibels	50–54	55–59	60–64	65–69	70–74	75–79	80–84	85–89
Frequency	2	2	4	4	4	6	8	8

Decibels	90–94	95–99	100–104	105–109	110–114	115–119
Frequency	10	12	14	10	8	4

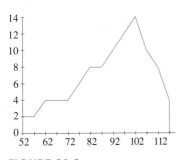

FIGURE 21.1

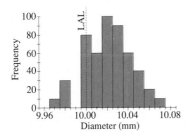

FIGURE 21.2

Graphs

Many times it is difficult for us to understand the numbers in a table such as a frequency distribution. We find that a chart or graph helps us to get a better idea of these numbers. One type of graph is the **histogram**. Another type of graph is the **frequency polygon** or **broken line graph**. A histogram for the data in the above frequency table showing the noise level along an interstate highway from the previous example is given in Figure 21.1. The numbers along the bottom mark the middle of each interval. A broken line graph for the same information is shown in Figure 21.2.

Sometimes a histogram can provide very important information. For example, the histogram in Figure 21.3 shows the results when inspectors measured the diameters in millimeters of 500 steel rods. The dotted line marked LAL is the lowest acceptable limit. Rods smaller than 10.00 mm are too loose in the bearing and would be thrown out or rejected in a later operation. When a rod is rejected, the company loses all the labor and material cost that went into making that rod.

The histogram in Figure 21.3 contains some very interesting information. We can see that 40 rods were rejected because they were below the LAL; but, there is a gap in the histogram at the interval just below the LAL. The gap was unexpected and is a result of inspectors passing rods that were barely below the LAL. This gap is one that the inspectors could have corrected.

FIGURE 21.3

Median, Quartiles, and Box Plots

Tables and graphs can give us some general ideas about the data, but we often need more information. Among this information are some indications for the location of the center of the distribution. We would also like some measure of how the data are spread. This gives us some numerical descriptions of the data and helps us compare groups of data.

The first types of information that we will examine are the ones that provide some indication of the center. These are known as the **measures of central tendency** and there are three of them: the median, the mean, and the mode.

The first measure of central tendency we will examine is the median.

> **Median**
>
> The **median** is the middle number of the numbers in the distribution. One-half of the values are larger than the median, and one-half are smaller. To find the median,
> 1. Arrange the numbers in increasing (or decreasing) order.
> 2. If there is an odd number of items, the median is the value of the middle item.
> 3. If there is an even number of items, the median is the number half-way between the two middle items.

EXAMPLE 21.8

Given the numbers 9, 8, 3, 2, 4, determine the median.

Solution   First arrange the numbers in increasing order: 2, 3, 4, 8, 9. There are five numbers, so the middle number is the third number. The third number is 4, so the median is 4.

EXAMPLE 21.9

Find the median of 11, 12, 15, 18, 20, 20.

Solution   These number are already in numerical order. There are six numbers. The median will be half-way between the third and fourth numbers. The third number is 15 and the fourth is 18. Midway between these is 16.5, thus the median is 16.5.

The median divides the items into two equally sized parts. In the same way, the **quartiles** Q_1, Q_2, and Q_3, divide the numbers into four equally sized parts when the numbers are arranged in increasing (or decreasing) order. As we will see, quartiles provide a quick way to graphically see how the numbers are distributed.

> **Finding Quartiles**
>
> 1. Arrange the numbers in increasing order.
> 2. Q_2 is the median. It divides the numbers into a lower half and an upper half.
> 3. Q_1 is the median of the lower half of the numbers.
> 4. Q_3 is the median of the upper half of the numbers.

EXAMPLE 21.10

Determine the quartiles of 12, 15, 42, 37, 61, 14, 14, 9, 25, 32, 27, and 30.

Solution   We begin by arranging the numbers in increasing order:

$$9, 12, 14, 14, 15, 25, 27, 30, 32, 37, 42, 61$$

There are 12 numbers, so the second quartile, Q_2 (or median), is midway between the sixth and seventh numbers. The sixth number is 25. The seventh number is 27.

$$Q_2 = \frac{25 + 27}{2} = 26$$

EXAMPLE 21.10 (Cont.)

Q_1 is the median of the lower half, and so it is the median of the smallest six numbers. Q_1 is midway between the third and fourth numbers. These are both 14, so $Q_1 = 14$.

Q_3 is the median of the upper half. The upper half of the items is 27, 30, 32, 37, 42, and 61. The median of these is midway between 32 and 37, so

$$Q_3 = \frac{32 + 37}{2} = 34.5$$

The three quartiles, the lowest number, and the highest number are used to make a diagram called a **box plot** or a **box and whisker diagram**. The following steps describe how to make a box and whisker diagram.

Making a Box and Whisker Diagram

1. Arrange the numbers in increasing order.
2. Find the lowest and highest scores, Q_1, Q_2, and Q_3.
3. Draw a scale that will include the lowest and highest numbers.
4. Draw a box with the ends of the box at Q_1 and Q_3.
5. Draw a line through the box at Q_2.
6. Draw a whisker from the lowest number to the box and draw another whisker from the highest number to the box.

≡ **Note**   The box of a box and whisker diagram contains the middle 50% of the data, one whisker shows the bottom 25% of the data, and the other whisker shows the top 25% of the data.

Application

EXAMPLE 21.11

At an automobile engine plant, a quality control technician pulls crankshafts from the assembly line at regular intervals. The technician measures a critical dimension on each of these crankshafts. Even though the dimension is supposed to be 182.000 mm, some variation will occur during production. Here are the measurements for one morning:

182.120	182.005	182.025	181.987	181.898	182.034
181.960	181.940	182.055	181.897	181.935	182.063
182.015	182.026	181.965	181.985	182.362	181.998
182.107	181.934	181.991	182.005	182.012	181.984

Make a box plot for these measurements.

Solution   We begin by listing the numbers in numerical order.

181.897	181.898	181.934	181.935	181.940	181.960
181.965	181.984	181.985	181.987	181.991	181.998
182.005	182.005	182.012	182.015	182.025	182.026
182.034	182.055	182.063	182.107	182.120	182.362

EXAMPLE 21.11 (Cont.)

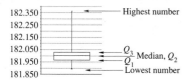

FIGURE 21.4

From this arrangement of the 24 numbers, we can see that the lowest is 181.897 mm and the highest is 182.362 mm. Notice that the numbers are arranged in four rows with six numbers in each row. Thus, Q_1 will be between the last number in the first row and the first number in the second row, or $Q_1 = \frac{181.960+181.965}{2} = 191.962\,5$. In a similar way, we find $Q_2 = \frac{181.998+182.005}{2} = 182.001\,5$; and $Q_3 = \frac{182.026+182.034}{2} = 182.03$. The box plot for these data is shown in Figure 21.4.

Mean and Mode

The second measure of central tendency is the **mean**, sometimes called the **arithmetic mean**.

Mean

To determine the mean, you add all the values and divide by the number of values. The symbol $\bar{x}$ is used to represent the mean. In symbols,

$$\bar{x} = \frac{\sum_{i=1}^{n} x_i}{n}$$

EXAMPLE 21.12

Find the mean of the values in **(a)** Example 21.8 and **(b)** Example 21.9.

Solutions

(a) In Example 21.8, the five numbers were $9, 8, 3, 2$, and 4, so

$$\bar{x} = \frac{9+8+3+2+4}{5} = \frac{26}{5} = 5.2$$

(b) In Example 21.9, the six numbers were $11, 12, 15, 18, 20$, and 20. The mean of these six numbers is

$$\bar{x} = \frac{11+12+15+18+20+20}{6} = \frac{96}{6} = 16$$

If you want to find the mean of a large number of values, and if some appear more than once, then there is a quicker method to get the total. Consider the following frequency distribution for the 500 steel rods in Figure 21.3.

Diameter	9.97	9.98	9.99	10.00	10.01	10.02	10.03	10.04
Frequency	10	30	0	80	60	100	90	60

Diameter	10.05	10.06	10.07
Frequency	40	20	10

It would take a long time to add these 500 values. Instead, if we multiply each diameter by the frequency for that diameter, we reduce the 500 additions to ten multiplications and nine additions. Thus,

$$\bar{x} = \frac{\begin{array}{c}9.97(10) + 9.98(30) + 9.99(0) + 10.00(80) + 10.01(60) + 10.02(100) \\ + 10.03(90) + 10.04(60) + 10.05(40) + 10.06(20) + 10.07(10)\end{array}}{10 + 30 + 80 + 60 + 100 + 90 + 60 + 40 + 20 + 10}$$

$$= \frac{5{,}010.7}{500}$$

$$= 10.0214$$

Using the sigma notation from Chapter 19, the mean is given by the following formula.

Mean

$$\bar{x} = \frac{\displaystyle\sum_{i=1}^{n} x_i f_i}{\displaystyle\sum_{i=1}^{n} f_i}, \text{ where } x_i \text{ represents the } i\text{th value and } f_i \text{ the frequency for that value.}$$

In Example 21.12, there were 11 values: $x_1 = 9.97$ and $f_1 = 10$, $x_2 = 9.98$ and $f_2 = 30$, and so on until $x_{11} = 10.07$ and $f_{11} = 10$.

Your calculator can be a great help in calculating the mean. For example, to determine the denominator for this example, you could proceed as follows. (Some middle steps are left out, but the process is the same as for those before and after.)

PRESS	DISPLAY
9.97 $\boxed{\times}$ 10 $\boxed{+}$	99.7
9.98 $\boxed{\times}$ 30 $\boxed{+}$	399.1
10 $\boxed{\times}$ 80 $\boxed{+}$	1199.1
⋮	⋮
10.06 $\boxed{\times}$ 20 $\boxed{+}$	4910
10.07 $\boxed{\times}$ 10 $\boxed{=}$	5010.7
$\boxed{\div}$ 500	10.0214

Thus, we have the mean $\bar{x} = 10.0214$. At the end, we divided by 500 because we knew that the frequency was 500. If we had not known the total frequencies, we should have stored the value for the numerator, 5010.7, by using $\boxed{\text{STO}}$ 1 and then totaled the frequencies.

There is another method for determining the mean with a calculator. This method uses the $\boxed{\sum +}$ key. We will show this method using the values from Example 21.8: 9, 8, 3, 2, 4. (Some calculators must be placed in the "statistics" or "stat" mode before the $\boxed{\sum +}$ key can be used; on others, it is done automatically.)

PRESS	DISPLAY
9 $\boxed{\sum +}$	1
8 $\boxed{\sum +}$	2
3 $\boxed{\sum +}$	3
2 $\boxed{\sum +}$	4
4 $\boxed{\sum +}$	5
$\boxed{2nd}$ $\boxed{Mean}$	5.2

Notice that until the last step the display shows the total number of values that have been entered. (On some calculators, you would use $\boxed{2nd}$ $\boxed{\bar{x}}$ instead of $\boxed{2nd}$ $\boxed{Mean}$.)

This method would not be very fast if you were calculating the mean of a lot of numbers, since it does not use the frequency of the values. Some calculators have a $\boxed{2nd}$ $\boxed{frq}$ key. This key allows you to enter the frequency for each value before you enter the $\boxed{\sum +}$ key. If the frequency is one, you skip the $\boxed{2nd}$ $\boxed{frq}$ key. We will use this on the 500 diameter measures we calculated earlier. Again, we will omit some of the middle steps.

PRESS	DISPLAY
9.97 $\boxed{2nd}$ $\boxed{frq}$	Fr 00
10 $\boxed{\sum +}$	10
9.98 $\boxed{2nd}$ $\boxed{frq}$	Fr 00
30 $\boxed{\sum +}$	40
10 $\boxed{2nd}$ $\boxed{frq}$ 80 $\boxed{\sum +}$	120
⋮	⋮
10.06 $\boxed{2nd}$ $\boxed{frq}$ 20 $\boxed{\sum +}$	490
10.07 $\boxed{2nd}$ $\boxed{frq}$ 10 $\boxed{\sum +}$	500
$\boxed{2nd}$ $\boxed{Mean}$	10.0214

There was one problem: You cannot enter a frequency larger than 99. The value 10.02 had a frequency of 100 so we had to enter that value twice, thus the frequencies totaled 100. We chose to use 10.02 $\boxed{2nd}$ $\boxed{frq}$ 50 $\boxed{\sum +}$ both times.

The use of the $\boxed{\sum +}$ key is not as fast as calculating the mean without it. But, as we will see in Section 21.3, the use of the $\boxed{\sum +}$ key allows us to get some additional information.

The third, and final, measure of central tendency is the mode.

Mode

The **mode** is the value that has the greatest frequency. A set of numbers can have more than one mode. If there are two modes, the data are said to be **bimodal**.

EXAMPLE 21.13

What is the mode for the data in Example 21.9: 11, 12, 15, 18, 20, and 20?

Solution   The mode is 20, because that value occurs twice and all the other values occur once.

Application

| EXAMPLE 21.14 | What is the mode for the sample of 500 steel rods in Figure 21.3? |

Solution The 500 steel rods had a mode of 10.20, because the 100 measures for that diameter were more than for any other.

Exercise Set 21.2

Use the following four sets of numbers for Exercises 1–20.

A : 4, 2, 6, 3, 7, 4, 6, 2, 4, 4

B : 50, 52, 52, 54, 54, 54, 54, 56, 58, 58

C : 80, 77, 82, 73, 92, 89, 100, 96, 96, 94, 74, 94, 94, 96, 83, 84, 96, 87, 84, 96

D : 100, 98, 96, 94, 93, 90, 89, 85, 82, 78, 76, 66, 64, 64, 78, 89, 93, 96, 98, 96, 93, 64, 96

In Exercises 1–4, set up a frequency distribution table for the numbers in the indicated set.

1. Set A

2. Set B

3. Set C

4. Set D

In Exercises 5–8, set up a frequency distribution table for each of the given intervals in the indicated sets.

5. 1–2, 3–4, 5–6, 7–8 in Set A

6. 50–52, 53–55, 56–58 in Set B

7. 71–75, 76–80, 81–85, 86–90, 91–95, 96–100 in Set C

8. 61–65, 66–70, 71–75, 76–80, 81–85, 86–90, 91–95, 96–100 in Set D

In Exercises 9–12, draw histograms for the data in the given exercises.

9. Exercise 1

10. Exercise 2

11. Exercise 7

12. Exercise 8

In Exercises 13–16, draw broken line graphs for the data in the indicated exercises.

13. Exercise 1

14. Exercise 2

15. Exercise 7

16. Exercise 8

In Exercises 17–20, determine the mean, median, mode, and quartiles of the given set. Then draw a box plot of the data.

17. Set A

18. Set B

19. Set C

20. Set D

Solve Exercises 21–32.

21. *Energy technology*   A group of 100 batteries were selected from the day's production for a machine. The batteries were tested to see how long they would operate a flashlight with these results:

Hours	211–215	216–220	221–225	226–230
Frequency	4	9	19	23

Hours	231–235	236–240	241–245	246–250
Frequency	16	14	10	5

Form a histogram for this data.

22. *Energy technology*   Form a histogram for the data in Exercise 21, using the intervals 211–220, 221–230, etc.

23. *Industrial technology*   A technician was measuring the thickness of a plastic coating on some pipe and obtained the following data:

Thickness (mm)	0.01	0.02	0.03	0.04	0.05
Frequency	1	5	40	50	36

Thickness (mm)	0.06	0.07	0.08	0.09
Frequency	30	25	10	3

Form a histogram of this data.

24. *Industrial technology*   Draw a broken-line graph for the data in Exercise 23.

25. *Industrial technology*   Form a histogram for the data in Exercise 23 over the intervals 0.01–0.02, 0.03–0.04, etc.

26. *Industrial technology*   Determine the mean, median, mode, and quartiles for the data in Exercise 23.

27. *Industrial technology*   Draw a box plot for the data in Exercise 23.

28. *Electrical technology*   A technician tested an electric circuit and found the following values in milliamperes on successive trials:

5.24, 5.31, 5.42, 5.26, 5.31, 5.47, 5.27, 5.29, 5.35, 5.44,

5.35, 5.31, 5.45, 5.46, 5.39, 5.34, 5.35, 5.46, 5.26, 5.27,

5.47, 5.34, 5.28, 5.39, 5.34, 5.42, 5.43, 5.46, 5.34, 5.29

Form a frequency distribution for this data.

29. *Electrical technology*   For the data in Exercise 28, draw a histogram for the intervals 5.21–5.25, 5.26–5.30, etc.

30. *Electrical technology*   Determine the mean, median, and mode for the data in Exercise 28.

31. *Electrical technology*   Determine the quartiles and draw a box plot for the data in Exercise 28.

32. Write a computer program that will allow you to enter individual values and their frequencies and to determine the mean of the data.

☰ 21.3
MEASURES OF DISPERSION

In Section 21.2, we looked at the three measures of central tendency. In this section, we will learn a technique that will tell us how close together the information is distributed. **Measures of dispersion** tell how close together or how spread out the data are.

Variance and Standard Deviation ▬▬▬▬▬

There are several measures of dispersion. We will look at two of them: variance and standard deviation.

Variance and Standard Deviation

The **variance** v of a set of numbers is given by the formula

$$v = \frac{\sum\limits_{i=1}^{n} (x_i - \overline{x})^2}{n}$$

and the **standard deviation** s is the square root of the variance. Thus,

$$s = \sqrt{v}$$

EXAMPLE 21.15

Find the variance and standard deviation of the numbers $9, 8, 3, 2, 4$.

Solution   This is the same set of numbers we used in Example 21.8. In Example 21.12, we determined that the mean was 5.2. We can set up a table to help with the calculations.

x	$x - \overline{x}$	$(x - \overline{x})^2$
9	3.8	14.44
8	2.8	7.84
3	−2.2	4.84
2	−3.2	10.24
4	−1.2	1.44
		38.80

$$v = \frac{38.8}{5} = 7.76$$

$$s = \sqrt{7.76} \approx 2.79$$

In some sets of data, some values occur more than once. The formula for the variance becomes

$$v = \frac{\sum f_i (x_i - \overline{x})^2}{n}$$

What you would do is add two more columns to your table, one for f_i and the other for $f_i(x_i - \overline{x})^2$, as is shown by the next example.

EXAMPLE 21.16

Determine the mean and standard deviation for the following values: $39, 41, 39, 44, 39, 40, 39, 40, 37, 42, 37, 43, 44, 38, 38, 38, 43, 38, 41, 39$.

EXAMPLE 21.16 (Cont.)

Solution   We first make a frequency distribution table and then add columns for $x_i f_i$, $x_i - \bar{x}$, $(x_i - \bar{x})^2$, and $(x_i - \bar{x})^2 f_i$.

x_i	f_i	$x_i f_i$	$x_i - \bar{x}$	$(x_i - \bar{x})^2$	$(x_i - \bar{x})^2 f_i$
37	2	74	−2.95	8.7025	17.4050
38	4	152	−1.95	3.8025	15.2100
39	5	195	−0.95	0.9025	4.5125
40	2	80	0.05	0.0025	0.0050
41	2	82	1.05	1.1025	2.2050
42	1	42	2.05	4.2025	4.2025
43	2	86	3.05	9.3025	18.6050
44	2	88	4.05	16.4025	32.8050
Totals	20	799			94.95

The mean is $\bar{x} = \dfrac{\sum x_i f_i}{\sum f_i} = \dfrac{799}{20} = 39.95$ and the variance is $v = \dfrac{\sum (x_i - \bar{x})^2 f_i}{\sum f_i} = \dfrac{94.95}{20} = 4.7475$. The standard deviation is $s = \sqrt{4.7475} \approx 2.18$.

Standard Deviation on a Calculator

Again, a calculator will do much of the work for you. If you use the $\boxed{\sum +}$ key to compute the mean, by pushing two more buttons you can determine the standard deviation. We will demonstrate this with the same data from Example 21.16.

	PRESS	DISPLAY
37	$\boxed{\text{2nd}}$ $\boxed{\text{frq}}$ 2 $\boxed{\sum +}$	2
38	$\boxed{\text{2nd}}$ $\boxed{\text{frq}}$ 4 $\boxed{\sum +}$	6
39	$\boxed{\text{2nd}}$ $\boxed{\text{frq}}$ 5 $\boxed{\sum +}$	11
40	$\boxed{\text{2nd}}$ $\boxed{\text{frq}}$ 2 $\boxed{\sum +}$	13
41	$\boxed{\text{2nd}}$ $\boxed{\text{frq}}$ 2 $\boxed{\sum +}$	15
42	$\boxed{\sum +}$	16
43	$\boxed{\text{2nd}}$ $\boxed{\text{frq}}$ 2 $\boxed{\sum +}$	18
44	$\boxed{\text{2nd}}$ $\boxed{\text{frq}}$ 2 $\boxed{\sum +}$	20

To get the mean, you enter $\boxed{\text{2nd}}$ $\boxed{\text{Mean}}$ and get 39.95. To get the standard deviation enter $\boxed{\text{2nd}}$ $\boxed{\sigma_n}$ and get 2.1788759, which rounds off to 2.18. If you want the variance, then press the $\boxed{x^2}$ and get 4.7475, the variance.

The symbol σ, a lower case $\sum$, is the mathematical symbol for the theoretical standard deviation. Some calculators use $\boxed{\text{2nd}}$ $\boxed{\text{S.Dev}}$ or $\boxed{\text{2nd}}$ $\boxed{s}$ instead of $\boxed{\text{2nd}}$ $\boxed{\sigma_n}$. But this uses a different formula and will give you a different value. On these calculators, use $\boxed{\text{2nd}}$ $\boxed{\text{Var}}$ to get the variance and then $\boxed{\sqrt{x}}$ to get the standard deviation.

There are two slightly different versions for the standard deviation. One version, the one we are using, has a denominator of n. The other version has a denominator of $n - 1$. To get our version, you must use the $\boxed{\text{2nd}}$ $\boxed{\sigma_n}$ or $\boxed{\text{Var}}$ keys. To get the other version, you use the 2nd σ_{n-1} or S.Dev. keys. For n larger than 30, both versions give approximately the same value.

The significance of different values of the standard deviation are obtained mostly from experience. But a small standard deviation indicates that the values are closely clustered about the mean. On the other hand, a large standard deviation indicates that the values are spread out widely from the mean.

Normal Curve

When we make a large number of measurements, we usually expect that a majority of them are near the mean. One of the most important frequency curves is called the **normal curve** or the **normal distribution curve**. The curve is bell shaped and symmetric about the mean. The normal curve was first discovered by DeMoivre, whose theorem for complex numbers you used in Chapter 14.

The standard form of the normal curve considers the mean to be the origin ($\overline{x} = 0$) and the standard deviation as 1, and has the equation

$$y = \frac{1}{\sqrt{2\pi}} e^{-x^2/2}$$

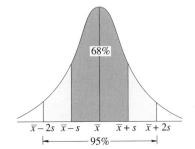

FIGURE 21.5

Some results of this curve are that 68% of the values should be within one standard deviation of the mean. (While 68.26% is closer, we will use 68%.) Thus, 68% of the values are in the interval from $\overline{x} - s$ to $\overline{x} + s$. A total of about 95% of the values (it is actually closer to 95.44%) are supposed to be within two standard deviations of the mean or from $\overline{x} - 2s$ to $\overline{x} + 2s$. (See Figure 21.5.) You may want to round off the values $\overline{x} - s$, $\overline{x} + s$, $\overline{x} - 2s$, and $\overline{x} + 2s$ before you count the values in each interval.

EXAMPLE 21.17

In Example 21.16, there were 20 values. The mean was 39.95 and the standard deviation was 2.18. Thus, the interval from $\overline{x} - s$ to $\overline{x} + s$ is from 37.77 to 42.13. This rounds off to the interval from 38 to 42 and there are 14 values in this interval. This is 70% of the values and is the closest you can get to 68% with a sample of 20 values.

When a distribution is described in intervals, and you have no other information, then you must make some decisions before counting or not counting the points in an interval. In computing the mean and standard deviation, the midpoint of each interval is used as the "value" for that interval. The same criteria are used when counting the number of points within one or more standard deviations of the mean. For example, if $\overline{x} + s$ is at or past the midpoint of an interval, then include all the points of that interval. On the other hand, if $\overline{x} + s$ does not reach the midpoint of the interval, then do not count any of the points in the interval.

Exercise Set 21.3

In Exercises 1–12, use the following sets of numbers.

A : 4, 2, 6, 3, 7, 4, 6, 2, 4, 4

B : 50, 52, 52, 54, 54, 54, 54, 56, 58, 58

C : 80, 77, 82, 73, 92, 89, 100, 96, 96, 94, 74, 94, 94, 96, 83, 84, 96, 87, 84, 96

D : 100, 98, 96, 94, 93, 90, 89, 85, 82, 78, 76, 66, 64, 64, 78, 89, 93, 96, 98, 96, 93, 64, 96

In Exercises 1–4 find the variance v for the indicated set of numbers.

1. Set A

2. Set B

3. Set C

4. Set D

In Exercises 5–8, find the standard deviation for the indicated set of numbers.

5. Set A

6. Set B

7. Set C

8. Set D

In Exercises 9–12, determine the number of values within (a) one standard deviation and (b) two standard deviations of the mean.

9. Set A

10. Set B

11. Set C

12. Set D

In Exercises 13–16, assume the given data are normally distributed, (a) find the standard deviation for the indicated sets of numbers, (b) determine the number of data points within one standard deviation of the mean, and (c) those within two standard deviations of the mean.

13. *Machine technology*   The 500 steel rods in Figure 21.3. (See frequency distribution following Example 21.12.)

14. *Energy technology*   The data from the frequency distribution for the batteries in Exercise 21 of Exercise Set 21.2. (Use the midpoint of each interval as the value for that interval.)

15. *Machine technology*   The data from the frequency distribution for the plastic pipe coating in Exercise 23 of Exercise Set 21.2.

16. *Electrical technology*   The data from the electric circuit tests in Exercise 28, Exercise Set 21.2.

 17. Write a computer program to calculate the mean, variance, and standard deviation. For the variance, you might want to use the formula

$$v = \frac{n \sum f_i x^2{}_i - \left(\sum x^2{}_i\right)}{n^2}$$

 18. Modify your program in Exercise 17 to have the program determine the total number of values and the number and percentage within one and two standard deviations of the mean.

☰ **21.4**
FITTING A LINE TO DATA

Until now, when we have dealt with the relationship between two variables we have been given the functional relationship between them. But, many times we do not know what the relationship is or even if it exists. Sometimes we have some pairs of values that were obtained from an experiment or from observation. In Sections 21.4 and 21.5, we will take these pairs of values and plot them. Then we will try to find a function $y = f(x)$ that is suggested by the data.

In this section, we will show how to "fit" a line to a given set of points. As so often happens, when we are dealing with information about people or with the results of an experiment, the results do not exactly match what theory says they should. When you are conducting an experiment, you are often trying to determine the type of curve, and its equation, that fits the data. In this section, we will be interested in fitting a straight line to the data. In order to fit a straight line, you are going to find the line that best fits these data.

We will assume that a relationship does exist between the variables in each of the examples and in all of the exercises. Our job is to find the straight line that comes the closest to fitting these data.

Application

EXAMPLE 21.18

Below are the heights, in inches, and the weights, in pounds, for 12 students.

Student	Height	Weight
A	66	130
B	67	140
C	68	180
D	69	160
E	70	185
F	71	190
G	72	200
H	70	195
I	72	195
J	71	210
K	72	210
L	73	180

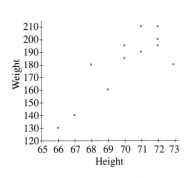

FIGURE 21.6

A graph of these 12 points is shown in Figure 21.6. A quick glance at this graph tells you that these points do not all lie on the same straight line. But, in general, the taller a person is, the more he or she weighs. Thus, we think that it is possible that there might be some straight line that would come close to these points and from which none of the points differ by very much.

A graph like the one in Figure 21.6, which plots individual data points on a coordinate system, is called a **scatter diagram**.

Method of Least Squares

There are several ways to determine the line that best fits these points. We will use a method known as the **method of least squares**. The idea behind this method is that

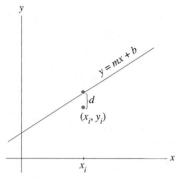

FIGURE 21.7

the sum of the squares of the deviation of all points from the best line is a minimum. The deviation we refer to is the difference between the y-value of the line and the y-value for a given value of x.

Suppose that the line $y = mx + b$ "passes through" these data points and that the point (x_1, y_1) is a point from the data. As the line passes through these points, it may or may not intersect any of them. But, some points will lie on one side of the line and other points on the other side of the line. The difference between the observed y-value and the one on the line is the deviation. In Figure 21.7, this is indicated by the letter d.

Now by finding these deviations for all the points of observation and squaring them, we will be finding the minimum of the sum of the squares. If

$$y = mx + b$$

is the equation of the least squares line, then it can be shown that

$$m = \frac{\dfrac{\sum x_i \sum y_i}{N} - \sum x_i y_i}{\dfrac{\left(\sum x_i\right)^2}{N} - \sum x_i^2}$$

and $\quad b = \bar{y} - m\bar{x}$

The formula for the slope looks very complicated. Algebraically it can be simplified as

$$m = \frac{\overline{xy} - \bar{x}\,\bar{y}}{s_x^2}$$

where $\overline{xy}$ is the mean of the xys, $\bar{x}$ is the mean of the xs, $\bar{y}$ is the mean of the ys, and s_x^2 is the variance of the x-values.

Normal Equations

There is another method that allows us to use our ability to solve a system of simultaneous equations. There are two equations, known as the **normal equations**, that involve m and b. These are

$$\sum_{i=1}^{n} y_i = bn + m \sum_{i=1}^{n} x_i$$

and $\quad \displaystyle\sum_{i=1}^{n} x_i y_i = b \sum_{i=1}^{n} x_i + m \sum_{i=1}^{n} x_i^2$

Here, n represents the number of points in the sample.

We will now determine the least square line for Example 21.18.

EXAMPLE 21.19

Determine the least square line for the data in Example 21.18.

Solutions The heights will be the x-scores and the weights the y-scores.

Student	Height x	Weight y	xy	x^2
A	66	130	8,580	4,356
B	67	140	9,380	4,489
C	68	180	12,240	4,624
D	69	160	11,040	4,761
E	70	185	12,950	4,900
F	71	190	13,490	5,041
G	72	200	14,400	5,184
H	70	195	13,650	4,900
I	72	195	14,040	5,184
J	71	210	14,910	5,041
K	72	210	15,120	5,184
L	73	180	13,140	5,329
Totals	841	2175	152,940	58,993

Method 1:

$$\bar{x} = \frac{841}{12} = 70.0833$$

$$\bar{y} = \frac{2,175}{12} = 181.25$$

$$\overline{xy} = \frac{152,940}{12} = 12,745$$

$$s_x = 2.10 \quad s^2_x = 4.41$$

$$m = \frac{\overline{xy} - \bar{x}\bar{y}}{s^2_x}$$

$$= \frac{12,745 - (70.0833)(181.25)}{4.41}$$

$$= \frac{12,745 - 12,702.6}{4.41}$$

$$= \frac{42.4}{4.41} \approx 9.61$$

$$b = \bar{y} - m\bar{x}$$

$$= 181.25 - (9.61)(70.08)$$

$$\approx -492.22$$

Thus, the equation of the line is $y = 9.61x - 492.22$.

EXAMPLE 21.19 (Cont.)

Method 2: Using the normal equations, we have the system

$$2,175 = 12b + 841m$$

$$152,940 = 841b + 58,993m$$

Using Cramer's rule and the techniques of Section 17.5, we find that $m = \frac{6,105}{635} \approx$ 9.61 and $b = \frac{-312,765}{635} \approx -492.54$. These are approximately the same results that we obtained using Method 1. The results using the second method are more accurate, because we did not round off until the end.

Using a Calculator

Again, the calculator can give you a great deal of help. In fact, it can be used to calculate the slope and intercept of the least squares line. You will use some new keys. The $\boxed{x \text{ş} y}$ key is the "x exchange y" key. It is used to swap x- and y-values. You will also use the $\boxed{2nd}$ $\boxed{CSR}$ keys. Pressing $\boxed{2nd}$ $\boxed{CSR}$ clears the statistical register. You should use it before you begin any statistical problem. There are some other new keys that we will introduce at the end of this example. The data in the example is the same set we have been using.

EXAMPLE 21.20

Use a calculator to determine the least square line for the data from Example 21.18.

Solution

PRESS	DISPLAY	
$\boxed{CLR}$ $\boxed{CLR}$ $\boxed{2nd}$ $\boxed{CSR}$	0	Clears display, memory, and statistical registers.
66 $\boxed{x \text{ş} y}$ 130 $\boxed{\sum +}$	1	
67 $\boxed{x \text{ş} y}$ 140 $\boxed{\sum +}$	2	
$\vdots$	$\vdots$	
72 $\boxed{x \text{ş} y}$ 210 $\boxed{\sum +}$	11	
73 $\boxed{x \text{ş} y}$ 180 $\boxed{\sum +}$	12	

To determine the y-intercept, press either $\boxed{2nd}$ $\boxed{b/a}$ or press $\boxed{2nd}$ $\boxed{Intcp}$, and to find the slope, next press $\boxed{x \text{ş} y}$.

$\boxed{2nd}$ $\boxed{b/a}$	-492.54331	y-intercept
$\boxed{x \text{ş} y}$	9.6141732	slope

EXAMPLE 21.20 (Cont.)

On an RPN calculator, you would enter this same data using this procedure:

PRESS	DISPLAY	
2nd CLEAR $\sum$	0	Clears memory and statistical registers.
130 ENTER 66 $\sum+$	1	
140 ENTER 67 $\sum+$	2	
⋮	⋮	
180 ENTER 73 $\sum+$	12	
2nd L.R.	−492.5433071	
$x \gtrless y$	9.614173228	

Correlation Coefficient

Another statistical value that is often used is the **correlation coefficient** r, where $-1 \leq r \leq 1$. A value of r close to +1 indicates that the relationship between x and y is close to being linear and both variables are increasing. A value close to -1 also indicates a nearly linear relationship, but one where one variable increases as the other decreases. If $-0.75 < r < 0.75$, the correlation is considered to be poor. Graphs of points with different values of r are shown in Figures 21.8a–d.

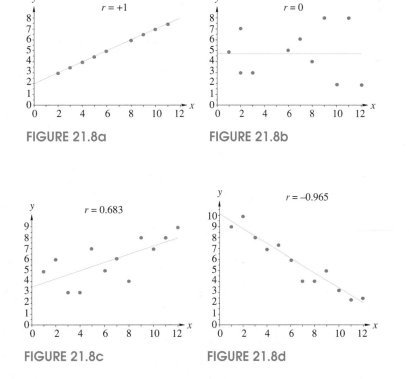

FIGURE 21.8a

FIGURE 21.8b

FIGURE 21.8c

FIGURE 21.8d

One formula for the correlation coefficient is given by $r = m\frac{s_x}{s_y}$, where m is the slope of the least square line, s_x is the standard deviation of the x-values, and s_y is the standard deviation of the y-values.

Application

EXAMPLE 21.21

Determine the correlation coefficient for the data on the students in Example 21.18.

Solution In Example 21.19, we determined that $m = 9.61$, and $s_x = 2.10$. We can calculate $s_y = 24.59$, so

$$r = m\frac{s_x}{s_y}$$

$$= 9.61\left(\frac{2.10}{24.59}\right)$$

$$\approx 0.82$$

It is also possible to determine the correlation coefficient with the help of some calculators. Once you have used the calculator procedures for determining the slope and intercept of the least square line, enter 2nd Corr if you are using an algebraic calculator, or 2nd $\hat{y},r$, then $x \lessgtr y$ on an RPN calculator. If we had done this for the data in Example 21.18, we would have gotten $r = 0.8210168$.

Exercise Set 21.4

In Exercises 1–8, find the equation of the least square line and the correlation coefficient for the given data. For each problem, graph the given data and the least square line.

1.

x	5	6	8	9	12	14
y	2	3	7	7	8	10

2.

x	1	3	5	7	9
y	16	45	86	104	132

3.

x	2	4	6	8	10
y	30	55	93	122	132

4.

x	68.5	67.2	67.7	63.8	69.8	64.7	66.4
y	33.6	35.0	30.2	30.0	33.2	30.8	30.2

x	69.1	65.3	64.8
y	33.3	32.9	37.3

5. *Machine technology* A tensile ring is to be calibrated by measuring the deflection in thousandths of an inch at various loads in thousands of pounds. The following results were obtained by applying increasingly larger load forces from 1,000 to 12,000 lb.

load force (lb × 1,000)	1	2	3	4	5	6
deflection (in × 0.001)	16	34	44	76	84	98

load force	7	8	9	10	11	12
deflection	108	126	137	158	165	185

6. *Business* A vending machine company studied the relationship between maintenance cost and dollar sales for their machines. Here are the results from eight machines:

maintenance cost	95	105	85	130	145
sales	1,100	1,250	600	890	1,450

maintenance cost	125	90	110	90
sales	1,500	800	1,300	870

7. *Energy technology* An engineering company studied the relationship between the air velocity and the evaporation of burning fuel droplets in an impulse engine. The following results were obtained:

air velocity (mm/sec)	200	600	1,000
evaporation coefficient (mm²/s)	0.18	0.37	0.39

air vel.	1,400	1,800	2,200	2,600	3,000
evap. coeff.	0.75	0.82	0.93	1.15	1.42

8. *Automotive engineering* An automobile company studied the relationship between highway mileage and automobile weight.

highway mpg	42	37	45	40	36	35
weight (lb × 100)	21	23	24	22	25	24

highway mpg	45	43	30	25
weight (lb × 100)	19	20	25	26

9. Write a computer program that will allow you to input pairs of values and then compute the slope and y-intercept of the least square line and the correlation coefficient. (For much of this, you can use the program for solving systems of linear equations that was developed in Section 17.5.)

≡ 21.5
FITTING NONLINEAR CURVES TO DATA

In Section 21.4, we studied the case where the regression curve of y on x is linear. Thus, the curve was of the form $y = mx + b$ and, if we knew the value of x, we could predict a value of y. In the same manner, we could predict the value of x for any value of y.

In this section, we will investigate cases where the regression curve is not linear, but where we can still use the methods we learned in Section 21.4. We will then look at some examples of polynomial regression.

Engineers and technicians often plot data on different kinds of graph paper. In Section 13.6, we plotted data on semilogarithmic and logarithmic graph paper. We learned that if the graph of data is a straight line with slope m on log-log (logarithmic) paper, then the data are related by a power function, which is of the form $y = ax^m$. If the graph of the data is a straight line on semilog (semilogarithmic) paper, then the data are related by a function of the form $y = ab^x$.

Suppose that you plot several pairs of points on regular, semilog, and log-log graph paper. If the points "straighten out" on one of them, say the semilog paper, you then have an idea of the nature of the curve. If it is "straight" on the semilog paper, you know it is of the form $y = ab^x$. If you take the logarithm of both sides, you get

$$\log y = \log a + x \log b$$

We can get estimates of $\log a$ and $\log b$ using the methods from Section 21.4. Of course, once we know $\log a$ and $\log b$ we can find a and b.

Application

EXAMPLE 21.22

A tire company collected the following data on the percentage of radial tires that are still usable after having been driven any given number of miles. The data are

EXAMPLE 21.22 (Cont.)

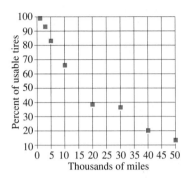

FIGURE 21.9a

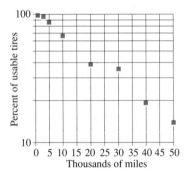

FIGURE 21.9b

shown in the following table.

Miles driven ($\times 1{,}000$)	1	2	5	10	20	30	40	50
Percent usuable	98.3	92.1	82.1	64.9	38.0	35.2	19.3	13.5

Solution The plot of these data are shown in Figure 21.9a. As you can see, when semilog graph paper was used in Figure 21.9b, the points appear to be in straight line.

Since we think that we have a straight line of the form $\log y = \log a + x \log b$, we expand the data table to include the values for $\log y$.

Miles (x)	Percent (y)	$\log y$
1	98.3	1.99255
2	92.1	1.96426
5	82.1	1.91434
10	64.9	1.81224
20	38.0	1.57978
30	35.2	1.54654
40	19.3	1.28556
50	13.5	1.13033

Using a calculator, or the technique from Section 21.4, we find that $\log a = 1.994676$ and $\log b = -0.0172898$, so $a \approx 98.78$ and $b \approx 0.96$, and we see that $y = (98.78)(0.96)^x$. The correlation of these data are $r = -0.99$ (which indicates a strong validity for the equation).

If you use a calculator to compute the slope and intercept, you do not need the $\log y$ column. You can enter the first line by pressing 1 $\boxed{x \S y}$ 98.3 $\boxed{\log}$ $\boxed{\sum +}$.

If the graph appeared to be a straight line on log-log paper, the function would then be of the form $y = ax^m$. Again, you would take the logarithm of both sides. This time you would find the least square line that fits

$$\log y = \log a + m \log x$$

If we have reason to believe that the data would best fit a polynomial of degree p, then the least square curve is of the form

$$y = A_0 + A_1 x + A_2 x^2 + \cdots + A_{p-1} x^{p-1} + A_p x^p$$

Advanced mathematics can be used to derive $p + 1$ **normal equations** that can be used to determine approximations for the A_i values. These normal equations are of the following form, where a_i represents the approximation for A_i that we will obtain.

$$\sum y = na_0 + a_1 \sum x + a_2 \sum x^2 + \cdots + a_p \sum x^p$$

$$\sum xy = a_0 \sum x + a_1 \sum x^2 + a_2 \sum x^3 + \cdots + a_p \sum x^{p+1}$$

$$\sum x^2 y = a_0 \sum x^2 + a_1 \sum x^3 + a_2 \sum x^4 + \cdots + a_p \sum x^{p+2}$$

$$\vdots$$

$$\sum x^p y = a_0 \sum x^p + a_1 \sum x^{p+1} + a_2 \sum x^{p+2} + \cdots + a_p \sum x^{2p}$$

Notice that this is a system of $p + 1$ linear equations in the $p + 1$ variables $a_0, a_1, a_2, \ldots, a_p$. Your choice is to select the degree of the polynomial that you think best approximates the plotting of the data.

EXAMPLE 21.23

A plastics company tested a new additive to determine the relation of the amount of additive to the time it takes the plastic to set. Find the polynomial that best fits the following data.

Amount of additive (g)	0	1	2	3	4	5	6	7	8
Setting time (hr)	7.5	5	3.5	2.5	2	2.5	2.5	4	5

Solution The plot of these points is shown in Figure 21.10. The points give the appearance of a parabola, so we will see which quadratic equation best fits these data. Calculating the sums needed for the normal equations, we get

$$\sum x = 36 \qquad \sum x^2 = 204 \qquad \sum x^3 = 1{,}296 \qquad \sum x^4 = 8{,}772$$
$$\sum y = 34.5 \qquad \sum xy = 123 \qquad \sum x^2 y = 742$$

With this information, we can solve the following system of three linear equations in three variables.

$$34.5 = 9a_0 + 36a_1 + 204a_2$$

$$123 = 36a_0 + 204a_1 + 1{,}296a_2$$

$$742 = 204a_0 + 1{,}296a_1 + 8{,}772a_2$$

Solving these, we get $a_0 = 7.258$, $a_1 = -2.328$, and $a_2 = 0.260$, and we obtain the least squares quadratic :

$$y = 7.258 - 2.328x + 0.260x^2$$

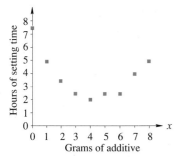

FIGURE 21.10

If you think you know which curve the data will fit, you can take a slightly different approach. Suppose the data seems to fit a function of the type $f(x)$. Then you can solve for the least squares line

$$y = mf(x) + b$$

Here, you first calculate $f(x)$ and then treat the problem as the least square line was treated in Section 21.4.

EXAMPLE 21.24

The resonant frequency of an electric circuit containing a 4 μF capacitor was measured as a function of an inductance in the circuit, and the following data were found.

f (Hz)	490	370	260	200	175	150
L (H)	1.0	2.0	4.0	6.0	8.0	10.0

Find the least square curve of $f = m\left(\dfrac{1}{\sqrt{L}}\right) + b$.

Solution   We are given $f(x) = \dfrac{1}{\sqrt{L}}$. We will solve the system of normal equations from Section 21.4:

$$\sum y = nb + m \sum f(x)$$

$$\sum f(x)y = b \sum f(x) + m \sum [f(x)]^2$$

Here we use $f(x)$ instead of x. Since $f(x) = \dfrac{1}{\sqrt{L}}$, we let y represent the frequency and get the following sums for our system of equations.

$$\sum y = 1{,}645 \qquad\qquad \sum f(x) = 3.285$$

$$\sum f(x)y = 1{,}072.585 \qquad\qquad \sum [f(x)]^2 = 2.142$$

Thus, we have the system of equations

$$1{,}645 = 6b + 3.285m$$

$$1{,}072.585 = 3.285b + 2.142m$$

Solving these, we find that

$$m = 500.630 \qquad \text{and} \qquad b = 0.0720$$

and so the most accurate curve is

$$f = \frac{500.630}{\sqrt{L}} + 0.0720$$

Exercise Set 21.5

1. *Chemistry*   The following data gives the amount of chlorine residual in a swimming pool at different times after it has been treated with chemicals.

Number of hours (x)	12	24	36	48	60
Chlorine residual (y) [parts per million (ppm)]	1.80	1.47	1.25	1.10	1.01

(a) Plot the data on semilog graph paper.
(b) Use the method of least squares to fit a curve of the form $y = ab^x$.

2. *Biology*   The following data represents the growth of a colony of bacteria in a culture medium:

Days since inoculation (x)	3	6	9	12
Bacteria count (×1,000) (y)	115	147	240	354

Days since inoculation (x)	15	18
Bacteria count (×1,000) (y)	580	866

(a) Plot the data on semilog graph paper.
(b) Use the method of least squares to fit a curve of the form $y = ab^x$.

3. *Business*   A company compiled the following data on the numbers of pipes in a shipment and the number of defective pipes in the shipment.

Pipes shipped (× 100) (x)	5	10	4	10	7
Defective pipe (y)	30	51	26	52	40

(x)	8	8	5	10	5	12	6
(y)	43	45	31	52	30	59	36

(a) Plot these points on regular graph paper.
(b) Find the best fit quadratic equation for these points.
(c) Graph the curve you found in (b) on the same graph you used for (a).

4. *Machine technology*   The following data pertains to the amount of hydrogen present in core drillings made at 100-mm intervals along the length of a vacuum-cast ingot.

Core location (mm from base) (x)	1	2	3
Hydrogen present (ppm) (y)	1.28	1.47	1.00

(x)	4	5	6	7	8	9	10
(y)	0.85	0.72	0.63	0.89	0.91	1.05	1.08

(a) Plot these points.
(b) Find a parabola by the method of least squares.
(c) Graph the parabola on your graph from (a).

5. *Physics*   The following data were gathered on the atmospheric pressure, in atmospheres, at various altitudes, in meters.

altitude (×1,000 m), x	1	2	3	4
pressure (atm), y	0.879	0.782	0.687	0.601

altitude (×1,000 m), x	5	6	7	8
pressure (atm), y	0.540	0.475	0.420	0.370

(a) Plot these points on semilog paper.
(b) Find the least square curve of $y = ae^{-kx}$.
(c) Plot the curve from (b) on your graph from (a).

6. *Electrical technology*   The following data for the resistance of a copper wire and a cross-sectional diameter were gathered.

diameter (mm), x	0.1	0.2	0.3	0.4	0.5
resistance (Ω), y	90	22	10	5.6	3.8

diameter (mm), x	0.6	0.7	0.8	0.9	1.0
resistance (Ω), y	2.5	1.8	1.4	1.1	0.8

(a) Plot these points on log-log paper.
(b) Fit the least square curve of $yx^2 = k$.
(c) Plot the curve from (b) on your graph from (a).

7. *Physics*   The following are data for the pressure and volume of a gas at a constant temperature.

pressure	1	2	3	4	5	6	7
volume	5.3	2.7	1.8	1.4	0.9	0.8	0.7

pressure	8	9	10
volume	0.6	0.5	0.5

(a) Plot these points on log-log paper.
(b) Find the least square curve of $pv = k$.
(c) Plot the curve from (b) on your graph for (a).

8. *Energy technology*   An experimenter is testing a new windmill design. She gathers the data in the following table showing the power P generated by the windmill for various wind velocities v.

v	10	20	30	40	50
P	18	78	390	2,050	9,900

(a) Sketch the data on semilog graph paper.
(b) Sketch the data on log-log graph paper.
(c) Find the least square curve that relates P as a function of v.
(d) Graph your curve from (c) on the graph from (a).

▤ CHAPTER 21 REVIEW

Important Terms and Concepts

Conditional probability

Correlation

Dependent events

Event

Frequency distribution

Histogram

Independent events

Mean

Median

Method of least squares

Mode

Normal curve

Normal equations

Outcome

Probability

Sample space

Standard deviation

Trial

Variance

Review Exercises

In Exercises 1–8, find the probability of the event.

1. Getting a "tail" when a coin is tossed.

2. Drawing a joker from a deck of 54 cards.

3. Drawing a green ball, blindfolded, from a bag containing 6 red and 10 green balls.

4. Not drawing a queen from a deck of 52 cards.

5. Rolling a 5 or higher with a single die.

6. Rolling a sum of 4 or less with a pair of die.

7. Rolling a 5 and then a 2 with a single die.

8. Drawing two aces from a well-shuffled deck of cards, if the first card is not replaced.

Solve Exercises 9–22.

9. *Quality control*   A quality control engineer inspects a random sample of three disc drives from each lot of 50 that is ready to be shipped. If such a lot contains four drives with slight defects, find the probability that the inspector's sample will include only drives with no defects.

10. *Quality control*   An inspector at a computer disc-drive manufacturing company has collected the following information on the revolutions per minute of the disc drives: 300.1, 300.3, 298.9, 299.6, 300.1, 300.4, 300.5, 297.8, 299.8, 300.7, and 301.2. Determine the median of these numbers.

11. *Quality control*   Determine the quartiles of the numbers in Exercise 10.

12. *Quality control*   Draw a box plot of the data in Exercise 10.

13. *Quality control*   Determine the mode of the numbers in Exercise 10.

14. *Quality control*   Determine the mean of the data in Exercise 10.

15. *Quality control*   Determine the standard deviation of the data in Exercise 10.

16. *Quality control*   What percent of the data is in the interval $\bar{x} - s.d., \bar{x} + s.d.$ for the data in Exercise 10?

17. *Quality control*   What percent of the data in Exercise 10 is in the interval $\bar{x} - 2s.d., \bar{x} + 2s.d$?

18. *Machine technology*   Some forged alloy bars were twisted until they broke. The number of twists for the bars tested were 33, 24, 37, 48, 26, 52, 37, 45, 23, 43, 39, 42, 32, 50, and 38. Find **(a)** the mean, **(b)** the median, **(c)** the quartiles, and **(d)** the standard deviation for the data.

19. *Business*   A microcomputer leasing company conducted periodic maintenance visits to the places where they rented their microcomputers. They collected the following data on the number of microcomputers at each location and the total time,

in minutes, needed to complete the maintenance.

No. of microcomputers	4	6	2	5
Time for maintenance (min)	205	282	93	237

No. of microcomputers	7	7	6	3
Time for maintenance (min)	324	336	277	153

No. of microcomputers	8	5	3	1
Time for maintenance (min)	368	242	140	75

(a) Graph these data.
(b) Find the equation of the least square line for the given data.
(c) Graph the least square line.
(d) What is the correlation coefficient?

20. *Physics*   An experiment was conducted to determine the specific heat ratio α for a certain gas. In the experiment, the gas was compressed adiabatically to several predetermined volumes V and the corresponding pressure P was measured. The results of the experiment are shown in the following table.

P (lb/in.^{2})	16.3	22.4	33.5	59.3	157.2	414.3
V (in.^{3})	50	40	30	20	10	5

(a) Plot these points on log-log graph paper.
(b) Find the least square curve of $pV^\alpha = K$.
(c) Plot the curve from (b) on your graph for (a).

21. *Business*   A company finds that when the price of its product is p dollars per unit, the number of units sold, n, is indicated in the following table.

price, p	10	30	40	50	60	70
number sold, n	70	68	63	50	43	30

(a) Plot these data.
(b) Find the equation of the least square line and the correlation coefficient for the given data.
(c) Graph the line from (b) on the graph for (a)

22. For the points in the following table, find the least square curve $y = m\sqrt{x} + b$.

x	0	2	4	6	8	10	12	14
y	6	8.3	9.1	9.5	10.0	11.0	10.6	11.3

x	16	18	20
y	11.5	11.9	12.3

≡ CHAPTER 21 TEST

1. What is the probability of drawing a red queen from a deck of 52 cards?

2. A quality control engineer inspects a random sample of 5 carburetors from each lot of 100 that is ready to be shipped. If such a lot of 100 carburetors contains 2 with a defect, find the probability that the engineer's sample will include only carburetors with no defects.

Use the following information to solve Exercises 3–10.

An inspector at a disc-drive manufacturing company has collected the following information on the revolutions per minute of 20 hard disc drives: 3,600.1, 3,600.4, 3,599.9, 3,599.2, 3,598.9, 3,601.2, 3,600.8, 3,598.7, 3,599.6, 3,600.4, 3,598.2, 3,600.1, 3,599.6, 3,600.4, 3,599.2, 3,601.6, 3,598.6, 3,600.4, 3,600.2, and 3,601.4.

3. Determine the median of these numbers.
4. Determine the quartiles for these numbers.
5. Draw a box plot for these data.
6. Determine the mode of these numbers.
7. Determine the mean of these data.

8. Determine the standard deviation.

9. What percent of the data is in the interval $\bar{x} - s.d., \bar{x} + s.d.$?

10. What percent of the data is in the interval $\bar{x} - 2s.d., \bar{x} + 2s.d.$?

11. An automobile rental agency wanted to see if there was a relationship between the minimum cost to rent a car and the number of cars rented. The following data were collected on the minimum cost and the number

of cars rented each week.

Minimum cost ($)	19.95	20.95	18.95	19.95
Number of cars rented	265	232	293	237

Minimum cost ($)	19.95	21.95	22.95	18.95
Number of cars rented	224	186	167	346

Minimum cost ($)	19.95	20.95	22.95	19.95
Number of cars rented	318	242	140	275

(a) Graph these data.
(b) Find the equation of the least square line for the given data.
(c) Graph the least square line.
(d) What is the correlation coefficient?

The Electronic Hand-Held Calculator

Until recently you would have had to work all the problems in this book in your head, with pencil and paper, or with a slide rule. There were some calculators available but they were large and bulky—almost the size of a typewriter—and they were very limited in the operations that you could perform with them.

The invention of the integrated circuit made it possible to process and store large amounts of information in a very small space and with very little energy. Later advances in integrated circuits allowed for even more information to be stored on an integrated circuit "chip." The advances in the chip plus the light emitting diode (LED) provided the basis for small, light, inexpensive electronic calculators. More recent calculators use a liquid crystal display (LCD) instead of the LEDs.

≡ A.1
INTRODUCTION

Electronic calculators have quietly and quickly become common tools. People use them at work, at home, and in school. Because we expect that you will use a calculator in your work as a technician, we include this section showing how calculators can and should be used. You are encouraged and expected to use a calculator and/or computer when working problems in this book.

Types of Calculators

What kind of calculator should you get? For the work in this book, you need a "scientific" calculator. This type of calculator naturally has keys for the ten digits, $\boxed{0}$ through $\boxed{9}$, and for the basic operations ($\boxed{+}$, $\boxed{-}$, $\boxed{\times}$, and $\boxed{\div}$). But, the calculator should also have the keys $\boxed{\sin}$, $\boxed{\cos}$, $\boxed{\tan}$, $\boxed{y^x}$ or $\boxed{x^y}$, $\boxed{\log}$, and $\boxed{\ln x}$. Make sure you have the instruction manual for your calculator. Because not all calculators work the same, the methods described in this book may not work on your calculator. If you have this problem, the instruction manual for your calculator and your teacher should help you learn how to work these problems.

All calculators work in one of two basic ways. Some use **algebraic notation** (**AN**) and the others use **reverse Polish notation** (**RPN**). In the text, most calculator operations are shown using algebraic notation. If it is likely that confusion might result for users of RPN calculators, then an example showing how to use an RPN calculator is provided. It does not matter whether your calculator is an AN or RPN type. Just make sure you read your instruction manual carefully and practice using your calculator.

There are other variations among calculators. Some have one or more addressable memories. Others have the ability to draw graphs.

Another major difference is the order of operations performed by the calculator. The order of operations is explained in Section 1.2. A calculator that performs operations according to the order of operations is said to have an **algebraic operating system** (**AOS**). If you need to specify that certain operations are performed before others, then you will need to use parentheses.

Consider the problem $9 + 5 \times 6$. Using the order of operations in Section 1.2, you should get an answer of 39. Work the problem on your calculator. Do you get 39 or 84? Some calculators will give one answer; some will give the other. If your calculator gives the answer 84, then you must be very careful to perform the operations in the correct order and either write the intermediate results or, if your calculator has a memory, store the intermediate results in memory. Some calculators have parenthesis keys that can be used to work problems like the one just described.

If you have any doubts as to which sequence of operations your calculator will follow when you use combined operations, press the $\boxed{=}$ key after each individual operation and record, or store in the calculator's memory, the intermediate results. Practice and experience will allow you to become familiar enough with your calculator to use it correctly and to take advantage of its built-in capabilities.

Most calculators display 8 to 10 significant digits and store 2 or 3 more digits for rounding off purposes. However, this is not always the case. Some calculators employ a method called **truncation** in which any digits not displayed are discarded. Thus, 489.781 truncated to tenths is 489.7. This same number rounded off to tenths is 489.8. While you will seldom use numbers with enough significant digits for calculator truncation to be a problem, you should be aware of this possibility.

A **graphics calculator** has a larger screen and can draw and display graphs. One advantage of a graphics calculator is that it displays the numbers and operations and allows you to edit or change them.

One other difference between calculators is how they handle the decimal point. Some calculators are designed to show exactly two decimal places with every computation. These calculators have a **fixed decimal point**. Other calculators only show as many decimal places as result from the computation. These calculators have a **floating decimal point**. For example, a fixed decimal point calculator would show 34.50 when a floating decimal point calculator would show 34.5. Similarly, a floating decimal point calculator would show 47.368 when a fixed decimal point calculator would show 47.37. Most scientific calculators have a $\boxed{\text{Fix}}$ key that allows you to set the number of digits to the right of the decimal point that will be displayed.

☰ A.2
BASIC OPERATIONS WITH A CALCULATOR

In Section A.1, we examined the differences between calculators. In this section, we will learn how to use the calculator to solve problems such as those in the first few chapters of the text.

Entering Positive and Negative Numbers

Data are entered by pressing the digits in order from left to right. Thus, the number 48.732 would be entered into the calculator by pressing the keys in this order: $\boxed{4}$ $\boxed{8}$ $\boxed{\cdot}$ $\boxed{7}$ $\boxed{3}$ $\boxed{2}$. If your calculator has a floating decimal point the display should read 48.732; if it has a fixed decimal point, it would read 48.73.

Your calculator probably has a $\boxed{+/-}$, $\boxed{\text{CHS}}$, or $\boxed{(-)}$ key. This key allows you to change the sign of the number in the display. Enter 59.638 in your calculator and press the $\boxed{+/-}$ or $\boxed{\text{CHS}}$ key. The display should now read −59.638. Press the $\boxed{+/-}$ or $\boxed{\text{CHS}}$ key again and you should see the calculator return to the original display of 59.638. Notice that the $\boxed{+/-}$ or $\boxed{\text{CHS}}$ key is pressed *after* the number is entered in the calculator.

Some calculators, such as graphics calculators, use a $\boxed{(-)}$ key to enter a negative number. The use of the parentheses is to help you tell it apart from the $\boxed{-}$ key which is used for subtraction. Here the $\boxed{(-)}$ key is pressed *before* the number is entered. Thus, you would press $\boxed{(-)}$ 59.638 in order to enter the number −59.638.

☰ Note

The $\boxed{+/-}$ or $\boxed{\text{CHS}}$ key should be pressed after the number has been entered. The $\boxed{(-)}$ key should be pressed before the number is entered.

Dual Function Key

Some calculators have a **dual function** key or $\boxed{\text{2nd}}$ key. In order to save space, calculator makers decided to let some keys perform more than one function. The first function is printed on the key; the second is printed just above the key; and a third function may be printed above or below the key.

To use the first function of a key, all you need to do is press the key. To use the second function, you first press the $\boxed{\text{2nd}}$ key and then the key below the desired function. For example, on some calculators there is a key labeled $\boxed{\text{CLR}}$ with CA above it much like this: $\dfrac{\text{CA}}{\boxed{\text{CLR}}}$. In order to get the CA function, press the $\boxed{\text{2nd}}$ key and then press the $\boxed{\text{CLR}}$ key. We will write this as $\boxed{\text{2nd}}$ $\boxed{\text{CA}}$ to prevent any possible confusion about which function to use. Some calculators indicate the $\boxed{\text{2nd}}$ key by $\boxed{\text{F}}$ or $\boxed{\text{2nd F}}$ or a blank key $\boxed{}$ printed in a solid color such as gold or blue.

Graphics calculators have both a second function key and an alpha function key. For example, on the Texas Instruments *TI-81*, the $\boxed{x^2}$ and the $\boxed{\text{LN}}$ keys each have two symbols above them as shown here:

$$\dfrac{\sqrt{}\ \text{I}}{\boxed{x^2}} \qquad \dfrac{e^x\text{S}}{\boxed{\text{LN}}}$$

To the left printed in light blue above a key is an operation. For example, to the left above the $\boxed{x^2}$ key is the operation $\sqrt{}$ and to the left above the $\boxed{\text{LN}}$ key is the operation e^x. To use an operation, press the $\boxed{\text{2nd}}$ key and then the key you require.

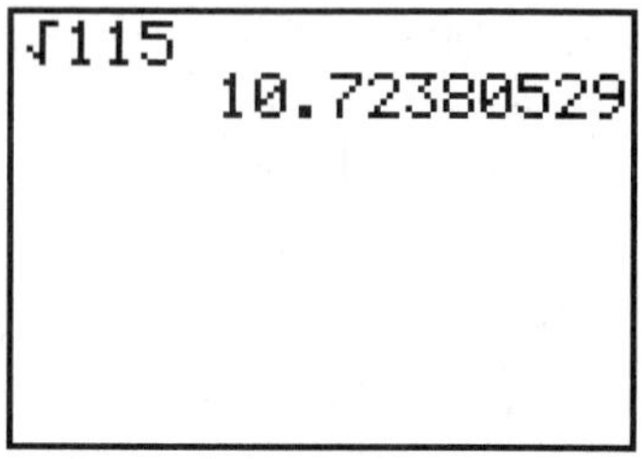

```
√115
        10.72380529
```

FIGURE A.1

For example, to calculate $\sqrt{115}$, press 2nd √ 115 ENTER . The screen displays $\sqrt{115}$ on one line and the result, 10.72380529 on the next line, as shown in Figure A.1. Notice that the 2nd key is colored a light blue to help you recognize that to use an operation printed in light blue you need to first press the 2nd key.

To the right above a key is a letter or other symbol printed in grey. For example, to the right above the x^2 key is the letter I and to the right above the LN key is the letter S. To use these keys, you first press the ALPHA key. So, to enter the letter I you press the ALPHA x^2 key. In this book we will indicate that you need to use the ALPHA key by placing the letter inside a square. So, if we want you to use the variable I, we will write ALPHA I . Remember, to use anything printed in grey that is above a key, you must first press the ALPHA key.

Basic Arithmetic Operations

The basic arithmetic operations are performed using the operation keys +, −, ×, and ÷, and the =, ENTER, or EXE key. The problem $4 \times 8 + 5 \times 6$ would be worked by keying into the calculator as follows:

Algebraic Calculator   RPN Calculator

4 × 8 + 5 × 6 =   4 Enter 8 × 5 Enter 6 × +

In each case, the final display is 62.

To find the sum of 14.32 and 9.37 you press

Algebraic Calculator   RPN Calculator

14.32 + 9.37 =   14.32 ENTER 9.37 +

and the result 23.69 is displayed.

Subtraction, multiplication, and division are handled in much the same manner.

≡ **Note**   From now on, we will indicate what you should enter and what is displayed in the following manner.

PRESS	DISPLAY
14.32 + 9.37 =	23.69

There will be times when a third column will be added, so we can include some comments or explanations.

EXAMPLE A.1

Use an algebraic or graphing calculator to evaluate **(a)** $72.1 - 83.12$, **(b)** 4.3×6.1, and **(c)** $7.8 \div 2.4$

Solution

	PRESS	DISPLAY
(a)	72.1 − 83.12 =	−11.09
(b)	4.3 × 6.1 =	26.23
(c)	7.8 ÷ 2.4 =	3.25

EXAMPLE A.2

Use an RPN calculator to work the problem $43.7 + 56.2$.

Solution

	PRESS	DISPLAY
	43.7 ENTER	43.7
	56.2 +	99.9

EXAMPLE A.3

Evaluate **(a)** $9.3 + 84.12 - 20.5 - 123.1$ and **(b)** $7.4 \times 2.5 \div 37 \times 8$.

Solution

	PRESS	DISPLAY
(a)	9.3 + 84.12 − 20.5 − 123.1 =	-50.18
(b)	7.4 × 2.5 ÷ 37 × 8 =	4.

These same two problems would be worked on an RPN calculator as follows:

	PRESS	DISPLAY
(a)	9.3 ENTER 84.12 + 20.5 − 123.1 −	−50.18
(b)	7.4 ENTER 2.5 × 37 ÷ 8 ×	4.0

Parentheses

Even if your calculator uses the algebraic operating system (AOS) described earlier, there are times when you will need to specify the exact order in which expressions are to be evaluated. There are several different symbols used for grouping—parentheses ( ), brackets [ ], and braces { }—but calculators and computers primarily use parentheses. The parentheses tell the calculator to perform the operations inside the parentheses first. If you are not sure how your calculator will handle an expression, you should use parentheses.

A calculator does not know about implied multiplication. People often write $(7 + 3)(8 - 5)$ when they mean $(7 + 3) \times (8 - 5)$. You must tell the calculator when quantities are being multiplied, as shown in Example A.4.

EXAMPLE A.4

Multiply $(7 + 3) \times (8 - 5)$.

Solution

	PRESS	DISPLAY
	(7 + 3) × (8 − 5) =	30.

A slightly more involved use of parentheses is given in Example A.5. Here you want the calculator to evaluate the entire numerator and then divide by the entire denominator. To make sure that this is what the calculator does, extra sets of parentheses must be added.

EXAMPLE A.5

Evaluate $\dfrac{(8+3) \times (7-21)}{(6+15 \div 3) \times 2}$ with your calculator.

Solution Add another set of parentheses around the numerator and the denominator. The problem now looks like $\dfrac{((8+3) \times (7-21))}{((6+15 \div 3) \times 2)}$. Now use your calculator to evaluate this expression.

PRESS	DISPLAY	COMMENT
((8 + 3) × (7 − 21)))	−154	(This is the value of the numerator.)
÷	−154	(The bar between the numerator and denominator means to divide.)
((6 + 15 ÷ 3) × 2)	22	(This is the value of the denominator)
=	−7	(The final result)

RPN Calculator

Add parentheses to aid in the correct order of operations. It is easiest to work from the inside of parentheses outward, much as you would if you were doing the calculations on paper. Since division is performed before addition, a set of parentheses should be added in the denominator.

$$\frac{(8+3) \times (7-21)}{(6+(15 \div 3) \times 2)}$$

You would then use an RPN calculator in the following manner:

PRESS	DISPLAY	COMMENT
8 ENTER 3 + 7 ENTER 21 − ×	−154	(The numerator)
ENTER	−154	(Stores the numerator)
15 ENTER 3 ÷ 6 + 2 ×	22	(The denominator)
÷	−7	(The result)

Note Remember that parentheses come in pairs. Whenever you use a left parenthesis, (, you must also use a right parenthesis,), in that same problem.

Example A.6 provides another chance for you to use parentheses with your calculator.

EXAMPLE A.6

Evaluate $\dfrac{9 \times 3 + 7 \times (-8) + (-16)}{((3 + 6 \times 5) + 4.5 - 7.5) \div 1.5}$.

Solution   Notice that there are no parentheses in the numerator. If your calculator does not use the AOS, then you need to place parentheses so that you get the right answer for the numerator. Parentheses should be placed as such:

$$(9 \times 3) + (7 \times -8) + -16$$

If your calculator has AOS, then this step is not necessary. You need to rewrite the problem by placing a set of parentheses around the numerator and a set around the denominator. The problem now looks like

$$\frac{(9 \times 3 + 7 \times -8 + (-16))}{(((3 + 6 \times 5) + 4.5 - 7.5) \div 1.5)}$$

Use your calculator to evaluate this expression.

PRESS	DISPLAY	COMMENT
(9 × 3 + 7 × 8 +/− +		
16 +/−)	−45.	(This is the value of the numerator.)
÷	−45.	(The bar separating the numerator and denominator means divide.)
(((3 + 6 × 5) +		
4.5 − 7.5) ÷ 1.5)	20.	(This is the value of the denominator.)
=	−2.25	(The final result)

Reciprocals

Many calculators have a $\boxed{1/x}$ or $\boxed{x^{-1}}$ key that is used to find the reciprocal of a number on display. Remember, your calculator will give the decimal value of the reciprocal. In Example A.7, we use the calculator to evaluate the reciprocals of three numbers.

EXAMPLE A.7

What are the reciprocals of **(a)** 3, **(b)** −19, and **(c)** $\frac{4}{5}$?

Solution	PRESS	DISPLAY
(a)	3 1/x	0.3333333
(b)	19 +/− 1/x	−0.0526316
(c)	Algebraic calculator: 4 ÷ 5 = 1/x	1.25
	RPN calculator: 4 ENTER 5 ÷ 1/x	1.25
	Graphics calculator: (4 ÷ 5) x^{-1} ENTER	1.25

Notice that in **(c)**, if you knew that the decimal equivalent of $\frac{4}{5}$ was 0.8, then you could have entered 0.8 $\boxed{1/x}$ or 0.8 $\boxed{x^{-1}}$ and the display would have read 1.25. But be careful when you try to use a short cut to solve a problem. Remember that the reciprocal of a reciprocal is the original number. We know that the reciprocal of 3 is $\frac{1}{3}$, so the reciprocal of $\frac{1}{3}$ is 3. Now work Example A.8.

EXAMPLE A.8	Use a calculator to **(a)** find the reciprocal of the reciprocal of 3, and **(b)** find the reciprocal of 0.3333333.

Solution

	PRESS	DISPLAY
(a)	3 $\boxed{1/x}$	0.3333333
	$\boxed{1/x}$	3.
(b)	0.3333333 $\boxed{1/x}$	3.0000003

As you can see, the reciprocal of a reciprocal is supposed to return the original number. But, if you enter the decimal equivalent of a number that does not terminate before the eighth place to the right of the decimal point you may not get the exact reciprocal that you want. Sometimes it is worth the few extra steps it takes to enter 1 $\boxed{\div}$ 3 instead of 0.3333333.

Exercise Set A.2

Use a calculator to evaluate Exercises 1–24.

1. $28 + 45$
2. $48 + 27$
3. $93 - 47$
4. $86 - 39$
5. $43 - 192$
6. $57 - 216$
7. 21×15
8. 47×12
9. $18 \div 5$
10. $38 \div 4$
11. $15.32 + -6.71$
12. $18.71 + -7.38$
13. $-14.67 - 8.932$

14. $-21.93 - 6.091$
15. 12.41×-3.4
16. 23.36×-517
17. $-8.35 \div 2.5$
18. $53.72 \div -6.32$
19. $16.37 + 81.4 - 6.437 + -8.5 - -16.372$
20. $-12.4 - 81.37 + -6.9 - 108.4 + -6.43$
21. $8.4 \times 3.7 \div 6.2 \div -4.2 \times 124 \div -0.74$
22. $-165 \div 1.1 \times 2.6 \times -8.5 \div 13 \div -0.015$
23. $\dfrac{(8+7) \times 3 \div ((6.2-5) \times 5))}{18 \div 6 - 3 \times 4.5 - 1.25 \times -8}$
24. $\dfrac{(17.3 + 6.82) \times 4.1 \div (-27.3 \div 3 - 8.9)}{(81.73 - 16.38 \times 220 \div 11 + 40.37)}$
$\div (97.6 - 15.4)$

Use a calculator to determine the reciprocal of the numbers in Exercises 25–30.

25. 4

26. 0.5

27. 4.5

28. 0.1111111

29. 215.3

30. 0.00025

≡ A.3
SOME SPECIAL CALCULATOR KEYS

In Section A.2, we looked at one special key—the $1/x$ or reciprocal key and at the grouping keys $(\;)$. In this section you will learn to use eight more special keys.

Keys for Powers or Roots

The first four special keys are used to find decimal values of powers and roots. Two of these should look familiar. These are the x^2 and $\sqrt{\;}$ or $\sqrt{x}$ keys. The x^2 will calculate the square (or second power) of any number you enter as shown in Example A.9.

EXAMPLE A.9

Evaluate **(a)** 3^2, **(b)** 5^2, **(c)** 4.2^2, **(d)** 135^2, and **(e)** $(-7)^2$.

Solution

	PRESS	DISPLAY
(a)	3 x^2	9.
(b)	5 x^2	25.
(c)	4.2 x^2	17.64
(d)	135 x^2	18225.
(e)	Algebraic: 7 $+/-$ x^2	49.
	RPN: 7 CHS x^2	49.
	Graphic: $($ $(-)$ 7 $)$ x^2 $ENTER$	49.

The $\sqrt{\;}$ key is used to determine the square root of a number. On many calculators you first enter the number and then press the $\sqrt{\;}$ key. (This may require pressing the $2nd$ key and then the $\sqrt{\;}$ key.) On a graphics calculator the $\sqrt{\;}$ key is pressed before the numbers.

EXAMPLE A.10

Evaluate **(a)** $\sqrt{25}$, **(b)** $\sqrt{2}$, and **(c)** $\sqrt{797}$.

Solution

	PRESS	DISPLAY
(a)	Algebraic or RPN: 25 $\sqrt{\;}$	5.
	Graphics: $\sqrt{\;}$ 25	5.
(b)	Algebraic or RPN: 2 $\sqrt{\;}$	1.4142136
	Graphics: $\sqrt{\;}$ 2	1.4142136
(c)	Algebraic or RPN: 797 $\sqrt{\;}$	28.231188
	Graphics: $\sqrt{\;}$ 797	28.231188

In Example A.10(a), the displayed number is the exact principal square root of the number. The numbers displayed for $\sqrt{2}$ and $\sqrt{797}$ are approximate values. Both of these are irrational numbers. If you evaluate $(1.4142136)^2$ and $(28.231188)^2$, you may not get 2 and 797. For example, on one calculator $(1.4142136)^2 = 2.0000001$ and $(28.231188)^2 = 796.99998$.

Remember, you can only take square roots of positive real numbers. What happens when you enter 2 $\boxed{+/-}$ $\boxed{\sqrt{}}$ into your calculator? Some calculators display 1.4142136 in the flashing mode. The flashing indicates that you have made an error. Other calculators display an "E" or the word "Error." A number, such as 0, may be displayed with the "E".

If you want powers larger than two, use the $\boxed{y^x}$, $\boxed{x^y}$, or $\boxed{\wedge}$ key. This discussion will be written for a calculator with a $\boxed{y^x}$ key. If your calculator has an $\boxed{x^y}$ or $\boxed{\wedge}$ key, then you should make any necessary changes.

Suppose you want to find 2^3. We know that $2^3 = 2 \times 2 \times 2 = 8$. To get this same answer with your algebraic calculator, enter 2 $\boxed{y^x}$ 3 $\boxed{=}$. On an RPN calculator, press 2 $\boxed{\text{ENTER}}$ 3 $\boxed{y^x}$ and on some graphing calculators press 2 $\boxed{\wedge}$ 3 $\boxed{\text{ENTER}}$.

To evaluate $4^{5/6}$, you need to use parentheses. You should enter 4 $\boxed{y^x}$ $\boxed{(}$ 5 $\boxed{\div}$ 6 $\boxed{)}$ $\boxed{=}$. The result is approximately 3.1748. Now work the problems in Example A.11 to see how to use your calculator.

EXAMPLE A.11			

Evaluate: **(a)** 4^3, **(b)** $4.3^{2.7}$, **(c)** $(\sqrt{3})^{\sqrt{2}}$, **(d)** 2^{-3}, **(e)** $\sqrt[4]{2}$, and **(f)** $9^{5/3}$.

Solution

	PRESS	DISPLAY
(a)	4 $\boxed{y^x}$ 3 $\boxed{=}$	64.
(b)	4.3 $\boxed{y^x}$ 2.7 $\boxed{=}$	51.32924
(c)	3 $\boxed{\sqrt{}}$ $\boxed{y^x}$ 2.7 $\boxed{\sqrt{}}$ $\boxed{=}$	2.466015575
	TI-81: $\boxed{\sqrt{}}$ 3 $\boxed{\wedge}$ $\boxed{\sqrt{}}$ 2.7 $\boxed{\text{ENTER}}$	2.466015575
(d)	2 $\boxed{y^x}$ 3 $\boxed{+/-}$ $\boxed{=}$	0.125
	Graphics: 2 $\boxed{y^x}$ $\boxed{(-)}$ 3 $\boxed{\text{ENTER}}$	0.125
(e)	2 $\boxed{y^x}$ 0.25 $\boxed{=}$	1.1892071
(f)	2 $\boxed{x^y}$ $\boxed{(}$ 5 $\boxed{\div}$ $\boxed{)}$ $\boxed{=}$	38.9407384

Memory Keys

Memory keys allow you to store numbers that you plan to use later. These numbers are stored in special places in a calculator. The number of memories available in your calculator may range from 1–26.

To STOre the number on display, you press the $\boxed{\text{STO}}$ key. Some calculators use $\boxed{\text{M}}$ for the Memory key. In calculators with more than one memory, you must tell the calculator in which memory you want to store the number. Calculator memories are numbered. A calculator with ten memories numbers the locations 0–9. If you want to place a displayed number in memory location six, press $\boxed{\text{STO}}$ $\boxed{6}$. To erase the numbers from all memories press $\boxed{\text{2nd}}$ $\boxed{\text{CA}}$.

Calculators that have alphabetical keys have a memory for each letter of the alphabet. For example, to place a number displayed on the *TI-81* in memory location B, you press $\boxed{\text{STO}\triangleright}$ and then the $\boxed{\text{B}}$ key. Do not press the $\boxed{\text{ALPHA}}$ key. On a Casio *fx7700-G*, the $\boxed{\rightarrow}$ key acts like the $\boxed{\text{STO}\triangleright}$ key, except that you need to press the $\boxed{\text{ALPHA}}$ key. For example, to store a number in memory location B, you press $\boxed{\rightarrow}$ $\boxed{\text{ALPHA}}$ $\boxed{\text{B}}$.

Once you have stored a number in memory, you need to be able to get it back so you can use it. To do this, use the ReCalL key, $\boxed{\text{RCL}}$. On some calculators, this is the Memory Recall, $\boxed{\text{MR}}$, key. If your calculator has more than one memory, then you must indicate which memory location contains the number. To recall the number in memory location five, press $\boxed{\text{RCL}}$ $\boxed{5}$. To recall the number in memory location B, press $\boxed{\text{ALPHA}}$ $\boxed{\text{B}}$.

≡ Note

Recalling a number from memory does not clear the memory. The number is still stored in that memory, so you can use it again.

Instructions for more special calculator keys are included at various places in the text.

Exercise Set A.3

Use a calculator to evaluate Exercises 1–44.

1. 7^2

2. 11^2

3. -25^2

4. 142^2

5. 3.1^2

6. 7.95^2

7. -18.43^2

8. -26.4^2

9. $/\sqrt{36}$

10. $\sqrt{169}$

11. $\sqrt{1296}$

12. $\sqrt{45796}$

13. $\sqrt{10.24}$

14. $\sqrt{34.81}$

15. $\sqrt{151.5361}$

16. $\sqrt{2631.69}$

17. 9^3

18. 7^4

19. 21^5

20. 53^4

21. 4.7^3

22. 0.24^8

23. 14.1^{22}

24. 23.4^{36}

25. $\sqrt[4]{6561}$

26. $\sqrt[5]{32768}$

27. $\sqrt[7]{62748517}$

28. $\sqrt[6]{11390625}$

29. $\sqrt[5]{340.48254}$

30. $\sqrt[5]{656.75808}$

31. $\sqrt[23]{241}$

32. $\sqrt[62]{11562}$

33. 8^{-3}

34. 9^{-7}

35. $1.1^{4.2}$

36. $2.3^{5.7}$

37. 7^3

38. 12^5

39. 25^{-10}

40. 47^{-8}

41. $\sqrt[3.1]{23}$

42. $\sqrt[2.45]{62}$

43. $\sqrt[3]{\sqrt[4]{9}}$

44. $\sqrt[3]{\sqrt[9]{56}}$

Enter each of the numbers in Exercises 45–52 into your calculator in scientific notation.

45. 6.21×10^{14}

46. 3.781×10^{23}

47. 8.92×10^{-21}

48. 7.631×10^{-46}

49. $8,437,100,000,000$

50. $62,790,000,000$

51. 0.00000000000271

52. 0.0000000000472

Evaluate Exercises 53–60 with the help of a calculator.

53. $4.8 \times 10^{12} \times 5.23 \times 10^{25}$

54. $7.912 \times 10^{35} \times 6.39 \times 10^{24}$

55. $1.73 \times 10^{18} \times 8.42 \times 10^{-21}$

56. $2.71 \times 10^{25} \times 4.32 \times 10^{-16}$

57. $4.31 \times 10^{14} \div 2.31 \times 10^{12}$

58. $8.92 \times 10^{19} \div 4.62 \times 10^{23}$

59. $5.42 \times 10^{21} \div 2.47 \times 10^{-5}$

60. $4.72 \times 10^{31} \div 4.93 \times 10^{-34}$

Use the $\boxed{\text{STO}}$ **and** $\boxed{\text{RCL}}$ **keys on a calculator to solve Exercises 61–68.**

61. $\dfrac{45.37 - 82.91}{6.4 \times 3.7 - 8.9 \times 4}$

62. $\dfrac{(9.31 + 8.72)5.4}{7.3 \times 4.6 - 4 \times 9.1}$

63. $\dfrac{3.7 - 91.4}{(18.32 - 6.7)9.2}$

64. $\dfrac{4.7 \times 3.1 - 5.4}{3.4 \div 4.3 - 6.2}$

65. $A = (4.3 + 8.1) \div 7.2, B = 5.7A - 34.9, C = \dfrac{B}{A} - A,$ Find C

66. $A = 12.4 \div (3.7 - 8.9), B = \sqrt{-A}, C = 4B + A,$ Find C

67. $A = 4.3 \times 8.7 - 9.1, B = A^2 + 7, C = (\sqrt{B})A,$ Find C

68. $A = (2.7 + 8.9)^2, B = \sqrt[3]{4A} - \sqrt{A}, C = \dfrac{B^5 + A^3}{AB},$ Find C

≡ A.4
GRAPHING CALCULATORS AND COMPUTER-AIDED GRAPHING

The introduction of microcomputers has allowed people to use inexpensive computers for writing and performing calculations quickly and accurately; but people discovered that words and numbers were not enough, and so computer graphics were introduced.

Some of these graphing capabilities are now available on calculators. Use of a graphing calculator or a computer removes some of the drudgery of plotting equations by hand. In this text, we show how you can use computer and calculator graphing capabilities. However, we will also try to show you how to interpret the information they give you.

The graphics capabilities of computers and calculators vary widely. Graphics examples in this text were run on either a Casio *fx-7700G* or a Texas Instruments *TI-81* graphing calculator or on a Macintosh computer.

Section 4.5 is an optional section that presents the basics of using a graphing calculator. Additional examples that explain more advanced features of graphing calculators are spread throughout the text.

Review Exercises

Use a calculator to evaluate Exercises 1–20.

1. $43 + -85$

2. $16 - 34$

3. 25×-81

4. $-62 \div 16$

5. 8.32×4.15

6. $-9.41 + 18.301$

7. $4.32 - 1.011$

8. $15.42 \div 0.002$

9. $8.3 + 9.1 - 4.3 \times 4 + 3 \div 5$

10. $-162 \div 1.2 + 4.3 \div 0.4 - 7$

11. $\dfrac{(3+5) \times -2 \div ((6-4.3) \times 7)}{9 \div 3 - 4 \times 2.3 - 5 \times -7}$

12. $\dfrac{(16.3 - 4) \times 6.2 \div (3.4 \div 18.1 + 3)}{8.1 \div 9 + 10}$

13. 35^2

14. $\sqrt{5}$

15. $(-4)^5$

16. 14^6

17. $\sqrt[7]{813}$

18. $\sqrt[4]{216}$

19. $\sqrt[2.3]{42.1}$

20. $\sqrt[1.3]{25.4}$

In Exercises 21–24, enter the number into your calculator in scientific notation.

21. 4.31×10^{23}

22. 6.42×10^{-16}

23. $34,700,000,000,000,000,000,000,000,000,000$

24. $0.00000000000000000000427$

Evaluate Exercises 25–26 with the help of a calculator.

25. $2.7 \times 10^{18} \times 3.4 \times 10^{23}$

26. $3.8 \times 10^{-15} \times 4.5 \times 10^{13}$

B

The Metric System

The metric system* is becoming more important to workers. Almost every major country uses the metric system. Many industries are converting their manufacturing processes and products to metric specifications. In Canada, and every other major industrial country except the United States, the SI metric system is the official measurement system. Competition in the form of international trade has forced many U.S. companies to convert their products to the metric system. The U.S. Congress passed legislation requiring federal agencies to use the metric system in their business activities.

Older technical service publications were printed using the U.S. Customary system, while the latest publications are either in metric or in a combination of both measurement systems. This section gives a brief introduction to the metric system and the different units and their symbols, and tells how to convert within the metric system or between the metric and U.S. Customary system.

Units of Measure

There are seven base units in the metric system and two supplemental units in the SI metric system. All seven base units and the two supplemental units are shown in Table B.1.

All other units are formed from these seven. The base units that you will use most often are those for length (meter), mass or weight (kilogram), time (second), and electric current (ampere). The base unit for temperature is Kelvin, but we will mostly use a variation called the degree Celsius. In addition to the seven base units, there are two supplemental units. We use the supplemental unit for a plane angle (radian).

Each unit can be divided into smaller units or made into larger units. To show that a unit has been made smaller or larger, a prefix is placed in front of the base unit. The prefixes are based on powers of 10. The most common prefixes are shown in Table B.2. For example, 1 kilometer (km) is 1 000 m, 1 milligram (mg) is 0.001 g, 1 megavolt (MV) is 1 000 000 V=10^6 V, and 1 nanosecond (ns) is 0.000 000 001 s =10^{-9} s.

---

* The official name for the metric system is The International System of Units (or, in French, Le Système International d'Unités). The official abbreviation for the metric system throughout the world is "SI."

TABLE B.1   SI Metric System Base and Supplemental Units

BASE UNITS		
Quantity	**Unit**	**Symbol**
Length	meter	m
Mass	kilogram	kg
Time	second	s
Electric current	ampere	A
Temperature	Kelvin	K
Luminous intensity	candela	cd
Amount of substance	mole	mol
SUPPLEMENTAL UNITS		
Quantity	**Unit**	**Symbol**
Plane angle	radian	rad
Solid angle	steradian	sr

TABLE B.2   Metric Prefixes

Multiple		**Prefix**	**Symbol**
1 000 000 000	$= 10^9$	giga	G
1 000 000	$= 10^6$	mega	M
1 000	$= 10^3$	kilo	k
1			
0.01	$= 10^{-2}$	centi	c
0.001	$= 10^{-3}$	milli	m
0.000 001	$= 10^{-6}$	micro	μ
0.000 000 001	$= 10^{-9}$	nano	n

Many quantities, such as area, volume, speed, velocity, acceleration, and force require a combination of two or more fundamental units. Combinations of these units are referred to as **derived units** and are listed in Table B.3

Table B.4 shows some of the metric units most commonly used and their uses in various trade and technical areas. This list is not intended to be exhaustive, but to provide you with some idea as to where the different units are used.

Writing Metrics

You should remember some important rules when you use the metric system.

1. The unit symbols that are used are the same in all languages. This means that the symbols used on Japanese cars are the same as those used on German or American cars.
2. The unit symbols are not abbreviations and a period is not put at the end of the symbol.

TABLE B.3   Derived Units for Common Physical Quantities

Quantity	Derived Unit	Symbol	Alternate Symbol
Acceleration	meter per second squared	m/s^2	
Angular acceleration	radian per second squared	rad/s^2	
Angular velocity	radian per second	rad/s	
Area	square meter	m^2	
Concentration, mass (density)	kilogram per cubic meter	kg/m^3	
	gram per liter	g/L	
Capacitance	farad	F	C/V
Electric current	ampere	A	V/Ω
Electric field strength	volt per meter	V/m	
Electric resistance	ohm	Ω	V/A
Electromotive force (emf)	volt	V	J/C
Force	newton	N	$kg \cdot m/s^2$
Frequency	hertz	Hz	s^{-1}
Illuminance	lux	lx	lm/m^2
Inductance	henry	H	$V \cdot s/A$
Light exposure	lumen second	$lm \cdot s$	
Luminance	candela per square meter	cd/m^2	
Magnetic flux	weber	Wb	$V \cdot s$
Moment of force	newton meter	$n \cdot m$	
Moment of inertia, dynamic	kilogram meter squared	$kg \cdot m^2$	
Momentum	kilogram meter per second	$kg \cdot m/s$	
Power	watt	W	J/s
Pressure	pascal	Pa	N/m^2
Quantity of electricity	coulomb	C	
Speed, velocity	meter per second	m/s	
Volume	cubic meter	m^3	
Volume flow rate	cubic meter per second	m^3/s	
	liter per minute	L/min	
Work, energy, quantity of heat	joule	J	$N \cdot m$

3. Unit symbols are shown in lowercase letters except when the unit is named for a person. Examples of unit symbols that are written in capital letters are joule (J), newton (N), pascal (Pa), watt (W), ampere (A), and coulomb (C). The only exception is the use of L for liter. The symbol L is often used for liter to eliminate any confusion between the lowercase unit symbol (l) and the numeral (1). Some people prefer to use a script ℓ for liter.

4. The symbol is the same for both singular and plural (e.g., 1 m and 12 m).

5. Numbers with four or more digits are written in groups of three separated by a space instead of a comma. (The space is optional on four-digit numbers.) This is done because some countries use the comma as a decimal point.

| **EXAMPLE B.1** | Use 2 473 or 2473 instead of 2,473; 45 689 instead of 45,689; 47 398 254.263 72 instead of 47,398,254.26372. |

TABLE B.4   Uses of Metric Units in Technical Areas

Quantity	Unit	Symbol	Use
Length	micrometer	μm	paint thickness; surface texture or finish
	millimeter	mm	motor vehicle dimensions; wood; hardware, bolt, and screw dimensions; tool sizes; floor plans
	centimeter	cm	bearing size; length and width of fabric or window; length of weld, channel, pipe, I-beam, or rod
	meter	m	braking distance; turning circle; room size; wall covering; landscaping; architectural drawings; length of pipe or conduit; highway width
	kilometer	km	land distances; maps; odometers
Area	square centimeter	cm^2	piston head surfaces; brake and clutch contact area; glass, tile, or wall covering; area of steel plate
	square meter	m^2	fabric, land, roof, and floor area; room sizes; carpeting; window and/or wall covering
Volume or capacity	cubic centimeter	cm^3	cylinder bore; small engine displacement; tank or container capacity
	cubic meter	m^3	work or storage space; truck body; room or building volume; trucking or shipping space; tank or container capacity; ordering concrete; earth removal
	milliliter	mL	chemicals; lubricant; oils; small liquids; paint;
	liter	L	fuel; large engine displacement; gasoline
Temperature	degree Celsius	°C	thermostats; engine operating temperature; oil or liquid temperature; melting points; welds
Mass or weight	gram	g	tire weights; mailing and shipping packages
	kilogram	kg	batteries; weights; mailing and shipping packages
	metric ton	t	vehicle and load weight; construction material such as sand or cement; crop sales
Bending force, torque, moment of force	newton meter	N · m	torque specifications; fasteners
Pressure/vacuum	kilopascal	kPa	gas, hydraulic, oxygen, tire, air, or air hose pressure; manifold pressure compression; tensile strength
Velocity	kilometers per hour	km/h	vehicle speed; wind speed
	meters per second	m/s	speed of air or liquid through a system
Force, thrust, drag	newton	N	pedal; spring; belt; drawbar
Power	watt	W	air conditioner; heater; engine; alternator
Illumination	lumens per square meter	lm/m^2	intensity of light on a given area

TABLE B.4 (continued)

Quantity	Unit	Symbol	Use
Density	milligrams per cubic meter	mg/m^3	industrial hygiene standards for fumes, mists, and dusts
Flow	cubic meters per second	m^3/s	measure of air exchange in a region; exhaust and air exchange system ratings

6. A zero is placed to the left of the decimal point if the number is less than one (0.52 L, not .52 L).
7. Liter and meter are often spelled litre and metre.
8. The units of area and volume are written by using exponents.

EXAMPLE B.2

5 square centimeters are written as 5 cm^2, not as 5 sq cm;

37 cubic meters are written as 37 m^3, not as 37 cu m

Changing Within the Metric System

Changing units of measurement within a system is called **reduction**. A change from one system to another is called **conversion**. We will now discuss reduction in the metric system, and then conversion between the metric and U.S. customary system afterward.

The metric prefixes in Table B.2 provide a method to help with metric reduction. Using the information in Table B.2, we construct a metric reduction diagram like the one shown in Figure B.1. Starting on the left with the largest prefix shown in Table B.2, we mark each multiple of 10 until we get to the smallest prefix. Notice that not all multiples are labeled. While there is a prefix for each multiple of 10, we have given you only those that you will need.

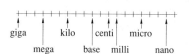

FIGURE B.1

Reduction in the SI Metric System

To change from one metric system unit to another:

1. Mark each unit on the reduction scale in Figure B.1.
2. Move the decimal point as many places as you move along the reduction scale according to the following rules:
   a. To change to a unit farther to the right, move the decimal point to the right.
   b. To change to a unit farther to the left, move the decimal point to the left.
3. If you are changing between square units, then move the decimal point *twice* as far as indicated in Step 2.
4. If you are changing between cubic units, then move the decimal point *three times* as far as indicated in Step 2.

EXAMPLE B.3

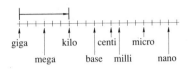

FIGURE B.2

Change 7.35 GV to kilovolts.

Solution   Since 7.35 GV represents 7.35 gigavolts, we put a mark at the giga point on the reduction scale in Figure B.2. We want to convert GV to kV, so an arrow is drawn from the first mark to the mark labeled "kilo." The arrow points to the right and is 6 units long, so we move the decimal point 6 units to the right. To do this, we need to insert some zeros, with the result

$$7.35 \text{ GV} = 7\ 350\ 000 \text{ kV}$$

EXAMPLE B.4

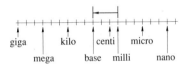

FIGURE B.3

Change 9.6 mm to meters.

Solution   Since 9.6 mm represents 9.6 millimeters, we put a mark at the milli point on the reduction scale in Figure B.3. We want to convert mm to m, the base unit, so an arrow is drawn from the first mark to the mark labeled "base." The arrow points to the left and is 3 units long, so we move the decimal point 3 units to the left. Again, we need to insert some zeros, with the result

$$9.6 \text{ mm} = 0.009\ 6 \text{ m}$$

EXAMPLE B.5

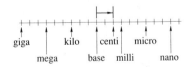

FIGURE B.4

Change 3.7 m^2 to square centimeters.

Solution   Since 3.7 m^2 represents 3.7 square meters, we put a mark at the base point on the reduction scale in Figure B.4. We want to convert m^2 to cm^2, so the arrow is drawn from the first mark to the mark labeled "centi." The arrow points to the right and is 2 units long. Since we are converting between square units, we double this number, so we move the decimal point 4 units to the right. Again, we need to insert some zeros, with the result

$$3.7 \text{ m}^2 = 3\ 700 \text{ cm}^2$$

To change one set of units into another set of units, you should perform algebraic operations with units to form new units for the derived quantity.

EXAMPLE B.6

Convert a speed of 88.00 km/h to meters per second.

EXAMPLE B.6 (Cont.)

Solution We will use two relationships that will give four possible conversion factors.

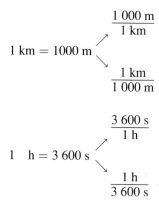

$$1 \text{ km} = 1000 \text{ m} \begin{cases} \dfrac{1\ 000 \text{ m}}{1 \text{ km}} \\[2mm] \dfrac{1 \text{ km}}{1\ 000 \text{ m}} \end{cases}$$

$$1 \quad \text{h} = 3\ 600 \text{ s} \begin{cases} \dfrac{3\ 600 \text{ s}}{1 \text{ h}} \\[2mm] \dfrac{1 \text{ h}}{3\ 600 \text{ s}} \end{cases}$$

We write the quantity to be changed, then choose the appropriate conversion factors so that all but two of the units "cancel," leaving the units m/s.

$$88\, \frac{\text{km}}{\text{h}} \times \frac{1\ 000 \text{ m}}{1\ \text{km}} \times \frac{1\ \text{h}}{3\ 600 \text{ s}} \approx 24.44 \text{ m/s}$$

Changing Between the Metric and Customary Systems ▬

Many technicians are often called upon to use both the SI metric and the U.S. Customary measurement systems. While it is best not to change from either the metric or the customary system to the other, sometimes a worker has to do so. For example, it is possible that the measuring tools available may be different than the measuring system in which the specifications are given. In such a case, the technician will be required to convert from one system to another. We will show one way to change between the U.S. Customary system and the SI metric system. Table B.5 should help you do this.

EXAMPLE B.7

Express 4.3 kg in pounds.

Solution You are changing from the metric (kg) to the customary system.

Look at Table B.5 under the heading, "From Metric to Customary." You are changing a measurement from kg to lb. Look in the column labeled "To change from" until you find the symbol for kilogram (kg). The symbol in the next right-hand column is the symbol for pound (lb).

Now look in the "Multiply by" column just to the right of these two symbols. The number there is 2.205. Multiply the given number of kilograms (4.3) by this number (2.205).

$$4.3 \times 2.205 = 9.4815$$

This means that 4.3 kg is equivalent to 9.4815 lb (or 4.3 kg ≈ 9.4815 lb).

TABLE B.5   Changing Units Between the Metric and Customary Systems

Quantity	FROM METRIC TO CUSTOMARY			FROM CUSTOMARY TO METRIC		
	To change from	To	Multiply by	To change from	To	Multiply by
Length	μm	mil	0.039 37	mil	μm	25.4
	mm	in.	0.039 37	in.	mm	25.4
	cm	in.	0.393 7	in.	cm	2.54
	m	ft	3.280 8	ft	m	0.304 8
	km	mile	0.621 37	mile	km	1.609 3
Area	cm^2	$in.^2$	0.155	$in.^2$	cm^2	6.451 6
	m^2	ft^2	10.763 9	ft^2	m^2	0.092 9
Volume	cm^3	$in.^3$	0.061	$in.^3$	cm^3	16.387
	m^3	yd^3	1.308	yd^3	m^3	0.764 6
	m^3	gal	264.172	gal	m^3	0.003 785
	mL	fl oz	0.033 8	fl oz	mL	29.574
	L	fl oz	33.814	fl oz	L	0.029 6
	L	pt	2.113	pt	L	0.473 2
	L	qt	1.056 7	qt	L	0.946 4
	L	gal	0.264 2	gal	L	3.785 4
Mass or weight	g	oz	0.035 3	oz	g	28.349 5
	kg	lb	2.205	lb	kg	0.453 6
	t	lb	2205	ton	kg	907.2
Bending moment, torque, moment of force	N · m	lbf · in.	8.850 7	lbf · in.	N · m	0.113
	N · m	lbf · ft	0.737 6	lbf · ft	N · m	1.355 8
Pressure, vacuum	kPa	psi	0.145	psi	kPa	6.894 8
Velocity	km/h	mph	0.621 4	mph	km/h	1.609 3
Force, thrust, drag	N	lbf	0.224 8	lbf	N	4.448 2
Power	W	W	1	W	W	1
Temperature	°C	°F	$\frac{9}{5}$ (°C) + 32	°F	°C	$\frac{5}{9}$ (°F − 32)

EXAMPLE B.8

Express 6.5 fluid ounces in liters.

Solution   You are changing from the customary (fluid ounces) to the metric (liters) systems.

Look in Table B.5 under the heading "From Customary to Metric." You are changing a measurement from fluid ounces to liters. Look under the column labeled "To change from" until you find the symbol for fluid ounces—fl oz. The fl oz symbol appears twice. Now look in the next right-hand column. Do you see the symbol for liters (L)? It is opposite the second fl oz symbol.

EXAMPLE B.8 (Cont.)

If you look in the "Multiply by" column to the right of these two symbols (fl oz and L), you find 0.029 6. Multiply the number of fluid ounces (6.5) by this number (0.029 6).

$$6.5 \times 0.029\ 6 = 0.192\ 4$$

This means that 6.5 fluid ounces is equivalent to 0.192 4 L (or 6.5 fl oz ≈ 0.192 4 L).

Exercise Set

In Exercises 1–16, reduce the given unit to the indicated unit.

1. 347 g to kilograms
2. 0.26 km to meters
3. 7.92 kW to watts
4. 2.3 μs to seconds
5. 0.023 85 Ω to milliohms
6. 0.000 235 47 MW to milliwatts

7. 9.72 mm to centimeters
8. 0.35 mm to nanometers
9. 835 000 cm^2 to square meters
10. 2.34 km^2 to m^2
11. 4.35 mm^3 to cubic centimeters
12. 91.52 m^3 to cubic millimeters

Solve Exercises 13–16.

13. *Machine technology*   A die is 14 mm long. Express this in centimeters.

14. *Law enforcement*   A male suspect is 1.97 m tall. Express this in centimeters.

15. *Environmental science*   A wastewater treatment plant has 92 600 kg of sludge. Express this in metric tons.

16. *Medical technology*   A technician needs 1 125 mL of a sterile solution. Express this in liters.

In Exercises 17–24, change the given units to the indicated unit. (You may want to consult Table B.3.)

17. 45 m/s to km/h
18. 27 mm/s to km/h
19. *Physics*   If a force of 126 N gives a body an acceleration of 9 m/s^2, then the body has a mass of $\dfrac{126\ \text{N}}{9\ \text{m/s}^2}$. Convert this to kilograms.
20. *Electricity*   An ac electric current that is generated by a source of 120 V and with an impedance of 125 Ω has an effective current of $\dfrac{120\ \text{V}}{125\ \Omega}$. Convert this to amperes.
21. *Electricity*   The resistance in a heater is $\dfrac{84\ \text{V}}{8\ \text{A}}$. Convert this to ohms.
22. *Physics*   A technician determines that the absolute pressure in a tank is 112 cm of mercury. The following computation will convert this pressure to kilopascals: (13 600 kg/m^3)(9.8 m/s^2)(112 cm). Convert this to kilopascals.

23. *Automotive technology*   If a 3 250-g piston is held 300 mm above a work surface, then its potential energy relative to the work surface is

$$(3\ 250\ \text{g})(9.8\ \text{m/s}^2)(300\ \text{mm})$$

Express this potential energy in joules.

24. *Construction*   At the instant a 4-kg sledgehammer reaches a velocity of 26 m/s, it has a kinetic energy of $\frac{1}{2}$(4 kg)(26 m/s)^2. Express this kinetic energy in joules.

In Exercises 25–40, convert the given measurement to the indicated unit.

25. 4 m = _____ ft

26. 16 L = _____ pt

27. 24 kg = _____ lb

28. 27.5 N · m = _____ lbf · ft

29. 105 km/h = _____ mph

30. 17 in. = _____ cm

31. 4.5 qt = _____ L

32. 21 ft^2 = _____ m^2

33. 5.8 lb = _____ kg

34. 32 psi = _____ kPa

35. 97 W = _____ W

36. 98.6°F = _____ °C

37. *Fire safety*   A fire extinguisher contains $2\frac{1}{2}$ gal. Convert this to liters.

38. *Law enforcement*   An adult female weighs 120 lb. Convert this to kilograms.

39. *Environmental science*   A flow-measuring meter at a wastewater treatment plant recorded 9,660,000 gal in one day. Convert this to liters.

40. *Environmental science*   A wastewater treatment plant has a daily flow of 59 500 m^3. Convert this to gallons.

Table of Integrals

Basic Forms

1. $\displaystyle\int u\,dv = uv - \int v\,du$

2. $\displaystyle\int u^n\,du = \frac{1}{n+1}\,u^{n+1} + C,\ n \neq -1$

3. $\displaystyle\int \frac{du}{u} = \ln|u| + C$

4. $\displaystyle\int e^u\,du = e^u + C$

5. $\displaystyle\int a^u\,du = \frac{1}{\ln a}\,a^u + C$

6. $\displaystyle\int \sin u\,du = -\cos u + C$

7. $\displaystyle\int \cos u\,du = \sin u + C$

8. $\displaystyle\int \sec^2 u\,du = \tan u + C$

9. $\displaystyle\int \csc^2 u\,du = -\cot u + C$

10. $\displaystyle\int \sec u \tan u\,du = \sec u + C$

11. $\displaystyle\int \csc u \cot u\,du = -\csc u + C$

12. $\displaystyle\int \tan u\,du = \ln|\sec u| + C$

$\qquad\qquad = -\ln|\cos u| + C$

13. $\displaystyle\int \cot u\,du = \ln|\sin u| + C$

14. $\displaystyle\int \sec u\,du = \ln|\sec u + \tan u| + C$

15. $\displaystyle\int \csc u\,du = \ln|\csc u - \cot u| + C$

16. $\displaystyle\int \frac{du}{\sqrt{a^2 - u^2}} = \sin^{-1}\frac{u}{a} + C$

17. $\displaystyle\int \frac{du}{a^2 + u^2} = \frac{1}{a}\tan^{-1}\frac{u}{a} + C$

18. $\displaystyle\int \frac{du}{u\sqrt{u^2 - a^2}} = \frac{1}{a}\sec^{-1}\frac{u}{a} + C$

19. $\displaystyle\int \frac{du}{a^2 - u^2} = \frac{1}{2a}\ln\left|\frac{u+a}{u-a}\right| + C$

20. $\displaystyle\int \frac{du}{u^2 - a^2} = \frac{1}{2a}\ln\left|\frac{u-a}{u+a}\right| + C$

Forms Involving $\sqrt{a^2 - u^2}$

21. $\displaystyle\int \sqrt{a^2 - u^2}\,du = \frac{u}{2}\sqrt{a^2 - u^2} + \frac{a^2}{2}\sin^{-1}\frac{u}{a} + C$

22. $\displaystyle\int u^2\sqrt{a^2 - u^2}\,du = \frac{u}{8}\left(2u^2 - a^2\right)\sqrt{a^2 - u^2} +$

$\qquad \dfrac{a^4}{8}\sin^{-1}\dfrac{u}{a} + C$

23. $\displaystyle\int \frac{\sqrt{a^2 - u^2}}{u}\,du = \sqrt{a^2 - u^2} - a\ln\left|\frac{a + \sqrt{a^2 - u^2}}{u}\right| + C$

24. $\displaystyle\int \frac{\sqrt{a^2 - u^2}}{u^2}\,du = -\frac{1}{u}\sqrt{a^2 - u^2} - \sin^{-1}\frac{u}{a} + C$

25. $\displaystyle\int \frac{u^2\,du}{\sqrt{a^2 - u^2}} = -\frac{u}{2}\sqrt{a^2 - u^2} + \frac{a^2}{2}\sin^{-1}\frac{u}{a} + C$

26. $\displaystyle\int \frac{du}{u\sqrt{a^2 - u^2}} = -\frac{1}{a}\ln\left|\frac{a + \sqrt{a^2 - u^2}}{u}\right| + C$

27. $\displaystyle\int \frac{du}{u^2\sqrt{a^2 - u^2}} = -\frac{1}{a^2 u}\sqrt{a^2 - u^2} + C$

28. $\int (a^2 - u^2)^{3/2}\, du = -\dfrac{u}{8}(2u^2 - 5a^2)\sqrt{a^2 - u^2} +$
$\dfrac{3a^4}{8}\sin^{-1}\dfrac{u}{a} + C$

29. $\int \dfrac{du}{(a^2 - u^2)^{3/2}} = \dfrac{u}{a^2\sqrt{a^2 - u^2}} + C$

Forms Involving $\sqrt{u^2 \pm a^2}$

30. $\int \sqrt{u^2 \pm a^2}\, du = \dfrac{u}{2}\sqrt{u^2 \pm a^2} \pm$
$\dfrac{a^2}{2}\ln\left|u + \sqrt{u^2 \pm a^2}\right| + C$

31. $\int u^2\sqrt{u^2 \pm a^2}\, du = \dfrac{u}{8}(2u^2 \pm a^2)\sqrt{u^2 \pm a^2} -$
$\dfrac{a^4}{8}\ln\left|u + \sqrt{u^2 \pm a^2}\right| + C$

32. $\int \dfrac{\sqrt{u^2 + a^2}}{u}\, du = \sqrt{u^2 + a^2} - a\ln\left|\dfrac{a + \sqrt{u^2 + a^2}}{u}\right| + C$

33. $\int \dfrac{\sqrt{u^2 - a^2}}{u}\, du = \sqrt{u^2 - a^2} - a\cos^{-1}\dfrac{a}{u} + C$

34. $\int \dfrac{\sqrt{u^2 \pm a^2}}{u^2}\, du = -\dfrac{\sqrt{u^2 \pm a^2}}{u} + \ln\left|u + \sqrt{u^2 \pm a^2}\right| + C$

35. $\int \dfrac{du}{\sqrt{u^2 \pm a^2}} = \ln\left|u + \sqrt{u^2 \pm a^2}\right| + C$

36. $\int \dfrac{du}{u\sqrt{u^2 + a^2}} = -\dfrac{1}{a}\ln\left|\dfrac{a + \sqrt{u^2 + a^2}}{u}\right| + C$

37. $\int \dfrac{du}{u\sqrt{u^2 - a^2}} = \dfrac{1}{a}\cos^{-1}\left|\dfrac{a}{u}\right| + C$
$= \dfrac{1}{a}\sec^{-1}\left|\dfrac{u}{a}\right|$

38. $\int \dfrac{u^2\, du}{\sqrt{u^2 \pm a^2}} = \dfrac{u}{2}\sqrt{u^2 \pm a^2} \mp$
$\dfrac{a^2}{2}\ln\left|u + \sqrt{u^2 \pm a^2}\right| + C$

39. $\int \dfrac{du}{u^2\sqrt{u^2 \pm a^2}} = \mp\dfrac{\sqrt{u^2 \pm a^2}}{a^2 u} + C$

40. $\int (u^2 \pm a^2)^{3/2}\, du = \dfrac{u}{4}(u^2 \pm a^2)^{3/2} \pm$
$\dfrac{3a^2 u}{8}\sqrt{u^2 \pm a^2} + \dfrac{3a^4}{8}\ln\left|u + \sqrt{u^2 \pm a^2}\right| + C$

41. $\int \dfrac{(u^2 + a^2)^{3/2}}{u}\, du = \dfrac{1}{3}(u^2 + a^2)^{3/2} + a^2\sqrt{u^2 + a^2} -$
$a^3\ln\left|\dfrac{a + \sqrt{u^2 + a^2}}{u}\right| + C$

42. $\int \dfrac{(u^2 - a^2)^{3/2}}{u}\, du = \dfrac{1}{3}(u^2 - a^2)^{3/2} - a^2\sqrt{u^2 - a^2} -$
$a^3\sec^{-1}\dfrac{u}{a} + C$

43. $\int \dfrac{du}{(u^2 \pm a^2)^{3/2}} = \dfrac{\pm u}{a^2\sqrt{u^2 \pm a^2}} + C$

44. $\int \dfrac{du}{u(u^2 + a^2)^{3/2}} = \dfrac{1}{a^2\sqrt{u^2 + a^2}} -$
$\dfrac{1}{a^3}\ln\left|\dfrac{a + \sqrt{u^2 + a^2}}{u}\right| + C$

45. $\int \dfrac{du}{u(u^2 - a^2)^{3/2}} = \dfrac{-1}{a^2\sqrt{u^2 - a^2}} - \dfrac{1}{|a^3|}\sec^{-1}\dfrac{u}{a} + C$

Forms Involving $a + bu$

46. $\int \dfrac{u\, du}{a + bu} = \dfrac{1}{b^2}(a + bu - a\ln|a + bu|) + C$

47. $\int \dfrac{u^2\, du}{a + bu} =$
$\dfrac{1}{2b^3}\left[(a + bu)^2 - 4a(a + bu) + 2a^2\ln|a + bu|\right] + C$

48. $\int \dfrac{du}{u(a + bu)} = \dfrac{1}{a}\ln\left|\dfrac{u}{a + bu}\right| + C$

49. $\int \dfrac{u\, du}{(a + bu)^2} = \dfrac{1}{b^2}\left(\dfrac{a}{a + bu} + \ln|a + bu|\right) + C$

50. $\int \dfrac{du}{u(a + bu)^2} = \dfrac{1}{a(a + bu)} - \dfrac{1}{a^2}\ln\left|\dfrac{a + bu}{u}\right| + C$

51. $\int u\sqrt{a + bu}\, du = \dfrac{2}{15b^2}(3bu - 2a)(a + bu)^{3/2} + C$

52. $\int \dfrac{u\, du}{\sqrt{a + bu}} = \dfrac{1}{3b^2}(bu - 2a)\sqrt{a + bu} + C$

53. $\int \dfrac{u^2\, du}{\sqrt{a + bu}} = \dfrac{2}{15b^3}(8a^2 + 3b^2 u^2 - 4abu)\sqrt{a + bu} + C$

54. $\int \dfrac{du}{u\sqrt{a + bu}} = \dfrac{1}{\sqrt{a}}\ln\left|\dfrac{\sqrt{a + bu} - \sqrt{a}}{\sqrt{a + bu} + \sqrt{a}}\right| + C,\ \text{if}\, a > 0$
$= \dfrac{2}{\sqrt{-a}}\tan^{-1}\sqrt{\dfrac{a + bu}{-a}} + C,\ \text{if}\, a < 0$

55. $\int \dfrac{\sqrt{a + bu}}{u}\, du = 2\sqrt{a + bu} + a\int \dfrac{du}{u\sqrt{a + bu}}$

Trigonometric Forms

56. $\int \sin^2 u \, du = \frac{1}{2} u - \frac{1}{4} \sin 2u + C$

$\qquad = \frac{1}{2} u - \frac{1}{2} \sin u \cos u + C$

57. $\int \cos^2 u \, du = \frac{1}{2} u + \frac{1}{4} \sin 2u + C$

$\qquad = \frac{1}{2} u + \frac{1}{2} \sin u \cos u + C$

58. $\int \tan^2 u \, du = \tan u - u + C$

59. $\int \cot^2 u \, du = - \cot u - u + C$

60. $\int \sin^3 u \, du = - \cos u + \frac{1}{3} \cos^3 u + C$

61. $\int \cos^3 u \, du = \sin u - \frac{1}{3} \sin^3 u + C$

62. $\int \tan^3 u \, du = \frac{1}{2} \tan^2 u + \ln |\cos u| + C$

63. $\int \cot^3 u \, du = -\frac{1}{2} \cot^2 u - \ln |\sin u| + C$

64. $\int \sin^n u \, du = -\frac{1}{n} \sin^{n-1} u \cos u + \frac{n-1}{n} \int \sin^{n-2} u \, du$

65. $\int \cos^n u \, du = \frac{1}{n} \cos^{n-1} u \sin u + \frac{n-1}{n} \int \cos^{n-2} u \, du$

66. $\int \tan^n u \, du = \frac{1}{n-1} \tan^{n-1} u - \int \tan^{n-2} u \, du$

67. $\int \cot^n u \, du = \frac{-1}{n-1} \cot^{n-1} u - \int \cot^{n-2} u \, du$

68. $\int \sec^n u \, du = \frac{1}{n-1} \tan u \sec^{n-2} u +$
$\frac{n-2}{n-1} \int \sec^{n-2} u \, du$

69. $\int \csc^n u \, du = \frac{-1}{n-1} \cot u \csc^{n-2} u +$
$\frac{n-2}{n-1} \int \csc^{n-2} u \, du$

70. $\int \sin au \sin bu \, du = \frac{\sin(a-b)u}{2(a-b)} - \frac{\sin(a+b)u}{2(a+b)} + C$

71. $\int \cos au \cos bu \, du = \frac{\sin(a-b)u}{2(a-b)} + \frac{\sin(a+b)u}{2(a+b)} + C$

72. $\int \sin au \cos bu \, du = - \frac{\cos(a-b)u}{2(a-b)} - \frac{\cos(a+b)u}{2(a+b)} + C$

73. $\int u \sin u \, du = \sin u - u \cos u + C$

74. $\int u \cos u \, du = \cos u + u \sin u + C$

75. $\int u^n \sin u \, du = -u^n \cos u + n \int u^{n-1} \cos u \, du$

76. $\int u^n \cos u \, du = u^n \sin u - n \int u^{n-1} \sin u \, du$

77. $\int \sin^n u \cos^m u \, du = - \frac{\sin^{n-1} u \cos^{m+1} u}{n+m}$

$\qquad + \frac{n-1}{n+m} \int \sin^{n-2} u \cos^m u \, du$

$\qquad = \frac{\sin^{n+1} u \cos^{m-1} u}{n+m}$

$\qquad + \frac{m-1}{n+m} \int \sin^n u \cos^{m-2} u \, du$

Inverse Trigonometric Forms

78. $\int \sin^{-1} u \, du = u \sin^{-1} u + \sqrt{1 - u^2} + C$

79. $\int \cos^{-1} u \, du = u \cos^{-1} u - \sqrt{1 - u^2} + C$

80. $\int \tan^{-1} u \, du = u \tan^{-1} u - \frac{1}{2} \ln(1 + u^2) + C$

81. $\int u^n \sin^{-1} u \, du = \frac{1}{n+1} \left[u^{n+1} \sin^{-1} u - \int \frac{u^{n+1} \, du}{\sqrt{1 - u^2}} \right],$
$n \neq -1$

82. $\int u^n \cos^{-1} u \, du = \frac{1}{n+1} \left[u^{n+1} \cos^{-1} u + \int \frac{u^{n+1} \, du}{\sqrt{1 - u^2}} \right],$
$n \neq -1$

83. $\int u^n \tan^{-1} u \, du = \frac{1}{n+1} \left[u^{n+1} \tan^{-1} u - \int \frac{u^{n+1} \, du}{1 + u^2} \right],$
$n \neq -1$

Exponential and Logarithmic Forms

84. $\displaystyle\int ue^{au}\,du = \frac{1}{a^2}(au-1)e^{au} + C$

85. $\displaystyle\int u^2 e^{au}\,du = \frac{1}{a^3}(a^2u^2 - 2au + 2)e^{au} + C$

86. $\displaystyle\int u^n e^{au}\,du = \frac{1}{a}u^n e^{au} - \frac{n}{a}\int u^{n-1}e^{au}\,du$

87. $\displaystyle\int e^{au}\sin bu\,du = \frac{e^{au}}{a^2+b^2}(a\sin bu - b\cos bu) + C$

88. $\displaystyle\int e^{au}\cos bu\,du = \frac{e^{au}}{a^2+b^2}(a\cos bu + b\sin bu) + C$

89. $\displaystyle\int \ln u\,du = u\ln u - u + C$

90. $\displaystyle\int u^n \ln u\,du = u^{n+1}\left(\frac{\ln u}{n+1} - \frac{1}{(n+1)^2}\right) + C,\, n \neq -1$

91. $\displaystyle\int \frac{1}{u\ln u}\,du = \ln|\ln u| + C$

Hyperbolic Forms

92. $\displaystyle\int \sinh u\,du = \cosh u + C$

93. $\displaystyle\int \cosh u\,du = \sinh u + C$

94. $\displaystyle\int \tanh u\,du = \ln\cosh u + C$

95. $\displaystyle\int \coth u\,du = \ln|\sinh u| + C$

96. $\displaystyle\int \operatorname{sech} u\,du = \tan^{-1}|\sinh u| + C$

97. $\displaystyle\int \operatorname{csch} u\,du = \ln\left|\tanh \tfrac{1}{2}u\right| + C$

98. $\displaystyle\int \operatorname{sech}^2 u\,du = \tanh u + C$

99. $\displaystyle\int \operatorname{csch}^2 u\,du = -\coth u + C$

100. $\displaystyle\int \operatorname{sech} u\tanh u\,du = -\operatorname{sech} u + C$

101. $\displaystyle\int \operatorname{csch} u\coth u\,du = -\operatorname{csch} u + C$

Answers to Odd-Numbered Exercises

ANSWERS FOR CHAPTER 1

Exercise Set 1.1

1. Natural number, whole number, integer, rational number, and real number **3.** Irrational number, real number
5. 15 **7.** $\frac{\sqrt{7}}{8}$ **9.**

11.

13. < **15.** <

17. > **19.** < **21.** $-\frac{1}{5}$ **23.** $\frac{3}{17}$ **25.** $-5, -|4|,$ $-\frac{2}{3}, \frac{-1}{3}, \frac{16}{3}, |-8|$ **27.** (a) $\frac{1}{32}$ inch; (b) $\frac{1}{8}$ inch; (c) 0.0625 inch **29.** 2.6 V

Exercise Set 1.2

1. Commutative law for addition **3.** Commutative law for multiplication **5.** Associative law for addition
7. Identity element for multiplication **9.** Additive inverse
11. Associative law for multiplication **13.** -91
15. $-\sqrt{2}$ **17.** 2 **19.** $\frac{2}{\sqrt{2}} = \sqrt{2}$ **21.** 14 **23.** 13
25. 1 **27.** 31 **29.** 1

Exercise Set 1.3

1. 50 **3.** 14 **5.** -9 **7.** -24 **9.** 12
11. 15 **13.** $-\frac{38}{4} = -\frac{19}{2}$ **15.** $\frac{1}{8}$ **17.** $\frac{8}{15}$ **19.** $\frac{3}{20}$
21. $\frac{5}{8}$ **23.** $-\frac{47}{30} = -1\frac{17}{30}$ **25.** $\frac{1}{32}$ **27.** $-\frac{8}{15}$
29. $\frac{3}{32}$ **31.** $-\frac{10}{3} = -3\frac{1}{3}$ **33.** $\frac{6}{5}$ **35.** $-\frac{3}{20}$
37. $\frac{49}{25} = 1\frac{24}{25}$ **39.** $\frac{7}{2}$ **41.** $78\frac{1}{2}$ inches $= 6'6\frac{1}{2}''$
43. 57 V **45.** (a) $7\frac{33}{40}$ mi; (b) $11\frac{1}{5}$ mi; (c) $3\frac{3}{10}$ mi
47. $84'9\frac{1}{2}''$ **49.** $\frac{7}{8}''$

Exercise Set 1.4

1. 125 **3.** $\frac{3}{2}$ **5.** 16 **7.** $\frac{1}{49}$ **9.** 3^6 **11.** 2^{12}
13. 2^2 **15.** 2^6 **17.** x^{20} **19.** $\frac{a^6}{b^3}$ **21.** $\frac{x^3}{4^3}$
23. $\frac{a^8 b^4}{c^{12}}$ **25.** $\frac{1}{x^7}$ **27.** $\frac{1}{5^3}$ **29.** $\frac{1}{7^5}$ **31.** $\frac{1}{a^3 y^4}$

33. $\frac{y^2}{a^4}$ **35.** $\frac{1}{pr^2}$ **37.** $\frac{5^2}{4y^3}$ **39.** $\frac{y^{15}}{2^3 b^6}$ **41.** b^{24}
43. 51.96 Ω

Exercise Set 1.5

1. Exact **3.** Approximate **5.** Approximate **7.** 3
9. 3 **11.** 1 **13.** 4 **15.** (a) 6.05, (b) 6.05
17. (a) 5.01, (b) 0.027 **19.** (a) 27,0̃00, (b) 27,0̃00
21. (a) 86, (b) 0.2 **23.** (a) 140.070, (b) 140.070
25. (a) 10, (b) 14, (c) 14.4 **27.** (a) 7, (b) 7.0, (c) 7.04
29. (a) 400, (b) 4̃00, (c) 403 **31.** (a) 300, (b) 310, (c) 305
33. (a) 10, (b) 14, (c) 14.4 **35.** (a) 7, (b) 7.0, (c) 7.04
37. (a) 400, (b) 4̃00, (c) 403 **39.** (a) 300, (b) 3̃00, (c) 305
41. (a) 90, (b) 89.9, (c) 89.899 **43.** (a) 240, (b) 237.3, (c) 237.302 **45.** (a) 440, (b) 438.0, (c) 437.998
47. (a) 80, (b) 78.7, (c) 78.671
49.

	Absolute error	Relative error	Percent error
Length	-0.28 mm	-0.0117	-1.17%
Width	0.35 mm	0.0438	4.38%
Thickness	-0.02 mm	-0.0067	-0.67%

51. 99.37 **53.** 1618 **55.** 1020 **57.** 17
59. 351.65 m **61.** 924.6 mm

Exercise Set 1.6

1. 4.2×10^4 **3.** 3.8×10^{-4} **5.** $9.807\,00 \times 10^9$
7. 9.70×10^{-5} **9.** 4.3×10^0 **11.** 4500
13. 40500000 **15.** 0.000063 **17.** 72
19. 1.5504×10^{16} **21.** 2.66×10^{-11} **23.** 2.94×10^5
25. 2.2528×10^{-1} **27.** 1.2×10^5 **29.** 2.5×10^{-3}
31. 2×10^4 **33.** 1.8×10^{-3}
35. $1.3791744 \times 10^{25} \approx 1.38 \times 10^{25}$ **37.** 2.5×10^{22}
39. 3.9×10^{14} **41.** 675 (Answers may vary.)
43. 1.11×10^6 mm **45.** 2.789808×10^{-19} kg **47.** One neutron; approximately 1.84×10^3 times heavier

Exercise Set 1.7
1. 5   **3.** 12   **5.** 2   **7.** −3   **9.** 2   **11.** $\frac{2}{3}$
13. 0.2   **15.** −0.1   **17.** 3   **19.** 8.32   **21.** 5
23. $\sqrt[4]{72}$ **25.** 5   **27.** $\frac{1}{2}$ **29.** 7   **31.** $\frac{1}{3}$ **33.** $5^{1/3}$
35. 2   **37.** 8   **39.** 4   **41.** $\frac{1}{4}$ **43.** 81   **45.** $\frac{9}{0.2\sqrt{2}}$
47. 0.25   **49.** 2.091 m   **51.** 1 449.1377 m/s

Review Exercises
1. (a) Integers, rational numbers, real numbers, (b) Rational
numbers, real numbers, (c) Irrational numbers, real
numbers   **3.** (a) $\frac{3}{2}$, (b) $\frac{-1}{8}$, (c) −5
5. (a) Commutative law for addition, (b) Distributive law,

(c) Additive identity, (d) Multiplicative inverse   **7.** −44
9. 98   **11.** $-\frac{1}{6}$ **13.** −3   **15.** $-\frac{2}{5}$ **17.** $\frac{8}{3}$
19. $\frac{10}{13}$ **21.** 32   **23.** −64   **25.** 2   **27.** 2^8 **29.** 2^2
31. 4^{15} **33.** 4   **35.** $\frac{a}{b}$ **37.** $\frac{1}{a^2x^4}$ **39.** (a) 2.37,
(b) 2.37   **41.** (a) both, (b) 0.7   **43.** (a) 7.4, (b) 7.35,
(c) 7.4, (d) 7.35   **45.** (a) 2.1, (b) 2.05, (c) 2.1, (d) 2.05
47. 3.71×10^{11} **49.** 2.4×10^{-11} **51.** 12   **53.** 5

Chapter 1 Test
1. (a) $\frac{4}{3} = 1\frac{1}{3}$; (b) $\frac{1}{8}$; (c) 6   **3.** $\frac{5}{8}$ **5.** −112
7. $\frac{22}{3} = 7\frac{1}{3}$ **9.** $-\frac{21}{2} = -10\frac{1}{2}$ **11.** $\frac{14}{29}$ **13.** 8
15. 25   **17.** $\frac{b^3}{a}$ **19.** 0.000 51   **21.** $\frac{7}{5}$

ANSWERS FOR CHAPTER 2

Exercise Set 2.1
1. $11x$ **3.** $2z$ **5.** $6x + 9x^2$ or $9x^2 + 6x$
7. $10w - 7w^2$ or $-7w^2 + 10w$ **9.** $2ax^2 + a^2x$
11. $11xy^2 - 5x^2y$ **13.** $12b$ **15.** $2a^2 + 5b + 4a$
17. $2x^2 + 6x$ **19.** $12y^2 + 14x$ **21.** $-12b + 6c$
23. $7a + 7b$ **25.** $x + y$ **27.** $5a + 5b - c$ **29.** $6x + 6y$
31. $5a + 5b$ **33.** $6a - 2b + 4c$ **35.** $9x + 7y$
37. $-6x + 2y - 13z$ **39.** $-4x + 3y + 25z$ **41.** $6x + 2y$
43. $3y + z$ **45.** $x + 6y$ **47.** $a - 3b$ **49.** $-70a - 8b$
51. $\frac{13}{6}p = 2\frac{1}{6}p$

Exercise Set 2.2
1. a^3x^3 **3.** $6a^2x^3$ **5.** $-6x^3w^3z$ **7.** $-24x^4ab$
9. $10y - 12$ **11.** $35 - 20w$ **13.** $21xy + 12x$
15. $15t - 5t^2$ **17.** $2a^2 - a$ **19.** $6x^3 - 2x^2 + 8x$
21. $-20y^4 + 8y^3 - 20y^2 + 12y - 24$ **23.** $a^2 + ab + ac + bc$
25. $x^3 + 5x^2 + 6x + 30$ **27.** $6x^2 + xy - y^2$
29. $6a^2 - 7ab + 2b^2$ **31.** $2b^2 + 3b - 5$
33. $56a^4b^2 + 3a^2bc - 9c^2$ **35.** $x^2 - 16$ **37.** $p^2 - 36$
39. $a^2x^2 - 4$ **41.** $4r^4 - 9x^2$ **43.** $25a^4x^6 - 16d^2$
45. $\frac{9}{16}t^2b^6 - \frac{4}{9}p^2a^4f^2$ **47.** $x^2 + 2xy + y^2$
49. $x^2 - 10x + 25$ **51.** $a^2 + 6a + 9$ **53.** $4a^2 + 4ab + b^2$
55. $9x^2 - 12xy + 4y^2$ **57.** $12x^3 + 40x^2 - 32x$
59. $x^2 - y^2 - z^2 + 2yz$ **61.** $2n^2 + 6n$

Exercise Set 2.3
1. x^4 **3.** $2x^2$ **5.** $3y^2$ **7.** $-3b$ **9.** $11y$
11. $-6ay$ **13.** $18c^2d^2$ **15.** $\frac{-3p}{5n^2}$ **17.** $\frac{4bcx}{7y}$
19. $2a^2 + a$ **21.** $4b^3 - 2b$ **23.** $6x^2 + 4x$ **25.** $2x^3 - 3$
27. $-6x^3 + 2x$ **29.** $5x + 5y$ **31.** $2x + 3y$ **33.** $ap - 2$
35. $a + 1$ **37.** $-3xy + z$ **39.** $-b^2x^2 - b^2$ **41.** $x + 1 - y$
43. $-\frac{3}{2}xy + 2y^2$ **45.** $x + 4$ **47.** $x - 1$ **49.** $x - 1$
51. $2a + 1$ **53.** $4y + 2$ **55.** $x - 2$ **57.** $2a - 1$
59. $2x^2 + x - 1 + \frac{3}{2x-1}$ **61.** $r^2 - 3r + \frac{5}{r+2}$
63. $x^3 + 3x^2 + 9x + 27$ **65.** $4x^3 - 3x^2 - x + 6$ **67.** $x + y$
69. $w^2 + wz + z^2$ **71.** $x^2 + y^2$ **73.** $c^2d^2 + 2cd + 4$
75. $x - y$ **77.** $p^2r - 2p + 3r^2$ **79.** $a + d + 4 + \frac{-4}{a-3d-1}$

81. af **83.** $a + b - c$ **85.** $a^2 - a + 1$
87. $\frac{1}{R_1} + \frac{1}{R_2} + \frac{1}{R_3}$

Exercise Set 2.4
1. 39   **3.** 12   **5.** −15   **7.** 4.4   **9.** $\frac{9}{2}$
11. −8   **13.** $\frac{3}{2}$ **15.** $-\frac{2}{3}$ **17.** 15   **19.** −24
21. 2   **23.** −4   **25.** $-\frac{17}{2}$ **27.** 4   **29.** 4   **31.** −1
33. 7   **35.** $\frac{1}{2}$ **37.** $\frac{24}{5}$ **39.** 5   **41.** −6   **43.** 9
45. −6   **47.** $-\frac{40}{3}$ **49.** 6   **51.** 6   **53.** $-\frac{11}{6}$
55. $\frac{13}{4}$ **57.** 14   **59.** −18   **61.** 13   **63.** $\frac{24}{7}$
65. 78   **67.** $\frac{b}{2a}$ **69.** $-\frac{x}{x-8}$ or $\frac{x}{8-x}$ **71.** $\frac{7}{3}$ **73.** 1
75. −3   **77.** −15   **79.** −2   **81.** $C = \frac{5}{9}(F - 32)$

Exercise Set 2.5
1. 82   **3.** 60, since no score can be below 60
5. $1150; $230   **7.** $6130   **9.** $2200 at 7.5% and
$2300 at 6%   **11.** 7 hours   **13.** 266 miles
15. 320 km   **17.** 2.4 hours or 2 hours 24 minutes
19. $\frac{4}{3}$ hours or 1 hour 20 minutes   **21.** 57.5 mL
23. 468.75 kg of 75% copper; 281.25 kg of 35% copper
25. 350 lb; 8.24 feet from the 500 lb end   **27.** 4.5 ft from the
left end   **29.** 5 in. from right end

Review Exercises
1. $3y$ **3.** $5x - 8$ **5.** $7x^2 + 11$ **7.** $16x + 8$
9. $1 - 3x$ **11.** $9x^2 + x - 2$ **13.** $3a + 9b$ **15.** a^3x^3
17. $27ax^3$ **19.** $20x - 24$ **21.** $8x^2 - 10x$ **23.** $a^2 - 16$
25. $6a^2 - 5ab + b^2$ **27.** $6x^4 - 5x^2 - 6$ **29.** $x^2 + 4x + 4$
31. $30x^3 - 25x^2 - 20x$ **33.** a^3 **35.** $4a$ **37.** $-9ax^2$
39. $18x - 8$ **41.** $x - 4$ **43.** $x^2 + 3x + 9$ **45.** $x - y$
47. $x - 2y$ **49.** 38   **51.** $\frac{15}{2}$ **53.** 36   **55.** 5
57. 4.5   **59.** $-\frac{3}{2}$ **61.** −7   **63.** −12   **65.** $-\frac{1}{5}$
67. 2   **69.** 76   **71.** 3.6 hours or 3 hours 36 minutes
73. 35 kg of lead; 63.64% lead   **75.** 1538.6 N on the left
cable, 1107.4 N on the right cable

Chapter 2 Test

1. $5x^2 - 3x$ **3.** $2x + 8y - 6$ **5.** $4b^2 - 9$

7. $3x^3 + 2x - \frac{1}{2x^2}$ **9.** $\frac{y^2 - 13y - 8}{(3y+2)(y-1)}$ **11.** $\frac{14}{5}$ **13.** 1.8 qt

▤ ANSWERS FOR CHAPTER 3

Exercise Set 3.1

1. $\frac{\pi}{12} \approx 0.2617994$ **3.** $\frac{7\pi}{6} \approx 3.6651914$

5. $\frac{427\pi}{900} \approx 1.4905112$ **7.** $\frac{109\pi}{120} \approx 2.8536133$ **9.** $240°$

11. $234°$ **13.** $14.324°$ **15.** (a) $145°$; (b) 2.5307274
$\approx \frac{29\pi}{36}$ **17.** $\angle A = 147°$; $\angle B = 33°$ **19.** $\angle A = \frac{3\pi}{8}$;
$\angle B = \frac{5\pi}{8}$ **21.** $\angle A = 109°30'$; $\angle B = 70°30'$

23. $\frac{24}{5} = 4.8$ **25.** $47.5°$ **27.** 120π rad/s

Exercise Set 3.2

1. $70°$ **3.** $x = 55°, y = 70°$ **5.** $x = y = 60°$

7. $36.87°$ **9.** $A = 23.4$ units^2; $P = 23.4$ units

11. $A = 126$ units^2; $P = 84$ units **13.** (a) 133.3 m;
(b) 478.5 m^2 **15.** $\sqrt{200} = 10\sqrt{2}$ m ≈ 14.1421 m

17. $\sqrt{5869} \approx 76.6$ ft **19.** (a) 5262 mm and 8123 mm;
(b) 9231 mm **21.** $\sqrt{119} \approx 10.9$ m **23.** (a) 212 m;
(b) 182.8 m

Exercise Set 3.3

1. $P = 60$ cm; $A = 225$ cm^2 **3.** $P = 96.5$ in.;
$A = 379.5$ in.^{2} **5.** $P = 69$ in.; $A = 343.6$ in.^{2}

7. $P = 44$ cm; $A = 168$ cm^2 **9.** $P = 73.8$ mm;
$A = 279.84$ mm^2 **11.** 444 ft^2 **13.** $10\,000$ mm^2

15. $P = 3.031$ in.; $A = 0.663$ in.^{2} **17.** (a) $A = 750$ ft^2;
(b) 94.76 ft **19.** 2.1875 in.^{2} or 2.19 in.^{2}

21. $18\,638$ mm^2

Exercise Set 3.4

1. $A = 16\pi$ cm^2; $C = 8\pi$ cm **3.** $A = 25\pi$ in.^{2};
$C = 10\pi$ in. **5.** $A = 201.64\pi$ mm^2; $C = 28.4\pi$ mm

7. $A = 146.41\pi$ mm^2; $C = 24.20\pi$ mm **9.** (a) 576π in.^{2};
(b) $48\pi \approx 151$ in. **11.** 32 **13.** (a) 163.2 mm;
(b) $6\,935.5$ mm^2 **15.** (a) 86.643 in.; (b) 481.52 in^2

17. (a) $A = 2529.876$ in.^{2}; (b) 197.1 in. **19.** 11.91 mm

Exercise Set 3.5

1. $L = 216\pi \approx 678.6$ in.^{2}, $T = 288\pi \approx 904.8$ in.^{2},
$V = 648\pi \approx 2,035.8$ in.^{3} **3.** $L = 136\pi \approx 427.3$ mm^2,
$T = 200\pi \approx 628.3$ mm^2, $V = 320\pi \approx 1,005.3$ mm^3

5. $L = 735$ in.^{2}, $T = 816.8$ in.^{2}, $V = 859.1$ in.^{3} **7.** A
sphere has no lateral surface area, $T = 256\pi \approx 804.25$ cm^2,
$V = 682.7\pi \approx 2\,144.7$ cm^3 **9.** 5866.7 yd^3

11. (a) 3600 ft^3; (b) 1560 ft^2

13. (a) $103\,455\pi \approx 325\,013.5$ mm^3; (b) $20\,172.79$ mm^2 of
paper **15.** $1333.3\pi \approx 4188.8$ m^3 **17.** 136.1 yd^3

19. $57\,172$ mm^3 **21.** $66\pi \approx 207.3$ in.^{2} **23.** 1208.23 m^3

25. $48\pi \approx 150.8$ ft^3

Review Exercises

1. $\frac{3}{20}\pi \approx 0.4712$ **3.** $198°$ **5.** $43°$

7. $\angle A = 152°$, $\angle B = 28°$ **9.** 3.3 **11.** $30°$ **13.** 33.36

15. $A = 294$ units^2; $P = 73$ units **17.** $A = 81\pi \approx 254.5$
units^2; $C = 18\pi \approx 56.55$ units **19.** $A = 924$ units^2;
$P = 154$ units **21.** $A = 186.75$ units^2; $P = 67.1$ units

23. $L = 232.32\pi \approx 729.85$ units^2, $T = 525.14\pi \approx 1649.78$
units^2, $V = 1405.54\pi \approx 4415.62$ units^3 **25.** $L = 2320$
units^2, $T = 3920$ units^2, $V = 11,200$ units^3

27. $L = 520\pi \approx 1633.6$ units^2, $T = 930\pi \approx 2921.68$ units^2,
$V = 2896\pi \approx 9098.05$ units^3 **29.** Spheres have no
lateral surface area, $T = 324\pi \approx 1017.88$ sq units,
$V = 972\pi \approx 3053.63$ units^3

Chapter 3 Test

1. $\frac{7\pi}{36}$ **3.** $104°$ **5.** 12.6 **7.** 101.4 cm

9. 280.8 cm^2 **11.** about 144.8 m **13.** 1945.944 cm^3

15. ≈ 201.59 cm^2

▤ ANSWERS FOR CHAPTER 4

Exercise Set 4.1

1. Function **3.** Not a function **5.** Function for $x \geq 0$

7. Not a function **9.** Yes, because for each value of x
there is just one value of y. **11.** No, because when $x = 4$
there are two values of y: -2 and 2, or when $x = 1$, y is 1 or
-1. **13.** Domain: $\{-3, -2, -1, 0, 1, 2\}$; Range:
$\{-7, -5, -3, -1, 1, 3\}$ **15.** Domain:
$\{-3, -2, -1, 0, 1, 2, 3\}$; Range: $\{-25, -6, 1, 2, 3, 10, 25\}$

17. Domain is all real numbers except 5. Range is all real
numbers except 1. **19.** Domain is all real numbers.
Range is all non-negative real numbers. **21.** -2

23. -11 **25.** $3b + 7$ **27.** 24 **29.** $x^2 + 5x$ **31.** 1

33. $\frac{2}{17} \approx 0.1176471$ **35.** $\frac{\frac{1-x}{2x^2+1}}{\frac{2-x}{x^2+2}} = \frac{(1-x)(x^2+2)}{(2x^2+1)(2-x)}$ **37.** 37

39. 22.222222 **41.** implicit **43.** explicit **45.** 2

47. 16 **49.** -4 **51.** -46 **53.** 22 **55.** $y = x + 5$

57. $y = \frac{x+8}{2}$ **59.** $y = \frac{9x-5}{3}$ **61.** $y = \sqrt[3]{x-7}$

63. (a) 4.91 m; (b) 19.64 m; (c) 122.75 m.

Exercise Set 4.2

1. $x^2 + 3x + 4$ **3.** $x^2 - 3x - 6$ **5.** $3x - x^2 + 6$
7. $3x^3 + 5x^2 - 3x - 5$ **9.** $\frac{x^2-1}{3x+5}$ **11.** $\frac{3x+5}{x^2-1}$ **13.** The
domains of both f and g are all real numbers.
15. $9x^2 + 30x + 24$ **17.** $3x^2 + 2$ **19.** $3x^2 + 4x - 1$
21. $2x - 3x^2 - 1$ **23.** $3x^2 - 2x + 1$ **25.** $9x^3 - x$

27. $\frac{3x-1}{3x^2+x}$ **29.** $\frac{3x^2+x}{3x-1}$ **31.** $9x^2 + 3x - 1$
33. $27x^2 - 15x + 2$ **35.** The domains of both f and g are all
real numbers. **37.** (a) $C(n) = 15n + 7500$, (b) $9000,
(c) $22,500 **39.** (a) $P(n) = R(n) - C(N) = 60n - 275$
dollars, (b) $2725
41. $(P \circ n)(a) = P(n(a)) = 300(7a + 4)^2 - 50(7a + 4)$

Exercise Set 4.3

1.

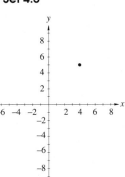

3.

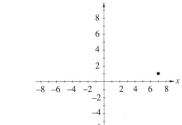

5.

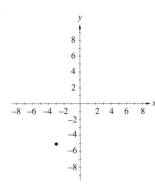

7.

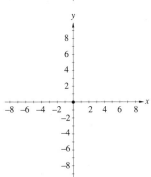

9.

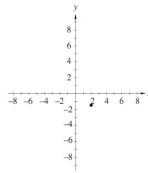

11. $(-1, -4)$

13. They are on a horizontal line and
have the same y-coordinate.

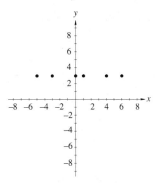

15. The points all lie on the same straight line.

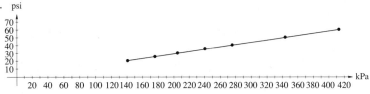

17. On the horizontal line $y = -2$.
19. In the fourth quadrant.

21. (a)

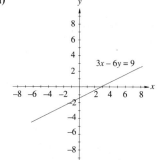

(b) $228°F$; **(c)** $236°F$

Exercise Set 4.4

1. (a)

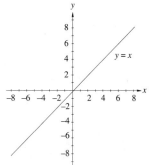

(b) x-intercept is 0; y-intercept is 0;
(c) $m = 1$

3. (a)

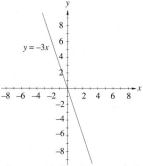

(b) x-intercept is 0; y-intercept is 0;
(c) $m = -3$

5. (a)

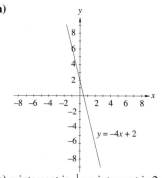

(b) x-intercept is $\frac{1}{2}$; y-intercept is 2;
(c) $m = -4$

7. (a)

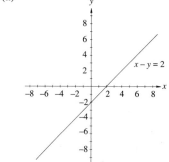

(b) x-intercept is 2; y-intercept is
-2; **(c)** $m = 1$

9. (a)

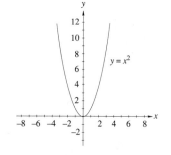

(b) x-intercept is 3; y-intercept is
$-\frac{3}{2}$; **(c)** $m = \frac{1}{2}$

11.

x	-3	-2	-1	0	1	2	3
y	9	4	1	0	1	4	9

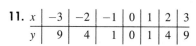

13.

x	-3	-2	-1	0	1	2	3
y	7	2	-1	-2	-1	2	7

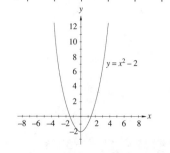

15.

x	-3	-2	-1	0	1	2	3	4	5	6
y	36	25	16	9	4	1	0	1	4	9

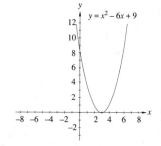

17.

x	-3	-2	-1	$-\frac{1}{2}$	$-\frac{1}{4}$	0	$\frac{1}{4}$	$\frac{1}{2}$	1	2	3
y	$-\frac{1}{3}$	$-\frac{1}{2}$	-1	-2	-4	Undefined	4	2	1	$\frac{1}{2}$	$\frac{1}{3}$

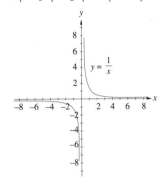

19.

x	-3	-2	-1	0	1	2	3
y	-27	-8	-1	0	1	8	27

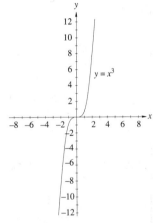

21.

x	0	0	5	-5	3	-3	3	-3	4	4	-4	-4
y	5	-5	0	0	4	4	-4	-4	3	-3	3	-3

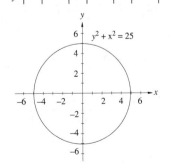

23.

x	-5	-4	-3	-2	-1	0	1	2
y	3	2	1	0	1	2	3	4

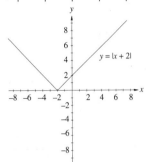

25.

x	0	1	-1	2	-2	3	-3	4	-4
y	Undefined	0	0	±3.35	±3.35	±5.2	±5.2		

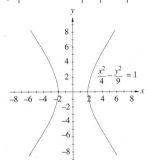

27. Function **29.** Function

31. (a)

k	0	1	2	3	4
	5	6	7	8	9
$p(k)$	25.0	8.33	5.00	3.57	2.78
	2.27	1.92	1.67	1.47	1.32

(b) yes

33. **(a)** $0 < p < \$4.65$;

(b)

p	0.25	0.50	0.75	1.00
	1.25	1.50	1.75	2.00
$d(p)$	10.56	11	11.31	11.50
	11.56	11.50	11.31	11

(d) $\$1.25$

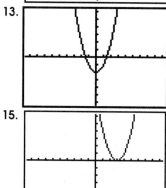

$$d(p) = -p^2 + 2.5p + 10$$

Exercise Set 4.5

1.

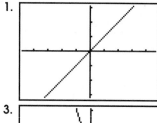

3.

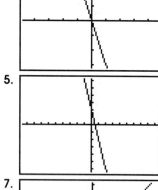

5.

7.

9.

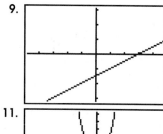

11.

13.

15.

17.

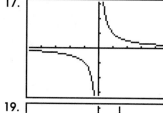

19.

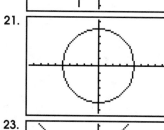

21.

23.

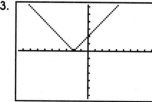

25.

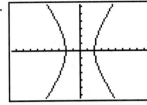

Exercise Set 4.6

1. The root is $x = -\frac{5}{2}$

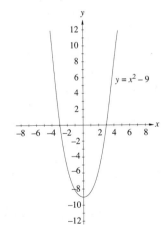

$y = 2x + 5$

3. The roots are $x = -3$ and $x = 3$

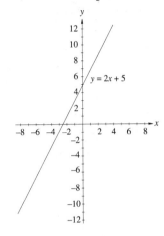

$y = x^2 - 9$

5. The roots are $x = 0$ and $x = 5$

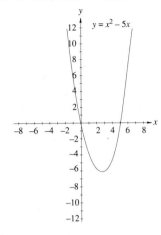

$y = x^2 - 5x$

7. The roots are
$x = -\frac{5}{2} - \frac{\sqrt{37}}{2} \approx -5.5414$ and
$x = -\frac{5}{2} + \frac{\sqrt{37}}{2} \approx 0.5414$

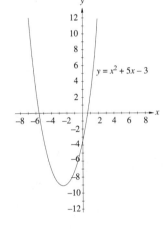

$y = x^2 + 5x - 3$

9. The roots are $x = -\frac{9}{4}$ and $x = \frac{6}{5}$

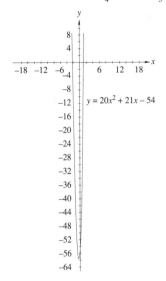

$y = 20x^2 + 21x - 54$

11. Has an inverse function
13. Does not have an inverse function
15.

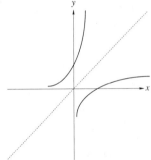

17.

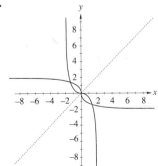

19. (a) 30%; (b) 60%; (c) 73%
21. No

Review Exercises

1. (a)

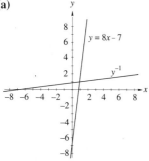

(b) Domain: all real numbers;
Range: all real numbers; x-intercept
$\frac{7}{8}$; y-intercept -7; **(c)** function;
(d) has an inverse function

3. (a)

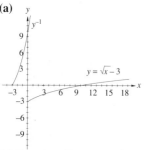

(b) Domain: all non-negative real
numbers, $x \geq 0$; Range: all real
numbers greater than or equal to -3;
x-intercept 9; y-intercept -3;
(c) function; **(d)** has an inverse
function

5. (b) Domain: all real numbers;
Range: all real numbers greater than
or equal to -1; x-intercepts 0, 2;
y-intercept 0; **(c)** function; **(d)** does
not have an inverse function

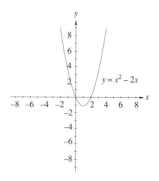

7. -12 **9.** 0   **11.** $4a - 20$
13. -1 **15.** 0   **17.** $\frac{7}{25} = 0.28$
19. The zeros are -3 and 3.

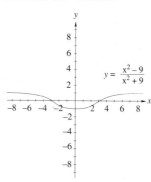

21. 0   **23.** $-20 + \frac{5}{13} = -\frac{255}{13}$
25. 12   **27.** 17   **29.** $\frac{5}{52}$
31. $-\frac{272}{25}$ **33.** -1
35. $y = x^2 - 20$; roots:
$x = -\sqrt{20} \approx -4.4721$ and
$x = \sqrt{20} \approx 4.4721$

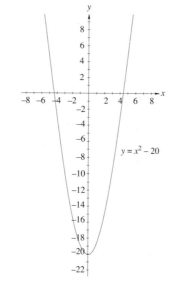

37. $y = 8x^3 - 20x^2 - 34x + 21$; roots:
$x = -1.5, x = 0.5,$ and $x = 3.5$

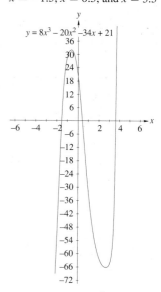

39. (a) $P(n) = 20n - \frac{n^2}{20} - 550$; **(b)** \$5,
\$950, \$1325, \$950, $-$\$550 (the store
will lose \$550 if it sells 400
videotapes)

41. (a)

v	0	10	20	30	40	50
$s(v)$	0	14	36	66	104	150

v	60	70
$s(v)$	204	266

(b) about 69.8 mph;

(c)

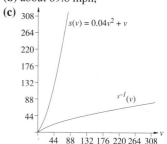

Chapter 4 Test
1. -19 **3.** -5

5. (a)

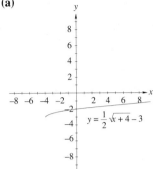

$y = \frac{1}{2}\sqrt{x+4} - 3$

(b) $x \geq -4$; **(c)** $y \geq -3$; **(d)** $x = 32$;
(e) $y = -2$ **7.** $\frac{3x^2 - x - 70}{x+5}$
9. $3(x+5) = 3x + 15$ **11.** $\frac{3x-20}{3x-10}$

≡ ANSWERS FOR CHAPTER 5

Exercise Set 5.1
1. 3   **3.** $-\frac{5}{4}$ **5.** $\frac{10}{7}$
7. $y - 3 = 4(x - 5)$
9. $y + 5 = \frac{2}{3}(x - 1)$
11. $y = -\frac{5}{3}(x - 2)$
13. $y - 2 = \frac{3}{4}(x + 3)$ or
     $y - 5 = \frac{3}{4}(x - 1)$ **15.** $y = 2x + 4$
17. $y = 5x - 3$

19. $y = 3x + 6$; $m = 3$, $b = 6$

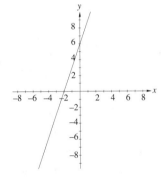

21. $y = \frac{5}{2}x + 4$; $m = \frac{5}{2}$, $b = 4$

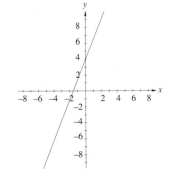

23. $y = \frac{5}{2}x - 5$; $m = \frac{5}{2}$, $b = -5$

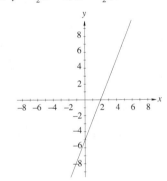

25. $y = -\frac{1}{3}x + \frac{7}{3}$; $m = -\frac{1}{3}$, $b = \frac{7}{3}$

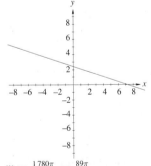

27. $m = \frac{1\,780\pi}{60} = \frac{89\pi}{3}$
29. (a) $F = kL - kL_0$;
     (b) $F = 4.5L - 27$;

(c)

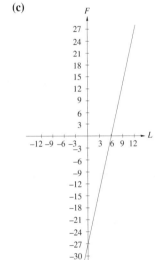

Exercise Set 5.2
1. $(4, 2)$

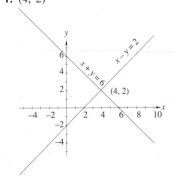

3. $(4.5, 2)$

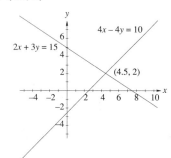

5. $(0, \frac{32}{5}) = (0, 6.4)$

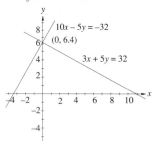

7. $(2.2, -1.1)$

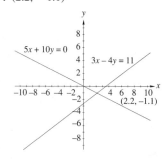

9. $(3, 5)$ **11.** $(-4, 6)$
13. $(8, -2)$ **15.** $(3.3, -1.2)$
17. $(7, 2)$ **19.** $(-2, 1)$
21. $(-3, 3)$ **23.** $(1.5, -6.5)$

25. $(4, -1)$

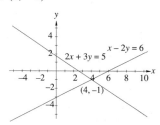

27. $(-1, 7)$

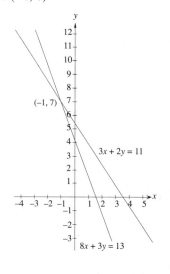

29. $(0.608, -1.324)$

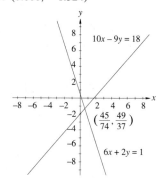

31. $(3, \frac{1}{3})$

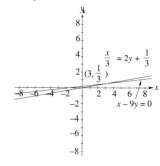

33. length $= 13$ km, width $= 5$ km
35. length $= 16.875$ km, width
    $= 5.625$ km   **37.** 6 250 L of 5%;
    3 750 L of 13% gasohol
39. 6.4 kg, 4 m   **41.** $I_1 = 1$ A,
    $I_2 = 1$ A, $I_3 = 0$ A
43. $I_1 = 0.224$ A, $I_2 = 1.052$ A,
    $I_3 = 0.828$ A

Exercise Set 5.3
1. $x = 2, y = -1, z = 4$ **3.** $x = -3, y = -2, z = 4$
5. $x = 2, y = -1, z = 4$ **7.** $x = -3, y = -2, z = 4$
9. $x = 1.3, y = 2.7, z = -2$ **11.** $x = 1.25, y = 3.75,$
    $z = -5.5$ **13.** $x = -6, y = 2, z = 5$
15. $I_1 = 0.875$ A, $I_2 = 2.125$ A, $I_3 = 1.25$ A
17. $2a + 6b - c = 40; a + 7b - c = 50; a = -4, b = 6,$
    $c = -12$ **19.** 4 large trucks, 2 medium, 3 small

Exercise Set 5.4
1. -14 **3.** 13 **5.** 25 **7.** -10 **9.** (a) 2,
    (b) $\begin{vmatrix} 3 & -4 \\ -2 & 1 \end{vmatrix} = -5$, (c) 5   **11.** (a) 1, (b) -13, (c) 13
13. (a) -2, (b) -1, (c) -1 **15.** $3(-3) + 7(2) + 4(8) = 37$
17. 0, rows 1 and 3 are identical   **19.** 0, row 3 is a multiple
    of row 1   **21.** 501   **23.** -120 **25.** 252.865 83

27. $-0.057 525$ **29.** 5   **31.** 30   **33.** 0   **35.** See
    Computer Programs   **37.** See *Computer Programs*

Exercise Set 5.5
1. $x = 1, y = 5$ **3.** $x = -2, y = 4$ **5.** $x = 3.522 8,$
    $y = 1.316 9$ **7.** no solution, inconsistent system
9. $x = -7.876 8, y = -17.689$ **11.** no solution,
    inconsistent system   **13.** $x = -4, y = 6$
15. $x = 1.282 5, y = -1.015, z = 0.14$ **17.** See *Computer
    Programs*   **19.** $s_0 = 200, v_0 = 15.8, a = -9.8$
21. $A = 530$ km, $B = 620$ km, $C = 750$ km

Review Exercises
1. 27   **3.** -67 **5.** 10   **7.** -32 **9.** (a) $\frac{-2}{7}$,
    (b) $y - 1 = \frac{-2}{7}(x - 5)$ or $y - 3 = \frac{-2}{7}(x + 2)$

11. (a) $\frac{-5}{11}$, (b) $y - 4 = \frac{-5}{11}(x + 2)$ or $y + 1 = \frac{-5}{11}(x - 9)$
13. $y = 2x - 3$ **15.** $y = \frac{4}{5}x + \frac{8}{5}$
17. $x = 6, y = -14$

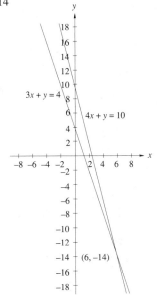

19. $x = 3, y = -4$ **21.** $x = \frac{19}{11} \approx 1.7273,$
$y = -\frac{1}{11} \approx -0.0909$ **23.** $x = 3, y = 5$ **25.** $x = 1,$
$y = 2$ **27.** $x = 5.59375, y = 3.71875, z = -10.40625$
29. $x = 2, y = 1$ **31.** $x = \frac{57}{49} \approx 1.1633,$
$y = \frac{-71}{49} \approx -1.4490$ **33.** $x = 2, y = -1, z = 1$
35. 30 pounds of Colombian Supreme and 20 pounds of Mocha
Java **37.** 24 large, 32 middle size, and 10 small offices

Chapter 5 Test
1. (a) $\frac{4}{9}$; (b) $y - 2 = \frac{4}{9}(x + 4)$ or $y - 6 = \frac{4}{9}(x - 5)$ or, in
slope-intercept form, $y = \frac{4}{9}x + \frac{34}{9}$ **3.** $x = -2, y = \frac{5}{3}$
5. -49 **7.** $x = -\frac{3}{2}, y = 5$ **9.** $30, $21, and $9

ANSWERS FOR CHAPTER 6

Exercise Set 6.1
1. $1.38 : 16$ **3.** $86 : 1$ **5.** $4.5 : 1$ **7.** $3.2857 : 1$
9. equal **11.** equal **13.** 28 **15.** 79 **17.** 4.267
19. 14 **21.** 1.78 **23.** $x = 5.6, z = 25.2$ **25.** $54 : 13$
27. $28 : 1$ **29.** $1.67 \ \mu F$ **31.** $11.03 \ cm^3$
33. 62,800,000 calories **35.** 3.5 in.

Exercise Set 6.2
1. $a = 1.5, b = 3.75$ **3.** $a = 17, b = 42.71, c = 47.17$
5. $a = 2.55, b = 3.4, c = d = 4.25$ **7.** $184{,}320 \ in.^2, 1280$
ft^2 **9.** 5.5 kg **11.** 2143.75 L **13.** $145.56 \ mm^3$
15. (a) $50.27 \ cm^2$; (b) $33.51 \ cm^3$

Exercise Set 6.3
1. $R = kl$ **3.** $A = kd^2$ **5.** $IN = k$
7. $r = kt, k = \frac{2}{3}$ **9.** $d = kr^2, k = \frac{1}{2}$ **11.** 9
13. 180 m/s **15.** 435.6 neutrons **17.** (a) 0.085,
(b) $5.54 \ m^3$ **19.** $106.67 \ \Omega$ **21.** 160 psi

23. 0.075 A **25.** 142.22 lb

Exercise Set 6.4
1. $K = kmv^2$ **3.** $f = \frac{kv}{l}$ **5.** $E = \frac{klI}{d^2}$ **7.** $a = kbc;$
$k = 2.5$ **9.** $u = \frac{kv}{w}; k = 8$ **11.** $r = kst; k = \frac{1}{3},$
$r = 4$ **13.** $a = kxy^2; k = 0.12, a = 162$
15. $p = \frac{kr\sqrt{t}}{w}; k = 1, p = 36$ **17.** (a) $\frac{1}{2}$, (b) 25
19. 22.36 V **21.** 17.14 m **23.** 77.5 Btu/hr
25. $1.25 \ ft^3$

Review Exercises
1. 4.5 **3.** 27 **5.** 38 **7.** $x = 36.75, y = 28$
9. 160 **11.** 90 mm; 112.5 mm **13.** $8 : 9$ or 88.9%
15. 0.2 m/s **17.** 8.99×10^9 **19.** 1.24×18^8 m/s

Chapter 6 Test
1. $22\frac{2}{3} \approx 22.67$ **3.** 16.8 **5.** $x = 14; y = 10.5$
7. $E = kd^2$ **9.** 27.5 and 66 **11.** $162 \ m^2$

ANSWERS FOR CHAPTER 7

Exercise Set 7.1
1. $3p + 3q$ **3.** $15x - 3xy$ **5.** $p^2 - q^2$
7. $4x^2 - 36p^2$ **9.** $r^2 + 2rw + w^2$ **11.** $4x^2 + 4xy + y^2$

13. $\frac{4}{9}x^2 + \frac{16}{3}xb + 16b^2$ **15.** $4p^2 - 3pr + \frac{9}{16}r^2$
17. $a^2 + 5a + 6$ **19.** $x^2 - 3x - 10$ **21.** $6a^2 + 5ab + b^2$
23. $6x^2 - 7x - 20$ **25.** $a^3 + 3a^2b + 3ab^2 + b^3$

27. $x^3 + 12x^2 + 48x + 64$ **29.** $8a^3 - 12a^2b + 6ab^2 - b^3$
31. $27x^3 - 54x^2y + 36xy^2 - 8y^3$ **33.** $m^3 + n^3$ **35.** $r^3 - t^3$
37. $8x^3 + b^3$ **39.** $27a^3 - d^3$ **41.** $3a^2 + 12a + 12$
43. $5rt + 2r^2 + \frac{r^3}{5t}$ **45.** $x^4 - 36$ **47.** $9x^4 - y^4$ **49.** 0
51. $x^3 - 3x^2 - 9x + 27$ **53.** $r^3 - 3r^2t + 3rt^2 - t^3$
55. $125 + 27x^3$ **57.** $(x + y)^2 - 2(x + y)(w + z) + (w + z)^2$
59. $(x + y)^2 - z^2 = x^2 + 2xy + y^2 - z^2$
61. $z^2 = R^2 + x_L^2 - 2x_Lx_C + x_C^2$
63. $a_c = \frac{(2t^2 - t)^2}{r} = \frac{4t^4 - 4t^3 + t^2}{r}$

Exercise Set 7.2

1. $6(x + 1)$ **3.** $6(2a - 1)$ **5.** $2(2x - y + 4)$
7. $5(x^2 + 2x + 3)$ **9.** $5(2x^2 - 3)$ **11.** $2x(2x + 3)$
13. $7b(by + 4)$ **15.** $ax(1 + 6x)$ **17.** $2ap(2p + 3aq + 4q^2)$
19. $(a - b)(a + b)$ **21.** $(x - 2)(x + 2)$ **23.** $(y - 9)(y + 9)$
25. $(2x - 3)(2x + 3)$ **27.** $(3a^2 - b)(3a^2 + b)$
29. $(5a - 7b)(5a + 7b)$ **31.** $(12 - 5b^2)(12 + 5b^2)$
33. $5(a - 5)(a + 5)$ **35.** $7(2a - 3b^2)(2a + 3b^2)$
37. $(a - 3)(a + 3)(a^2 + 9)$ **39.** $16(x - 2y)(x + 2y)(x^2 + 4y^2)$
41. $\pi r(r + \sqrt{h^2 + r^2})$ **43.** $\frac{1}{2}d(v_2 - v_1)(v_2 + v_1)$
45. $\frac{(\omega_f - \omega_0)(\omega_f + \omega_0)}{2\theta}$

Exercise Set 7.3

1. $b^2 - 4ac = 113$; $\sqrt{113} \approx 10.6$; does not factor using rational numbers   **3.** $b^2 - 4ac = 196$; $\sqrt{196} = 14$; factors   **5.** $b^2 - 4ac = 169$; $\sqrt{169} = 13$; factors
7. $(x + 2)(x + 5)$ **9.** $(x - 3)(x - 9)$ **11.** $(x - 2)(x - 25)$
13. $(x - 2)(x + 1)$ **15.** $(x - 5)(x + 2)$ **17.** $(r + 5)^2$
19. $(a + 11)^2$ **21.** $(f - 15)^2$ **23.** $(6y - 1)(y - 1)$
25. $(7t + 2)(t + 1)$ **27.** $(7b + 1)(b - 5)$ **29.** $(4e - 1)(e + 5)$
31. $(3u + 4)(u + 2)$ **33.** $(9t + 2)(t - 3)$
35. $(3x - 1)(2x + 5)$ **37.** $(5a + 3)(3a - 5)$
39. $(5e + 3)(3e + 5)$ **41.** $(5x - 2)(2x - 3)$
43. $3(r - 7)(r + 1)$ **45.** $7t^2(7t - 1)(t - 2)$
47. $(3x + 2y)(2x - 5y)$ **49.** $(2a + b)(4a - 9b)$
51. $(a - b)(a^2 + ab + b^2)$ **53.** $(2x - 3)(4x^2 + 6x + 9)$
55. $i = 0.7(t^2 - 3t - 4) = 0.7(t - 4)(t + 1)$
57. $(0.0001n^2 - 3)(n - 2,000) = 0.0001(n^2 - 30,000)(n - 2,000)$
59. (a) $4x^2 + 32x - 36 = 0$; (b) $4(x + 9)(x - 1) = 0$

Exercise Set 7.4

1. $\frac{35}{40}$ **3.** $\frac{ax}{ay}$ **5.** $\frac{3ax^3y}{3a^2x}$ **7.** $\frac{4(x+y)}{x^2-y^2} = \frac{4x+4y}{x^2-y^2}$

9. $\frac{(a+b)^2}{a^2-b^2} = \frac{a^2+2ab+b^2}{a^2-b^2}$ **11.** $\frac{19}{12}$ **13.** $\frac{x}{4}$ **15.** $\frac{4}{x-3}$
17. $\frac{x-4}{x+4}$ **19.** $\frac{x}{3}$ **21.** $\frac{x+3}{x^2+5}$ **23.** $\frac{2m-m^2}{3+6m^2}$ **25.** $\frac{x}{x-3}$
27. $\frac{2b}{3(b+5)}$ **29.** $\frac{z+3}{z-3}$ **31.** $\frac{x+1}{x+4}$ **33.** $\frac{2x+1}{x+5}$ **35.** $\frac{y(2y+1)}{y-1}$
37. $\frac{y^2+xy+x^2}{2}$ **39.** $x - y$

Exercise Set 7.5

1. $\frac{10}{xy}$ **3.** $\frac{12x^2}{5y^3}$ **5.** $\frac{3y}{7x}$ **7.** $\frac{8x^3}{21y}$ **9.** $\frac{5}{6xy}$
11. $\frac{5a^3d}{2b}$ **13.** $\frac{8y^2}{25x^2}$ **15.** $\frac{y^3}{5p^2}$ **17.** $4y$ **19.** $5(a - b)$
21. $\frac{3(x^2-100)}{4(x+5)}$ **23.** $\frac{(2x-1)(x+3)}{-3x}$ **25.** $a + 2$ **27.** $\frac{1}{4a}$
29. $\frac{1}{x+1}$ **31.** $\frac{4}{y}$ **33.** $\frac{x-1}{3(x+2)}$ **35.** $\frac{(x-2y)(x-5y)(x-4y)}{(x+4y)(x+2y)(x-7y)}$
37. $\frac{3x-5}{x+3}$ **39.** $\frac{x+y}{x-y}$

Exercise Set 7.6

1. 1 **3.** $\frac{2}{3}$ **5.** $\frac{5}{6}$ **7.** $\frac{2}{15}$ **9.** $\frac{6}{x}$ **11.** $\frac{1}{a}$
13. $\frac{5x}{y}$ **15.** $-\frac{3r}{2t}$ **17.** $\frac{3+x}{x+2}$ **19.** $\frac{t-2}{t+1}$ **21.** $\frac{2y}{x+2}$
23. $\frac{7}{a+b}$ **25.** $\frac{2y+3x}{xy}$ **27.** $\frac{ad-4b}{bd}$ **29.** $\frac{7x-3}{x(x^2-1)}$
31. $\frac{6-2x}{(x^2-1)(x+1)}$ **33.** $\frac{x^2+8x-10}{(x^2-36)(x-5)}$ **35.** $\frac{11-3x}{(x-3)(x^2-4)}$
37. $\frac{1-x-8x^2}{x(3x-1)(x-4)}$ **39.** $\frac{3x^2-8x-5}{(x^2-1)(x+4)}$ **41.** $\frac{6y+13-y^2}{(y-2)(y+1)(y+4)}$
43. $\frac{2x^4-x^3+13x^2+2x-4}{(x^2+3)(x-1)^2(x+2)}$ **45.** $\frac{x+2}{x-3}$ **47.** $\frac{x(x-1)}{x+1}$
49. $\frac{x^2-2xy-y^2}{x^2+y^2}$ **51.** $\frac{x+3}{x+2}$ **53.** $\frac{t(t-1)}{t^2+1}$ **55.** $\frac{x^2+y^2}{2x}$
57. $\frac{R_1+R_2}{R_1R_2}$ **59.** $\frac{C_1C_2+C_1C_3+C_2C_3}{C_1C_2C_3}$

Review Exercises

1. $5x^2 - 5xy$ **3.** $x^3 - 6x^2y + 12xy^2 - 8y^3$
5. $2x^2 - 9x - 18$ **7.** $x^4 - 25$ **9.** $8 + 12x + 6x^2 + x^3$
11. $9(1 + y)$ **13.** $7(x - 3)(x + 3)$ **15.** $(x - 6)(x - 5)$
17. $(x + 8)(x - 2)$ **19.** $(2x + 3)(x - 3)$ **21.** $\frac{x}{3y}$
23. $\frac{x-3}{x+3}$ **25.** $\frac{x^2-xy+y^2}{x+y}$ **27.** $\frac{3xy}{7}$ **29.** $\frac{4}{x}$ **31.** $\frac{7x}{y}$
33. $\frac{7x^2-27x+2}{(x+2)^2(x-5)^2}$ **35.** $\frac{-2x^2}{y^2-x^2}$ **37.** $\frac{2(x^2+36)(x+1)}{(x+2)(x^2-36)}$
39. $\frac{(x-6)^2}{(x+6)^2}$ **41.** $\frac{y-x}{y+x}$ **43.** -1 **45.** $2x^2 - 1$

Chapter 7 Test

1. $x^2 + 2x - 15$ **3.** $2(x - 8)(x + 8)$ **5.** $\frac{x-5}{x+1}$
7. $\frac{2(x+2)}{3}$ **9.** $\frac{x^2-11x-12}{(x+3)(x-2)} = \frac{x^2-11x-12}{x^2+x-6}$
11. $\frac{R_2R_3+R_1R_3+R_1R_2}{R_1R_2R_3}$

≡ ANSWERS FOR CHAPTER 8

Exercise Set 8.1

1. 0.3 **3.** 10 **5.** $-\frac{1}{7}$ **7.** 9 **9.** -3 **11.** $\frac{15}{8}$
13. $-\frac{1}{3}$ **15.** 15 **17.** -9 **19.** $\frac{2}{5}$ **21.** no solution, since x cannot equal 1.   **23.** $\frac{30}{37}$ **25.** $-\frac{3}{5}$ **27.** -6
29. no solution   **31.** $\frac{rt}{r-t}$ **33.** $\frac{R_1R_2R_3}{R_1R_2+R_1R_3+R_2R_3}$

35. $\frac{V-2\pi r^2}{2\pi r}$ **37.** $\frac{R_1f(n-1)}{R_1-f(n-1)}$ **39.** 3 hours
41. $h = \frac{28}{5} = 5$ hours 36 minutes; total $= h + 4 = 9$ hours, 36 minutes

Exercise Set 8.2

1. ± 3 **3.** $-3, 2$ **5.** $-1, 12$ **7.** $2, -4$

9. 0, 5   **11.** 3, 4   **13.** $-2, \frac{7}{2}$ **15.** $\frac{3}{2}$, 4
17. 3, $-\frac{1}{3}$ **19.** $\frac{5}{2}, \frac{7}{2}$ **21.** $\frac{5}{3}, -\frac{7}{2}$ **23.** $\frac{1}{5}, \frac{3}{2}$
25. $-\frac{3}{2}, \frac{20}{3}$ **27.** $-1, 3$ **29.** $\frac{6}{5}, -\frac{2}{5}$ **31.** $2, -\frac{1}{2}$
33. $-1, 24$ **35.** 4 s   **37.** width 8 cm, length 13 cm
39. 5 m and 12 m

Exercise Set 8.3
1. $-2, -4$ **3.** $-1, 11$ **5.** $-3 \pm \sqrt{6}$ **7.** $\frac{5 \pm \sqrt{5}}{2}$
9. $\frac{3 \pm \sqrt{29}}{2}$ **11.** $\frac{3 \pm \sqrt{27}}{2}$ **13.** $-k \pm \sqrt{k^2 - c}$ **15.** 3.5 s
17. either 20 or 370 objects were sold   **19.** There are 2 possible answers. In one, each field is 300 ft × 450 ft; and in the other, each field measures 600 ft × 225 ft.

Exercise Set 8.4
1. $-4, 1$ **3.** $2, -\frac{1}{3}$ **5.** $-1, \frac{1}{7}$ **7.** $-1, \frac{7}{2}$
9. $-2, \frac{4}{3}$ **11.** $-\frac{2}{3}, -\frac{2}{3}$ **13.** $\frac{3 \pm \sqrt{17}}{4}$ **15.** $\frac{-5 \pm \sqrt{17}}{2}$
17. $\frac{-3 \pm \sqrt{15}}{2}$ **19.** $1 \pm \sqrt{8} = 1 \pm 2\sqrt{2}$ **21.** $\pm \frac{\sqrt{6}}{2}$
23. no real roots because the discriminant is -48 **25.** no real roots because the discriminant is $-\frac{647}{81}$
27. $\frac{-0.2 \pm \sqrt{0.064}}{0.02}$ **29.** $5 \pm \sqrt{13}$ **31.** $\frac{2 \pm \sqrt{1.6}}{2.4}$

33. $\frac{-\sqrt{3} \pm \sqrt{87}}{6}$ **35.** no real root because the discriminant is negative   **37.** $\frac{-2 \pm \sqrt{13}}{3}$ **39.** approximately 9.16 s
41. approximately 11.42 s   **43.** approximately 16.38 cm wide and 24.57 cm long   **45.** 2.25 inches wide, 7.5 inches long, or 3.75 inches wide and 4.5 inches deep
47. $20   **49.** 1062.1 Ω   **51.** See *Computer Programs*

Review Exercises
1. 6   **3.** -2 **5.** $\frac{2 \pm \sqrt{10}}{3}$ **7.** $\frac{A - 2lw}{2(l+w)}$ **9.** $\frac{pq}{p+q}$
11. 1, 7   **13.** 1, 10   **15.** $-7, -8$ **17.** $1, \frac{3}{2}$ **19.** $-2, \frac{5}{6}$ **21.** $\pm \frac{3}{2}$ **23.** $-5, 1$ **25.** $\frac{-19 \pm 9\sqrt{5}}{2}$
27. $\frac{-5 \pm \sqrt{193}}{6}$ **29.** $4 \pm \sqrt{11}$ **31.** $1, -\frac{5}{3}$ **33.** $\frac{-1 \pm \sqrt{13}}{3}$
35. $\frac{4 \pm \sqrt{6}}{5}$ **37.** no answer because the discriminant is negative   **39.** $\frac{3 \pm \sqrt{17}}{4}$ **41.** 0.0899 A   **43.** 3 sec

Chapter 8 Test
1. 4   **3.** $-1, 5$ **5.** double root, $\frac{3}{4}$ **7.** $\frac{-3 \pm \sqrt{5}}{2}$
9. 8.5 m long and 5.5 m wide

▤ ANSWERS FOR CHAPTER 9

Exercise Set 9.1
1. 1.5705   **3.** 1.396   **5.** 2.70475   **7.** 3.75175
9. 114.592°   **11.** 85.944°   **13.** 60°   **15.** $-45°$

17. (c) 510°, $-210°$

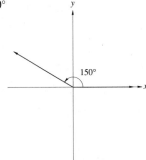

19. (c) $-495°$, 225°

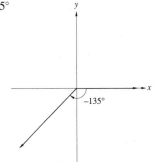

21. $\sin\theta = \frac{3}{5}; \cos\theta = \frac{4}{5}; \tan\theta = \frac{3}{4}; \csc\theta = \frac{5}{3}; \sec\theta = \frac{5}{4}; \cot\theta = \frac{4}{3}$ **23.** $\sin\theta = \frac{-15}{17}; \cos\theta = \frac{8}{17}; \tan\theta = \frac{-15}{8}; \csc\theta = \frac{-17}{15}; \sec\theta = \frac{17}{8}; \cot\theta = \frac{-8}{15}$ **25.** $\sin\theta = \frac{2}{\sqrt{5}}; \cos\theta = \frac{1}{\sqrt{5}}; \tan\theta = 2; \csc\theta = \frac{\sqrt{5}}{2}; \sec\theta = \sqrt{5}; \cot\theta = \frac{1}{2}$ **27.** $\sin\theta = -\frac{4}{\sqrt{41}}; \cos\theta = \frac{5}{\sqrt{41}}; \tan\theta = \frac{-4}{5}; \csc\theta = -\frac{\sqrt{41}}{4}; \sec\theta = \frac{\sqrt{41}}{5}; \cot\theta = \frac{-5}{4}$
29. $\sin\theta = \frac{5}{6}; \cos\theta = \frac{\sqrt{11}}{6}; \tan\theta = \frac{5}{\sqrt{11}}; \csc\theta = \frac{6}{5}; \sec\theta = \frac{6}{\sqrt{11}}; \cot\theta = \frac{\sqrt{11}}{5}$ **31.** $\sin\theta = 0; \cos\theta = 1; \tan\theta = 0; \csc\theta = $ Does not exist; $\sec\theta = 1; \cot\theta = $ Does not exist.   **33.** $\sin\theta = 0; \cos\theta = -1; \tan\theta = 0; \csc\theta = $ Does not exist; $\sec\theta = -1; \cot\theta = $ Does not exist.
35. $\sin\theta = \frac{4}{5}; \cos\theta = \frac{3}{5}; \tan\theta = \frac{4}{3}; \csc\theta = \frac{5}{4}; \sec\theta = \frac{5}{3}; \cot\theta = \frac{3}{4}$ **37.** $\sin\theta = \frac{21}{29}; \cos\theta = \frac{-20}{29}; \tan\theta = \frac{-21}{20}; \csc\theta = \frac{29}{21}; \sec\theta = \frac{-29}{20}; \cot\theta = \frac{-20}{21}$
39. $\sin\theta = -\frac{\sqrt{15}}{8}; \cos\theta = \frac{-7}{8}; \tan\theta = \frac{\sqrt{15}}{7}; \csc\theta = -\frac{8}{\sqrt{15}}; \sec\theta = \frac{-8}{7}; \cot\theta = \frac{7}{\sqrt{15}}$

Exercise Set 9.2
1. $\sin\theta = \frac{5}{13}; \cos\theta = \frac{12}{13}; \tan\theta = \frac{5}{12}; \csc\theta = \frac{13}{5}; \sec\theta = \frac{13}{12}; \cot\theta = \frac{12}{5}$ **3.** $\sin\theta = \frac{3}{5}; \cos\theta = \frac{4}{5}; \tan\theta = \frac{3}{4}; \csc\theta = \frac{5}{3}; \sec\theta = \frac{5}{4}; \cot\theta = \frac{4}{3}$
5. $\sin\theta = \frac{1.4}{\sqrt{7.25}}; \cos\theta = \frac{2.3}{\sqrt{7.25}}; \tan\theta = \frac{14}{23}; \csc\theta = \frac{\sqrt{7.25}}{1.4}; \sec\theta = \frac{\sqrt{7.25}}{2.3}; \cot\theta = \frac{23}{14}$ **7.** $\sin\theta = \frac{15}{17}; \cos\theta = \frac{8}{17}; \tan\theta = \frac{15}{8}; \csc\theta = \frac{17}{15}; \sec\theta = \frac{17}{8}; \cot\theta = \frac{8}{15}$

9. $\sin\theta = \frac{3}{5}$; $\cos\theta = \frac{4}{5}$; $\tan\theta = \frac{3}{4}$; $\csc\theta = \frac{5}{3}$; $\sec\theta = \frac{5}{4}$; $\cot\theta = \frac{4}{3}$ **11.** $\sin\theta = \frac{20}{29}$; $\cos\theta = \frac{21}{29}$; $\tan\theta = \frac{20}{21}$; $\csc\theta = \frac{29}{20}$; $\sec\theta = \frac{29}{21}$; $\cot\theta = \frac{21}{20}$
13. $\sin\theta = 0.866$; $\cos\theta = 0.5$; $\tan\theta = 1.732$; $\csc\theta = 1.155$; $\sec\theta = 2$; $\cot\theta = 0.577$ **15.** $\sin\theta = 0.085$; $\cos\theta = 0.996$; $\tan\theta = 0.085$; $\csc\theta = 11.765$; $\sec\theta = 1.004$; $\cot\theta = 11.718$ **17.** 0.3189593
19. 0.3307184 **21.** 4.1760011 **23.** 0.1298773
25. 0.247404 **27.** 0.7291147 **29.** 0.1417341
31. 4.7970857 **33.** 22.61 m **35.** 4.92 V **37.** See *Computer Programs*

Exercise Set 9.3
1. 47.05° **3.** 77.92° **5.** 32.97° **7.** 73.00°
9. 61.68° **11.** 12.12° **13.** $B = 73.5°$, $b = 24.6$, $c = 25.7$ **15.** $B = 17.4°$, $a = 19.1$, $b = 6.0$
17. $B = 47°$, $a = 32.3$, $c = 47.3$ **19.** $B = 0.65$, $b = 4.9$, $c = 8.2$ **21.** $B = 1.42$, $a = 2.7$, $b = 17.8$
23. $B = 0.16$, $a = 246.6$, $c = 249.8$ **25.** $A = 0.54$, $B = 1.03$, $c = 17.5$ **27.** $A = 1.03$, $B = 0.54$, $a = 15.4$
29. $A = 0.73$, $B = 0.84$, $b = 22.4$ **31.** 32° **33.** 1147.3 m
35. $F_y = 5$ N, $F_x = 8.7$ N **37.** 9.62° **39.** 117.2 m from the intersection; 276.2 m from service station

Exercise Set 9.4
1. 87° **3.** 43° **5.** $\frac{\pi}{8}$ **7.** 1.36 rad **9.** 158°
11. 209° **13.** 1.02 rad **15.** 4.11 rad **17.** Quadrant II
19. IV **21.** III **23.** II **25.** II and IV **27.** II and III
29. IV **31.** positive **33.** positive **35.** negative
37. negative **39.** 0.6820 **41.** −2.3559 **43.** −0.2830
45. 0.6755 **47.** −0.1853 **49.** 4.9131 **51.** 0.8241
53. −0.0454 **55.** −1.0491 **57.** 0.2151 **59.** −1.8479
61. 0.9090

Exercise Set 9.5
1. I, II **3.** II, IV **5.** I, IV **7.** III, IV
9. 30.0°; 150.0° **11.** 153.4°; 333.4° **13.** 76.6°; 283.4°
15. 189.4°; 350.6° **17.** 0.85; 2.29 **19.** 1.95; 5.09
21. 0.23; 2.91 **23.** 1.19; 5.09 **25.** 57.1° **27.** 76.6°
29. 18.7° **31.** −32.6° **33.** 1.91 rad **35.** 1.28 rad
37. 1.25 rad **39.** 0.24 rad **41.** See *Computer Programs*
43. See *Computer Programs*

Exercise Set 9.6
1. 861.520 m **3.** 2262 in./min or 37.7 in./s **5.** 29.45 cm **7.** 220.89 cm² **9.** 0.168 m
11. 24776.5 mm/min or 412.9 mm/sec; 24776.5 mm
13. 9.515478°; 657.66 miles **15.** 83 (The actual answer is 83.77, but this would be 83 bytes.) **17.** 5100°/s
19. 8.80 m/sec **21.** 7393.3 km

Review Exercises
1. $\frac{\pi}{3} \approx 1.047$ **3.** 5.672 **5.** −2.007 **7.** 135°
9. 123.19° **11.** −246.94° **13.** (a) I, (b) 60°, (c) 420° and −300° **15.** (a) IV, (b) 35°, (c) 685° and −35°
17. (a) III, (b) 65°, (c) 245° and −475° **19.** (a) II, (b) $\frac{\pi}{4}$, (c) $\frac{11\pi}{4}$ and $\frac{-5\pi}{4}$ **21.** (a) II, (b) 0.99, (c) 8.43 and −4.13
23. (a) II, (b) 1.17, (c) 1.97 and −10.59 **25.** $\sin\theta = -4/5$; $\cos\theta = 3/5$; $\tan\theta = -4/3$; $\csc\theta = -5/4$; $\sec\theta = 5/3$; $\cot\theta = -3/4$ **27.** $\sin\theta = 21/29$, $\cos\theta = -20/29$, $\tan\theta = -21/20$, $\csc\theta = 29/21$, $\sec\theta = -29/20$, $\cot\theta = -20/21$ **29.** $\sin\theta = 1/\sqrt{50}$, $\cos\theta = 7/\sqrt{50}$, $\tan\theta = 1/7$, $\csc\theta = \sqrt{50}$, $\sec\theta = \sqrt{50}/7$, $\cot\theta = 7$
31. $\sin\theta = 8/17$, $\cos\theta = 15/17$, $\tan\theta = 8/15$, $\csc\theta = 17/8$, $\sec\theta = 17/15$, $\cot\theta = 15/8$ **33.** $\sin\theta = 4/5$, $\cos\theta = 3/5$, $\tan\theta = 4/3$, $\csc\theta = 5/4$, $\sec\theta = 5/3$, $\cot\theta = 3/4$ **35.** $\sin\theta = 84/91$, $\cos\theta = 35/91$, $\tan\theta = 84/35$, $\csc\theta = 91/84$, $\sec\theta = 91/35$, $\cot\theta = 35/84$
37. $\cos\theta = 24/27.2$, $\tan\theta = 12.8/24$, $\csc\theta = 27.2/12.8$, $\sec\theta = 27.2/24$, $\cot\theta = 24/12.8$ **39.** $\sin\theta = 3.12/4$, $\cos\theta = 2.5/4$, $\tan\theta = 3.12/2.5$, $\csc\theta = 4/3.12$, $\cot\theta = 2.5/3.12$ **41.** $\tan\theta = 0.577$, $\csc\theta = 2$, $\sec\theta = 1.155$, $\cot\theta = 1.732$ **43.** 0.7071 **45.** 0.6619
47. −0.6663 **49.** −0.0585 **51.** 30.0°; 150.0°
53. 138.6°; 221.4° **55.** 2.09; 4.19 **57.** 2.38; 5.52
59. 60.0° **61.** −45.0° **63.** 22.6° **65.** 22.1°
67. 7.41 A **69.** $\theta = 41.3°$; $V = 66.6$ V **71.** 581.2 ft
73. 66,700 mph **75.** 23.0° **77.** 204.2 feet
79. $F_y = 2891.3$ lb; $F_x = 1972.3$ lb.

Chapter 9 Test
1. $\frac{5\pi}{18} \approx 0.8725$ **3.** (a) III; (b) 57°
5. (a) $\frac{9}{23.4} \approx 0.38462$; (b) $\frac{9}{21.6} \approx 0.41667$
7. (a) 0.79864; (b) 2.47509; (c) −1.66164; (d) −0.49026
9. 179.15 ft

ANSWERS FOR CHAPTER 10

Exercise Set 10.1

1.

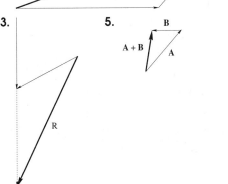

3.   **5.**

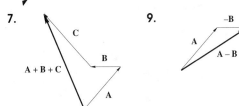

7.   **9.**

11.   **13.**

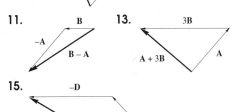

15.

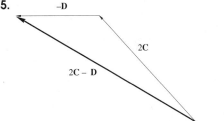

17.   **19.**

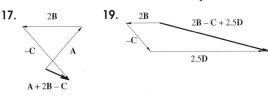

21. $P_x = 5.176; P_y = 19.319$ **23.** $P_x = 4.688;$
$P_y = -17.793$ **25.** $V_x = 9.372; V_y = 2.687$
27. $A = 15, \theta = 126.87°$ **29.** $C = 17, \theta = 61.93°$
31. 13 km/h   **33.** $V_x \approx 295$ N, $V_x \approx 930$ N   **35.** parallel

to the ramp ≈ 31.8 lb, perpendicular to the ramp ≈ 149.7 lb
37. $V = 17.5$ V, $\phi \approx 30.96°$

Exercise Set 10.2

1. $R = 12.5, \theta_R = 180°$ **3.** $R = 34.8, \theta_R = 270°$
5. $R = 73, \theta_R = 131.1°$ **7.** $R = 89.7, \theta_R = 24.79°$
9. $A = 65, \theta_A = 59.49°$ **11.** $C = 12.5, \theta_C = 20.61°$
13. $E = 6.5, \theta_E = 14.25°$ **15.** $G = 15.1, \theta_G = 56.31°$
17. $R = 12.33, \theta_R = 32.03°$ **19.** $R = 53, \theta_R = 178.11°$
21. $R = 88.50, \theta_R = 223.83°$ **23.** $R = 22.44, \theta_R = 348.90°$
25. $R = 13.04, \theta_R = 150.30°$ **27.** $R = 63.58, \theta_R = 4.62$ rad
29. $R = 13.13, \theta_R = 64.41°$ **31.** ground speed
≈ 488.3 mph, course $\approx 68.2°$, drift angle $\approx 5.2°$
33. See *Computer Programs*

Exercise Set 10.3

1. $R = 114.60, \theta = 14.49°$ if 70 lb force has direction $0°$
3. **(a)** 20.48 lb, **(b)** 14.34 lb, **(c)** vertical $= 16.07$,
horizontal $= 19.15$ **5.** 1 616.8 N forward, 525.33 N
sideward   **7.** 372.16 mi/hr in the compass direction
173.83°   **9.** 1 112.62 kg perpendicular, 449.53 kg
parallel   **11.** 119.11 lb horizontal, 154.40 lb vertical
13. 6 A, 38.66°   **15.** 10.68 A, $-59.46°$ **17.** 31.24 A,
$\theta = -39.81°$ **19.** 55.84, $\theta = 6.52°$ **21.** 126.24 m/s
at an angle of $-18.09°$

Exercise Set 10.4

1. $a = 7.59, b = 22.77, C = 75.3°$ **3.** $A = 67.1°$,
$C = 15.9°, c = 4.22$ **5.** $C = 73.31°, B = 20.37°$,
$b = 6.69$ **7.** no solution, not a triangle
9. $B = 57.31°, C = 77.69°, c = 22.52$; or $B = 122.69$,
$C = 12.31°, c = 4.91$ **11.** $A = 148.8°, b = 20.17$,
$c = 23.74$ **13.** $B = 64.62°, A = 79.78°, a = 21.13$; or
$B = 115.38°, A = 29.02°, a = 10.42$ **15.** no solution
17. $B = 0.44, A = 1.33, a = 19.52$ **19.** $B = 1.35, a = 90.7$,
$c = 194$ **21.** From B to C is 305.5 m; From A to C is
372.0 m   **23.** 34.3 miles   **25.** 294.77 m

Exercise Set 10.5

1. $c = 11.31, A = 33.6°, B = 104.1°$ **3.** $b = 68.02$,
$A = 40.1°, C = 26.2°$ **5.** $A = 45°, B = 23.2°$,
$C = 111.8°$ **7.** $b = 103.25, A = 38.21°, C = 49.35°$
9. $A = 30.34°, B = 44.11°, C = 105.55°$ **11.** $c = 11.77$,
$A = 0.56, B = 1.76$ **13.** $b = 74.52, A = 0.70, C = 0.46$
15. $A = 0.96, B = 0.59, C = 1.59$ **17.** $b = 68.56$,
$A = 0.924, C = 0.593$ **19.** $A = 1.59, B = 0.83$,
$C = 0.72$ **21.** 10 426 km   **23.** 92.25 ft
25. 30.98 N, $\theta = 39.36°$

Review Exercises

1.

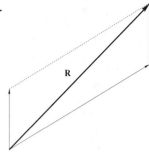

3.

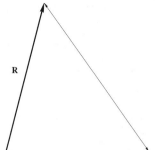

5. $P_x = 13.68, P_y = 32.22$ **7.** $P_x = -23.19, P_y = 3.09$
9. $R = 17.89, \theta_R = 333.43°$ **11.** $A_x = 36.71, A_y = 9.84$
13. $C_x = 6.12, C_y = 18.41$ **15.** $R = 50.41, \theta_R = 20.69°$
17. $R = 72.77, \theta_R = 3.13$ rad   **19.** $A = 25.7°, B = 97.5°,$
 $C = 56.8°$ **21.** $A = 145.94°, a = 146.14, C = 14.49°$
23. $B = 0.12, C = 2.90, c = 254.51$ **25.** $A = 0.52,$

$B = 0.41, c = 109.3$ **27.** $2\,831.17$ kg at $83.97°$
29. 66.47 lb parallel; 107.63 lb perpendicular   **31.** 195.47 m

Chapter 10 Test
1. $V_x = -21.34, V_y = 41.88$ **3.** $R = 59.70, \theta_R = 93.86°$
5. $a = 9.38$ **7.** 49.59 in.

≡ ANSWERS FOR CHAPTER 11

Exercise Set 11.1
1. Period $= \pi$, Amplitude $= 3$,
 Frequency $= \frac{1}{\pi}$
3. Period $= 2\pi$, Amplitude $= 2$,
 Frequency $= \frac{1}{2\pi}$
5. Period $= 1$, Amplitude $= 8$,
 Frequency $= 1$ **7.** Period $= \frac{\pi}{2}$
 Amplitude $= \frac{1}{2}$, Frequency $= \frac{2}{\pi}$
9. Period $= 4\pi$, Amplitude $= \frac{1}{3}$,
 Frequency $= \frac{1}{4\pi}$ **11.** Period 8π,
 Amplitude $= 3$, Frequency $= \frac{1}{8\pi}$
13. Period 2π, Amplitude $= \frac{1}{2}$,
 Frequency $= \frac{1}{2\pi}$

15.

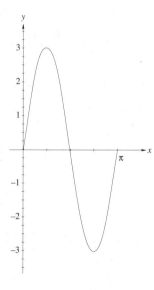

17.

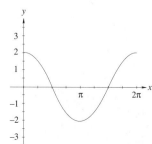

19.

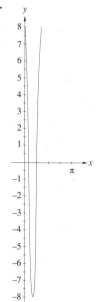

21.

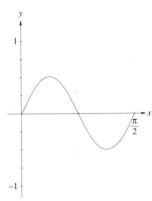

23.

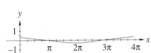

25.

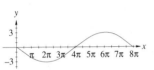

27.

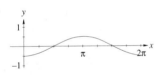

29. (a) amplitude: 6.5 A, Period: $\frac{1}{60}$,
Frequency: 60;

(b)

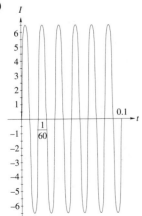

5. Amplitude = 6,
Period = 240° = $\frac{4\pi}{3}$, Phase
Shift = $-\pi/1.5 = -\frac{2\pi}{3}$

7. Amplitude = 2,
Period = 120° = $2\pi/3$, Phase
Shift = $\pi/6 = 30°$

9. Amplitude = 4.5, Period = $\pi/3$,
Phase Shift = 4/3

11. Amplitude = 0.5, Period = 2, Phase
Shift = $-1/8$

13. Amplitude = 0.75, Period = 2,
Phase Shift = $-\pi/3$

15.

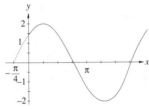

17.

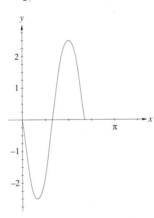

19.

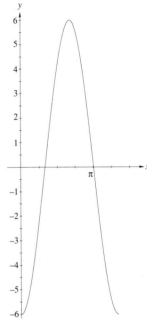

21.

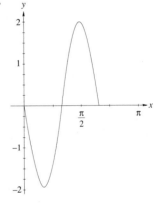

Exercise Set 11.2

1. Amplitude = 2, Period = 2π, Phase
Shift = $-\pi/4$

3. Amplitude = 2.5, Period = $2\pi/3$,
Phase Shift = $\pi/3$

23.

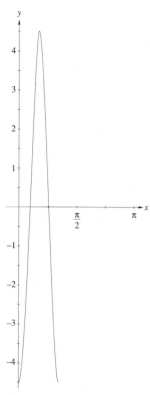

25.

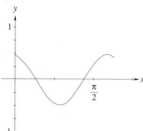

27.

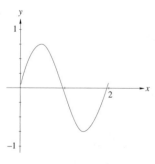

29. (a) Amplitude: 65; Period $\frac{1}{60}$; phase shift: $-\frac{1}{240}$; (b).

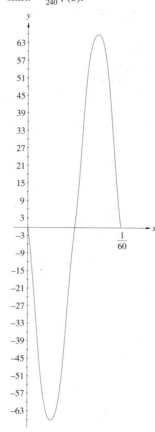

Exercise Set 11.3

1.

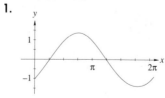

3.

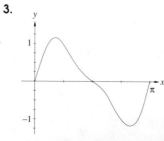

5.

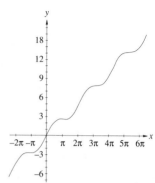

7.

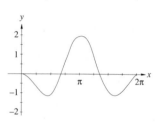

9.

11.

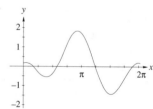

13.

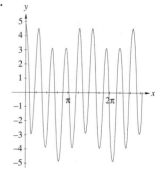

15.

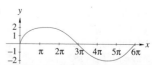

17.

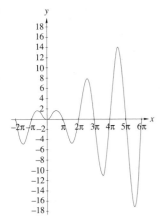

Exercise Set 11.4

1.

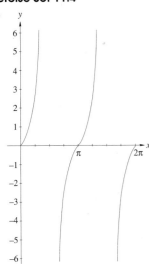

3.

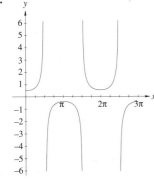

5.

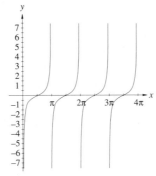

7.

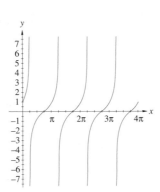

9.

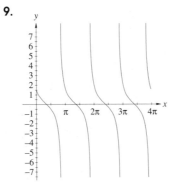

11.

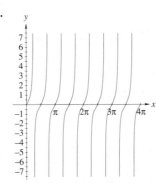

13.

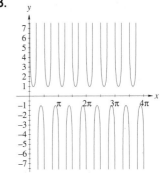

15.

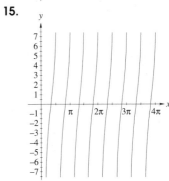

17.

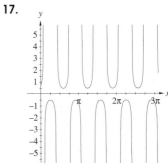

Exercise Set 11.5

1. $y = 10 \sin 8\pi t$ **3.** $y = 0.8 \sin \frac{\pi}{3} t$
5. -2.79 ft/s^2

7. (a) 8.5 cm; (b) $\frac{\pi}{1.4}$; (c) $\frac{1.4}{\pi}$

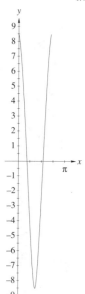

9. amplitude 10 A, period $\frac{1}{60}$ sec/cycle, $f = 60$ Hz, $\omega = 120\pi$ rad/s

11. $I = 6.8 \sin(160\pi t + \pi/3)$

13. $V = 220 \sin(80\pi t - \frac{\pi}{3})$

15. $V = 220 \cos(80\pi t + \frac{\pi}{6})$

17. $f = 10^{23}$, amp $= 10^{-12}$

3.

t	-3	-2	-1	0	1	2	3; $y = \frac{1}{x}$
x	-3	-2	-1	0	1	2	3
y	$-\frac{1}{3}$	$-\frac{1}{2}$	-1	*	1	$\frac{1}{2}$	$\frac{1}{3}$

*Not defined

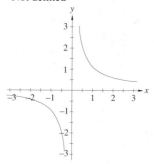

19.

1.

t	-3	-2	-1	0	1	2	3;
x	-3	-2	-1	0	1	2	3
y	-9	-6	-3	0	3	6	9

$y = 3x$

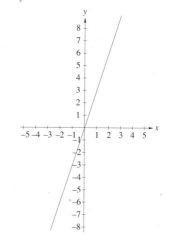

5.

t	-5	-4	-3	-2	-1	0	1	2	3	4	5;
x	8	7	6	5	4	3	2	1	0	-1	-2
y	16	7	0	-5	-8	-9	-8	-5	0	7	16

$y = (x - 3)^2 - 9 = x^2 - 6x$

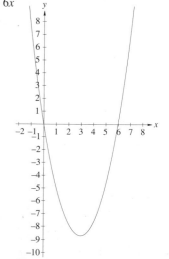

7.

t	0	$\frac{\pi}{4}$	$\frac{\pi}{2}$	$\frac{3\pi}{4}$	π	$\frac{5\pi}{4}$	$\frac{3\pi}{2}$	$\frac{7\pi}{4}$
x	0	$\sqrt{2}$	2	$\sqrt{2}$	0	$-\sqrt{2}$	-2	$-\sqrt{2}$
y	2	$\sqrt{2}$	0	$-\sqrt{2}$	-2	$-\sqrt{2}$	0	$\sqrt{2}$

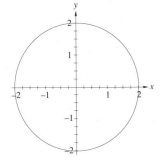

9.

t	0	$\frac{\pi}{4}$	$\frac{\pi}{2}$	$\frac{3\pi}{4}$	π	$\frac{5\pi}{4}$	$\frac{3\pi}{2}$	$\frac{7\pi}{4}$	2π
x	2	$\frac{5\sqrt{2}}{2}$	5	$\frac{5\sqrt{2}}{2}$	0	$-\frac{5\sqrt{2}}{2}$	-5	$-\frac{5\sqrt{2}}{2}$	0
y	0	$\frac{3\sqrt{2}}{2}$	3	$\frac{3\sqrt{2}}{2}$	0	$-\frac{3\sqrt{2}}{2}$	-3	$-\frac{3\sqrt{2}}{2}$	0

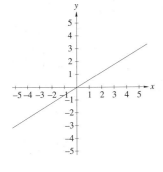

11.

t	-5	-4	-3	-2	-1	0	1
x	-5.96	-4.76	-2.86	-1.09	-0.16	0	0.16
y	0.72	1.65	1.99	1.41	0.46	0	0.46

t	2	3	4	5
x	1.09	2.86	4.76	5.96
y	1.41	1.99	1.65	0.72

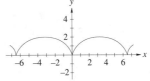

13.

t	0	$\frac{\pi}{6}$	$\frac{\pi}{4}$	$\frac{\pi}{3}$	$\frac{\pi}{2}$	$\frac{2\pi}{3}$	$\frac{3\pi}{4}$	$\frac{5\pi}{6}$
x	1	$\frac{2}{\sqrt{3}}$	$\sqrt{2}$	2	$*$	-2	$-\sqrt{2}$	$-\frac{2}{\sqrt{3}}$
y	$*$	4	$2\sqrt{2}$	$\frac{4}{\sqrt{3}}$	2	$\frac{4}{\sqrt{3}}$	$2\sqrt{2}$	4

t	π	$\frac{7\pi}{6}$	$\frac{5\pi}{4}$	$\frac{4\pi}{3}$	$\frac{3\pi}{2}$	$\frac{5\pi}{3}$	$\frac{7\pi}{4}$	$\frac{11\pi}{6}$	2π
x	-1	$-\frac{2}{\sqrt{3}}$	$-\sqrt{2}$	-2	$*$	2	$\sqrt{2}$	$\frac{2}{\sqrt{3}}$	1
y	$*$	-4	$-2\sqrt{2}$	$-\frac{4}{\sqrt{3}}$	-2	$-\frac{4}{\sqrt{3}}$	$-2\sqrt{2}$	-4	$*$

*Not defined

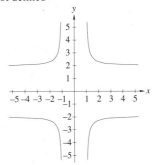

15. (a) $1 : 1$;

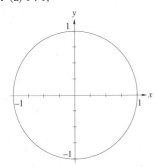

17. (a) $2 : 1$;

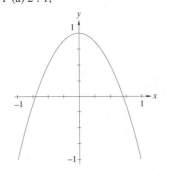

19. (a) $1 : 4$;

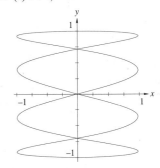

21. (a) 2 : 3;

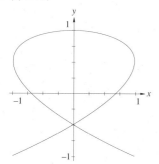

23. (a) 2 : 5;

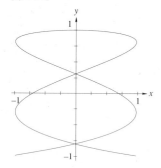

25.

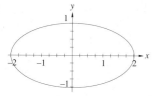

27.

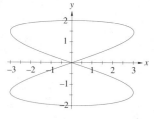

29.

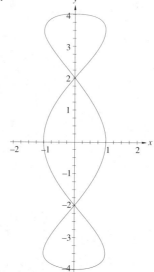

Exercise Set 11.7

1.

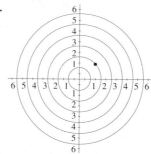

3.

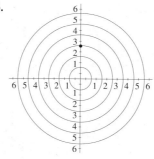

5.

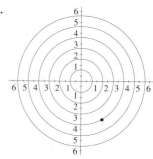

7.

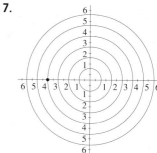

9.

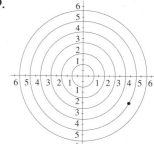

11.

13. (2, 3.46) **15.** (−1.414, 1.414) or
(−√2, √2) **17.** (−5.64, −2.05)
19. (0.80, 1.83) **21.** (−2.95, −0.52)
23. (4.81, 3.59) **25.** (5.66, 45°) ≈
(4√2, 45°) = (4√2, $\frac{\pi}{4}$)
27. (5, 36.87°) = (5, 0.64)

29. $(29,\ 133.60°) = (29,\ 2.33)$
31. $(5,\ 126.87°) = (5,\ 2.21)$
33. $(12.21,\ 235.01°) = (12.21,\ 4.102)$
35. $(9.22,\ 77.47°) = (9.22,\ 1.352)$
37.

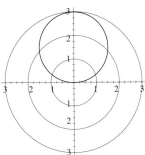

39.

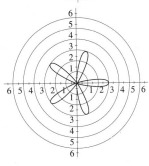

41.

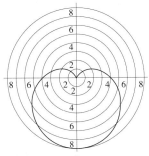

43.

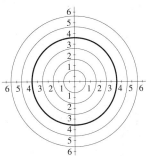

45.

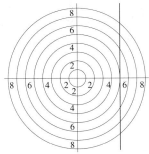

47.

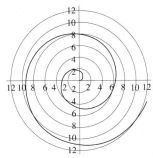

49.

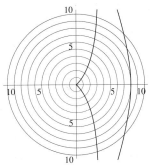

51.

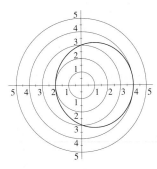

53.

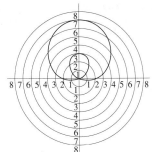

55.

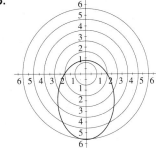

57.

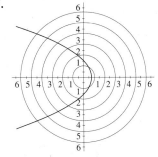

Review Exercises

1. period $\frac{\pi}{2}$, amplitude 8, frequency $\frac{2}{\pi}$, displacement 0

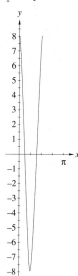

3. period $\frac{\pi}{3}$, amplitude ∞, frequency $\frac{3}{\pi}$, displacement 0

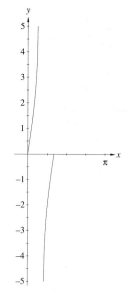

5. period $\frac{2\pi}{3}$, amplitude $\frac{1}{2}$, frequency $\frac{3}{2\pi}$, displacement $\frac{\pi}{9}$

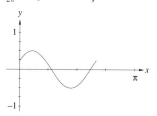

7. period π, amplitude ∞, frequency $\frac{1}{\pi}$, displacement $-\frac{\pi}{4}$

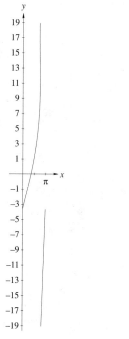

9. period 3π, amplitude 2, frequency $\frac{1}{3\pi}$, displacement $-\frac{\pi}{4}$

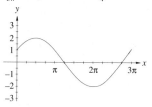

11.

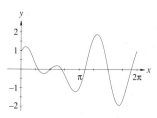

13.

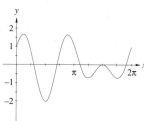

15.

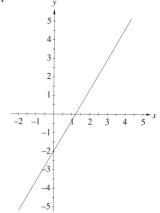

17.

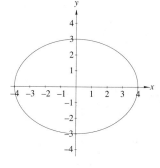

19.

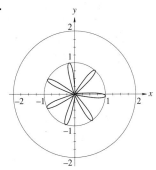

21. $y = 85 \sin 10\pi t$

23. $I = 4.8 \sin(120\pi t + \frac{\pi}{2})$

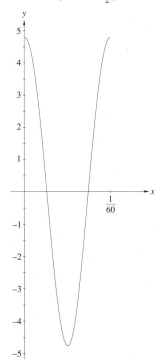

Chapter 11 Test

1. (a) period $\frac{2\pi}{5}$, amplitude 3, frequency $\frac{5}{2\pi}$, displacement 0
(b) period $\frac{2\pi}{3}$, amplitude 2.4, frequency $\frac{3}{2\pi}$, displacement $\frac{\pi}{12}$
(c) period $\frac{\pi}{2}$, amplitude ∞ , frequency $\frac{2}{\pi}$, displacement $-\frac{\pi}{10}$

3.

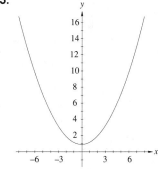

5. (1.147, 1.638)
7. $y = 5.7 \sin 120\pi t$

☰ ANSWERS FOR CHAPTER 12

Exercise Set 12.1

1. 5 **3.** 4 **5.** 1/5 **7.** 1/2 **9.** 9 **11.** 25
13. 1/8 **15.** $-1/2$ **17.** 1/2 **19.** 32 **21.** 3^7
23. 7^4 **25.** x^{10} **27.** y^2 **29.** 9^{10} **31.** x^{21}
33. $x^5 y^5$ **35.** $1/a^5 b^5$ **37.** x^8 **39.** $1/x^6$ **41.** x^2
43. $r^{7/4}$ **45.** $a^{5/6}$ **47.** $d^{5/12}$ **49.** b^3/a^3 **51.** r^2/s^3
53. y^6 **55.** a^5/b^{12} **57.** $1/yb^{13/2} = \frac{1}{yb^{13/2}}$ **59.** p^3/x^2
61. 4.0993852 **63.** 25.368006 **65.** 3.5162154
67. 4.5606226 **69.** 221.1125 m $\approx$ 221 m
71. $p = 905\,146.3$ N/m^2 $T = 91\,238.747$ K

Exercise Set 12.2

1. $2\sqrt[3]{2}$ **3.** $3\sqrt{5}$ **5.** y^4 **7.** $a\sqrt[5]{a^2}$ **9.** $xy^3\sqrt{y}$
11. $a\sqrt[4]{ab^3}$ **13.** $2x\sqrt[3]{x}$ **15.** $3x\sqrt{3xy}$ **17.** -2
19. ab^3 **21.** $pq^2 r\sqrt[4]{r^3}$ **23.** $2x/3$ **25.** xy^2/z
27. $\frac{2x\sqrt[3]{2y^2}}{z^2}$ **29.** $8xy^2\sqrt{x}/3z^2 p = \frac{8xy^2\sqrt{x}}{3z^2 p}$
31. $4\sqrt{3}/3 = \frac{4\sqrt{3}}{3}$ **33.** $3\sqrt[3]{2}/2 = \frac{3\sqrt[3]{2}}{2}$
35. $5\sqrt{2x}/2x = \frac{5\sqrt{2x}}{2x}$ **37.** $3\sqrt[4]{8z^2}/4z = \frac{3\sqrt[4]{8z^2}}{4z}$
39. $2\sqrt[3]{2y}/x = \frac{2\sqrt[3]{2y}}{x}$ **41.** $2x\sqrt[3]{by}/3bz = \frac{2x\sqrt[3]{by}}{3bz}$

43. $2 \times 10^2 = 200$ **45.** $50\sqrt{10}$
47. $2\sqrt{10} \times 10^3 = 2000\sqrt{10}$ **49.** $\sqrt[3]{12.5} \times 10^3$
51. $\sqrt{\frac{x^2+y^2}{xy}} = \frac{\sqrt{x^3 y+xy^3}}{xy}$ **53.** $|a+b|$
55. $\sqrt{\frac{b+a^2}{a^2 b}} = \frac{\sqrt{b^2+a^2 b}}{ab}$ **57.** $f = \frac{\sqrt{T\mu}}{2L\mu} = \frac{1}{2L\mu}\sqrt{T\mu}$
59. $z = \frac{1}{\sqrt{\frac{R^2+x^2}{R^2 x^2}}} = \sqrt{\frac{R^2 x^2}{R^2+x^2}} = \frac{Rx\sqrt{R^2+x^2}}{R^2+x^2}$

Exercise Set 12.3

1. $7\sqrt{3}$ **3.** $5\sqrt[3]{9}$ **5.** $8\sqrt{3}+4\sqrt{2}$ **7.** $3\sqrt{5}$
9. $-\sqrt{7}$ **11.** $5\sqrt{15}/3 = \frac{5}{3}\sqrt{15}$ **13.** $-\sqrt{2}$
15. $3x\sqrt{xy}$ **17.** $(2q+p^2)\sqrt[3]{3p^2 q}$ **19.** $\frac{x^2-y^2}{x^2 y^2}\sqrt{xy}$
21. $2\sqrt{3ab}/3$ **23.** $\sqrt{40} = 2\sqrt{10}$ **25.** $x\sqrt{15}$
27. $8x\sqrt{x}$ **29.** 3/4 **31.** $\sqrt{2x}+2$ **33.** $x+2\sqrt{xy}+y$
35. $\frac{\sqrt[3]{5\cdot 7^2}}{7} = \frac{\sqrt[3]{245}}{7}$ **37.** $|a|-|b|$ **39.** $\sqrt[6]{x^5}$ **41.** 4
43. $\sqrt[3]{2b}/2$ **45.** $\frac{x-\sqrt{5}}{x^2-5}$ **47.** $\frac{(\sqrt{5}-\sqrt{3})^2}{2} = 4-\sqrt{15}$
49. $2x\sqrt{x^2-1}/(x^2-1), x \neq \pm 1$
51. $-\frac{\sqrt{x^2-y^2}+\sqrt{x^2+xy}}{y}, y \neq 0$ **53.** $-\frac{b}{a}$ **55. (a)** $R = \frac{x^{3/2}}{x+1}$,
(b) 4.259 Ω **57.** $\frac{c}{a}$

Exercise Set 12.4

1. 22   **3.** 22.5   **5.** no solution   **7.** 15   **9.** 16
11. 10   **13.** 2   **15.** $\frac{2829}{196} \approx 14.43$ **17.** no solution
19. no solution   **21.** $R = \frac{v^2}{\mu_s g}$ **23.** $I_A = 1.419$;
$I_B = 4.691$

Review Exercises

1. 7   **3.** 125   **5.** 1/27   **7.** 5^{13} **9.** 2^{32}
11. $x^8 y^{12}$ **13.** $y^5 a^5$ **15.** $x^{5/6}$ **17.** $xy^8/a = \frac{xy^8}{a}$
19. -3 **21.** $ab^2 \sqrt{b}$ **23.** $\frac{-2xy^2}{z} \sqrt[3]{z^2}$ **25.** $\frac{\sqrt{a^3 - b^3}}{ab}$

27. $8\sqrt{5}$ **29.** $8\sqrt{6} - 2\sqrt{3}$ **31.** $(2 + 3a)\sqrt[4]{a^3 b}$
33. $\frac{(a^2 - b^2)\sqrt{7ab}}{7ab}$ **35.** $\sqrt{15}$ **37.** $x\sqrt{14}$ **39.** $a - 2b$
41. $\frac{\sqrt[3]{a}}{2}$ **43.** 82   **45.** no solution   **47.** 18.14   **49.** 2
51. **(a)** 28.72 mg; **(b)** 16.49 mg   **53.** $v = \sqrt{25t^2 + 4t + 5}$

Chapter 12 Test

1. 8^{-2} **3.** $\frac{x^5}{y^{15}}$ **5.** -4 **7.** $x^2 y \sqrt{y}$ **9.** $x + 3$
11. $2\sqrt{3}x$ **13.** $\frac{\sqrt[3]{2y^2}}{y}$ **15.** 12   **17.** 27

☰ ANSWERS FOR CHAPTER 13

Exercise Set 13.1

1. 4.7288   **3.** 31.0063   **5.** 3.4154

7.

x	−3	−2	−1	0	1	2	3	4	5
f(x)	$\frac{1}{64}$	$\frac{1}{16}$	$\frac{1}{4}$	1	4	16	64	256	1024

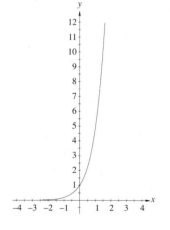

9.

x	−3	−2	−1	0	1	2	3	4	5
h(x)	$\frac{8}{27}$	$\frac{4}{9}$	$\frac{2}{3}$	1	1.5	2.25	3.375	5.0625	7.59375

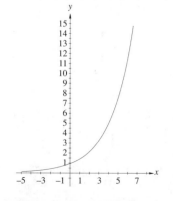

11.

x	−3	−2	−1	0	1	2	3	4	5
f(x)	27	9	3	1	$\frac{1}{3}$	$\frac{1}{9}$	$\frac{1}{27}$	$\frac{1}{81}$	$\frac{1}{243}$

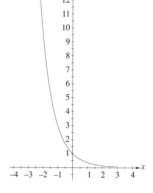

13.

x	−3	−2	−1	0	1	2	3
h(x)	13.824	5.760	2.4	1	0.417	0.174	0.072

x	4	5
h(x)	0.030	0.0126

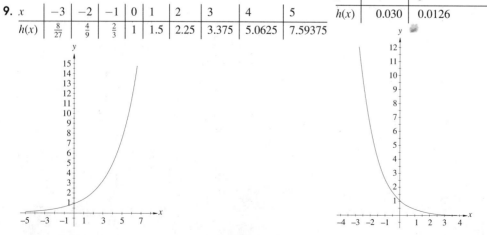

15.

x	-3	-2	-1	0	1	2
$f(x)$	0.064	0.192	0.577	1.732	5.196	15.588

x	3	4
$f(x)$	46.765	140.296

11.

x	-4	-3	-2	-1	0
$h(x)$	218.3926	80.3421	29.5562	10.8731	4

x	1	2
$h(x)$	1.4715	0.5413

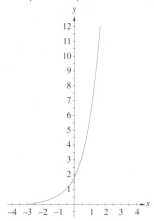

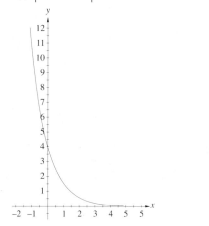

17. (a) \$1338.23; (b) \$1343.92; (c) \$1346.86; (d) \$1348.85
19. \$4.93

Exercise Set 13.2
1. 20.085 537   **3.** 104.584 986   **5.** 0.018 316
7. 0.063 928

9.

x	-2	-1	0	1	2
$f(x)$	0.5413	1.4715	4	10.8731	29.5562

x	3	4
$f(x)$	80.3421	218.3926

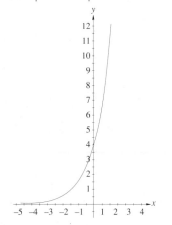

13. (a) 125 mg; (b) 85.91 mg; (c) 0.31 mg   **15.** (a) 8660;
(b) 25,981; (c) 260,980   **17.** (a) 280,510; (b) 393,430
19. 110.6°F   **21.** 110.26 min or about 1 hour 50 minutes
23. 0.0066 C

Exercise Set 13.3
1. $6^3 = 216$ **3.** $4^2 = 16$ **5.** $(\frac{1}{7})^2 = \frac{1}{49}$
7. $2^{-5} = \frac{1}{32}$ **9.** $9^{7/2} = 2{,}187$ **11.** $\log_5 625 = 4$
13. $\log_2 128 = 7$ **15.** $\log_7 343 = 3$ **17.** $\log_5 \frac{1}{125} = -3$
19. $\log_4 128 = \frac{7}{2}$ **21.** 1.609 437 912   **23.** 1.558 355 122
25. 0.602 059 991 3   **27.** 1.102 776 615   **29.** 1.292 029 674
31. 1.125 708 821

33.

x	0.1	0.5	1	2	4	6	8	10
$\ln x$	-2.30	-0.69	0	0.69	1.39	1.79	2.08	2.30

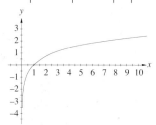

35.

x	0.1	0.5	1	2	4	6	8	10
$\log_2 x$	-3.32	-1	0	1	2	2.58	3	3.32

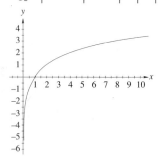

37.

x	0.1	0.5	1	2	4	6
$\log_{12} x$	-0.93	-0.28	0	0.28	0.56	0.72

x	8	10
$\log_{12} x$	0.84	0.93

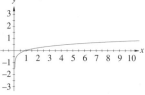

39.

x	0.1	0.5	1	2	4	6	8	10
$\log_{1/4} x$	1.66	0.5	0	-0.5	-1	-1.29	-1.5	-1.66

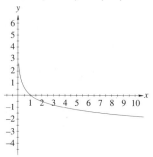

41. (a) $I = I_0 10^{\beta/10}$; (b) 50 dB; (c) 90 dB

Exercise Set 13.4
 1. $\log 2 - \log 3$ **3.** $\log 2 + \log 7$
 5. $\log 2 + \log 2 + \log 3$
 7. $\log 2 + \log 3 + \log 5 + \log 5 - \log 7$ **9.** $\log 2 + \log x$
 11. $\log 2ax - \log 3y$ **13.** $\log 22$ **15.** $\log \frac{11}{3}$
 17. $\log 12$ **19.** $\log \frac{4x}{y}$ **21.** $\log(2^5 \cdot 5^3) = \log 4{,}000$
 23. $\log \frac{2}{3} \cdot \frac{6}{7} = \log \frac{4}{7}$ **25.** 0.9030   **27.** 1.0791
 29. 1.1761   **31.** 1.5741   **33.** 2.3010   **35.** 3.6990
 37. 1.5476   **39.** 0.2851   **41.** 6

Exercise Set 13.5
 1. 2.092   **3.** 0.575   **5.** -2.476 **7.** 2.651
 9. 5.5   **11.** 2.609   **13.** $10^{2.3} \approx 199.53$ **15.** $10^{17} + 5$
 17. $\sqrt{\frac{1}{2} e^9} \approx 63.6517$ **19.** $\frac{1}{18} \approx 0.056$ **21.** 101

23. 4.040404   **25.** 11.55 yr   **27.** 138.63 yr   **29.** 5.5%
31. 8.2%   **33.** 5.6   **35.** 4418.07 m   **37.** 0.104 s
39. 0.1354 seconds

Exercise Set 13.6
 1.

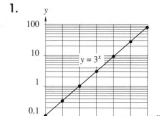

3.

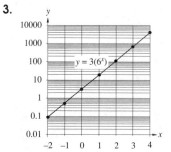

5.

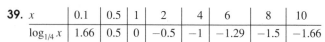

7.

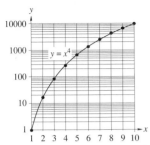

9.

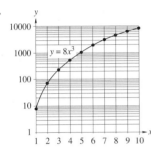

11.

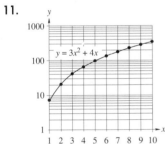

13.

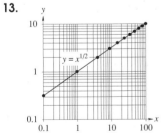

15.

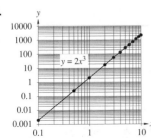

17.

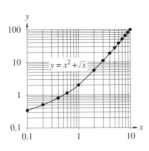

19.

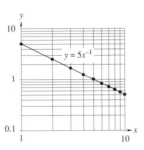

21.

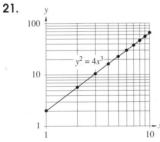

23.

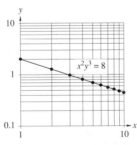

25.

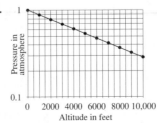

27.

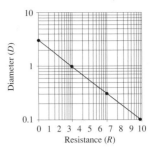

29.

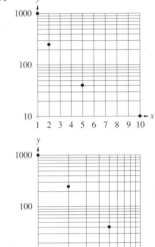

$y = 1000x^{-2}$

Review Exercises

1. 148.413 16   **3.** 106.697 74
5. 0.903 09   **7.** 4.398 146

9.

x	−2	−1	0	1	2	3	4
$5e^x$	0.677	1.839	5	13.59	36.95	100.43	272.99

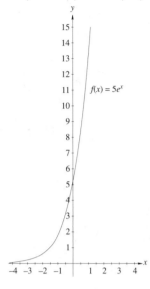

11.

x	−3	−2	−1	0	1	2
5^x	0.008	0.04	0.2	1	5	25

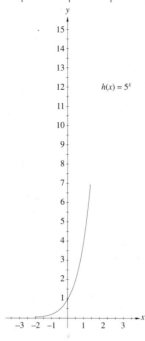

13.

x	0	1	2	3	4	5	6	7	8	9	10
$\log x$	not defined	0	0.301	0.477	0.602	0.699	0.778	0.845	0.903	0.959	1

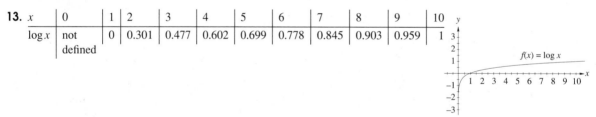

15.

x	$-\frac{1}{2}$	0	1	2	3	4	5	6	7	8	9	10
$\ln(2x+1)$	not defined	0	1.099	1.609	1.946	2.197	2.398	2.565	2.708	2.833	2.944	3.045

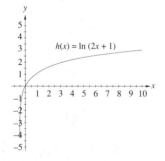

17. $\log 3 - \log 4$ **19.** $\log 5 + \log x$ **21.** $3(\ln 4 + \ln x)$

23. $\frac{1}{2}(\log 6 + \log 8)$ **25.** $\log 45$ **27.** $\log x^5$ **29.** $\log x^8$

31. 2.404   **33.** 0.480   **35.** $e^{9.1} \approx 8955.29$

37. $0.5 \times 10^{50} - 0.5$

39.

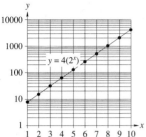

41.

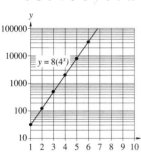

43.

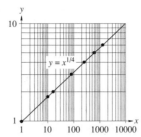

45.

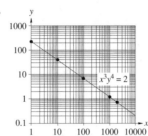

47. **(a)** $2088.15 in the first bank (compounded semiannually); **(b)** $2112.06 in the second bank (compounded monthly); **(c)** $2117.00 in the third bank (compounded continuously); **(d)** 9.24 years **49. (a)** 10,000;

(b) 156,250
51. $y = 0.4(6.2764)^x$
53. (a) $I = I_0 \cdot 10^{(6-M)/2.5}$;
  (b) $10^{5.8} \approx 630,957.34$ times

Chapter 13 Test

1. 200.33681 **3.** 4.281 515 453
5. $\log 5 + \log 13 + \log x$
7. $\frac{1}{5}(\log 5 + 3\log x) - \log 9 - \log x$
9. $\log \frac{17}{x^2}$ **11.** $e^{17.2} \approx 29\,502\,925.92$
13. 3.236 **15.** $10^5 - 2 \approx 99\,998$
17.

x	1	2	3	4
$0.4(3^x)$	1.2	3.6	10.8	32.4

x	5	6	7
$0.4(3^x)$	97.2	291.6	874.8

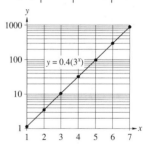

ANSWERS FOR CHAPTER 14

Exercise Set 14.1

1. $5j$ **3.** $0.2j$ **5.** $5j\sqrt{3}$ **7.** $-6j\sqrt{5}$ **9.** $\frac{3}{4}j$
11. $-3j$ **13.** -11 **15.** -18 **17.** -10 **19.** -42
21. $-5j$ **23.** $0.5j$ **25.** $-2j\sqrt{8.1} = -6j\sqrt{0.9}$
27. $-2j\sqrt{15}$ **29.** $-14j$ **31.** $x = 7, y = -2$
33. $x = 10, y = -3$ **35.** $x = 2, y = -2$ **37.** $8 + 10j$
39. $-10 - 5j$ **41.** $2 + 3j$ **43.** $-6 + 8j$ **45.** $3 - 4j$
47. $1 - 2j$ **49.** $2 + j\sqrt{2}$ **51.** $2 - \frac{1}{2}j\sqrt{6}$ **53.** $7 - 2j$
55. $6 + 5j$ **57.** 19 **59.** $8j$ **61.** $-\frac{1}{2} + \frac{3}{2}j, -\frac{1}{2} - \frac{3}{2}j$
63. $3j, -3j$ **65.** $-\frac{3}{4} + \frac{j\sqrt{47}}{4}, -\frac{3}{4} - \frac{j\sqrt{47}}{4}$ **67.** $-\frac{1}{5} \pm \frac{2j\sqrt{6}}{5}$

Exercise Set 14.2

1. $-1 + 7j$ **3.** $5 - 2j$ **5.** $7 - j$ **7.** $2 - \sqrt{5} + 11j$
9. $8 + 2j$ **11.** $3 - 6j$ **13.** $7 - 2j$ **15.** $-3 + 6j$
17. $10 - 45j$ **19.** $7 + 11j$ **21.** $36 + 8j$ **23.** $-126 + 48j$
25. 9 **27.** $-3 + 4j$ **29.** $48 - 14j$ **31.** 29
33. $1 - 5j$ **35.** $-3j$ **37.** $2j$ **39.** $\frac{1}{13} + \frac{5}{13}j$
41. $-\sqrt{2} - j$ **43.** $4.059 - 0.765j$ **45.** -4
47. $-0.5 + \frac{j\sqrt{3}}{2}$ **49.** $(a + bj) + (a - bj) = 2a$, which is a real
  number. **51.** $(a + bj)(a - bj) = a^2 + b^2$, which is a real
  number. **53.** $10.53 - 4.21j$ V **55.** $5.08 + 2.88j$ V
57. $5.84 + 1.16j\ \Omega$ **59.** $15 - 5j\ \Omega$ **61.** $0.74 + 1.12j$ A
63. See *Computer Programs* **65.** See *Computer Programs*

Exercise Set 14.3

1. $2\sqrt{3} + 2j$

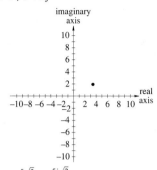

3. $-\frac{5\sqrt{2}}{2} + \frac{5j\sqrt{2}}{2} \approx -3.5355 + 3.5355j$

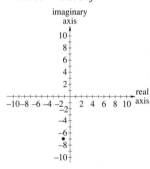

5. $-1.2155 - 6.8937j$

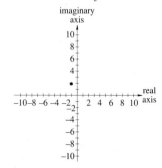

7. $-0.8452 + 1.8126j$

9. $2.7189 + 1.2679j$

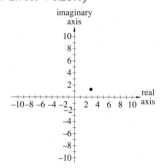

11. $4.6985 - 1.7101j$

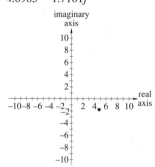

13. $-1.9018 - 4.0784j$

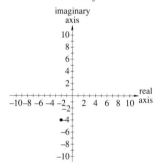

15. -2.5

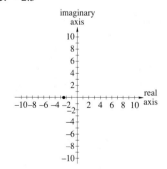

17. $7.071 \underline{/45°}$

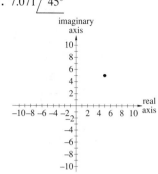

19. $8.944 \underline{/296.6°}$

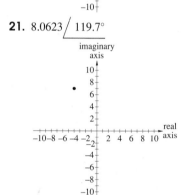

21. $8.0623 \underline{/119.7°}$

23. $6.3246 \underline{/198.4°}$

25. $6\underline{/\,0°}$

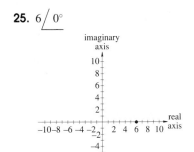

27. $7.5717\underline{/\,303.7°}$

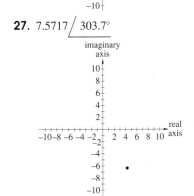

29. $5.8034\underline{/\,178.0°}$

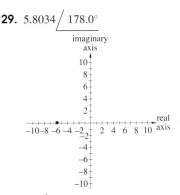

31. $2.7\underline{/\,180°}$

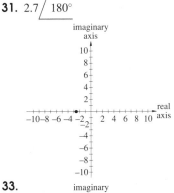

33.

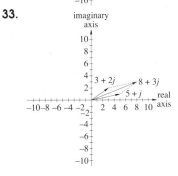

35.

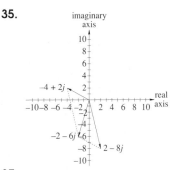

37.

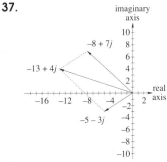

39.

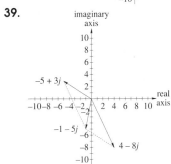

41. See *Computer Programs*

Exercise Set 14.4

1. $3e^{3\pi/2j} \approx 3e^{4.7j}$ **3.** $2e^{1.05j}$ **5.** $1.3e^{5.7j}$
7. $3.1e^{0.44j}$ **9.** $10e^{0.64j}$ **11.** $15e^{2.21j}$
13. $5\text{ cis } 0.5 = 4.3879 + 2.3971j$
15. $2.3\text{ cis } 4.2 = -1.1276 - 2.0046j$ **17.** $12e^{5j}$
19. $2.4e^{5.9j}$ **21.** $28e^{3.72j}$ **23.** $4e^{2j}$ **25.** $4.25e^{1.5j}$
27. $81e^{8j}$ *or* $81e^{1.72j}$ **29.** $39.0625e^{6j}$ **31.** $2e^{3j}$
33. $2.5e^{2.1j}$ **35.** $61.4e^{0.4031j}\text{V}$ **37.** In exponential form
$Z = 135e^{-0.9163j}\,\Omega$ or $135e^{5.37j}$; in rectangular form
$Z = 82.2 - 107.1j\,\Omega$ **39.** $V = 80e^{0.7029j}$ V
41. $I = 46e^{5.49j}$ A

Exercise Set 14.5

1. $15(\cos 69° + j\sin 69°)$ **3.** $10(\cos 4.1 + j\sin 4.1)$
5. $4(\cos 60° + j\sin 60°)$ **7.** $4.5\underline{/\,111°}$ **9.** $12\underline{/\,8.0}$

11. $15625(\cos 144° + j\sin 144°)$
13. $1124.864(\cos 10.26 + j\sin 10.26)$
15. $3.4^4\underline{/\,21.2} = 133.6\underline{/\,21.2}$ **17.** $12e^{3.8j}$
19. $0.125e^{0.9j}$ **21.** $1, -0.5 + 0.866j, -0.5 - 0.866j$
23. $1.732 - j, 2j, -1.732 - j$ **25.** $1.8478 - 0.7654j,$
$0.7654 + 1.8478j, -1.8478 + 0.7654j, -0.7654 - 1.8478j$
27. $1.0586 + 0.1677j, 0.1677 + 1.0586j, -0.9550 + 0.4866j,$
$-0.7579 - 0.7579j, 0.4866 - 0.9550j$ **29.** $0.866 - 0.5j,$
$j, -0.866 - 0.5j$ **31.** $1.9319 + 0.5176j,$
$0.5176 + 1.9319j, -1.4142 + 1.4142j, -1.9319 - 0.5176j,$
$-0.5176 - 1.9319j, 1.4142 - 1.4142j$ **33.** $108\underline{/\,19°}$

35. See *Computer Programs*   **37.** See *Computer Programs*

Exercise Set 14.6

1. (a) $3 - 2j$, (b) $5 + j$ **3.** (a) $1 + 2j$, (b) $1.8 - 0.6j$

5. (a) $6.803 \,\underline{/\,53.26°}$, (b) $1.6611 \,\underline{/\,32.44°}$

7. (a) $8.208 \,\underline{/\,0.449}$, (b) $1.226 \,\underline{/\,0.45}$ **9.** (a) $12 + 5j$,

   (b) $1.9207 + 0.2387j$ **11.** (a) $6.595 \,\underline{/\,-2.95°}$,

   (b) $0.916 \,\underline{/\,29.32°}$ **13.** $-13 - 84j$ V   **15.** 2 V

17. $2.8 \,\underline{/\,23.7°}$ Ω   **19.** $X_L = 75.40$ Ω, $X_C = 66.31$ Ω,

   $Z = 39.07$ Ω, $ø = 13.45°$ **21.** $X_L = 150.80$ Ω,

   $X_C = 44.21$ Ω, $Z = 108.45$ Ω, $\phi = 79.37°$

23. $X_L = 12.5$ Ω, $X_C = 100$ Ω, $Z = 91.87$ Ω, $\phi = -72.26°$

25. (a) $77.62 \,\underline{/\,14.93°}$, (b) 271.67 V   **27.** (a) $\sqrt{10} \,\underline{/\,18.43°}$,

   (b) 9.01 V   **29.** $2 - 9j$ **31.** 22.51 Hz

33. $0.16 - 0.12j$ **35.** See *Computer Programs*

Review Exercises

1. $7j$ **3.** $3\sqrt{6}j$ or $3j\sqrt{6}$ **5.** -6 **7.** $9 - 3j$

9. $-1 + 4j$ **11.** $-36 + 8j$ **13.** 25

15. $10.82 \,\underline{/\,-33.69°}$ *or* $10.82 \,\underline{/\,326.31°}$

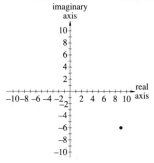

17. $5.657 \,\underline{/\,-45°}$

19. $-4.331 + 4.847j$

21. -10 **23.** $-4{,}782{,}969j$ **25.** $-26.080 + 6.988j$

27. 4096   **29.** $0.866 - 0.5j, j, -0.866 - 0.5j$

31. $2 + 3.464j, -2 - 3.464j$ **33.** $18.385e^{0.391j}$

35. $6740.6e^{2.702j}$ **37.** magnitude: 7.433, direction: $-10.86°$

39. $X_L = 113.10$ Ω, $X_C = 53.05$ Ω, $Z = 81.43$, $\phi = 47.51°$

Chapter 14 Test

1. $4j\sqrt{5}$ **3.** $-4\sqrt{3} + 4j$ **5.** $-13 + 8j$

7. $-\frac{38}{25} + \frac{9}{25}j = -1.52 + 0.36j$ **9.** $\frac{4}{3}j$

11. $2.1941 + 2.0460j$ **13.** (a) $9 - j$ Ω,

   (b) $\frac{118}{41} - \frac{4}{41}j$ Ω $\cong 2.878 + 0.098j$

■ ANSWERS FOR CHAPTER 15

Exercise Set 15.1

1. $\sqrt{194}$ **3.** $\sqrt{65}$ **5.** $\sqrt{277}$ **7.** 7

9. $(\frac{9}{2}, -\frac{5}{2})$ **11.** $(1, -\frac{11}{2})$ **13.** $(\frac{15}{2}, -6)$

15. $(\frac{3}{2}, 5)$ **17.** $-\frac{13}{5} = -2.6$ **19.** $-\frac{1}{8}$ **21.** $\frac{14}{9}$

23. 0   **25.** $y - 4 = -2.6(x - 2)$ or $y = -2.6x + 9.2$

27. $y + 6 = -\frac{1}{8}(x - 5)$ or $y = -\frac{1}{8}x - 5\frac{3}{8}$

29. $y - 1 = \frac{14}{9}(x - 12)$ or $y = \frac{14}{9}x - 17\frac{2}{3}$ **31.** $y = 5$

33. 1   **35.** approximately 0.1511352   **37.** 68.19859°

39. 153.43495°   **41.** $-\frac{1}{3}$ **43.** 2   **45.** $y + 5 = 6(x - 2)$

   or $y = 6x - 17$ **47.** $y + 2 = -\frac{5}{2}(x - 4)$ or $y = -\frac{5}{2}x + 8$

49. $y + 4 = 1.732(x + 2)$ or $y = 1.732x - 0.536$

51. $y = 0.364x + 3$ **53.** $2x - 3y + 11 = 0$

55. $5x - 2y = 22$ **57.** $m = -\frac{3}{2}$, y-intercept = 6,

   x-intercept = 4   **59.** $m = \frac{1}{3}$, y-intercept = -3,

   x-intercept = 9   **61.** $v = 2.6 + 0.4t$ **63.** 0.8Ω

Exercise Set 15.2

1. $(x - 2)^2 + (y - 5)^2 = 9$; $x^2 + y^2 - 4x - 10y + 20 = 0$

3. $(x + 2)^2 + y^2 = 16$; $x^2 + y^2 + 4x - 12 = 0$

5. $(x + 5)^2 + (y + 1)^2 = \frac{25}{4}$; $x^2 + y^2 + 10x + 2y + 19.75 = 0$

7. $(x - 2)^2 + (y + 4)^2 = 1$; $x^2 + y^2 - 4x + 8y + 19 = 0$

9. $C = (3, 4)$, $r = 3$

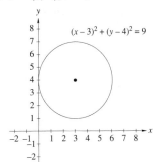

11. $C = (-\frac{1}{2}, -\frac{13}{4})$, $r = \sqrt{7}$

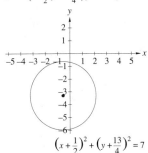

13. $C = (0, \frac{7}{3})$, $r = \sqrt{6}$

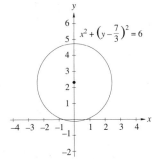

15. Circle, $C = (-2, 3)$, $r = 3$
17. Circle, $C = (-5, 3)$, $r = 9$
19. Not a circle since $r^2 = -1$
21. Circle, $C = (-3, 0)$, $r = 5$
23. Circle, $C = (-\frac{5}{2}, \frac{9}{2})$, $r = 6$
25. Not a circle since $r^2 = -1.3$
27. $x^2 + y^2 = 330.7^2$ or
$x^2 + y^2 \approx 1.094 \times 10^5$ (Assume
circle is at the center of a coordinate
system.)
29. $(x - 1.495 \times 10^8)^2 + y^2 = 1.478 \times 10^{11}$
31. $x^2 + (y - 0.9)^2 = 0.55^2$
33. **(a)** $x^2 + (y - 5)^2 = 13^2$; **(b)** 28 cm

Exercise Set 15.3

1. $F = (0, 1)$, directrix: $y = -1$, opens upward

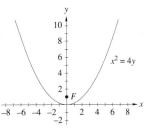

3. $F = (-1, 0)$, directrix: $x = 1$, opens left

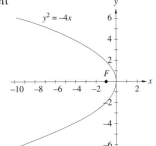

5. $F = (0, -2)$, directrix: $y = 2$, opens downward

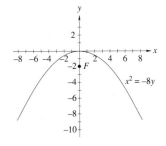

7. $F = (\frac{5}{2}, 0)$, directrix: $x = -\frac{5}{2}$, opens right

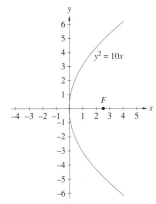

9. $F = (0, \frac{1}{2})$, directrix: $y = -\frac{1}{2}$, opens upward

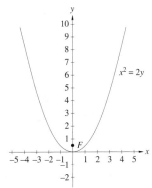

11. $F = (-\frac{21}{4}, 0)$, directrix: $x = \frac{21}{4}$, opens left

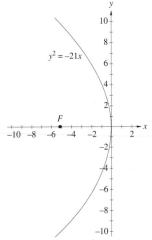

13. $x^2 = 16y$ **15.** $y^2 = -24x$

17. $y^2 = 8x$ **19.** $y^2 = 6x$
21. 27.78 m   **23.** $\frac{6.25}{6} \approx 1.042$ m
from vertex

Exercise Set 15.4

1. $\frac{x^2}{36} + \frac{y^2}{20} = 1$ **3.** $\frac{x^2}{12} + \frac{y^2}{16} = 1$
5. $\frac{x^2}{25} + \frac{y^2}{16} = 1$ **7.** $\frac{x^2}{16} + \frac{y^2}{9} = 1$
9. $V = (0, 3), V' = (0, -3), F = (0, \sqrt{5}), F' = (0, -\sqrt{5}),$
$M = (2, 0), M' = (-2, 0)$

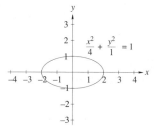

11. $V = (2, 0), V' = (-2, 0), F = (\sqrt{3}, 0), F' = (-\sqrt{3}, 0),$
$M = (0, 1), M' = (0, -1)$

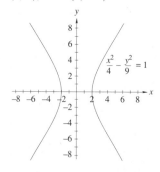

13. $V = (0, 2), V' = (0, -2), F = (0, \sqrt{3}), F' = (0, -\sqrt{3}), M = (1, 0), M' = (-1, 0)$

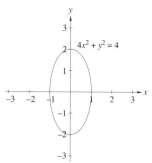

15. $V = (6, 0), V' = (-6, 0), F = (\sqrt{11}, 0), F' = (-\sqrt{11}, 0),$
$M = (0, 5), M' = (0, -5)$

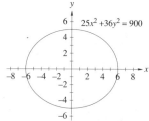

17. 2.992×10^8 km   **19.** major
axis $= \sqrt{200} \approx 14.14$; minor
axis $= 10$ **21.** 0.9999253

Exercise Set 15.5

1. $\frac{x^2}{16} - \frac{y^2}{20} = 1$ **3.** $\frac{y^2}{9} - \frac{x^2}{16} = 1$
5. $\frac{x^2}{9} - \frac{y^2}{16} = 1$ **7.** $\frac{x^2}{7} - \frac{y^2}{9} = 1$
9. $V = (2, 0), V' = (-2, 0), F = (\sqrt{13}, 0), F' = (-\sqrt{13}, 0), M = (0, 3), M' = (0, -3)$

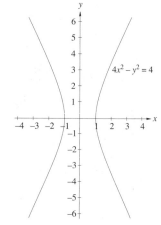

11. $V = (0, 2), V' = (0, -2),$
$F = (0, \sqrt{5}), F' = (0, -\sqrt{5}),$
$M = (1, 0), M' = (-1, 0)$

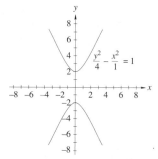

13. $V = (1, 0), V' = (-1, 0),$
$F = (\sqrt{5}, 0), F' = (-\sqrt{5}, 0),$
$M = (0, 2), M' = (0, -2)$

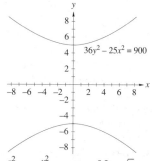

15. $V = (0, 5), V' = (0, -5),$
$F = (0, \sqrt{61}), F' = (0, -\sqrt{61}),$
$M = (6, 0), M' = (-6, 0)$

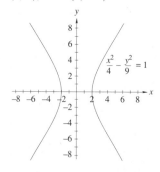

17. $\frac{x^2}{49} - \frac{y^2}{61.25} = 1$ **19.** $\sqrt{2}$

Exercise Set 15.6

1. ellipse;

	$x'y'$-system	xy-system
center	$(0', 0')$	$(4, -3)$
vertices	$(\pm 3', 0')$	$(7, -3)$, $(1, -3)$
foci	$(\pm\sqrt{5}', 0')$	$(\sqrt{5}+4, -3)$, $(-\sqrt{5}+4, -3)$
endpoints of minor axis	$(0', \pm 2')$	$(4, -1)$, $(4, -5)$

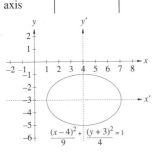

$$\frac{(x-4)^2}{9} + \frac{(y+3)^2}{4} = 1$$

3. hyperbola;

	$x'y'$-system	xy-system
center	$(0', 0')$	$(-4, 3)$
vertices	$(0', \pm 6)$	$(-4, 9)$, $(-4, -3)$
foci	$(0', \pm\sqrt{37}')$	$(-4, \sqrt{37}+3)$, $(-4, -\sqrt{37}+3)$
endpoints of conjugate axis	$(\pm 1', 0')$	$(-3, 3)$, $(-5, 3)$

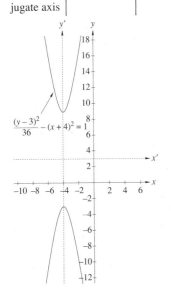

$$\frac{(y-3)^2}{36} - (x+4)^2 = 1$$

5. hyperbola;

	$x'y'$-system	xy-system
center	$(0', 0')$	$(-5, 0)$
vertices	$(\pm 2', 0')$	$(-3, 0)$, $(-7, 0)$
foci	$(\pm 2\sqrt{26}', 0')$	$(2\sqrt{26}-5, 0)$, $(-2\sqrt{26}-5, 0)$
endpoints of conjugate axis	$(0', \pm 10')$	$(-5, 10)$, $(-5, -10)$

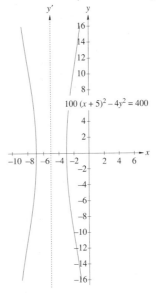

$$100(x+5)^2 - 4y^2 = 400$$

7. ellipse;

	$x'y'$-system	xy-system
center	$(0', 0')$	$(-2, \frac{3}{2})$
vertices	$(0', \pm 2')$	$(-2, \frac{7}{2})$, $(-2, -\frac{1}{2})$
foci	$(0', \pm\sqrt{3}')$	$(-2, \sqrt{3}+\frac{3}{2})$, $(-2, -\sqrt{3}+\frac{3}{2})$
endpoints of minor axis	$(\pm 1', 0')$	$(-1, \frac{3}{2})$, $(-3, \frac{3}{2})$

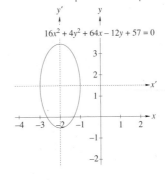

$$16x^2 + 4y^2 + 64x - 12y + 57 = 0$$

9. ellipse;

	$x'y'$-system	xy-system
center	$(0', 0')$	$(5, 2)$
vertices	$(0', \pm 5')$	$(5, 7),\quad (5, -3)$
foci	$(0', \pm\sqrt{21}')$	$(5, 2 + \sqrt{21}),\quad (5, 2 - \sqrt{21})$
endpoints of minor axis	$(\pm 2', 0')$	$(7, 2),\quad (3, 2)$

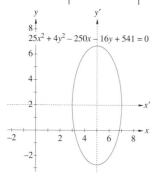

$25x^2 + 4y^2 - 250x - 16y + 541 = 0$

11. hyperbola;

	$x'y'$-system	xy-system
center	$(0', 0')$	$(4, 2)$
vertices	$(\pm\sqrt{2}', 0')$	$(4 + \sqrt{2}, 2),\quad (4 - \sqrt{2}, 2)$
foci	$(\pm\sqrt{6}', 0')$	$(4 + \sqrt{6}, 2),\quad (4 - \sqrt{6}, 2)$
endpoints of conjugate axis	$(0', \pm 2')$	$(4, 0),\quad (4, 4)$

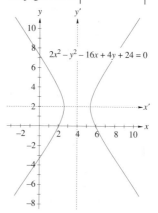

$2x^2 - y^2 - 16x + 4y + 24 = 0$

13. $(x - 2)^2 = 32(y + 3)$

15. $\frac{(x-4)^2}{36} + \frac{(y+3)^2}{20} = 1$

17. $\frac{(y-2)^2}{9} - \frac{(x+3)^2}{16} = 1$

19. $(y - 1)^2 = -16(x + 5)$

21. $\frac{(x+4)^2}{36} + \frac{(y-1)^2}{100} = 1$

23. This is a parabola with the equation $-4.9(t - 3)^2 = s - 44.1$. Maximum height is when $t = 3$ and

$s = 44.1$ m.

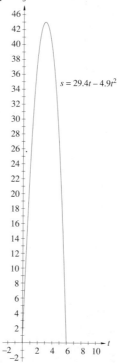

$s = 29.4t - 4.9t^2$

25. If the transverse axis passes through A and B and the conjugate axis

passes through the midpoint of $\overline{AB}$, then A is at $(-500, 0)$ and B at $(500, 0)$. The plane is at $(216, 1073.054)$ and lies on the hyperbola $\frac{x^2}{8\,100} - \frac{y^2}{241\,900} = 1$

27. (a) $x^2 = -200(y - 0.02)$;
(b) 2.00 cm

Exercise Set 15.7

1. (a) discriminant is 1; curve is a hyperbola; (b) $45°$;
(d) $\frac{y''^2}{18} - \frac{x''^2}{18} = 1$
(e)

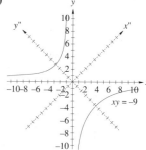

$xy = -9$

3. (a) discriminant is 32; curve is a hyperbola; (b) $45°$;
(d) $\frac{y''^2}{2} - \frac{x''^2}{4} = 1$

(e)

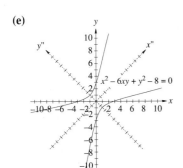

$x^2 - 6xy + y^2 - 8 = 0$

5. (a) discriminant is $-10,000$; curve is an ellipse; **(b)** $36.870°$;
(d) $\frac{x''^2}{4} + y''^2 = 1$
(e)

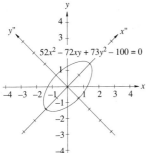

$52x^2 - 72xy + 73y^2 - 100 = 0$

7. (a) discriminant is 0; curve is a parabola; **(b)** $45°$; **(d)** $y''^2 = -\frac{\sqrt{2}}{2}x''$
(e)

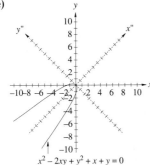

$x^2 - 2xy + y^2 + x + y = 0$

9. (a) discriminant is 168; curve is a hyperbola; **(b)** $33.690°$;
(d) $\frac{x''^2}{7} - \frac{y''^2}{6} = 1$

(e)

$2x^2 + 12xy - 3y^2 - 42 = 0$

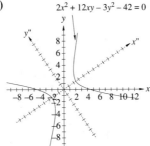

11. (a) discriminant is -119; curve is an ellipse; **(b)** $45°$;
(d) $7x''^2 + 17y''^2 = 52$
(e)

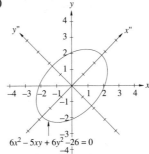

$6x^2 - 5xy + 6y^2 - 26 = 0$

13.

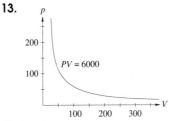

$PV = 6000$

15. Parabola

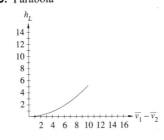

Exercise Set 15.8

1. hyperbola

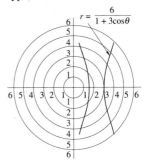

$r = \dfrac{6}{1 + 3\cos\theta}$

3. ellipse

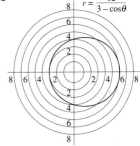

$r = \dfrac{12}{3 - \cos\theta}$

5. parabola

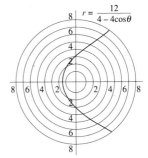

$r = \dfrac{12}{4 - 4\cos\theta}$

7. ellipse

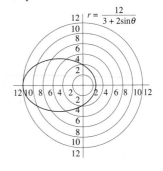

$r = \dfrac{12}{3 + 2\sin\theta}$

9. hyperbola rotated $\frac{\pi}{3}$ radians

$$r = \frac{12}{2 - 4\cos\left(\theta + \frac{\pi}{3}\right)}$$

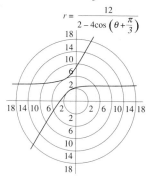

11. $r = \frac{6}{1 - \frac{3}{2}\cos\theta} = \frac{12}{2 - 3\cos\theta}$

13. $r = \frac{5}{1 - \sin\theta}$

15. $r = \frac{\frac{2}{3}}{1 - \frac{2}{3}\cos\theta} = \frac{2}{3 - 2\cos\theta}$

17. Greatest distance: 4.335×10^7 miles; shortest distance: 2.854×10^7 miles

19.

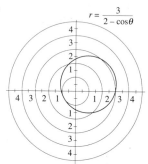

$$r = \frac{3}{2 - \cos\theta}$$

Review Exercises

1. (a) 5; (b) $(\frac{1}{2}, 7)$; (c) $-\frac{4}{3}$;
   (d) $y - 5 = -\frac{4}{3}(x - 2)$ or
   $4x + 3y = 23$

3. (a) $\sqrt{104} = 2\sqrt{26}$; (b) (2, 1); (c) 5;
   (d) $y + 4 = 5(x - 1)$ or $y - 5x = -9$

5. line in #1: $\frac{3}{4}$; line in #2: $\frac{12}{5}$; line in
   #3: $-\frac{1}{5}$; line in #4: 1

7. $y - 5 = -2(x + 3)$ or $y + 2x = -1$

9.

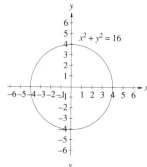

$x^2 + y^2 = 16$

11.

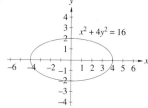

$x^2 + 4y^2 = 16$

13.

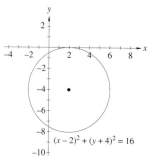

$(x - 2)^2 + (y + 4)^2 = 16$

15.

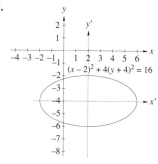

$(x - 2)^2 + 4(y + 4)^2 = 16$

17.

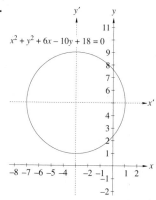

$x^2 + y^2 + 6x - 10y + 18 = 0$

19.

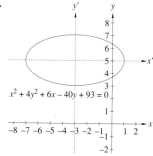

$x^2 + 4y^2 + 6x - 40y + 93 = 0$

21.

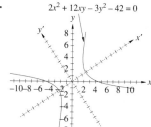

$2x^2 + 12xy - 3y^2 - 42 = 0$

23.

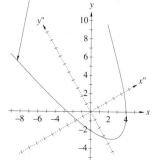

$3x^2 + 2\sqrt{3}\,xy + y^2 + 8x - 8\sqrt{3}\,y = 32$

25.

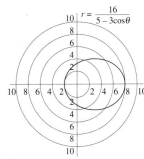

$$r = \frac{16}{5 - 3\cos\theta}$$

27.

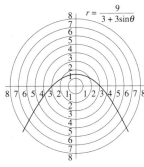

$$r = \frac{9}{3 + 3\sin\theta}$$

29. $\frac{x^2}{50^2} + \frac{y^2}{195.84} = 1$ **31.** If the x-axis passes through A and B and the y-axis through the midpoint of $\overline{AB}$, then B has the coordinates $(200, 0)$ and $A = (-200, 0)$. The plane is at $(82.92, -50)$. The plane lies on the hyperbola

$$\frac{x^2}{6\,400} - \frac{y^2}{33\,600} = 1.$$

Chapter 15 Test

1. Focus $(-4.5, 0)$; directrix $x = 4.5$

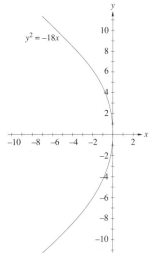

3. $\sqrt{149}$ **5.** $y = \frac{\sqrt{3}}{3}x - \sqrt{3}$
7. **(a)** $\frac{\pi}{4}$; **(b)** an ellipse **9.** 75 ft above its lowest point.

☰ ANSWERS FOR CHAPTER 16 ▬▬▬▬▬

Exercise Set 16.1
1. $(7, 1)$

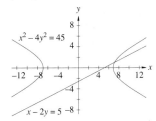

3. $(4, -2), (-4, 2)$

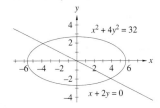

5. $(\sqrt{3}, 1), (-\sqrt{3}, 1), (j\sqrt{5}, -3),$
$(-j\sqrt{5}, -3)$

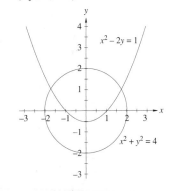

7. $(1, \sqrt{6}), (1, -\sqrt{6}), (-7, j\sqrt{42}),$
$(-7, -j\sqrt{42})$

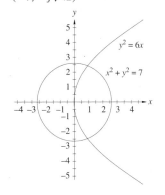

9. $(3, 1), (-3, -1), (j\sqrt{\frac{3}{2}}, -3j\sqrt{\frac{2}{3}}),$
$(-j\sqrt{\frac{3}{2}}, 3j\sqrt{\frac{2}{3}})$

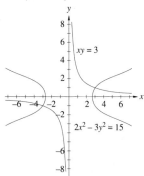

11. 42.08 mph for the first part of the trip; 32.08 mph for the last part.
13. One is 40 Ω and the other is 60 Ω
15. Either $v_1 = 3$ m/s and $v_2 = 5.5$ m/s or $v_1 = 6$ m/s and $v_2 = 3.5$ m/s
17. $F = 67.80; \ell = 2.36$

Exercise Set 16.2

1. $x + 5 < 15$ **3.** $3x < 30$
5. $\frac{-x}{2} > -5$ **7.** $x^2 < 100$
9. $x > 2$

11. $x > -6$

13. $x < 5$

15. $x \geq 6$

17. $x < 18$

19. $x \leq 3\frac{1}{2}$

21. $x \geq -\frac{9}{5}$

23. $-6 < x < 4$

25. $x < -10$ or $x > 2$

27. This is a contradictory inequality, and so there are no solutions.
29. $-4 \leq x < 7$

31. This is a contradictory inequality and so there are no solutions.
33. 6357 km $\leq r \leq$ 6378 km
35. $+0.10° \leq c \leq +1.10°$
37. 266.67 $\Omega < R <$ 1333.33 Ω
39. 43 or more

Exercise Set 16.3

1. $x < -1$ or $x > 3$
3. $-4 \leq x \leq 1$ **5.** $-1 < x < 1$
7. $-3 \leq x \leq 5$ **9.** $x \leq -1$ or $x \geq 2$ **11.** $x < 2$ or $x > 3$
13. $-3 < x < -\frac{1}{2}$ **15.** $-\frac{1}{2} < x < 1$
17. Absolute inequality—all real numbers satisfy this inequality.
19. $x < -3$ or $-1 < x < 2$
21. $-4 < x < -3$ or $x > \frac{5}{2}$
23. $x < -4$ **25.** $-2 < x < 0$ or $x > 2$ **27.** Contradictory inequality. No real numbers solve this inequality. **29.** $x < -1$ or $2 < x < 5$ **31.** $-1 < x < 0$ or $x > 2$ **33.** $x \leq -\frac{1}{3}$ or $\frac{1}{2} < x \leq 3$ **35.** $x > 9$ or $-1 < x < 1$ **37.** $x > 1$
39. $\frac{-3-\sqrt{17}}{2} \leq x \leq \frac{-3+\sqrt{17}}{2}$ **41.** 1 m $< d < \sqrt{6}$ m **43.** $x < 0.123$ or $x > 0.977$

Exercise Set 16.4

1.

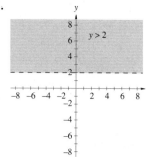

3.

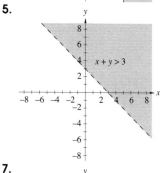

5.

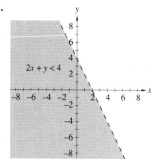

7.

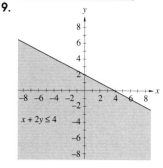

9.

11.
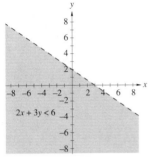
$2x + 3y < 6$

13.
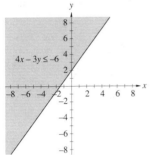
$4x - 3y \leq -6$

15.
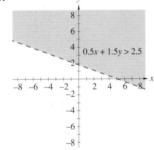
$0.5x + 1.5y > 2.5$

17.
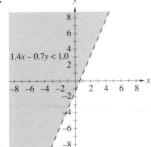
$1.4x - 0.7y < 1.0$

19.
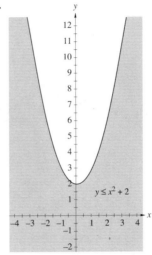
$y \leq x^2 + 2$

21.
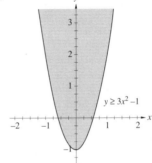
$y \geq 3x^2 - 1$

23.
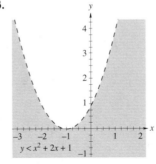
$y < x^2 + 2x + 1$

25.
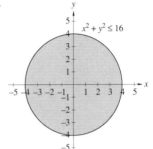
$x^2 + y^2 \leq 16$

27.
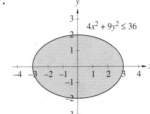
$4x^2 + 9y^2 \leq 36$

29.
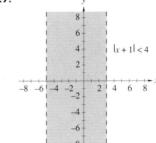
$|x + 1| < 4$

31.
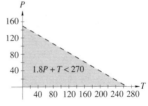
$1.8P + T < 270$

Exercise Set 16.5

1.
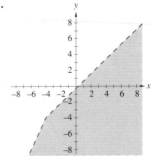

3.

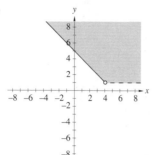

11.

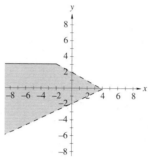

9.

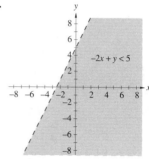

−2x + y < 5

5.

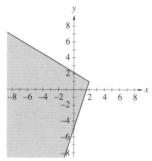

13.

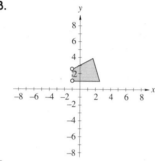

11.

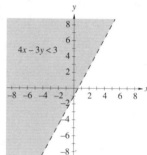

4x − 3y < 3

7.

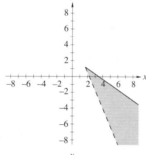

15. (4, 3) max is 11 at (4, 3)

17. max is 400 at any point on
$3x + 2y = 80$ where $20 \leq x \leq \frac{80}{3}$.

19. max is 216 at (12, 12)   **21.** 0 of x
and 400 of y will produce the most
profit: $3,200.   **23.** 120 of B and
none of A for a profit of $1140.00.

Review Exercises

1. $x < -4$

3. $x \geq 11$

5. $x > 1.9$

$\frac{19}{10}$

7. $x < -3$ or $x > \frac{5}{3}$

$\frac{5}{3}$

9.

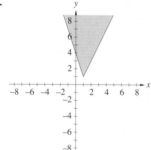

13.

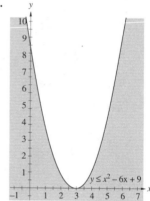

$y \leq x^2 - 6x + 9$

15.

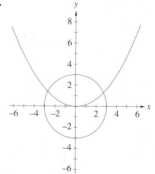

17.

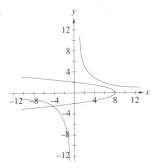

19.

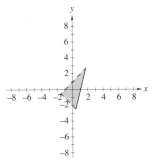

21.

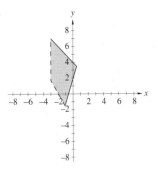

23. min is $-6\frac{1}{3}$ at $(1\frac{1}{6}, 3\frac{2}{3})$
25. $R_1 + R_2 \le 5\,\Omega$ **27.** 6 of C and 8 of S for a profit of \$20,460.

Chapter 16 Test
1. $x > 6$

3. $x \ge -\frac{11}{2}$

5. $-1 \le x \le -\frac{1}{3}$

7. (a) 9 of brand X and none of brand Y; **(b)** 26¢

☰ ANSWERS FOR CHAPTER 17 ━━━━━

Exercise Set 17.1
1. 2×3 **3.** 3×6 **5.** 4×2 **7.** $a_{11} = 1; a_{24} = 13;$
$a_{21} = 8; a_{32} = 6$ **9.** $x = 12, y = -6, z = 2, w = 5$

11. $\begin{bmatrix} -3 & 6 & 6 & -8 \\ 12 & 9 & 17 & 0 \\ 9 & 0 & 5 & 10 \end{bmatrix}$ **13.** $\begin{bmatrix} 2 & -1 & 4 \\ 5 & 9 & -2 \end{bmatrix}$

15. $\begin{bmatrix} -1 & 7 & 4 \\ 8 & 1 & 15 \end{bmatrix}$ **17.** $\begin{bmatrix} 12 & 9 & 6 \\ 15 & 0 & 21 \end{bmatrix}$

19. $\begin{bmatrix} -5 & 16 & 8 \\ 12 & 0 & 25 \end{bmatrix}$ **21.** $\begin{bmatrix} -22 & -1 & -2 \\ -9 & 2 & -5 \end{bmatrix}$

23. $\begin{bmatrix} -3 \\ -7 \\ 3 \\ -5 \end{bmatrix}$ **25.** $\begin{bmatrix} 21 \\ 27 \\ -12 \\ 30 \end{bmatrix}$ **27.** $\begin{bmatrix} 26 \\ 24 \\ -11 \\ 35 \end{bmatrix}$ **29.** $\begin{bmatrix} 0 \\ -22 \\ 9 \\ 25 \end{bmatrix}$

31. $\begin{bmatrix} 27 & 12 & 15 \\ 18 & 15 & 16 \end{bmatrix}$ or 27 of computer chip A, 12 of computer
chip B, 15 EPROMS, 18 keyboards, 15 motherboards, and
16 disk drives. **33.** $\begin{bmatrix} 135 & 82 \\ 65 & 118 \end{bmatrix}$

35.
Chip type:	286	386	486	586
Warehouse A	297	2343	3531	292
Warehouse B	1232	4653	3436	83
Warehouse C	352	3439	3017	1113

37. See *Computer Programs* **39.** See *Computer Programs*

Exercise Set 17.2
1. $\begin{bmatrix} 26 & 9 \\ -20 & -23 \end{bmatrix}$ **3.** $[-23]$ **5.** $\begin{bmatrix} -3 & 13 & 8 \\ -2 & 16 & 14 \end{bmatrix}$

7. $\begin{bmatrix} 5 & 17 & 5 \\ 44 & 67 & 37 \\ -9 & -6 & -7 \end{bmatrix}$ **9.** $\begin{bmatrix} -12 & 14 & 11 \\ 88 & 18 & 23 \end{bmatrix}$

11. $\begin{bmatrix} 7 & 16 \\ 17 & 38 \end{bmatrix}$ **13.** $\begin{bmatrix} -8 & 2 \\ -12 & 2 \end{bmatrix}$ **15.** $\begin{bmatrix} -68 & 18 \\ -160 & 42 \end{bmatrix}$

17. $\begin{bmatrix} -1 & 18 \\ 5 & 40 \end{bmatrix}$ **19.** $\begin{bmatrix} 42 & 96 \\ 102 & 228 \end{bmatrix}$ **21.** $\begin{bmatrix} 0 & 0 \\ 0 & 0 \end{bmatrix}$, No

23. Warehouse A: \$283,025, warehouse B: \$354,995, and warehouse C: \$364,415

25. $A^2 = \begin{bmatrix} 0 & 1 \\ 1 & 0 \end{bmatrix}\begin{bmatrix} 0 & 1 \\ 1 & 0 \end{bmatrix} = \begin{bmatrix} 0+1 & 0+0 \\ 0+0 & 1+0 \end{bmatrix} = \begin{bmatrix} 1 & 0 \\ 0 & 1 \end{bmatrix} =$
I. Similarly, $B^2 = C^2 = I$. **27.** a) The commutator of A
and B is $AB - BA = \begin{bmatrix} i & 0 \\ 0 & -i \end{bmatrix} - \begin{bmatrix} -i & 0 \\ 0 & i \end{bmatrix} =$
$2\begin{bmatrix} i & 0 \\ 0 & -i \end{bmatrix} = 2i\begin{bmatrix} 1 & 0 \\ 0 & -1 \end{bmatrix} = 2iC$. Similar methods are
used for **(b)** and **(c)**. **29.** See *Computer Programs*

Exercise Set 17.3
1. $\begin{bmatrix} 1 & -1 \\ -5 & 6 \end{bmatrix}$ **3.** $\begin{bmatrix} -1.5 & 2 \\ 4 & -5 \end{bmatrix}$ **5.** $\begin{bmatrix} -1 & -1.5 \\ -1.5 & -2 \end{bmatrix}$

7. $\begin{bmatrix} 0.6 & -2 \\ -0.8 & 3 \end{bmatrix}$ **9.** $\begin{bmatrix} 0 & 1 & 0 \\ -\frac{1}{3} & 0 & 0 \\ 0 & 0 & \frac{1}{4} \end{bmatrix}$

11. Singular matrix because column 2 is twice column 1.

13. $\begin{bmatrix} 1 & 3 & -2 \\ -1 & -4 & 3 \\ 0 & -4 & 5 \end{bmatrix}$

15. Singular matrix because row 2 is -2 times row 1

17. $\begin{bmatrix} 0.5 & 0 & 0 \\ -0.5 & 0.5 & 0 \\ 0 & -0.5 & 0.5 \end{bmatrix}$ **19.** $\begin{bmatrix} 3 & 3 & -1 \\ -2 & -2 & 1 \\ -4 & -5 & 2 \end{bmatrix}$

21. All answers should check.

Exercise Set 17.4

1. $x = -1, y = 2$ **3.** $x = -1.5, y = 2.5$ **5.** No solution **7.** $x = 1.5, y = -4, z = 3.5$ **9.** $x = 5.5,$ $y = -3.5, z = -6.5$ **11.** $x = -2, y = 5$
13. $x = -5, y = 7$ **15.** $x = 4.2, y = -2.4$ **17.** $x = -2,$ $y = 4, z = 1$ **19.** $x = -3.4, y = 2.2, z = 1.5$
21. $I_A = 6.05, I_B = 5.26, I_C = 0.79$
23. $I_1 = 1.22, I_2 = 2.37, I_3 = 1.64$ **25.** See *Computer Programs*

Review Exercises

1. $\begin{bmatrix} 4 & 4 & 2 & 7 \\ 9 & 7 & 3 & 4 \\ 9 & 5 & -8 & -3 \end{bmatrix}$ **3.** $\begin{bmatrix} 1 & 5 & 1 & 5 \\ 1 & 8 & -3 & 4 \\ 5 & 7 & -6 & 3 \end{bmatrix}$

5. $\begin{bmatrix} 8 & 3 & 4 & 4 \\ 3 & 14 & -14 & 8 \\ 18 & 35 & -16 & 24 \end{bmatrix}$ **7.** $\begin{bmatrix} 3 & 8 \\ 4 & 13 \\ 1 & 2 \end{bmatrix}$

9. $\begin{bmatrix} 12 & 15 & 3 \\ 8 & 10 & 2 \\ -4 & -5 & -1 \end{bmatrix}$ **11.** $\begin{bmatrix} -2.5 & -1.5 \\ 2 & 1 \end{bmatrix}$

13. $\begin{bmatrix} -0.5 & 0.125 & 0 \\ 0 & 0.25 & 0 \\ 0.5 & -0.125 & 1 \end{bmatrix}$ **15.** $x = 2.5, y = -6.4$

17. $x = 2, y = -3, z = 5$ **19.** $T = 86.47, a = 1.23$
21. $x'' = -x, y'' = -y$

23. $M = \begin{bmatrix} 1 - \frac{d}{f_2} & \frac{d}{f_1 f_2} - \frac{1}{f_1} - \frac{1}{f_2} \\ d & 1 - \frac{d}{f_1} \end{bmatrix}$

Chapter 17 Test

1. (a) $\begin{bmatrix} 7 & 5 & -7 \\ 19 & -6 & 6 \end{bmatrix}$, and **(b)** $\begin{bmatrix} 26 & -10 & -6 \\ 42 & -18 & -2 \end{bmatrix}$

3. (a) $CD = \begin{bmatrix} -7 & 24 \\ -11 & 16 \end{bmatrix}$, **(b)** $DC = \begin{bmatrix} 2 & 3 \\ -46 & 7 \end{bmatrix}$

5. $x = 1, y = -2, z = 3$

☰ ANSWERS FOR CHAPTER 18

Exercise Set 18.1

1. 9 **3.** 3 **5.** 130 **7.** 0 **9.** 0 **11.** 11
13. 0 **15.** 79 **17.** yes **19.** yes **21.** yes **23.** yes
25. $Q(x) = x^4 + 3x^3 - 8x^2 - 24x + 3; R(x) = 18$
27. $Q(x) = 5x^2 - 8x + 24; R(x) = -63$
29. $Q(x) = 8x^4 + 4x^3 - 2x^2 + 6x + 1; R(x) = \frac{1}{2}$
31. $Q(x) = 2x^3 - 3x^2 - 4; R(x) = 0$

Exercise Set 18.2

1. $1, \frac{-5+\sqrt{85}}{10}, \frac{-5-\sqrt{85}}{10}$ **3.** $\frac{1}{3}, 3, -3$ **5.** $2, -2, j, -j$
7. $3, -3, -1 + 2j, -1 - 2j$ **9.** $-1, -1, 2, 3$ **11.** $-1,$ $-1, -3, 1$ **13.** $-\frac{1}{2}, \frac{1}{3}, -2 + j, -2 - j$ **15.** $1, \frac{2}{3},$
$\frac{-1+j\sqrt{3}}{2}, \frac{-1-j\sqrt{3}}{2}$ **17.** $1, 1, 1, -1 + j\sqrt{3}, -1 - j\sqrt{3}$
19. $3, -2, \frac{2}{3}, \frac{-1+\sqrt{5}}{2}, \frac{-1-\sqrt{5}}{2}$ **21.** $j, -j, 3j, -3j, 7, -1$

Exercise Set 18.3

1. Number of Positive Roots: 1; Negative Roots: 0 or 2; Possible Rational Roots: $\pm 1, \pm 2$ **3.** Number of Positive Roots: 0 or 2; Negative Roots: 1; Possible Rational Roots: $\pm 1, \pm 2, \pm 3, \pm 6$ **5.** Number of Positive Roots: 0 or 2; Negative Roots: 1; Possible Rational Roots: $\pm 1, \pm 2,$ $\pm 3, \pm 6$ **7.** Number of Positive Roots: 0 or 2; Negative Roots: 0 or 2; Possible Rational Roots: $\pm 1, \pm 3, \pm \frac{1}{2}, \pm \frac{3}{2}$
9. Number of Positive Roots: 0 or 2; Negative Roots: 0 or 2; Possible Rational Roots: $\pm 1, \pm 2, \pm 4, \pm \frac{1}{2}, \pm \frac{1}{3}, \pm \frac{2}{3}, \pm \frac{4}{3},$ $\pm \frac{1}{6}$ **11.** $1, 3, 5$ **13.** $-1, -1, 2$
15. $-\frac{3}{4}, 1 + j\sqrt{3}, 1 - j\sqrt{3}$ **17.** $2, 2, j\sqrt{3}, -j\sqrt{3}$
19. $-\frac{1}{2}, -\frac{1}{2}, 3 + \sqrt{3}, 3 - \sqrt{3}$ **21.** no rational roots
23. $\frac{1}{2}, 3, 3, j, -j$ **25.** $1, 3, -2, \frac{3}{2} + j\frac{\sqrt{11}}{2}, \frac{3}{2} - j\frac{\sqrt{11}}{2}$
27. 1 in. or 2 in. **29.** 9 feet **31.** $R = 2$ **33.** 3.5 cm *or* $6 - \sqrt{21} \cong 1.417$

Exercise Set 18.4

1. 0.56 **3.** 0.52 **5.** 0.48 **7.** $-1, 1.35$ (the other two roots are complex numbers) **9.** $-1, 2.09$
11. $1, -1.66$ **13.** $-1, 0.24$ **15.** $-1.41, 1.41, -\frac{1}{2}, -\frac{1}{4}$
17. 6.22731 years or about 6 years 83 days
19. (a) 63.87006 cm; **(b)** 521 101π cm^3 **21.** The approximate roots are $\pm 0.3536, \pm 0.8654, \pm 1.4639,$ $\pm 1.7321,$ and ± 2.2323

Exercise Set 18.5

1. vertical asymptote at $x = -2$; horizontal asymptote at $y = 0$

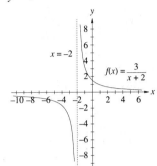

3. vertical asymptote at $x = -4$; horizontal asymptote at $y = 2$

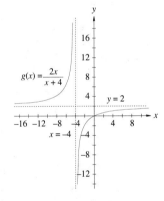

19. vertical asymptotes at $x = 2$ and $x = -2$; horizontal asymptote at $y = 0$; solution $x = 18$ **21.** 1 (100%)

Chapter 18 Test

1. 5 **3.** Quotient: $x^3 - 2x^2 - 9x - 10$; Remainder: 15
5. $1, -2, -\frac{2}{3}$ **7.** vertical asymptotes: $x = -3$ and

5. vertical asymptotes at $x = -4$ and $x = -2$; horizontal asymptote at $y = 3$

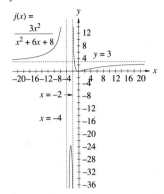

7. vertical asymptote at $x = -2$; horizontal asymptote at $y = 1$

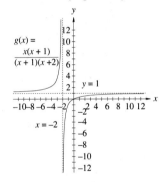

$x = -2$; horizontal asymptote: $y = 3$; solutions: $x = 1$ and $x = -\frac{1}{3}$

9. vertical asymptotes at $x = 4$; and at $x = 2$ horizontal asymptote at $y = 4$

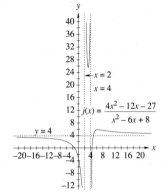

11. $x = -2$ or $x = 3$ **13.** 1
15. -5 **17.** $\frac{21}{8}$ **19.** 3
21. 4.49 hours

Review Exercises

1. 30, degree $= 2$ **3.** -13, degree $= 3$ **5.** $Q(x) = 4x - 18$; $R(x) = 51$
7. $Q(x) = 9x^3 + 4x^2 + 6x - 1$; $R(x) = 0$
9. 1, 1, 1 **11.** $-1, -1, -1, 3$
13. $-3.044, -0.328, 0.548, 1.824$
15. $-1, -1, j, j, -j, -j$ **17.** vertical asymptotes at $x = -4$ and at $x = 1$; horizontal asymptote at $y= 0$; solution: $x = 0$

ANSWERS FOR CHAPTER 19

Exercise Set 19.1

1. $\frac{1}{2}$; $\frac{1}{3}$; $\frac{1}{4}$; $\frac{1}{5}$; $\frac{1}{6}$; $\frac{1}{7}$ **3.** 4; 9; 16; 25; 36; 49
5. -1; 2; -3; 4; -5; 6 **7.** $\frac{1}{3}$; $\frac{3}{5}$; $\frac{5}{7}$; $\frac{7}{9}$; $\frac{9}{11}$; $\frac{11}{13}$
9. 1; $\frac{2}{3}$; $\frac{4}{9}$; $\frac{8}{27}$; $\frac{16}{81}$; $\frac{32}{243}$ **11.** 0; $\frac{1}{9}$; $\frac{1}{4}$; $\frac{9}{25}$; $\frac{4}{9}$; $\frac{25}{49}$
13. 1; 2; 6; 24; 120; 720 **15.** 5; 8; 11; 14; 17; 20
17. 2; 2; 4; 64; 16,777,216; 1.329228×10^{36}
19. 1; 2; 9; 262,144; $5^{262,144}$; $6^{(5^{262,144})}$
21. 1; $\frac{1}{2}$; $\frac{1}{2}$; $\frac{1}{4}$; $\frac{1}{8}$; $\frac{1}{32}$ **23.** 0; 2; 2; 0; -2; -2
25. 18.75 ft *or* 18 ft 9 in. **27.** **(a)** \$30,400; **(b)** \$35,900

Exercise Set 19.2

1. Arithmetic sequence with $d = 8$; $a_9 = 65$
3. Geometric sequence with $r = \frac{-1}{2}$; $a_{10} = \frac{-1}{64}$
5. Geometric sequence with $r = -4$; $a_6 = (-1)(-4)^5 = 1024$
7. Arithmetic sequence with $d = -5$; $a_8 = -32$
9. Neither; can't tell from the given information what a_{10} will be **11.** Arithmetic sequence with $d = 0.4$; $a_9 = 3.6$
13. Arithmetic sequence with $d = \frac{-1}{6}$; $a_8 = -\frac{1}{2}$
15. Geometric sequence with $r = 0.1$; $a_6 = 0.00005$
17. Arithmetic sequence with $d = 2$; $a_9 = 17 + \sqrt{3}$

19. Neither; $a_6 = 1.2, a_n = (-1)^{n-1}[4.3 - (1.1)(n-1)]$
21. -51 **23.** 243 **25.** 20 **27.** 9 **29.** -3.5
31. $\frac{1}{8}$ mile **33.** 12.3 cm; 51 **35. (a)** about 14.68 miles; **(b)** about 38.44 miles

Exercise Set 19.3
1. $\frac{3}{2} + \frac{3}{4} + \frac{3}{8} + \frac{3}{16}$ **3.** $1 - \frac{3}{2} + 3 - \frac{27}{4}$ **5.** 0
7. 91 **9.** 225 **11.** $\frac{1,593,342}{67,925} \approx 23.457$
13. Arithmetic; $S_{10} = 165$ **15.** Geometric; $S_8 = 255$
17. Arithmetic; $S_{12} = 192$ **19.** Arithmetic; $S_{20} = 57.5$
21. Geometric; $S_{16} \approx 0.5625$ **23.** Arithmetic; $S_{14} = 114.8$
25. Geometric; $S_{15} = 17,920.253$ **27.** Geometric;
$S_8 = 9,840$ **29.** 392.32m when it hits the fourth time,
634.10m when it hits the tenth time. **31.** 104 ft
33. \$2,540.47; \$3.99 **35. (a)** \$3,662,000; **(b)** \$7,682,000

Exercise Set 19.4
1. Converges; $\frac{1}{3}$ **3.** Converges; 0.6 **5.** Converges; 4
7. Converges; about 1.389 **9.** Converges; $0.0333\ldots$
11. Converges; $0.555\ldots$ **13.** Diverges; $r = 5/4$
15. Diverges; $r = 10$ **17.** Converges; 2 **19.** Converges;
2.4142 **21.** Diverges; $r = \sqrt{5}$ **23.** 4/9 **25.** $\frac{19}{33}$
27. $\frac{13,520}{9,999}$ **29.** $\frac{20,986}{3,330} = \frac{10,493}{1,665}$ **31.** 45 m
33. 1,250 revolutions

Exercise Set 19.5
1. $a^4 + 4a^3 + 6a^2 + 4a + 1$ **3.** $81x^4 - 108x^3 + 54x^2 - 12x + 1$
5. $\frac{x^6}{64} + \frac{3}{16}x^5d + \frac{15}{16}x^4d^2 + \frac{5}{4}x^3d^3 + \frac{15}{4}x^2d^4 + 3xd^5 + d^6$
7. $\frac{a^5}{32} - \frac{5a^4}{4b} + 20\frac{a^3}{b^2} - 160\frac{a^2}{b^3} + 640\frac{a}{b^4} - \frac{1,024}{b^5}$
9. $a^7 + 7a^6b + 21a^5b^2 + 35a^4b^3 + 35a^3b^4 + 21a^2b^5 + 7ab^6 + b^7$

11. $t^8 - 8t^7a + 28t^6a^2 - 56t^5a^3 + 70t^4a^4 - 56t^3a^5 + 28t^2a^6 - 8ta^7 + a^8$
13. $64a^6 - 192a^5 + 240a^4 - 160a^3 + 60a^2 - 12a + 1$
15. $x^{14}y^7 + \frac{7}{2}x^{12}y^6a + \frac{21}{4}x^{10}y^5a^2 + \frac{35}{8}x^8y^4a^3 + \frac{35}{16}x^6y^3a^4 +$
$\frac{21}{32}x^4y^2a^5 + \frac{7}{64}x^2ya^6 + \frac{a^7}{128}$
17. $x^{12} + 12x^{11}y + 66x^{10}y^2 + 220x^9y^3$ **19.** $1 + 2a + 3a^2 + 4a^3$
21. $1 - \frac{b}{3} + \frac{2b^2}{9} - \frac{14b^3}{81}$ **23.** 1.464 **25.** 1.049
27. 1.008 **29.** 0.983 **31.** $3,003x^{10}y^5$
33. $126,720x^8y^4$ **35.** $1,001a^{10}b^4$ **37. (a)** $1 + \frac{v^2}{2c^2} + \frac{3v^4}{8c^4}$
(b) $mc^2 + \frac{mv^2}{2} + \frac{3mv^4}{8c^2}$ **39.** $r^3 + \frac{3}{2}\ell^2r + \frac{3\ell^4}{8r}$

Review Exercises
1. $\frac{1}{3}; \frac{1}{4}; \frac{1}{5}; \frac{1}{6}; \frac{1}{7}; \frac{1}{8}$ **3.** $\frac{-1}{3}; \frac{1}{10}; \frac{-1}{21}; \frac{1}{36}; \frac{-1}{55}; \frac{1}{78}$
5. $1; \frac{1}{2}; \frac{1}{6}; \frac{1}{24}; \frac{1}{120}; \frac{1}{720}$ **7.** 1; 3; 4; 7; 11; 18
9. Arithmetic sequence, $d = -3, a_{10} = -17$
11. Arithmetic sequence, $d = 4, a_7 = 23$ **13.** Geometric
sequence, $r = -\frac{1}{6}, a_{10} = -\frac{1}{3,359,232} \approx -0.0000003$.
15. -24 **17. (a)** $\frac{4}{3} + \frac{4}{9} + \frac{4}{27} + \frac{4}{81}$ **(b)** $2 + \frac{3}{2} + \frac{4}{3} + \frac{5}{4}$
19. 39 **21.** Arithmetic series; $S_{10} = 265$ **23.** Geometric
series; $S_{14} = 1.4999997$ **25.** Converges; $S = \frac{2}{3}$
27. Diverges **29.** Converges; $S \approx 1.3660$ **31.** $\frac{185}{999} = \frac{5}{27}$
33. $a^5 + 10a^4 + 40a^3 + 80a^2 + 80a + 32$
35. $\frac{x^6}{64} - \frac{9x^5y^2}{16} + \frac{135x^4y^4}{2} - \frac{135x^3y^6}{2} + \frac{1215x^2y^8}{4} - 729xy^{10} + 729y^{12}$
37. $32,768x^{15} + 245,760x^{14}y + 860,160x^{13}y^2 + 1,863,680x^{12}y^3$
39. $1 + 5x + 15x^2 + 35x^3$ **41.** $2,480,640x^{14}y^6$ **43.** 1.004
45. \$727.76 **47.** 65.76 cm **49.** 19.834 sec

Chapter 19 Test
1. $-5, 5, \frac{5}{3}, 1, \frac{5}{7}, \frac{5}{9}$ **3.** arithmetic sequence, $d = -2.5$,
$a_{10} = -7.5$ **5.** 25 **7.** $\frac{435}{999} = \frac{145}{333}$ **9.** 123,000

☰ ANSWERS FOR CHAPTER 20

Exercise Set 20.1
All answers are proofs and will not be displayed in this section.

Exercise Set 20.2
1. $\frac{\sqrt{6}-\sqrt{2}}{4}$ **3.** $\frac{\sqrt{3}}{2}$ **5.** $\frac{\sqrt{6}-\sqrt{2}}{\sqrt{6}+\sqrt{2}} = \frac{\sqrt{3}-1}{\sqrt{3}+1} = 2 - \sqrt{3}$
7. $\frac{1}{2}$ **9.** $\cos x$ **11.** $-\sin x$ **13.** $-\cos x$
15. $\sin x$ **17.** $\frac{21+\sqrt{105}}{32} \approx 0.97647$
19. $\frac{21+\sqrt{105}}{7\sqrt{7}-3\sqrt{15}} \approx 4.52768$ **21.** $\frac{7\sqrt{7}+3\sqrt{15}}{32} \approx 0.94185$ **23.** I
25. $\frac{\sqrt{105}-21}{32} \approx -0.33603$ **27.** $\frac{\sqrt{105}-21}{7\sqrt{7}+3\sqrt{15}} \approx -0.35678$
29. $\frac{7\sqrt{7}-3\sqrt{15}}{32} \approx 0.21567$ **31.** IV **33.** $\sin 60° = \frac{\sqrt{3}}{2}$
35. $\cos 44°$ **37.** $\cos \alpha$ **39.** $\cos^2 x - \sin^2 x = \cos 2x$

Exercise Set 20.3
1. $\sqrt{\frac{2+\sqrt{3}}{4}} \approx 0.96593$ **3.** $\sqrt{\frac{2-\sqrt{3}}{4}} \approx 0.25882$
5. $\sqrt{\frac{2+\sqrt{3}}{4}} \approx 0.96593$

7. $\sqrt{\frac{1+\sqrt{\frac{2+\sqrt{3}}{4}}}{2}} = \sqrt{\frac{2+\sqrt{2+\sqrt{3}}}{4}} \approx 0.99145$
9. $\frac{\sqrt{2}}{2+\sqrt{2}} \approx 0.41421$ **11.** $\sqrt{\frac{2-\sqrt{3}}{4}} \approx 0.25882$
13. $\frac{1}{4}(\sqrt{3}\sqrt{2+\sqrt{2+\sqrt{3}}} - \sqrt{2-\sqrt{2+\sqrt{3}}}) \approx 0.79335$
15. $\sin 2x = -\frac{336}{625}; \cos 2x = \frac{527}{625}; \tan 2x = -\frac{336}{527};$
$\sin \frac{x}{2} = \frac{7\sqrt{2}}{10}; \cos \frac{x}{2} = \frac{\sqrt{2}}{10}; \tan \frac{x}{2} = 7$
17. $\sin 2x = \frac{840}{841}; \cos 2x = -\frac{41}{841}; \tan 2x = -\frac{840}{41};$
$\sin \frac{x}{2} = \sqrt{\frac{9}{58}}; \cos \frac{x}{2} = \sqrt{\frac{49}{58}}; \tan \frac{x}{2} = \frac{3}{7}$
19. $\sin 2x = \frac{840}{1369}; \cos 2x = \frac{-1081}{1369}; \tan 2x = -\frac{840}{1081};$
$\sin \frac{x}{2} = \sqrt{\frac{49}{74}}; \cos \frac{x}{2} = -\sqrt{\frac{25}{74}}; \tan \frac{x}{2} = -\frac{7}{5}$

Exercise Set 20.4
1. 90°, 270° **3.** 30°, 210° **5.** 228.59°, 311.41°
7. 51.34°, 231.34° **9.** 90°, 270° **11.** 180° **13.** 0°,
90°, 180° **15.** 0°, 90°, 180°, 270° **17.** 30°, 150°,

210°, 330°   **19.** 15°, 75°, 195°, 255°   **21.** 45°, 225°
23. 0°, 40°, 60°, 80°, 120°, 160°, 180°, 200°, 240°, 280°, 300°,
320°   **25.** 45°, 135°, 225°, 315°   **27.** 0°, 135°, 180°,
315°   **29.** 30°, 90°, 150°, 270°   **31.** 20.87°
33. 0.007 005 s   **35.** 0, π

Review Exercises
Answers 1 through 9 are proofs and will not be displayed in this section.

11. 139.11°, 319.11°   **13.** 90°, 120°, 240°, 270°   **15.** 0°,
90°, 180°, 270°   **17.** 0°, 180°   **19.** 218.17°, 321.83°
21. $-\frac{120}{169}$ **23.** $\sqrt{25/26} \approx 0.98058$ **25.** $\frac{-21}{221} \approx -0.09502$
27. $\frac{220}{221} \approx 0.99548$ **29.** $\frac{140}{221} \approx 0.63348$ **31.** 90°, 210°,
330°

Chapter 20 Test
1. $-\frac{33}{65}$ **3.** $-\frac{63}{65}$ **5.** $-\frac{24}{25}$ **7.** $\frac{2}{\sqrt{5}}$
9. $4\sin 12x$

≡ ANSWERS FOR CHAPTER 21

Exercise Set 21.1
1. $\frac{1}{4}$ **3.** $\frac{1}{13}$ **5.** $\frac{3}{13}$ **7.** $\frac{1}{4}$ **9.** $\frac{13}{51}$ **11.** $\frac{1}{6}$
13. $\frac{1}{2}$ **15.** $\frac{1}{36}$ **17.** $\frac{11}{36}$

19.

Die 1 \ Die 2	1	2	3	4	5	6
1	(1, 1)	(1, 2)	(1, 3)	(1, 4)	(1, 5)	(1, 6)
2	(2, 1)	(2, 2)	(2, 3)	(2, 4)	(2, 5)	(2, 6)
3	(3, 1)	(3, 2)	(3, 3)	(3, 4)	(3, 5)	(3, 6)
4	(4, 1)	(4, 2)	(4, 3)	(4, 4)	(4, 5)	(4, 6)
5	(5, 1)	(5, 2)	(5, 3)	(5, 4)	(5, 5)	(5, 6)
6	(6, 1)	(6, 2)	(6, 3)	(6, 4)	(6, 5)	(6, 6)

21. $\frac{1}{36}$ **23.** 0   **25.** $\frac{1}{6}$ **27.** $\frac{1}{6}$ **29.** $\frac{1}{1,296}$ **31.** $\frac{4}{49}$
33. $\frac{4}{7}$ **35.** (a) $\frac{9}{16}$ =56.25% (b) $\frac{1}{16}$ =6.25% (c) $\frac{3}{8}$ =37.5%
37. (a) 0.56 (b) 0.06 (c) 0.38   **39.** 0.04875

Exercise Set 21.2
1.

Number	2	3	4	6	7
Frequency	2	1	4	2	1

3.

Number	73	74	77	80	82	83	84	87	89
Frequency	1	1	1	1	1	1	2	1	1

Number	92	94	96	100
Frequency	1	3	5	1

5.

Interval	1 − 2	3 − 4	5 − 6	7 − 8
Frequency	2	5	2	1

7.

Interval	71 − 75	76 − 80	81 − 85	86 − 90
Frequency	2	2	4	2

Interval	91 − 95	96 − 100
Frequency	4	6

9.

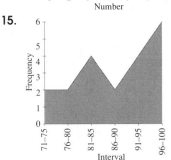

11.

13.

15.

17. Mean: 4.2; Median: 4; Mode: 4;
$Q_1 = 3; Q_2 = 4; Q_3 = 6$

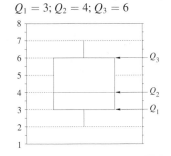

19. Mean: 88.35; Median: 90.5; Mode: 96; $Q_1 = 82.5$; $Q_2 = 90.5$; $Q_3 = 96$

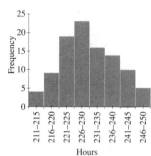

21.

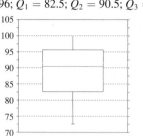

23.

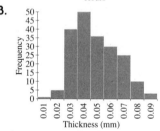

25.

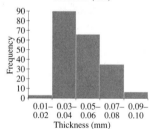

27.

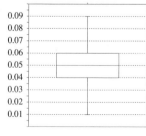

29.

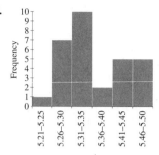

31. $Q_1 = 5.29$; $Q_2 = 5.345$; $Q_3 = 5.43$

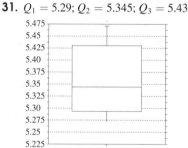

Exercise Set 21.3

1. 2.56　**3.** 64.3275　**5.** 1.6
7. 8.02　**9.** (a) 5, (b) 10
11. (a) 15, (b) 20　**13.** (a) 0.0217,
(b) 390, (c) 480　**15.** (a) 0.0164,
(b) 156, (c) 197　**17.** See
Computer Programs

Exercise Set 21.4

1. $y = 0.833x - 1.333$, $r = 0.943$

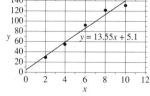

3. $y = 13.55x + 5.1$, $r = 0.986$

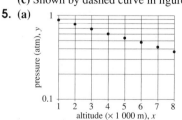

5. $y = 14.955x + 5.379$, $r = 0.996$

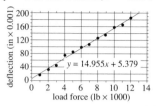

7. $y = 0.000425x + 0.072$, $r = 0.986$

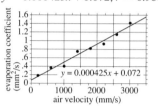

9. See *Computer Programs*

Exercise Set 21.5

1. (a)

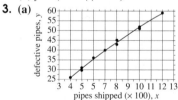

(b) $y = (2.002)(0.989)^x$

3. (a)

(b) $y = 4.084 + 5.848x + 0.107x^2$;
(c) Shown by dashed curve in figure

5. (a)

(b) $y = 0.9958e^{-0.12356x}$; (c) Shown
by dashed curve in figure

7. (a)

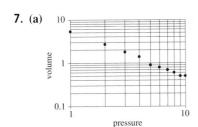

(b) $v = \frac{5.41}{p} - 0.064$; **(c)** Shown by dashed curve in figure

Review Exercises

1. $\frac{1}{2}$ **3.** $\frac{5}{8}$ **5.** $\frac{1}{3}$ **7.** $\frac{1}{36}$

9. $(\frac{23}{25})^3 = \frac{12167}{15625}$

11. $Q_1 = 299.7$; $Q_2 = 300.1$; $Q_3 = 300.45$

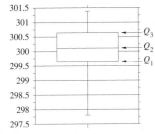

13. 299.945 **15.** 72.7%

17. 81.8%

19. (a)

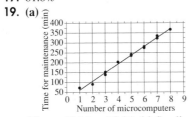

(b) $y = 43.94x + 18.94$; **(c)** See line in graph; **(d)** $r = 0.996$

21. (a)

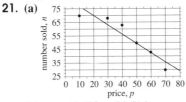

(b) $n = -0.686p + 83.714$;

(c) Shown by curve in figure

Chapter 21 Test

1. $\frac{1}{26}$ **3.** 3600.1

5.

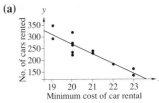

7. 3599.945 **9.** 65%

11. (a)

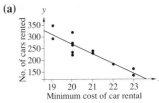

(b) $y = -39.774x + 1063.761$;

(c) See line in graph;

(d) $r = -0.8997$

Computer Programs and Additional Exercises

This appendix consists of programs written in GWBASIC that serve as answers to each of the computer programming exercises found in this text. Just as the textbook increases in sophistication and complexity as you progress from beginning to end, the computer programs also increase in sophistication and complexity.

The name of each program corresponds to its location in the text. Each name consists of the three letters PRG followed immediately by five digits. The first two digits represent the chapter in the text. Thus, Chapter 9 is 09 and chapter 25 is 25. The next (third) digit denotes the section in the chapter; and the last two digits, the exercise in that section where the program was assigned.

Using the above description, the program for Chapter 9, Section 5, Exercise 43, has the name PRG09543. Similarly, the computer program for Chapter 17, Section 3, Exercise 22, is named PRG17322.

* * * * *

You may wish to write your own versions of these programs. If you do, then use the following programs as guidelines. In some cases, we have added some additional steps to the programs to make them easier to use. You might have some ideas for making the programs even more user-friendly.

However, it is important that anyone programming a computer remember that a computer must be told what to do and how it is done. If the directions are wrong, the computer will give a wrong answer.

There are several special commands, statements, and functions that are used in BASIC. In fact, most of these, with some variation such as SQRT rather than SQR, are used in other computer programming languages. This is not intended as a programming course, and so we assume that you are familiar with BASIC.

Commands, such as LOAD, RUN, LIST, and SAVE are not part of a program, but allow you to tell the computer to do something with a program.

Program *statements* used in these programs include PRINT, PRINT USING, INPUT, REM, GOTO (or GO TO), DEF FN, DIM, FOR ... NEXT, IF ... THEN, GOSUB ... RETURN, CLS, CLEAR, and END.

Functions used in these programs include ABS, ATN, COS, EXP, INT, LEFT$, LOG, SGN, SIN, GQR, TAB, and TAN. You should remember that a computer evaluates trigonometric functions expressed in radians and not degrees. You will need to include conversion factors in any programs where you will enter, or have the computer print, angles given in degrees.

You may also have noticed that BASIC includes only the sine, cosine, tangent, and arctangent functions. If your programs include any other trigonometric formulas, you will have to define them in that particular program.

Naturally, since these are programs involving mathematics, you will need to use the symbols for the mathematical *operations*. The computer uses +, −, *, and / for addition, subtraction, multiplication, and division, respectively. For exponentiation, these BASIC programs use $^\wedge$. (Some computers and some languages use either ** or ↑.)

* * * * *

Not everyone wishes to take the time to write a computer program. Thus, each of these programs is available on a 3.5-inch IBM formatted disk.

Included with these program listings are 30 exercises. Some of the exercises ask you to modify a program so that it will do a better job of performing the desired task or so that it can be used in some additional situations. The remaining exercises ask you to use one of the programs with a given set of numbers or equations.

A solution to each of these additional exercises is given in the *Instructor's Solutions Manual*; solutions to the odd numbered exercises are in the *Student's Solutions Manual*. Whenever possible, each solution contains a copy of the exact wording and format used when the computer printed each solution.

Chapter 5

Section 5.2, Exercise 44

```
10 1 REM *** PRG05244 ***
20 DIM I(20), X(20), Y(20)
30 CLS
40 REM THIS PROGRAM USES THE GAUSS-SEIDEL
   METHOD FOR SOLVING SYSTEMS OF TWO
   LINEAR EQUATIONS
50 PRINT "THIS PROGRAM USES THE GAUSS-
   SEIDEL METHOD FOR SOLVING SYSTEMS OF
   TWO LINEAR EQUATIONS."
60 REM EQUATION 1 IS OF THE FORM
   AX+BY+C=0
70 REM EQUATION 2 IS OF THE FORM
   DX+EY+F=0
80 PRINT"ENTER THE COEFFICIENTS FOR
   EQUATION 1: AX+BY+C=0"
90 PRINT:INPUT"A = ",A:INPUT"B =
   ",B:INPUT "C = ",C
100 PRINT:PRINT:PRINT"ENTER THE
    COEFFICIENTS FOR EQUATION 2:
    DX+EY+F=0"
110 PRINT:INPUT"D = ",D:INPUT"E =
    ",E:INPUT"F = ",F
120 PRINT:PRINT:PRINT"IF YOU WANT TO
    GUESS A VALUE FOR X ENTER A (1)
    ":PRINT"IF YOU WANT TO GUESS A VALUE
    FOR Y";:INPUT" ENTER A (2) ",Q
130 IF Q <> 1 AND Q <> 2 THEN GOTO 120
140 I = 1
150 IF Q = 2 THEN GOTO 260
160 PRINT:INPUT"ENTER YOUR GUESS FOR X:
    ",X(I)
170 PRINT "I","X(I)","Y(I)"
180 Y(I) = -(D*X(I)+F)/E
190 PRINT:PRINT I,X(I),Y(I)
200 IF I = 1 GOTO 220
210 IF ABS(X(I) - X(I-1))<.01 AND
    ABS(Y(I)-Y(I-1)) < .01 GOTO 400
220 I = I + 1
230 IF I > 20 GOTO 360
240 X(I) = - (B*Y(I-1)+C)/A
250 GOTO 180
260 PRINT:INPUT"ENTER YOUR GUESS FOR Y:
    ",Y(1)
270 PRINT "I","X(I)","Y(I)"
280 X(I) =-(B*Y(I) + C)/A
290 PRINT:PRINT I,X(I),Y(I)
300 IF I = 1 THEN GOTO 320
```

```
310 IF ABS(X(I)-X(I-1))<.01 AND ABS(Y(I)-
    Y(I-1))<.01 THEN GOTO 400
320 I = I + 1
330 IF I > 20 GOTO 360
340 Y(I)=-(D*X(I-1)+F)/E
350 GOTO 280
360 PRINT "YOUR GUESS WAS NOT CLOSE
    ENOUGH." : PRINT
370 INPUT "DO YOU WANT TO TRY ANOTHER
    GUESS? ", Q2$
380 IF Q2$ = "Y" THEN GOTO 120
390 IF Q2$ = "N" THEN GOTO 420
400 PRINT : PRINT : PRINT "THE ANSWER IS"
    : PRINT : PRINT "X = "X(I)", Y= "Y(I)
410 PRINT
420 INPUT "IF YOU WANT TO WORK THIS SAME
    PROBLEM TYPE A 99 ", QU
430 IF QU = 99 THEN GOTO 120
440 END
```

 Use this program (PRG05244) to employ the Gauss–Seidel method to find the solution of the linear system

$$\begin{cases} 2x + 5y - 2 = 0 \\ 3x - 8y + 15 = 0 \end{cases} \text{ near } (-2,\ 1).$$

Section 5.4, Exercise 35

```
 1 REM ***PRG05435***
10 REM THIS PROGRAM WILL ALLOW YOU TO
   INPUT ELEMENTS OF A 2 x 2 DETERMINANT
   OF THE FORM:
20 REM
30 REM    A    B
40 REM
50 REM    C    D
60 REM
70 REM AND EVALUATE IT.
80 CLS
90 PRINT " THIS PROGRAM DETERMINES THE
   VALUE OF A 2 X 2 DETERMINANT"
91 PRINT "OF THE FORM:"
92 PRINT
93 PRINT" A B"
94 PRINT
95 PRINT" C D"
96 PRINT
98 INPUT "GIVE VALUE FOR A: ",A
100 INPUT "GIVE VALUE FOR B: ",B
110 INPUT "GIVE VALUE FOR C: ",C
120 INPUT "GIVE VALUE FOR D: ",D
130 DET = A*D-C*B
```

```
140 PRINT
150 PRINT "THE DETERMINANT IS: ";DET
160 END
```

 Use this program (PRG05435) to evaluate each of the following determinants.

(a) $\begin{vmatrix} 2 & -\dfrac{1}{2} \\ \sqrt{2} & \dfrac{1}{\sqrt{7}} \end{vmatrix}$ **(b)** $\begin{vmatrix} \pi & 0.002 \\ -5 & -\sqrt{19} \end{vmatrix}$

Section 5.4, Exercise 37

```
  1 REM PRG05437
 10 REM THIS PROGRAM DETERMINES THE VALUE
    OF A 3 X 3 DETERMINANT
 20 REM OF THE FORM
 30 REM A1 B.P. C1
 40 REM A2 B2 C2
 50 REM A3 B3 C3
 60 REM
 70 CLS
 80 PRINT"THIS PROGRAM DETERMINES THE
    VALUE OF A 3 X 3 DETERMINANT"
 90 PRINT"OF THE FORM:"
100 PRINT"A1 B.P. C1"
110 PRINT"A2 B2 C2"
120 PRINT"A3 B3 C3"
130 PRINT
140 PRINT
150 INPUT"GIVE THE VALUE FOR A1: ",A1
160 INPUT"GIVE THE VALUE FOR B1: ",B.P.
170 INPUT"GIVE THE VALUE FOR C1: ",C1
180 PRINT
190 PRINT
200 INPUT"GIVE THE VALUE FOR A2: ",A2
210 INPUT"GIVE THE VALUE FOR B2: ",B2
220 INPUT"GIVE THE VALUE FOR C2: ",C2
230 PRINT
240 PRINT
250 INPUT"GIVE THE VALUE FOR A3: ",A3
260 INPUT"GIVE THE VALUE FOR B3: ",B3
270 INPUT"GIVE THE VALUE FOR C3: ",C3
275 PRINT
276 PRINT
280 DET = A1*B2*C3 + B.P.*C2*A3 + C1*A2*B3
    - C1*B2*A3 - A1*C2*B3 - B.P.*A2*C3
290 PRINT USING"THE VALUE OF THIS
    DETERMINANT IS: ##########.######";DET
300 END
```

Use this program (PRG05437) to evaluate each of the following determinants.

(a) $\begin{vmatrix} 1 & 2 & 3 \\ 0 & 2 & 4 \\ 0 & 0 & -6 \end{vmatrix}$ (b) $\begin{vmatrix} \pi & \dfrac{4}{3} & -\dfrac{17}{6} \\ 1+\sqrt{2} & 7 & 0.4 \\ 2 & \sqrt{11} & 0 \end{vmatrix}$

Section 5.5, Exercise 17

```
1 REM PRG05517
10 CLS
20 PRINT "THIS PROGRAM WILL USE CRAMER'S
   RULE TO SOLVE A SYSTEM OF LINEAR
   EQUATIONS"
30 PRINT "TYPE A '2' IF YOU WANT TO SOLVE
   A SYSTEM OF TWO EQUATIONS OR TYPE A
   '3' IF YOU WANT TO SOLVE A SYSTEM OF
   THREE EQUATIONS.": INPUT Q
40 IF Q <> 2 AND Q <> 3 AND Q <> 0 THEN
   GOTO 30
50 IF Q = 0 THEN END
60 IF Q = 3 THEN GOTO 240
70 PRINT : PRINT : PRINT "THE TWO
   EQUATIONS ARE OF THE FORM:"
80 PRINT "AX + BY = H"
90 PRINT "CX + DY = K"
100 PRINT : PRINT "ENTER THE VALUES FOR
    EQUATION 1:"
110 INPUT" A = ",A
120 INPUT" B = ",B
130 INPUT" H = ",H
140 PRINT : PRINT "NOW ENTER THE VALUES
    FOR EQUATION 1:"
150 INPUT" A = ",A
160 INPUT" B = ",B
170 INPUT" K = ",K
180 DET = A * D - B * C
190 IF DET = 0 THEN GOTO 450
200 DX = H * D - K * B:DY = A * K - C * H
210 X = DX/DET:Y = DY / DET
220 PRINT "THE SOLUTIONS ARE:"
230 PRINT : PRINT "X = ";X : PRINT "Y =
    ";Y:GOTO 520
240 PRINT " THIS PROGRAM WILL USE CRAMER'S
    RULE TO SOLVE A SYSTEM OF THREE
    EQUATIONS OF THE FORM: "
250 PRINT "A(1)X + B(1)Y + C(1)Z = K(1)"
260 PRINT "A(2)X + B(2)Y + C(2)Z = K(2)"
270 PRINT "A(1)X + B(3)Y + C(3)Z = K(3)"
280 PRINT : PRINT "ENTER THE VALUES FOR
    EQUATION 1: "
290 INPUT" A(1) = ",A1
300 INPUT" B(1) = ",B.P.
310 INPUT" C(1) = ",C1
320 INPUT" K(1) = ",K1
330 PRINT : PRINT "ENTER THE VALUES FOR
    EQUATION 2: "
340 INPUT" A(2) = ",A2
350 INPUT" B(2) = ",B2
360 INPUT" C(2) = ",C2
370 INPUT" K(2) = ",K2
380 PRINT : PRINT "ENTER THE VALUES FOR
    EQUATION 3: "
390 INPUT" A(3) = ",A3
400 INPUT" B(3) = ",B3
410 INPUT" C(3) = ",C3
420 INPUT" K(3) = ",K3
430 DET = A1*B2*C3 + B.P.*C2*A3 + C1*A2*B3
    - C1*B2*A3 - A1*C2*B3 - B.P.*A2*C3
440 IF DET <> 0 THEN GOTO 460
450 PRINT : PRINT : PRINT "THE DETER-
    MINANT IS ZERO. THE SYSTEM HAS NO
    SOLUTION!!!":GOTO 520
460 DX = K1*B2*C3 + B.P.*C2*K3 + C1*K2*B3
    - C1*B2*K3 - K1*C2*B3 - B.P.*K2*C3
470 DY = A1*K2*C3 + K1*C2*A3 + C1*A2*K3 -
    C1*K2*A3 - A1*C2*K3 - K1*A2*C3
480 DZ = A1*B2*K3 + B.P.*K2*A3 + K1*A2*B3
    - K1*B2*A3 - A1*K2*B3 - B.P.*A2*K3
490 X = DX/DET:Y=DY/DET:Z=DZ/DET
500 PRINT "THE SOLUTIONS ARE: "
510 PRINT "X = ";X : PRINT "Y = ";Y :
    PRINT "Z = ";Z
520 END
```

 1. Change this program (PRG05517) to print out either "THIS SYSTEM IS INCONSISTENT" or "THIS SYSTEM IS DEPENDENT", whichever is correct, rather than the statement on line 450.

2. After you have revised PRG05517, as in Exercise 1 above, use it to solve each of the following systems.

(a) $\begin{cases} 2x + y - 3z = 1 \\ x - 4y + z = 6 \\ 4x - 16y + 4z = 24 \end{cases}$

(b) $\begin{cases} 2x + y + z = 0 \\ x + y - z = 0 \\ x + 2y + 2z = 0 \end{cases}$

(c) $\begin{cases} x + 2y - z = 0 \\ 2x - y + z = 0 \\ 8x + y + z = 2 \end{cases}$

Chapter 8

Section 8.4, Exercise 51

```
10 REM ***PRG08451***
20 PRINT"THIS PROGRAM WILL DETERMINE THE
   ROOTS OF A QUADRATIC EQUATION OF THE
   FORM AX^2 + BX + C = 0."
30 PRINT:INPUT"A= ",A:INPUT "B=
   ",B:INPUT"C= ",C
40 D = B * B - 4 *A*C
50 IF ABS(D) <= .00001 THEN D = 0:GOTO
   120
60 IF D > 0 THEN GOTO 80
70 PRINT"THE DISCRIMINANT IS LESS THAN
   ZERO. THERE ARE NO REAL ROOTS.":GOTO
   140
80 Y = SQR(D)
90 X1 = ( - B + Y)/(2*A)
100 X2 = ( - B - Y)/(2*A)
110 PRINT : PRINT"THE ROOTS ARE: ";X1;"
    AND ";X1
120 X1 = -B/(2*A)
130 PRINT : PRINT"THE ROOTS ARE BOTH THE
    SAME: ";X1;" AND ";X2
140 PRINT : PRINT : PRINT"TYPE A 99 IF
    YOU WANT TO SOLVE ANOTHER QUADRATIC
    EQUATION. ":INPUT Q
150 IF Q = 99 THEN GOTO 30
160 END
```

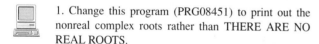 1. Change this program (PRG08451) to print out the nonreal complex roots rather than THERE ARE NO REAL ROOTS.

2. Use your new version of PRG08451 from Exercise 1 above to find the roots of
(a) $4x^2 - 9x + 3.5 = 0$,
(b) $\sqrt{2}x^2 + 4x - \sqrt{12} = 0$,
(c) $1.2x^2 - 5.52x + 6.348 = 0$,
(d) $3.2x^2 - 7.1x + 5.3 = 0$.

Chapter 9

Section 9.2, Exercise 36

```
10 REM *** PRG09236 ***
20 REM THIS PROGRAM COMPUTES AND PRINTS
   THE SINES AND COSINES OF THE ANGLES
30 REM FROM: 0 TO 1.6 RADIANS EVERY 0.1
   RADIAN.
```

```
40 PRINT"THIS PROGRAM COMPUTES AND PRINTS
   THE SINES AND COSINES OF THE ANGLES"
50 PRINT"FROM: 0 TO 1.6 RADIANS EVERY 0.1
   RADIAN."
60 PRINT"ANGLE","SIN","COS" : PRINT
70 FOR X = 0 TO 1.6 STEP 0.1
80 SX = X - (X^3)/6 + (X^5) / 120 - (X^7)
   / 5040
90 CX = 1 - (X^2)/2 + (X^4) / 24 - (X^6)
   / 720 + (X^8)/ 40320
100 PRINT USING"##.# ##.######
    ##.######";X;SX;CX
110 NEXT X
120 END
```

Section 9.2, Exercise 37

```
10 REM *** PRG09237 ***
20 REM THIS PROGRAM COMPUTES AND PRINTS
   THE SINE AND COSINE OF ANGLES
30 REM*FROM 0 TO 90 DEGREES EVERY 1
   DEGREE
40 PRINT "ANGLE","SIN","COS"
50 PRINT
60 PI = 3.141592653559#
70 FOR I = 1 TO 90
80 X = I * PI / 180
90 SX = X - (X ^ 3) / 6 + (X ^ 5) / 120 -
   (X ^ 7) / 5040
100 CX = 1 - (X ^ 2) / 2 + (X ^ 4) / 24 -
    (X ^ 6) /720 + (X ^ 8) / 40320!
110 PRINT USING" ##   ##.####
    ##.####";I;SX;CX
120 NEXT I
130 END
```

 Change these programs (PRG09236 and PRG09237) to print the sines, cosines, and tangents from 0 to 2π radians every 0.1 radian and from $0°$ to $180°$ every $1°$, respectively.

Section 9.5, Exercise 41

```
10 REM *** PRG09541 ***
20 CLS
30 REM THIS PROGRAM WILL DETERMINE THE
   ANGLE, IN RADIANS, FROM THE VALUE
40 REM FOR THE TANGENT OF THE ANGLE.
50 PRINT"THIS PROGRAM WILL DETERMINE AN
   ANGLE MEASURE, IN RADIANS, FROM THE
   VALUE"
```

```
 60 PRINT"ENTERED IN FOR THE ANGLE'S
    TANGENT."
 70 INPUT"WHAT IS THE TANGENT OF THE
    ANGLE: ",X
 80 Y = ATN(X)
 90 PRINT : PRINT"IF ATN(Y) = ";X
100 PRINT"THEN Y IS: ";Y;" RADIANS."
110 END
```

Section 9.5, Exercise 42

```
 10 REM PRG09542
 20 CLS
 30 REM THIS PROGRAM WILL DETERMINE THE
    ANGLE, IN RADIANS, FROM THE VALUE FOR
 40 REM THE COSINE OF THE ANGLE.
 50 PRINT "THIS PROGRAM WILL DETERMINE THE
    ANGLE, IN RADIANS, FROM THE VALUE FOR"
 60 PRINT "THE COSINE OF THE ANGLE."
 70 PRINT : PRINT:INPUT"WHAT IS THE COSINE
    OF THE ANGLE: ",X
 80 IF X = 1 THEN Y = 1:GOTO 140
 90 IF X = - 1 THEN Y = 3.1415927#:GOTO
    140
100 IF ABS(X) < 1 GOTO 130
110 PRINT : PRINT : PRINT X;" IS NOT AN
    ACCEPTABLE VALUE. PLEASE ENTER A VALUE
    BETWEEN"
120 PRINT " - 1 AND + 1.":GOTO 70
130 Y = - ATN(X/SQR( - X*X + 1)) +
    1.57079633#
140 PRINT : PRINT"IF ARC COS(Y) = ";X:
    PRINT"THEN Y IS: ";Y;" RADIANS."
150 END
```

Section 9.5, Exercise 43

```
  1 REM RPG09543
 10 CLS
 20 REM THIS PROGRAM WILL ALLOW YOU TO
    ENTER A VALUE AND PRINT ALL POSSIBLE
    INVERSE TRIGONOMETRY FUNCTIONS FOR
    THAT VALUE.
 30 PRINT"THIS PROGRAM WILL ALLOW YOU TO
    ENTER A VALUE AND PRINT ALL POSSIBLE
    INVERSE TRIGONOMETRY FUNCTIONS FOR
    THAT VALUE."
 40 PI = 3.14159265359#
 50 INPUT "WHAT IS THE VALUE OF X: ",X
 60 IF ABS(X) > 1 THEN GOTO 140
 70 IF X = 1 THEN Y1 = PI /2:Y2 = 0:GOTO
    120
```

```
 80 IF X = - 1 THEN Y1 = - PI / 2: Y2 =
    PI:GOTO 120
 90 Y1 = ATN(X/SQR( - X*X + 1))
100 Y2 = - ATN(X/SQR( - X*X + 1)) + PI /2
110 IF ABS(X) < 1 THEN GOTO 160
120 IF X = 1 THEN Y3 = PI / 2:Y4 = PI:GOTO
    160
130 IF X = - 1 THEN Y3 = - PI/2:Y4 =
    0:GOTO 160
140 Y3 = ATN(1/SQR(X*X - 1)) + (SGN(X) -
    1)* PI/2
150 Y4 = ATN(SQR(X*X - 1)) + (SGN(X) - 1)
    *PI/2
160 Y5 = ATN(X)
170 Y6 = - ATN(X) + PI/2
180 PRINT : PRINT"IF THE VALUE OF X IS
    ";X" THEN "
190 IF ABS(X) > 1 GOTO 220
200 PRINT : PRINT"ARC SIN(X) = ";Y3;" RAD"
    : PRINT"ARC COS(X) = ";Y2;" RAD"
210 IF ABS(X) < 1 THEN GOTO 240
220 PRINT : PRINT"ARC SIN(X) AND ARC
    COS(X) DO NOT EXIST OR VALUE LARGER
    THAN 1 OR SMALLER THEN - 1"
230 PRINT : PRINT"ARC CSC(X) = ";Y3;" RAD"
    : PRINT"ARC SEC(X) = ";Y4;" RAD":GOTO
    250
240 PRINT : PRINT"ARC CSC(X) AND ARC
    SEC(X) DO NOT EXIST OR VALUES BETWEEN
    - 1 AND + 1."
250 PRINT : PRINT"ARC TAN(X) = ";Y5;" RAD"
    : PRINT"ARC COT(X) = ";Y6;" RAD"
260 END
```

Section 9.5, Exercise 44

Add the following lines to each program:

To Exercise 41 add   43 Y = Y * 180/3.14159254359
                45 PRINT "OR"; Y; "DEGREES"

To Exercise 42 add   83 Y = Y * 180/3.1415927
                85 PRINT "OR"; Y; DEGREES."

To Exercise 43 multiply each value of Y1, Y2, etc. by 180/PI

Chapter 10

Section 10.2, Exercise 33

```
  1 REM *** PRG10233 ***
 10 CLS
```

```
20 REM THIS PROGRAM WILL DETERMINE
   THE COMPONENT VECTOR FOR A POSITION
   VECTOR.
30 PRINT "THIS PROGRAM WILL DETERMINE
   THE COMPONENT VECTOR FOR A POSITION
   VECTOR."
40 PI = 3.14159265359#
50 PRINT : PRINT"IF THE DIRECTION OF THE
   POSITION VECTOR IS IN RADIANS A '99'.
   IF THEY ARE GIVEN IN DEGREES THEN TYPE
   A ' - 1'.":INPUT Q
60 PRINT: INPUT"ENTER THE MAGNITUDE OF
   THE VECTOR ",M: PRINT :INPUT" AND ITS
   DIRECTION : ",D
70 IF Q = 99 THEN GOTO 90
80 D = D * PI / 180
90 AX = M * COS(D):AY = M * SIN(D)
100 TX = 0:TY = PI / 2
110 IF AX < 0 THEN TX = PI
120 IF AY < 0 THEN TY = TY + PI
130 PRINT : PRINT : PRINT"THE HORIZON-
    TAL VECTOR HAS " : PRINT"MAGNITUDE:
    ";ABS(AX):PRINT" AND DIRECTION: ";TX;"
    RADIANS" : PRINT " OR ";TX * 180 /
    PI;" DEGREES"
140 PRINT : PRINT"THE VERTICAL VECTOR
    HAS " : PRINT"MAGNITUDE: ";ABS(AY) :
    PRINT"AND DIRECTION ";TY;" RADIANS" :
    PRINT "OR ";TY * 180 / PI;" DEGREES."
150 END
```

Use this program (PRG10233) to determine the horizontal and vertical components of the given vectors.

(a) Magnitude 20, $\theta = 75°$
(b) Magnitude 16, $\theta = 212°$
(c) Magnitude 18.4, $\theta = 4.97$ rad
(d) Magnitude 23.7, $\theta = 2.22$ rad

Section 10.2, Exercise 34

```
10 REM *** PRG10234***
20 CLS
30 REM THIS PROGRAM WILL DETERMINE THE
   RESULTANT VECTORS WHEN TWO OR MORE
   VECTORS ARE ADDED.
40 PRINT "THIS PROGRAM WILL DETERMINE
   THE RESULTANT VECTORS WHEN TWO OR MORE
   VECTORS ARE ADDED."
50 AX = 0:AY = 0:PI = 3.14159265359#
60 PRINT : PRINT "IF THE DIRECTION FOR
   THE VECTORS ARE IN RADIANS TYPE A
   '99'. IF THEY ARE GIVEN IN DEGREES
   THEN TYPE ' -1'." : INPUT Q
70 PRINT : PRINT:INPUT "ENTER THE NUMBER
   OF VECTORS: ";N
80 PRINT : PRINT : PRINT "FOR THE FIRST
   VECTOR ENTER THE":INPUT" MAGNITUDE:
   ";A(1):INPUT" AND THE DIRECTION:
   ";TA(1)
90 FOR I = 2 TO N
100 PRINT : PRINT "FOR VECTOR NUMBER
    ";I" ENTER THE":INPUT" MAGNITUDE:
    ";A(I):INPUT" AND THE DIRECTION:
    ";TA(I)
110 NEXT I
120 IF Q = 99 THEN GOTO 140
130 FOR I = 1 TO N:TA(I) = TA(I) * PI/
    180:NEXT I
140 FOR I = 1 TO N
150 AX(I)=A(I) * COS(TA(I)):AY(I) = A(I) *
    SIN(TA(I)):AX = AX + AX(I):AY = AY +
    AY(I)
160 NEXT I
170 R = SQR(AX^2 + AY^2)
180 IF AX <> 0 THEN GOTO 220
190 IF AY < 0 THEN GOTO 210
200 ANG = PI/2:GOTO 230
210 ANG = 3 * PI/2:GOTO 230
220 ANG = ATN(AY/AX)
230 IF AX < 0 THEN GOTO 260
240 IF AY < 0 THEN ANG = ANG + 2 * PI
250 GOTO 270
260 ANG = ANG + PI
270 PRINT : PRINT : PRINT "THE RESULTANT
    VECTOR HAS" : PRINT : PRINT "MAGNI-
    TUDE = ";R : PRINT" AND DIRECTION =
    ";ANG;"RADIANS": PRINT "OR ";ANG * 180
    / PI;" DEGREES"
280 END
```

Use this program (PRG10234) to determine the resultant vector when the given vectors are added.

(a) $A = 109$, $\theta_A = 33.4°$, $B = 125$, $\theta_B = 69.4°$
(b) $C = 137$, $\theta_C = 30°$, $D = 89.4$, $\theta_D = -72.5°$, $D = 164.6$, $\theta_E = 159.8°$
(c) $F = 37.5$, $\theta_F = 0.25$ rad, $G = 49.3$, $\theta_G = 1.92$ rad

Chapter 14

Section 14.2, Exercise 63

```
1 1 REM *** PRG14263 ***
10 CLS
20 PRINT"THIS PROGRAM WILL ADD OR SUB-
   TRACT TWO COMPLEX NUMBERS. THE NUMBERS
   MUST BE IN THE FORM: A + BJ"
30 '
40 '
50 PRINT" ENTER AN 'A' IF YOU WANT TO ADD
   TWO NUMBERS, ENTER AN 'S' IF YOU WANT
   TO SUBTRACT THEM":INPUT A$
60 '
70 '
80 REM **** ENTER COMPLEX NUMBER IN A +
   BJ FORM ****
90 '
100 '
110 CLS : PRINT"ENTER EACH OF THE COMPLEX
    NUMBERS IN THE FORM: A + BJ"
120 IF A$ = "A" THEN GOTO 140
130 PRINT"ENTER THE NUMBER BEING
    SUBTRACTED AS THE SECOND NUMBER"
140 FOR I = 1 TO 2 : PRINT : PRINT"FOR
    NUMBER ";I : PRINT:INPUT"A= ";A(I - 1)
    : PRINT:INPUT"AND B = ";B(I - 1):NEXT
    I
150 IF A$ = "S" THEN GOTO 280
160 '
170 '
180 REM *** SUBTRACT COMPLEX NUMBERS ***
190 '
200 '
210 A = A(0) + A(1):B = B(0) + B(1)
220 GOTO 350
230 '
240 '
250 REM *** SUBTRACTING COMPLEX NUMBERS
    ***
260 '
270 '
280 A = A(0) - A(1):B = B(0) - B(1)
290 PRINT : PRINT A;" + ";B;"J"
300 '
310 '
320 REM *** PRINTING ANSWERS ***
330 '
340 '
350 PRINT : PRINT : PRINT"THE ANSWER IS "
360 PRINT : PRINT A;" + ";B;"J"
```

```
370 END
```

 Use this program (PRG14263) to add or subtract the following complex numbers.

(a) $(7.51 - 2.36j) + (-12.47 + 4.61j)$
(b) $(15.59 - 6.34j) - (-21.3 + 7.21j)$

Section 14.2, Exercise 64

```
1 REM *** PRG14264 ***
10 CLS : PRINT"THIS PROGRAM WILL MULTIPLY
   TWO COMPLEX NUMBERS. THE NUMBERS MUST
   BE ENTERED IN THE FORM: A + BJ."
80 '
90 '
100 REM *** ENTER COMPLEX NUMBER IN A + BJ
    FORM ***
110 '
120 '
130 PRINT"ENTER EACH OF THE NUMBERS IN THE
    FORM: A + BJ.
140 FOR I = 1 TO 2 : PRINT : PRINT"FOR
    NUMBER ";I: INPUT"GIVE A =",A(I - 1) :
    PRINT:INPUT"GIVE B = ";B(I - 1):NEXT I
150 GOTO 220
170 '
180 '
190 REM *** MULTIPLYING COMPLEX NUMBERS
    ***
200 '
210 '
220 A = A(0) * A(1) - B(0)*B(1):B = A(0) *
    B(1) + A(1) * B(0)
280 '
290 '
300 REM *** PRINT PRODUCT ***
310 '
320 '
330 PRINT : PRINT : PRINT"THE PRODUCT IS:
    "
340 PRINT : PRINT A;" + ";B;"J"
350 END
```

 Use this program (PRG14264) to multiply the following complex numbers.

(a) $(5.72 - 4.36j)(-35.87 + 1.61j)$
(b) $(0.09 - 1.74j)(-1.3 + 0.11j)$

Section 14.2, Exercise 65

```
10 REM *** PRG14265 ***
20 CLS : PRINT" THIS PROGRAM WILL DIVIDE
   TWO COMPLEX NUMBERS. THE NUMBER MUST
   BE ENTERED IN THE FORM: A + BJ"
30 '
40 '
50 REM *** ENTER COMPLEX NUMBERS IN THE
   FORM: A + BJ ***
60 '
70 '
80 PRINT"ENTER EACH OF THE COMPLEX
   NUMBERS IN THE FORM A + BJ"
90 PRINT : PRINT"ENTER THE NUMERATOR
   FIRST"
100 PRINT"THE NUMERATOR IS:":INPUT"A =
    ";A(0):PRINT:INPUT"AND B = ";B(0)
110 PRINT:PRINT"AND THE DENOMINATOR
    IS:":INPUT"A = ";A(1):PRINT:INPUT"AND
    B = "; B(1)
120 '
130 ,
140 REM *** DIVIDING COMPLEX NUMBERS ***
150 '
160 DENOM = A(1)*A(1) + B(1)*B(1)
170 A = (A(0) * A(1) + B(0) * B(1))
    / DENOM:B = (B(0) * A(1) - A(0) *
    B(1))/DENOM
180 '
190 PRINT : PRINT : PRINT"THE QUOTIENT IS:
    "
200 PRINT : PRINT A;" + ";B;"J"
210 END
```

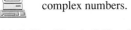 Use this program (PRG14265) to divide the following complex numbers.

(a) $(0.25 + 7j) \div (-9.12 - 0.21j)$
(b) $4j \div (7 + 2j)$

Section 14.3, Exercise 41

```
10 REM *** PRG14341 ***
20 PRINT"THIS PROGRAM WILL CHANGE A
   COMPLEX NUMBER FROM RECTANGULAR FORM
   TO POLAR FORM OR FROM POLAR FORM TO
   RECTANGULAR."
30 PRINT : PRINT : PRINT"IF THE NUMBERS
   YOU ARE ENTERING ARE IN RECTANGULAR
```

```
   FORM, THEN TYPE A '1'. IF THEY ARE IN
   POLAR FORM, THEN TYPE A '9'."
40 INPUT A$
50 PI = 3.14159265359#
60 IF A$ = "9" THEN GOTO 200
70 '
80 '
90 REM *** ENTER COMPLEX NUMBERS IN
   RECTANGULAR FORM ***
100 '
110 '
120 CLS:PRINT"ENTER THE COMPLEX NUMBER IN
    THE FORM: A + BJ."
130 INPUT"A = ";A:PRINT:INPUT" AND B = ";B
140 GOTO 420
150 '
160 '
170 REM *** ENTER THE COMPLEX NUMBERS IN
    POLAR FORM ***
180 '
190 '
200 PRINT:PRINT"ENTER THE COMPLEX NUMBER
    IN THE FORM R CIS (AN)"
210 PRINT:PRINT"IF ANGLES ARE IN DEGREES
    TYPE '55' IF IN RADIANS TYPE '11'":IN-
    PUT DR:IF DR <> 11 AND DR <> 55 THEN
    GOTO 210
220 INPUT"R = ";R:PRINT:INPUT "AND THE
    ANGLE IS ";AN
230 '
240 '
250 REM *** CONVERT DEGREES TO RADIANS ***
260 '
270 '
280 IF DR = 11 THEN GOTO 350
290 AD = AN:AN = AN * PI / 180
300 '
310 '
320 REM *** CONVERT TO RECTANGULAR FORM
    ***
330 '
340 '
350 A = R * COS(AN):B = R * SIN(AN)
360 GOTO 490
370 '
380 '
390 REM *** PRINTING ANSWER ***
400 '
410 '
420 R = SQR(A*A + B*B)
```

```
430 IF A<> 0 THEN GOTO 460
440 IF B> 0 THEN AN = PI / 2:GOTO 490
450 IF B< 0 THEN AN = 3 * PI / 2:GOTO 490
460 AN = ATN(B/A)
470 IF A = 0 THEN AN = AN - PI
480 IF AN < 0 THEN AN = AN + PI
490 IF AN >= 2 * PI THEN AN = AN - 2 *
    PI:GOTO 490
500 AD = AN * 180/PI
510 '
520 '
530 REM *** PRINTING ANSWER ***
540 '
550 '
560 PRINT:PRINT: PRINT"THE NUMBER IS "
570 PRINT:PRINT A;" + ";B;"J"
580 PRINT:PRINT" OR ":PRINT R;" CIS
    ";AN;" RADIANS" :PRINT R;" CIS ";AD; "
    DEGREES"
590 END
```

 Use this program (PRG14341) to convert the following complex numbers from their given form to the indicated form.

(a) Change $-12.4 + 3.7j$ from rectangular form to polar form
(b) Change 4.2 cis 247° from polar form to rectangular form
(c) Change 11.25 cis 5.13 from polar form to rectangular form

Section 14.5, Exercise 35

```
10 REM *** PRG14535 ***
20 PRINT" THIS PROGRAM WILL MULTIPLY A
   GROUP OF COMPLEX NUMBERS. THE NUMBERS
   CAN BE ENTERED IN RECTANGULAR FORM OR
   IN POLAR FORM"
30 PRINT:PRINT:PRINT"IF THE NUMBERS YOU
   ARE ENTERING ARE IN RECTANGULAR FORM
   TYP E A '1' IF IN POLAR FORM TYPE
   '9'."
40 INPUT A$
50 PI = 3.14159265359#
60 N = 2
70 PRINT:PRINT:INPUT"ENTER THE NUMBER OF
   COMPLEX NUMBERS YOU ARE MULTIPLYING
   ";N
80 IF A$ = "9" THEN GOTO 220
90 '
100 '
110 REM *** ENTER COMPLEX NUMBERS IN
    RECTANGULAR FORM ***
```

```
120 '
130 '
140 CLS:PRINT"ENTER EACH OF THE ";N;"
    COMPLEX NUMBERS IN THE FORM A + BJ"
150 FOR I = 1 TO N:PRINT:PRINT"FOR NUMBER
    ";I:PRINT:INPUT"A = ";A(I-1):PRINT:INP
    UT"AND B = ";B(I-1):NEXT I
160 GOTO 430
170 '
180 '
190 REM *** ENTER COMPLEX NUMBER IN POLAR
    FORM ***
200 '
210 '
220 PRINT:PRINT"ENTER THE COMPLEX NUMBER
    IN THE FORM R CIS(AN)"
230 PRINT:PRINT"IF ANGLES ARE IN DEGREES
    TYPE '55' IF IN RADIANS TYPE '11'":IN-
    PUT DR:IF DR <> 11 AND DR <> 55 THEN
    GOTO 230
240 FOR I = 1 TO N:PRINT:PRINT"FOR
    NUMBER ";I:PRINT:INPUT"R = ";R(I-
    1):PRINT:INPUT "AND THE ANGLE IS
    ";AN(I-1):NEXT I
250 '
260 '
270 REM *** CONVERT DEGREES TO RADIANS ***
280 '
290 '
300 IF DR = 11 THEN GOTO 370
310 FOR I = 0 TO N-1:AD(I) = AN(I):AN(I) =
    AN(I) * PI /180:NEXT I
320 '
330 '
340 REM *** CONVERT TO RECTANGULAR FORM
    ***
350 '
360 '
370 FOR I = 0 TO N-1:A(I) = R(I)
    * COS(AN(I)):B(I) = R(I) *
    SIN(AN(I)):NEXT I
380 '
390 '
400 REM *** MULTIPLY COMPLEX NUMBERS ***
410 '
420 '
430 A = A(0) * A(1) - B(0) * B(1):B = A(0)
    * B(1) + A(1) * B(0)
440 IF A$ = "9" THEN AN = AN(0) + AN(1)
450 IF N = 2 THEN GOTO 490
```

```
460 FOR I = 3 TO N:A(N) = A:B(N) = B:A
    = A(I-1) * A(N) - B(I-1)* B(N):B =
    A(I-1) * B(N) + A(N) * B(I-1)
470 IF A$ = "9" THEN AN = AN + AN(I-1)
480 NEXT I
490 R = SQR(A*A+B*B)
500 IF A$ = "9" THEN GOTO 560
510 IF A<> 0 THEN GOTO 540
520 IF B> 0 THEN AN = PI/2:GOTO 570
530 IF B< 0 THEN AN = 3 * PI / 2:GOTO 570
540 AN = ATN(B/A)
550 IF A$ = "1" AND AN < 0 THEN AN = AN +
    PI
560 IF AN < 0 THEN AN = AN + 2*PI:GOTO 560
570 IF AN >= 2*PI THEN AN = AN - 2 *
    PI:GOTO 570
580 AD = AN * 180 / PI
590 PRINT:PRINT:PRINT"THE TOTAL IS "
600 PRINT:PRINT A;" + ";B;"J"
610 PRINT:PRINT"OR":PRINT R;"CIS ";AN;"
    RADIANS":PRINT R;" CIS ";AD;" DEGREES"
620 END
```

 Use this program (PRG14535) to multiply the following complex numbers.

(a) $(5.72 - 4.36j)(-35.87 + 1.61j)$
(b) $(4 \text{ cis } 321°)(2.7 \text{ cis } 217°)(2 \text{ cis } 153°)$
(c) $(6 \text{ cis } 2.75)(4.5 \text{ cis } 1.25)$

Section 14.5, Exercise 36

```
10 REM *** PRG14536 ***
20 PRINT" THIS PROGRAM WILL DIVIDE TWO
   COMPLEX NUMBERS. THE NUMBERS CAN BE
   ENTERED IN RECTANGULAR FORM OR IN
   POLAR FORM"
30 PRINT:PRINT:PRINT"IF THE NUMBERS YOU
   ARE ENTERING ARE IN RECTANGULAR FORM
   TYPE A '1' IF IN POLAR FORM TYPE '9'."
40 INPUT A$
50 PI = 3.14159265359#
60 IF A$ = "9" THEN GOTO 210
70 '
80 '
90 REM *** ENTER COMPLEX NUMBERS IN
   RECTANGULAR FORM ***
100 '
110 '
120 CLS : PRINT"ENTER THE NUMERATOR FIRST"
130 PRINT"THE NUMERATOR IS:":INPUT"A =
    ";A(0):PRINT:INPUT"AND B = ";B(0)
140 PRINT:PRINT"AND THE DENOMINATOR
    IS:":INPUT"A = "; A(1):PRINT:INPUT"AND
    B = "; B(1)
150 GOTO 420
160 '
170 '
180 REM *** ENTER COMPLEX NUMBER IN POLAR
    FORM ***
190 '
200 '
210 PRINT:PRINT"ENTER THE COMPLEX NUMBER
    IN THE FORM R CIS(AN)"
220 PRINT:PRINT"IF ANGLES ARE IN DEGREES
    TYPE '55' IF IN RADIANS TYPE '11'":IN-
    PUT DR:IF DR <> 11 AND DR <> 55 THEN
    GOTO 220
230 FOR I = 1 TO 2:PRINT: PRINT"FOR NUMBER
    ";I: PRINT: INPUT"R = "; R(I-1):
    PRINT: INPUT "AND THE ANGLE IS ";
    AN(I-1):NEXT I
240 '
250 '
260 REM *** CONVERT DEGREES TO RADIANS ***
270 '
280 '
290 IF DR = 11 THEN GOTO 360
300 FOR I = 0 TO 1:AD(I) = AN(I):AN(I) =
    AN(I) * PI /180:NEXT I
310 '
320 '
330 REM *** CONVERT TO RECTANGULAR FORM
    ***
340 '
350 '
360 FOR I = 0 TO 1:A(I) = R(I)
    * COS(AN(I)):B(I) = R(I) *
    SIN(AN(I)):NEXT I
370 '
380 '
390 REM *** DIVIDING COMPLEX NUMBERS ***
400 '
410 '
420 DENOM = A(1)*A(1) + B(1)*B(1)
430 A = (A(0) * A(1) + B(0) * B(1)) /
    DENOM:B = (B(0) * A(1) - A(0) * B(1))/
    DENOM
440 '
450 '
```

```
460 '
470 R = SQR(A*A+B*B)
480 IF A$ = "9" THEN AN = AN(0) -
    AN(1):GOTO 540
490 IF A<> 0 THEN GOTO 520
500 IF B> 0 THEN AN = PI/2:GOTO 550
510 IF B< 0 THEN AN = 3 * PI / 2:GOTO 550
520 AN = ATN(B/A)
530 IF A$ = "1" AND AN < 0 THEN AN = AN +
    PI
540 IF AN < 0 THEN AN = AN + 2*PI:GOTO 540
550 IF AN >= 2*PI THEN AN = AN - 2 *
    PI:GOTO 550
560 AD = AN * 180 / PI
570 PRINT:PRINT:PRINT"THE TOTAL IS "
580 PRINT:PRINT A;" + ";B;"J"
590 PRINT:PRINT"OR":PRINT R;"CIS ";AN;"
    RADIANS":PRINT R;" CIS ";AD;" DEGREES"
600 END
```

 Use this program (PRG14536) to divide the following complex numbers.

(a) $(0.25 + 7j) \div (-9.12 - 0.21j)$
(b) $7.5 \text{ cis } 36° \div 2.5 \text{ cis } 108°$
(c) $2.8 \text{ cis } 5.34 \div 0.7 \text{ cis } 1.65$

Section 14.5, Exercise 37

```
  1 REM *** PRG14537 ***
 10 PRINT "THIS PROGRAM WILL USE DE
    MOIVRE'S FORMULA TO FIND THE N-TH
    POWER OF A COMPLEX NUMBER OR THE N
    N-TH ROOTS OF A COMPLEX NUMBER."
 20 PRINT : PRINT : PRINT "IF YOU ARE
    FINDING A ROOT TYPE A '2' IF YOU ARE
    FINDING A POWER TYPE A '9'."
 30 INPUT RP: IF RP <> 2 AND RP <> 9 THEN
    GOTO 20
 40 PRINT : PRINT : PRINT "IF THE NUMBER
    YOU ARE ENTERING IS IN RECTANGULAR
    FORM TYPE '1' IF IT IS IN POLAR FORM
    TYPE A '9'."
 50 INPUT A$
 60 PI = 3.14159265359#
 70 IF A$ = "9" THEN GOTO 330
 80 '
 90 '
100 REM *** ENTER COMPLEX NUMBER IN
    RECTANGULAR FORM ***
110 '
```

```
120 '
130 CLS: PRINT "ENTER THE COMPLEX NUMBER
    IN THE FORM A + BJ"
140 PRINT : PRINT : INPUT "A= ";A: PRINT :
    PRINT : INPUT "AND B = ";B
150 '
160 '
170 REM *** CONVERT FROM RECTANGULAR TO
    POLAR FORM ***
180 '
190 '
200 R = SQR(A*A + B*B)
210 IF A <> 0 THEN GOTO 240
220 IF B > 0 THEN AN = PI/2:GOTO 260
230 IF B< 0 THEN AN = 3 * PI/2:GOTO 260
240 AN = ATN(B/A)
250 IF A < 0 THEN AN = AN + PI
260 AD = AD * 180/PI
270 GOTO 480
280 '
290 '
300 REM ENTER COMPLEX NUMBER IN COMPLEX
    FORM ***
310 '
320 '
330 PRINT : PRINT "ENTER THE COMPLEX
    NUMBER IN THE FORM R CIS(AN)"
340 PRINT : PRINT "IF THE ANGLES ARE IN
    DEGREES TYPE '55' IF": INPUT " IN
    RADIANS TYPE '11'";DR:IF DR <> 11 AND
    DR <> 55 GOTO 340
350 PRINT : PRINT : PRINT : INPUT "R =
    ";R: PRINT : INPUT " AND THE ANGLE IS
    ";AN
350 PRINT : PRINT : PRINT : INPUT "R =
    ";R: PRINT : INPUT " AND THE ANGLE IS
    ";AN
360 '
370 '
380 REM CONVERT DEGREES TO RADIANS ***
390 '
400 '
410 IF DR = 11 THEN GOTO 480
420 AD = AN:AN = AN * PI / 180
430 '
440 '
450 REM *** FIND THE N - TH POWER OF THE
    NUMBER ***
460 '
470 '
```

```
480 IF RP = 2 GOTO 630
490 PRINT : PRINT : INPUT "ENTER THE
    POWER";N
500 AN = N * AN
510 R = R^N
520 IF AN >= 0 AND AN < 2*PI THEN GOTO 570
530 IF AN < 0 THEN GOTO 560
540 AN = AN - 2*PI:IF AN >= 2 * PI GOTO
    540
550 GOTO 570
560 AN = AN + 2 * PI: IF AN <0 GOTO 560
570 PRINT : PRINT : PRINT "THE TOTAL IS "
580 A = R * COS(AN):B = R * SIN(AN)
590 AD = AN * 180/PI
600 PRINT : PRINT A;" + ";B;"J"
610 PRINT : PRINT "OR": PRINT R;" CIS
    ";AN;" RADIANS": PRINT : PRINT R;" CIS
    ";AD;" DEGREES"
620 GOTO 850
630 PRINT : PRINT : INPUT "ENTER THE
    NUMBER OF ROOTS ";N
640 R = R^(1/N)
650 AG = AN
660 FOR I = O TO N - 1
670 AN = (AG + 2 * PI * I)/N
680 IF I = 0 THEN GOTO 740
690 IF I = 1 THEN GOTO 730
700 IF I = 2 THEN GOTO 720
710 PRINT : PRINT : PRINT "THE ";I + 1;"TH
    ROOT IS":GOTO 750
720 PRINT : PRINT : PRINT "THE 3RD ROOT
    IS":GOTO 750
730 PRINT : PRINT : PRINT "THE 2ND ROOT
    IS":GOTO 750
740 PRINT : PRINT : PRINT "THE FIRST ROOT
    IS"
750 IF AN >= 0 AND AN < 2*PI THEN GOTO 800
760 IF AN < 0 THEN GOTO 790
770 AN = AN - 2 * PI: IF AN >= 2*PI THEN
    GOTO 770
780 GOTO 800
790 AN = AN + 2 * PI:IF AN < 0 THEN GOTO
    790
800 A = R * COS(AN):B = R * SIN(AN)
810 AD = AN * 180/PI
820 PRINT : PRINT A ;" + ";B;"J"
830 PRINT : PRINT "OR": PRINT R;" CIS
    ";AN;" RADIANS": PRINT : PRINT R;" CIS
    ";AD;" DEGREES"
840 NEXT I
850 END
```

 Use this program (PRG14537) to evaluate the following complex numbers.

(a) $(1.5 + 3.2j)^4$
(b) the four fourth roots of $2 - 2j$

Section 14.6, Exercise 35

```
  1 REM RPG14635
 10 PRINT" THIS PROGRAM WILL COMPUTE
    THE TOTAL IMPEDANCE OF A GROUP OF N
    IMPEDANCES IN AN AC CIRCUIT THAT ARE
    CONNECTED IN SERIES OR PARALLEL"
 20 PRINT:PRINT:PRINT"THE IMPEDANCE MAY BE
    ENTERED EITHER IN RECTANGULAR OR IN
    POLAR FORM."
 30 FOR I = 1 TO 4000:NEXT I
 40 PI = 3.14159265359#
 50 N = 2
 51 PRINT:INPUT"HOW MANY IMPEDANCES ARE IN
    THE CIRCUIT: ",N
 60 DIM A(N),B(N)
 70 PRINT:PRINT:INPUT"IF THE IMPEDANCES
    ARE IN RECTANGULAR FORM TYPE A '1' ;
    IF THEY ARE IN POLAR FORM THEN TYPE A
    '9':",FR:IF FR <> 1 AND FR <> 9 THEN
    GOTO 70
 80 IF FR = 9 THEN GOTO 220
 90 '
100 '
110 REM **** ENTER RECTANGULAR COORDINATES
    ****
120 '
130 '
140 PRINT:PRINT"ENTER EACH OF THE ";N;"
    IMPEDANCES IN THE FORM A + BJ"
150 FOR I = 1 TO N:PRINT"FOR IMPEDANCE
    #";I:PRINT:INPUT"A = ",A(I -
    1):PRINT:INPUT "AND B = ",B(I -
    1):NEXT I
160 GOTO 390
170 '
180 '
190 REM *** ENTER POLAR COORDINATES ***
200 '
210 '
220 DIM R(N),AN(N)
230 PRINT:PRINT"ENTER EACH OF THE ";N;"
    IMPEDANCES IN THE FORM R CIS(A)"
240 PRINT:PRINT"IF ANGLES ARE IN DEGREES
    TYPE '55'; IF":INPUT"IN RADIANS TYPE
    '11'",DR:IF DR <> 55 AND DR <> 11 THEN
    GOTO 240
250 FOR I = 1 TO N :PRINT:PRINT"FOR
    IMPEDANCE NUMBER ";I:PRINT:PRINT:
    PRINT:INPUT "R = ",R(I - 1):PRINT
    :INPUT" AND THE ANGLE IS ",AN(I - 1)
260 '
270 '
280 REM *** CONVERT DEGREES TO RADIANS ***
290 '
300 '
310 IF DR = 11 GOTO 380
320 AN(I - 1) = AN(I - 1) * PI/180
330 '
340 '
350 REM *** CONVERT TO RECTANGULAR FORM
    ***
360 '
370 '
380 A(I - 1) = R(I - 1) * COS(AN(I -
    1)):B(I - 1) = R(I - 1) * SIN(AN(I -
    1)):NEXT I
390 PRINT:PRINT:PRINT"THE TOTAL IMPEDANCE
    FOR A"
400 '
410 '
420 REM *** DETERMINE THE SERIES IMPEDANCE
    ***
430 '
440 '
450 FOR I=1 TO N:A(N) = A(N) + A(N -
    1):B(N) = B(N) + B(N - 1):NEXT I
460 RT = SQR(A(N) * A(N) + B(N) * B(N)):TA
    = ATN(B(N)/A(N)):DA=TA*180/PI
470 PRINT"SERIES CIRCUIT IS" :PRINT:PRINT
    A(N);" + ";B(N);"J,":PRINT:PRINT RT;"
    CIS ";TA;" RADIANS OR ":PRINT:PRINT
    RT;"CIS ";DA;" DEGREES."
480 '
490 '
500 REM *** DETERMINE THE PARALLEL
    IMPEDANCE ***
510 '
520 '
530 ZA = 0:ZB = 0
540 FOR I = 1 TO N:B = A(I - 1) * A(I - 1)
    + B(I - 1)*B(I - 1):ZA = ZA + A(I -
    1)/B:ZB = ZB - B(I - 1) /B:NEXT I
550 B = ZA * ZA + ZB * ZB:PA=ZA/B:PB = -
    ZB/B
560 RP = SQR(PA * PA + PB * PB):AR =
    ATN(PB/PA):AD = AR * 180/PI
570 PRINT :PRINT"AND FOR A":PRINT:PRINT
    "PARALLEL CIRCUIT IS" :PRINT:PRINT
    PA;" + ";PB;"J,":PRINT:PRINT RP;"CIS
    ";AR;"RADIANS OR ":PRINT:PRINT RP;"CIS
    ";AD;" DEGREES."
```

```
580 '
590 '
600 REM *** ASK FOR ANOTHER PROBLEM ***
610 '
620 '
630 PRINT:PRINT:PRINT:INPUT"IF YOU WANT TO
    WORK ANOTHER PROBLEM TYPE A '1'; IF
    NOT TYPE A '5'.",QU:IF QU <> 1 AND QU
    <> 5 THEN GOTO 630
640 IF QU = 5 THEN GOTO 670
650 CLEAR
660 GOTO 40
670 END
```

Use this program (PRG14635) to find the total impedances if the given impedances are connected **(a)** in series and **(b)** in parallel.

(a) $Z_1 = 2 + 3j$, $Z_2 = 1 - 5j$
(b) $Z_1 = -2 + j$, $Z_2 = 3 - 4j$, $Z_3 = 9 - 2j$
(c) $Z_1 = 2.19\underline{/18.4°}$, $Z_2 = 5.16\underline{/67.3°}$
(d) $Z_1 = 2.57\underline{/0.25}$, $Z_2 = 1.63\underline{/1.38}$

Chapter 17

Section 17.1, Exercise 37

```
  1 REM *** PRG17137 ***
  5 CLS
 10 REM ***********************
 20 REM ***********************
 30 REM **                  **
 40 REM **    MATRIX ADDITION **
 50 REM **    AND SUBTRACTION **
 60 REM **                  **
 70 REM ***********************
 80 REM ***********************
100 PRINT : PRINT : PRINT : PRINT
110 PRINT "THIS PROGRAM ADDS AND SUBTRACTS
    TWO": PRINT : PRINT "M X N MATRICES":
    PRINT :
120 GOTO 140
130 PRINT "PLEASE PRINT AN A OR AN S":
    PRINT: PRINT" 140 '
150 INPUT "DO YOU WANT TO ADD OR SUBTRACT?
    ";QU$
160 IF LEFT$ (QU$,1) < > "A" AND LEFT$
    (QU$,1) < > "S" THEN GOTO 130
170 PRINT : PRINT : PRINT : INPUT "ENTER
    THE NUMBER OF ROWS: ";M: PRINT : PRINT
```

```
    : INPUT "AND THE NUMBER OF COLUMNS:
    ";N
180 DIM A(M-1, N-1), B(M-1, N-1), C(M-1,
    N-1)
190 CLS : PRINT "ENTER THE ELEMENTS IN
    MATRIX A"
200 PRINT : PRINT
210 FOR I = 0 TO M - 1
220 FOR J = 0 TO N - 1
230 PRINT "A(";I;",";J;") = ";: INPUT
    A(I,J)
240 NEXT J
250 PRINT : PRINT : NEXT I
260 CLS : PRINT "ENTER THE ELEMENTS IN
    MATRIX B"
270 PRINT : PRINT
280 FOR I = 0 TO M - 1
290 FOR J = 0 TO N - 1
300 PRINT "B(";I;",";J;") = ";: INPUT
    B(I,J)
310 NEXT J
320 PRINT : PRINT : NEXT I
330 CLS : PRINT "THE ANSWER IS": PRINT :
    PRINT
340 FOR I = 0 TO M - 1
350 FOR J = 0 TO N - 1
360 IF LEFT$ (QU$,1) = "S" THEN GOTO 390
370 C(I,J) = A(I,J) + B(I,J)
380 GOTO 400
390 C(I,J) = A(I,J) - B(I,J)
400 Z = 8 * J + 1
410 PRINT USING "########.####"; C(I,J);
420 NEXT J
430 PRINT : PRINT : NEXT I
440 GOTO 460
450 PRINT : PRINT "PLEASE REENTER AND TYPE
    A Y OR N": PRINT : PRINT
460 INPUT "DO YOU WANT TO WORK ANOTHER
    PROBLEM? "; QU$
470 IF LEFT$ (QU$,1) < > "Y" AND LEFT$
    (QU$,1) < > "N" THEN GOTO 450
480 IF QU$ = "Y" THEN GOTO 150
490 END
```

Use this program (PRG17137) to find the indicated sum or difference **(a)** $\begin{bmatrix} 2 & 4 \\ -5.3 & 1 \\ 3.1 & 4 \end{bmatrix} + \begin{bmatrix} 5.7 & -6 \\ 2.4 & 9 \\ 8 & 2 \end{bmatrix}$ and

(b) $\begin{bmatrix} 3.4 & 2.1 & -1.5 \\ 9.7 & 12.1 & 4 \end{bmatrix} - \begin{bmatrix} 1 & 3.4 & -5 \\ 4 & 3.9 & 6 \end{bmatrix}$

Section 17.1, Exercise 38

```
  1 REM *** PRG17138 ***
  5 CLS
 10 REM *****************************
 20 REM *****************************
 30 REM **                       **
 40 REM **   SCALAR MULTIPLICATION **
 50 REM **        OF MATRICES      **
 60 REM **                       **
 70 REM *****************************
 80 REM *****************************
100 PRINT : PRINT : PRINT : PRINT
110 PRINT "THIS PROGRAM MULTIPLIES A
    SCALAR": PRINT : PRINT " BY AN M X N
    MATRIX : PRINT : PRINT
115 CLEAR
120 PRINT : PRINT : INPUT "ENTER THE
    SCALAR: "; K: PRINT : PRINT
130 PRINT : PRINT : INPUT "ENTER
    THE NUMBER OF ROWS:";M: PRINT : PRINT
    : INPUT "AND THE NUMBER OF COLUMNS:
    ";N
140 DIM A(M - 1,N - 1),B(M - 1,N - 1),C(M
    - 1,N - 1)
150 CLS : PRINT "ENTER THE ELEMENTS IN THE
    MATRIX"
160 PRINT : PRINT
170 FOR I = 0 TO M - 1
180 FOR J = 0 TO N - 1
190 PRINT "A(";I;",";J;") = ";: INPUT
    A(I,J)
200 NEXT J
210 PRINT : PRINT : NEXT I
220 CLS : PRINT "THE ANSWER IS": PRINT :
    PRINT
230 FOR I = 0 TO M - 1
240 FOR J = 0 TO N - 1
250 C(I,J) = K * A(I,J)
250 C(I,J) = K * A(I,J)
260 Z = 8 * J + 1
270 PRINT USING "########.####"; C(I,J);
280 NEXT J
290 PRINT : PRINT : NEXT I
300 GOTO 320
310 PRINT : PRINT"PLEASE REENTER AND TYPE
    A Y OR N"; PRINT : PRINT
320 INPUT "DO YOUR WANT TO WORK ANOTHER
    PROBLEM";QU$
330 IF LEFT$ (QU$,1) < > "Y" AND LEFT$
    (QU$,1)< > "N" THEN GOTO 310
```

```
340 IF QU$ = "Y" THEN GOTO 115
350 END
```

Section 17.1, Exercise 39

```
  1 REM *** PRG17139 ***
  5 REM *****************************
 10 REM *****************************
 20 REM **                       **
 30 REM **   SCALAR MULTIPLICATION **
 40 REM **     AND THEN ADDITION    **
 50 REM **      OR SUBTRACTION      **
 60 REM **       OF TWO MATRICES    **
 70 REM **                       **
 80 REM *****************************
 90 REM *****************************
100 PRINT : PRINT : PRINT : PRINT
110 PRINT "THIS PROGRAM ALLOWS YOU TO
    MULTIPLY TWO M X N MATRICES BY A
    SCALAR" : PRINT : PRINT
112 PRINT "AND THEN IT ADDS OR SUBTRACTS
    THE NEW MATRICES."
115 CLEAR
120 PRINT : PRINT : INPUT "ENTER THE
    SCALAR FOR THE FIRST MATRIX: "; K1:
    PRINT : PRINT
130 PRINT : PRINT : INPUT "ENTER THE
    SCALAR FOR THE SECOND MATRIX: "; K2:
    PRINT : PRINT
220 GOTO 240
230 PRINT "PLEASE PRINT AN A OR AN S":
    PRINT: PRINT"
240 '
250 INPUT "DO YOU WANT TO ADD OR SUBTRACT?
    ";QU$
260 IF LEFT$ (QU$,1) < > "A" AND LEFT$
    (QU$,1) < > "S" THEN GOTO 230
270 PRINT : PRINT : PRINT : INPUT "ENTER
    THE NUMBER OF ROWS: ";M: PRINT : PRINT
    : INPUT "AND THE NUMBER OF COLUMNS:
    ";N
290 CLS : PRINT "ENTER THE ELEMENTS IN
    MATRIX A"
300 PRINT : PRINT
310 FOR I = 0 TO M - 1
320 FOR J = 0 TO N - 1
330 PRINT "A(";I;",";J;") = ";: INPUT
    A(I,J)
340 NEXT J
350 PRINT : PRINT : NEXT I
```

```
360 CLS : PRINT "ENTER THE ELEMENTS IN
    MATRIX B"
370 PRINT : PRINT
380 FOR I = 0 TO M - 1
390 FOR J = 0 TO N - 1
400 PRINT "B(";I;",";J;") = ";: INPUT
    B(I,J)
410 NEXT J
420 PRINT : PRINT : NEXT I
430 CLS : PRINT "THE ANSWER IS": PRINT :
    PRINT
440 FOR I = 0 TO M - 1
450 FOR J = 0 TO N - 1
460 IF LEFT$ (QU$,1) = "S" THEN GOTO 490
470 C(I,J) = K1*A(I,J) + K2*B(I,J)
480 GOTO 500
490 C(I,J) = A(I,J) - B(I,J)
500 Z = 8 * J + 1
510 PRINT USING "########.####"; C(I,J);
520 NEXT J
530 PRINT : PRINT : NEXT I
540 GOTO 560
550 PRINT : PRINT "PLEASE REENTER AND TYPE
    A Y OR N": PRINT : PRINT
560 INPUT "DO YOU WANT TO WORK ANOTHER
    PROBLEM? "; QU$
570 IF LEFT$ (QU$,1) < > "Y" AND LEFT$
    (QU$,1) < > "N" THEN GOTO 550
580 IF QU$ = "Y" THEN GOTO 115
590 END
```

Use this program (PRG17139) to determine

$$3\begin{bmatrix} 2 & 4 \\ -5.3 & 1 \\ 3.1 & 4 \end{bmatrix} + \frac{1}{2}\begin{bmatrix} 5.7 & -6 \\ 2.4 & 9 \\ 8 & 2 \end{bmatrix}$$

Section 17.2, Exercise 29

```
  1 REM *** PRG17229 ***
  5 CLS
 10 REM *****************************
 20 REM *****************************
 30 REM **                       **
 40 REM **   MATRIX MULTIPLICATION   **
 50 REM **                       **
 60 REM *****************************
 70 REM *****************************
100 PRINT : PRINT : PRINT : PRINT
110 PRINT "THIS PROGRAM MULTIPLIES TWO
    MATRICES": PRINT : PRINT "MATRIX A AND
    MATRIX B": PRINT :
```

```
120 CLEAR
130 PRINT : PRINT : PRINT "FOR MATRIX A:"
    : PRINT : PRINT : INPUT "ENTER THE
    NUMBER OF ROWS: " ;M : PRINT : PRINT :
    INPUT "AND THE NUMBER OF COLUMNS: ";N
140 PRINT : PRINT : PRINT "FOR MATRIX B:"
    : PRINT : PRINT : INPUT "ENTER THE
    NUMBER OF ROWS: " ;P: PRINT : PRINT :
    INPUT "AND THE NUMBER OF COLUMNS: ";R
150 IF N = P THEN GOTO 210
160 CLS : PRINT : PRINT "YOU HAVE MADE A
    MISTAKE.": PRINT : PRINT "THE NO. OF
    COLUMNS IN MATRIX A MUST BE THE SAME
    AS THE NO. OF ROWS IN MATRIX B"
170 FOR X = 1 TO 4000: NEXT X
180 PRINT: PRINT : PRINT "YOU HAD ";N;"
    COLUMNS IN MATRIX A": PRINT : PRINT "
    AND ";P;" ROWS IN MATRIX B."
190 FOR X = 1 TO 4000: NEXT X
200 PRINT : PRINT : PRINT "PLEASE REEN-
    TER": FOR X = 1 TO 2000: NEXT X: CLS :
    GOTO 130
210 DIM A(M - 1,N - 1),B(P - 1,R - 1),C(M
    - 1,R - 1)
220 CLS : PRINT "ENTER THE ELEMENTS IN
    MATRIX A":
230 PRINT : PRINT :
240 FOR I = 0 TO M - 1
250 FOR J = 0 TO N - 1
260 PRINT "A(";I;",";J;") = ";: INPUT
    A(I,J)
270 NEXT J
280 PRINT : PRINT : NEXT I
290 CLS : PRINT "ENTER THE ELEMENTS IN
    MATRIX B"
300 PRINT : PRINT
310 FOR I = 0 TO P - 1
320 FOR J = 0 TO R - 1
330 PRINT "B(";I;",";J;") = ";: INPUT
    B(I,J)
340 NEXT J
350 PRINT : PRINT : NEXT I
360 CLS : PRINT "THE ANSWER IS": PRINT :
    PRINT
370 FOR I = 0 TO M - 1
380 FOR K = 0 TO R - 1
390 FOR J = 0 TO N - 1
400 C(I,K) = C(I,K) + A(I,J) * B(J,K)
410 NEXT J
420 Z = 8 * J + 1
430 PRINT USING "########.####"; C(I,K);
```

```
440 NEXT K
450 PRINT : PRINT : NEXT I
460 GOTO 480
470 PRINT : PRINT "PLEASE REENTER AND TYPE
    A Y OR N": PRINT : PRINT
480 INPUT "DO YOU WANT TO WORK ANOTHER
    PROBLEM? "; QU$
490 IF LEFT$ (QU$,1) < > "Y" AND LEFT$
    (QU$,1) < > "N" THEN GOTO 470
500 IF QU$ = "Y" THEN GOTO 120
510 END
```

 Use this program (PRG17229) to find the indicated product:

$$\begin{bmatrix} 3.7 & 4.3 & -2.9 & -17.4 \\ -7.1 & 8.3 & 10 & -5.4 \end{bmatrix} \begin{bmatrix} 0 & 4.2 & 1.7 \\ 1.2 & 0.4 & 2 \\ 8 & -1.9 & 0 \\ -3.5 & -4.1 & 12.7 \end{bmatrix}$$

Section 17.3, Exercise 22

```
  1 REM *** PRG17322 ***
  5 CLS
 10 REM *************************
 20 REM *************************
 30 REM **                     **
 40 REM **    MATRIX INVERSION  **
 50 REM **                     **
 60 REM *************************
 70 REM *************************
100 PRINT : PRINT : PRINT : PRINT
110 PRINT "THIS PROGRAM INVERTS AN N X N
    MATRIX":
120 FOR X = 1 TO 1000: NEXT X
130 CLEAR
140 PRINT : PRINT : INPUT "ENTER THE SIZE
    OF THE SQUARE MATRIX ";X
150 DIM A(X - 1,X - 1), B(X - 1,X - 1)
160 CLS : PRINT "ENTER THE ELEMENTS IN THE
    MATRIX"
170 FOR I = 0 TO X - 1
180 FOR J = 0 TO X - 1
190 PRINT "A(";I;",";J;") = ";: INPUT
    A(I,J)
200 NEXT J
210 B(I,I) = 1
220 PRINT : PRINT : NEXT I
230 FOR J = 0 TO X - 1
240 FOR I = J TO X - 1
250 IF A(I,J) < > 0 THEN 280
260 NEXT I
270 PRINT : PRINT "THIS MATRIX IS SINGU-
    LAR. IT DOES NOT HAVE AN INVERSE.":
    GOTO 530
280 FOR H = 0 TO X - 1
290 T = A(J,H):A(J,H) = A(I,H):A(I,H) = T
300 M = B(J,H):B(J,H) = B(I,H):B(I,H) = M
310 NEXT H
320 D = A(J,J)
330 FOR H = 0 TO X - 1
340 A(J,H) = A(J,H) / D:B(J,H) = B(J,H) /
    D
350 NEXT H
360 FOR H = 0 TO X - 1
370 IF J = H THEN 420
380 D = A(H,J)
390 FOR K = 0 TO X - 1
400 A(H,K) = A(H,K) - A(J,K) * D:B(H,K) =
    B(H,K) - B(J,K) * D
410 NEXT K
420 NEXT H
430 NEXT J
440 CLS : PRINT : PRINT : PRINT "THE
    INVERSE MATRIX IS": PRINT : PRINT
450 FOR I = 0 TO X - 1
460 FOR J = 0 TO X - 1
470 B(I,J) = INT ( B(I,J) * 10000 + .5) /
    10000
480 PRINT USING "######.####"; B(I,J);
490 NEXT J
500 PRINT : PRINT : NEXT I
510 GOTO 530
520 PRINT : PRINT "PLEASE REENTER AND TYPE
    A Y OR N": PRINT : PRINT
530 INPUT "DO YOU WANT TO INVERT ANOTHER
    MATRIX"; QU$
540 IF LEFT$ (QU$,1) < > "Y" AND LEFT$
    (QU$,1) < > "N" AND LEFT$ (QU$,1) <
    > "y" AND LEFT$ (QU$,1) < > "n" THEN
    GOTO 530
550 IF QU$ = "Y" OR QU$ = "y" THEN GOTO
    130
560 END
```

 Use this program (PRG17322) to find the inverse of the given matrix (if it exists).

(a) $\begin{bmatrix} -20 & 5.1 \\ -7.5 & 1.5 \end{bmatrix}$, (b) $\begin{bmatrix} 1 & 2 & -1 \\ 3 & 7 & -10 \\ 7 & 16 & -21 \end{bmatrix}$, and

(c) $\begin{bmatrix} 1.5 & -2.2 & -1 & -2.8 \\ 3.6 & -5.4 & -2 & -7 \\ 2 & -5 & -1 & -5 \\ -1 & 6 & 4 & 10 \end{bmatrix}$

Section 17.4, Exercise 25

```
  1 REM *** PRG17425 ***
  5 CLS
 10 REM ******************************
 20 REM ******************************
 30 REM **                        **
 40 REM **   USING MATRICES TO SOLVE  **
 50 REM **SYSTEMS OF LINEAR EQUATIONS**
 60 REM **                        **
 70 REM ******************************
 80 REM ******************************
100 CLS
110 PRINT "THIS PROGRAM SOLVES A SYSTEM OF
    N EQUATIONS WITH N VARIABLES":
120 FOR X = 1 TO 1000: NEXT X
130 CLEAR
140 CLS : INPUT "ENTER THE NUMBER OF
    EQUATIONS ";X
150 DIM A(X - 1,X - 1),B(X - 1,X - 1), C(X
    - 1),D(X - 1)
160 CLS: PRINT "ENTER THE COEFFICIENTS OF
    THE SYSTEM OF EQUATIONS"
170 FOR I = 0 TO X - 1
180 FOR J = 0 TO X - 1
190 PRINT "A(";I;",";J;") = ";: INPUT
    A(I,J)
200 NEXT J
210 B(I,I) = 1
220 PRINT : PRINT : NEXT I
230 PRINT : PRINT "NOW ENTER THE
    CONSTANTS.": PRINT
240 FOR I = 0 TO X - 1
250 PRINT "C(";I;") = ";: INPUT C(I)
260 NEXT I
270 FOR J = 0 TO X - 1
280 FOR I = J TO X - 1
290 IF A(I,J) < > 0 THEN 320
300 NEXT I
310 PRINT : PRINT "THE MATRIX IS SINGULAR.
    THIS SYSTEM HAS NO SOLUTION.": GOTO
    650
320 FOR H = 0 TO X - 1
330 T = A(J,H):A(J,H) = A(I,H):A(I,H) = T
340 M = B(J,H):B(J,H) = B(I,H):B(I,H) = M
350 NEXT H
360 D = A(J,J)
370 FOR H = 0 TO X - 1
380 A(J,H) = A(J,H) / D:B(J,H) = B(J,H) /
    D
390 NEXT H
400 FOR H = 0 TO X - 1
410 IF J = H THEN 460
420 D = A(H,J)
430 FOR K = 0 TO X - 1
440 A(H,K) = A(H,K) - A(J,K) * D:B(H,K) =
    B(H,K) - B(J,K) * D
450 NEXT K
460 NEXT H
470 NEXT J
480 CLS : PRINT : PRINT : PRINT "THE
    INVERTED MATRIX IS ": PRINT : PRINT
490 FOR I = 0 TO X - 1
500 FOR J = 0 TO X - 1
510 Y = INT ( B(I,J) * 1000 + .5) / 1000
520 PRINT Y,;
530 NEXT J
540 PRINT : PRINT : NEXT I
550 PRINT : PRINT "THE ANSWER IS": PRINT
560 FOR I = 0 TO X - 1
570 FOR J = 0 TO X - 1
580 D(I) = D(I) + B(I,J) * C(J)
600 NEXT J
610 PRINT "X(";I + 1;") = ";D(I)
620 PRINT : PRINT : NEXT I
630 GOTO 650
640 PRINT : PRINT "PLEASE REENTER AND TYPE
    A Y OR N"; PRINT : PRINT
650 INPUT "DO YOU WANT TO SOLVE ANOTHER
    SYSTEM "; QU$
660 IF LEFT$ (QU$,1) < > "Y" AND LEFT$
    (QU$,1) < > "N" AND LEFT$ (QU$,1) <
    > "y" AND LEFT$ (QU$,1) < > "n" THEN
    GOTO 650
670 IF QU$ = "Y" OR QU$ = "y" THEN GOTO
    130
680 END
```

Use this program (PRG17425) to solve each of the following systems

(a) $\begin{cases} 2x + y - 3z = 1 \\ x - 4y + z = 6 \\ 4x - 16y + 4z = 24 \end{cases}$

(b) $\begin{cases} -x + y + 3z = 4 \\ 2x + 3y + z = -2 \\ 6x + 4y + 2z = 6 \end{cases}$

Chapter 18

Section 18.4, Exercise 22

```
  1 *** PRG18422 ***
 10 REM ****************************
 20 REM ****************************
 30 REM **                      **
 40 REM **   LINEAR INTERPOLATION  **
 50 REM **                      **
 60 REM ****************************
 70 REM ****************************
 80 REM
 90 REM
100 REM
110 REM **  DEFINE THE FUNCTION  **
120 REM
130 CLS
140 INPUT "DO YOU WANT TO CHANGE THE
    FUNCTION? (Y/N) "; QU$
150 IF QU$ < > "Y" AND QU$ < > "N" THEN
    GOTO 140
160 IF QU$ = "N" GOTO 220
170 CLS : PRINT : PRINT : PRINT "TO CHANGE
    THE FUNCTION TYPE "
180 PRINT : PRINT "1050 DEF FN F(X) = . .
    . "
190 PRINT : PRINT "WHERE YOU PUT THE NEW
    FUNCTION AFTER THE EQUAL SYMBOL."
200 PRINT : PRINT "AFTER YOU HAVE ENTERED
    THE FUNCTION YOU WILL HAVE TO TYPE
    'RUN' TO GET THE PROGRAM TO START.":
    PRINT
210 END
220 REM
230 REM
240 REM ** SEARCH FOR ROOTS **
250 REM
260 REM
270 GOSUB 1050
280 GOTO 300
290 PRINT : PRINT : PRINT "YOU MADE A
    MISTAKE. PLEASE ANSWER THE QUESTION
    WITH A Y OR N.": PRINT : PRINT
300 INPUT "DO YOU WANT TO SEARCH FOR
    ROOTS? "; QU$
310 IF LEFT$ (QU$,1) < > "Y" AND LEFT$
    (QU$,1) < > "N" GOTO 290
320 IF LEFT$ (QU$,1) = "N" GOTO 560
330 PRINT : PRINT : INPUT "ENTER THE LOWER
    LIMIT: ";A
340 PRINT : INPUT "NOW ENTER THE UPPER
    LIMIT: ";B
350 CLS : PRINT "THE LOWER LIMIT IS ";A;"
    AND THE" : PRINT "UPPER LIMIT IS ";B
360 PRINT : PRINT : PRINT "X-1"; TAB(
    8);"F(X-1)"; TAB(21);"X"; TAB(30);
    "F(X) "
370 FOR X = A TO B - 1
380 P = FN F(X)
390 Q = FN F(X + 1)
400 IF SGN (P) < > SGN (Q) GOTO 420
410 GOTO 440
420 PRINT : PRINT X; TAB( 6);P; TAB( 21);
    X + 1;TAB(26);Q
430 PRINT " THERE IS A REAL ROOT BETWEEN
    ";X;" AND ";X + 1
440 NEXT X
450 PRINT "SEARCH COMPLETED.": PRINT
460 GOTO 480
470 PRINT : PRINT : "PLEASE ANSWER AGAIN.
    TYPE A Y OR N."
480 PRINT : INPUT "DO YOU WANT TO SEARCH
    OVER ANOTHER INTERVAL? ";QU$
490 IF LEFT$ (QU$,1) < > "Y" AND LEFT$
    (QU$,1) < > "N" GOTO 470
500 IF LEFT$ (QU$,1) = "Y" GOTO 330
510 REM
520 REM
530 REM ** USE OF LINEAR INTERPOLATION **
540 REM
550 REM
560 PRINT : PRINT : PRINT "WE WILL NOW
    USE LINEAR INTERPOLATION TO DETERMINE
    APPROXIMATE ROOTS OF F(X).": PRINT
570 PRINT : PRINT "ENTER THE ENDPOINTS OF
    THE INTERVAL": PRINT " WHERE YOU WANT
    THE ROOT.": PRINT
580 INPUT "LEFT ENDPOINT = ";A: INPUT
    "RIGHT ENDPOINT = ";B
590 IF SGN (FN F(A)) = SGN (FN F(B)) GOTO
    610
600 GOTO 620
610 PRINT : PRINT "THERE IS NO REAL ROOT
    BETWEEN ";A;" AND ";B;".": PRINT
    "PLEASE TRY A DIFFERENT INTERVAL.":
    PRINT : GOTO 570
620 M = (FN F(B) - FN F(A))/(B - A)
630 C = A - (FN F(A)/M)
640 IF SGN (FN F(A)) = SGN (FN F(C)) GOTO
    660
650 B = C: GOTO 670
```

```
660 A = C
670 PRINT : PRINT "THE ROOT IS APPROXI-
    MATELY ";C : PRINT "THE ACTUAL ROOT IS
    BETWEEN ";A;" AND ";B
680 CHK = ABS (FN F(C))
690 IF CHK < .001 GOTO 710
700 GOTO 620
710 PRINT : PRINT : PRINT "INTERPOLATION
    COMPLETED.": PRINT : PRINT "THE ROOT
    IS APPROXIMATELY ";C: PRINT " AND F(C)
    = "; FN F(C)
720 GOTO 740
730 PRINT : PRINT "PLEASE REANSWER WITH A
    Y OR N."
740 PRINT : INPUT "DO YOU WANT TO SEARCH
    OVER ANOTHER INTERVAL? ";QU$
750 IF LEFT$ (QU$,1) < > "Y" AND LEFT$
    (QU$,1) < > "N" GOTO 730
760 IF LEFT$ (QU$,1) = "Y" GOTO 570
770 END
1000 REM
1010 REM
1020 REM ** SUBROUTINE **
1030 REM
1040 REM
1050 DEF FN F(X) = X ^ 3 + X ^ 2 − 7 * X
1060 PRINT : RETURN
```

 Use this program (PRG18422) to approximate the root of $f(x) = \cos(x^2) - x$

Chapter 21

Section 21.2, Exercise 32

```
 1 *** PRG21232 ***
15 REM THIS PROGRAM LETS YOU ENTER INDI-
   VIDUAL SCORES AND THEIR FREQUENCIES TO
   FIND THE MEAN.
20 CLS
30 PRINT"ENTER EACH SCORE AND THEN" :
   PRINT : PRINT"ITS FREQUENCY" : PRINT
   : PRINT"WHEN DONE ENTER A NEGATIVE
   NUMBER FOR A SCORE."
40 T = 0:FR = 0
50 PRINT:INPUT "INPUT A SCORE: ";S
60 IF S < 0 THEN GOTO 100
70 PRINT:INPUT "AND IT'S FREQUENCY: ";F :
   PRINT
80 T = T + F*S:FR = FR + F
90 GOTO 50
```

```
100 MEAN = T / FR
110 PRINT : PRINT"THE MEAN IS ";MEAN
120 END
```

Section 21.3, Exercise 17

```
 10 REM ** PRG21317 **
 20 CLS
 30 PRINT"THIS PROGRAM GIVES THE MEAN,
    VARIANCE, AND STANDARD DEVIATIONS"
 40 PRINT"OF A NUMBER OF SCORES."
 50 PRINT"ENTER EACH SCORE AND THEN" :
    PRINT : PRINT"WHEN YOU HAVE FIN-
    ISHED, ENTER A NEGATIVE NUMBER FOR THE
    SCORE."
 60 T = 0:FR = 0:A = 0
 70 PRINT:INPUT "HOW MANY SCORES ARE YOU
    GOING TO ENTER? ";N
 80 DIM S(N),F(N)
 90 FOR I = 0 TO N-1
100 PRINT:INPUT"SCORE = ";S(I)
110 PRINT:INPUT "FREQUENCY = ";F(I) :
    PRINT
120 T = T + F(I)*S(I):FR = FR + F(I):A = A
    + F(I) * S(I) * S(I)
130 NEXT I
140 MEAN = T/FR
150 VAR = (FR * A - T * T)/(FR*FR)
160 SD = SQR(VAR)
170 PRINT : PRINT"THE MEAN IS: ";MEAN
180 PRINT : PRINT"THE VARIANCE IS: ";VAR
190 PRINT : PRINT"AND THEN STANDARD
    DEVIATION IS: ";SD
200 END
```

Section 21.3, Exercise 18

```
 1 REM *** PRG21318 ***
10 CLS
20 PRINT "THIS PROGRAM GIVES THE MEAN,
   VARIANCE, STANDARD DEVIATION, AND THE
   TOTAL NUMBER AND PERCENTAGE OF SCORES
   WITHIN ONE AND TWO STANDARD DEVIATIONS
   OF THE MEAN."
30 PRINT:PRINT "ENTER EACH SCORE AND THEN
   ITS FREQUENCY" :PRINT
40 T = 0:FR = 0:A = 0:N1 = 0:N2 = 0
50 PRINT : INPUT "HOW MANY SCORES ARE
   THERE TO ENTER? ",N
60 DIM S(N),F(N)
70 FOR I = 0 TO N - 1
80 PRINT : INPUT "SCORE = ";S(I)
```

```
 90 PRINT : INPUT "FREQUENCY = ";F(I):
    PRINT
100 T = T + F(I)*S(I):FR = FR + F(I):A = A
    + F(I) * S(I)*S(I)
110 NEXT I
120 MEAN = T / FR
130 VAR = (FR * A - T * T) / (FR * FR)
140 SD = SQR(VAR)
150 S1 = MEAN - SD : S2 = MEAN + SD:S3 =
    S1 - SD:S4 = S2 + SD
160 FOR I = 0 TO N
170 IF S(I) > S1 AND S(I) < S2 THEN N1 =
    N1 + F(I)
180 IF S(I) > S3 AND S(I) < S4 THEN N2 =
    N2 + F(I)
190 NEXT I
200 PRINT "THE TOTAL NUMBER IN THE SAMPLE
    IS: ";FR
210 PRINT "THE MEAN IS: ";MEAN
220 PRINT "THE VARIANCE IS: ";VAR
230 PRINT : PRINT "AND THE STANDARD
    DEVIATION IS: ";SD
240 PRINT : PRINT "THE NUMBER OF SCORES
    WITHIN ONE STANDARD DEVIATION OF THE
    MEAN I S: ";N1;" OR "
250 PRINT N1 * 100 / FR;" PERCENT OF THE
    TOTAL VALUE."
260 PRINT : PRINT "AND THE NUMBER WITHIN
    TWO S.D. OF THE MEAN IS: ";N2;" OR
    ";N2 *1 00/FR;"PERCENT."
270 END
```

 Use this program (PRG21318) on the data in the following table to determine the mean, variance, the standard deviation, the number within one standard deviation of the mean, and those within two standard deviations of the mean.

Score	1	2	3	4	5	6	7	8
Frequency	23	51	31	27	12	4	1	1

Section 21.4, Exercise 9

```
  1 REM *** PRG21409 ***
  5 CLS
 10 PRINT "THIS PROGRAM DETERMINES THE
    SLOPE, M, AND Y - INTERCEPTS OF THE
    LEAST SQUARES LINE AND THE CORRELATION
    COEFFICIENT OF A SET OF DATA."
 30 PRINT : PRINT : PRINT :INPUT "ENTER
    THE NUMBER OF PAIRS OF POINTS (X,Y) IN
    THE SAMPLE ";N
 40 FOR I = 0 TO N - 1
 50 PRINT : PRINT : PRINT "FOR POINT
    NO.";I + 1
 60 PRINT : INPUT "X = ";X
 70 PRINT : INPUT "Y = ";Y
 80 XX=XX + X : YY=YY + Y : XY=XY + X*Y :
    X2=X2 + X*X : Y2=Y2 + Y*Y
 90 NEXT I
100 XV = (N * X2 - XX * XX)/(N*N) : YV =
    (N * Y2 - YY*YY)/(N*N)
110 XS = SQR(XV):YS = SQR(YV)
120 PRINT : PRINT : PRINT "THE FOLLOWING
    ARE THE TOTALS"
130 M = (XY/N - XX*YY/(N*N))/XV
140 B = YY/N - M*XX/N
150 PRINT : PRINT "M = ";M: PRINT : PRINT
    "B = ";B
160 R = M * XS/YS
170 PRINT : PRINT "R = ";R
180 END
```

Use this program (PRG21409) to determine the slope and y-intercept of the least squares line and the correlation coefficient for the data in the following table.

x	17	21	15	16	15	16	24	27	23	18
y	73	66	64	61	70	71	90	86	84	72

Index of Applications

This index categorizes the applications problems in this book alphabetically. After each topic, the index lists the text page number, followed by the problem number in parentheses. For example, if you are looking for a problem on acoustical engineering, you would turn to page 248 and look for problem number 45.

Acoustical engineering, 248(45), 335(11, 12), 453(41), 705(32)
Acoustics, 47(51)
Aeronautical engineering, 61(62), 657(34)
Aeronautics, 712(36)
Agricultural technology, 126(23), 225(10), 566(12)
Agriculture, 293(18, 19), 635(29), 703(34)
Architectural technology, 225(7)
Architecture, 110(20), 115(14, 15), 116(16), 577(43, 44), 636(34), 652(28)
Astronomy, 42(44), 469(53), 524(28, 29), 537(17), 547(24), 548(26), 557(17), 558(32), 572(33, 34)
Automotive, 83(8)
Automotive technician, 6(27, 28), 22(42), 23(46)
Automotive technology, 35(58), 126(24), 151(15, 21, 22), 159(34), 171(19), 173(41), 187(40), 212(21), 221(27, 31, 32), 339(79), 399(8), 462(32), 572(35, 36), 704(35), 705(31)

Biology, 448(15, 17), 468(49), 469(54), 735(2)
Business, 82(40), 83(6, 11, 12), 86(70), 143(64), 148(37, 38, 39, 40), 159(33), 173(38, 39), 214(35, 36, 37), 256(57), 293(16, 17), 299(47), 572(39), 580(32), 586(21, 22, 23, 24), 588(27), 598(34, 35, 36), 603(23), 614(24), 652(27), 679(50), 730(6), 735(3), 736(19), 737(21)

Chemistry, 83(21, 23), 84(24), 86(73), 230(17), 423(60), 580(31), 640(21)
Civil engineering, 56(51), 84(25), 96(26), 109(14, 17), 110(19, 21), 116(18), 124(9, 10), 125(17), 368(21, 22, 24), 369(21, 22, 24, 25), 374(28), 375(31), 434(21), 532(21), 537(18), 543(20), 558(29, 30)
Computer science, 35(49), 61(61), 280(40), 335(15, 16)
Computer technology, 288(40), 339(72)
Construction, 22(44), 23(48, 49), 35(48, 60), 56(52), 83(18), 103(18, 19), 104(20, 23), 109(11, 12), 125(18), 126(25), 171(21), 225(11), 298(44), 299(50), 321(34), 339(74, 77, 80), 349(33, 34), 361(11), 372(23), 603(24), 652(26), 665(31, 35), 665(35),
Construction technology, 212(22)

Drafting, 635(28)

Dynamics, 143(63), 256(58), 288(35, 36, 38), 298(39, 40, 41), 301(43), 315(33, 34)

Ecology, 159(31, 32), 256(59)
Economics, 236(10)
Electrical engineering, 104(21, 22)
Electrical technician, 23(47)
Electrical technology, 720(28, 29, 30, 31), 724(16), 735(6, 8)
Electricity, 6(29), 22(43), 28(43), 35(59), 115(10), 221(29, 34), 231(19, 23, 24), 235(18, 19, 20, 21, 22), 236(9, 17, 18), 245(61), 272(57, 59), 280(41), 299(48, 49), 301(41), 315(35), 321(31), 338(67, 69), 350(36, 37), 361(12, 13, 14, 15, 16), 362(17, 18, 19), 375(30), 399(9, 10, 11, 12, 13, 14, 15, 18, 19, 20), 423(59), 430(55), 448(22), 467(27), 486(59, 60, 61, 62), 496(35, 36, 37, 38, 39, 40, 41, 42), 503(33, 34), 39, 40), 552(14), 566(13, 14), 587(25), 613(21, 22), 614(23), 636(31), 691(58), 698(39)
Electronics, 42(48), 47(49), 68(87), 96(27), 116(19), 187(41, 42, 43), 194(15, 16), 212(18), 255(55), 382(29), 385(43), 411(23, 24), 462(37, 39), 508(30, 31, 32, 33, 34), 510(38, 39, 40), 519(63, 64), 532(23), 703(33)
Energy, 83(20), 125(12, 15), 235(23), 236(16), 245(64), 335(19), 532(24)
Energy technology, 335(19), 436(54), 572(40), 720(21, 22), 724(14), 731(7)
Environmental science, 83(15), 148(41), 171(20), 172(22), 225(12), 462(33, 34, 36), 652(25), 658(35)
Environmental technology, 836(4)

Finance, 83(9, 10), 443(17, 18, 19, 20), 448(16), 461(25, 26), 462(31), 468(47), 665(32, 33, 34)
Forestry, 339(76)

Industrial design, 115(11, 12), 116(17), 301(44), 524(30), 639(18),
Industrial technology, 335(4), 636(33), 712(38, 39, 40), 720(23, 24, 25, 26, 27)
Insurance, 712(37)
Interior design, 110(18), 115(9)

Land management, 186(33, 34, 35, 36)

Landscape architecture, 103(13)
Light, 237(19)
Lighting technology, 434(23, 24), 576(41)

Machine technology, 83(17), 84(28, 29), 109(15, 16), 180(27), 236(13), 338(68, 70), 572(38), 670(33), 724(13, 15), 730(5), 735(4), 736(18)
Mechanical engineering, 103(16), 109(13), 335(14), 335(17, 18), 537(19, 20), 548(27), 557(18, 19), 640(20)
Mechanical technology, 96(27)
Mechanics, 125(14), 221(25), 399(1, 2, 7), 411(21), 705(33)
Medical technology, 173(40), 350(35), 399(16), 448(18), 598(33), 645(21), 647(21, 22), 657(31), 712(35)
Metalurgy, 212(20)
Metalworking, 35(61)
Meteorology, 75(81), 148(42), 180(28), 462(35), 467(25, 26)
Music, 423(57, 58)

Navigation, 231(18), 321(33), 335(13), 349(31, 32), 358(31, 32), 361(5, 6, 7, 8), 362(20, 21, 22), 369(23), 372(22), 436(53), 542(18), 548(25), 558(31)
Nuclear physics, 676(37, 38)
Nuclear technology, 418(72), 436(51, 52), 448(13, 14), 462(27, 28, 29, 30), 468(48), 639(19)

Oceanography, 385(44), 399(3, 4), 411(22), 430(54)
Optics, 221(30), 272(58), 338(65, 66), 691(57), 698(37, 38), 703(31, 32), 704(36)

Package design, 299(46)
Petroleum engineering, 83(22), 143(62), 186(37), 187(38), 195(20), 236(15), 636(30), 639(17)

Physics, 42(43, 45, 46, 47), 75(82), 84(27, 30), 86(74, 75), 104(24), 180(29, 30), 187(39), 212(18), 221(33), 222(36), 231(20, 21, 22), 235(25), 236(12, 14), 245(62, 63), 293(15), 335(2, 3, 9, 20), 360(1, 2, 3, 4), 361(9, 10), 372(24, 25, 26), 374(27, 29), 399(5, 6, 17), 418(69, 71), 434(22), 467(28), 510(37), 519(61, 62), 524(27), 547(23), 552(13), 566(15, 17, 18), 576(42), 604(25, 26, 27), 657(32, 33), 665(29, 30), 670(31, 32, 34), 676(39), 679(49), 691(56), 698(40), 735(5, 7), 737(20)

Quality control, 736(9, 10, 11, 12, 13, 14, 15, 16, 17)

Recreation, 22(41, 45)

Seismology, 453(40), 468(50)
Sheet metal technology, 115(13), 125(16), 126(19, 20, 21, 22), 225(13), 226(14, 15), 256(56), 298(42, 43), 299(45), 635(27)
Solar energy, 288(39)
Sound, 235(24), 430(56)
Space technology, 86(72), 116(20), 231(25), 335(10), 339(73), 372(21), 566(16)
Surveying, 375(32)

Thermodynamics, 248(42), 249(46), 448(19, 20, 21), 462(38), 552(15),
Transportation, 83(13, 14), 86(71), 125(11), 194(19), 272(60), 280(42), 565(11), 588(26)
Transportation engineering, 84(26), 103(17), 322(37, 38, 39), 335(1), 339(71)

Wastewater technology, 47(50, 52), 225(8), 248(43)

Index

Absolute
  error, **30**
  inequality, 567
  values, 4
AC circuits, complex numbers and, 503–9
Acute
  angle, 92
  triangle, 97
Addition
  algebra and, 52–56
  complex numbers, 477–79
  fractions, 264–71
  integers, 13–14
  linear equations,
    three variables, 189–93
    two variables, 182–85
  matrices, 593–94
  multinomials, 54–55
  radicals, 424–25
  rational numbers, 18
  vectors and, 350–57
Admittance, AC circuits and, 508
Algebra
  addition/subtraction and, 52–56
  binomials, multiplication of, 58–60
  constants in, 52
  division and, 62–67
  equations,
    applications of, 76–82
    solving, 69–75
  expressions in, 52
  fundamental theorem of, 624
  logic, calculators use of, **9**
  monomials and, 53–54
    multiplication of, 56–57
  multinomials and, 53–54
    adding/subtracting, 54–55
    multiplication of, 57–58
  terms in, 52–53
  variables in, 52
Alternate angles
  exterior, 93
  interior, 93
Alternating current, trigonometric graphs and,
  396–98
Altitude, cones and, 119
Amplitude, 378
Analytic geometry, plane. See Plane analytic
  geometry
Angles, 89
  complementary, trigonometric functions and,
    310
  coterminal, 305–6
  of depression, 319–20
  of elevation, 319–20
  negative, trigonometric functions and, 303

positive, trigonometric functions and, 303
  quadrantal, 305
  reference, trigonometric functions, 322–27
  types of, 92–94
Approximate, numbers, 31
Arc length, 112–14
Argand plane, complex planes and, 486
Arithmetic
  rational numbers, 17–21
  sequences, 652–54
    sum of first n term, **660**
    terms of a, **653**
  series, 660–61
Associative laws, numbers and, 7–8
Asymptotes
  horizontal, **643**
  vertical, **642**
Augmented matrices, 592–93
Axes
  rotation of, second-degree equation, 548–53
  translation of, plane analytic geometry and,
    543–48

Base
  cones and, 119
  triangles and, 98
Binomial formula, 671–74
  series, 674–76
  theorem, 670–79
    binomial formula and, 671–74
    Pascal's triangle and, 670–71
Binomials
  difference of squares, **60**
  multiplication of, FOIL method, 58–60
  square of, **60**
  two, same first term, **240**
Blasius' formula, 45
Business functions, exponential functions and,
  441–42

Calculators
  cosine curves and, 393–94
  exponents and, evaluating, 416
  finding logs with, 450–51
  graphing, 159–63
  inverse trigonometric functions and, 327–28
  parametric equations, graphing with, 401–3
  rounding off numbers with, 33–34
  scientific notation on, 37–38
  sine/cosine curves and, 389–91
  standard deviation on, 722–23
  trigonometric values on, 312–13
Cartesian coordinate system, 149
Center, ellipse and, 532
Centroids, triangles and, 98
Charles' law, 227–28

Circles, 110–16
  arc length, 112–14
  circumference/area of, 111–12
  equations of,
    general, **521**
    standard, **520**
    plane analytic geometry, 519–25
Circumference, circle, 111–12
Clearing the equation, 276
Coefficient matrix, 592
Cofactors, linear equations and, 200–202
Combined variation, 232–34
Common
  factors, 245–46
  logs, 450
Commutative laws, numbers and, 7
Complementary angles, 93
  trigonometric functions and, **310**
Complex fractions, 268–71
Complex numbers, 471–510
  absolute value of, **487**
  AC circuits and, 503–9
  addition of, 477–79
  changing,
    polar to rectangular, **488**
    rectangular to polar, **489**
  complex plane and, 486–87
  conjugates of, 475–76
  DeMoivre's formula, 497–98
    powers/roots and, 499–503
  division of, 480–82
    polar form, 498
  equality of, **474**
  exponential form of, 492–96
    multiplying/dividing, 494–95
  imaginary unit and, 472–74
  multiplication of, 479–80
  polar form of, 487–92
    multiplication of, 497–98
  rectangular form of, **474**
  roots of, **500**
  subtraction of, 477–79
  using calculators with, 482–84
Complex plane, complex numbers and, 486–87
Complex roots theorem, **626**
Component vectors, 343
  finding, 343–47
Composite functions, 145–47
Compound
  inequality, 566
  interest, **442, 663**
Computer-aided graphing, 159–63
Computers, inverse trigonometric functions and,
  329–30
Conditional
  inequality, 567

probability, 710
Cones, 119
Conic sections
　defined, **553**
　polar coordinates and, 553–57
　polar equations of, **554**
Conjugate axis, hyperbola and, 538
Constants, algebraic, 52
Continued proportion, 220
Contradictory inequality, 567
Corresponding segments, 94
Cosecant
　functions, graphs of, 392
Cosine
　curves, 379–81
　　composite, 385–91
　　displacement and, 382–85
　law of, oblique triangles and, 369–71
Cot, graphs of, 392–93
Cotangent
　graphs of, 392
Coterminal angles, 305–6
Cramer's rule, linear equations and, 205–12
Csc, graphs of, 392–93
Cube prism, 117
Cubes
　perfect, **242**
　two,
　　difference of, **242**, **254**
　　sum of, **242**, **254**
Curves
　cosine, 379–81
　　composite, 385–91
　　displacement and, 382–85
　fitting nonlinear curves to, 725–31
　normal, probability and, 723–24
　sine, 377
　　composite, 385–91
　　displacement and, 382–85
Cylinders, 117

Data, fitting a line to, 725–31
Decimals, 3–4
Degrees
　conversion, between radians, **90**
　converting radians to, **304**
　geometry and, 89–91
　radian conversions, **303**
　trigonometry and, 303–5
DeMoivre's formula, 497–98
　powers/roots and, 499–503
Denominators, rationalizing, 419–21
Dependent variables, 136–38
Depression, angles of, 319–20
Descartes' rule of signs, 630
Determinants
　evaluating, **200**
　　using a calculator, 202–3
　linear equations and, 195–205
　properties of, 197–99
　of the second order, 195
Difference
　identities, 685–91
Dimensions
　matrices and, 590
Direct variation, 226–28
Directrix, 525
Discriminant

plane analytic geometry, 551–53
　trinomial, **249**
Dispersion, measures of, 720–24
Distance formula, plane analytic geometry, 512
Distributive law, **239**
　numbers and, 8
Division
　algebra and, 62–67
　complex numbers, 480–82
　　exponential form, 494–95
　　polar form, 498
　fractions, 261–63
　monomial, by a monomial, 62–63
　multinominal,
　　by a monomial, 63–64
　　by a multinominal, 64–67
　rational numbers, 20–21
　real numbers, 16–17
Domain, 1046
　defined, 134–35
Double-angle identities, 691–93

Elevation, angles of, 319–20
Elimination methods, linear equations and, 180
Ellipse
　major axis, 532
　　horizontal, **533**
　　vertical, **534**
　plane analytic geometry and, 532–38
　reflective properties of, 535–36
Ellipsoid, of revolution, 535
Equal ratios, **219**
Equations
　algebraic,
　　applications of, 76–82
　　solving, 69–75
　equivalent, 69–72
　exponential, 457–59
　fractional, 276–80
　linear, 176–80
　　matrices and, 609–14
　logarithms, 459–60
　nonlinear systems, solutions of, 561–66
　parametric, trigonometric functions and,
　　400–401
　polar, 407–8
　　using calculators with, 408–9
　quadratic, 281–82
　　formula, 293–98
　　roots of, 282
　radicals, 430–34
　roots of, 624–28
　solving, using graphs, 163–72
　trigonometric, 699–702
Equilateral
　hyperbola, 549
　triangles, 97
Equivalent equations, 69–72
Errors
　concept of, 30–31
　types of, **30**
Estimates, scientific notation for, 40–41
Estimating, 33
Event, probability and, 708
Explicit functions, 139
Exponential
　equations, 457–59
　functions, 439–43

business functions, 441–42
　e[x] [superscript], 443–49
　graphing, asymptotes, 439–41
　growth/decay and, 445–47
Exponents, 23
　evaluating, using a calculator, 416
　fractional, 44–46, 414–16
　rules of, 23–27
Expressions, algebraic, 52
Exterior angles, 93

Faces, prisms, 117
Factor theorem, 620
　remainder, 618–24
Factoring, 245–49
　finding roots of, 282–87
　trinomials, 249–56
Factors, common, 245–46
Finite sequence, 649
Focus, 525
Fractional
　equations, 276–80
　exponents, 44–46, 414–16
Fractions, 256–59
　addition of, 264–71
　complex, 268–71
　division of, 261–63
　fundamental principle of, **256**
　least common denominator, 266–68
　multiplication of, 260
　subtraction of, 264–71
Frequency
　distribution, probability, 712–13
　polygon, 713
Frustums, 120–22
Functions, 135
　composite, 145–47
　defined, 133
　explicit/implicit, 139
　inverse, 139–41
　operations on, 143–45
　variables and, 136–38
　see also specific type of function

Geometric
　sequences, 654–55
　　terms of a, **654**
　series, 661–65
　　application of, 662–63
　　infinite, 665–70
　shapes,
　similar, 222–23
　　areas of, 223–24
　　scale drawings of, 224–25
　　volumes of, 224
Geometry
　angles, 89
　　types of, 92–94
　circles and, 110–16
　degrees, 89–91
　lines, 89
　radians, 89–91
　rays, 89
　segments, 89
　solids, 117–26
　　cones, 119
　　cylinders, 117
　　frustums, 120–22

prisms, 117–19
pyramids, 119–20
spheres, 122
triangles and, 97–104
Graph paper
logarithmic, 464–66
semilogarithmic, 463–66
Graphing calculators, 159–63
Graphs
equation for, **152**
horizontal line test, **167**
intercepts and, 152–53
other than trigonometric, 391–94
probability and, 713
slope and, 153–54
solving equations with, 163–72
tangent function, 391–92
that are not lines, 154–56
trigonometric. See Trigonometric functions
vertical line test, 156–57

Half-angle identities, 694–97
Harmonic motion, simple, 395–96
Heron's formula, 98
Hero's formula, 98
Histogram, 713
Horizontal asymptotes, **643**
Hyperbola, 538–43
equilateral, 549
major axis,
horizontal, **539**
vertical axis, **539**
rectangular, 549
reflective properties of, 541
Hypotenuse, 100

Identities
defined, 681
double-angle, 691–93
half-angle, 694–97
proving, 681–84
Pythagorean, 681
quotient, **310**
reciprocal, **310**
sum/difference, 685–91
Identity elements, numbers, 10
Imaginary
axis, complex planes and, 486
numbers, 472–74
Impedance, defined, 503
Implicit
functions, 139
Independent variables, 136–38
Inductive reactance, defined, 503
Inequalities
graphing, 580
linear, 566–72
programming and, 580–86
solving, 569–71
nonlinear, 572–77
numbers and, 4–5
properties of, 567–69
two variables in, 577–80
Infinite geometric series, 665–70
sum of an, **667**
sequence, 649
Integers, 2
addition of, 13–14

subtraction of, 15–16
Intercepts, graphs and, 152–53
Interest, compound, **442**, **663**
Interior angles, 93
Inverse
functions, 139–41
trigonometric, 327–31
matrices, 604–9
numbers, 11–12
variation, 228–30
Irrational
numbers, 2–3
roots, 636–40
Isosceles trapezoid, 104

Joint variation, 231–32

Laplace transforms
partial fractions and, 659–60
Lateral area
prism, 117, **118**
pyramid, **119**
Lateral edges
prisms, 117
pyramids, 119
Lateral face
prisms, 117
pyramids, 119
Laws of radicals, 418–23
Least common denominator, 266–68
Least squares, method of, 725–26
Line
equations of,
different forms of, **516**
point-slope, 176–77
slope-intercept, 177–78
fitting data to, 725–31
horizontal/vertical, **178**
slope of, **513**
Linear
equations, 176–80
cofactors and, 200–202
determinants and, 195–205
matrices and, 609–14
minors and, 199–200
solving, Cramer's rule and, 205–12
three variables, algebraic methods for
solving, 188–95
two variables, graphical/algebraic methods
of solving, 180–88
factorization theorem, **625**
inequalities, 566–72
solving, 569–71
interpolation, irrational roots and, 637–39
programming, inequalities and, 580–86
Lines, geometry and, 89
Lissajous figures, 401
Locator theorem, irrational roots and, 636–37
Logarithmic
functions, 449–53
graph paper, 464–66
Logarithms
business/finance and, 460–61
common/natural, 450
equations, 459–60
finding with a calculator, 450–51
properties of, 453–57

Malus's law, 684
Matrices
addition of, 593–94
augmented, 592–93
coefficient, 592
dimension and, 590
inverses of, 604–9
linear equations and, 609–14
multiplication of, 599–602
properties of, **596**
square, 591–92
subtraction of, 593–94
vectors, row/column, 590–91
zero, 591
Mean, probability and, 716–19
Measures of central tendency, 712–20
Median, 713–16
triangles and, 98
Method of least squares, 725–26
Midpoint formula, plane analytic geometry, 513
Minors, linear equations and, 199–200
Mixture problems, 80–81
Mode, probability and, 716–19
Monomials
algebra and, 53–54
dividing,
by a monomial, 62–63
multinominal by a, 63–64
multiplication of, 56–57
Motion problems, uniform, 77–80
Multinomials
algebra and, 53–54
adding, 54–55
dividing,
by a monomial, 63
by a multinominal, 64–67
multiplication of, 57–58
Multiplication
binomials, FOIL method, 58–60
complex numbers, 479–80
exponential form, 494–95
polar form, 497–98
fractions, 260
matrices, 599–602
monomials, 56–57
multinomials, 57–58
radicals, 425–28, **473**
rational numbers, 19–20
real numbers, 16–17
scalars, 595

Natural logs, 450
Nonlinear
curves, fitting to data, 731–35
inequalities, 572–77
systems,
defined, 561
equations, solutions of, 561–66
Normal curve, probability and, 723–24
Numbers
absolute values, 4
approximate, 31
decimals, 3–4
error, concept of, 30–31
estimation and, 33
exponents, 23
rules of, 23–27
identity elements of, 10